D0845443

SPONSORING EDITOR: Steve Quigley
PRODUCTION EDITOR: Mary Cafarella
DESIGNER: Marshall Henrichs
ILLUSTRATOR: ANCO (Boston)
COVER DESIGN: Richard Hannus
CHAPTER OPENING PHOTOGRAPHS: Marshall Henrichs

Library of Congress Cataloging in Publication Data

Bittinger, Marvin L
  Mathematics.

  Includes index.
1. Mathematics—1961-   2. Mathematical models.
I. Crown, J. Conrad, joint author. II. Title.
QA37.2.B58   510   80-19988
ISBN 0-201-03116-7

Copyright © 1982 by Addison-Wesley Publishing Company, Inc. Philippines copyright 1982 by Addison-Wesley Publishing Company, Inc.

All rights reserved. No part of this publication may be reproduced, stored in a retrieval system, or transmitted, in any form or by any means, electronic, mechanical, photocopying, recording, or otherwise, without the prior written permission of the publisher. Printed in the United States of America. Published simultaneously in Canada.

ISBN 0-201-03116-7
ABCDEFGHIJ-MA-8987654321

Mathematics:

A MODELING APPROACH

MARVIN L. BITTINGER
J. CONRAD CROWN

Indiana University—
Purdue University at Indianapolis

ADDISON-WESLE
Reading, Massachusett
London • Amsterdam • Do

Preface

The material in this book continues on from basic algebra and introduces the student to areas of Finite Mathematics and Calculus which have applications in business, economics, management, the social and behavioral sciences, and biology. The basic material can be covered in two semester courses.

While much of the material in this book has been taken from the second editions of *Finite Mathematics: A Modeling Approach* by Bittinger and Crown and *Calculus: A Modeling Approach* by Bittinger, many sections have been rewritten after class-testing the original texts, incorporating suggestions from both instructors and students to present the ideas more clearly. Also, by request, a section on "Mathematics of Finance" has been included in the present text.

1. Intuitive approach

While this word has many meanings and interpretations, its use here, for the most part, means "experience-based." That is, when a concept is being taught, the learning is based on the student's prior experience or new experience given before the concept is formalized. For example, in a maximum–minimum problem a function is usually derived which is to be maximized or minimized. Instead of forging ahead with the standard calculus solution, the student is asked to stop and compute some function values. This experience provides the student with more insight into the problem. Not only does the student discover that different dimensions yield different volumes, if volume is to be maximized, but the dimensions which yield the maximum volume might even be conjectured as a result of the calculations. Provision for use of the hand calculator also provides for an intuitive approach (see later comments).

2. Design and format

Each page has an outer margin which is used in several ways. (1) In the margin are sample, developmental, and exploratory exercises, placed with the text material so that the student can become actively involved in the development of the topic. These margin exercises have proved to be extremely beneficial. (2) For each section the objectives are stated in behavioral terms at the top of the page. These can be easily spotted by the student, and when the typical question arises "What material am I responsible for?" these objectives provide an answer. They may also help take the fear out of the word "mathematics."

3. The hand calculator

Exercises in this text can be done with or without a hand calculator. Most students, we find, not only have calculators but assume that calculators are *always* helpful to them. While there are many types of problems for which the calculator can reduce the work of computation, there are also many problems where there are naturally occurring fractions, and automatic conversion of all fractions to decimals may bring more distress than relief. For example, in the solution of systems of linear equations, fractions such as ⅓ have no exact decimal equivalent, and consequently conversion to decimals introduces the problem of "round-off" error. In general we feel that calculators should *not* be used automatically but rather should be reserved for cases where they are necessary (or where tables would be required), or where they reduce the tedium of computation.

4. Applications

Relevant and factual applications are included throughout the text to maintain interest and motivation. Problems in linear programming are of particular interest to students in business and management curricula. Problems in natural growth and decay (involving exponential and logarithmic functions) have applications in almost all areas ranging from birth rate to salvage value. Population growth is considered in the context of several mathematical models, not just exponential (see Chapter 10). When the exponential model is studied, other applications such as continuously compounded interest and the demand for natural resources are also considered. The notions of total revenue, cost, and profit, together with their related derivatives (marginal functions), are threads which run through the text, providing continued reinforcement and unification.

5. Tests

Each chapter ends with a chapter test. We have found that such sample tests supplement the Chapter Objectives and take much of the anxiety out of mathematics since the students know what is expected of them. All the answers to these tests are in the back of the book. Additional alternate forms of these tests appear, classroom-ready, in the *Instructor's Manual,* which is now in 8½″ × 11″ format to facilitate copying.

6. Exercises

Great care has been given to constructing the exercises, most of which support the behavioral objectives. Many of the Linear Programming exercises have been reworked to simplify the calculations and minimize the occurrence of fractions. The first exercises in each set are quite easy, while later ones become progressively more difficult. Most of the exercises are similar to Examples worked out in the preceding section. The exercises are also arranged in matching pairs; that is, each odd-numbered exercise is very much like the one immediately following. The odd-numbered exercises have answers in the book, while the even-numbered exercises have answers in the *Instructor's Manual.* All Margin Exercises have answers in the book. It is recommended that the student do all of these, stopping to do them when the text so indicates.

The authors wish to acknowledge the assistance of the following, whose professional reviews were extremely valuable in the preparation of the book: A. Astromoff (San Francisco State University), Walter A. Coole (Skagit Valley College), Richard M. Crownover (University of Missouri—Columbia), Matt Hassett (Arizona State University), Carl D. Meyer (North Carolina State University), Ann O'Connell (Providence College), John W. Toole (University of Maine at Orono), Francis A. Varrichio (Saint Peter's College), Samuel J. Wiley (LaSalle College), and Thomas J. Woods (Central Connecticut State College).

Indianapolis, Indiana   M.L.B.
July 1981   J.C.C.

Contents

CHAPTER 1 ALGEBRA REVIEW, FUNCTIONS, AND MODELING   1

1.1 Exponents, multiplying, and factoring   2
1.2 Equations, inequalities, and interval notation   10
1.3 Graphs and functions   19
1.4 Straight lines and linear functions   32
1.5 Systems of equations in two variables—Graphical solution   47
1.6 Systems of equations in two variables—Algebraic solution   51
1.7 Other types of functions   56
1.8 Mathematical modeling   69
     Chapter test   77

CHAPTER 2 EQUATIONS AND THE ECHELON METHOD— MATRICES   79

2.1 The echelon method—Unique solutions   80
2.2 The echelon method—Special cases   89
2.3 Basic matrix properties   95
2.4 Matrix multiplication   101
*2.5 (Optional) Computation of the matrix inverse   114
     Chapter test   117

CHAPTER 3 LINEAR PROGRAMMING   119

3.1 Graphing a system of linear constraints (inequalities) in two
     variables   121
3.2 Determining the optimum value   130

*Indicates an optional section in the sense that subsequent material is not dependent on it.

3.3 Formulating maximum-type linear programs and solving graphically 137

3.4 Formulating minimum-type linear programs and solving graphically 142

3.5 The simplex algorithm—Maximum-type linear programs 148

3.6 The simplex algorithm—Duality and minimum-type linear programs 164

Chapter test 178

CHAPTER 4 SETS AND COUNTING TECHNIQUES 179

4.1 Sets 180

4.2 Union and intersection of sets and Venn diagrams 186

4.3 Experiments, trees, and distinguishability of outcomes 198

4.4 Counting techniques—Permutations 202

4.5 Counting techniques—Combinations 211

*4.6 (Optional) The Binomial Theorem 220

Chapter Test 226

CHAPTER 5 PROBABILITY 227

5.1 Introduction to probability 228

5.2 Compound events 236

5.3 Independent events—Multiplication Theorem 242

5.4 The Addition Theorem 257

5.5 Conditional probability—Multiplication Theorem—Dependent and independent events 264

5.6 Conditional probability—Bayes' Theorem 275

Chapter test 282

CHAPTER 6 RANDOM VARIABLES—STATISTICS 283

6.1 Random variables and probability functions 284

6.2 Average and expected value 288

6.3 Variance and standard deviation 297

6.4 Bernoulli trials—Binomial probability 300

Chapter test 308

CHAPTER 7 DECISION THEORY—MARKOV CHAINS AND GAMES 309

7.1 Transition matrices and probability vectors 310

7.2 Regular and irregular Markov chains 316

7.3 Fixed points and "The Long Run" 321
7.4 Pure strategies and matrix games 327
7.5 Mixed strategies: $m \times n$ and 2×2 games 334
7.6 Voting coalitions and characteristic form 339
     Chapter test 345

CHAPTER 8 MATHEMATICS OF FINANCE **347**

8.1 Arithmetic sequences 348
8.2 Geometric sequences 353
*8.3 (Optional) Depreciation 360
8.4 Simple and compound interest 367
8.5 Annual percentage rate 374
8.6 Annuities and sinking funds 376
8.7 Present value of an annuity and amortization 380
     Chapter test 385

CHAPTER 9 DIFFERENTIAL CALCULUS **387**

9.1 Continuity and limits 388
9.2 Average rates of change 400
9.3 Differentiation using limits 409
9.4 Differentiation techniques: Power rule and sum–difference
     rule 419
9.5 Applications and rates of change 426
9.6 Differentiation techniques: Product and quotient rules 433
9.7 The extended power rule /The chain rule 437
9.8 Higher derivatives 443
9.9 The shape of a graph: finding maximum and minimum
     values 445
9.10 Maximum-minimum problems 460
*9.11 (Optional) Business applications: Minimizing inventory
      costs 474
      Chapter test 479

CHAPTER 10 EXPONENTIAL AND LOGARITHMIC FUNCTIONS **481**

10.1 Exponential and logarithmic functions 482
10.2 The exponential function, base e 494
10.3 The natural logarithm function 504
10.4 Applications: The uninhibited growth model $\frac{dP}{dt} = kP$ 512
10.5 Applications: Decay 522
      Chapter test 528

CHAPTER 11 INTEGRATION   **529**

11.1 The antiderivative   530
11.2 The definite integral: Area   537
11.3 Integration on an interval: The definite integral   547
11.4 Integration techniques: Substitution   556
11.5 Integration techniques: Integration by parts—Tables   561
*11.6 (Optional) The definite integral as a limit of sums   565
11.7 Improper integrals   571
11.8 Probability   574
11.9 Probability: Expected value; the normal distribution   585
       Chapter test   595

CHAPTER 12 FUNCTIONS OF SEVERAL VARIABLES   **597**

12.1 Partial derivatives   598
12.2 Higher-order partial derivatives   608
12.3 Maximum–minimum problems   610
12.4 Application: The least squares technique   618
12.5 Constrained maximum and minimum values—Lagrange multi-
       pliers   624
       Chapter test   632

TABLES   **633**

ANSWERS   **643**

INDEX   **I–1**

CHAPTER ONE

World running records can be modeled by linear functions. (*Marshall Henrichs*)

Algebra Review, Functions, and Modeling

OBJECTIVES

You should be able to

a) Rename an exponential expression without exponents.
b) Multiply exponential expressions by adding exponents.
c) Divide exponential expressions by subtracting exponents.
d) Raise a power to a power by multiplying exponents.
e) Multiply algebraic expressions.
f) Factor algebraic expressions.
g) Solve applied problems involving the comparison of a power like $(3.1)^2$ with 3^2.
h) Solve applied problems involving compound interest.

Rename without exponents.

1. 3^4 **2.** $(-3)^2$

3. $(1.02)^3$ **4.** $(\frac{1}{4})^2$

Rename without exponents.

5. $(5t)^0$ **6.** $(5t)^1$

7. k^0 **8.** m^1

9. $(\frac{1}{4})^1$ **10.** $(\frac{1}{4})^0$

1.1 EXPONENTS, MULTIPLYING, AND FACTORING

Exponential Notation

The set of integers is as follows:

$$\ldots, -3, -2, -1, 0, 1, 2, 3, \ldots$$

Let us review the meaning of an expression

$$a^n,$$

where n is an integer. The number a above is called the *base* and n is called the *exponent*. When n is larger than 1, then

$$a^n = \underbrace{a \cdot a \cdot a \cdots a}_{n \text{ factors}}.$$

In other words, a^n is the product of n factors, each of which is a.

Examples Rename without exponents.

a) $4^3 = 4 \cdot 4 \cdot 4$, or 64
b) $(-2)^5 = (-2)(-2)(-2)(-2)(-2)$, or -32
c) $(1.08)^2 = 1.08 \times 1.08$, or 1.1664
d) $(\frac{1}{2})^3 = \frac{1}{2} \cdot \frac{1}{2} \cdot \frac{1}{2}$, or $\frac{1}{8}$

DO EXERCISES 1 THROUGH 4. (EXERCISES ARE IN THE MARGIN.)

We define an exponent of 1 as follows:

$$a^1 = a, \quad \text{for any number } a.$$

That is, any number to the first power is that number itself. We define an exponent of 0 as follows:

$$a^0 = 1, \quad \text{for any nonzero number } a.$$

That is, any nonzero number a to the 0 power is 1.

Examples Rename without exponents.

a) $(-2x)^0 = 1$ b) $(-2x)^1 = -2x$ c) $(\frac{1}{2})^0 = 1$
d) $e^0 = 1$ e) $e^1 = e$ f) $(\frac{1}{2})^1 = \frac{1}{2}$

DO EXERCISES 5 THROUGH 10.

The meaning of a negative integer as an exponent is as follows:

$$a^{-n} = \frac{1}{a^n}, \quad \text{for any nonzero number } a.$$

That is, any nonzero number a to the $-n$ power is the reciprocal of a^n.

Examples   Rename without negative exponents.

a) $2^{-5} = \dfrac{1}{2 \cdot 2 \cdot 2 \cdot 2 \cdot 2} = \dfrac{1}{32}$

b) $10^{-3} = \dfrac{1}{10 \cdot 10 \cdot 10} = \dfrac{1}{1000}$, or 0.001

c) $(\frac{1}{4})^{-2} = \dfrac{1}{(\frac{1}{4})^2} = \dfrac{1}{\frac{1}{4} \cdot \frac{1}{4}} = \dfrac{1}{\frac{1}{16}} = 1 \cdot \dfrac{16}{1} = 16$ d) $x^{-5} = \dfrac{1}{x^5}$

e) $e^{-k} = \dfrac{1}{e^k}$ f) $t^{-1} = \dfrac{1}{t^1} = \dfrac{1}{t}$

DO EXERCISES 11 THROUGH 17.

Properties of Exponents

Note the following:

$$b^5 \cdot b^{-3} = (b \cdot b \cdot b \cdot b \cdot b) \frac{1}{b \cdot b \cdot b}$$

$$= \frac{b \cdot b \cdot b \cdot b \cdot b}{b \cdot b \cdot b}$$

$$= \frac{b \cdot b \cdot b}{b \cdot b \cdot b} \cdot b \cdot b$$

$$= 1 \cdot b \cdot b$$

$$= b^2.$$

The result could have been obtained by adding the exponents. This is true in general.

For any number a, and any integers n and m,

$$a^n \cdot a^m = a^{n+m}.$$

(To multiply when the bases are the same, add the exponents.)

Rename without negative exponents.

11. 2^{-4} **12.** 10^{-2}

13. $(\frac{1}{4})^{-3}$ **14.** t^{-7}

15. e^{-t} **16.** M^{-1}

17. $(x + 1)^{-2}$

Multiply.

18. $t^4 \cdot t^5$

19. $t^{-4} \cdot t$

20. $10e^{-4} \cdot 5e^{-9}$

21. $t^{-3} \cdot t^{-4} \cdot t$

22. $4b^5 \cdot 6b^{-2}$

Divide.

23. $\dfrac{x^6}{x^2}$

24. $\dfrac{x^2}{x^6}$

25. $\dfrac{e^t}{e^t}$

26. $\dfrac{e^2}{e^k}$

27. $\dfrac{e^5}{e^{-7}}$

28. $\dfrac{e^{-5}}{e^{-7}}$

Examples Multiply.

a) $x^5 \cdot x^6 = x^{5+6} = x^{11}$

b) $x^{-5} \cdot x^6 = x^{-5+6} = x$

c) $2x^{-3} \cdot 5x^{-4} = 10x^{-3+(-4)} = 10x^{-7}$

d) $r^2 \cdot r = r^{2+1} = r^3$

DO EXERCISES 18 THROUGH 22.

Note the following:

$$b^5 \div b^2 = \frac{b^5}{b^2} = \frac{b \cdot b \cdot b \cdot b \cdot b}{b \cdot b} = \frac{b \cdot b}{b \cdot b} \cdot b \cdot b \cdot b = 1 \cdot b \cdot b \cdot b = b^3.$$

The result could have been obtained by subtracting the exponents. This is true in general.

For any nonzero number a and any integers n and m,

$$\frac{a^n}{a^m} = a^{n-m}.$$

(To divide when the bases are the same, subtract the exponents.)

Examples Divide.

a) $\dfrac{a^3}{a^2} = a^{3-2} = a^1 = a$

b) $\dfrac{x^7}{x^7} = x^{7-7} = x^0 = 1$

c) $\dfrac{e^3}{e^{-4}} = e^{3-(-4)} = e^{3+4} = e^7$

d) $\dfrac{e^{-4}}{e^{-1}} = e^{-4-(-1)} = e^{-4+1} = e^{-3}$, or $\dfrac{1}{e^3}$

DO EXERCISES 23 THROUGH 28.

Note the following:

$$(b^2)^3 = b^2 \cdot b^2 \cdot b^2 = b^{2+2+2} = b^6.$$

The result could have been obtained by multiplying the exponents. This is true in general.

For any number a, and any integers n and m,

$$(a^n)^m = a^{nm}.$$

(To raise a power to a power, multiply the exponents.)

Examples Simplify.

a) $(x^{-2})^3 = x^{-2 \cdot 3} = x^{-6}$, or $\dfrac{1}{x^6}$

b) $(e^x)^2 = e^{2x}$

c) $(3x^3y^4)^2 = 3^2(x^3)^2(y^4)^2 = 9x^6y^8$

d) $(2x^4y^{-5}z^3)^{-3} = 2^{-3}(x^4)^{-3}(y^{-5})^{-3}(z^3)^{-3}$

$$= \frac{1}{2^3}x^{-12}y^{15}z^{-9}, \text{ or}$$

$$= \frac{y^{15}}{8x^{12}z^9}$$

DO EXERCISES 29 THROUGH 33.

Multiplication

The distributive laws are important in multiplying. The laws are as follows:

For any numbers a, b, and c,

$$a(b + c) = ab + ac, \quad \text{and} \quad a(b - c) = ab - ac.$$

Examples Multiply.

a) $3(x-5) = 3 \cdot x - 3 \cdot 5 = 3x - 15$

b) $P(1+i) = P \cdot 1 + P \cdot i = P + Pi$

c) $(x - 5)(x + 3) = (x - 5)x + (x - 5)3$
$$= x \cdot x - 5x + 3x - 5 \cdot 3$$
$$= x^2 - 2x - 15$$

d) $(a + b)(a + b) = (a + b)a + (a + b)b$
$$= a \cdot a + ba + ab + b \cdot b$$
$$= a^2 + 2ab + b^2$$

DO EXERCISES 34 THROUGH 38.

The following formulas, which are obtained using the distributive laws, are useful in multiplying.

$$(a + b)^2 = a^2 + 2ab + b^2 \qquad (1)$$
$$(a - b)^2 = a^2 - 2ab + b^2 \qquad (2)$$
$$(a - b)(a + b) = a^2 - b^2 \qquad (3)$$

Simplify.

29. $(x^{-4})^3$ **30.** $(e^2)^2$

31. $(e^x)^3$ **32.** $(5x^3y^5)^2$

33. $(4x^{-5}y^{-6}z^2)^{-4}$

Multiply.

34. $2(x + 7)$ **35.** $P(1 - i)$

36. $(x - 4)(x + 7)$ **37.** $(a - b)(a - b)$

38. $(a - b)(a + b)$

Multiply.

39. $(x - h)^2$ **40.** $(3x + t)^2$

41. $(5t - m)(5t + m)$

Factor.

42. $P - Pi$

43. $x^2 + 10xy + 25y^2$

44. $4x^2 + 28x + 40$

45. $25c^2 - d^2$

46. $3x^2h + 3xh^2 + h^3$

47. How close is $(5.1)^2$ to 5^2?

Examples Multiply.

a) $(x + h)^2 = x^2 + 2xh + h^2$
b) $(2x - t)^2 = (2x)^2 - 2(2x)t + t^2 = 4x^2 - 4xt + t^2$
c) $(3c + d)(3c - d) = (3c)^2 - d^2 = 9c^2 - d^2$

DO EXERCISES 39 THROUGH 41.

Factoring

Factoring is the reverse of multiplication. That is, to factor an expression, we find an equivalent expression that is a product. Always remember to look first for a common factor.

Examples Factor.

a) $P + Pi = P \cdot 1 + P \cdot i = P(1 + i)$ (We used a distributive law.)
b) $2xh + h^2 = h(2x + h)$
c) $x^2 - 6xy + 9y^2 = (x - 3y)^2$
d) $x^2 - 5x - 14 = (x - 7)(x + 2)$ (Here we looked for factors of -14 whose sum is -5.)
e) $x^2 - 9t^2 = (x - 3t)(x + 3t)$ (We used $(a - b)(a + b) = a^2 - b^2$.)

DO EXERCISES 42 THROUGH 46.

In later work we will consider expressions like

$$(x + h)^2 - x^2.$$

To simplify this, first note that

$$(x + h)^2 = x^2 + 2xh + h^2.$$

Subtracting x^2 on both sides of this equation, we get

$$(x + h)^2 - x^2 = 2xh + h^2.$$

Factoring out an h on the right side we get

$$(x + h)^2 - x^2 = h(2x + h). \tag{4}$$

Let us now use this result to compare two squares.

Example How close is $(3.1)^2$ to 3^2?

Solution Substituting $x = 3$ and $h = 0.1$ in equation (4) we get

$$(3.1)^2 - 3^2 = 0.1(2 \cdot 3 + 0.1) = 0.1(6.1) = 0.61.$$

So $(3.1)^2$ differs from 3^2 by 0.61.

DO EXERCISE 47.

*(Optional) Compound Interest

Suppose we invest P dollars at interest rate i, compounded annually. The amount A_1 in the account at the end of 1 year is given by

$$A_1 = P + Pi = P(1 + i) = Pr,$$

where, for convenience,

$$r = 1 + i.$$

Going into the second year we have Pr dollars, so by the end of the second year we would have the amount A_2 given by

$$A_2 = A_1 \cdot r = (Pr)r = Pr^2.$$

Going into the third year we have Pr^2 dollars, so by the end of the third year we would have the amount A_3 given by

$$A_3 = A_2 \cdot r = (Pr^2)r = Pr^3.$$

In general,

If an amount P is invested at interest rate i, compounded annually, in t years it will grow to the amount A given by

$$A = P(1 + i)^t.$$

Example 1   Suppose \$1000 is invested at 8% compounded annually. How much is in the account at the end of 2 years?

Solution   We substitute into the equation $A = P(1 + i)^t$ and get

$$A = 1000(1 + 0.08)^2 = 1000(1.08)^2 = 1000(1.1664) = \$1166.40.$$

DO EXERCISE 48.

If interest is compounded quarterly, we can find a formula like the one above as follows:

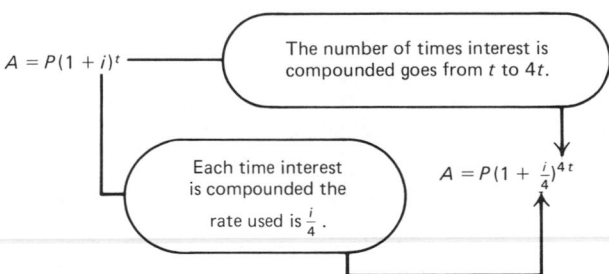

$A = P(1 + i)^t$ — The number of times interest is compounded goes from t to $4t$.

Each time interest is compounded the rate used is $\frac{i}{4}$.

$A = P(1 + \frac{i}{4})^{4t}$

---

48. Suppose \$1000 is invested at 7% compounded annually. How much is in the account at the end of 2 years?

---

* This will be reconsidered in Chapter 8, "Mathematics of Finance."

49. Suppose $1000 is invested at 9% compounded semiannually ($n = 2$). How much is in the account at the end of 3 years?

In general,

If a principal P is invested at interest rate i, compounded n times a year, in t years it will grow to the amount A given by

$$A = P\left(1 + \frac{i}{n}\right)^{nt}.$$

Example 2   Suppose $1000 is invested at 8% compounded quarterly. How much is in the account at the end of 2 years?

Solution   We substitute into the equation $A = P\left(1 + \dfrac{i}{n}\right)^{nt}$ and get

$$A = 1000\left(1 + \frac{0.08}{4}\right)^{4 \times 2} = 1000(1 + 0.02)^8 = 1000(1.02)^8,$$

$$\approx 1000(1.171659381)$$

$$= 1171.659381$$

$$\approx \$1171.66^*$$

DO EXERCISE 49.

---

* A CALCULATOR NOTE: A calculator with a y^x key and a ten-digit readout was used to find $(1.02)^8$ in Example 2. This power could also be found on a calculator without a y^x key by multiplying $1.02 \times 1.02 \times 1.02 \times 1.02 \times 1.02 \times 1.02 \times 1.02 \times 1.02$. In this case, the answer has 17 digits, and is approximated by the calculator to fit its readout, ten digits. The number of places on the calculator may affect the accuracy of the answer. Thus, you may occasionally find your answers do not agree with those in the key, which were found on a calculator with a ten-digit readout. In general, if you are using a calculator, do all your computations, and round only at the end, as in Example 2. Usually, your answer should agree to at least four digits. It might be wise to consult with your instructor on the accuracy required.

EXERCISE SET 1.1

Rename without exponents.

1. 5^3 **2.** 7^2 **3.** $(-7)^2$ **4.** $(-5)^3$ **5.** $(1.01)^2$

6. $(1.01)^3$ **7.** $(\frac{1}{2})^4$ **8.** $(\frac{1}{4})^3$ **9.** $(6x)^0$ **10.** $(6x)^1$

11. t^1 **12.** t^0 **13.** $(\frac{1}{3})^0$ **14.** $(\frac{1}{3})^1$

Rename without negative exponents.

15. 3^{-2} **16.** 4^{-2} **17.** $\left(\frac{1}{2}\right)^{-3}$ **18.** $\left(\frac{1}{2}\right)^{-2}$ **19.** 10^{-1}

20. 10^{-4} **21.** e^{-b} **22.** t^{-k} **23.** b^{-1} **24.** h^{-1}

Multiply.

25. $x^2 \cdot x^3$ **26.** $t^3 \cdot t^4$ **27.** $x^{-7} \cdot x$ **28.** $x^5 \cdot x$ **29.** $5x^2 \cdot 7x^3$

30. $4t^3 \cdot 2t^4$ **31.** $x^{-4} \cdot x^7 \cdot x$ **32.** $x^{-3} \cdot x \cdot x^3$ **33.** $e^{-t} \cdot e^t$ **34.** $e^k \cdot e^{-k}$

Divide.

35. $\dfrac{x^5}{x^2}$ **36.** $\dfrac{x^7}{x^3}$ **37.** $\dfrac{x^2}{x^5}$ **38.** $\dfrac{x^3}{x^7}$ **39.** $\dfrac{e^k}{e^k}$

40. $\dfrac{t^k}{t^k}$ **41.** $\dfrac{e^t}{e^4}$ **42.** $\dfrac{e^k}{e^3}$ **43.** $\dfrac{t^6}{t^{-8}}$ **44.** $\dfrac{t^5}{t^{-7}}$

45. $\dfrac{t^{-9}}{t^{-11}}$ **46.** $\dfrac{t^{-11}}{t^{-7}}$

Simplify.

47. $(t^{-2})^3$ **48.** $(t^{-3})^4$ **49.** $(e^x)^4$ **50.** $(e^x)^5$ **51.** $(2x^2y^4)^3$

52. $(2x^2y^4)^5$ **53.** $(3x^{-2}y^{-5}z^4)^{-4}$ **54.** $(5x^3y^{-7}z^{-5})^{-3}$ **55.** $(-3x^{-8}y^7z^2)^2$ **56.** $(-5x^4y^{-5}z^{-3})^4$

Multiply.

57. $5(x-7)$ **58.** $4(x-3)$ **59.** $x(1-t)$ **60.** $x(1+t)$

61. $(x-5)(x-2)$ **62.** $(x-4)(x-3)$ **63.** $(x+5)(x-2)$ **64.** $(x+4)(x-3)$

65. $(2x+5)(x-1)$ **66.** $(3x+4)(x-1)$ **67.** $(a-2)(a+2)$ **68.** $(3x-1)(3x+1)$

69. $(5x+2)(5x-2)$ **70.** $(t-1)(t+1)$ **71.** $(a-h)^2$ **72.** $(a+h)^2$

73. $(5x+t)^2$ **74.** $(7a-c)^2$ **75.** $5x(x^2+3)^2$ **76.** $-3x^2(x^2-4)(x^2+4)$

Use the following equation (Eq. 1) for Exercises 77 through 80.

$$(x+h)^3 = (x+h)(x+h)^2 = (x+h)(x^2+2xh+h^2)$$
$$= (x+h)x^2 + (x+h)2xh + (x+h)h^2$$
$$= x^3 + x^2h + 2x^2h + 2xh^2 + xh^2 + h^3$$
$$= x^3 + 3x^2h + 3xh^2 + h^3. \tag{1}$$

77. $(a+b)^3$ **78.** $(a-b)^3$ **79.** $(x-5)^3$ **80.** $(2x+3)^3$

Factor.

81. $x-xt$ **82.** $x+xh$ **83.** $x^2+6xy+9y^2$ **84.** $x^2-10xy+25y^2$

85. $x^2-2x-15$ **86.** $x^2+8x+15$ **87.** x^2-x-20 **88.** $x^2-9x-10$

89. $49x^2-t^2$ **90.** $9x^2-b^2$ **91.** $36t^2-16m^2$ **92.** $25y^2-9z^2$

93. $a^3b-16ab^3$ **94.** $2x^4-32$ **95.** a^8-b^8 **96.** $36y^2+12y-35$

97. $10a^2x-40b^2x$ **98.** $x^3y-25xy^3$ **99.** $2-32x^4$ **100.** $2xy^2-50x$

Use the following for Exercises 101 and 102: $(x + h)^2 - x^2 = h(2x + h)$

101. a) How close is $(4.1)^2$ to 4^2?
  b) How close is $(4.01)^2$ to 4^2?
  c) How close is $(4.001)^2$ to 4^2?

102. a) How close is $(2.1)^2$ to 2^2?
  b) How close is $(2.01)^2$ to 2^2?
  c) How close is $(2.001)^2$ to 2^2?

From Eq. (1) it follows that $(x + h)^3 - x^3 = h(3x^2 + 3xh + h^2)$. Use this for Exercises 103 and 104.

103. a) How close is $(2.1)^3$ to 2^3?
  b) How close is $(2.01)^3$ to 2^3?
  c) How close is $(2.001)^3$ to 2^3?

104. a) How close is $(4.1)^3$ to 4^3?
  b) How close is $(4.01)^3$ to 4^3?
  c) How close is $(4.001)^3$ to 4^3?

The symbol ▦ indicates an exercise designed to be done using a calculator.

Business — Compound interest.

105. Suppose $1000 is invested at 8%. How much is in the account at the end of 1 year, if interest is compounded
  a) annually?
  b) semiannually?
  c) quarterly?
  d) daily? (▦ with y^x key.)
  e) hourly?

106. Suppose $1000 is invested at 10%. How much is in the account at the end of 1 year, if interest is compounded
  a) annually?
  b) semiannually?
  c) quarterly?
  d) daily? (▦ with y^x key.)
  e) hourly?

Business—Determining monthly payment on a loan. P dollars are borrowed on a home mortgage. The monthly payment, M, is given by

$$M = P \cdot \left[\frac{\dfrac{i}{12}\left(1 + \dfrac{i}{12}\right)^n}{\left(1 + \dfrac{i}{12}\right)^n - 1} \right],$$

where i is the interest rate and n is the total number of monthly payments.

107. ▦ The mortgage on a house is $43,000. The interest rate is $11\frac{3}{4}\%$. The loan period is 25 years. What is the monthly payment?

108. ▦ The mortgage on a house is $80,000. The interest rate is $12\frac{1}{2}\%$. The loan period is 30 years. What is the monthly payment?

OBJECTIVES

You should be able to

a) Solve equations like
  $-5x + 7 = 8x + 4$, and $2t^2 = 9 + t$.
b) Solve inequalities like
  $-5x + 7 < 8x + 4$.
c) Solve applied problems.

d) Write interval notation for a given graph or inequality.

1.2 EQUATIONS, INEQUALITIES, AND INTERVAL NOTATION

Equations

Basic to the solution of many equations are these two simple principles.

THE ADDITION PRINCIPLE. If an equation $a = b$ is true, then the equation $a + c = b + c$ is true for any number c.

THE MULTIPLICATION PRINCIPLE. If an equation $a = b$ is true, then the equation $ac = bc$ is true for any number c.

Example 1 Solve $-\frac{5}{6}x + 10 = \frac{1}{2}x + 2$.

Solution We first multiply on both sides by 6 to clear of fractions.

$6(-\frac{5}{6}x + 10) = 6(\frac{1}{2}x + 2)$ (Multiplication Principle)

$6(-\frac{5}{6}x) + 6 \cdot 10 = 6(\frac{1}{2}x) + 6 \cdot 2$ (Distributive Law)

$-5x + 60 = 3x + 12$ (Simplifying)

$60 = 8x + 12$ (Addition Principle: We add $5x$ to get the variable by itself on one side of the equation.)

$48 = 8x$ (We add -12.)

$\frac{1}{8} \cdot 48 = \frac{1}{8} \cdot 8x$ (We multiply by $\frac{1}{8}$.)

$6 = x$

The number 6 checks when it is substituted into the original equation; thus it is the solution.

DO EXERCISE 50.

To solve applied problems, we first translate to mathematical language, usually an equation. Then we solve the equation and check to see if the solution of the equation is a solution of the problem.

Example 2 After a 5% gain in weight an animal weighs 693 lb. What was its original weight?

Solution We first translate to an equation

$\underbrace{(\text{Original weight})}_{w} + 5\% \underbrace{(\text{Original weight})}_{w} = 693$

$+ 5\% \cdot \qquad = 693$

Now we solve the equation.

$w + 5\%w = 693$

$1 \cdot w + 0.05w = 693$

$(1 + 0.05)w = 693$

$1.05w = 693$

$w = \dfrac{693}{1.05} = 660$

Check: $660 + 5\% \times 660 = 660 + 0.05 \times 660 = 660 + 33 = 693$

DO EXERCISE 51.

The third principle for solving equations is the *Principle of Zero Products*.

50. Solve

$$-\frac{7}{8}x + 5 = \frac{1}{4}x - 2.$$

51. An investment is made at 8%, compounded annually. It grows to $783 at the end of 1 year. How much was invested originally?

52. Solve $5x(x + 2)(2x - 3) = 0$.

THE PRINCIPLE OF ZERO PRODUCTS. **For any numbers a and b, if $ab = 0$, then $a = 0$ or $b = 0$; and if $a = 0$ or $b = 0$, then $ab = 0$.**

To solve an equation using this principle, there *must* be a 0 on one side of the equation and a product on the other. The solutions are then obtained by setting each of the factors equal to 0 and solving the resulting equations.

Example 3 Solve $3x(x - 2)(5x + 4) = 0$.

Solution $3x(x - 2)(5x + 4) = 0$

$3x = 0$ or $x - 2 = 0$ or $5x + 4 = 0$ (Principle of Zero Products)

$\frac{1}{3} \cdot 3x = \frac{1}{3} \cdot 0$ or $x = 2$ or $5x = -4$ (Solve each separately.)

$x = 0$ or $x = 2$ or $x = -\frac{4}{5}$

The solutions are $0, 2$, and $-\frac{4}{5}$.

DO EXERCISE 52.

53. Solve $x^2 + x = 12$.

Example 4 Solve $x^2 - x = 20$.

Solution $x^2 - x = 20$ (Adding -20)

$x^2 - x - 20 = 0$ (Factoring)

$(x - 5)(x + 4) = 0$ (Principle of Zero Products)

$x - 5 = 0$ or $x + 4 = 0$

$x = 5$ or $x = -4$

The solutions are 5 and -4.

DO EXERCISE 53.

Example 5 Solve $4x^3 = x$.

Solution $4x^3 = x$

$4x^3 - x = 0$ (Adding $-x$)

$x(4x^2 - 1) = 0$

$x(2x - 1)(2x + 1) = 0$ (Factoring)

$x = 0$ or $2x - 1 = 0$ or $2x + 1 = 0$ (Principle of Zero Products)

$x = 0$ or $2x = 1$ or $2x = -1$

$x = 0$ or $x = \frac{1}{2}$ or $x = -\frac{1}{2}$

The solutions are $0, \frac{1}{2}$, and $-\frac{1}{2}$.

DO EXERCISE 54.

Inequalities

Principles for solving inequalities are similar to those for solving equations. We can add the same number on both sides of an inequality. We can also multiply on both sides by the same nonzero number; but if that number is negative, we must reverse the inequality sign. Let us see why this is necessary. Consider the true inequality

$$5 < 9. \tag{1}$$

Let us multiply both members by 2. We get another true inequality

$$10 < 18.$$

Let us multiply both members in (1) by -3.

$$-15 < -27$$

This time the inequality is false. However, if we reverse the inequality symbol (use $>$ instead of $<$), we will get a true inequality

$$-15 > -27.$$

The following is a reformulation of the inequality-solving principles.

If the inequality $a < b$ is true, then

> **i) $a + c < b + c$ is true, for any c,**
> **ii) $a \cdot c < b \cdot c$, for any *positive* c,**
> **iii) $a \cdot c > b \cdot c$, for any *negative* c.**

Similar principles hold when $<$ is replaced by $\leqslant$, and $>$ is replaced by $\geqslant$.

Example 6   Solve $5x > 12 - 3x$.

Solution
$$
\begin{aligned}
5x &> 12 - 3x \\
5x + 3x &> 12 \qquad \text{(Adding } 3x) \\
8x &> 12 \\
\tfrac{1}{8} \cdot 8x &> \tfrac{1}{8} \cdot 12 \qquad \text{(Multiplying by } \tfrac{1}{8}) \\
x &> \tfrac{3}{2}
\end{aligned}
$$

Any number greater than $\frac{3}{2}$ is a solution.

DO EXERCISE 55.

54. Solve $x^3 = x$.

55. Solve $3x < 11 - 2x$.

56. Solve $16 - 7x \leqslant 10x - 4$.

57. In Example 8, determine the number of suits the firm must sell so that its total revenue will be more than $40,000.

Example 7 Solve $17 - 8x \geqslant 5x - 4$.

Solution

$$
\begin{aligned}
17 - 8x &\geqslant 5x - 4 \\
-8x &\geqslant 5x - 21 \quad &\text{(Adding } -17) \\
-13x &\geqslant -21 \quad &\text{(Adding } -5x) \\
-\tfrac{1}{13}(-13x) &\leqslant -\tfrac{1}{13}(-21) \quad &\text{(Multiplying by } -\tfrac{1}{13}, \text{ and} \\
& &\textit{reversing} \text{ the inequality sign)} \\
x &\leqslant \tfrac{21}{13}
\end{aligned}
$$

Any number less than or equal to $\frac{21}{13}$ is a solution.

DO EXERCISE 56.

Example 8 Raggs, Ltd., a clothing firm, determines that its total revenue, in dollars, from the sale of x suits is

$$2x + 50.$$

Determine the number of suits the firm must sell so that its total revenue will be more than $70,000.

Solution We translate to an inequality and solve

$$
\begin{aligned}
2x + 50 &> 70{,}000 \\
2x &> 69{,}950 \quad &\text{(Adding } -50) \\
x &> 34{,}975. \quad &\text{(Multiplying by } \tfrac{1}{2})
\end{aligned}
$$

Thus the company's total revenue will exceed $70,000 when it sells more than 34,975 suits.

DO EXERCISE 57.

Interval Notation

The set of real numbers corresponds to the set of points on a line.

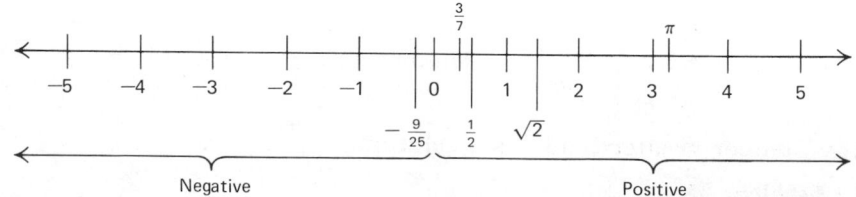

For real numbers a and b such that $a < b$ (a is to the left of b on a number line), we define the *open interval* (a, b) to be the set of numbers between, but not including, a and b. That is,

(a, b) = the set of all numbers x such that $a < x < b$.

The graph of (a, b) is shown above. The open circles and the parentheses indicate that a and b are *not* included. The numbers a and b are called *endpoints*.

DO EXERCISES 58 AND 59.

The *closed interval* $[a, b]$ is the set of numbers between and including a and b. That is,

$[a, b]$ = the set of all numbers x such that $a \leq x \leq b$.

The graph of $[a, b]$ is shown above. The solid circles and the brackets indicate that a and b are included.

There are two kinds of *half-open intervals* defined as follows:

$(a, b]$ = the set of all numbers x such that $a < x \leq b$.

The open circle and the parenthesis indicate that a is not included. The solid circle and the bracket indicate that b is included. Also,

$[a, b)$ = the set of all numbers x such that $a \leq x < b$.

The solid circle and the bracket indicate that a is included. The open circle and the parenthesis indicate that b is not included.

DO EXERCISES 60 AND 61.

58. Write interval notation for each graph.

a)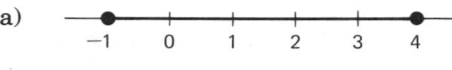

b)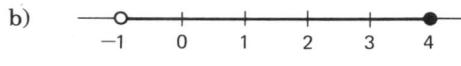

59. Write interval notation for
a) the set of all numbers x such that $-1 < x < 4$;
b) the set of all numbers x such that $-\frac{1}{4} < x < \frac{1}{4}$.

60. Write interval notation for each graph.

a)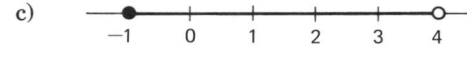

b)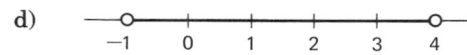

c)

d)

61. Write interval notation for
a) the set of all numbers x such that $-\sqrt{2} < x < \sqrt{2}$,
b) the set of all numbers x such that $0 \leq x < 1$,
c) the set of all numbers x such that $-6.7 < x \leq -4.2$,
d) the set of all numbers x such that $3 \leq x \leq 7\frac{1}{2}$.

62. Write interval notation for each graph.

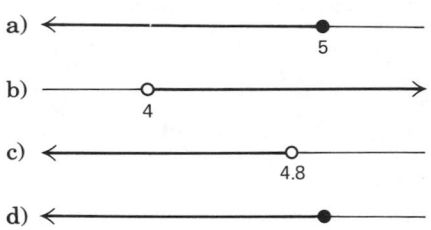

a) (solid dot at 5)

b) (open dot at 4)

c) (open dot at 4.8)

d) (solid dot at 5)

63. Write interval notation for
a) the set of all numbers x such that $x \geqslant 8$,
b) the set of all numbers x such that $x < -7$,
c) the set of all numbers x such that $x > 10$,
d) the set of all numbers x such that $x \leqslant -0.78$.

Some intervals are of unlimited extent in one or both directions. In such cases we use the infinity symbol ∞. For example,

$$[a, \infty) = \text{the set of all numbers } x \text{ such that } x \geqslant a.$$

Note that ∞ is not a number.

$$(a, \infty) = \text{the set of all numbers } x \text{ such that } x > a.$$

$$(-\infty, b] = \text{the set of all numbers } x \text{ such that } x \leqslant b.$$

$$(-\infty, b) = \text{the set of all numbers } x \text{ such that } x < b.$$

We can name the entire set of real numbers using $(-\infty, \infty)$.

DO EXERCISES 62 AND 63.

Any point in an interval which is not an endpoint is an *interior* point.

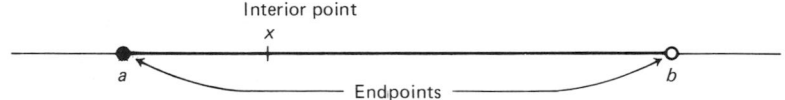

Note that all of the points in an open interval are interior points.

EXERCISE SET 1.2

Solve.

1. $-7x + 10 = 5x - 11$

2. $-8x + 9 = 4x - 70$

3. $5x - 17 - 2x = 6x - 1 - x$

4. $5x - 2 + 3x = 2x + 6 - 4x$

5. $x + 0.8x = 216.$

6. $x + 0.5x = 210$

7. $x + 0.08x = 216$

8. $x + 0.05x = 210$

Applied Problems

Biology

9. After a 6% gain in weight an animal weighs 508.8 lb. What was its original weight?

10. After a 7% gain in weight an animal weighs 363.8 lb. What was its original weight?

Business.

11. An investment is made at 9%, compounded annually. It grows to $708.50 at the end of 1 year. How much was invested originally?

12. An investment is made at 7%, compounded annually. It grows to $856 at the end of 1 year. How much was invested originally?

Sociology.

13. After a 2% increase, the population of a city is 826,200. What was the former population?

14. After a 3% increase, the population of a city is 741,600. What was the former population?

Solve.

15. $2x(x + 3)(5x - 4) = 0$

16. $7x(x - 2)(2x + 3) = 0$

17. $x^2 + 1 = 2x + 1$

18. $2t^2 = 9 + t^2$

19. $t^2 - 2t = t$

20. $6x - x^2 = x$

21. $6x - x^2 = -x$

22. $2x - x^2 = -x$

23. $9x^3 = x$

24. $16x^3 = x$

25. $(x - 3)^2 = x^2 + 2x + 1$

26. $(x - 5)^2 = x^2 + x + 3$

Solve.

27. $3 - x \leqslant 4x + 7$

28. $x + 6 \leqslant 5x - 6$

29. $5x - 5 + x > 2 - 6x - 8$

30. $3x - 3 + 3x > 1 - 7x - 9$

31. $-7x < 4$

32. $-5x \geqslant 6$

33. $5x + 2x \leqslant -21$

34. $9x + 3x \geqslant -24$

35. $2x - 7 < 5x - 9$

36. $10x - 3 \geqslant 13x - 8$

37. $8x - 9 < 3x - 11$

38. $11x - 2 \geqslant 15x - 7$

39. $8 < 3x + 2 < 14$

40. $2 < 5x - 8 \leqslant 12$

41. $3 \leqslant 4x - 3 \leqslant 19$

42. $9 \leqslant 5x + 3 < 19$

43. $-7 \leqslant 5x - 2 \leqslant 12$

44. $-11 \leqslant 2x - 1 < -5$

Applied Problems

45. *Business.* A firm determines that the total revenue, in dollars, from the sale of x units of a product is

$$3x + 1000.$$

Determine the number of units that must be sold so that its total revenue will be more than $22,000.

46. *Business.* A firm determines that the total revenue, in dollars, from the sale of x units of a product is

$$5x + 1000.$$

Determine the number of units that must be sold so that its total revenue will be more than $22,000.

47. To get a B in a course a student's average must be greater than or equal to 80% (at least 80%) and less than 90%. On the first three tests the student scores 78%, 90%, and 92%. Determine the scores on the 4th test that will yield a B.

48. To get a C in a course a student's average must be greater than or equal to 70% and less than 80%. On the first three tests the student scores 65%, 83%, and 82%. Determine the scores on the 4th test that will yield a C.

Write interval notation for each graph in Exercise 49 through 56.

49.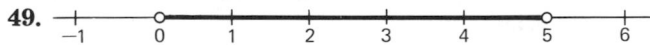

50.

51.

52.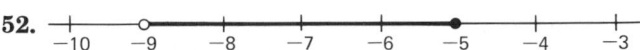

53.

54.

55.

56.

Write interval notation for Exercises 57 through 62.

57. The set of all numbers x such that $-3 \leq x \leq 3$.

58. The set of all numbers x such that $-4 < x < 4$.

59. The set of all numbers x such that $-14 \leq x < -11$.

60. The set of all numbers x such that $6 < x \leq 20$.

61. The set of all numbers x such that $x \leq -4$.

62. The set of all numbers x such that $x > -5$.

1.3 GRAPHS AND FUNCTIONS

Graphs

Each point in the plane corresponds to an ordered pair of numbers. Note that the pair $(2, 5)$ is different from the pair $(5, 2)$. This is why we call $(2, 5)$ an *ordered pair*. The first member 2 is called the *first coordinate* of the point, and the second member 5 is called the *second coordinate*. Together these are called the *coordinates of the point*. The vertical line is called the *y-axis* and the horizontal line is called the *x-axis*.

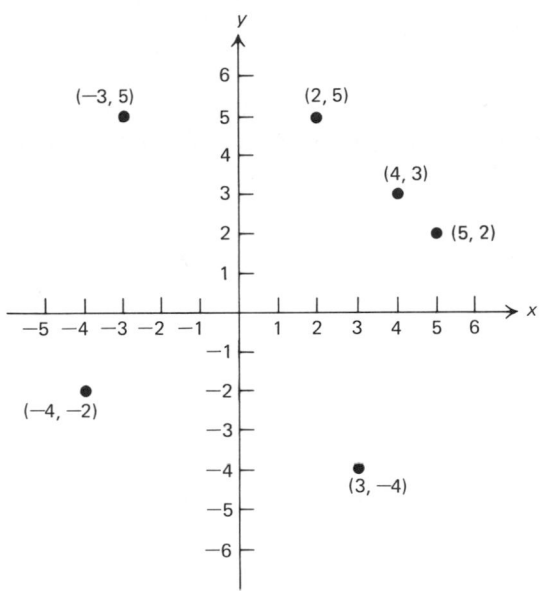

DO EXERCISE 64.

Graphs of Equations

A *solution* of an equation in two variables is an ordered pair of numbers that, when substituted alphabetically for the variables, gives a true sentence. For example, $(-1, 2)$ is a solution of the equation $3x^2 + y = 5$, because when we substitute -1 for x and 2 for y we get a true sentence:

$$\begin{array}{c|c} 3x^2 + y & = 5 \\ \hline 3(-1)^2 + 2 & 5 \\ 3 + 2 & \\ 5 & \end{array}$$

DO EXERCISE 65.

OBJECTIVES

You should be able to

a) Given a function and several inputs, find the outputs.
b) Graph a given function.
c) Decide if a graph is that of a function.
d) Tell where a function is increasing, decreasing, or neither, given its graph.

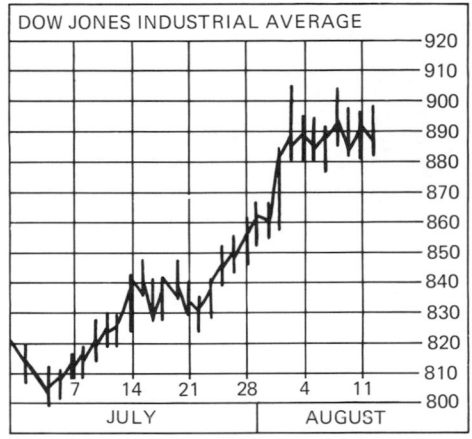

A graph of the Dow Jones Industrial Average.

64. Plot these ordered pairs: $(2, 0)$, $(0, 2)$, $(-1, 3)$, $(4, 3)$, and $(-2, -3)$.

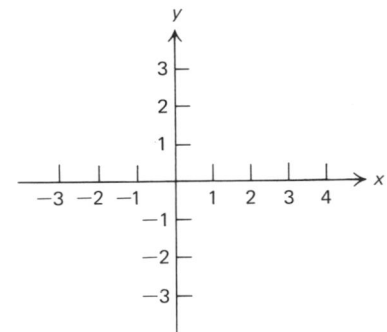

65. Decide whether each pair is a solution of
$$x^2 - 2y = 6.$$

a) $(-2, -1)$ b) $(3, 0)$

66. Graph $y = -2x + 1$.

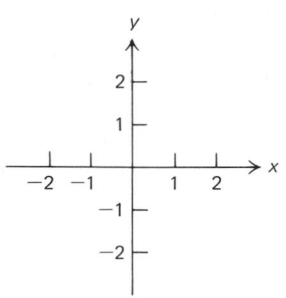

The *graph* of an equation is a geometric representation of all of its solutions. It is obtained by plotting enough ordered pairs (which are solutions) to see a pattern. The graph could be a line, curve (or curves), or some other configuration.

Example 1. Graph $y = 2x + 1$.

x	0	-1	-2	1	2
y	1	-1	-3	3	5

We choose these numbers at random (since y is expressed in terms of x).

We find these numbers by substituting in the equation.

For example, when $x = -2$, $y = 2(-2) + 1 = -3$. This yields the pair $(-2, -3)$. We plot all the pairs from the table and, in this case, draw a line to complete the graph.

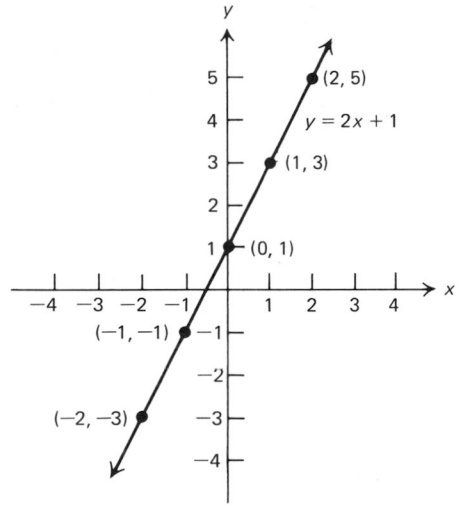

DO EXERCISE 66.

Example 2. Graph $y = x^2 - 1$.

x	0	1	2	-1	-2
y	-1	0	3	0	3

We choose these numbers at random (since y is expressed in terms of x).

We find these numbers by substituting in the equation.

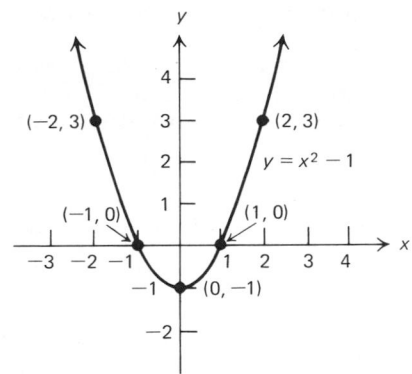

DO EXERCISE 67.

Example 3 Graph $x = y^2$.

x	0	1	4	1	4
y	0	1	2	-1	-2

We find these numbers by substituting in the equation.

This time we choose numbers at random since x is expressed in terms of y.

We plot these points, keeping in mind that x is still the first coordinate and y the second.

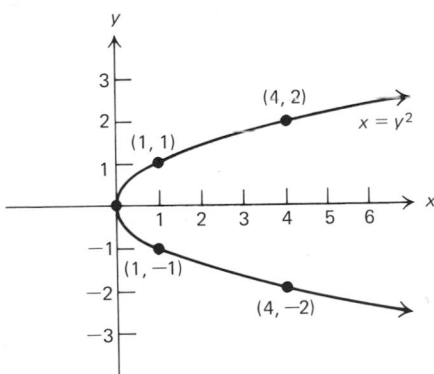

DO EXERCISE 68.

Functions

A *function* is a special kind of relation between two or more variables. Such relations are of fundamental importance in calculus.

67. Graph $y = x^2 - 3$.

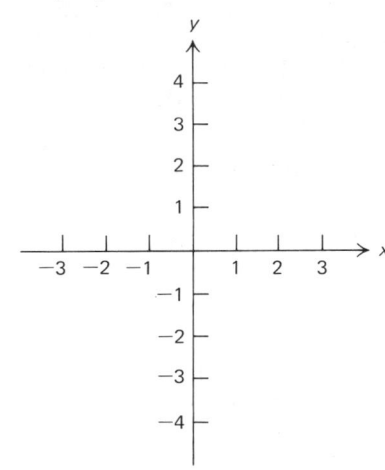

68. Graph $x = y^2 + 1$.

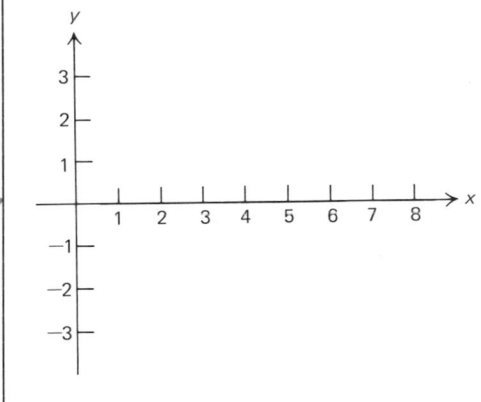

69. The operation of "taking the reciprocal" is a function. That is, the operation of going from x to $1/x$ is a function defined for all numbers except 0. Thus the domain is the set of all non-zero real numbers.

Complete this table.

Inputs	Outputs
5	
$-\frac{2}{3}$	
$\frac{1}{4}$	
$\frac{1}{a}$	
k	
$1 + t$	

A Function as an Input–Output Relation

DEFINITION. A *function* is a relation that assigns to each "input" number a unique "output" number. The set of all input numbers is called the *domain*. The set of all output numbers is called the *range*.

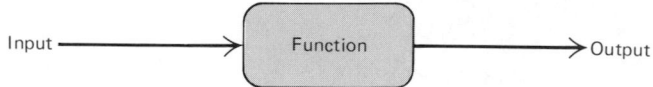

Example 4   Squaring numbers is a function. We can take any number as an input. We square that number to find the output, x^2.

Inputs	Outputs
-3	9
1.73	2.9929
k	k^2
$\sqrt{a}$	a
$1 + t$	$(1 + t)^2$, or $1 + 2t + t^2$

The domain of this function is the set of all real numbers, because any real number can be squared.

DO EXERCISE 69.

It is customary to use letters such as f and g to represent functions. Suppose f is a function and x is a number in its domain. For the input x, we can name the output as

$$f(x), \text{ read "}f \text{ of } x\text{," or "the value of } f \text{ at } x\text{."}$$

If f is the squaring function, then $f(3)$ is the output for the input 3. Thus $f(3) = 3^2 = 9$.

Example 5   The squaring function is given by

$$f(x) = x^2.$$

Find $f(-3)$, $f(1)$, $f(k)$, $f(\sqrt{k})$, $f(1 + t)$, and $f(x + h)$.

Solution

$$f(-3) = (-3)^2 = 9, \qquad f(1) = 1^2 = 1,$$
$$f(k) = k^2, \qquad f(\sqrt{k}) = (\sqrt{k})^2 = k,$$
$$f(1 + t) = (1 + t)^2 = 1 + 2t + t^2,$$
$$f(x + h) = (x + h)^2 = x^2 + 2xh + h^2$$

To find $f(x + h)$, remember what the function does—it squares the input. Thus $f(x + h) = (x + h)^2 = x^2 + 2xh + h^2$. This amounts to replacing x on both sides of $f(x) = x^2$, by $x + h$.

DO EXERCISE 70.

Example 6 A function f subtracts the square of an input from the input. A description of f is given by

$$f(x) = x - x^2.$$

Find $f(4)$ and $f(x + h)$.

Solution   We replace the x's on both sides by the inputs. Thus

$$f(4) = 4 - 4^2 = 4 - 16 = -12,$$
$$f(x + h) = (x + h) - (x + h)^2 = x + h - (x^2 + 2xh + h^2)$$
$$= x + h - x^2 - 2xh - h^2.$$

DO EXERCISE 71.

Taking square roots is *not* a function. This is because an input can have more than one output. For example, the input 4 has two outputs 2 and -2.

Example 7 Taking principal square roots (nonnegative roots) is a function. Let g be this function. Then g can be described as

$$g(x) = \sqrt{x}.$$

(Recall from algebra that the symbol "$\sqrt{a}$" represents the nonnegative square root of a.) The domain of this function is the set of nonnegative real numbers. Find $g(0)$, $g(2)$, $g(a)$, $g(16)$, and $g(t + h)$.

Solution

$$g(0) = \sqrt{0} = 0, \qquad g(2) = \sqrt{2},$$
$$g(a) = \sqrt{a}, \qquad g(16) = \sqrt{16} = 4,$$
$$g(t + h) = \sqrt{t + h}$$

DO EXERCISE 72.

70. The reciprocal function is given by

$$f(x) = \frac{1}{x}.$$

Find $f(5)$, $f(-2)$, $f(\frac{1}{4})$, $f\left(\dfrac{1}{a}\right)$, $f(k)$,

$f(1 + t)$,   and $f(x + h)$.

71. A function t is given by

$$t(x) = x + x^2.$$

Find $t(5)$, $t(-5)$,   and $t(x + h)$.

72. Subtracting 3 from a number and then taking the reciprocal is a function f given by

$$f(x) = \frac{1}{x - 3}.$$

a) What is the domain of this function? Explain.
b) Find $f(5)$, $f(4)$, $f(2.5)$, and $f(x + h)$.

A Function as a Mapping

Another way of thinking of a function is as a "mapping" of one set to another.

A *function* is a mapping that associates with each number x in one set (called the domain) a unique number y in another set.

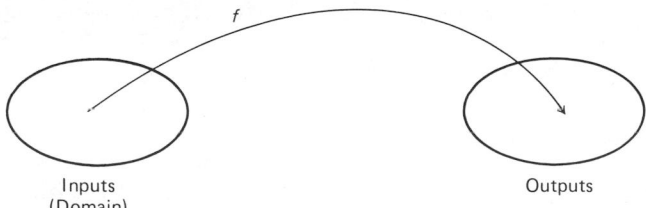

For example, the squaring function maps members of the set of real numbers to members of the set of nonnegative numbers.

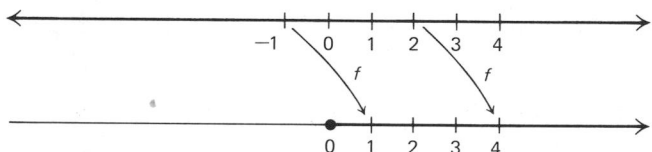

The statement

$$y = f(x)$$

means that the number x is mapped to the number y by the function f. Functions are often implicit in certain equations. For example, consider

$$xy = 2.$$

For any nonzero x there is a unique number y satisfying the equation. This yields a function which is given explicitly by

$$y = f(x) = \frac{2}{x}.$$

On the other hand, consider the equation

$$x = y^2.$$

A number x would be related to two values of y, namely $\sqrt{x}$ and $-\sqrt{x}$. Thus, this equation is not an implicit description of a function which maps inputs x to outputs y.

Graphs of Functions

Consider again the squaring function. The input 3 is associated with the output 9. The input–output pair $(3, 9)$ is one point on the *graph* of this function.

The *graph* of a function f is a geometric representation of all of its input–output pairs $(x, f(x))$. In cases where the function is given by an equation, the graph of a function is the graph of the equation $y = f(x)$.

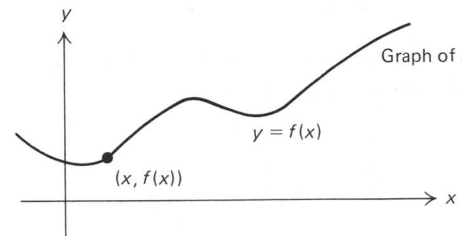

It is customary to locate input values (the domain) on the horizontal axis and output values on the vertical axis.

Example 8   Graph $f(x) = 2x + 1$.

Solution

x	0	-1	-2	1	2
$f(x)$	1	-1	-3	3	5

We choose these inputs at random.

We compute these outputs.

Next we plot the input–output pairs from the table and, in this case, draw a line to complete the graph.

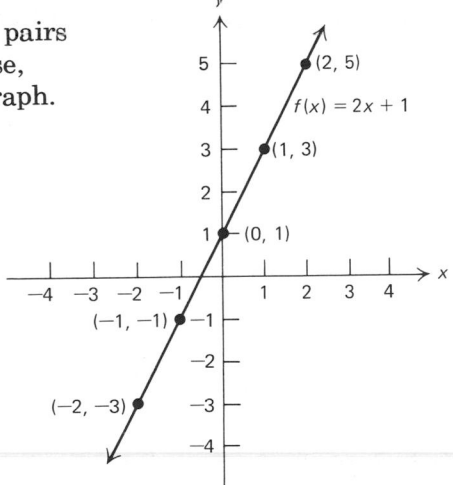

73. Graph $f(x) = -2x + 1$.

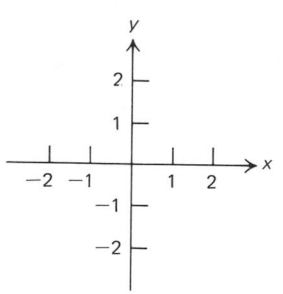

Example 9 Graph $f(x) = x^2 - 1$.

Solution

x	0	1	2	−1	−2
$f(x)$	−1	0	3	0	3

We choose these inputs at random.

We compute these outputs.

Next we plot the input–output pairs from the table and, in this case, draw a curve to complete the graph.

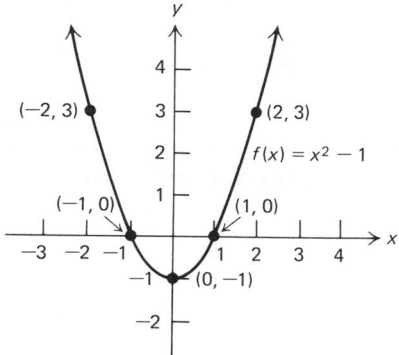

74. Graph $g(x) = x^2 - 3$.

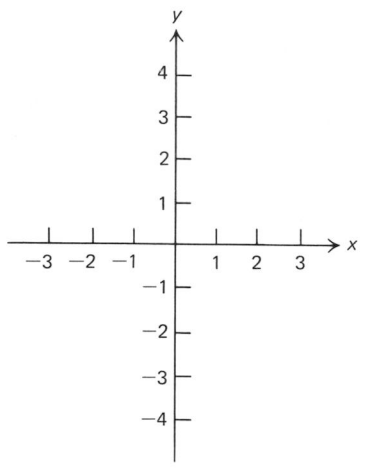

DO EXERCISES 73 AND 74

The following figure illustrates how the idea of a mapping is connected with the graph of a function.

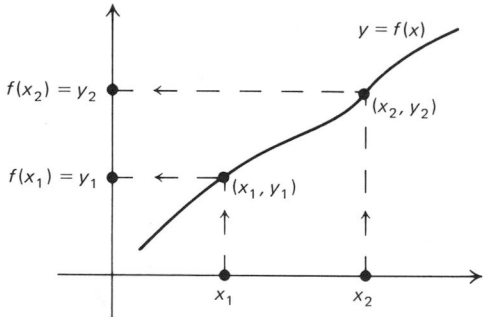

Let us now determine how we can look at a graph and decide whether it is a graph of a function. We already know

$$x = y^2$$

does not yield a function that maps a number x to a unique number y. Look at its graph.

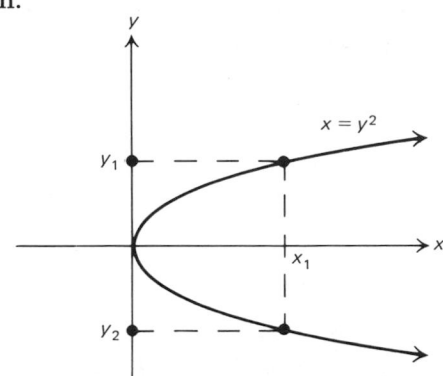

Note that there is a point x_1 that has two outputs. Equivalently, we have a vertical line that meets the graph more than once.

VERTICAL LINE TEST.   A graph is that of a function provided no vertical line meets the graph more than once.

Examples   Which of the following are graphs of functions?

a)

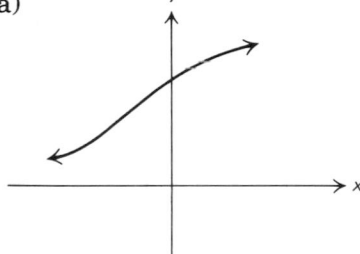

b)

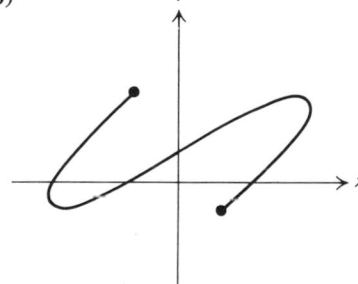

c)

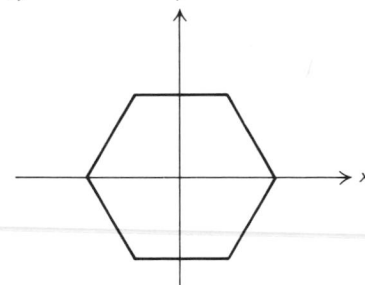

d)

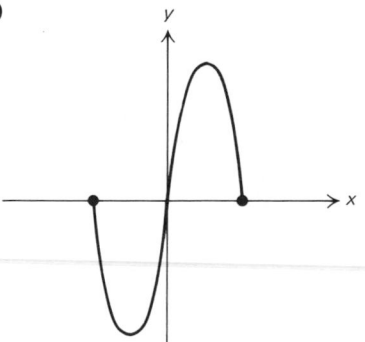

75. Which of the following are graphs of functions?

a)
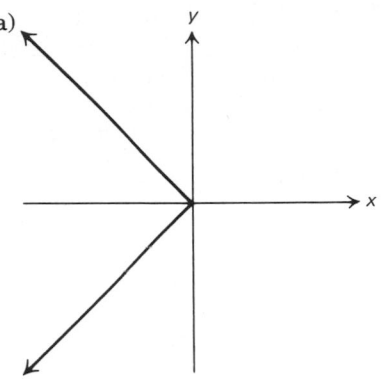

b)
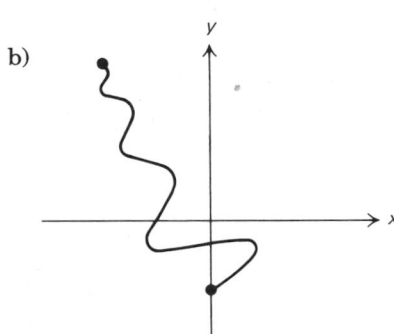

c)
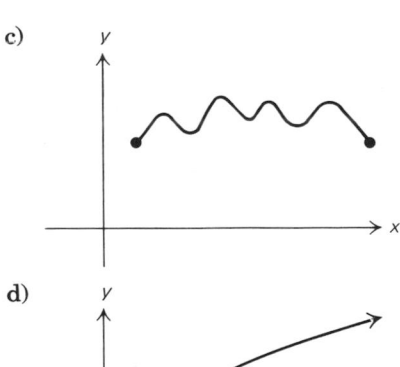

d)

Solution

a) A function. No vertical line meets the graph more than once.
b) Not a function. A vertical line (in fact many) meets the graph more than once.
c) Not a function.
d) A function.

DO EXERCISE 75.

Increasing and Decreasing Functions

If the graph of a function rises from left to right, it is said to be *increasing*. If the graph drops from left to right, it is said to be *decreasing*.

Examples

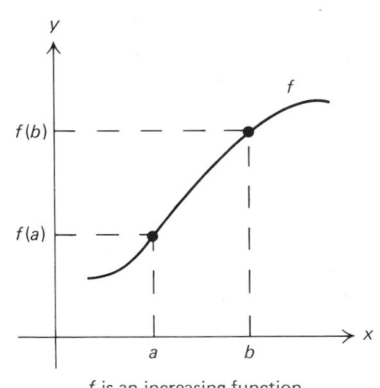

f is an increasing function

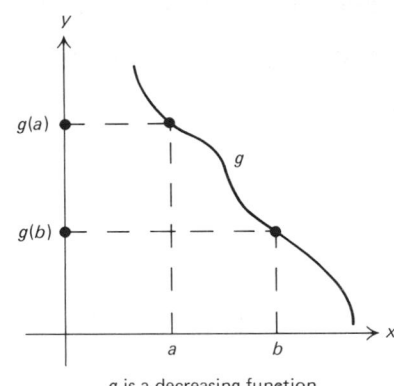

g is a decreasing function

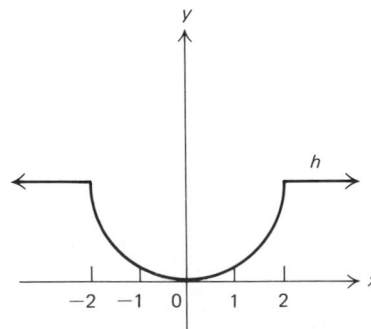

h is neither increasing nor decreasing

Note that while the function *h* is neither increasing nor decreasing it is decreasing on the interval $[-2, 0]$ and increasing on the interval $[0, 2]$. More formally:

DEFINITION. **The function *f* is increasing on an interval *l* in its domain if $f(a) < f(b)$ whenever $a < b$ in *l*. Similarly, *f* is** *decreasing* **on *l* if $f(a) > f(b)$ whenever $a < b$ in *l*.**

DO EXERCISES 76 AND 77.

Some Final Remarks

Almost all of the functions in this text can be described by equations. Some functions, however, cannot. We sometimes use the terminology *y is a function of x.* This means that *x* is an input and *y* is an output. We sometimes refer to *x* as the *independent* variable when it represents inputs, and *y* as the *dependent* variable when it represents outputs. We may refer to "the function, $y = x^2$," without naming it with a letter *f*. We may simply refer to x^2 (alone) as a function.

The height attained is a function of the force applied by the hammer. (*David Krasnor: Photo Researchers, Inc.*)

76. Which are (a) increasing? (b) decreasing? (c) neither?

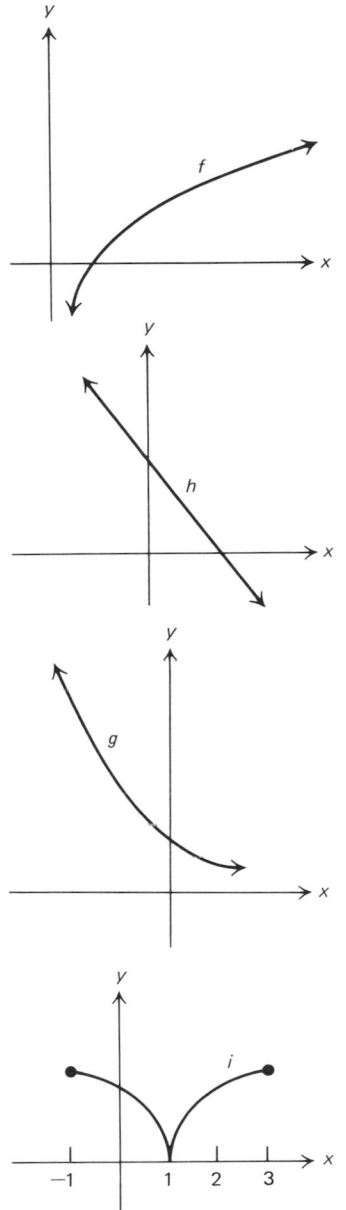

77. For the function *i* of Margin Exercise 76, on what interval is the function (a) increasing? (b) decreasing?

EXERCISE SET 1.3

1. A function f is given by

$$f(x) = 2x + 3.$$

This function takes a number x, multiplies it by 2 and adds 3.

a) Complete this table. b) Find $f(5)$, $f(-1)$, $f(k)$, $f(1 + t)$, and $f(x + h)$.

Inputs	Outputs
4.1	
4.01	
4.001	
4	

2. A function f is given by

$$f(x) = 3x - 1.$$

This function takes a number x, multiplies it by 3 and subtracts 1.

a) Complete this table. b) Find $f(4)$, $f(-2)$, $f(k)$, $f(1 + t)$, and $f(x + h)$.

Inputs	Outputs
5.1	
5.01	
5.001	
5	

3. A function g is given by

$$g(x) = x^2 - 3.$$

This function takes number x, squares it, and subtracts 3. Find $g(-1)$, $g(0)$, $g(1)$, $g(5)$, $g(u)$, $g(a + h)$, and $g(1 - h)$.

4. A function g is given by

$$g(x) = x^2 + 4.$$

This function takes a number x, squares it, and adds 4. Find $g(-3)$, $g(0)$, $g(-1)$, $g(7)$, $g(v)$, $g(a + h)$, and $g(1 - t)$.

5. A function f is given by

$$f(x) = (x - 3)^2.$$

This function takes a number x, subtracts 3 from it, and squares the result.

a) Find $f(4)$, $f(-2)$, $f(0)$, $f(a)$, $f(t + 1)$, $f(t + 3)$, and $f(x + h)$.
b) Note that f could also be given by

$$f(x) = x^2 - 6x + 9.$$

Explain what this does to an input number x.

6. A function f is given by

$$f(x) = (x + 4)^2.$$

This function takes a number x, adds 4 to it, and squares the result.

a) Find $f(3)$, $f(-6)$, $f(0)$, $f(k)$, $f(t - 1)$, $f(t - 4)$, and $f(x + h)$.
b) Note that f could also be given by

$$f(x) = x^2 + 8x + 16.$$

Explain what this does to an input number x.

Graph the following functions.

7. $f(x) = 2x + 3$ **8.** $f(x) = 3x - 1$ **9.** $g(x) = -4x$ **10.** $g(x) = -2x$

11. $f(x) = x^2 - 1$ **12.** $f(x) = x^2 + 4$ **13.** $g(x) = x^3$ **14.** $g(x) = \frac{1}{2}x^3$

Which of the following are graphs of functions?

15.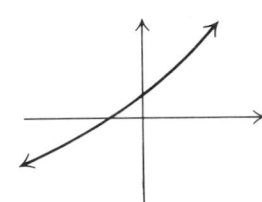

16.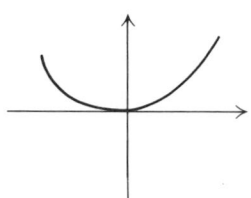

17.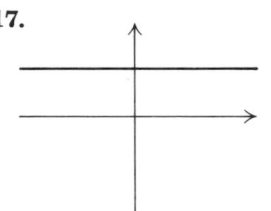

18.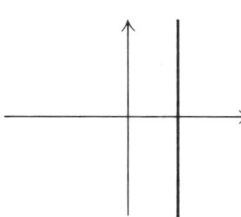

19.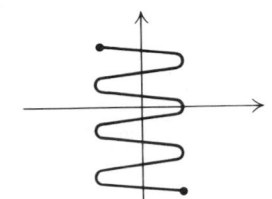

20.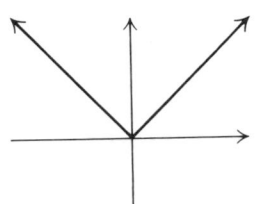

21.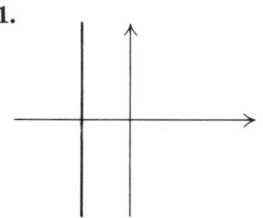

22.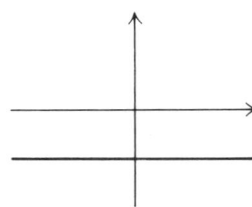

23. a) Graph $x = y^2 - 1$.
   b) Is this a function?

24. a) Graph $x = y^2 - 3$.
   b) Is this a function?

25. For $f(x) = x^2 - 3x$, find $f(x + h)$.

26. For $f(x) = x^2 + 4x$, find $f(x + h)$.

27. *Business—Revenue.* Raggs, Ltd., a clothing firm, determines that its total revenue (money coming in) from the sale of x suits is given by the function

$$R(x) = 2x + 50,$$

where $R(x)$ is the revenue, in dollars, from the sale of x suits. Find $R(10)$ and $R(100)$.

28. *Business—Compound interest.* The amount of money in a savings account at 8% compounded annually depends on the initial investment x and is given by the function

$$A(x) = x + 8\%x,$$

where $A(x)$ = amount in the account at the end of one year. Find $A(100)$ and $A(1000)$.

Decide whether the following graphs of functions are increasing, decreasing, or neither.

29.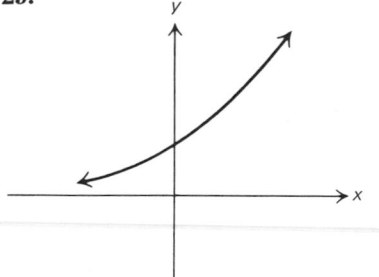

30.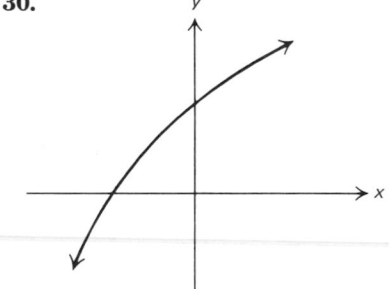

31.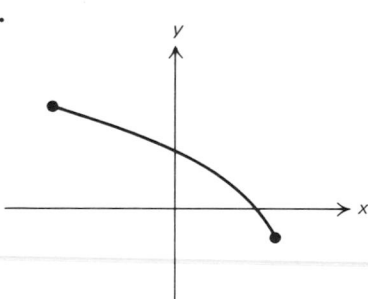

32.

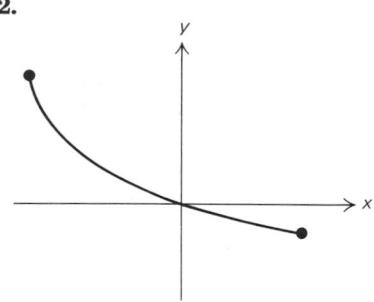

33.

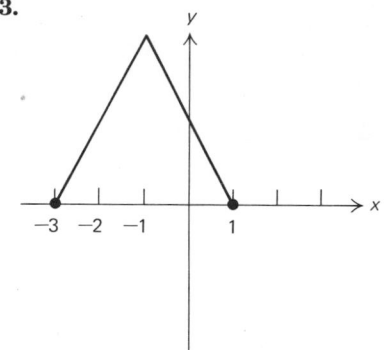

34.

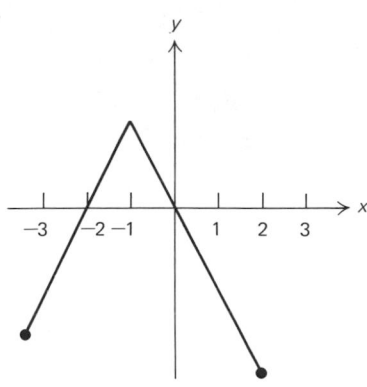

35. On what interval in Exercise 33 is the function increasing? decreasing?

36. On what interval in Exercise 34 is the function increasing? decreasing?

▶────────────────────────────────

Solve each of the following for y in terms of x. Decide whether each of the resulting equations represents a function.

37. $2x + y - 16 = 4 - 3y + 2x$ **38.** $2y^2 + 3x = 4x + 5$ **39.** $(4y^{2/3})^3 = 64x$ **40.** $(3y^{3/2})^2 = 72x$

OBJECTIVES

You should be able to

a) Graph equations of the type $y = b$ and $x = a$.
b) Graph linear functions.
c) Find an equation of a line given its slope and one point on the line.
d) Find the slope of the line containing a given pair of points.
e) Find an equation of the line containing a given pair of points.
f) Given a linear function, tell whether it is increasing, decreasing, or neither, without graphing.

1.4 STRAIGHT LINES AND LINEAR FUNCTIONS

Horizontal and Vertical Lines

Let us consider graphs of equations $y = b$ and $x = a$.

Example 1

a) Graph $y = 4$.
b) Decide if the relation is a function.

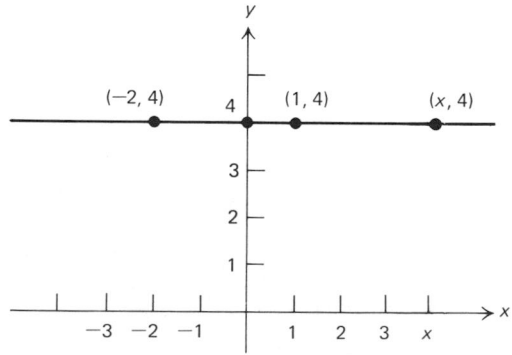

Solution

a) The graph consists of all ordered pairs whose second coordinate is 4. To see how a pair such as $(-2, 4)$ could be a solution we can consider the above equation in the form

$$0x + y = 4.$$

Then $(-2, 4)$ is a solution because

$$0(-2) + 4 = 4 \text{ is true.}$$

b) The vertical line test holds. Thus, this is a function.

DO EXERCISE 78.

Example 2

a) Graph $x = -3$.
b) Decide if it is a function.

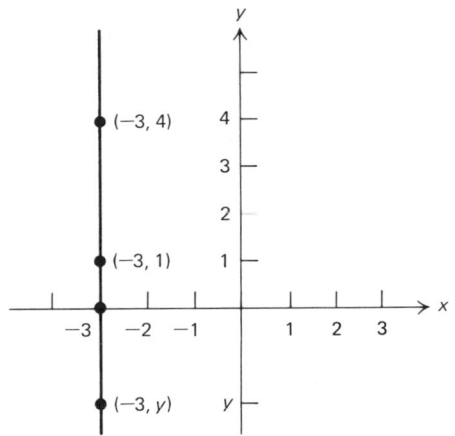

Solution

a) The graph consists of all ordered pairs whose first coordinate is -3.

b) This is *not* a function. It fails the vertical line test. The line itself meets the graph more than once, in fact, infinitely many times more.

DO EXERCISE 79.

In general,

The graph of $y = b$, a horizontal line, is the graph of a function.
The graph of $x = a$, a vertical line, is not the graph of a function.

78. a) Graph $y = 3$.
   b) Decide if it is a function.

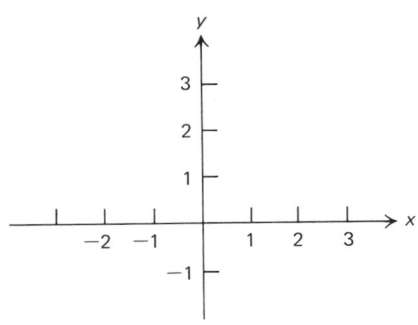

79. a) Graph $x = 1$.
   b) Decide if it is a function.

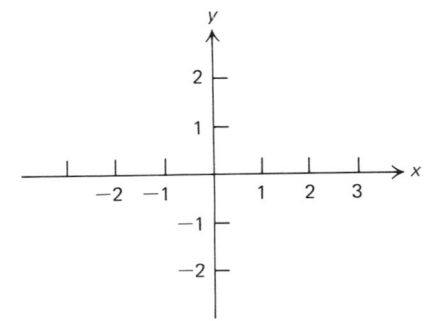

80. a) Graph $y = -2x$.

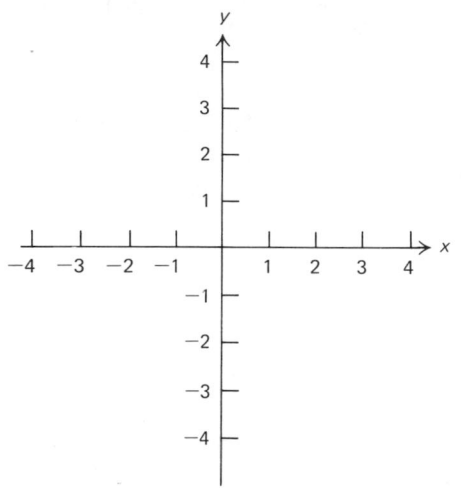

b) Is this a function?
c) What is the slope?

The Equation $y = mx$

Consider the following table of numbers and look for a pattern.

x	1	-1	$-\frac{1}{2}$	2	-2	3	-7	5
y	3	-3	$-\frac{3}{2}$	6	-6	9	-21	15

Note that the ratio of the bottom number to the top one is 3. That is,

$$\frac{y}{x} = 3, \quad \text{or} \quad y = 3x.$$

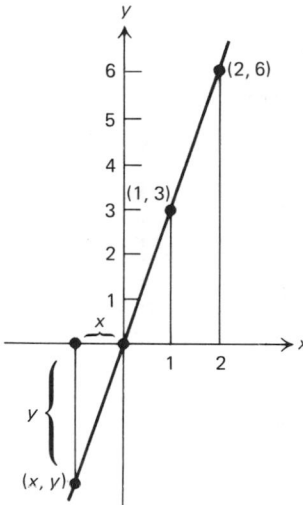

Ordered pairs from the table can be used to graph the equation $y = 3x$. Note that this is a function.

The function f given by

$$y = mx \quad \text{or} \quad f(x) = mx$$

is the straight line through the origin $(0, 0)$ and the point $(1, m)$. The constant m is called the *slope* of the line.

DO EXERCISE 80.

Various graphs of $y = mx$ for positive m are shown below. Note that such graphs rise from left to right. A line with large positive slope rises faster than a line with smaller positive slope.

a)

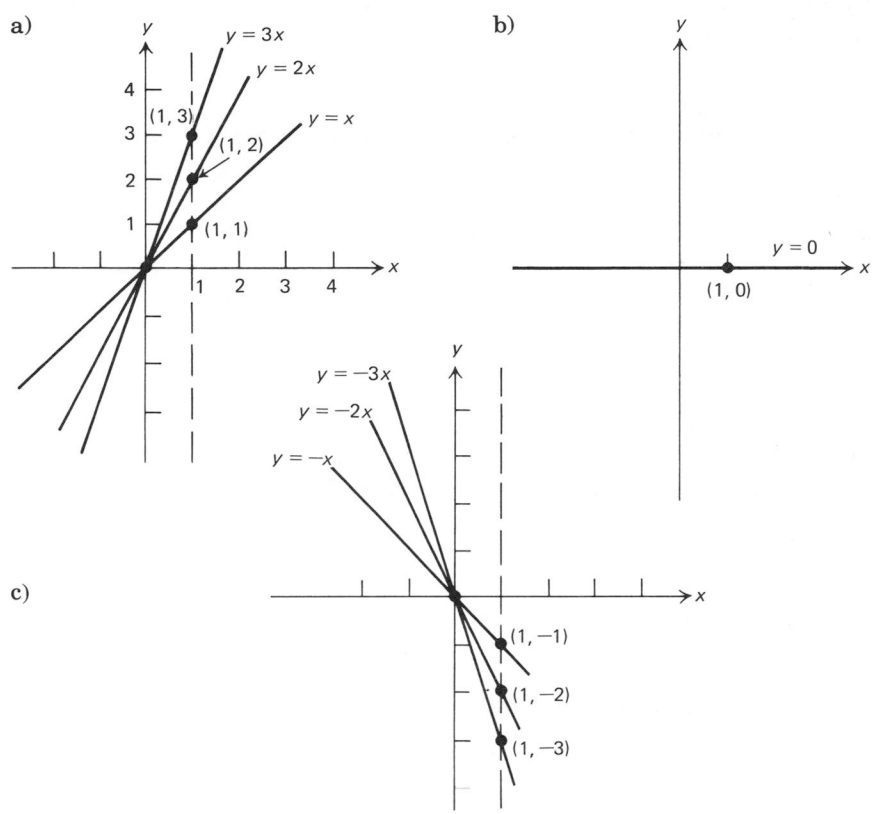

b)

c)

81. Consider these lines.

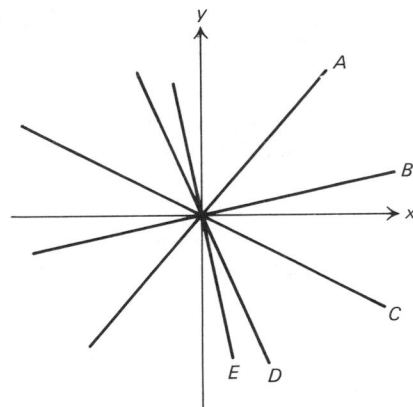

a) Which line has the largest positive slope?
b) Which line has the smallest positive slope?
c) Which line has the largest negative slope?
d) Which line has the smallest negative slope?
e) Which line has the largest slope?
f) Which line has the smallest slope?

When $m = 0$, $y = 0x$, or $y = 0$. Graph (b) is a graph of $y = 0$. Note that this is the x-axis and is a horizontal line. Graphs of $y = mx$ for negative m are shown in (c). Note that such graphs fall from left to right. A line with small negative slope is steeper than a line with larger negative slope.

DO EXERCISE 81.

Direct Variation

There are many applications involving equations like $y = mx$, where m is some positive number. In such situations we say we have *direct variation*, and m (the slope) is called the *variation constant*, or *constant of proportionality*. Usually only positive values of x and y are considered.

The variable *y varies directly* as x if there is some positive constant m such that $y = mx$. We also say that y is *directly proportional* to x.

82. *Ecology—Newspaper recycling.* The number T of trees saved by recycling is directly proportional to the height h of a stack of recyclable newspaper.

a) It is known that a stack of newspaper 36 in. high will save 1 tree. Find an equation of variation expressing T as a function of h.

b) How many trees are saved by a stack of paper 162 in. (13.5 ft) high?

Example 3   *Biomedical—Hair growth.* The number N of inches that human hair will grow is directly proportional to the time t in months. Hair will grow 6 inches in 12 months.

a) Find an equation of variation.

b) How many months does it take for hair to grow 10 inches?

Solution

a) $N = mt$, so $6 = m(12)$ and $\frac{1}{2} = m$. Thus $N = \frac{1}{2}t$.

b) To find how many months it takes for hair to grow 10 inches we solve

$$10 = \tfrac{1}{2}t$$

and get

$$20 = t.$$

Thus it takes 20 months for hair to grow 10 inches.

DO EXERCISE 82.

The Equation $y = mx + b$

Compare the graphs of the equations

$$y = 3x \quad \text{and} \quad y = 3x - 2.$$

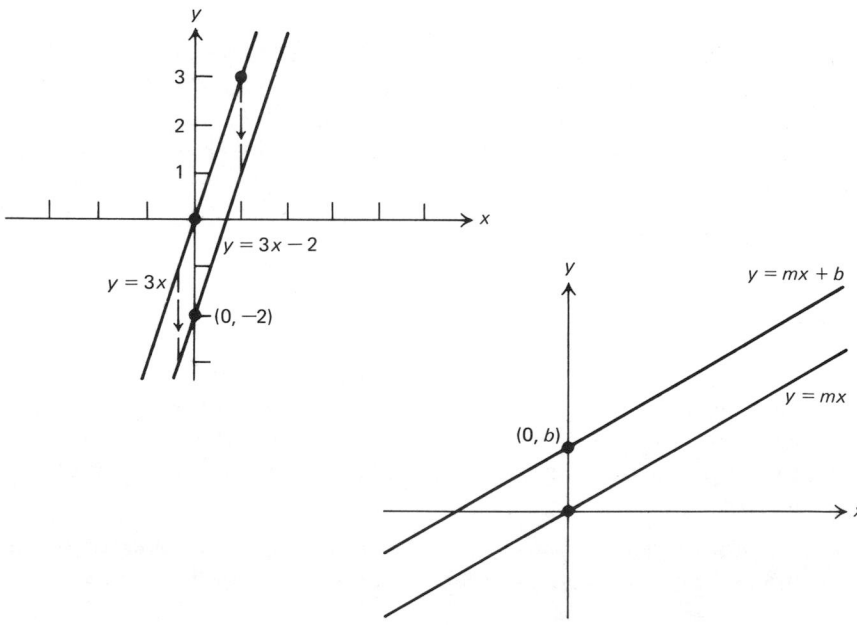

Note that $y = 3x - 2$ is a shift downward 2 units of the graph of $y = 3x$ and $y = 3x - 2$ has y-intercept $(0, -2)$. Note also that $y = 3x - 2$ is a graph of a function.

DO EXERCISE 83.

A *linear function* is given by

$$y = mx + b \quad \text{or} \quad f(x) = mx + b$$

and has a graph which is the straight line parallel to $y = mx$ with y-intercept $(0, b)$. The constant m is called the *slope*.

When $m = 0$, $y = 0x + b = b$, and we have what is known as a *constant function*. The graph of such a function is a horizontal line.

The Slope-Intercept Equation

Any nonvertical line l is uniquely determined by its slope m and its y-intercept $(0, b)$. In other words, the slope describes the "slant" of the line and the y-intercept is the point where it crosses the y-axis. Accordingly,

$y = mx + b$ is called the *slope-intercept* equation of a line.

Example 4   Find the slope and y-intercept of $2x - 4y - 7 = 0$.

Solution   We solve for y: $-4y = -2x + 7$

$$y = \tfrac{1}{2}x - \tfrac{7}{4}$$

Slope: $\tfrac{1}{2}$ y-intercept: $(0, -\tfrac{7}{4})$

DO EXERCISE 84.

The Point-Slope Equation

Suppose we know the slope of a line and some point of the line other than the y-intercept. We can still find an equation of the line.

Example 5   Find an equation of the line with slope 3 containing the point $(-1, -5)$.

Solution   From the slope intercept equation we have

$$y = 3x + b,$$

so we must determine b. Since $(-1, -5)$ is on the line, it follows that

$$-5 = 3(-1) + b,$$

so $-2 = b$ and $y = 3x - 2.$

DO EXERCISE 85.

---

83. a) Using the same axes, graph

$$y = 3x$$

and

$$y = 3x + 1.$$

b) How can the graph of $y = 3x + 1$ be obtained from the graph of $y = 3x$?

84. Find the slope and y-intercept of $2x + 3y - 6 = 0$.

85. Find an equation of the line with slope -4 containing the point $(2, -7)$.

If a point (x_1, y_1) is on the line

$$y = mx + b, \tag{1}$$

it must follow that

$$y_1 = mx_1 + b. \tag{2}$$

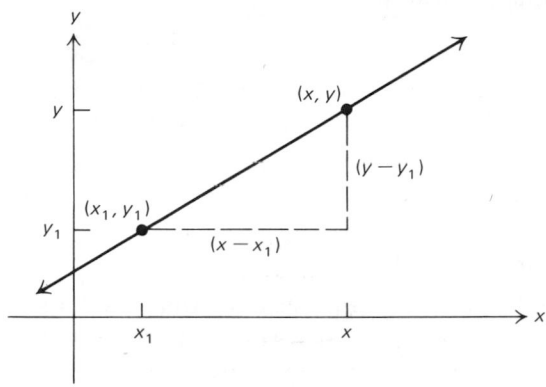

Subtracting equation (2) from (1) gets rid of the b's and we have

$$\begin{aligned}
y - y_1 &= (mx + b) - (mx_1 + b) \\
&= mx + b - mx_1 - b \\
&= mx - mx_1 = m(x - x_1).
\end{aligned}$$

Now

$y - y_1 = m(x - x_1)$ **is called the *point-slope* equation**

of a line L. This allows us to write an equation of a line given its slope and the coordinates of *any* point on it.

Example 6   Find an equation of the line with slope 3 containing the point $(-1, -5)$.

Solution   Substituting in

$$y - y_1 = m(x - x_1)$$

we get

$$y - (-5) = 3[x - (-1)].$$

Simplifying and solving for y we get the slope-intercept equation as found in Example 5.

$$\begin{aligned}
y + 5 &= 3(x + 1) \\
y &= 3x + 3 - 5 \\
y &= 3x - 2
\end{aligned}$$

DO EXERCISE 86.

86. Find an equation of the line with slope -4 containing the point $(2, -7)$.

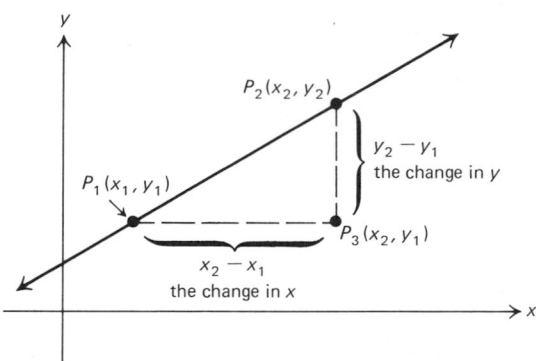

We now determine a way to compute the slope of a line when we know the coordinates of two of its points. Suppose (x_1, y_1) and (x_2, y_2) are the coordinates of two different points, P_1 and P_2, on a line that is not parallel to an axis. Consider a right triangle as shown, with legs parallel to the axes. The point P_3 with coordinates (x_2, y_1) is the third vertex of the triangle. As we move from P_1 to P_2, y changes from y_1 to y_2. The change in y is $y_2 - y_1$. Similarly, the change in x is $x_2 - x_1$. The ratio of these changes is the slope. To see this, consider the point-slope equation

$$y - y_1 = m(x - x_1).$$

Since (x_2, y_2) is on the line, it must follow that

$$y_2 - y_1 = m(x_2 - x_1).$$

Since the line is not vertical, the two x coordinates must be different, so $x_2 - x_1$ is nonzero and we can divide by it to get

$$m = \frac{y_2 - y_1}{x_2 - x_1} = \frac{\text{change in } y}{\text{change in } x} = \begin{array}{l}\text{slope of line containing points} \\ (x_1, y_1) \text{ and } (x_2, y_2).\end{array}$$

Example 7 Find the slope of the line containing the points $(-2, 6)$ and $(-4, 9)$.

Solution

$$m = \frac{y_2 - y_1}{x_2 - x_1} = \frac{6 - 9}{-2 - (-4)} = \frac{-3}{2} = -\frac{3}{2}$$

Find the slope of the line containing each pair of points.

87. $(1, 3), (2, 5)$

88. $(-6, 4), (2, 5)$

89. $(4, 7), (6, -10)$

90. $(3, 5), (-1, 5)$

Find the slope, if it exists, of the line containing each pair of points.

91. $(4, -7), (-2, -7)$

92. $(4, -7), (4, -9)$

Note that it does not matter which point is taken first, as long as we subtract coordinates in the same order. In this example we can also find m as follows:

$$m = \frac{9 - 6}{-4 - (-2)} = \frac{3}{-2} = -\frac{3}{2}.$$

DO EXERCISES 87 THROUGH 90.

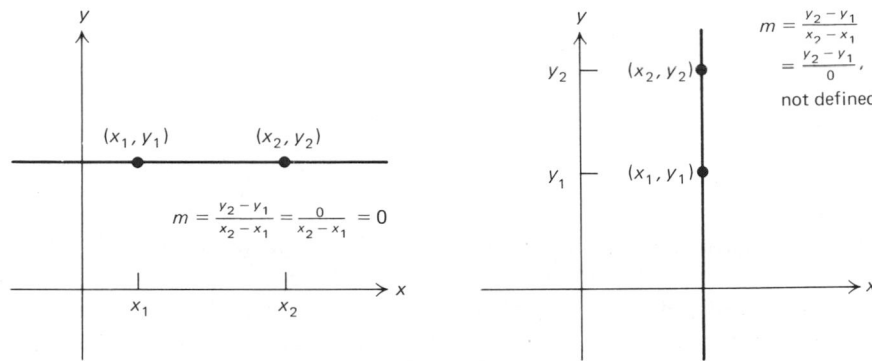

If a line is horizontal the change is y for any two points is 0. Thus a horizontal line has slope 0. If a line is vertical, the change in x for any two points is 0. Thus the slope is not defined because we cannot divide by 0. A vertical line has no slope. Thus "0 slope" and "no slope" are two very distinct concepts.

DO EXERCISES 91 AND 92.

Increasing and Decreasing Linear Functions

We do not need to graph a linear function $f(x) = mx + b$ to determine whether it is increasing, decreasing, or neither. We merely look at the slope m. If $m > 0$, the function is increasing. If $m < 0$, the function is decreasing. If $m = 0$, the function is constant, so is neither increasing nor decreasing.

Example 8 Determine whether increasing, decreasing, or neither:

$$g(x) = -2x + 3, \qquad f(x) = \tfrac{2}{3}x + 2, \qquad h(x) = 4.$$

Solution The graphs are provided for illustration and should not be needed when doing the exercises.

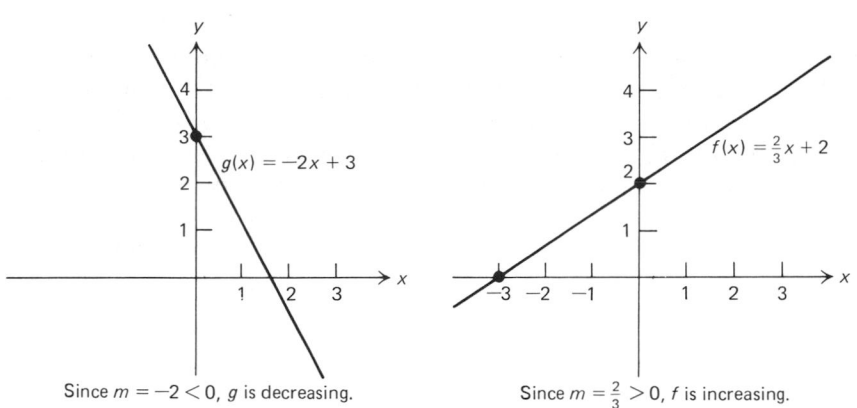

Since $m = -2 < 0$, g is decreasing.

Since $m = \frac{2}{3} > 0$, f is increasing.

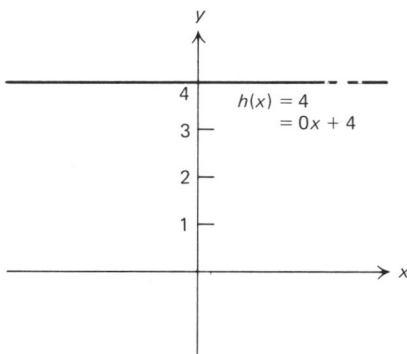

Since $m = 0$, h is neither increasing nor decreasing.

93. Without graphing, decide whether each function is increasing, decreasing, or neither.

a) $f(x) = -\frac{1}{4}x - 9$

b) $g(x) = x - 9$

c) $h(x) = -9$

DO EXERCISE 93.

Applications of Linear Functions

Many applications are modeled by linear functions.

Example 9 *Business—Total cost.* Raggs, Ltd., a clothing firm, has *fixed costs* of $10,000 per year. These are costs such as rent, maintenance, and so on, which must be paid no matter how much the company produces. To produce x units of a certain kind of suit it costs $20 per unit in addition to the fixed costs. That is, the *variable costs* are $20x$ dollars. These are costs which are directly related to production, such as material, wages, fuel, and so on. Then the *total cost*, $C(x)$, of producing x suits in a year is given by a function C:

$$C(x) = (\text{variable costs}) + (\text{fixed costs}) = 20x + 10{,}000$$

a) Graph the variable cost, fixed cost, and total cost functions.

b) What is the total cost of producing 100 suits? 400 suits?

c) How much more does it cost to produce 400 suits than 100 suits?

94. Rework Example 9, where variable costs = 30x, fixed costs = $15,000, and total costs = $C(x) = 30x + 15,000$.

Solution

a) The variable cost and fixed cost functions appear below left; the total cost function, below right. From a practical standpoint, the domains of these functions are nonnegative integers 0, 1, 2, 3, and so on. This is because it does not make sense to make a negative number of suits or a fractional number of suits. Nevertheless, it is common practice to draw the graphs as if the domains were the entire set of nonnegative real numbers.

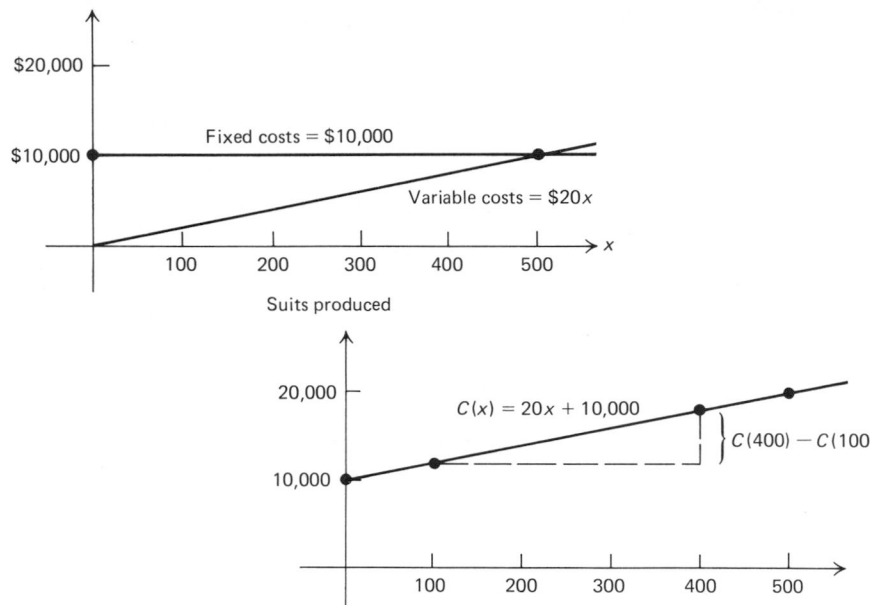

b) The total cost of producing 100 suits is

$$C(100) = 20 \cdot 100 + 10,000 = \$12,000.$$

The total cost of producing 400 suits is

$$C(400) = 20 \cdot 400 + 10,000 = \$18,000.$$

c) The extra cost of producing 400 rather than 100 suits is given by

$$C(400) - C(100) = \$18,000 - \$12,000 = \$6000.$$

DO EXERCISE 94.

Example 10   *Business—Profit and loss analysis.* In reference to Example 9, Raggs, Ltd. determines that its total revenue (money coming in) from the sale of x suits is $80 per suit. That is, total revenue $R(x)$ is given by the function

$$R(x) = 80x.$$

a) Graph $R(x)$ and $C(x)$ using the same axes.
b) Total profit $P(x)$ is given by a function P:

$$P(x) = \text{(total revenue)} - \text{(total costs)} = R(x) - C(x).$$

Determine $P(x)$ and draw its graph using the same axes.
c) The company will *break even* at that value of x for which $P(x) = 0$ (that is, no profit and no loss). This is where $R(x) = C(x)$. Find the break-even value of x.

Solution

a) The graphs of $R(x) = 80x$ and $C(x) = 20x + 10,000$ are shown here.

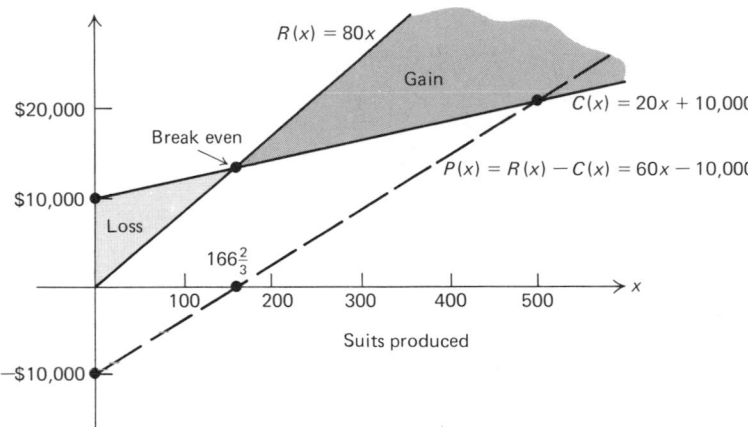

b) $P(x) = R(x) - C(x) = 80x - (20x + 10,000) = 60x - 10,000$. The graph of $P(x)$ is shown.
c) To find the break-even value we solve $R(x) = C(x)$.

$$80x = 20x + 10,000,$$
$$60x = 10,000,$$
$$x = 166\tfrac{2}{3}.$$

How do we interpret the fractional answer, since it is not possible to produce $\tfrac{2}{3}$ of a suit? One would simply round to 167. Estimates of break-even points are usually sufficient since companies want to operate well above break-even points where profit is maximized.

DO EXERCISE 95.

95. Rework Example 10, where

$$C(x) = 30x + 15,000$$

and

$$R(x) = 90x.$$

EXERCISE SET 1.4

Graph.

1. $y = -4$ **2.** $y = -3.5$ **3.** $x = 4.5$ **4.** $x = 10$

Graph. Find the slope and y-intercept.

5. $y = -3x$ **6.** $y = -0.5x$ **7.** $y = 0.5x$ **8.** $y = 3x$

9. $y = -2x + 3$ **10.** $y = -x + 4$ **11.** $y = -x - 2$ **12.** $y = -3x + 2$

Find the slope and y-intercept.

13. $2x + y - 2 = 0$ **14.** $2x - y + 3 = 0$ **15.** $2x + 2y + 5 = 0$ **16.** $3x - 3y + 6 = 0$

Find an equation of the line:

17. with $m = -5$, containing $(1, -5)$. **18.** with $m = 7$, containing $(1, 7)$.

19. with $m = -2$, containing $(2, 3)$. **20.** with $m = -3$ containing $(5, -2)$.

21. with y-intercept $(0, -6)$ and slope $\frac{1}{2}$. **22.** with y-intercept $(0, 7)$ and slope $\frac{4}{3}$.

23. with slope 0, containing $(2, 3)$. **24.** with slope 0, containing $(4, 8)$.

Find the slope of the line containing each pair of points.

25. $(-4, -2), (-2, 1)$ **26.** $(-2, 1), (6, 3)$ **27.** $(2, -4), (4, -3)$.

28. $(-5, 8), (5, -3)$ **29.** $(3, -7), (3, -9)$ **30.** $(-4, 2), (-4, 10)$

31. $(2, 3), (-1, 3)$ **32.** $(-6, \frac{1}{2}), (-7, \frac{1}{2})$ **33.** $(x, 3x), (x + h, 3(x + h))$

34. $(x, 4x), (x + h, 4(x + h))$ **35.** $(x, 2x + 3), (x + h, 2(x + h) + 3)$

36. $(x, 3x - 1), (x + h, 3(x + h) - 1)$

THE TWO-POINT EQUATION. **An equation of the nonvertical line containing the points (x_1, y_1) and (x_2, y_2) is given by**

$$y - y_1 = \frac{y_2 - y_1}{x_2 - x_1}(x - x_1). \quad \textit{Two-point equation}$$

This can be proved by replacing m in the point-slope equation $y - y_1 = m(x - x_1)$ by $\dfrac{y_2 - y_1}{x_2 - x_1}$.

37.–48. Find an equation of the line containing each pair of points in Exercises 25 through 36.

Without graphing, decide whether each function is increasing, decreasing, or neither.

49. $f(x) = 43$ **50.** $f(x) = -27$ **51.** $f(x) = -4x + 3$ **52.** $f(x) = -3x - 7$

53. $f(x) = 0.2x + 170$ **54.** $f(x) = 0.1x + 50,000$

Applied Problems

55. *Ecology—Energy conservation.* The R-factor of home insulation is directly proportional to its thickness T.

   a) Find an equation of variation where $R = 12.51$ when $T = 3$ in.

   b) What is the R-factor for insulation which is 6 inches thick?

56. *Biomedical—Nerve impulse speed.* Impulses in nerve fibers travel at a speed of 293 ft/sec. The distance D traveled in t sec is given by $D = 293t$. How long would it take an impulse to travel from the brain to the toes of a person who is 6 ft tall?

57. *Biomedical—Brain weight.* The weight, B, of a human's brain is directly proportional to its body weight, W.

  a) It is known that a person who weighs 200 lb has a brain which weighs 5 lb. Find an equation of variation expressing B as a function of W.
  b) Express the variation constant as a percent, and interpret the resulting equation.
  c) What is the weight of the brain of a person who weighs 120 lb?

59. *Business—Investment.* A person makes an investment of P dollars at 8%. After 1 year it grows to an amount, A.

  a) Show that A is directly proportional to P.
  b) Find A when $P = \$100$.
  c) Find P when $A = \$259.20$.

60. *Sociology—Urban population.* The population of a town is P. After a growth of 2% its new population is N.

  a) Assuming that N is directly proportional to P, find an equation of variation.
  b) Find N when $P = 200,000$.
  c) Find P when $N = 367,200$.

61. *Stopping distance on glare ice.* The stopping distance (at some fixed speed) of regular tires is given by a linear function of the air temperature, F:

$$D(F) = 2F + 115,$$

where $D(F)$ = stopping distance, in feet, when the air temperature is F, in degrees Fahrenheit.

  a) Find $D(0°)$, $D(-20°)$, $D(10°)$, and $D(32°)$.
  b) Graph $D(F)$.
  c) Explain why the domain should be restricted to the interval $[-57.5°, 32°]$.

63. *Biology—Spread of an organism.* A certain kind of organism is released over an area of 2 sq. mi. It grows and spreads over more area. The area covered by the organism after time t is given by a linear function

$$A(t) = 1.1t + 2,$$

where $A(t)$ = area covered, in square miles, after time t, in years.

  a) Find $A(0)$, $A(1)$, $A(4)$, and $A(10)$.
  b) Graph $A(t)$.
  c) Why should the domain be restricted to interval $[0, \infty)$?

58. *Biomedical—Muscle weight.* The weight M of the muscles in a human is directly proportional to body weight, W.

  a) It is known that a person who weighs 200 lb has 80 lb of muscles. Find an equation of variation expressing M as a function of W.
  b) Express the variation constant as a percent and interpret the resulting equation.
  c) What is the muscle weight of a person who weighs 120 lb?

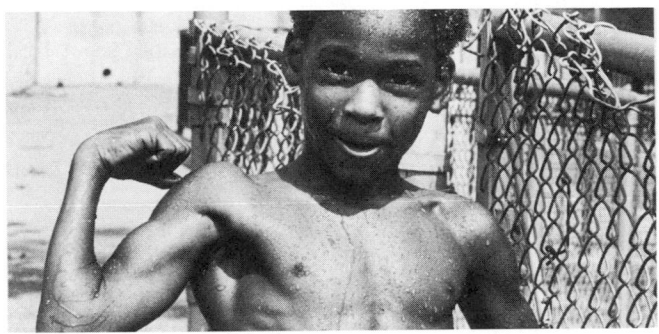

The muscle weight is directly proportional to body weight.
(*Anna Kaufman Moon: Stock, Boston*)

62. *Sociology—Percentage of the population in college.* The percentage of the population in college is given by a linear function

$$P(t) = 1.25t + 15,$$

where $P(t)$ = percentage in college the tth year after 1940. Thus $P(0)$ is the percentage in college in 1940, $P(30)$ is the percentage in college in 1970, and so on.

  a) Find $P(0)$, $P(1)$, $P(30)$, and $P(40)$.
  b) What percentage of the population will be in college in 1980?
  c) Graph $P(t)$.

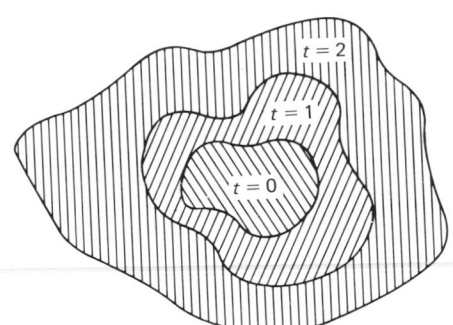

64. *Business—Profit and loss analysis.* A ski manufacturer is planning a new line of skis. For the first year, the fixed costs for setting up the new production line are $22,500. Variable costs for producing each pair of skis are estimated to be $40. The sales department projects that 3000 pairs can be sold during the first year at a price of $85 per pair.

a) Formulate a function $C(x)$ for the total cost of producing x pairs of skis.

b) Formulate a function $R(x)$ for the total revenue from the sale of x pairs of skis.

c) Formulate a function $P(x)$ for the total profit from the production and sale of x pairs of skis.

d) What profit or loss will the company realize if expected sales of 3000 pairs occur?

e) How many pairs must the company sell to break even?

66. *Business—Sales commissions.* A person applying for a sales position is offered alternative salary plans:

Plan A: A base salary of $600 per month plus a commission of 4% of the gross sales for the month.

Plan B: A base salary of $700 per month plus a commission of 6% of the gross sales for the month in excess of $10,000.

a) For each plan formulate a function that expresses monthly earnings as a function of gross sales x.

b) For what gross sales values is Plan B preferable?

65. *Business—Profit and loss analysis.* Boxowitz, Inc., a computer firm, is planning to sell a new minicalculator. For the first year, the fixed costs for setting up the new production line are $100,000. Variable costs for producing each calculator are estimated to be $20. The sales department projects that 150,000 calculators can be sold during the first year at a price of $45 each.

a) Formulate a function $C(x)$ for the total cost of producing x calculators.

b) Formulate a function $R(x)$ for the total revenue from the sale of x calculators.

c) Formulate a function $P(x)$ for the total profit from the production and sale of x calculators.

d) What profit or loss will the firm realize if the expected sales of 150,000 calculators occur?

e) How many calculators must the firm sell to break even?

67. *Anthropology—Estimating heights.* An anthropologist can use certain linear functions to estimate the height of a male or female, given the length of certain bones. A *humerus* is the bone from the elbow to the shoulder. Let x = length of the humerus in centimeters. Then the height, in centimeters, of a male with a humerus of length x is given by

$$M(x) = 2.89x + 70.64.$$

The height, in centimeters, of a female with a humerus of length x is given by

$$F(x) = 2.75x + 71.48.$$

A 45-cm humerus was uncovered in a ruins.

a) Assuming it was from a male, how tall was he?

b) Assuming it was from a female, how tall was she?

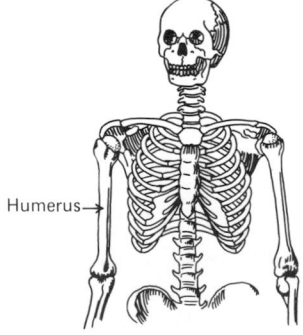

Humerus→

1.5 SYSTEMS OF EQUATIONS IN TWO VARIABLES—GRAPHICAL SOLUTION

Variables with Subscripts

When working with linear equations, it is convenient to use a variable x_1 to represent the x, or first coordinate. The "1" is called a *subscript*. Similarly, we use x_2 to represent the y, or second coordinate. Note the connection: x_1 is the "first" coordinate and x_2 is the "second" coordinate. An equation like

$$4x - 6y = -10$$

can now be written

$$4x_1 - 6x_2 = -10.$$

Graphing is done the same way as before, except that the axes are labeled x_1 and x_2.

Systems of Equations in Two Variables

A pair of linear equations

$$a_1x_1 + b_1x_2 = c_1$$
$$a_2x_1 + b_2x_2 = c_2$$

is called a *system* of two linear equations in two variables. A *solution* of a system is an ordered pair which is a solution of *both* equations.

Example 1   Decide whether $(2, 3)$ is a solution of the system

$$x_1 - x_2 = -1$$
$$4x_1 + 2x_2 = 14.$$

Solution   We substitute 2 for x_1 and 3 for x_2 in each equation.

$x_1 - x_2 = -1$		$4x_1 + 2x_2 =$	14
$2 - 3$	-1	$4 \cdot 2 + 2 \cdot 3$	14
-1		$8 + 6$	
		14	

We see that $(2, 3)$ is a solution of both equations, so it is a solution of the system.

DO EXERCISES 96 AND 97.

OBJECTIVES

You should be able to

a) Decide whether an ordered pair is a solution of a system of equations.
b) Graph a system of equations and classify it as consistent or inconsistent, dependent or independent.
c) Solve a system of equations graphically.

Decide whether $(-2, 1)$ is a solution of each system.

96.

$$x_1 + x_2 = -1$$
$$-3x_1 - x_2 = 5$$

97.

$$2x_1 - x_2 = -5$$
$$3x_1 + 2x_2 = 3$$

The graph of each equation in a system is a line. Given two lines, the following can happen:

1. The lines have no point in common—they are parallel.

2. The lines have exactly one point in common.

3. The lines are the same—they have infinitely many points in common.

A system of two linear equations in two variables is:

Consistent **if it has one or more solutions.**
Inconsistent **if it has** *no* **solution.**
Linearly dependent **if it is possible to multiply one equation by a constant and obtain the other.**
Linearly independent **if it is not possible to multiply one equation by a constant and obtain the other equation.**

Let us look at the three possibilities for lines to intersect, and describe the system in terms of the preceding terminology.

Example 2   The graph of the system

$$x_2 = 2x_1 - 1$$
$$x_2 = 2x_1 + 1$$

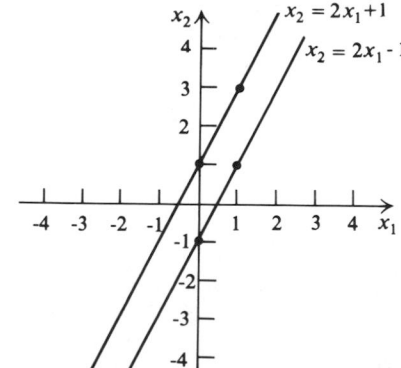

is on page 48. Note that the lines are parallel. Thus the system has no solution—it is inconsistent. There is no way to obtain one equation from the other by multiplying by a constant, so the system is independent.

Example 3 The graph of the system

$$x_2 = 2x_1 - 1$$
$$3x_2 = 6x_1 - 3$$

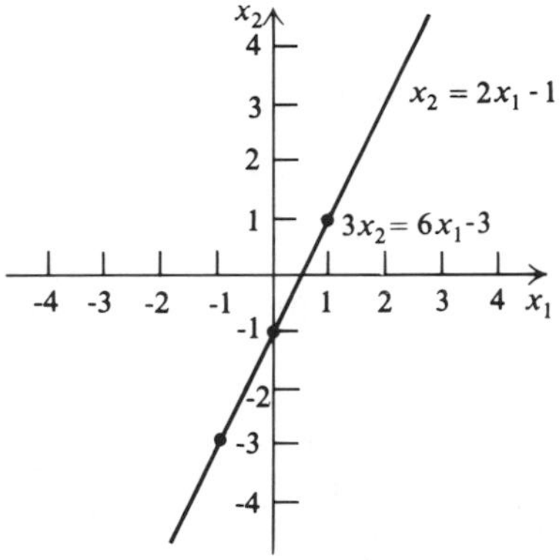

is shown above. Note that the lines are the same. Thus the system has infinitely many solutions—it is consistent. We obtain the second equation from the first by multiplying by 3. Thus the system is dependent.

Example 4 The graph of the system

$$x_1 - 2x_2 = 0$$
$$-2x_1 + x_2 = 2$$

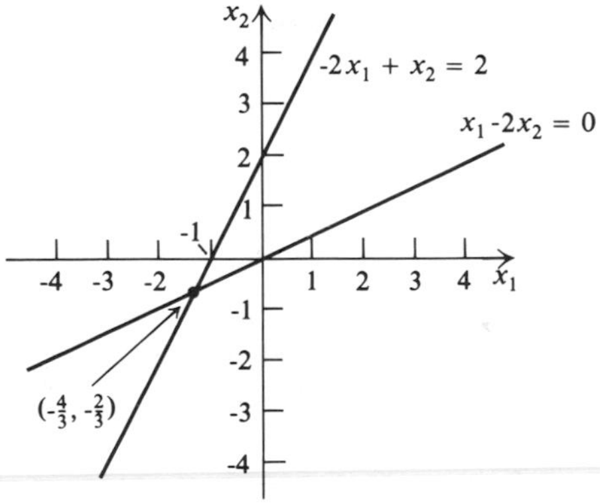

Graph each system and classify it as consistent or inconsistent, dependent or independent.

98. $2x_1 - x_2 = 1$
$-6x_1 + 3x_2 = -3$

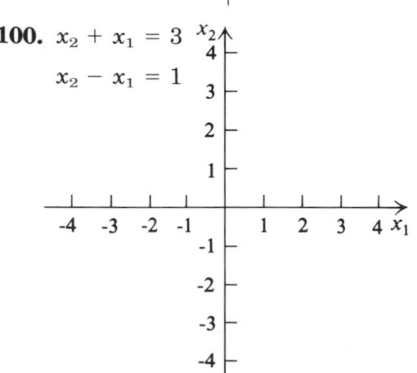

99. $x_1 - 4x_2 = -4$
$-x_1 + 4x_2 = 8$

100. $x_2 + x_1 = 3$
$x_2 - x_1 = 1$

is on page 49. Note that the lines intersect at exactly one point—it is consistent. There is no way to obtain one equation from the other by multiplying by a constant, so the system is independent.

DO EXERCISES 98 THROUGH 100.

Graphical Solution of Systems of Linear Equations

We can solve systems of equations graphically.

Example 5 Solve graphically.

$$x_2 - x_1 = 1$$
$$x_2 + x_1 = 5$$

Solution We graph the two equations.

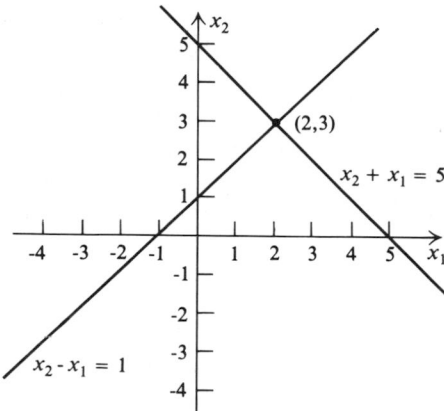

The point of intersection appears to be $(2, 3)$. We can check this as follows.

$x_2 - x_1 = 1$		$x_2 + x_1 = 5$	
$3 - 2$	1	$3 + 2$	5
1		5	

Thus, the solution is $(2, 3)$. Note that this procedure is subject to error, especially when fractional solutions are involved (see Example 4). Algebraic procedures will be developed in the next section and in Chapter 2 for obtaining exact answers.

DO EXERCISE 101.

101. a) Solve the system in Margin exercise 100, graphically.

b) Solve graphically

$$x_2 + x_1 = 0$$
$$2x_1 - x_2 = -6$$

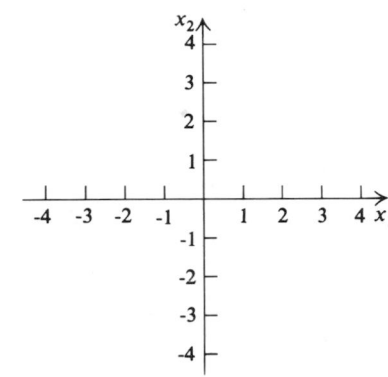

EXERCISE SET 1.5

Decide whether $(3, -2)$ is a solution of each system

1. $x_1 + x_2 = 1$
$\quad x_1 - x_2 = 6$

2. $2x_1 + x_2 = 4$
$\quad x_1 - 2x_2 = 7$

Graph each system and classify it as consistent or inconsistent, dependent or independent.

3. $x_1 + x_2 = 1$
$\quad x_1 - x_2 = 6$

4. $x_2 - 2x_1 = 1$
$\quad x_2 - 2x_1 = 3$

5. $2x_1 + 3x_2 = 1$
$\quad -x_1 - 1.5x_2 = -\frac{1}{2}$

6. $2x_1 - 4x_2 = 8$
$\quad -\frac{1}{2}x_1 + x_2 = -2$

7. $x_1 + 3x_2 = 4$
$\quad x_1 + 3x_2 = 6$

8. $2x_1 + x_2 = 4$
$\quad x_1 - 2x_2 = 7$

Solve graphically

9. $x_2 + 3x_1 = 5$
$\quad 2x_2 - x_1 = -4$

10. $2x_1 - x_2 = 4$
$\quad 5x_1 - x_2 = 13$

11. $2x_1 - 4x_2 = 7$
$\quad x_1 - 2x_2 = 5$

12. $3x_2 - 6x_1 = 10$
$\quad x_2 - 2x_1 = -1$

13. $x_2 - 4 = 0$
$\quad x_1 + 5 = 0$

14. $x_1 + 3 = 0$
$\quad x_2 - 1 = 0$

1.6 SYSTEMS OF EQUATIONS IN TWO VARIABLES—ALGEBRAIC SOLUTION

Here we consider algebraic procedures for finding exact solutions to systems of equations.

OBJECTIVE

You should be able to solve a system of equations using the addition method.

102. Solve, using the substitution method

$$-2x_1 + x_2 = 2$$
$$x_1 + x_2 = 6$$

103. Solve using the substitution method

$$4x_2 + 3x_1 = 4$$
$$x_2 + 2x_1 = 6$$

The Substitution Method

SUBSTITUTION METHOD. **Solve one of the equations for one of the variables. Then substitute the resulting expression in the other equation and solve for the second variable.**

Example 1 Solve

$$x_1 - 2x_2 = 0 \tag{1}$$
$$-2x_1 + x_2 = 2 \tag{2}$$

Solution Since Eq. (2) has x_2 with coefficient 1, it is easiest to solve that equation for x_2.

$$x_2 = 2 + 2x_1 \tag{3}$$

Now we substitute $2 + 2x_1$ for x_2 in Eq. (1).

$$x_1 - 2(2 + 2x_1) = 0$$

We now have an equation in one variable, x_1. We solve for x_1 using the addition and multiplication principles.

$$x_1 - 4 - 4x_1 = 0$$
$$-4 - 3x_1 = 0$$
$$-3x_1 = 4$$
$$x_1 = -\tfrac{4}{3}$$

We now substitute $-\tfrac{4}{3}$ for x_1 in Eq. (3) to find x_2. We could also substitute in either of the original equations, but it is faster to use Eq. (3) since x_2 has coefficient 1 on one side.

$$x_2 = 2 + 2(-\tfrac{4}{3}) = 2 - \tfrac{8}{3} = -\tfrac{2}{3}$$

The solution is $(-\tfrac{4}{3}, -\tfrac{2}{3})$. The reader should check this.

Always check! It is very easy to do with systems of linear equations.

DO EXERCISES 102 and 103.

The Addition Method

The *addition method* for solving systems of equations makes use of the *addition* and *multiplication principles* for solving equations. The idea, just as with substitution, is to obtain an equation with one variable.

Example 2 Solve

$$3x_2 + 5x_1 = 17 \tag{1}$$
$$2x_2 - 5x_1 = 3 \tag{2}$$

Solution   We add the equations as follows.

$$3x_2 + 5x_1 = 17$$

$$2x_2 - 5x_1 = 3$$

$$5x_2 \qquad = 20 \qquad \text{(Adding the "sides" of the equa-}$$
$$5x_2 = 20 \qquad\qquad \text{tions, using the addition principle)}$$

$$x_2 = 4$$

We substitute in Eq. (1) to find x_1 (we could use Eq. (2) also).

$$3 \cdot 4 + 5x_1 = 17$$

$$12 + 5x_1 = 17$$

$$5x_1 = 5$$

$$x_1 = 1$$

The solution is $(1, 4)$. The reader should check this.

DO EXERCISE 104.

Note in Example 2 that the term $5x_1$ of Eq. (1) and the term $-5x_1$ of Eq. (2) add to 0. Thus when we added we eliminated x_1. In the following examples, we first multiply one or both of the equations in order to create a situation like Example 2.

Example 3   Solve

$$9x_1 - 2x_2 - -4 \qquad\qquad (1)$$

$$3x_1 + 4x_2 = 1 \qquad\qquad (2)$$

Solution   We first multiply Eq. (1) by 2, then add.

$$18x_1 \quad 4x_2 = -8 \qquad \text{(Multiplying by 2, using the}$$
$$3x_1 + 4x_2 = 1 \qquad\qquad \text{multiplication principle)}$$

$$21x_1 \qquad = -7 \qquad \text{(Adding, using the addition}$$
$$21x_1 = -7 \qquad\qquad \text{principle)}$$

$$x_1 = \frac{-7}{21} = -\frac{1}{3}$$

We substitute $-\frac{1}{3}$ for x_1 in Eq. (2) and solve for x_2.

$$3(-\tfrac{1}{3}) + 4x_2 = 1$$

$$-1 + 4x_2 = 1$$

$$4x_2 = 2$$

$$x_2 = \tfrac{2}{4} = \tfrac{1}{2}$$

The solution is $(-\frac{1}{3}, \frac{1}{2})$.

104. Solve

$$-3x_2 + 2x_1 = 0$$

$$3x_2 - 4x_1 = -1$$

105. Solve

$$8x_1 - 3x_2 = -31$$
$$2x_1 + 6x_2 = 26$$

106. Solve

$$3x_1 + 2x_2 = -1$$
$$4x_1 + 3x_2 = 2$$

DO EXERCISE 105.

Example 4   Solve

$$3x_1 + 5x_2 = 7 \tag{1}$$
$$5x_1 + 3x_2 = -23 \tag{2}$$

Solution   We multiply Eq. (1) by 5 and Eq. (2) by -3.

$$15x_1 + 25x_2 = 35 \qquad \text{(Multiplying by 5)}$$
$$\underline{-15x_1 - 9x_2 = 69} \qquad \text{(Multiplying by -3)}$$
$$16x_2 = 104$$
$$x_2 = \tfrac{104}{16} = \tfrac{13}{2}$$

We substitute $\tfrac{13}{2}$ for x_2 in Eq. (1) and solve for x_1.

$$3x_1 + 5(\tfrac{13}{2}) = 7$$
$$3x_1 + \tfrac{65}{2} = 7$$
$$3x_1 = 7 - \tfrac{65}{2}$$
$$3x_1 = \tfrac{14}{2} - \tfrac{65}{2}$$
$$3x_1 = -\tfrac{51}{2}$$
$$x_1 = (-\tfrac{51}{2})\tfrac{1}{3} = -\tfrac{17}{2}$$

The solution is $(\tfrac{13}{2}, -\tfrac{17}{2})$.

DO EXERCISE 106.

Example 5   Solve

$$4x_1 + 6x_2 = -8 \tag{1}$$
$$-2x_1 - 3x_2 = 4 \tag{2}$$

Solution   We multiply Eq. (2) by 2 and add

$$4x_1 + 6x_2 = -8$$
$$\underline{-4x_1 - 6x_2 = 8} \qquad \text{(Multiplying by 2)}$$
$$0 = 0 \qquad \text{(Adding)}$$

We get the true equation $0 = 0$. This will happen for any ordered pair that is a solution of one of the equations. Thus we have an infinite number of solutions. If we had multiplied Eq. (2) by -2, we would have gotten Eq. (1), which is another way of verifying that we have an infinite number of solutions, because the system is linearly dependent.

Example 6   Solve

$$3x_2 + x_1 = 10$$
$$6x_2 + 2x_1 = 23$$

Solution   We multiply Eq. (1) by -2 and add

$$-6x_2 - 2x_1 = -20 \qquad \text{(Multiplying by } -2)$$
$$\underline{6x_2 + 2x_1 = 23}$$
$$0 = 3 \qquad \text{(Adding)}$$

Since we get the false equation $0 = 3$, the system has no solution. We could check this by graphing the system—the lines would be *parallel*.

DO EXERCISES 107 AND 108.

The substitution and addition methods, when applied correctly, will yield the solution(s). Overall, the fastest method is usually the addition method, which is the basis for other procedures that we will develop later in the text. For this reason, it is better to practice using it more than the substitution method.

107. Solve

$$5x_2 + 3x_1 = 14$$
$$10x_2 + 6x_1 = 29$$

108. Solve

$$2x_1 - 6x_2 = -10$$
$$3x_1 - 9x_2 = -15$$

EXERCISE SET 1.6

Solve, using the substitution method

1. $x_2 + 4x_1 = 5$
$-3x_2 + 2x_1 = 13$

2. $4x_2 + x_1 = 8$
$5x_2 + 3x_1 = 3$

3. $5x_1 + x_2 = 8$
$3x_1 - 4x_2 = 14$

4. $2x_1 - 3x_2 = 8$
$4x_1 + x_2 = 2$

Solve, using the addition method

5. $3x_1 + 5x_2 = 28$
$5x_1 - 3x_2 = 24$

6. $4x_1 + 3x_2 = 17$
$6x_1 + 5x_2 = 27$

7. $5x_1 - 4x_2 = -3$
$7x_1 + 2x_2 = 6$

8. $-2x_1 + 4x_2 = 3$
$3x_1 - 7x_2 = 1$

9. $4x_1 + 2x_2 = 11$
$3x_1 - x_2 = 2$

10. $5x_1 - 3x_2 = -2$
$4x_1 + 2x_2 = 5$

11. $9x_1 - 2x_2 = 5$
$3x_1 - 3x_2 = 11$

12. $3x_1 + 4x_2 = 7$
$-5x_1 + 2x_2 = 10$

13. $3x_2 - 6x_1 = 15$
$4x_2 - 8x_1 = 20$

14. $8x_1 + 4x_2 = 20$
$6x_1 + 3x_2 = 14$

15. $5x_1 + 10x_2 = 20$
$2x_1 + 4x_2 = 9$

16. $2x_1 - 4x_2 = 8$
$5x_1 - 10x_2 = 20$

17. Eight times a certain number added to five times a second number is 184. The first number minus the second number is -3. Find the numbers.

18. One number is 4 times another number. Their sum is 175. Find the numbers.

19. *Business.* One day a business sold 20 pairs of gloves. The cloth gloves brought $4.95 per pair and the pigskin gloves sold for $7.50 per pair. The business took in $137.25. How many of each kind were sold?

20. *Business.* A store sold 30 sweatshirts. They sold white ones for $8.95 and red ones for $9.50. They took in $272.90. How many of each color did they sell?

21. *Biomedical.* Solution A is 2% alcohol. Solution B is 6% alcohol. A lab technician wants to mix the two to get 60 liters of a solution that is 3.2% alcohol. How many liters of each should the owner use?

22. *Agriculture.* A gardener has two kinds of solutions containing weedkiller and water. One is 5% weedkiller and the other is 15% weedkiller. The gardener needs 100 liters of a 12% solution and wants to make it by mixing. How much of each solution should be used?

Recall the formula for simple interest $I = Prt$, where I is interest, P is principal, r is rate, and t is time in years.

23. *Business.* Two investments are made totaling $4800. In the first year they yield $604 in simple interest. Part of the money is invested at 12% and the rest at 13%. Find the amount invested at each rate of interest.

24. *Business.* For a certain year $9500 is received in interest from two investments. A certain amount is invested at 13% and $10,000 more than this is invested at 14%. Find the amount invested at each rate.

25. A boat travels 46 km downstream in 2 hr. It travels 51 km upstream in 3 hr. Find the speed of the boat and the speed of the stream.

26. An airplane travels 3000 km with a tail wind in 3 hr. It travels 3000 km with a head wind in 4 hr. Find the speed of the airplane and the speed of the wind.

▶ ───

Solve for (x, y)

27. $\sqrt{2}x + \pi y = 3$
$\pi x - \sqrt{2}y = 1$

28. $ax - by = a^2$
$bx + ay = ab$

The symbol ▦ indicates an exercise or problem facilitated by use of a calculator.

Solve

29. ▦ $4.026x - 1.448y = 18.32$
$0.724y = -9.16 + 2.013x$

30. ▦ $4.83x + 9.06y = -39.42$
$-1.35x + 6.67y = -33.99$

───

OBJECTIVES

You should be able to

a) Graph a given function.
b) Convert from radical notation to fractional exponents, and from fractional exponents to radical notation.
c) Determine the domain of a rational function.
d) Find the equilibrium point, given a demand and supply function.

1.7 OTHER TYPES OF FUNCTIONS

Quadratic Functions

A *quadratic function f* is given by

$$f(x) = ax^2 + bx + c, \text{where } a \neq 0.$$

We have already considered some such functions, for example $f(x) = x^2$ and $g(x) = x^2 - 1$. Graphs of quadratic functions are always cup-shaped, like those in Example 1. Each has a dashed line of symmetry.

Example 1 Graph $y = x^2 - 2x - 3$ and $y = -2x^2 + 4x + 1$.

Solutions $y = x^2 - 2x - 3$ $y = -2x^2 + 4x + 1$

x	0	1	2	3	4	-1	-2
y	-3	-4	-3	0	5	0	5

x	0	1	2	3	-1
y	1	3	1	-5	-5

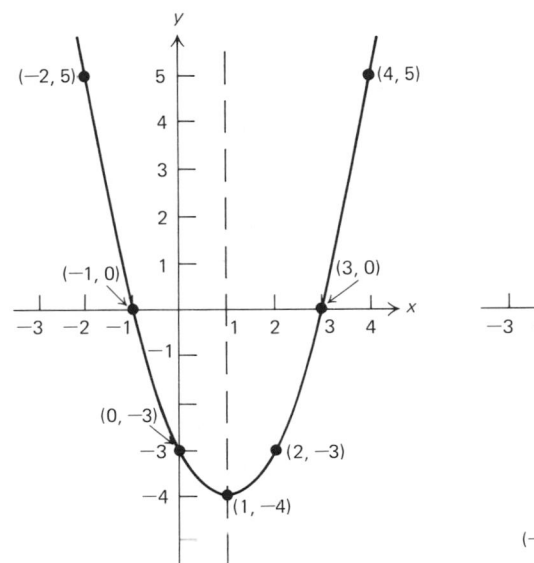

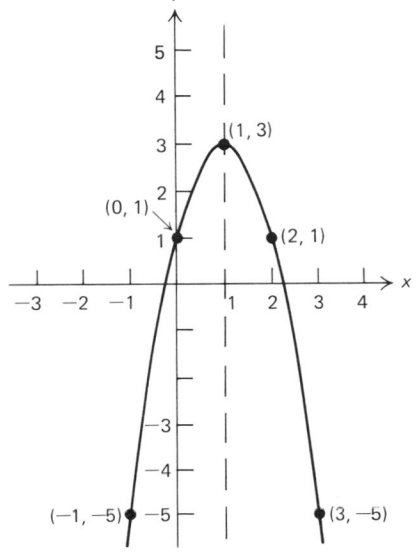

If the coefficient a is positive, the graph opens upward. If a is negative, the graph opens downward.

DO EXERCISES 109 AND 110.

First coordinates of points where a quadratic function intersects the x-axis (x-intercepts), if they exist, can be found by solving the quadratic equation $ax^2 + bx + c = 0$. If real number solutions exist, they can be found using the quadratic formula.

THE QUADRATIC FORMULA. The solutions of any quadratic equation $ax^2 + bx + c = 0$, $a \neq 0$, are given by

$$x = \frac{-b \pm \sqrt{b^2 - 4ac}}{2a}$$

Example 2   Solve. $3x^2 - 4x = 2$

109. Using the same axes, graph $y = x^2$ and $y = -x^2$. [*Note*: $-x^2$ means $-1 \cdot x^2$].

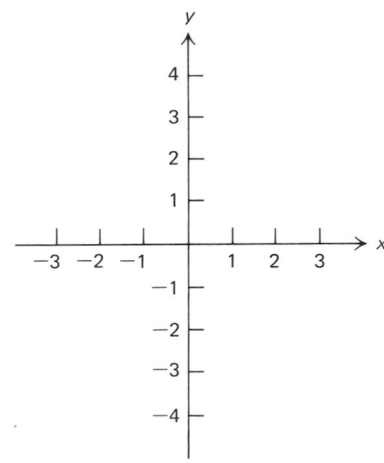

110. a) Using the same axes, graph $y = x^2$ and $y = (x - 6)^2$.

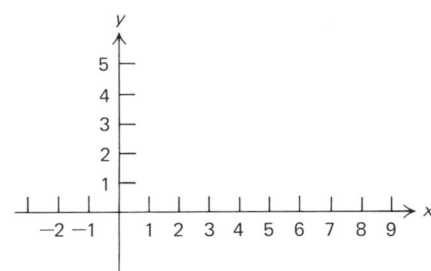

b) How could the graph of $y = (x - 6)^2$ be obtained from the graph of $y = x^2$?

111. Solve. $3x^2 = 7 - 2x$

Solution First find standard form $ax^2 + bx + c = 0$, and determine a, b, and c.

$$3x^2 - 4x - 2 = 0,$$
$$a = 3, \quad b = -4, \quad c = -2$$

Then use the quadratic formula

$$x = \frac{-b \pm \sqrt{b^2 - 4ac}}{2a} = \frac{-(-4) \pm \sqrt{(-4)^2 - 4(3)(-2)}}{2 \cdot 3}$$

$$= \frac{4 \pm \sqrt{16 + 24}}{6} = \frac{4 \pm \sqrt{40}}{6}$$

$$= \frac{4 \pm \sqrt{4 \cdot 10}}{6} = \frac{4 \pm 2\sqrt{10}}{6} = \frac{2(2 \pm \sqrt{10})}{2 \cdot 3} = \frac{2 \pm \sqrt{10}}{3}.$$

The solutions are $\dfrac{2 + \sqrt{10}}{3}$ and $\dfrac{2 - \sqrt{10}}{3}$.

DO EXERCISE 111.

Polynomial Functions

Linear and quadratic functions are part of a general class of polynomial functions.

A *polynomial function f* is given by

$$f(x) = a_n x^n + a_{n-1} x^{n-1} + \cdots + a_2 x^2 + a_1 x^1 + a_0,$$

where n **is a nonnegative integer, and** a_n, a_{n-1}, ..., a_1, a_0, **are real numbers.**

The following are some examples:

$$\begin{array}{ll} f(x) = -5 & \text{(A constant function)} \\ f(x) = 4x + 3 & \text{(A linear function)} \\ f(x) = -x^2 + 2x + 3 & \text{(A quadratic function)} \\ f(x) = 2x^3 - 4x^2 + x + 1 & \text{(A cubic function)} \end{array}$$

In general, graphing polynomial functions other than linear and quadratic is difficult. We use calculus to sketch such graphs in Section 3.3. Some *power* functions, such as

$$y = ax^n,$$

are relatively easy to graph.

Example 3 Using the same set of axes, graph $y = x^2$ and $y = x^3$.

Solution

x	-2	-1	$-\frac{1}{2}$	0	$\frac{1}{2}$	1	2
x^2	4	1	$\frac{1}{4}$	0	$\frac{1}{4}$	1	4
x^3	-8	-1	$-\frac{1}{8}$	0	$\frac{1}{8}$	1	8

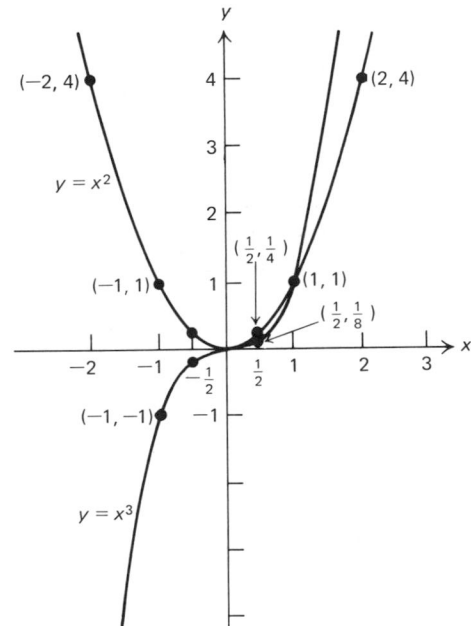

DO EXERCISE 112.

Rational Functions

Functions given by the ratio of two polynomials are called *rational*.

The following are examples of rational functions.

$$f(x) = \frac{x^2 - 9}{x - 3}, \qquad g(x) = \frac{x^2 - 16}{x + 4}, \qquad h(x) = \frac{x - 3}{x^2 - x - 2}$$

The domain of a rational function is restricted to those input values that do not result in division by 0. Thus for f the domain consists of all real numbers except 3. To determine the domain of h, we set the denominator equal to 0 and solve.

$$x^2 - x - 2 = 0,$$
$$(x + 1)(x - 2) = 0,$$
$$x = -1 \quad \text{or} \quad x = 2.$$

112. Using the same set of axes, graph $y = x^2$ and $y = x^4$.

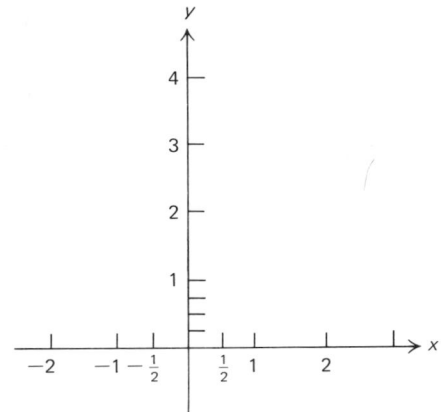

113. Determine the domain of each function.

a) $g(x) = \dfrac{x^2 - 16}{x + 4}$

b) $t(x) = \dfrac{x + 7}{x^2 + 4x - 5}$

c) $k(x) = \dfrac{1}{x - 5}$

114. Graph $y = \dfrac{-1}{x}$.

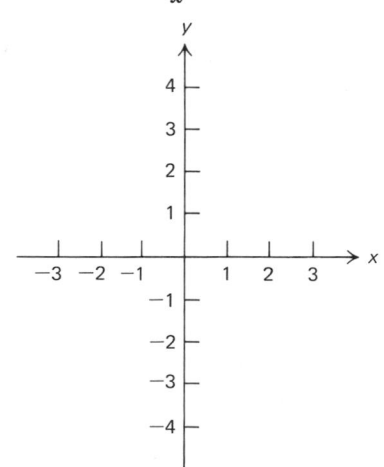

Thus -1 and 2 are not in the domain. The domain consists of all real numbers except -1 and 2.

DO EXERCISE 113.

One important class of rational functions is given by $y = \dfrac{k}{x}$.

Example 4 Graph $y = \dfrac{1}{x}$.

Solution

x	-3	-2	-1	$-\frac{1}{2}$	$-\frac{1}{4}$	$\frac{1}{4}$	$\frac{1}{2}$	1	2	3
y	$-\frac{1}{3}$	$-\frac{1}{2}$	-1	-2	-4	4	2	1	$\frac{1}{2}$	$\frac{1}{3}$

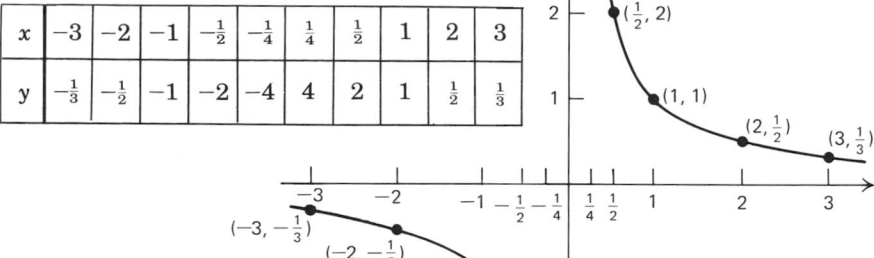

Note that 0 is not in the domain of this function since it would yield a 0 denominator. This function is decreasing over the intervals $(-\infty, 0)$ and $(0, \infty)$. It is an example of inverse variation. That is,

y varies inversely **as x if there is some positive number k such that $y = k/x$. We also say that y is *inversely proportional* to x.**

DO EXERCISE 114.

Absolute Value Functions

The following is an example of an absolute value function and its graph. The absolute value of a number is its distance from 0. We denote the absolute value of a number x as $|x|$.

Example 5 Graph $y = |x|$.

Solution

x	-3	-2	-1	0	1	2	3
y	3	2	1	0	1	2	3

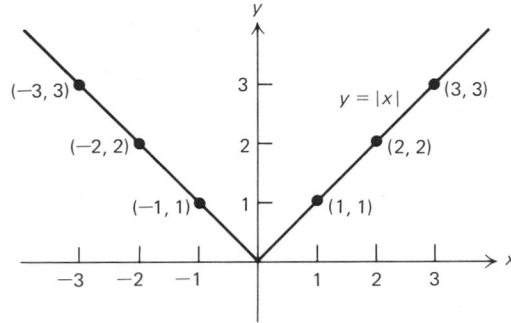

DO EXERCISE 115.

Square Root Functions

The following is an example of a square root function and its graph.

Example 6 Graph $y = -\sqrt{x}$.

Solution The domain of this function is just the nonnegative numbers—the interval $[0, \infty)$. Table 1 at the back of the book contains approximate values of square roots of certain numbers.

x	0	1	2	3	4	5	10
$-\sqrt{x}$	0	-1	-1.4	-1.7	-2	-2.2	-3.2

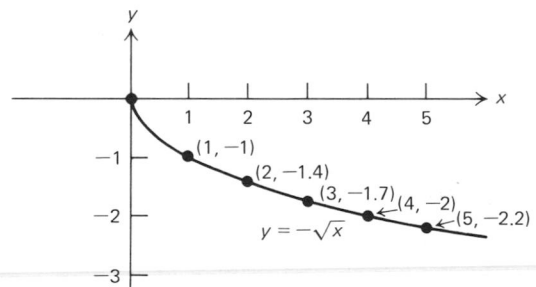

115. Graph $y = |x - 1|$. To find an output, take an input, subtract 1 from it, and then take the absolute value.

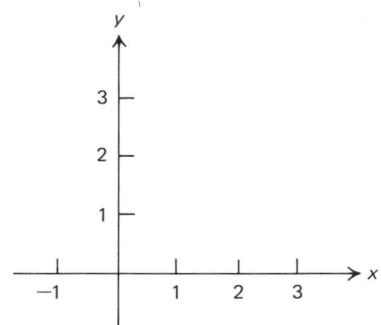

116. Graph $y = \sqrt{x}$.

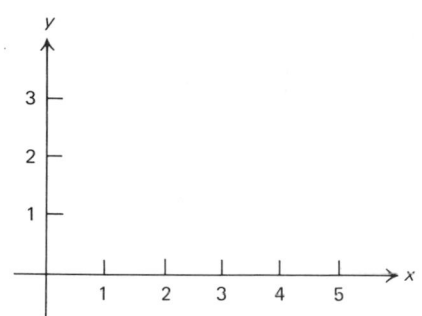

117. Find the domain of $y = \sqrt{2x + 3}$
[*Hint*: Solve $2x + 3 \geqslant 0$.]

118. Convert to fractional exponents.

a) $\sqrt[4]{t^3}$ b) $\sqrt[5]{y}$

c) $\dfrac{1}{\sqrt[5]{x^2}}$ d) $\dfrac{1}{\sqrt[3]{t}}$

e) $\sqrt{x^6}$ f) $\sqrt{x^7}$

119. Convert to radical notation.

a) $y^{1/7}$ b) $x^{3/2}$

c) $t^{-3/2}$ d) $b^{-1/2}$

DO EXERCISES 116 AND 117.

Power Functions with Fractional Exponents

We are motivated to define fractional exponents so that the same laws of Section 1.1 still hold. For example, if the laws of exponents are to hold, we would have

$$a^{1/2} \cdot a^{1/2} = a^{1/2+1/2} = a^1 = a.$$

Thus we are led to define $a^{1/2}$ to be $\sqrt{a}$. Similarly, we are led to define $a^{1/3}$ to be the cube root of a, $\sqrt[3]{a}$. In general,

$$a^{1/n} = \sqrt[n]{a}.$$

Again, if the laws of exponents are to hold, we would have

$$\sqrt[n]{a^m} = (a^m)^{1/n} = (a^{1/n})^m = a^{m/n}.$$

An expression $a^{-m/n}$ is defined by

$$a^{-m/n} = \frac{1}{a^{m/n}} = \frac{1}{\sqrt[n]{a^m}}.$$

Examples   Convert to fractional exponents.

a) $\sqrt[3]{x^2} = x^{2/3}$ b) $\sqrt[4]{y} = y^{1/4}$

c) $\dfrac{1}{\sqrt[3]{b^5}} = \dfrac{1}{b^{5/3}} = b^{-5/3}$ d) $\dfrac{1}{\sqrt{x}} = \dfrac{1}{x^{1/2}} = x^{-1/2}$

e) $\sqrt{x^8} = x^{8/2}$, or x^4

DO EXERCISE 118.

Examples   Convert to radical notation.

a) $x^{1/3} = \sqrt[3]{x}$ b) $t^{6/7} = \sqrt[7]{t^6}$

c) $x^{-2/3} = \dfrac{1}{x^{2/3}} = \dfrac{1}{\sqrt[3]{x^2}}$ d) $e^{-1/4} = \dfrac{1}{e^{1/4}} = \dfrac{1}{\sqrt[4]{e}}$

DO EXERCISE 119.

Examples   Simplify.

a) $8^{5/3} = (8^{1/3})^5 = (\sqrt[3]{8})^5 = 2^5 = 32$

b) $81^{3/4} = (81^{1/4})^3 = (\sqrt[4]{81})^3 = 3^3 = 27$

DO EXERCISES 120 AND 121.

Earlier when we graphed $y = \sqrt{x}$, we were also graphing $y = x^{1/2}$, or $y = x^{0.5}$. The power functions

$$y = ax^k, \quad k \text{ fractional,}$$

do arise in application. For example, the *home range* of an animal is defined as the region to which it confines its movements. It has been hypothesized in statistical studies* that the area H of that region can be approximated using the body weight W of an animal by the function

$$H = W^{1.41}.$$

W	0	10	20	30	40	50
H	0	26	68	121	182	249

Note that

$$H = W^{1.41} = W^{141/100} = \sqrt[100]{W^{141}}.$$

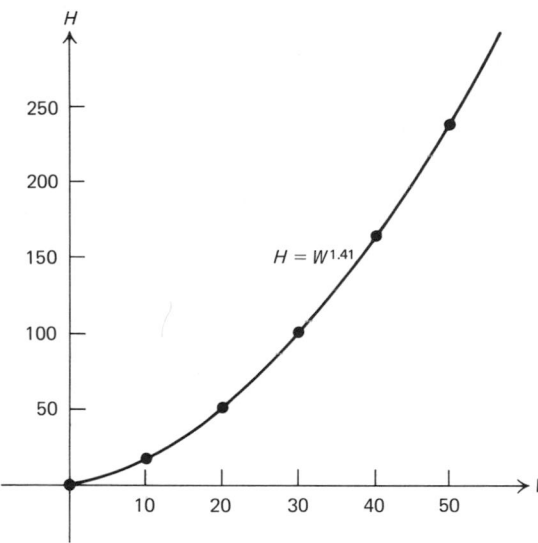

$H = W^{1.41}$

In the graph shown here, note that this is an increasing function. As body weight increases, the area over which the animal moves increases. It would be hard, however, to find accurate function values unless we used a calculator; we are simply illustrating that such functions do have application.

Supply and Demand Functions

Supply and demand in economics are modeled by increasing and decreasing functions. While specific scientific formulas for these concepts are not usually known, the notions of increasing and decreasing yield understanding of the ideas.

---

*See J. M. Emlen. *Ecology: An Evolutionary Approach*. Reading, Mass.: Addison-Wesley, 1973, p. 200.

Simplify.

120. $4^{5/2}$

121. $27^{2/3}$

Demand Functions. Look at the following table.

Demand Schedule	
Quantity (x) (number of 5-lb bags in millions)	Price (p) (per bag)
4	$5
5	4
7	3
10	2
15	1

The table shows the relationship between the price p per bag of sugar and the quantity x of 5-lb bags that the consumer will buy at that price. Note that as price per bag increases, the quantity demanded by the consumer decreases; and as price per bag decreases, the quantity demanded by the consumer increases. Thus it is natural to think of x as a function of p. In our later work it will be more convenient to think of p as a function of x. Thus, for a *demand* function, D, $D(x)$ is the price per unit of an item when x units are demanded by the consumer.

The following figure is the graph of a demand function for sugar (using the preceding table).

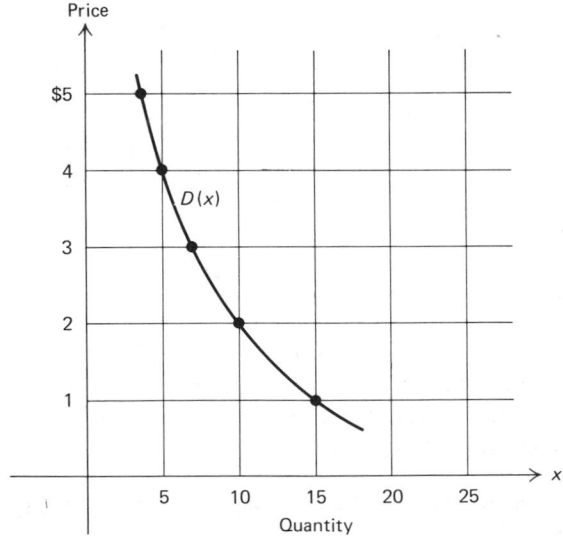

Supply Functions.   Look at the following table.

	Supply Schedule	
Quantity (x) (number of 5-lb bags in millions)	Price (p) (per bag)
24	$5
20	4
15	3
10	2
0	1

The table shows the relationship between the price p per bag of sugar and the quantity x of 5-lb bags that the seller is willing to supply at that price. Note that as the price per bag increases, the more the seller is willing to supply; and as the price per bag decreases, the less the seller is willing to supply. Again, it is natural to think of x as a function of p, but for our later work it is more convenient to think of p as a function of x. Thus, for a *supply* function S, $S(x)$ is the price per unit of an item at which the seller is willing to supply x units of a product to the consumer. The following figure is the graph of a supply function for sugar (using the preceding table).

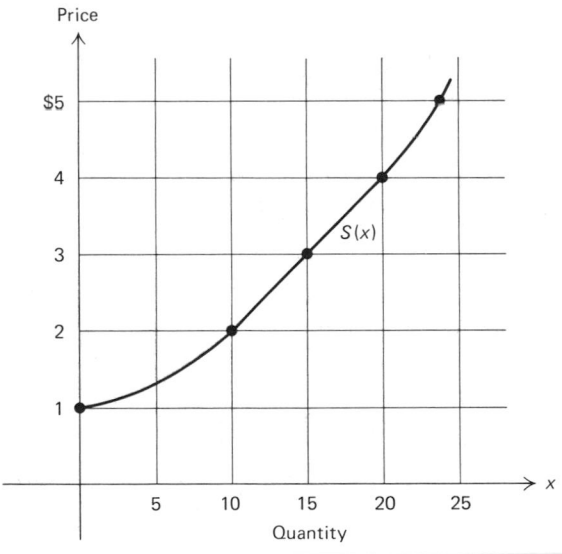

Let us now look at these curves together. Note that as supply increases demand decreases, and as supply decreases demand increases. The point of intersection of the two curves (x_E, p_E) is called the *equilibrium point*. The equilibrium price, p_E (in this case $2 per bag), is where the amount, x_E (in this case 10 million bags), which the seller willingly supplies is the same as the amount which the consumer willingly demands. The situation is analogous to a buyer and seller haggling over the sale of an item. The equilibrium point or selling price is what they finally agree on.

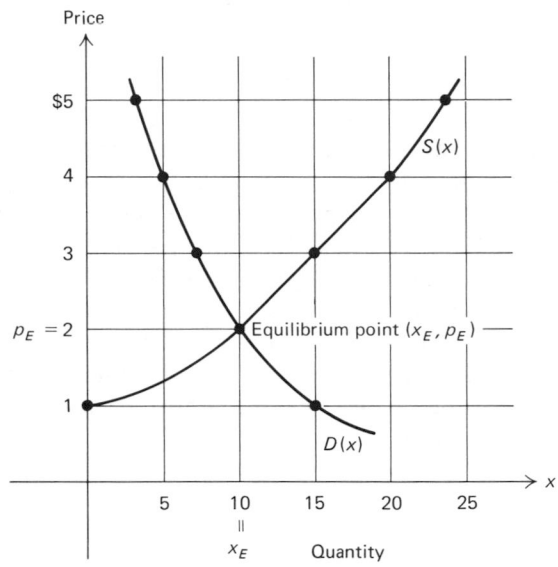

Example 7 Find the equilibrium point for the demand and supply functions

$$D(x) = (x - 6)^2 \qquad \text{and} \qquad S(x) = x^2 + x + 10.$$

Solution To find the equilibrium point we set $D(x) = S(x)$ and solve.

$$(x - 6)^2 = x^2 + x + 10,$$
$$x^2 - 12x + 36 = x^2 + x + 10,$$
$$-12x + 36 = x + 10,$$
$$-13x = -26,$$
$$x = \frac{-26}{-13},$$
$$x = 2.$$

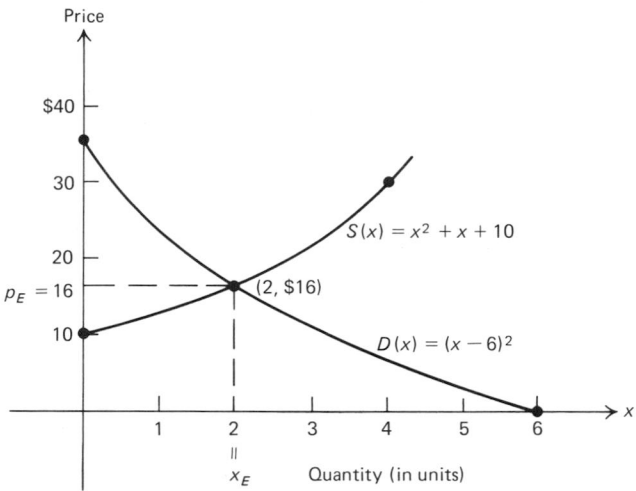

122. Given $D(x) = (x - 5)^2$ and $S(x) = x^2 + x + 3$, find the equilibrium point.

Thus $x_E = 2$(units). To find p_E we substitute x_E into either $D(x)$ or $S(x)$. We use $D(x)$. Then

$$p_E = D(x_E) = D(2) = (2 - 6)^2 = (-4)^2 = 16.$$

Thus the equilibrium price is \$16 per unit and the equilibrium point is (2, \$16).

DO EXERCISE 122.

EXERCISE SET 1.7

Using the same set of axes, graph each pair of equations.

1. $y = \frac{1}{2}x^2$, $y = -\frac{1}{2}x^2$

2. $y = \frac{1}{4}x^2$, $y = -\frac{1}{4}x^2$

3. $y = x^2$, $y = (x - 1)^2$

4. $y = x^2$, $y = (x - 3)^2$

5. $y = x^2$, $y = (x + 1)^2$

6. $y = x^2$, $y = (x + 3)^2$

7. $y = |x|$, $y = |x + 3|$

8. $y = |x|$, $y = |x + 1|$

9. $y = x^3$, $y = x^3 + 1$

10. $y = x^3$, $y = x^3 - 1$

11. $y = \sqrt{x}$, $y = \sqrt{x + 1}$

12. $y = \sqrt{x}$, $y = \sqrt{x - 2}$

Graph.

13. $y = x^2 - 4x + 3$

14. $y = x^2 - 6x + 5$

15. $y = -x^2 + 2x - 1$

16. $y = -x^2 - x + 6$

17. $y = \dfrac{2}{x}$

18. $y = \dfrac{3}{x}$

19. $y = \dfrac{-2}{x}$

20. $y = \dfrac{-3}{x}$

21. $y = \dfrac{1}{x^2}$

22. $y = \dfrac{1}{x - 1}$

23. $y = \dfrac{1}{|x|}$

24. $y = \sqrt[3]{x}$ [*Hint:* 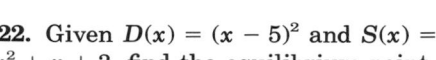 or use Table 1 at the end of the book]

Solve.

25. $x^2 - 2x = 2$

26. $x^2 - 2x + 1 = 5$

27. $x^2 + 6x = 1$

28. $x^2 + 4x = 3$

29. $4x^2 = 4x + 1$

30. $-4x^2 = 4x - 1$

31. $3y^2 + 8y + 2 = 0$

32. $2p^2 - 5p = 1$

Convert to fractional exponents.

33. $\sqrt{x^3}$

34. $\sqrt{x^5}$

35. $\sqrt[5]{a^3}$

36. $\sqrt[4]{b^2}$

37. $\sqrt[7]{t}$

38. $\sqrt[8]{c}$

39. $\dfrac{1}{\sqrt[3]{t^4}}$

40. $\dfrac{1}{\sqrt[5]{b^6}}$

41. $\dfrac{1}{\sqrt{t}}$

42. $\dfrac{1}{\sqrt{m}}$

43. $\dfrac{1}{\sqrt{x^2 + 7}}$

44. $\sqrt{x^3 + 4}$

Convert to radical notation.

45. $x^{1/5}$

46. $t^{1/7}$

47. $y^{2/3}$

48. $t^{2/5}$

49. $t^{-2/5}$

50. $y^{-2/3}$

51. $b^{-1/3}$

52. $b^{1/5}$

53. $e^{-17/6}$

54. $m^{-19/6}$

55. $(x^2 - 3)^{-1/2}$

56. $(y^2 + 7)^{-1/4}$

Simplify.

57. $9^{3/2}$

58. $16^{5/2}$

59. $64^{2/3}$

60. $8^{2/3}$

61. $16^{3/4}$

62. $25^{5/2}$

Determine the domain of each function.

63. $f(x) = \dfrac{x^2 - 25}{x - 5}$

64. $f(x) = \dfrac{x^2 - 4}{x + 2}$

65. $f(x) = \dfrac{x^3}{x^2 - 5x + 6}$

66. $f(x) = \dfrac{x^4 + 7}{x^2 + 6x + 5}$

67. $f(x) = \sqrt{5x + 4}$

68. $f(x) = \sqrt{2x - 6}$

Find the equilibrium point for the following demand and supply functions.

69. $D(x) = -2x + 8, S(x) = x + 2$

70. $D(x) = -\frac{5}{6}x + 10, S(x) = \frac{1}{2}x + 2$

71. $D(x) = (x - 3)^2, S(x) = x^2 + 2x + 1$

72. $D(x) = (x - 4)^2, S(x) = x^2 + 2x + 6$

73. $D(x) = (x - 4)^2, S(x) = x^2$

74. $D(x) = (x - 6)^2, S(x) = x^2$

75. (▦—with y^x key.) *Biology*—The *territory area* of an animal is defined to be its defended region, or exclusive region. For example, a lion has a certain region over which it is ruler. It has been hypothesized in statistical studies* that the area, T, of that region is approximated using body weight, W, by the power function

$$T = W^{1.31}$$

W	0	10	20	30	40	50	100	150
T	0	20						

Complete the table of approximate function values and graph the function.

* Emlen, *Op. cit.*, p. 57.

1.8 MATHEMATICAL MODELING

What Is a Mathematical Model?

When the essential parts of a problem situation are described in mathematical language, we say that we have a *mathematical model*. For example, the arithmetic of the natural numbers constitutes a mathematical model for situations in which counting is the essential ingredient. Situations in which calculus can be brought to bear often require the use of equations and functions, and typically there is concern with the way a change in one variable effects a change in another.

OBJECTIVES

You should be able to use curve fitting to find a model for a set of data, then use the model to make predictions. Use the model for the world record in the mile run.

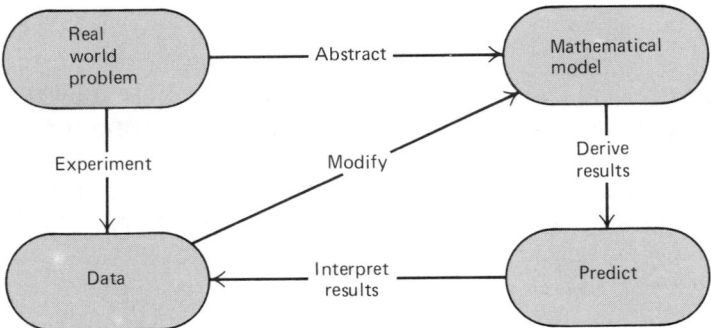

Mathematical models are abstracted from real-world situations. Procedures within the mathematical model then give results that allow one to predict what will happen in that real-world situation. To the extent that these predictions are inaccurate or the results of experimentation do not conform to the model, the model is in need of modification. This is shown in the diagram.

The diagram seems to indicate that mathematical modeling is an ongoing, possibly ever-changing, process. This is often the case. For example, finding a mathematical model that will enable accurate prediction of population growth is not a simple problem. Surely any population model one might devise will have to be altered as further relevant information is acquired.

While models can reveal worthwhile information, one must always be cautious in their use. An interesting case in point is a study* which showed that world records in *any* running race can be modeled by a linear function. In particular, for the mile run,

$$R = -0.00582x + 15.3476,$$

where R is the world record in minutes, and x is the year. Roger

---

* H. W. Ryder, H. J. Carr, and P. Herget, "Future Performance in Footracing," *Scientific American*, 234 (June 1976): 109–119.

123. What will the world record be in 1988?

124. When will the world record be 3:48.0? [*Hint:* 3:48.0 = 3.80]

Bannister shocked the world in 1954 by breaking the 4-minute mile. Had people been aware of this model they would not have been shocked, for when we substitute 1954 for x we get

$$R = -0.00582(1954) + 15.3476 = 3.97532 \approx 3{:}58.5$$

The actual record was 3:59.4. While this model will continue for 40 to 50 years to be worthwhile in predicting the world record in the mile, upon using some common sense, we see that we can't get meaningful answers to some questions. For example, we could use the model to find when the 1-minute mile will be broken. We set $R = 1$ and solve for x:

$$1 = -0.00582x + 15.3476$$
$$2465 = x$$

Most track people would assure us that the 1-minute mile is beyond human capability. In fact, at the time of this writing, experienced runners think it will never reach 3:40:0, the current world record being 3:49. Going to even further extreme, the model predicts that the 0-minute mile will be run in 2637. In conclusion, one must be careful in the use of any model. (You will develop this model in Exercise 7 of Exercise Set 12.4.)

DO EXERCISES 123 AND 124.

On p. 66 we saw an example of a mathematical model utilizing supply and demand functions. In general, the idea is to find a function that fits observations and theoretical reasoning (including common sense) as well as possible. Later in Section 12.4, we will see how calculus can be used to develop and analyze models. For now we will consider one type of modeling procedure using a somewhat over-simplified procedure that we call *curve fitting*.

Curve Fitting

The following four functions fit many situations.

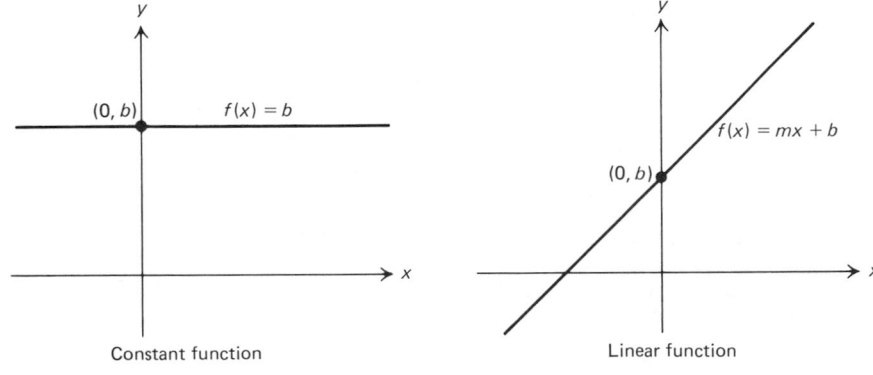

Constant function

Linear function

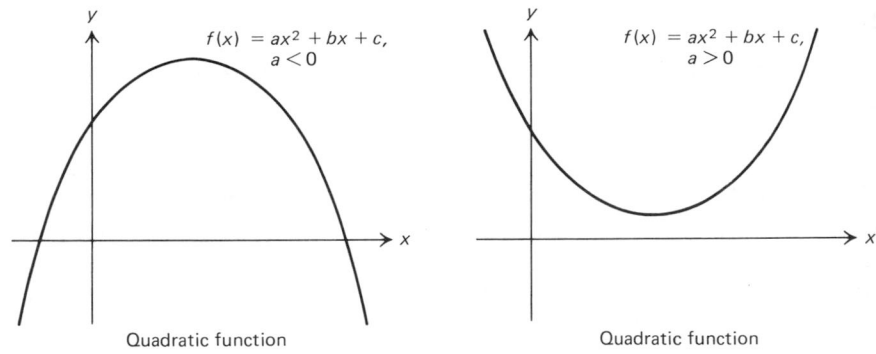

Quadratic function   Quadratic function

The following is a procedure that sometimes works for finding mathematical models.

CURVE FITTING.   **Given a set of data,**
1. Graph the data.
2. Look at the data and determine whether a known function seems to fit.
3. Find a function which fits the data by using data points to derive the constants.

The following problem is based on factual data.

Example 1 *Business—Taxes from each dollar earned.* From the given set of data, find a model; then use the model to find that part of each dollar earned in 1985 that will go for taxes.

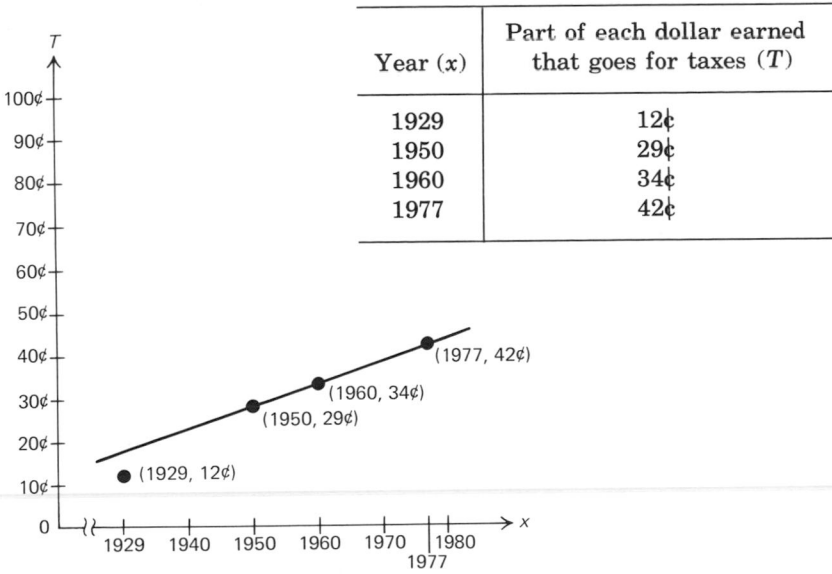

Year (x)	Part of each dollar earned that goes for taxes (T)
1929	12¢
1950	29¢
1960	34¢
1977	42¢

125. Rework Example 1, but this time use the data points $(1950, 29\cancel{c})$ and $(1960, 34\cancel{c})$. How does the answer to (b) compare to (b) in Example 1?

Solution

a) The graph is on page 71. It looks as though a linear function fits this data fairly well,

$$T = mx + b. \tag{1}$$

To derive the constants (or parameters) m and b, we pick two data points. This is a subjective matter, but, of course, you should not pick a point that deviates greatly from the general pattern. We pick the points $(1950, 29\cancel{c})$ and $(1977, 42\cancel{c})$. Since the points are to be solutions of Eq. (1) it follows that

$$42 = m \cdot 1977 + b \quad \text{or} \quad 42 = 1977m + b, \tag{2}$$

$$29 = m \cdot 1950 + b \quad \text{or} \quad 29 = 1950m + b. \tag{3}$$

This is a system of equations. We subtract Eq. (3) from Eq. (2) to get rid of b:

$$13 = 27m.$$

Then

$$\tfrac{13}{27} = m.$$

Substituting $\tfrac{13}{27}$ for m in Eq. (3), we get

$$29 = 1950 \cdot \tfrac{13}{27} + b$$
$$29 = 938.9 + b \quad \text{(We estimate } 1950 \cdot \tfrac{13}{27})$$
$$-909.9 = b.$$

Substituting these values of m and b in Eq. (1), we get the function (model) given by

$$T = \tfrac{13}{27}x - 909.9.$$

Since we are only interested in estimates, we use an approximation for $\tfrac{13}{27}$ and get

$$T = 0.481x - 909.9. \tag{4}$$

b) That part of the earned dollar that will go for taxes in 1985 is found by letting $x = 1985$ in Eq. (4).

$$T = 0.481(1985) - 909.9 \approx 45\cancel{c}.$$

DO EXERCISE 125.

To repeat, curve fitting is an oversimplified method of finding models. Other techniques such as Least Squares (Section 12.4) and computer simulations are used for more thorough research.

Example 2   For the given set of factual data, find a model, then use the model to determine the cost per mile to drive at 55 mph, 70 mph. How much more does it cost to drive a car at 70 mph than 55 mph?

Speed (x) of car (mph)	Operating cost (C) (cents per mile)
15	15.0
20	12.0
25	10.5
45	12.0

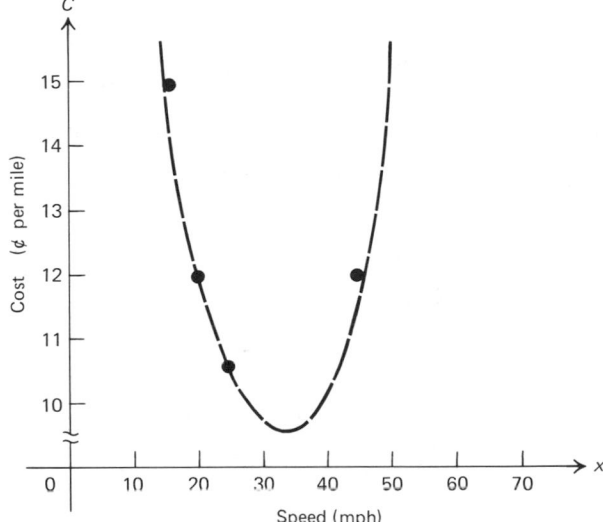

Solution

a) The graph is shown above. It looks as if a quadratic function might fit,

$$C = ax^2 + bx + c. \tag{1}$$

To derive the constants (or parameters) a, b, and c, we pick three data points $(15, 15)$, $(20, 12)$, and $(45, 12)$. Since these three points are to be solutions of Eq. (1), it follows that

$$15 = a \cdot 15^2 + b \cdot 15 + c, \quad \text{or} \quad 15 = 225a + 15b + c, \tag{2}$$
$$12 = a \cdot 20^2 + b \cdot 20 + c, \quad \text{or} \quad 12 = 400a + 20b + c, \tag{3}$$
$$12 = a \cdot 45^2 + b \cdot 45 + c, \quad \text{or} \quad 12 = 2025a + 45b + c. \tag{4}$$

126.

Age (x) of driver in years	Number (A) of daytime accidents committed by driver of age x
20	420
40	150
60	210
70	400

a) For the given set of data, graph the data and find a quadratic function that fits the data. Use the data points $(20, 420)$, $(40, 150)$, and $(70, 400)$. First give exact fractional values for $a, b,$ and c. Then give decimal values rounded to the nearest thousandth.

b) How many daytime accidents are committed by a driver of age 16?

This is a system of three equations in three unknowns. We subtract Eq. (3) from Eq. (4) to get rid of c:

$$0 = 1625a + 25b. \tag{5}$$

We subtract Eq. (3) from Eq. (2) which also gets rid of c:

$$3 = -175a - 5b. \tag{6}$$

Now we solve the system, Eqs. (5) and (6), which has just two unknowns. We multiply Eq. (6) by 5 and add.

$$0 = 1625a + 25b$$
$$\underline{15 = -875a - 25b}$$
$$15 = 750a$$

Thus

$$a = \frac{15}{750} = 0.02.$$

Then substituting 0.02 for a in Eq. (5) we solve for b.

$$0 = 1625(0.02) + 25b$$
$$0 = 32.5 + 25b$$
$$-32.5 = 25b$$
$$\frac{-32.5}{25} = b$$
$$-1.3 = b$$

Then substituting 0.02 for a and -1.3 for b in Eq. (3) we solve for c.

$$12 = 400(0.02) + 20(-1.3) + c$$
$$12 = 8 - 26 + c$$
$$30 = c$$

Substituting these values of $a, b,$ and c in Eq. (1) we get the function (model) given by

$$C = 0.02x^2 - 1.3x + 30. \tag{7}$$

b) The cost per mile at 55 mph is given by

$$C = 0.02(55^2) - 1.3(55) + 30 = 19¢.$$

The cost per mile at 70 mph is given by

$$C = 0.02(70^2) - 1.3(70) + 30 = 37¢.$$

Thus it costs 18¢ per mile more to drive at 70 mph than at 55 mph.

DO EXERCISE 126.

EXERCISE SET 1.8

For each exercise, (a) and (b) are as follows:
a) Graph the data given.
b) Look at the data and determine whether one of the six models discussed in the text of this section seems to fit.

1. *Business.* Raggs, Ltd. keeps track of its total costs of producing x items of a certain suit. These data are shown below.

Number of suits (x)	Total cost (C) of producing x suits
0	$10,000
1	10,030
2	10,059
3	10,094

c) Use the data points $(1, \$10,030)$ and $(3, \$10,094)$ to find a linear function that fits the data.
d) Predict the total cost of producing 4 suits, 10 suits.
e) Use the data points $(0, \$10,000)$ and $(2, \$10,059)$ to find a linear function which fits the data.
f) Use the model of (e) to predict the total cost of producing 4 suits, 10 suits.

2. *Business.* Pizza, Unltd. keeps track of its total costs of producing x pizzas. These data are shown below.

Number of pizzas (x)	Total cost (C) of producing x pizzas
0	$1000
1	1001
2	1001.80
3	1002.50

c) Use the data points $(1, \$1001)$ and $(3, \$1002.50)$ to find a linear function that fits the data.
d) Predict the total cost of producing 4 pizzas, 100 pizzas.
e) Use the data points $(0, \$1000)$ and $(2, \$1001.80)$ to find a linear function which fits the data.
f) Use the model of (e) to predict the total cost of producing 4 pizzas, 100 pizzas.

⊞ The problems in Exercises 3 and 4 are based on factual data.

3.

Travel speed (x) in mph	Number (D) of vehicles involved in an accident in daytime (for every 100 million miles of travel)
20	10,000
30	1,000
40	200
50	150
60	95
70	90
80	190

c) Use the data points $(30, 1000)$, $(50, 150)$, and $(70, 90)$ to find a quadratic function which fits the data.
d) Use the model to find the number of vehicles involved in an accident at 60 mph. Check this with the data.

4.

Travel speed (x) in mph	Number (N) of vehicles involved in an accident in nighttime (for every 100 million miles of travel)
20	10,000
30	2,000
40	400
50	250
60	250
70	350
80	1,500

c) Use the data points $(20, 10,000)$, $(50, 250)$, and $(80, 1500)$ to find a quadratic function fitting the data.
d) Use the model to find the number of vehicles involved in an accident at 80 mph. Check this with the data.

5. *Business.*

Year (t)	Total Sales (S) in dollars
1	$100,310
2	100,290
3	100,305
4	100,280

c) Use the data points (1, $100,310) and (2, $100,290) to find a linear function that fits the data.

d) This data set approximates a constant function. What procedure, apart from that of (c), could you use to find the constant?

7. *Biomedical.* This problem is based on a study by Dr. Harold J. Morowitz.

Average number of hours of sleep (x)	Death rate per 100,000 males (y)
5	1121
7	626
9	967

c) Use the given data points to find a quadratic function that fits the data.

d) Use the model to find the death rate of males who sleep 4 hr, 6 hr, and 10 hr.

6. *Business.*

Year (t)	Sales (S) in dollars
1	$10,000
2	21,000
3	27,000
4	37,000

c) Use the data points (1, $10,000) and (4, $37,000) to find a linear function that fits the data.

d) Predict the sales of the company in the 5th year.

8. *Biomedical.* This problem is based on factual data.

Year (x)	Estimated employment of medical assistants (E)
1972	200,000
1985	320,000

c) Use the given data points to find a linear function that fits the data.

d) Use the model to estimate the employment of medical assistants in 1990.

CHAPTER 1 TEST

1. Rename without a negative exponent. e^{-k}

2. Divide. $\dfrac{e^{-5}}{e^{8}}$

3. Multiply. $(x + h)^2$

4. Factor. $25x^2 - t^2$

5. A person makes an investment at 8% compounded annually. It grows to $993.60 at the end of 1 year. How much was originally invested?

6. Solve. $-3x < 12$.

7. A function is given by $f(x) = x^2 - 4$. Find

a) $f(-3)$.   b) $f(x + h)$.

8. What is the slope and y-intercept of $y = -3x + 2$?

9. Find an equation of the line with slope $\frac{1}{4}$, containing the point $(8, -5)$.

10. Find the slope of the line containing the points $(-2, 3)$ and $(-4, -9)$.

11. The weight, F, of fluids in a human is directly proportional to body weight, W. It is known that a person who weighs 180 lb has 120 lb of fluids. Find an equation of variation expressing F as a function of W.

12. A record company has fixed costs of $10,000 for producing a record master. Thereafter, the variable costs are $0.50 per record for duplicating from the record master. Revenue from each record is expected to be $1.30.

a) Formulate a function $C(x)$ for the total cost of producing x records.
b) Formulate a function $R(x)$ for the total revenue from the sale of x records.
c) Formulate a function $P(x)$ for the total profit from the production and sale of x records.
d) How many records must the company sell to break even?

13. Decide whether this function is increasing, decreasing, or neither.

$$f(x) = -0.2x + 7$$

14. Find the equilibrium point for the demand and supply functions

$$D(x) = (x - 7)^2 \quad \text{and} \quad S(x) = x^2 + x + 4.$$

15. Graph. $y = \dfrac{4}{x}$

16. Convert to fractional exponents. $\dfrac{1}{\sqrt{t}}$

17. Convert to radical notation. $t^{-3/5}$

Determine the domain of each function.

18. $f(x) = \dfrac{x^2 + 20}{(x - 2)(x + 7)}$.

19. $f(x) = \sqrt{5x + 10}$

20. Find a linear function that fits the data points $(1, 3)$ and $(2, 7)$.

21. Find a quadratic function that fits the data points $(1, 5)$, $(2, 9)$, and $(3, 4)$.

22. Write interval notation for this graph.

23. Graph this system and classify it as consistent or inconsistent, dependent or independent

$$5x_1 - 15x_2 = -10$$
$$3x_1 - 9x_2 = 6$$

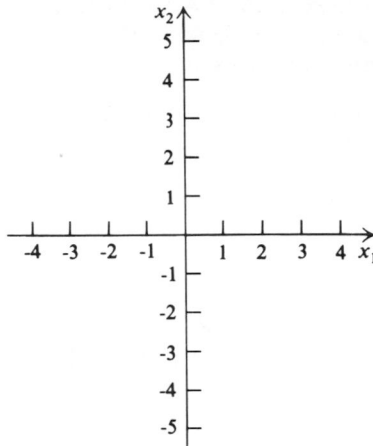

24. Decide whether $(-2, 3)$ is a solution of the system

$$x_1 + x_2 = 1$$
$$3x_1 + x_2 = -3$$

25. Solve graphically

$$x_1 - x_2 = 4$$
$$x_1 + x_2 = 2$$

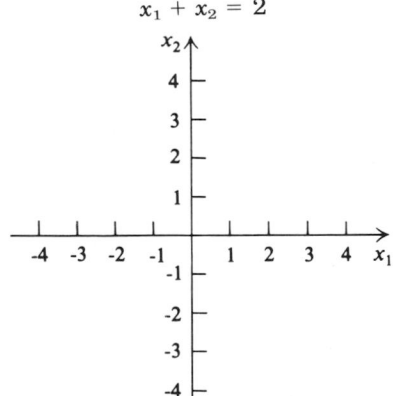

26. Solve, using the addition method

$$5x_1 + 4x_2 = 24$$
$$2x_1 - 3x_2 = 5$$

27. Solve, using the addition method

$$3x_1 - x_2 = 2$$
$$4x_1 + 2x_2 = 11$$

28. Solve, using the addition method

$$5x_1 + 10x_2 = 15$$
$$3x_1 + 6x_2 = 9$$

CHAPTER TWO

Equations
and the
Echelon Method—Matrices

OBJECTIVE

You should be able to solve systems of equations using the echelon method.

1. Translate this tableau to the corresponding system of equations.

x_1	x_2	1
2	−6	$\frac{1}{4}$
−3	1	8

2. Translate this system to the corresponding echelon tableau.

$$2x_1 + 5x_2 = -24$$
$$5x_1 - 2x_2 = -2$$

2.1 THE ECHELON METHOD—UNIQUE SOLUTIONS

In your experience with solving a system such as

$$2x_1 + 6x_2 = 26$$
$$8x_1 - 3x_2 = -31,$$

you may have noticed that one actually works with the coefficients, or constants, and not the variables. It is helpful to just list the constants in what is called an *echelon* tableau:*

x_1	x_2	1
2	6	26
8	−3	−31

The *rows* of a tableau are horizontal. The *columns* are vertical.

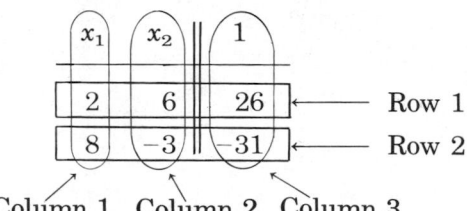

Column 1   Column 2   Column 3

Note that the first column is labeled with the variable x_1, the second with the variable x_2, and the third with a 1. The double vertical lines correspond to "=." To see how to translate from a tableau to the corresponding system, we multiply the elements of each row by the column headings and add:

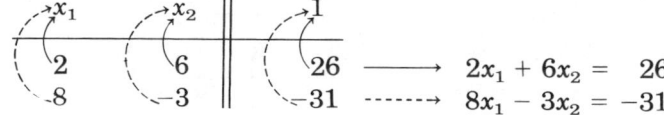

$$2x_1 + 6x_2 = 26$$
$$8x_1 - 3x_2 = -31$$

DO EXERCISES 1 AND 2.

The echelon method† for solving systems uses certain operations that correspond to operations on equations. We will carry out these operations in a way that may seem odd at first, but the reason for it will become apparent later. Before we formalize the method, let us consider an example. Compare each operation with the corresponding operation with the equations.

---

* The word "echelon" is a French word meaning "a series of steps."
† Short for *reduced row echelon method.*

Example 1 Solve

$$2x_1 + 6x_2 = 26$$
$$8x_1 - 3x_2 = -31$$

Solution

Addition Method   *Echelon Method*

x_1	x_2	1
2	6	26
8	-3	-31

$$2x_1 + 6x_2 = 26$$
$$8x_1 - 3x_2 = -31$$

Our first goal is to get a 1 in the first row and first column.

1. Multiply the first equation by $\frac{1}{2}$:

1. Multiply the first row by $\frac{1}{2}$:

$$x_1 + 3x_2 = 13$$
$$8x_1 - 3x_2 = -31$$

x_1	x_2	1
1*	3	13
8	-3	-31

In the tableau we put a star (*) on the 1. It is called a *pivot* element. Our goal will always be to get 0's in the rest of a column where a pivot element occurs. To do this here, we multiply the first (pivot) row by −8 and add.

2. Multiply the first equation by −8 and add to the second. We leave the first equation as is: we are simply adding a multiple of it to the second equation.

2. Multiply the first (pivot) row by −8 and add to the second. We leave the first row as is in the tableau. We are simply adding a multiple of the pivot row to the second row.

x_1	x_2	1
1	3	13
0	-27	-135

$$x_1 + 3x_2 = 13$$
$$-27x_2 = -135$$

Our next goal is to get a 1 in the second row and second column; this will be a new pivot element.

3. Multiply the second equation by $-\frac{1}{27}$:

3. Multiply the second row by $-\frac{1}{27}$:

$$x_1 + 3x_2 = 13$$
$$x_2 = 5$$

x_1	x_2	1
1	3	13
0	1*	5

We put a star on the 1 to indicate that this is the new pivot element. Remember that we always want to get 0's in the rest of a column where a pivot element occurs. To do this here, we multiply the second (pivot) row by -3 and add.

4. Multiply the second equation by -3 and add to the first:

$$x_1 = -2$$
$$x_2 = 5$$

4. Multiply the second (pivot) row by -3 and add to the first:

x_1	x_2	1
1	0	-2
0	1	5

The solution is $(-2, 5)$. We can obtain this directly from the tableau by translating to the corresponding system of equations shown on the left.

THE ECHELON METHOD. In carrying out the echelon method, we obtain pivot elements of 1 diagonally from upper left to lower right. Then we get 0's in the rest of each column by adding multiples of the pivot row to the other rows. We use any of these operations in carrying out this procedure:

  i) **Interchange any two rows.**
 ii) **Interchange any two variable columns, provided we interchange the headings.**
iii) **Multiply any row by a nonzero constant.**
 iv) **Add a multiple of the pivot row to another row.**

It is important to note that we can add *rows* only. We cannot add columns since this would mean adding unlike terms in the equations.

Example 2 Solve using the echelon method.

$$3x_1 - 4x_2 = -1$$
$$-3x_1 + 2x_2 = 0$$

Solution

1. We first obtain a tableau.

x_1	x_2	1
3	-4	-1
-3	2	0

2. We obtain the pivot element 1 in the first row and first column (upper left) by multiplying the first row by $\frac{1}{3}$.

x_1	x_2	1
1^*	$-\frac{4}{3}$	$-\frac{1}{3}$
-3	2	0

3. Next we *pivot*; that is, we obtain 0's in the rest of the first column. To do this we multiply the first (pivot) row by 3 and add to the second.

x_1	x_2	1
1	$-\frac{4}{3}$	$-\frac{1}{3}$
0	-2	-1

Based on your experience with the addition method, you may have been tempted at the outset to just add the first row to the second. While this is not incorrect, remember that we are proceeding in a special way to anticipate and ease later work.

4. We obtain the next pivot element by multiplying the second row by $-\frac{1}{2}$.

x_1	x_2	1
1	$-\frac{4}{3}$	$-\frac{1}{3}$
0	1^*	$\frac{1}{2}$

5. We pivot again, multiplying the second (pivot) row by $\frac{4}{3}$ and adding to the first.

x_1	x_2	1
1	0	$\frac{1}{3}$
0	1	$\frac{1}{2}$

The solution is $(\frac{1}{3}, \frac{1}{2})$, found by translating from the tableau to the system of equations $x_1 = \frac{1}{3}$ and $x_2 = \frac{1}{2}$.

DO EXERCISES 3 AND 4.

Now let us solve a system with three variables.

Example 3 Solve, using the echelon method.

$$2x_1 - x_2 + x_3 = -1$$
$$x_1 - 2x_2 + 3x_3 = 4$$
$$4x_1 + x_2 + 2x_3 = 4$$

Solution
1. We first write the tableau:

x_1	x_2	x_3	1
2	-1	1	-1
1	-2	3	4
4	1	2	4

3. Solve, using the addition and echelon methods together, as in Example 1.

$$2x_1 + 5x_2 = -24$$
$$5x_1 - 2x_2 = -2$$

4. Solve, using the echelon method.

$$9x_1 - 2x_2 = -4$$
$$3x_1 + 4x_2 = 1$$

2. We obtain a pivot element 1 in the first row and first column. We could do this by multiplying the first equation by $\frac{1}{2}$, but this would introduce fractions. An easier way to get the pivot element is to interchange the first and second rows.

x_1	x_2	x_3	1
1*	−2	3	4
2	−1	1	−1
4	1	2	4

3. Next we pivot to obtain 0's in the rest of the first column. We first multiply the first (pivot) row by −2 and add to the second row.

x_1	x_2	x_3	1
1*	−2	3	4
0	3	−5	−9
4	1	2	4

4. To complete the pivot, we multiply the first (pivot) row by −4 and add to the third row. (We usually work down the column in this way.)

x_1	x_2	x_3	1
1	−2	3	4
0	3	−5	−9
0	9	−10	−12

5. To obtain the next pivot, element we multiply the second row by $\frac{1}{3}$.

x_1	x_2	x_3	1
1	−2	3	4
0	1*	$-\frac{5}{3}$	−3
0	9	−10	−12

6. We pivot on the starred 1. We first multiply the second (pivot) row by 2 and add to the first.

x_1	x_2	x_3	1
1	0	$-\frac{1}{3}$	−2
0	1*	$-\frac{5}{3}$	−3
0	9	−10	−12

7. To complete the pivot, we multiply the second (pivot) row by -9 and add to the third.

x_1	x_2	x_3	1
1	0	$-\frac{1}{3}$	-2
0	1	$-\frac{5}{3}$	-3
0	0	5	15

8. To obtain the last pivot element, we multiply the third row by $\frac{1}{5}$.

x_1	x_2	x_3	1
1	0	$-\frac{1}{3}$	-2
0	1	$-\frac{5}{3}$	-3
0	0	1^*	3

9. We pivot on the starred 1. We first multiply the third (pivot) row by $\frac{1}{3}$ and add to the first.

x_1	x_2	x_3	1
1	0	0	-1
0	1	$-\frac{5}{3}$	-3
0	0	1^*	3

10. To complete the pivot, we multiply the third (pivot) row by $\frac{5}{3}$ and add to the second.

x_1	x_2	x_3	1
1	0	0	-1
0	1	0	2
0	0	1	3

The solution is found by translating from the tableau to the system of equations

$$x_1 = -1,$$
$$x_2 = 2,$$
$$x_3 = 3.$$

We can also say that the solution is the ordered triple $(-1, 2, 3)$.

In an actual computation we would record only the tableaux of Steps 1, 4, 7, and 10. The intermediate steps are recorded here so that the

5. Solve, using the echelon method.

$$3x_1 + 2x_2 + 2x_3 = 3$$
$$2x_1 - 4x_2 + x_3 = 0$$
$$x_1 + 2x_2 - x_3 = 5$$

student can follow the details. The actual computation would look like this:

x_1	x_2	x_3	1
2	-1	1	-1
1	-2	3	4
4	1	2	4
1^*	-2	3	4
0	3	-5	-9
0	9	-10	-12
1	0	$-\frac{1}{3}$	-2
0	1^*	$-\frac{5}{3}$	-3
0	0	5	15
1	0	0	-1
0	1	0	2
0	0	1^*	3

DO EXERCISE 5.

Application

Example 4 *Interest problem* Two investments are made that total $4800. For a certain year, these investments yield $412 in simple interest. Part of the $4800 is invested at 8% and the other part at 9%. Find the amount invested at each rate.

Solution Recall the formula for simple interest.

$$I = Pit.$$

Interest I is principal P times rate i times time t.

a) Let x_1 represent the amount invested at 8% and x_2 the amount invested at 9%. Then the interest from x_1 is $8\%x_1$, and the interest from x_2 is $9\%x_2$. Thus the $412 total interest is given by

$$8\%x_1 + 9\%x_2 = 412,$$

or

$$0.08x_1 + 0.09x_2 = 412.$$

b) Considering the total amount invested, we have

$$x_1 + x_2 = 4800.$$

c) We now have a system of equations:

$$0.08x_1 + 0.09x_2 = \ \ 412$$
$$x_1 + x_2 = 4800$$

We translate this system to an echelon tableau and solve.

x_1	x_2	1
0.08	0.09	412
1	1	4800

We first multiply the first row by 100 to clear the decimals.

x_1	x_2	1
8	9	41200
1	1	4800

To get the pivot element 1 in the first row, first column, we inter-change the rows.

x_1	x_2	1
1*	1	4800
8	9	41200

We multiply the first row by -8 and add to the second.

x_1	x_2	1
1	1	4800
0	1*	2800

We already have a pivot element 1 in the second row, second column. We complete the pivot by multiplying the second row by -1 and adding to the first row.

x_1	x_2	1
1	0	2000
0	1	2800

6. Two investments are made that total $3700. For a certain year, these investments yield $297 in simple interest. Part of the $3700 is invested at 7% and the other part at 9%. Find the amount invested at each rate.

Thus the solution is

$$x_1 = 2000,$$
$$x_2 = 2800,$$

so $2000 is invested at 8% and $2800 is invested at 9%.

DO EXERCISE 6.

EXERCISE SET 2.1

Solve, using the echelon method. Interchange rows and/or columns to avoid fractions when possible.

1.
$$x_1 + 4x_2 = 5$$
$$-3x_1 + 2x_2 = 13$$

2.
$$x_1 + 4x_2 = 8$$
$$3x_1 + 5x_2 = 3$$

3.
$$-x_1 + 3x_2 = 2$$
$$2x_1 - x_2 = 11$$

4.
$$9x_1 - 2x_2 = 5$$
$$3x_1 - 3x_2 = 11$$

5.
$$2x_1 - 5x_2 = 10$$
$$4x_1 + 3x_2 = 7$$

6.
$$5x_1 - 3x_2 = -2$$
$$4x_1 + 2x_2 = 5$$

7.
$$x_1 + x_2 + x_3 = 9$$
$$x_1 - x_2 - x_3 = -15$$
$$x_1 + x_2 - x_3 = -5$$

8.
$$x_1 + x_2 + x_3 = 1$$
$$x_1 + 2x_2 + 3x_3 = 4$$
$$x_1 + 3x_2 + 7x_3 = 13$$

9.
$$x_1 - x_2 + 2x_3 = 0$$
$$x_1 - 2x_2 + 3x_3 = -1$$
$$2x_1 - 2x_2 + x_3 = -3$$

10.
$$x_1 + 2x_2 - 3x_3 = 9$$
$$2x_1 - x_2 + 2x_3 = -8$$
$$3x_1 - x_2 - 4x_3 = 3$$

11.
$$3x_1 + 2x_2 + 2x_3 = 3$$
$$2x_1 + 4x_2 - x_3 = 8$$
$$2x_1 - 4x_2 + x_3 = 0$$

12.
$$4x_1 - x_2 - 3x_3 = 1$$
$$8x_1 + x_2 - x_3 = 5$$
$$2x_1 + x_2 + 2x_3 = 5$$

13.
$$2x_1 - 3x_2 + x_3 - x_4 = -8$$
$$x_1 + x_2 - x_3 - x_4 = -4$$
$$x_1 - x_2 - x_3 - x_4 = -14$$
$$x_1 + x_2 + x_3 + x_4 = 22$$

14.
$$3x_1 - 2x_2 + 2x_3 + x_4 = -6$$
$$x_1 - x_2 + 4x_3 + 3x_4 = -2$$
$$x_1 + x_2 + x_3 + x_4 = -5$$
$$2x_1 + 2x_2 - 2x_3 - 2x_4 = -10$$

15. *Business.* Two investments are made that total $8800. For a certain year, these investments yield $663 in simple interest. Part of the $8800 is invested at 7% and part at 8%. Find the amount invested at each rate.

16. *Business.* Two investments are made that total $15,000. For a certain year, these investments yield $1432 in simple interest. Part of the $15,000 is invested at 9% and part at 10%. Find the amount invested at each rate.

17. *Business.* For a certain year $3900 is received in interest from two investments. A certain amount is invested at 5%, and $10,000 more than this is invested at 6%. Find the amount invested at each rate. *Hint.* Express each equation in standard form $ax_1 + bx_2 = c$.

18. *Business.* For a certain year $876 is received in interest from two investments. A certain amount is invested at 7%, and $1200 more than this is invested at 8%. Find the amount invested at each rate.

19. *Business.* One day a campus bookstore sold 30 sweatshirts. White ones cost $9.95 and yellow ones cost $10.50. In dollars, $310.60 worth of sweatshirts were sold. How many of each color were sold?

20. *Business.* One week a business sold 40 scarves. White ones cost $4.95 and designed ones cost $7.95. In dollars, $282 worth of scarves were sold. How many of each kind were sold?

21. *Biology—Nutrition.* Soybean meal is 16% protein; corn meal is 9% protein. How many pounds of each should be mixed together to get 350 pounds of a mixture that is 12% protein?

22. *Chemistry.* A chemist has one solution of acid and water that is 25% acid and a second that is 50% acid. How many gallons of each should be mixed together to get 10 gallons of a solution that is 40% acid?

23. *Business.* A person receives $212 per year in simple interest from three investments totalling $2500. Part is invested at 7%, part at 8%, and part at 9%. There is $1100 more invested at 9% than at 8%. Find the amount invested at each rate.

24. *Business.* A person receives $306 per year in simple interest from three investments totalling $3200. Part is invested at 8%, part at 9%, and part at 10%. There is $1800 more invested at 10% than at 8%. Find the amount invested at each rate.

25. *Curve fitting.* Find numbers a, b, and c, such that the data points $(1, 4)$, $(-1, -2)$, and $(2, 13)$ are solutions of the quadratic function

$$y = ax^2 + bx + c.$$

26. *Curve fitting.* Find numbers a, b, and c, such that the data points $(1, 4)$, $(-1, 6)$, and $(-2, 16)$ are solutions of the quadratic function

$$y = ax^2 + bx + c.$$

27. *Business—Predicting earnings.* A business earns $1000 in its first month, $2000 in the second month, and $8000 in the third month.
a) Find a quadratic function $y = ax^2 + bx + c$, which fits the data (see Exercise 25), where y = earnings, and x = month;
b) Use the model in (a) to predict the earnings for the fourth month.

28. *Biomedical—Death rate as a function of sleep.* (This problem is based on a study by Dr. Harold J. Morowitz.)

Average number of hours of sleep, x	Death rate per 100,000 males, y
5	1121
7	626
9	967

a) Use the given data points to find a quadratic function $y = ax^2 + bx + c$ that fits the data.
b) Use the model to find the death rate of males who sleep 4 hr; 6 hr; and 10 hr.

2.2 THE ECHELON METHOD—SPECIAL CASES

In the preceding section, each system had exactly one solution and the final tableau had a form like the following:

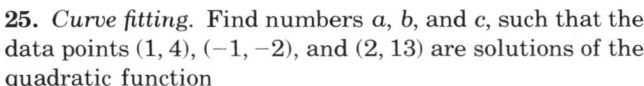

$$\begin{array}{ccc||c} x_1 & x_2 & x_3 & 1 \\ \hline 1 & 0 & 0 & p \\ 0 & 1 & 0 & q \\ 0 & 0 & 1 & r \end{array}$$

This is called the *reduced row echelon form*, or *echelon form*, for short. Here we consider special cases where systems have no solution, or infinitely many solutions, or the number of variables is not the same as the number of equations. All these cases can be analyzed in the same general way.*

* Homogeneous equations, that is, equations whose constant terms (righthand sides) are zero require *no* special consideration.

OBJECTIVE

You should be able to solve systems of equations using the echelon method for special cases where systems may have no solution, infinitely many solutions, or where the number of variables may not be the same as the number of equations.

7. Solve, using the echelon method.

$$4x_1 - 2x_2 = 2$$
$$2x_1 - x_2 = -8$$

Example 1 Solve

$$6x_1 - 3x_2 = 21$$
$$4x_1 - 2x_2 = 19$$

Solution We translate to the echelon tableau, multiply the first row by $\frac{1}{6}$, and obtain

x_1	x_2	1
1^*	$-\frac{1}{2}$	$\frac{7}{2}$
4	-2	19

We carry out the pivot by multiplying the pivot row by -4 and adding to the second row.

x_1	x_2	1
1	$-\frac{1}{2}$	$\frac{7}{2}$
0	0	5

The pivoting is considered to be complete here even though we do not have 1's down the main diagonal.

Earlier we found the solution by translating back to a system of equations. If we do that here, we obtain

$$x_1 - \tfrac{1}{2}x_2 = \tfrac{7}{2}$$
$$0 = 5$$

But the second equation is false. Thus the system is inconsistent. It has *no solution*.

DO EXERCISE 7.

Example 2 Solve

$$6x_1 - 3x_2 = 21$$
$$4x_1 - 2x_2 = 14$$

Solution We translate to the echelon tableau, multiply the first row by $\frac{1}{6}$, and obtain

x_1	x_2	1
1^*	$-\frac{1}{2}$	$\frac{7}{2}$
4	-2	14

We carry out the pivot by multiplying the pivot row by -4 and adding to the second row.

$$\begin{array}{cc|c} x_1 & x_2 & 1 \\ \hline 1 & -\frac{1}{2} & \frac{7}{2} \\ 0 & 0 & 0 \end{array}$$

Translating back to a system of equations we obtain

$$x_1 - \tfrac{1}{2}x_2 = \tfrac{7}{2}$$
$$0 = 0$$

We know from previous work that this system is consistent and the second equation is linearly dependent upon the first. The system has infinitely many solutions. Every point on the line $x_1 - \frac{1}{2}x_2 = \frac{7}{2}$ is a solution. We can describe this by solving the equation for x_1. Then the solutions can be described by

$$x_1 = \tfrac{7}{2} + \tfrac{1}{2}x_2, \qquad x_2 = \text{any number.}$$

By selecting arbitrary values of x_2 and computing x_1 we find the following as some solutions:

$$\begin{array}{lll} x_2 = 0 & x_2 = 1 & x_2 = -3 \\ x_1 = \frac{7}{2} & x_1 = \frac{7}{2} + \frac{1}{2} = 4 & x_1 = \frac{7}{2} - \frac{3}{2} = 2 \end{array}$$

DO EXERCISE 8.

Example 3 Suppose we are trying to solve a system with 3 equations and 4 variables, and we reach this stage in the tableau:

$$\begin{array}{cccc|c} x_1 & x_2 & x_3 & x_4 & 1 \\ \hline 1^* & 0 & -2 & 0 & 2 \\ 2 & 1 & 2 & 0 & 1 \\ 3 & 0 & -6 & 5 & 26 \end{array}$$

The pivot element is in the first row and first column. We carry out the pivoting by multiplying the first row by -2 and adding to the second and multiplying the first row by -3 and adding to the third. We obtain.

$$\begin{array}{cccc|c} x_1 & x_2 & x_3 & x_4 & 1 \\ \hline 1 & 0 & -2 & 0 & 2 \\ 0 & 1 & 6 & 0 & -3 \\ 0 & 0 & 0 & 5 & 20 \end{array}$$

Note that pivoting in the second column is also complete.

8. Solve using the echelon method.

$$4x_1 - 2x_2 = 6$$
$$-2x_1 + x_2 = -3$$

List the solutions for $x_2 = 0$, $x_2 = 1$, and $x_2 = -4$.

9. Consider the tableau

x_1	x_2	x_3	x_4	1
1	4	3	0	22
−2	−8	−1	0	−4
0	0	−2	2	−22

a) Find the reduced echelon tableau.

b) Describe the solutions.

c) List three specific solutions.
   Let $x_2 = 0, 1, -2$.

Consider now the third row and third column, where we have a 0. The only way to get a 1 there is to interchange the third and fourth columns and multiply by a constant. Instead of doing this we simply move on to the third row and fourth column. We multiply by $\frac{1}{5}$ and obtain

x_1	x_2	x_3	x_4	1
1	0	−2	0	2
0	1	6	0	−3
0	0	0	1*	4

Since the rest of the fourth column already has 0's, the pivoting is complete. This is a more general example of the *echelon form*. The solutions are found by translating back to a system of equations. We obtain

$$x_1 - 2x_3 = 2$$
$$x_2 + 6x_3 = -3$$
$$x_4 = 4$$

The solutions can be described by

$$x_1 = 2 + 2x_3$$
$$x_2 = -3 - 6x_3$$
$$x_3 = \text{any number}$$
$$x_4 = 4$$

The following are some solutions obtained by picking arbitrary values of x_3.

$x_1 = 2$	$x_1 = 4$	$x_1 = 0$
$x_2 = -3$	$x_2 = -9$	$x_2 = 3$
$x_3 = 0$	$x_3 = 1$	$x_3 = -1$
$x_4 = 4$	$x_4 = 4$	$x_4 = 4$

The echelon form has a general "staircase" description like any of the following, where # means that any type of number can be in that location.

x_1	x_2	x_3	x_4	x_5	1
1	0	#	0	#	#
0	1	#	0	#	#
0	0	0	1	#	#

x_1	x_2	x_3	1
1	0	0	#
0	1	0	#
0	0	1	#

x_1	x_2	x_3	1
1	0	0	#
0	1	0	#
0	0	0	#

These do not show all possibilities, but they give the general idea.

DO EXERCISE 9.

The following discussion concerns those situations where there is a row with all 0's to the left of the double vertical line.

i) **Any time we have a row with all 0's to the left of the double vertical line and a nonzero number to the right, the system has no solution.**

$$0 \quad 0 \quad 0 \quad 0 \,\|\, k \qquad \text{(where } k \neq 0\text{)}$$

ii) **For a row of *all* 0's, we cannot determine the nature of the solutions without further analysis.**

$$0 \quad 0 \quad 0 \quad 0 \,\|\, 0$$

This row corresponds to an equation that is linearly dependent. So up to this point the system would seem to be consistent. We shift that row to the bottom of the tableau and make further analysis of the upper part of the tableau. We may have exactly one solution, infinitely many solutions, or no solution, should Case (i) later occur.

Example 4 Suppose we are trying to solve a system of 4 equations in 3 variables, and we reach this stage in the tableau.

x_1	x_2	x_3	1
1	0	5	8
0	1	$-\frac{1}{4}$	-2
0	0	0	$-\frac{1}{2}$
0	0	-3	6

Because the third row has all 0's to the left of the double vertical line and a nonzero number to the right, the system has *no solution*. No further analysis is necessary.

DO EXERCISE 10.

Example 5 Suppose we are trying to solve a system of 4 equations in 3 variables and we reach this stage in the tableau.

x_1	x_2	x_3	1
1	0	5	8
0	1	$-\frac{1}{4}$	-2
0	0	0	0
0	0	-3	6

10. Suppose we are solving a system of 4 equations in 3 variables and we reach this stage in the tableau.

x_1	x_2	x_3	1
1	0	6	-9
0	0	0	4
0	-2	8	-6
0	5	10	20

Complete the solution, using the echelon method.

11. Suppose we are solving a system of 4 equations in 3 variables and we reach this stage in the tableau.

x_1	x_2	x_3	1
1	0	6	−9
0	0	0	0
0	−2	8	−6
0	5	10	20

Complete the solution, using the echelon method.

Since we have a row with all 0's, we interchange it with the fourth row.

x_1	x_2	x_3	1
1	0	5	8
0	1	$-\frac{1}{4}$	−2
0	0	−3	6
0	0	0	0

The pivot element is in the third row and third column. We multiply the third row by $-\frac{1}{3}$:

x_1	x_2	x_3	1
1	0	5	8
0	1	$-\frac{1}{4}$	−2
0	0	1^*	−2
0	0	0	0

Now we pivot. We multiply the third row by $\frac{1}{4}$ and add to the second, and we multiply the third row by −5 and add to the first, obtaining

x_1	x_2	x_3	1
1	0	0	18
0	1	0	$-\frac{5}{2}$
0	0	1	−2
0	0	0	0

We now find the solution by translating back to a system of equations:

$$x_1 = 18,$$
$$x_2 = -\frac{5}{2},$$
$$x_3 = -2,$$
$$0 = 0.$$

The last equation plays no role here. The solution can be stated as an ordered triple $(18, -\frac{5}{2}, -2)$.

In Example 5 the original system of equations was consistent and *linearly dependent*. If we have a system of more than two equations that is consistent and linearly dependent, then one of the equations is a multiple of the others or a sum of multiples of the others.

DO EXERCISE 11.

EXERCISE SET 2.2

Solve using the echelon method. Interchange rows and/or columns to avoid fractions when possible.

1. $x_1 - 3x_2 = 2$
$-2x_1 + 6x_2 = -4$

2. $3x_1 + 6x_2 = 9$
$x_1 + 2x_2 = 3$

3. $x_1 + 3x_2 = 2$
$-2x_1 + 6x_2 = -3$

4. $3x_1 + 6x_2 = 8$
$x_1 + 2x_2 = 3$

5. $4x_1 + 12x_2 + 16x_3 = 4$
$3x_1 + 4x_2 + 7x_3 = 3$
$x_1 + 8x_2 + 9x_3 = 1$

6. $2x_1 - 3x_2 + 7x_3 = 2$
$x_1 - 4x_2 + x_3 = 6$
$4x_1 - 16x_2 + 4x_3 = 24$

7. $4x_1 + 12x_2 + 16x_3 = 0$
$3x_1 + 4x_2 + 5x_3 = 0$
$x_1 + 8x_2 + 11x_3 = 0$

8. $2x_1 + x_2 - 3x_3 = 0$
$x_1 - 4x_2 + x_3 = 0$
$4x_1 - 16x_2 + 4x_3 = 0$

9. $x_1 + x_2 + 13x_3 = 0$
$x_1 - x_2 - 6x_3 = 0$

10. $x_1 + x_2 - x_3 = -3$
$x_1 + 2x_2 + 2x_3 = -1$

11. $3x_1 - 9x_3 = 3$
$2x_1 + x_2 - x_3 = 6$
$x_1 + 2x_2 + 7x_3 + x_4 = 7$

12. $x_1 + x_2 + 12x_3 + 2x_4 = 20$
$-2x_1 - x_2 - 20x_3 - 3x_4 = -31$
$3x_1 + 4x_2 + 40x_3 + 7x_4 = 69$

13. $2x_1 - 2x_2 + 18x_3 = -14$
$x_1 - 2x_2 + 13x_3 = -4$
$-2x_2 + 8x_3 = 4$
$2x_1 + x_2 + 36x_3 = 7$

14. $-2x_1 - 3x_2 - 4x_3 = -13$
$x_1 + 2x_3 = 3$
$x_1 + x_2 + 2x_3 = 8$
$x_1 + 3x_3 = 5$

Complete the solution using the echelon method.

15.

x_1	x_2	x_3	x_4	1
1	2	-4	8	7
0	-3	9	12	18
0	3	-9	-12	-18

16.

x_1	x_2	x_3	x_4	1
1	0	5	0	6
0	1	-3	0	4
0	0	0	-2	10

17.

x_1	x_2	x_3	1
1	-1	-2	-5
0	0	0	0
0	0	0	0
0	-2	4	-8

18.

x_1	x_2	x_3	1
1	-2	9	6
0	0	0	0
0	3	1	5

19.

x_1	x_2	x_3	x_4	x_5	1
1	0	8	-3	0	6
0	1	4	2	0	4
0	0	0	0	-2	10

20.

x_1	x_2	x_3	x_4	x_5	1
1	0	8	-3	0	6
0	1	4	2	0	4
0	0	0	0	3	-6

2.3 BASIC MATRIX PROPERTIES

A company makes two types of stereos. Type I requires 65 transistors, 50 capacitors, and 4 dials. Type II requires 85 transistors, 42 capacitors, and 6 dials. We can represent this information as follows.

	Transistors	Capacitors	Dials
Type I	65	50	4
Type II	85	42	6

OBJECTIVES

You should be able to

a) Find the dimensions of a matrix.
b) Find the sum and difference of two matrices of the same dimensions.
c) Find the scalar product of a matrix A and a constant c.
d) Find the *transpose* of a matrix.

Find the dimensions of each matrix.

12. $\begin{bmatrix} -2 & 0 & 3 \\ \frac{1}{2} & 16 & 3 \end{bmatrix}$

13. $[-4 \quad 8 \quad 7]$

14. $\begin{bmatrix} -9 \\ 4 \\ 5 \end{bmatrix}$

15. $[-5]$

16. $\begin{bmatrix} -3 & 6 \\ 4 & 0 \end{bmatrix}$

This table forms a rectangular array of numbers called a *matrix*.

A *matrix* is a rectangular array of numbers. The elements of a matrix are enclosed in brackets.

We can also think of a matrix being obtained from the coefficients and constants in a system of equations. Thus, from the system

$$-2x_1 + 8x_2 = 3,$$
$$\tfrac{1}{2}x_1 + 16x_2 = 5,$$

we get the matrix

$$\begin{bmatrix} -2 & 8 & 3 \\ \frac{1}{2} & 16 & 5 \end{bmatrix}$$

This matrix has 2 *rows* and 3 *columns*.

A matrix with *m* rows and *n* columns has *dimensions m × n, read "m by n."*

Example 1 Find the dimensions of each matrix.

a) $[-2 \quad 3 \quad 4 \quad \tfrac{1}{4}]$ b) $\begin{bmatrix} 6 \\ 7 \\ -3 \\ -\frac{1}{2} \end{bmatrix}$ c) $\begin{bmatrix} -2 & \frac{1}{4} & 8 \\ 0 & 1 & 5 \\ -8 & 6 & 4 \end{bmatrix}$ d) $[8]$

Solution

a) The dimensions are 1×4
b) The dimensions are 4×1
c) The dimensions are 3×3. Such a matrix is called a *square* matrix since it has the same number of rows as columns.
d) The dimensions are 1×1.

We will usually drop the brackets from a 1×1 matrix. That is,

$$[8] = 8.$$

DO EXERCISES 12 THROUGH 16.

We will use capital letters to represent matrices. The elements of a matrix will be denoted by lower-case letters with subscripts. For example, with

$$A = \begin{bmatrix} a_{11} & a_{12} & a_{13} \\ a_{21} & a_{22} & a_{23} \end{bmatrix},$$

the element in the ith row and jth column is given by a_{ij}. The above is a 2×3 matrix. We may also denote it

$$A = [a_{ij}]_{2\times3}, \qquad \text{or} \qquad [a_{ij}].$$

DO EXERCISE 17.

Two matrices are *equal* if and only if they have the same dimensions and corresponding elements are equal. Formally if $A = [a_{ij}]_{m\times n}$ and $B = [b_{ij}]_{m\times n}$, then $A = B$ if and only if $a_{ij} = b_{ij}$ for each i and j, where i ranges from 1 to m and j ranges from 1 to n.

Examples

a) $\begin{bmatrix} 2^3 & 0 \\ 1-5 & 9 \end{bmatrix} = \begin{bmatrix} 8 & 0 \\ -4 & 9 \end{bmatrix}$
b) $\begin{bmatrix} -2 & 4 & 7 \\ 0 & 1 & 5 \end{bmatrix} \neq \begin{bmatrix} -2 & 4 \\ 0 & 1 \end{bmatrix}$

c) $\begin{bmatrix} 8 & -9 \\ -6 & 7 \end{bmatrix} \neq \begin{bmatrix} 8 & -9 \\ 6 & 7 \end{bmatrix}$

Example 2 Solve for p and r.

$$\begin{bmatrix} -6 & 3r-5 \\ 0 & p \end{bmatrix} = \begin{bmatrix} -6 & 14 \\ 0 & -9 \end{bmatrix}$$

Solution Since the matrices are equal, $p = -9$, and $3r - 5 = 14$, or $r = \frac{19}{3}$.

DO EXERCISES 18 AND 19.

A $1 \times n$ matrix is often referred to as a *row vector*, and an $m \times 1$ matrix is often referred to as a *column vector*.

Example 3 Which are row vectors and which are column vectors?

$$A = [5 \quad -3 \quad 0], \quad B = \begin{bmatrix} 7 \\ -4 \\ 3 \end{bmatrix}, \quad C = [-8 \quad 9 \quad 10 \quad 0 \quad \tfrac{1}{4}], \quad D = \begin{bmatrix} -3 \\ 1 \end{bmatrix}$$

Solution The row vectors are A and C, and the column vectors are B and D.

DO EXERCISE 20.

---

17. Consider

$$[a_{ij}] = \begin{bmatrix} -3 & 0 & 6 \\ 4 & -6 & 7 \\ -1 & -2 & \tfrac{1}{2} \end{bmatrix}$$

a) Find a_{12}.

b) Find a_{22}.

c) Find a_{21}.

d) Find a_{32}.

18. Decide whether each pair of matrices is equal.

a) $\begin{bmatrix} 3^2 & -1 \\ 2-3 & 7 \end{bmatrix}, \begin{bmatrix} 6 & -1 \\ -1 & 7 \end{bmatrix}$

b) $[-2 \quad 9 \quad 8 \quad 10]$,
   $[\ 3 \quad -2 \quad 9 \quad 8]$

19. Solve for a and b.

$$\begin{bmatrix} a & 3 & -4 \\ 0 & 6 & -8 \end{bmatrix} = \begin{bmatrix} -6 & 3 & -4 \\ 4b-2 & 6 & -8 \end{bmatrix}$$

20. Which are row vectors and which are column vectors?

$$A = \begin{bmatrix} 5 \\ 4 \\ 6 \\ -1 \end{bmatrix}, \qquad B = [-2 \quad 4],$$

$$C = [-2 \quad 4 \quad 0],$$

$$D = \begin{bmatrix} 5 \\ 1 \\ 3 \\ 2 \end{bmatrix}.$$

21. Find the transpose of each matrix.

$$A = \begin{bmatrix} -8 & 1 & -2 \\ -4 & 0 & -1 \\ 6 & 7 & 8 \end{bmatrix},$$

$$B = [-7 \quad 9 \quad 10 \quad \tfrac{1}{4}],$$

$$C = \begin{bmatrix} -20 \\ 41 \end{bmatrix},$$

$$D = \begin{bmatrix} -4 & 5 \\ 1 & 0 \\ 0 & 1 \end{bmatrix}.$$

The Transpose of a Matrix

The *transpose* of a matrix A, denoted A^T, is found by interchanging the rows and columns of A. That is, if $A = [a_{ij}]$, then $A^T = [a_{ji}]$.

Example 4 Find A^T, B^T, C^T, and D^T.

$$A = \begin{bmatrix} 2 & 4 & 6 \\ 9 & 8 & -2 \\ 0 & -1 & 4 \end{bmatrix}, \qquad B = \begin{bmatrix} -3 & 0 & 4 \\ 7 & 1 & 6 \end{bmatrix},$$

$$C = \begin{bmatrix} -4 \\ 3 \\ 2 \end{bmatrix}, \qquad D = [-1 \quad 2 \quad 3 \quad 0]$$

Solution

$$A^T = \begin{bmatrix} 2 & 9 & 0 \\ 4 & 8 & -1 \\ 6 & -2 & 4 \end{bmatrix}, \qquad B^T = \begin{bmatrix} -3 & 7 \\ 0 & 1 \\ 4 & 6 \end{bmatrix},$$

$$C^T = [-4 \quad 3 \quad 2], \qquad D^T = \begin{bmatrix} -1 \\ 2 \\ 3 \\ 0 \end{bmatrix}$$

DO EXERCISE 21.

You have probably discovered in the margin exercises and Example 4 that if the dimensions of A are $m \times n$, then the dimensions of A^T are $n \times m$. Also, the transpose of a row vector is a column vector, and the transpose of a column vector is a row vector. The latter is a convenient method of saving space. That is, instead of writing

$$A = \begin{bmatrix} x_1 \\ x_2 \\ x_3 \end{bmatrix}$$

we may write

$$A^T = [x_1 \quad x_2 \quad x_3], \qquad \text{or} \qquad A = [x_1 \quad x_2 \quad x_3]^T.$$

Addition of Matrices

The *sum* of two matrices of the same dimensions is the matrix whose elements are the sums of corresponding elements of the given matrices. Formally, if $A = [a_{ij}]$ and $B = [b_{ij}]$, then $A + B = [a_{ij} + b_{ij}]$.

Note that matrix addition is defined only for matrices of the same dimensions.

Example 5 Find $A + B$ and $B + A$.

$$A = \begin{bmatrix} -4 & 0 \\ 3 & \frac{1}{4} \end{bmatrix}, \qquad B = \begin{bmatrix} 7 & -5 \\ 2 & \frac{1}{2} \end{bmatrix}$$

Solution

$$A + B = \begin{bmatrix} -4 + 7 & 0 + (-5) \\ 3 + 2 & \frac{1}{4} + \frac{1}{2} \end{bmatrix} = \begin{bmatrix} 3 & -5 \\ 5 & \frac{3}{4} \end{bmatrix},$$

$$B + A = \begin{bmatrix} 7 + (-4) & -5 + 0 \\ 2 + 3 & \frac{1}{2} + \frac{1}{4} \end{bmatrix} = \begin{bmatrix} 3 & -5 \\ 5 & \frac{3}{4} \end{bmatrix}.$$

DO EXERCISES 22 THROUGH 24.

For any matrices, *A*, *B*, and *C*, of the same dimensions,
     $A + B = B + A$ **(Addition of matrices is commutative)**
  $A + (B + C) = (A + B) + C$ **(Addition of matrices is associative)**

We give a proof of the commutative law,

$A + B = [a_{ij}] + [b_{ij}]$

  $= [a_{ij} + b_{ij}]$ (Definition of matrix addition)

  $= [b_{ij} + a_{ij}]$ (Addition of real numbers is commutative)

  $= [b_{ij}] + [a_{ij}]$ (Reverse of matrix addition)

  $= B + A.$

A *zero matrix* is denoted by the capital letter O. From Margin Exercise 24, we can conjecture that, if A and O have the same dimensions, then

$$A + O = O + A = A. \qquad \textbf{(\emph{O} is the \emph{additive identity})}$$

The product of a constant and a matrix, a *scalar product*, is defined as follows.

The *scalar product* of a number *c* (sometimes called a *scalar*) and a matrix *A* is the matrix obtained by multiplying the elements of *A* by *c*. Formally, if $A = [a_{ij}]$, then $cA = [ca_{ij}]$.

Example 6 Find $4A$ and $(-1)A$.

$$A = \begin{bmatrix} -2 & 0 \\ 1 & 5 \end{bmatrix}$$

22. Consider

$$A = \begin{bmatrix} 5 & -2 \\ 7 & -4 \end{bmatrix} \quad \text{and} \quad B = \begin{bmatrix} -8 & -4 \\ 2 & 4 \end{bmatrix}$$

a) Find $A + B$.

b) Find $B + A$.

c) Determine whether $A + B = B + A$.

23. Add:

$$\begin{bmatrix} -2 & 6 \\ 4 & 5 \\ 0 & \frac{1}{4} \end{bmatrix} + \begin{bmatrix} 2 & 9 \\ -3 & 7 \\ 1 & \frac{1}{4} \end{bmatrix}$$

24. Consider

$$A = \begin{bmatrix} 5 & -2 \\ 7 & -4 \end{bmatrix} \quad \text{and} \quad O = \begin{bmatrix} 0 & 0 \\ 0 & 0 \end{bmatrix}.$$

a) Find $A + O$.

b) Find $O + A$.

c) Determine whether $A + O = O + A$.

25. Consider

$$A = \begin{bmatrix} 6 & 1 & 0 \\ 2 & 1 & -5 \\ -3 & 9 & \frac{1}{2} \end{bmatrix}$$

Find:

a) $6A$ b) $(-1)A$

c) $A + (-1)A$ d) $(-1)A + A$

e) $-\dfrac{1}{30} A$ f) tA

26. Consider matrices A and B of Margin Exercise 22.

Find

a) $A - B$

b) $B - A$

c) Determine whether $A - B = B - A$.

Solution

$$4A = \begin{bmatrix} 4(-2) & 4 \cdot 0 \\ 4 \cdot 1 & 4 \cdot 5 \end{bmatrix} = \begin{bmatrix} -8 & 0 \\ 4 & 20 \end{bmatrix},$$

$$(-1)A = \begin{bmatrix} -1(-2) & -1 \cdot 0 \\ -1 \cdot 1 & -1 \cdot 5 \end{bmatrix} = \begin{bmatrix} 2 & 0 \\ -1 & -5 \end{bmatrix}$$

DO EXERCISE 25.

For real numbers we know that $-a$ represents the number we add to a to get 0. We call $-a$ the *additive inverse* of a. For matrices, $-A$ is the matrix we add to A to get the zero matrix O. You may have conjectured the following from Margin Exercise 25.

$$A + (-1)A = (-1)A + A = O$$

(The additive inverse of A, $- A$, is $(- 1)A$.)

We subtract as follows.

The difference $A - B = A + (-1)B$. We subtract B from A by adding the additive inverse of B.

Note that A and B must have the *same dimensions* in order to subtract one from the other.

Example 7 Find $A - B$.

$$A = \begin{bmatrix} -3 & 5 \\ 4 & 8 \end{bmatrix}, \qquad B = \begin{bmatrix} 11 & -2 \\ 6 & 7 \end{bmatrix}$$

$$A - B = A + (-1)B = \begin{bmatrix} -3 & 5 \\ 4 & 8 \end{bmatrix} + (-1)\begin{bmatrix} 11 & -2 \\ 6 & 7 \end{bmatrix}$$

$$= \begin{bmatrix} -3 & 5 \\ 4 & 8 \end{bmatrix} + \begin{bmatrix} -11 & 2 \\ -6 & -7 \end{bmatrix}$$

$$= \begin{bmatrix} -14 & 7 \\ -2 & 1 \end{bmatrix}$$

Note that $A - B$ can be found directly by subtracting corresponding elements of B from those of A.

DO EXERCISE 26.

EXERCISE SET 2.3

Consider

$$A = \begin{bmatrix} 1 & 3 \\ 4 & 2 \end{bmatrix}, \qquad B = \begin{bmatrix} -2 & 0 \\ -2 & -1 \end{bmatrix}, \qquad C = \begin{bmatrix} -1 & -2 & -3 \\ 3 & 2 & 1 \end{bmatrix}, \qquad \text{and} \qquad D = \begin{bmatrix} 0 & 8 & -4 \\ 1 & 0 & -1 \end{bmatrix}.$$

Find:

1. The dimensions of A **2.** The dimensions of B **3.** The dimensions of C **4.** The dimensions of D

5. $A + B$ **6.** $B + A$ **7.** $C + D$ **8.** $D + C$

9. $3A$ **10.** $3B$ **11.** $-5C$ **12.** $-6D$

13. $A - B$ **14.** $B - A$ **15.** $C - D$ **16.** $D - C$

17. $A + C$ **18.** $A + D$ **19.** kC **20.** kD

21. $A + O$ **22.** $O + D$ **23.** A^T **24.** B^T

25. C^T **26.** D^T **27.** $A^T + B^T$ **28.** $C^T + D^T$

29. For $[a_{ij}] = \begin{bmatrix} -4 & 5 \\ 0 & 9 \\ 1 & 3 \end{bmatrix}$, find a_{11}, a_{12}, a_{31}, a_{22}, a_{32}, and a_{21}.

30. For $[b_{ij}] = \begin{bmatrix} -2 & -3 & 0 \\ \frac{1}{4} & \frac{1}{2} & 1 \end{bmatrix}$, find b_{11}, b_{21}, b_{23}, b_{22}, b_{13}, and b_{12}.

31. Find X^T where $X = \begin{bmatrix} x_1 \\ x_2 \\ x_3 \\ x_4 \end{bmatrix}$.   **32.** Find Y^T where $Y = \begin{bmatrix} y_1 \\ y_2 \\ y_3 \\ y_4 \end{bmatrix}$.

2.4 MATRIX MULTIPLICATION

Summation Notation

Consider this sum:

$$a_1 + a_2 + a_3 + a_4$$

We can denote this sum using *summation notation* which utilizes the Greek capital letter $\sum$ (sigma),

$$\sum_{i=1}^{4} a_i \quad \text{or} \quad \sum_{i=1}^{4} a_i$$

This is read "the sum of the numbers a_i from $i = 1$ to $i = 4$." To recover the original sum, substitute the numbers 1 through 4 successively into a_i and write plus signs between the results.

Example 1   Write summation notation for $2 + 4 + 6 + 8 + 10$.

Solution

$$2 + 4 + 6 + 8 + 10 = \sum_{i=1}^{5} 2i$$

OBJECTIVES

You should be able to

a) Write summation notation for certain sums.

b) Express a sum without summation notation.

c) Find the product AB of two matrices A and B, where the number of columns in A is the same as the number of rows in B.

d) Given the inverse of a coefficient matrix, use that inverse to solve systems of 2 equations in 2 variables and 3 equations in 3 variables.

e) Given matrices A and B, find matrices like

$$[A \ \ B], \quad [A \ \ B]^T, \quad \text{and} \quad \begin{bmatrix} A \\ B \end{bmatrix}.$$

Write in summation notation.

27. $1 + 4 + 9 + 16 + 25 + 36$

28. $t + t^2 + t^3 + t^4$

29. $p_1 + p_2 + p_3 + \cdots + p_{38}$

Express without using summation notation.

30. $\displaystyle\sum_{i=1}^{3} 2^i$

31. $\displaystyle\sum_{i=1}^{20} p_i q_i$

32. $\displaystyle\sum_{i=1}^{5} it^i$

Example 2   Write summation notation for

$$a_1 + a_2 + a_3 + a_4 + \cdots + a_{19}$$

The three dots indicate that we are not writing all the terms in between.

Solution

$$a_1 + a_2 + a_3 + a_4 + \cdots + a_{19} = \sum_{i=1}^{19} a_i$$

DO EXERCISES 27 THROUGH 29.

Example 3   Express $\sum_{i=1}^{4} 3^i$ without using summation notation.

Solution

$$\sum_{i=1}^{4} 3^i = 3^1 + 3^2 + 3^3 + 3^4$$

Example 4 $\sum_{i=1}^{30} a_i b_i$ without using summation notation.

Solution

$$\sum_{i=1}^{30} a_i b_i = a_1 b_1 + a_2 b_2 + \cdots + a_{30} b_{30}$$

DO EXERCISES 30 THROUGH 32.

Matrix Multiplication

The product of two matrices A and B will *not* be defined as the matrix whose elements are products of corresponding elements of A and B. Some motivation for the definition of matrix multiplication is based on converting an equation such as

$$2x_1 - 4x_2 + 7x_3 = 8$$

to a *product of a row vector and a column vector:*

$$[2 \quad -4 \quad 7] \begin{bmatrix} x_1 \\ x_2 \\ x_3 \end{bmatrix} = [8], \quad \text{or} \quad 8$$

The *product* of a row vector A, a matrix of dimensions $1 \times n$, and a column vector B, a matrix of dimensions $n \times 1$, is the 1×1 matrix (or scalar) whose element is the sum of products of corresponding elements A and B. Formally, if

$$A = [a_1 \quad a_2 \quad a_3 \cdots a_n] \quad \text{and} \quad B = \begin{bmatrix} b_1 \\ b_2 \\ b_3 \\ \cdot \\ \cdot \\ \cdot \\ b_n \end{bmatrix},$$

then

$$AB = a_1 b_1 + a_2 b_2 + a_3 b_3 + \cdots + a_n b_n$$

or, using summation notation,

$$AB = \sum_{i=1}^{n} a_i b_i.$$

Example 5 Find each product.

a) $[2 \quad -4 \quad 7] \begin{bmatrix} -1 \\ 0 \\ 5 \end{bmatrix} = [2(-1) - 4 \cdot 0 + 7 \cdot 5] = [33], \quad$ or $\quad 33$

b) $[\frac{1}{4} \quad -8] \begin{bmatrix} 12 \\ 1 \end{bmatrix} = [\frac{1}{4} \cdot 12 - 8 \cdot 1] = [-5], \quad$ or $\quad -5$

c) $[-2 \quad 1 \quad 4 \quad -5] \begin{bmatrix} x_1 \\ x_2 \\ x_3 \\ x_4 \end{bmatrix} = [-2x_1 + x_2 + 4x_3 - 5x_4], \quad$ or

$$-2x_1 + x_2 + 4x_3 - 5x_4$$

To multiply the row vector A and the column vector B, the number of elements in A must be the same as the number of elements in B. The following illustration should help you remember this.

$$\begin{array}{ccc} A & & B \\ 1 \times n & & m \times 1 \end{array}$$

$$\underset{}{\underline{\qquad}} \; n = m \; \underset{}{\underline{\qquad}}$$

DO EXERCISES 33 THROUGH 35

Before we define multiplication of more general matrices, let us reconsider an example given earlier. A company makes two types of stereos. Type I requires 65 transistors, 50 capacitors, and 4 dials. Type II requires 85 transistors, 42 capacitors, and 6 dials. We can

Find each product.

33. $[4 \quad 5] \begin{bmatrix} a \\ b \end{bmatrix}$

34. $[4 \quad 5] \begin{bmatrix} -2 \\ 3 \end{bmatrix}$

35. $[3 \quad -2 \quad 4] \begin{bmatrix} -6 \\ 7 \\ -1 \end{bmatrix}$

36. Put in matrix form:

$$11x_1 + 2x_2 = -1$$
$$7x_1 - 13x_2 = 8$$

37. Write as separate equations:

$$\begin{bmatrix} 3 & -7 \\ -2 & 1 \end{bmatrix}\begin{bmatrix} y_1 \\ y_2 \end{bmatrix} = \begin{bmatrix} 4 \\ 5 \end{bmatrix}$$

represent this information using a matrix.

	Transistors	Capacitors	Dials
Type I	65	50	4
Type II	85	42	6

Suppose the company wanted to make 20 stereos of Type I and 30 stereos of Type II. It would determine the number of transistors needed as follows:

$$20 \cdot 65 + 30 \cdot 85 = 3850.$$

It would determine the number of capacitors needed as follows:

$$20 \cdot 50 + 30 \cdot 42 = 2260.$$

It would determine the dials needed as follows:

$$20 \cdot 4 + 30 \cdot 6 = 260.$$

The entire procedure could be done using matrices as follows:

$$[20 \quad 30] \cdot \begin{bmatrix} 65 & 50 & 4 \\ 85 & 42 & 6 \end{bmatrix}$$
$$= [20 \cdot 65 + 30 \cdot 85 \quad 20 \cdot 50 + 30 \cdot 42 \quad 20 \cdot 4 + 30 \cdot 6]$$
$$= [3850 \quad 2260 \quad 260]$$

Some further motivation for multiplication of matrices is based on converting a system of equations such as

$$2x_1 - 3x_2 = 7$$
$$4x_1 + 5x_2 = 9,$$

to matrix form:

$$\begin{bmatrix} 2 & -3 \\ 4 & 5 \end{bmatrix}\begin{bmatrix} x_1 \\ x_2 \end{bmatrix} = \begin{bmatrix} 7 \\ 9 \end{bmatrix}.$$

Multiplying these matrices, we obtain the original equations. Note the similarity with the echelon tableau, Section 2.1.

DO EXERCISES 36 AND 37.

The *product* of two matrices A and B is the matrix $C = [c_{ij}]$, where c_{ij} is obtained by multiplying the ith row of A (as a vector) by the jth column of B (as a vector).

Example 6 Multiply

$$[2 \quad 3]\begin{bmatrix} x & a \\ y & b \end{bmatrix} = [2x + 3y \quad 2a + 3b].$$

Mentally this product is found in two steps:

$$[① \quad ②]$$

where

① is $[\blacksquare\blacksquare] \begin{bmatrix} \blacksquare & \end{bmatrix}$

and

② is $[\blacksquare\blacksquare] \begin{bmatrix} & \blacksquare \end{bmatrix}$

Example 7 Multiply

$$[2 \quad 3]\begin{bmatrix} 1 & -2 \\ -8 & 4 \end{bmatrix} = [2 \cdot 1 + 3(-8) \quad 2(-2) + 3 \cdot 4] = [-22 \quad 8]$$

DO EXERCISE 38.

Example 8 Multiply

$$\begin{bmatrix} 2 & 3 \\ -6 & 7 \end{bmatrix}\begin{bmatrix} x & a \\ y & b \end{bmatrix} = \begin{bmatrix} 2x + 3y & 2a + 3b \\ -6x + 7y & -6a + 7b \end{bmatrix}$$

Mentally this product is found in four steps:

$$\begin{bmatrix} ① & ② \\ ③ & ④ \end{bmatrix}$$

① $\begin{bmatrix} \blacksquare \end{bmatrix}\begin{bmatrix} \blacksquare & \end{bmatrix}$ ② $\begin{bmatrix} \blacksquare \end{bmatrix}\begin{bmatrix} & \blacksquare \end{bmatrix}$

③ $\begin{bmatrix} \blacksquare \end{bmatrix}\begin{bmatrix} \blacksquare & \end{bmatrix}$ ④ $\begin{bmatrix} \blacksquare \end{bmatrix}\begin{bmatrix} & \blacksquare \end{bmatrix}$

Example 9 Multiply

$$\begin{bmatrix} 2 & 3 \\ -6 & 7 \end{bmatrix}\begin{bmatrix} 1 & -2 \\ -8 & 4 \end{bmatrix} = \begin{bmatrix} 2 \cdot 1 + 3(-8) & 2(-2) + 3 \cdot 4 \\ -6 \cdot 1 + 7(-8) & -6(-2) + 7 \cdot 4 \end{bmatrix} = \begin{bmatrix} -22 & 8 \\ -62 & 40 \end{bmatrix}$$

DO EXERCISE 39.

So that we can carry out the multiplication of the rows of A and the columns of B, the number of columns in A must be the same as the number of rows in B. That is, if A has dimensions $m \times n$ and B has dimensions $p \times q$, then in order to multiply we must have $n = p$. In such a case we say that the matrices are *conformable*. The product

38. Multiply.

a) $[5 \quad 6]\begin{bmatrix} c & g \\ d & h \end{bmatrix}$

b) $[5 \quad 6]\begin{bmatrix} -3 & 0 \\ 2 & -4 \end{bmatrix}$

39. Multiply.

a) $\begin{bmatrix} 5 & 6 \\ 3 & -1 \end{bmatrix}\begin{bmatrix} c & g \\ d & h \end{bmatrix}$

b) $\begin{bmatrix} 5 & 6 \\ 3 & -1 \end{bmatrix}\begin{bmatrix} -3 & 0 \\ 2 & -4 \end{bmatrix}$

matrix has dimensions $m \times q$. The following may help you remember this:

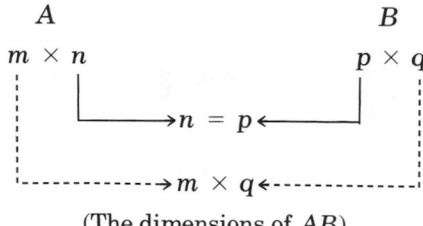

(The dimensions of AB)

Example 10 Find AB and BA, if possible.

$$A = \begin{bmatrix} -2 & 1 \\ 4 & 0 \\ -3 & -5 \end{bmatrix}, \qquad B = \begin{bmatrix} 2 & -1 & 0 & -7 \\ 4 & -3 & -1 & 0 \end{bmatrix}$$

Solution

$$AB = \begin{bmatrix} -2 \cdot 2 + 1 \cdot 4 & -2(-1) + 1(-3) & -2 \cdot 0 + 1(-1) & -2(-7) + 1 \cdot 0 \\ 4 \cdot 2 + 0 \cdot 4 & 4(-1) + 0(-3) & 4 \cdot 0 + 0(-1) & 4(-7) + 0 \cdot 0 \\ -3 \cdot 2 - 5 \cdot 4 & -3(-1) - 5(-3) & -3 \cdot 0 - 5(-1) & -3(-7) - 5 \cdot 0 \end{bmatrix}$$

$$= \begin{bmatrix} 0 & -1 & -1 & 14 \\ 8 & -4 & 0 & -28 \\ -26 & 18 & 5 & 21 \end{bmatrix}$$

BA cannot be found because B and A are not conformable; that is, the number of columns in B is not the same as the number of rows in A.

The products AB and BA of a row matrix A and a column matrix B are of special interest.

Example 11 Find AB and BA, if possible.

$$A = [-1 \quad 2 \quad -5], \qquad B = \begin{bmatrix} 4 \\ 0 \\ -2 \end{bmatrix}$$

Solution

A is a 1×3 matrix and B is a 3×1 matrix. Thus the product AB is a 1×1 matrix given by

$$AB = [-1 \quad 2 \quad -5] \begin{bmatrix} 4 \\ 0 \\ -2 \end{bmatrix} = [-1 \cdot 4 + 2 \cdot 0 - 5(-2)] = [6], \qquad \text{or} \qquad 6$$

B is a 3×1 matrix and A is a 1×3 matrix. Thus the product BA is a 3×3 matrix given by

$$BA = \begin{bmatrix} 4 \\ 0 \\ -2 \end{bmatrix} \begin{bmatrix} -1 & 2 & -5 \end{bmatrix} = \begin{bmatrix} 4(-1) & 4 \cdot 2 & 4(-5) \\ 0(-1) & 0 \cdot 2 & 0(-5) \\ -2(-1) & -2 \cdot 2 & -2(-5) \end{bmatrix}$$

$$= \begin{bmatrix} -4 & 8 & -20 \\ 0 & 0 & 0 \\ 2 & -4 & 10 \end{bmatrix}$$

Note that each element of BA is the product of two numbers.

DO EXERCISES 40 AND 41.

We have seen that matrix multiplication is not commutative. In some cases, there are matrices A and B such that $AB = BA$. In such cases we say that the matrices *commute*.

The notion of *conformability* is often used in many contexts with matrices. For example, matrices A and B are *equality conformable* if A and B have the same dimensions. A and B are *addition conformable* if A and B have the same dimensions. A and B are *multiplication conformable* if the number of columns in A is the same as the number of rows in B.

For any conformable matrices A, B, and C.

$(AB)C = A(BC) = ABC$ **(Multiplication of matrices is associative.)**

$A(B + C) = AB + AC$ **(Multiplication of matrices is distributive.)**

DO EXERCISE 42.

Identity Matrices

The letter I is used to represent square matrices such as

$$I = \begin{bmatrix} 1 & 0 \\ 0 & 1 \end{bmatrix}, \quad \text{and} \quad I = \begin{bmatrix} 1 & 0 & 0 \\ 0 & 1 & 0 \\ 0 & 0 & 1 \end{bmatrix}.$$

These square matrices have 1's extending from the upper left down to the lower right along what is called the *main diagonal*. The rest of the elements are 0.

40. Find AB and BA, if possible.

a) $A = \begin{bmatrix} -3 & 0 & 5 \\ -2 & 8 & 0 \end{bmatrix}$,

$B = \begin{bmatrix} 1 & -2 & -3 \\ 1 & 0 & 1 \\ -4 & 2 & 6 \end{bmatrix}$

b) $A = \begin{bmatrix} -2 & -3 & 1 & -1 \end{bmatrix}$,

$B = \begin{bmatrix} -6 \\ -8 \\ 0 \\ 5 \end{bmatrix}$

41. Consider

$$A = \begin{bmatrix} 3 & -8 \\ 1 & -5 \end{bmatrix}, \quad B = \begin{bmatrix} 0 & -1 \\ 4 & 0 \end{bmatrix}$$

a) Find AB.

b) Find BA.

c) Decide whether AB and BA are equal.

d) Is matrix multiplication commutative? Explain.

42. Given

$$A = \begin{bmatrix} 3 & -2 \\ 1 & 5 \end{bmatrix}, \quad B = \begin{bmatrix} 0 & -1 \\ 4 & 0 \end{bmatrix},$$

and

$$C = \begin{bmatrix} 5 & 2 \\ 1 & 1 \end{bmatrix}$$

a) Verify the associative law by matrix multiplication.

b) Verify the distributive law by matrix addition and multiplication.

43. Consider

$$A = \begin{bmatrix} x & a \\ y & b \end{bmatrix}, \qquad X = \begin{bmatrix} x_1 \\ x_2 \end{bmatrix}$$

and

$$I = \begin{bmatrix} 1 & 0 \\ 0 & 1 \end{bmatrix}.$$

a) Find AI.

b) Find IA.

c) Compare AI and IA.

d) Find IX.

e) Find XI.

f) Compare IX and X.

DO EXERCISE 43.

You have just proved the following for the square matrix I of dimensions 2×2.

For any square matrix A of dimensions $n \times n$,

$$AI = IA = A \qquad (I \text{ is a } multiplicative \ identity),$$

where I is the square matrix, described above, of dimensions $n \times n$.

We also have the following.

For any matrix A with exactly n rows,

$$IA = A,$$

where I is the square matrix, described above, of dimensions $n \times n$.

Matrix Inverses

The equation in real numbers, or *scalar* equation,

$$ax = b, \qquad \text{where } a \text{ is real,}$$

can be solved for x by multiplying both sides of the equation by a^{-1}, the inverse of a:

$$a^{-1}ax = a^{-1}b.$$

Since $a^{-1}a = aa^{-1} = 1$, the multiplicative identity element such that $1x = x$, we obtain

$$x = a^{-1}b.$$

For scalar numbers

$$a^{-1} = \frac{1}{a}, \qquad \text{for } a \neq 0,$$

so that

$$x = \frac{b}{a}.$$

Let us see how we can use matrices to represent and solve systems of equations. Consider the system

$$3x_1 + 2x_2 = 1,$$
$$5x_1 + 3x_2 = -2.$$

We first express this system as a matrix equation:

$$\begin{bmatrix} 3 & 2 \\ 5 & 3 \end{bmatrix} \begin{bmatrix} x_1 \\ x_2 \end{bmatrix} = \begin{bmatrix} 1 \\ -2 \end{bmatrix}$$

To see that this is correct, multiply the matrices on the left and use the fact that matrices are equal if corresponding elements are equal. We let

$$A = \text{the } \textit{coefficient matrix} = \begin{bmatrix} 3 & 2 \\ 5 & 3 \end{bmatrix}, \quad X = \begin{bmatrix} x_1 \\ x_2 \end{bmatrix}, \text{ and } B = \begin{bmatrix} 1 \\ -2 \end{bmatrix}.$$

Then

$$AX = B.$$

The solution of this *matrix* equation can*not* be written as the quotient

$$X = \frac{B}{A}.$$

Multiplying this equation by A on the *left*, yields

$$AX = B,$$

while multiplying the same equation by A on the *right*, yields

$$XA = B.$$

This implies that the two products AX and XA commute, that is

$$AX = XA.$$

This is not always true. As an example, consider

$$AX = B,$$

where

$$A = \begin{bmatrix} 3 & 2 \\ 5 & 3 \end{bmatrix}, \quad X = \begin{bmatrix} x_1 \\ x_2 \end{bmatrix}, \quad \text{and} \quad B = \begin{bmatrix} 1 \\ -2 \end{bmatrix}.$$

Then

$$AX = \begin{bmatrix} 3 & 2 \\ 5 & 3 \end{bmatrix}\begin{bmatrix} x_1 \\ x_2 \end{bmatrix} = \begin{bmatrix} 3x_1 + 2x_2 \\ 5x_1 + 3x_2 \end{bmatrix} = \begin{bmatrix} 1 \\ -2 \end{bmatrix}.$$

On the other hand

$$XA = \begin{bmatrix} x_1 \\ x_2 \end{bmatrix}\begin{bmatrix} 3 & 2 \\ 5 & 3 \end{bmatrix}$$

is not possible, since the matrices X and A in the product XA are not conformable.

Thus, **division by a matrix is not possible.**

However, we can replace division by multiplication by the *matrix inverse*. For example, consider solving

$$AX = B$$

44. Consider

$$A = \begin{bmatrix} 3 & 5 \\ 1 & 2 \end{bmatrix}, \qquad A^{-1} = \begin{bmatrix} 2 & -5 \\ -1 & 3 \end{bmatrix}$$

and

$$I = \begin{bmatrix} 1 & 0 \\ 0 & 1 \end{bmatrix}.$$

Find $A^{-1}A$ and AA^{-1} and compare to I.

45. Consider the system

$$3x_1 + 5x_2 = -1$$
$$x_1 + 2x_2 = 4$$

a) Express this system as a matrix equation.
b) What is the coefficient matrix A?
c) Given the inverse of the coefficient matrix A:

$$A^{-1} = \begin{bmatrix} 2 & -5 \\ -1 & 3 \end{bmatrix},$$

use it and the matrix equation to solve the system.

for X as we solved $ax = b$ for x. To do this we define the *multiplicative inverse*, or simply the *inverse*, of a square matrix A to be the square matrix A^{-1} with the property that (as for the scalar a)

$$A^{-1}A = AA^{-1} = I,$$

where I is the multiplicative identity element previously defined.

DO EXERCISE 44.

Now we multiply both sides of $AX = B$ by A^{-1} on the left and obtain

$$A^{-1}AX = A^{-1}B.$$

Since $A^{-1}A = I$ and $IX = X$, we obtain

$$X = A^{-1}B.$$

If we knew the inverse matrix A^{-1}, then we could find the solution of the system of equations by computing the product $A^{-1}B$.

In the present example, we give A^{-1} without explaining how we found it.

$$A^{-1} = \begin{bmatrix} -3 & 2 \\ 5 & -3 \end{bmatrix}$$

Compare the following:

$$AX = B \qquad \begin{bmatrix} 3 & 2 \\ 5 & 3 \end{bmatrix}\begin{bmatrix} x_1 \\ x_2 \end{bmatrix} = \begin{bmatrix} 1 \\ -2 \end{bmatrix}$$

$$X = A^{-1}B \qquad \begin{bmatrix} x_1 \\ x_2 \end{bmatrix} = \begin{bmatrix} -3 & 2 \\ 5 & -3 \end{bmatrix}\begin{bmatrix} 1 \\ -2 \end{bmatrix} = \begin{bmatrix} -7 \\ 11 \end{bmatrix}$$

From equality of matrices we can read off the solution. That is,

$$x_1 = -7 \qquad \text{and} \qquad x_2 = 11,$$

or, simply, the solution is $(-7, 11)$.

DO EXERCISE 45.

Let us relate this procedure to the echelon method. We consider the echelon tableau for the previous system:

x_1	x_2	1
3	2	1
5	3	-2

The entries to the left of the double vertical line make up the coefficient matrix A. The entries to the right make up the matrix B. When the echelon method is completed we have the tableau

$$
\begin{array}{cc||c}
x_1 & x_2 & 1 \\
\hline
1 & 0 & -7 \\
0 & 1 & 11
\end{array}
$$

The entries to the left of the double vertical line make up the identity matrix I. Those to the right make up the solution matrix. The steps of the echelon method have the same effect as multiplying by A^{-1} without actually knowing what it is.

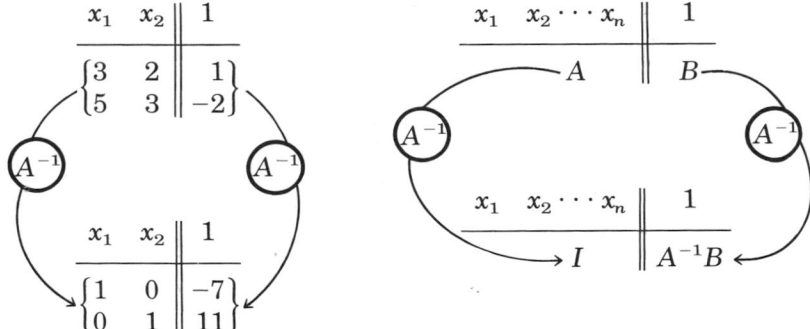

It is important to know the meaning and notation of matrix inverses. It is less important to know how to compute the inverse, although the preceding is the basis for such a method.

Augmented Matrices

For later purposes we need some additional notation. Let

$$X = [x_1 \quad x_2] \quad \text{and} \quad Y = [y_1 \quad y_2].$$

Then we can form the *augmented* matrix

$$[X \quad Y] = [x_1 \quad x_2 \vdots y_1 \quad y_2].$$

An augmented matrix is simply a particular way of putting two matrices together to form a new matrix. Sometimes, as above, a dashed line is used to indicate that $[X \quad Y]$ can be *partitioned* into X and Y.

Since the matrices X and Y have the same dimensions, we can also form the augmented matrix

$$\begin{bmatrix} X \\ Y \end{bmatrix} = \begin{bmatrix} x_1 & x_2 \\ \hdashline y_1 & y_2 \end{bmatrix}.$$

46. Consider

$$X = \begin{bmatrix} x_1 \\ x_2 \\ x_3 \end{bmatrix} \quad \text{and} \quad Y = \begin{bmatrix} y_1 \\ y_2 \\ y_3 \end{bmatrix}.$$

a) Find $[X \quad Y]$.

b) Find $\begin{bmatrix} X \\ Y \end{bmatrix}$.

47. Consider

$$A = \begin{bmatrix} -2 & 3 \\ 4 & 5 \end{bmatrix} \quad \text{and} \quad I = \begin{bmatrix} 1 & 0 \\ 0 & 1 \end{bmatrix}$$

a) Find $[A \quad I]$

b) Find $\begin{bmatrix} A \\ I \end{bmatrix}$.

48. Consider

$$A = \begin{bmatrix} 5 & 7 \\ -2 & 0 \end{bmatrix} \quad \text{and} \quad I = \begin{bmatrix} 1 & 0 \\ 0 & 1 \end{bmatrix}$$

a) Find $[AI]^T$.

b) Find $\begin{bmatrix} A \\ I \end{bmatrix}^T$.

When forming augmented matrices, the resulting array must be rectangular; that is, the matrices must conform. Thus, if $Z = [z_1 \quad z_2 \quad z_3]$, we can form the augmented matrix $[X \quad Z]$ but not $\begin{bmatrix} X \\ Z \end{bmatrix}$.

DO EXERCISE 46.

Suppose

$$A = \begin{bmatrix} 3 & 5 \\ 1 & -2 \end{bmatrix} \quad \text{and} \quad I = \begin{bmatrix} 1 & 0 \\ 0 & 1 \end{bmatrix}.$$

Then we can define an augmented matrix A' by

$$A' = [A \quad I] = \begin{bmatrix} 3 & 5 & \vdots & 1 & 0 \\ 1 & -2 & \vdots & 0 & 1 \end{bmatrix}$$

and another augmented matrix A'' by

$$A'' = \begin{bmatrix} A \\ I \end{bmatrix} = \begin{bmatrix} 3 & 5 \\ 1 & -2 \\ \hline 1 & 0 \\ 0 & 1 \end{bmatrix}.$$

DO EXERCISE 47.

Sometimes it is necessary to find the *transpose* of an augmented matrix. To do that, *first* form the augmented matrix and *then* form the transpose in the usual manner. Thus,

$$[X \quad Y]^T = \begin{bmatrix} x_1 \\ x_2 \\ y_1 \\ y_2 \end{bmatrix} \quad \text{and} \quad \begin{bmatrix} X \\ Y \end{bmatrix}^T = \begin{bmatrix} x_1 & \vdots & y_1 \\ x_2 & \vdots & y_2 \end{bmatrix}.$$

DO EXERCISE 48.

EXERCISE SET 2.4

Find AB and BA, if possible.

1. $A = [-2 \quad 1], \quad B = \begin{bmatrix} 3 \\ -4 \end{bmatrix}$

2. $A = [-1 \quad -2], \quad B = \begin{bmatrix} 6 \\ -7 \end{bmatrix}$

3. $A = [2 \quad 0 \quad -4], \quad B = \begin{bmatrix} 9 \\ -5 \\ \frac{1}{4} \end{bmatrix}$

4. $A = [-3 \quad 6 \quad 8], \quad B = \begin{bmatrix} 5 \\ \frac{1}{2} \\ 1 \end{bmatrix}$

5. $A = \begin{bmatrix} 1 & 2 & 0 \\ -1 & 0 & 4 \\ 2 & 5 & 6 \end{bmatrix}$, $\quad B = \begin{bmatrix} 3 & -4 & 1 \\ 2 & -1 & 0 \\ -3 & 2 & 1 \end{bmatrix}$

6. $A = \begin{bmatrix} -1 & 0 & 0 \\ 0 & -1 & 0 \\ 0 & 0 & -1 \end{bmatrix}$, $\quad B = \begin{bmatrix} 2 & -5 & 1 \\ -4 & 4 & 3 \\ 5 & 6 & 9 \end{bmatrix}$

7. $A = \begin{bmatrix} -4 & 1 & 3 \end{bmatrix}$, $\quad B = \begin{bmatrix} -4 & 2 \\ 1 & 0 \\ 6 & -9 \end{bmatrix}$

8. $A = \begin{bmatrix} -2 & 3 \\ 1 & 0 \\ -5 & 4 \end{bmatrix}$, $\quad B = \begin{bmatrix} 2 \\ 3 \end{bmatrix}$

9. $A = \begin{bmatrix} 1 & 0 & 2 \\ 5 & 0 & 1 \\ -1 & 0 & 4 \end{bmatrix}$, $\quad B = \begin{bmatrix} 0 & 0 & 0 \\ 1 & 3 & 7 \\ 0 & 0 & 0 \end{bmatrix}$. Find AB.

10. Find two matrices A and B each of dimensions 2×2, such that $A \neq 0$ and $B \neq 0$ but $AB = 0$. *Hint.* See Exercise 9.

11. Write the system of equations of Exercise 17 in matrix form.

12. Write the system of equations of Exercise 18 in matrix form.

13. Write the system of equations of Exercise 21 in matrix form.

14. Write the system of equations of Exercise 22 in matrix form.

Write as separate equations:

15. $\begin{bmatrix} 1 & 2 \\ 4 & -3 \end{bmatrix} \begin{bmatrix} x_1 \\ x_2 \end{bmatrix} = \begin{bmatrix} -1 \\ 2 \end{bmatrix}$

16. $\begin{bmatrix} 2 & 4 \\ 3 & 5 \end{bmatrix} \begin{bmatrix} x_1 \\ x_2 \end{bmatrix} = \begin{bmatrix} -2 \\ -4 \end{bmatrix}$

In Exercises 17 through 22, a system of equations is given, together with the inverse of the coefficient matrix. Use the matrix inverse to solve the system.

17. $\begin{aligned} 11x_1 + 3x_2 &= -4, \\ 7x_1 + 2x_2 &= 5 \end{aligned}$ $\quad A^{-1} = \begin{bmatrix} 2 & -3 \\ -7 & 11 \end{bmatrix}$

18. $\begin{aligned} 8x_1 + 5x_2 &= -6, \\ 5x_1 + 3x_2 &= 2 \end{aligned}$ $\quad A^{-1} = \begin{bmatrix} -3 & 5 \\ 5 & -8 \end{bmatrix}$

19. $\begin{aligned} 4x_1 - 3x_2 &= 2, \\ x_1 + 2x_2 &= -1 \end{aligned}$ $\quad A^{-1} = \frac{1}{11}\begin{bmatrix} 2 & 3 \\ -1 & 4 \end{bmatrix}$

20. $\begin{aligned} 3x_1 + 5x_2 &= -4, \\ 2x_1 + 4x_2 &= -2 \end{aligned}$ $\quad A^{-1} = \frac{1}{2}\begin{bmatrix} 4 & -5 \\ -2 & 3 \end{bmatrix}$

21. $\begin{aligned} 3x_1 + x_2 &= 2, \\ x_1 - x_2 + 2x_3 &= -4 \\ x_1 + x_2 + x_3 &= 5 \end{aligned}$ $\quad A^{-1} = \frac{1}{8}\begin{bmatrix} 3 & 1 & -2 \\ -1 & -3 & 6 \\ -2 & 2 & 4 \end{bmatrix}$

22. $\begin{aligned} x_1 + x_3 &= -4, \\ 2x_1 + x_2 &= -3 \\ x_1 - x_2 + x_3 &= 1 \end{aligned}$ $\quad A^{-1} = -\frac{1}{2}\begin{bmatrix} 1 & -1 & -1 \\ -2 & 0 & 2 \\ -3 & 1 & 1 \end{bmatrix}$

Consider

$$A = \begin{bmatrix} 0 & -1 \\ 1 & 2 \end{bmatrix} \quad \text{and} \quad B = \begin{bmatrix} -1 & 1 \\ 3 & 0 \end{bmatrix}.$$

23. Show that $(A + B)(A + B) \neq A^2 + 2AB + B^2$, where $AA = A^2$ and $BB = B^2$.

24. Show that $(A - B)(A + B) \neq A^2 - B^2$.

25. For $X = \begin{bmatrix} a \\ b \\ c \end{bmatrix}$ and $Y = \begin{bmatrix} e \\ f \\ g \end{bmatrix}$, find $\begin{bmatrix} X \\ Y \end{bmatrix}^T$.

26. For $A = \begin{bmatrix} -2 & 3 & 7 \end{bmatrix}$ and $O = \begin{bmatrix} 0 & 0 & 0 \end{bmatrix}$, find $\begin{bmatrix} A & O \end{bmatrix}$.

Given

$$A = \begin{bmatrix} 2 & -1 & 3 \\ 4 & 1 & 0 \end{bmatrix} \quad \text{and} \quad B = \begin{bmatrix} 0 & 1 & -2 \\ 1 & -3 & 7 \end{bmatrix}.$$

27. Find $[A \quad B]$ and $[A \quad B]^T$.

28. Find $\begin{bmatrix} A \\ B \end{bmatrix}$ and $\begin{bmatrix} A \\ B \end{bmatrix}^T$.

29. Find $\begin{bmatrix} A^T \\ B^T \end{bmatrix}$. Is $[A \quad B]^T = \begin{bmatrix} A^T \\ B^T \end{bmatrix}$?

30. Find $[A^T \quad B^T]$. Is $\begin{bmatrix} A \\ B \end{bmatrix}^T = [A^T \quad B^T]$?

OBJECTIVE

You should be able to compute the inverse of a square matrix.

*2.5 (Optional) COMPUTATION OF THE MATRIX INVERSE

We now describe a way to compute a matrix inverse. We do this with an example and in the abstract. Suppose we wanted to find the inverse of the matrix

$$A = \begin{bmatrix} 2 & -1 & 1 \\ 1 & -2 & 3 \\ 4 & 1 & 2 \end{bmatrix}.$$

We are going to use an echelon tableau but a bit differently. Proceeding as if A were the coefficient matrix of some system of equations, we write the matrix A in the left side as we normally do, but on the right side we write the identity matrix and use 1's as headings:

x_1	x_2	x_3	1	1	1
2	-1	1	1	0	0
1	-2	3	0	1	0
4	1	2	0	0	1

x_1	$x_2 \cdots x_n$	1	$1 \cdots 1$
	A		I

Now we proceed with the echelon method, but with one exception. We *never* interchange columns. In truth, we never did interchange columns in any of the examples of Sections 2.1 and 2.2, but we could have. Thus, we use only *row operations*, performing them on the entire augmented (or lengthened) rows. From the explanation in Section 2.4, we know that carrying out the echelon method has the same effect as multiplying by A^{-1}. But this time, when we multiply I on the right side of the tableau, we get A^{-1}.

Suppose we wanted to find the inverse of the given matrix A. The procedure is to perform row operations to obtain I, or

$$\begin{bmatrix} 1 & 0 & 0 \\ 0 & 1 & 0 \\ 0 & 0 & 1 \end{bmatrix}$$

on the *left side* of the tableau. The resulting matrix appearing on the *right side* is the inverse matrix A^{-1}. We illustrate this as follows.

1. We first obtain a 1 in the first row, first column, by interchanging the first and second rows.

x_1	x_2	x_3	1	1	1
1^*	-2	3	0	1	0
2	-1	1	1	0	0
4	1	2	0	0	1

2. Next we pivot to obtain 0's in the rest of the first column. We multiply the pivot row by -2 and **add** to the second row. We multiply the pivot row by -4 and add to the third row.

x_1	x_2	x_3	1	1	1
1	-2	3	0	1	0
0	3	-5	1	-2	0
0	9	-10	0	-4	1

3. Next we obtain a 1 in the second row, second column. We do this by multiplying the second row by $\frac{1}{3}$.

x_1	x_2	x_3	1	1	1
1	-2	3	0	1	0
0	1^*	$-\frac{5}{3}$	$\frac{1}{3}$	$-\frac{2}{3}$	0
0	9	-10	0	-4	1

4. We pivot to obtain 0's in the rest of the second column. We multiply the pivot row by 2 and add to the first. We multiply the pivot row by -9 and add to the third:

x_1	x_2	x_3	1	1	1
1	0	$-\frac{1}{3}$	$\frac{2}{3}$	$-\frac{1}{3}$	0
0	1	$-\frac{5}{3}$	$\frac{1}{3}$	$-\frac{2}{3}$	0
0	0	5	-3	2	1

49. Use the echelon method to find the inverse.

$$A = \begin{bmatrix} 2 & -3 \\ 4 & 5 \end{bmatrix}$$

5. To get a 1 in the third row, third column, we multiply by $\frac{1}{5}$:

x_1	x_2	x_3	1	1	1
1	0	$-\frac{1}{3}$	$\frac{2}{3}$	$-\frac{1}{3}$	0
0	1	$-\frac{5}{3}$	$\frac{1}{3}$	$-\frac{2}{3}$	0
0	0	1^*	$-\frac{3}{5}$	$\frac{2}{5}$	$\frac{1}{5}$

6. We pivot to obtain 0's in the rest of the third column. We multiply the pivot row by $\frac{1}{3}$ and add to the first. We multiply the pivot row by $\frac{5}{3}$ and add to the second:

x_1	x_2	x_3	1	1	1
1	0	0	$\frac{7}{15}$	$-\frac{1}{5}$	$\frac{1}{15}$
0	1	0	$-\frac{2}{3}$	0	$\frac{1}{3}$
0	0	1	$-\frac{3}{5}$	$\frac{2}{5}$	$\frac{1}{5}$

Thus

$$A^{-1} = \begin{bmatrix} \frac{7}{15} & -\frac{1}{5} & \frac{1}{15} \\ -\frac{2}{3} & 0 & \frac{1}{3} \\ -\frac{3}{5} & \frac{2}{5} & \frac{1}{5} \end{bmatrix}$$

The reader can check this by doing the multiplication $A^{-1}A$.

If we do not obtain the identity matrix on the left, as would be the case when the system has no solution or infinitely many solutions, then A^{-1} does not exist.

This procedure will work for any square matrix that has an inverse.

DO EXERCISE 49.

EXERCISE SET 2.5

Find A^{-1}.

1. $A = \begin{bmatrix} 3 & 2 \\ 5 & 3 \end{bmatrix}$
    2. $A = \begin{bmatrix} 3 & 5 \\ 1 & 2 \end{bmatrix}$
    3. $A = \begin{bmatrix} 11 & 3 \\ 7 & 2 \end{bmatrix}$
    4. $A = \begin{bmatrix} 8 & 5 \\ 5 & 3 \end{bmatrix}$

5. $A = \begin{bmatrix} 4 & -3 \\ 1 & 2 \end{bmatrix}$
    6. $A = \begin{bmatrix} 3 & 5 \\ 2 & 4 \end{bmatrix}$

7. $A = \begin{bmatrix} 3 & 1 & 0 \\ 1 & -1 & 2 \\ 1 & 1 & 1 \end{bmatrix}$
  8. $A = \begin{bmatrix} 1 & 0 & 1 \\ 2 & 1 & 0 \\ 1 & -1 & 1 \end{bmatrix}$
  9. $A = \begin{bmatrix} 1 & -1 & 2 \\ 0 & 1 & 3 \\ 2 & 1 & -2 \end{bmatrix}$
  10. $A = \begin{bmatrix} 1 & -1 & 2 \\ 0 & 1 & 2 \\ 1 & -3 & -4 \end{bmatrix}$

CHAPTER 2 TEST

Put in matrix form and solve using the echelon method.

1. $7x_1 + 4x_2 = -21,$
$\quad 3x_1 + x_2 = -9$

2. $3x_1 - 2x_2 + 3x_3 = 24,$
$\quad x_1 + x_2 - x_3 = -7,$
$\quad 2x_1 + 3x_2 - 5x_3 = -32$

Write as separate equations and solve using the echelon method.

3. $\begin{bmatrix} 4 & -8 \\ 3 & -6 \end{bmatrix}\begin{bmatrix} x_1 \\ x_2 \end{bmatrix} = \begin{bmatrix} -20 \\ -15 \end{bmatrix}$

4. $\begin{bmatrix} 8 & -4 \\ 6 & -3 \end{bmatrix}\begin{bmatrix} x_1 \\ x_2 \end{bmatrix} = \begin{bmatrix} 20 \\ 16 \end{bmatrix}$

Complete the solution using the echelon method.

5.

x_1	x_2	x_3	x_4	1
1	0	6	0	5
0	1	-2	0	3
0	0	0	-4	-8

6.

x_1	x_2	x_3	1
1	3	-2	6
0	4	-1	8
0	4	1	2
0	8	0	-10

For Questions 7 through 10, consider

$$A = \begin{bmatrix} -3 & 2 \\ -5 & 1 \end{bmatrix} \quad \text{and} \quad B = \begin{bmatrix} 0 & -1 \\ 1 & 0 \end{bmatrix}.$$

Find:

7. $A + B$

8. AB

9. $A - B$

10. $-4A$

11. For

$$C = \begin{bmatrix} 2 \\ -3 \\ 4 \end{bmatrix}, \text{ find } C^{\text{T}}.$$

12. Find AB and BA, if possible.

$$A = \begin{bmatrix} 1 \\ -1 \\ 2 \end{bmatrix}, \quad B = \begin{bmatrix} 2 & -3 & 0 \\ 1 & 2 & 4 \end{bmatrix}.$$

In Questions 13 and 14, a system of equations is given, together with the inverse of the coefficient matrix. Use the matrix inverse to solve the system. Show your work.

13. $8x_1 + 3x_2 = 6,$
$\quad 4x_1 - 6x_2 = -2$
$\qquad A^{-1} = \begin{bmatrix} \frac{1}{10} & \frac{1}{20} \\ \frac{1}{15} & -\frac{2}{15} \end{bmatrix}$

14. $x_1 + x_2 + x_3 = 10,$
$\quad x_1 + 2x_2 - x_3 = -5,$
$\quad 2x_1 - x_2 + 3x_3 = 20$
$\qquad A^{-1} = \begin{bmatrix} -1 & \frac{4}{5} & \frac{3}{5} \\ 1 & -\frac{1}{5} & -\frac{2}{5} \\ 1 & -\frac{3}{5} & -\frac{1}{5} \end{bmatrix}$

15. For $X = \begin{bmatrix} p \\ q \end{bmatrix}$ and $Y = \begin{bmatrix} t \\ u \end{bmatrix}$, find $\begin{bmatrix} X \\ Y \end{bmatrix}^{\text{T}}$.

16. For a certain year, $850 is received in interest from two investments. A certain amount is invested at 8%, and $1300 more than this is invested at 10%. Find the amount invested at each rate.

*17. Use the echelon method to find A^{-1}.

$$A = \begin{bmatrix} 1 & 0 & -1 \\ -1 & 1 & -1 \\ -1 & 0 & 2 \end{bmatrix}$$

CHAPTER THREE

Linear Programming

During World War II the Army began to formulate certain *linear optimization* problems. Their solutions were called plans or *programs*. Subsequently, many other problems, particularly economic, were found to have a similar mathematical formulation. Such problems are called linear-programming problems or *linear programs* for short.

For example, consider the manager of a department store who sends his buyer to the "market." The buyer has a budget and can spend no more than a given amount of money. Furthermore, the goods must be brought back to the store in a company truck with a maximum cargo volume and a maximum cargo weight. That is, the total *volume* of the goods bought cannot exceed the cargo *volume* limit of the truck, nor can the total *weight* of the goods bought exceed the cargo *weight* limit of the truck. Each type of item bought can be marked up a certain percent. What items should the buyer buy to *maximize* the total value of the goods bought subject to the given constraints?

Such problems are characterized by the following features:

i) The specifications of the problem are related by *inequalities*, or *constraints*, rather than by equations. For example,

$$2x_1 + 3x_2 \leqslant 6 \quad \text{is an inequality,}$$

while

$$2x_1 + 3x_2 = 6 \quad \text{is an equation.}$$

ii) The inequalities or constraints of the problem are linear. For example, $2x_1 + 3x_2 \leqslant 6$ is a linear inequality; that is, all variables are to the first power and there are no divisions by a variable. Note that

$$2x_1^2 + 9x_2^2 + \frac{1}{x_1} \leqslant 36 \quad \text{is } not \text{ a linear inequality.}$$

iii) The *solution* of a linear program must satisfy these constraints. In addition some quantity, a *linear* function of the variables, must be maximized or minimized. For example, maximize f where $f = 3x_1 + 2x_2$, and, for short, we have written f in place of $f(x_1, x_2)$.

In the following sections we shall consider various aspects of the formulation and solution of linear programs.

3.1 GRAPHING A SYSTEM OF LINEAR CONSTRAINTS (INEQUALITIES) IN TWO VARIABLES

The first stage in the solution of a linear program is graphing the constraints (inequalities). Consider an inequality like any of the following:

$$a_1x_1 + a_2x_2 \geqslant b,$$

or

$$a_1x_1 + a_2x_2 \leqslant b,$$

or

$$a_1x_1 + a_2x_2 > b,$$

or

$$a_1x_1 + a_2x_2 < b,$$

where a_1, a_2, and b are constants.

Corresponding to any of these inequalities there is a *related equation*

$$a_1x_1 + a_2x_2 = b.$$

Its graph is a line that divides the plane into two half-planes. We graph this equation first.

For inequalities with

a) $\geqslant$ or $\leqslant$, **use a solid line for the related equation;**
b) $>$ or $<$, **use a dashed line for the related equation.**

Example 1   Graph $2x_1 + 3x_2 \leqslant 6$.

Solution   There are two steps. First, graph the related equation $2x_1 + 3x_2 = 6$.

Since the equality is included ($\leqslant$) in the present example, the line is solid in Fig. 3.1. See page 122.

Second, having divided the plane into two half-planes, we must decide which half-plane contains the solutions of the inequality. To do this, we need consider only one test point. If *any* point on one side of the graph of the related equation satisfies the inequality, then *all* points in the same half-plane satisfy the inequality. If any point on one side of the graph of the related equation does *not* satisfy the inequality, then all points in the *opposite* half-plane satisfy the inequality.

OBJECTIVES

You should be able, given a system of constraints, to:

a) Graph the solution set of the system.
b) Decide if the solution set is empty? nonempty?
c) Decide if the solution set is bounded—unbounded.
d) Decide which constraints if any, are redundant.
e) Decide which corners, if any, are degenerate.

1. Graph $x_1 + 5x_2 \leqslant 10$

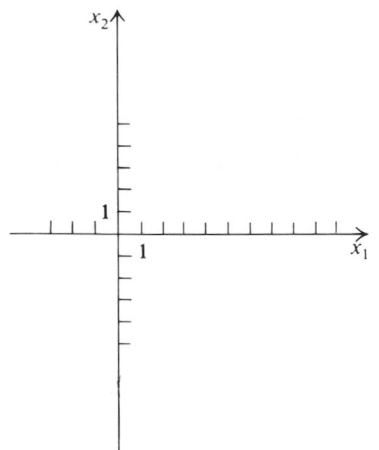

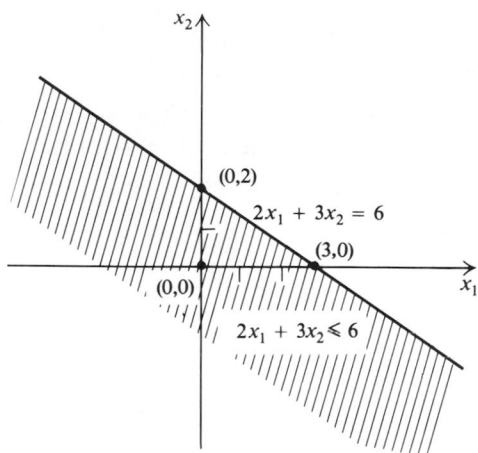

Figure 3.1

If the graph of the related equation does not include the origin $(0, 0)$, we use the origin as a test point. In the present example, we thus ask "Is $(0, 0)$ a solution of $2x_1 + 3x_2 \leqslant 6$?"

We replace x_1 by 0 and x_2 by 0:

$$\begin{array}{c|c} 2x_1 + 3x_2 & \leqslant 6 \\ \hline 2 \cdot 0 + 3 \cdot 0 & 6 \\ 0 + \quad 0 & \\ 0 & \end{array}$$

Because $0 \leqslant 6$ is true, $(0, 0)$ is a solution.

Since $(0, 0)$ is in the lower half-plane, *all* points in the lower half-plane are solutions. Thus, the shaded half-plane of Fig. 3.1 represents the solutions of the inequality.

DO EXERCISE 1.

Example 2 Graph $2x_1 - x_2 > 0$.

The related equation $2x_1 - x_2 = 0$ is graphed, using a dashed line since the line is not included in the inequality. See Fig. 3.2. Here the line passes through the origin $(0, 0)$, so we use either $(1, 0)$ or $(0, 1)$ as a test point. For $(1, 0)$ we ask "Is $(1, 0)$ a solution of $2x_1 - x_2 > 0$?"

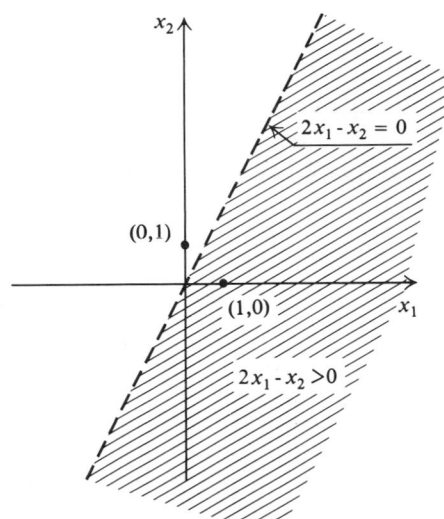

Figure 3.2

We replace x_1 by 1 and x_2 by 0.

$$2x_1 - x_2 > 0$$

$2 \cdot 1 - 0$	0
$2 - 0$	
2	

Because $2 > 0$ is true, $(1, 0)$ is a solution.

Since $(1, 0)$ is in the lower half-plane, all points in the lower half-plane are solutions of the inequality.

Alternatively, we could use $(0, 1)$ as a test point and ask "Is $(0, 1)$ a solution of $2x_1 - x_2 > 0$?"

We replace x_1 by 0 and x_2 by 1.

$$2x_1 - x_2 > 0$$

$2 \cdot 0 - 1$	0
$0 - 1$	
-1	

Since $-1 > 0$ is false, $(0, 1)$ is not a solution.

Since $(0, 1)$ in the *upper* half-plane is *not* a solution, all points in the *lower* half-plane are solutions of the inequality.

2. Graph $x_1 - 5x_2 > 0$.

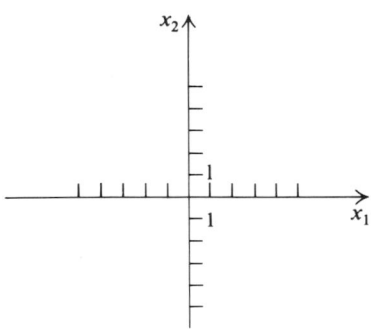

3. Graph $x_2 \geq 1$.

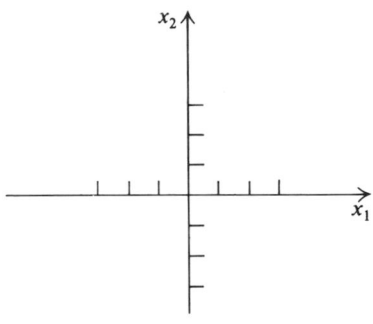

Either case tells us the solution set is the half-plane shaded in Fig. 3.2.

It should be noted that the inequality $2x_1 - x_2 > 0$ is equivalent to $x_2 - 2x_1 < 0$, if we recall that, when both sides of an inequality are multiplied by -1, the inequality sign is *reversed*.

DO EXERCISES 2 AND 3.

To graph a set, or *system*, of linear constraints, we first graph the solution set of each individual constraint using the same set of axes. The solution set of the *system* is that region, or set of ordered pairs, which satisfies *all* the constraints.

Example 3 Graph the solution set of the system of constraints

i) $x_1 - 2x_2 \leq 0$.
ii) $-2x_1 + \; x_2 \leq 2$.

These two constraints are each graphed in Fig. 3.3. A *pair of arrows* points in the direction of the half-plane representing solutions of each inequality. The region satisfying *both* constraints is indicated by the shading.

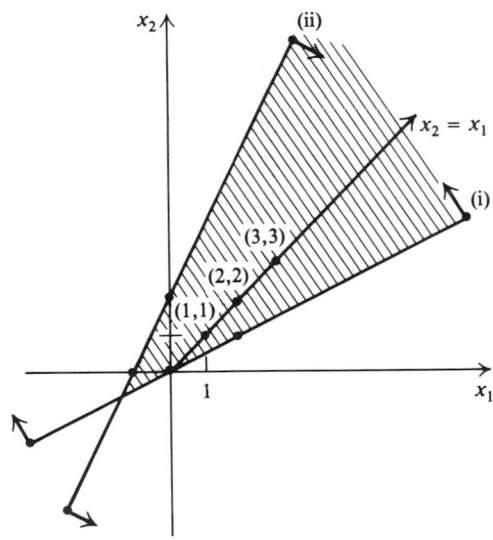

Figure 3.3

DO EXERCISE 4.

Example 4   Now let us add a third constraint to those of Example 3, so that we have

i) $x_1 - 2x_2 \leqslant 0,$
ii) $-2x_1 + x_2 \leqslant 2,$
iii) $x_1 + x_2 \leqslant 6.$

These are graphed in Fig. 3.4. The solution set of this system of *three* constraints is shaded. The solution set of Example 4 (Fig. 3.4) is *bounded*. This means simply that the solution set is confined to the boundary and interior of some polygon.

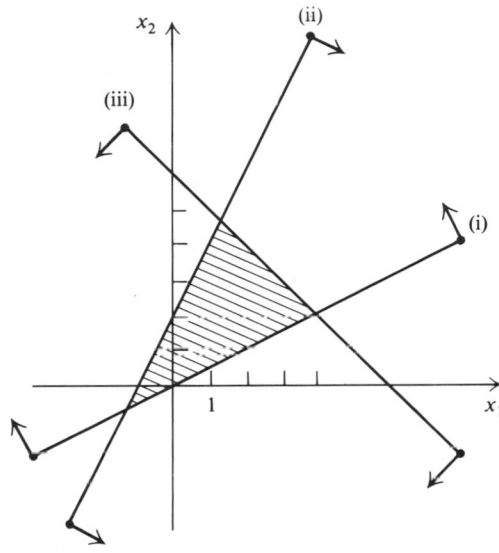

Figure 3.4

Not all solution sets are bounded. For example, the solution set of Example 3 (Fig. 3.3) is *unbounded*. This means that the solution set is *not* confined to the boundary and interior of some polygon. Note that in the direction of the arrow along the line $x_2 = x_1$, the boundary is *open*, so that in such a direction, given any solution one can find another solution farther out. In Example 3 any point (c, c) where $c \geqslant 0$ is a solution to the system of constraints (i) and (ii). The parameter c can become arbitrarily large and the point (c, c) will still be in the solution set.

DO EXERCISE 5.

4. Graph the solution set of the system of constraints

i) $x_1 + x_2 \geqslant 1$

ii) $-x_1 + x_2 \leqslant 2$

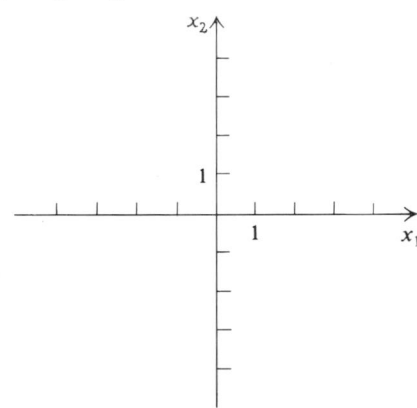

5. To the system of constraints of Margin Exercise 4, add

iii) $x_1 \leqslant 4$

and graph.

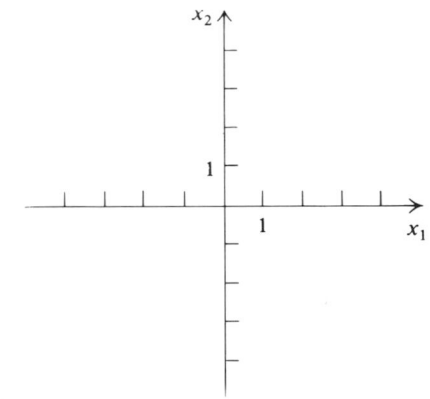

6. Add these constraints to those of Margin Exercise 5:

iv) $x_2 \geq 0$

v) $x_2 \leq 4$

and graph.

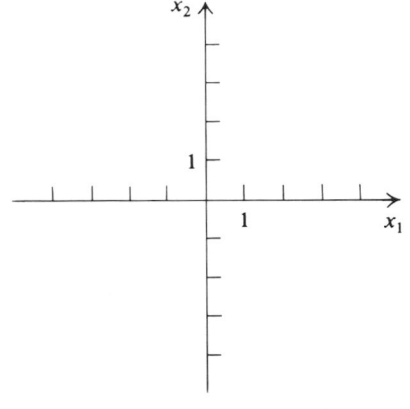

Example 5   Adding two more constraints, we have

i) $x_1 - 2x_2 \leq 0,$
ii) $-2x_1 +   x_2 \leq 2,$
iii) $x_1 +   x_2 \leq 6,$
iv) $x_1   \leq 2,$
v) $x_2 \leq 2.$

These are graphed in Fig. 3.5.

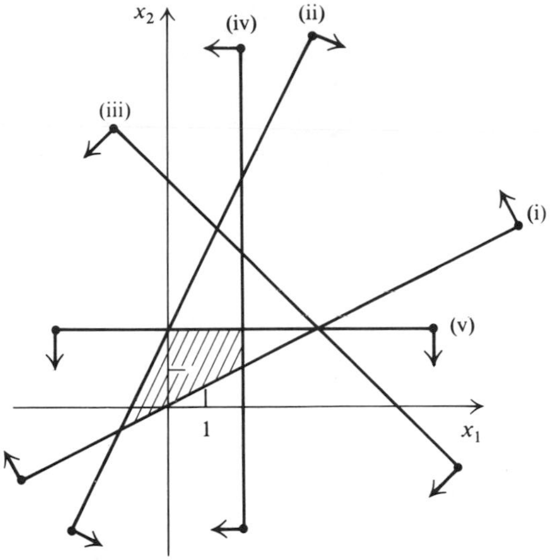

Figure 3.5

The solution set has now been sufficiently reduced so that constraint (iii) no longer affects the solution set and hence is considered *redundant*. Redundant constraints, when identified as such, may be disregarded for present purposes.

DO EXERCISE 6.

Example 6   Now let us replace constraint (iii) by (vi), so that we have:

i) $x_1 - 2x_2 \leq 0,$
ii) $-2x_1 +   x_2 \leq 2,$
iv) $x_1   \leq 2,$
v) $x_2 \leq 2,$
vi) $x_1 +   x_2 \leq 4.$

Thus, we obtain the graph in Fig. 3.6.

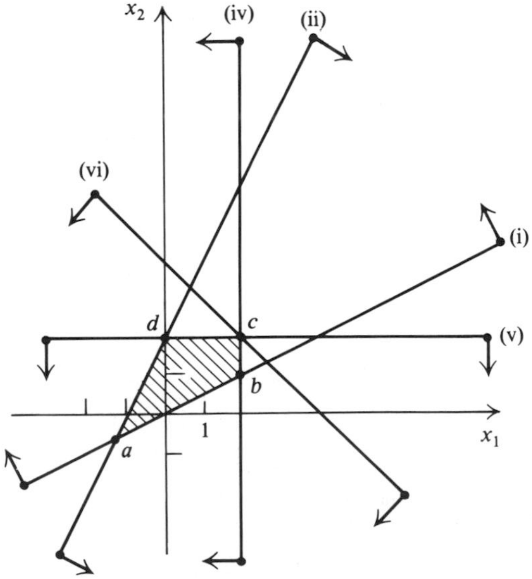

Figure 3.6

The *corners* (or vertices) of the solution set are each the intersection of two lines. Thus, the coordinates of the corners can be determined by the solution of the corresponding *two* related equations. For example, point a with coordinates $(-\frac{4}{3}, -\frac{2}{3})$ is the simultaneous solution of the equations related to the *two* constraints (i) and (ii). Similarly, point b with coordinates $(2, 1)$ corresponds to the intersection of the equations related to the *two* constraints (i) and (iv). Point c, on the other hand, with coordinates $(2, 2)$ is the simultaneous solution of the equations related to the *three* constraints (iv), (v), and (vi). Such points are called *degenerate* in linear programs. These degeneracies are important in later use but not here. For present purposes we consider constraint (vi) redundant. It may not be obvious from the graph which points are degenerate or which constraints are redundant. In this case, when we solved the pairs of related equations, the same solution would have occurred more than once, indicating a degeneracy.

DO EXERCISE 7.

Example 7   If constraint (vi) is replaced by (vii), so that we have:

i) $x_1 - 2x_2 \leq 0,$
ii) $-2x_1 + x_2 \leq 2,$
iv) $x_1 \leq 2,$
v) $x_2 \leq 2,$
vii) $x_1 + x_2 \geq 5,$

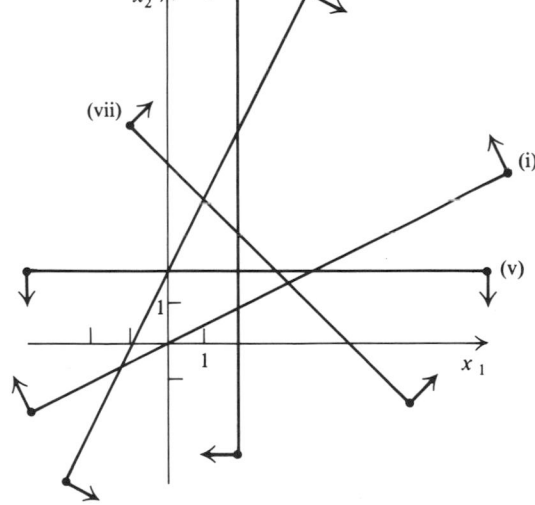

Figure 3.7

then we obtain the graph in Fig. 3.7. In this case there is *no* region common to all constraints (i), (ii), (iv), (v), and (vii). The solution set is then *empty*. This is not the same as saying that the solution set is $(0, 0)$, because if the point $(0, 0)$ were in the solution set, it would *not* be empty. Furthermore, the empty set is bounded.

DO EXERCISE 8.

7. Add the constraint

vi) $x_1 \leq 2$

to the preceding and graph.

a) Determine which points (if any) are degenerate.

b) Determine which constraints (if any) are redundant.

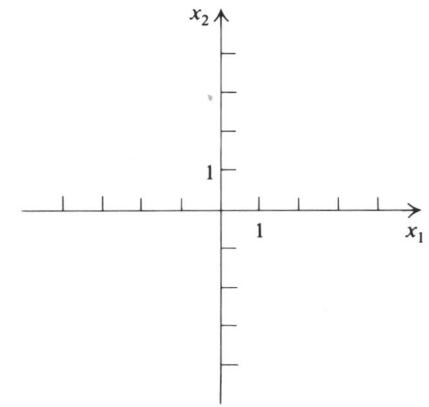

8. Replace constraint (vi) by

vii) $x_1 + 2 \leq 0$

and graph.

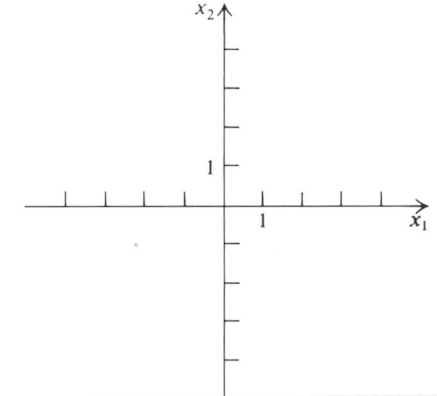

9. Which of the following areas are convex?

a)

b)

c)

d)

e)

f)

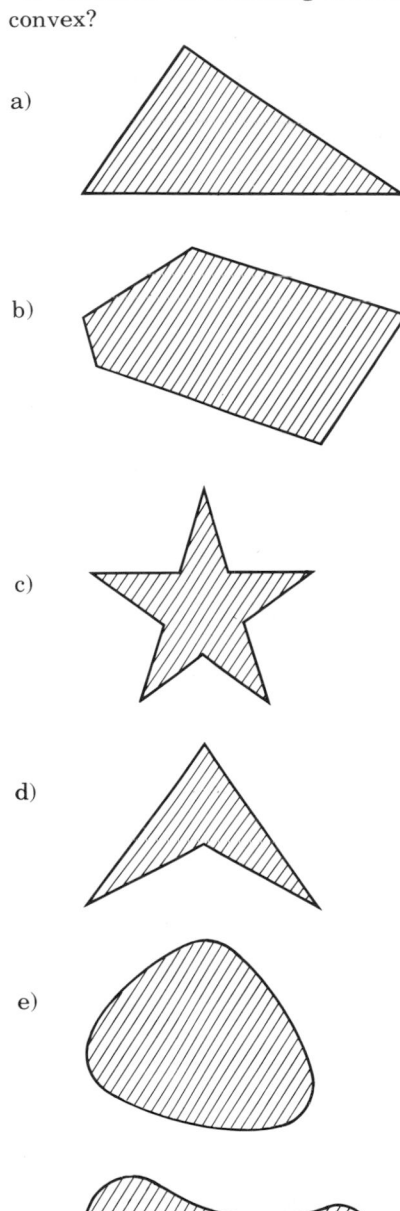

Consider two points in the solution set shown shaded in Fig. 3.8. Any point on the line segment between these two points is also in the solution set.*

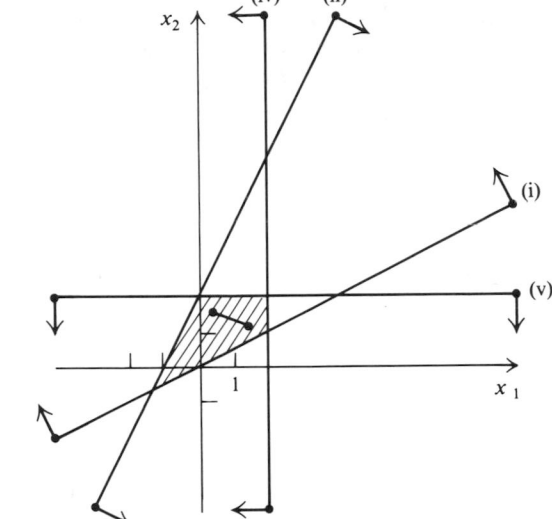

Figure 3.8

If the line segment between any pair of points in a set is also in the set, then such a set is called a *convex* set.

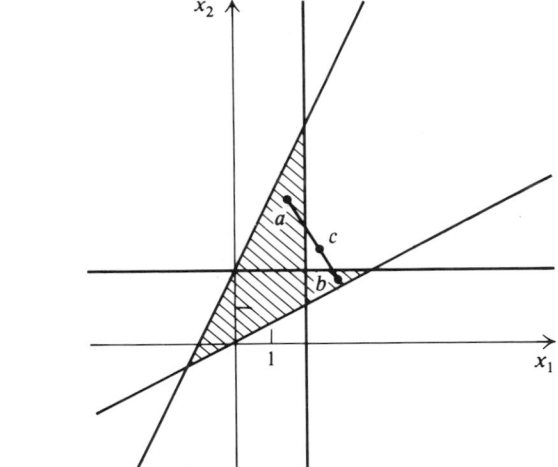

Figure 3.9

---

*Any point on the line segment *between* two points is called a *convex combination* of the two points.

An example of a set which is not convex is the shaded region of Fig. 3.9. In this case there are two points, such as a and b in the figure, such that the line segment *between* them includes points *outside* the given set, for example, point c. Thus, this set is *not* convex. It can be shown that

For any system of linear constraints, the corresponding solution set must be convex.

Knowing this is useful in determining the solution set of a graphed system of constraints. For example, the shaded region of Fig. 3.9 is not convex. Therefore, it could not possibly be a solution set for *any* system of linear constraints.

DO EXERCISE 9. (See preceding page.)

EXERCISE SET 3.1

Graph.

1. $5x_1 + x_2 > 10$

2. $3x_1 - 2x_2 \leqslant 12$

3. $3x_1 + 2x_2 \geqslant 6$

4. $x_1 - 4x_2 \leqslant 0$

5. $2x_1 + 5x_2 \leqslant 8$

6. $3x_1 + 7x_2 > 10$

Graph the following systems of constraints and shade the solution set.

7. $x_1 + x_2 \leqslant 6,$
$\quad\quad x_2 \leqslant 5,$
$\quad\quad x_1, x_2 \geqslant 0.$

8. $x_1 + 2x_2 \leqslant 8,$
$\quad\quad x_1 \leqslant 6,$
$\quad\quad x_1, x_2 \geqslant 0.$

9. $3x_1 + 2x_2 \leqslant 12,$
$\quad\quad x_1 + x_2 \leqslant 5,$
$\quad\quad x_1, x_2 \geqslant 0.$

10. $3x_1 + 5x_2 \leqslant 15,$
$\quad\quad 3x_1 + 2x_2 \leqslant 12,$
$\quad\quad x_1, x_2 \geqslant 0.$

11. $x_1 + 2x_2 \leqslant 14,$
$\quad\quad 4x_1 + 3x_2 \leqslant 26,$
$\quad\quad 2x_1 + x_2 \leqslant 12,$
$\quad\quad x_1, x_2 \geqslant 0.$

12. $x_1 + 3x_2 \leqslant 18,$
$\quad\quad 3x_1 + 2x_2 \leqslant 19,$
$\quad\quad 2x_1 + x_2 \leqslant 12,$
$\quad\quad x_1, x_2 \geqslant 0,$

13. $3y_1 + y_2 \geqslant 9,$
$\quad\quad y_1 + y_2 \geqslant 7,$
$\quad\quad y_1 + 2y_2 \geqslant 8,$
$\quad\quad y_1, y_2 \geqslant 0.$

14. $2y_1 + y_2 \geqslant 8,$
$\quad\quad 4y_1 + 3y_2 \geqslant 22,$
$\quad\quad 2y_1 + 5y_2 \geqslant 18,$
$\quad\quad y_1, y_2 \geqslant 0.$

15. $2y_1 + y_2 \geqslant 9,$
$\quad\quad 4y_1 + 3y_2 \geqslant 23,$
$\quad\quad y_1 + 3y_2 \geqslant 8,$
$\quad\quad y_1, y_2 \geqslant 0.$

16. $5y_1 + 3y_2 \geqslant 30,$
$\quad\quad 2y_1 + 3y_2 \geqslant 21,$
$\quad\quad 3y_1 + 6y_2 \geqslant 36,$
$\quad\quad y_1, y_2 \geqslant 0.$

17. $4y_1 + y_2 \geqslant 9,$
$\quad\quad 3y_1 + 2y_2 \geqslant 13,$
$\quad\quad 2y_1 + 5y_2 \geqslant 16,$
$\quad\quad y_1, y_2 \geqslant 0.$

18. $4y_1 + y_2 \geqslant 7,$
$\quad\quad y_1 + y_2 \geqslant 4,$
$\quad\quad 2y_1 + 5y_2 \geqslant 14,$
$\quad\quad y_1, y_2 \geqslant 0.$

For the following exercises

a) Graph the solution set of the system of constraints.
c) Is the solution set bounded? unbounded?
e) Which corners, if any, are degenerate?

b) Is the solution set empty? nonempty?
d) Which constraints, if any, are redundant?

Note. It may be difficult to answer (d) and (e) unless the equations are drawn carefully. In the next section, an alternative (algebraic) method will be used to determine the solution set more accurately.

Save your results for use at the end of the next section.

19. $x_1 + 2x_2 \leq 6,$
$\quad 0 \leq x_1 \leq 5,$
$\quad\quad x_2 \geq -2$

20. $x_1 - x_2 \geq -4,$
$\quad x_1 - x_2 \leq 6,$
$\quad -2 \leq x_2 \leq 2$

21. $\quad\quad x_1 \geq -3,$
$\quad x_1 - 2x_2 \leq 4,$
$\quad x_2 - 3x_1 \leq 9,$
$\quad 3x_1 + x_2 \leq 10,$

22. $\quad\quad 3x_1 \geq x_2,$
$\quad\quad 3x_2 \geq x_1,$
$\quad x_1 + x_2 \geq 5,$
$\quad 2x_1 + 3x_2 \leq 24$

▶

23. $-3x_1 + 2x_2 \geq 6,$
$\quad 2x_1 + x_2 \leq -2,$
$\quad x_1 + x_2 \geq 4,$
$\quad 2x_1 + 7x_2 \leq 21$

24. $x_1 + 4x_2 \geq -4,$
$\quad 2x_1 + x_2 \leq 2,$
$\quad\quad x_2 \geq 0,$
$\quad\quad x_1 \leq 5$

25. $\quad\quad x_1 \geq 0,$
$\quad\quad x_2 \geq 0,$
$\quad x_1 + x_2 \geq 2,$
$\quad x_1 - x_2 \leq 2,$
$\quad\quad x_2 \leq 6$

26. $-3x_1 + 4x_2 \leq 12,$
$\quad 3x_1 + 2x_2 \leq 24,$
$\quad\quad x_1 \geq 0,$
$\quad\quad x_2 \geq 0,$
$\quad\quad x_2 \geq 6$

27. $x_1 + x_2 \leq 0,$
$\quad 2x_1 - 3x_2 \leq 15,$
$\quad\quad x_2 \leq 5,$
$\quad\quad x_1 \geq 0,$
$\quad 2x_1 + x_2 \geq 3$

28. $3x_1 + 2x_2 \geq 6,$
$\quad\quad x_1 \geq 1,$
$\quad 0 \leq x_2 \leq 6,$
$\quad 2x_1 + 3x_2 \leq 24,$
$\quad 3x_1 + x_2 \leq 15$

29. $\quad\quad x_1 \geq 0,$
$\quad\quad x_2 \geq 0,$
$\quad 5x_2 - 3x_1 \leq 15,$
$\quad\quad x_1 \leq 4x_2$
$\quad 2x_1 - 5x_2 \leq 10$

30. $\quad\quad x_1 \geq 0,$
$\quad\quad x_2 \geq 0,$
$\quad -7x_1 + x_2 \leq 7,$
$\quad 4x_2 - 5x_1 \geq 20,$
$\quad x_1 + x_2 \leq 10,$
$\quad\quad x_2 \leq 3$

OBJECTIVES

You should be able to

a) Find the corners of a system of constraints.
b) Find the optimum values of an objective function and the points at which they are attained, subject to a system of constraints.

3.2 DETERMINING THE OPTIMUM VALUE

We now find optimum values of some linear function of the variables. By this we mean the largest or smallest values of the function. The function being optimized is called the *objective function*. Let us again consider the solution set corresponding to constraints (i), (ii), (iv), and (v) of Example 5 of Section 3.1.

i) $\quad x_1 - 2x_2 \leq 0,$
ii) $-2x_1 + x_2 \leq 2,$
iv) $\quad x_1 \quad\quad \leq 2,$
v) $\quad\quad x_2 \leq 2$

They are graphed in Fig. 3.8 and in Fig. 3.10.

The *corners* have been labeled a, b, c, and d.

We need to find the coordinates of the corners. We determine them algebraically. Point a (Fig. 3.10) is the intersection of the equations related to constraints (i) and (ii):

i') $\quad x_1 - 2x_2 = 0,$
ii') $-2x_1 + x_2 = 2.$

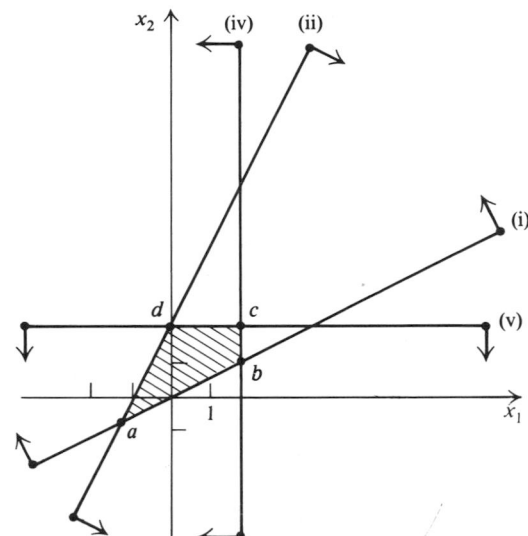

Figure 3.10

10. Use the echelon method and solve.
$$2x_1 - x_2 = 4,$$
$$3x_1 + 2x_2 = 12$$

We find the coordinates of point a, represented by (a_1, a_2), by solving the system of equations (i') and (ij'). We use the echelon method described in Section 2.1. Other methods of solution could be used, but we are practicing the echelon method for later use.

Example 1 Solve the system of equations (i') and (ii') and determine the coordinates of point a.

Solution The initial tableau is at the left.

x_1	x_2	1		x_1	x_2	1		x_1	x_2	1
1*	−2	0		1	−2	0		1	0	$-\frac{4}{3}$
−2	1	2		0	−3	2		0	1*	$-\frac{2}{3}$

Indicating the pivot by an * and pivoting (multiplying the first row by 2 and adding), we obtain the second tableau. Multiplying the second row by $-\frac{1}{3}$ to make the pivot element 1 and pivoting yields the final tableau. Thus, the coordinates of point a are $(a_1, a_2) = (-\frac{4}{3}, -\frac{2}{3})$.

DO EXERCISE 10.

By using the appropriate equations and solving algebraically (with the echelon method), we find the coordinates of the corners:

(i) and (ii) yield $(a_1, a_2) = (-\frac{4}{3}, -\frac{2}{3})$,
(i) and (iv) yield $(b_1, b_2) = (2, 1)$,
(iv) and (v) yield $(c_1, c_2) = (2, 2)$,
(ii) and (v) yield $(d_1, d_2) = (0, 2)$.

Example 2 Find the optimum (maximum and minimum) values and the points at which they are obtained, of the objective function[*] $f(x_1, x_2) = x_1 + x_2 = f$, for short, subject to the constraints:

i) $\quad x_1 - 2x_2 \leqslant 0,$
ii) $-2x_1 + \quad x_2 \leqslant 2,$
iv) $\quad x_1 \quad\quad \leqslant 2,$
v) $\quad\quad\quad x_2 \leqslant 2.$

Solution To do this, let f assume various values. Thus, $f = 6$ leads to the equation

$$6 = x_1 + x_2,$$

the graph of which is a straight line.

Similarly, $f = \quad$ 4 leads to the equation $\quad 4 = x_1 + x_2,$
$\quad\quad\quad f = \quad$ 2 leads to the equation $\quad 2 = x_1 + x_2,$
$\quad\quad\quad f = \quad$ 0 leads to the equation $\quad 0 = x_1 + x_2,$
$\quad\quad\quad f = -2$ leads to the equation $-2 = x_1 + x_2.$

The graphs of these lines are shown in Fig. **3.11** together with the solution set (shaded area).

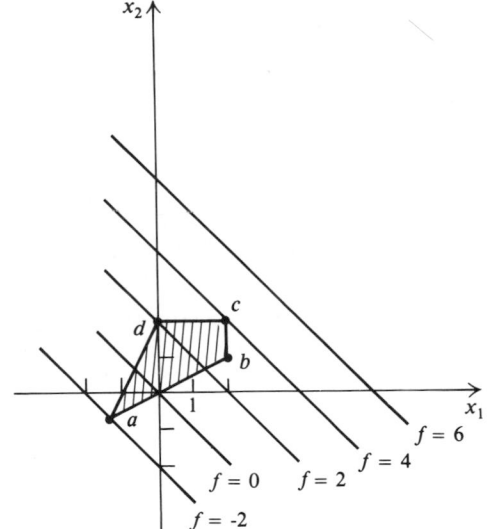

Figure 3.11

---

[*] A function of *two* variables is a relation f that assigns to each input pair (x_1, x_2) a unique output number $f(x_1, x_2)$. For example, for $f(x_1, x_2) = 4x_1 + 6x_2$, the function value $f(15, 20)$ is found by substituting 15 for x_1 and 20 for x_2:

$$f(15, 20) = 4(15) + 6(20) = 60 + 120 = 180.$$

From the figure we see that the line corresponding to $f = 6$ has *no* point in common with the solution set (shaded area). The lines corresponding to values of f between -2 and 4 $(-2 \leqslant f \leqslant 4)$ *do* have points in common with the solution set. Of all such lines $(-2 \leqslant f \leqslant 4)$, that corresponding to $f = 4$ has the maximum value of the objective function f. This maximum value is obtained at point c. Similarly, the minimum value of f is -2, obtained at point a.

It is important to note that the optima (maximum and minimum values) were obtained at the *boundary* of the solution set and, furthermore, at *corner points*. For linear programs, it can be shown that the optima will always be obtained at corner points.

To determine the optima we need evaluate the objective function *only at the corner points*. In the present example,

$$f(x_1, x_2) = x_1 + x_2$$

$$f(-\tfrac{4}{3}, -\tfrac{2}{3}) = (-\tfrac{4}{3}) + (-\tfrac{2}{3}) = -2 \quad \text{(minimum)},$$
$$f(2, 1) = 2 + 1 = 3,$$
$$f(2, 2) = 2 + 2 = 4 \quad \text{(maximum)},$$
$$f(0, 2) = 0 + 2 = 2.$$

Note. The notation $f(2, 1)$ represents the value of $f(x_1, x_2)$ when 2 is substituted for x_1 and 1 is substituted for x_2. Thus:

Maximum: $f(2, 2) = 4$,
Minimum: $f(-\tfrac{4}{3}, -\tfrac{2}{3}) = -2$.

Do Exercise 11.

Example 3 Find the optimum values of the objective function

$$f(x_1, x_2) = 2x_1 - x_2,$$

subject to the constraints:

i) $x_1 - 2x_2 \leqslant 0,$
ii) $-2x_1 + x_2 \leqslant 2,$
iv) $x_1 \leqslant 2,$
v) $x_2 \leqslant 2.$

Solution As before, let f assume various values, as indicated on Fig. 3.12. The maximum value of f occurs now at point b, while the minimum occurs all along the line segment from a to d. Evaluating f

11. Find the optima (maximum and minimum) values and the points at which they are obtained, for the objective function

$$f(x_1, x_2) = 2x_1 + 3x_2,$$

subject to the same constraints as in Example 2:

i) $x_1 - 2x_2 \leqslant 0$
ii) $-2x_1 + x_2 \leqslant 2$
iv) $x_1 \leqslant 2$
v) $x_2 \leqslant 2$

Remember you need consider only corner points.

12. Find the optimum values and the points at which they are obtained, for the objectives function

$$f(x_1, x_2) = 2x_2 - x_1,$$

subject to the same constraints as in Margin Exercise 11.

at the corner points, we have:

$$f(x_1, x_2) = 2x_1 \quad - x_2$$

$f(-\frac{4}{3}, -\frac{2}{3}) = 2(-\frac{4}{3}) - (-\frac{2}{3}) = -2$	(minimum),
$f(2, 1) \quad = 2 \cdot 2 - 1 \quad = 3$	(maximum),
$f(2, 2) \quad = 2 \cdot 2 - 2 \quad = 2,$	
$f(0, 2) \quad = 2 \cdot 0 - 2 \quad = -2$	(minimum).

The minimum value -2 can be seen to be attained at *more than one point*, specifically at the points a and d. It turns out that -2 is also attained for any *convex combination* of the coordinates of these two points, that is, for any point on the line segment between them.

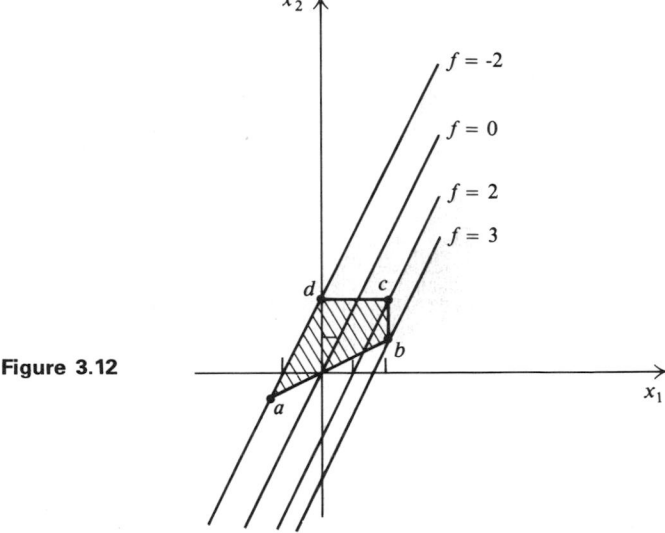

Figure 3.12

DO EXERCISE 12.

EXERCISE SET 3.2

Find the optimum value (maximum or minimum, as indicated) for each objective function and the point at which it is obtained for each set of constraints. Save your results for later use.

1. Maximize $f(x_1, x_2) = x_1 + 2x_2$, subject to the constraints (Exercise 7, Set 3.1):

$$x_1 + x_2 \leq 6,$$
$$x_2 \leq 5,$$
$$x_1, x_2 \geq 0.$$

2. Maximize $f(x_1, x_2) = x_1 + x_2$, subject to the constraints (Exercise 8, Set 3.1):

$$x_1 + 2x_2 \leq 8,$$
$$x_1 \leq 6,$$
$$x_1, x_2 \geq 0.$$

3. Maximize $f(x_1, x_2) = 5x_1 + 4x_2$, subject to the constraints (Exercise 9, Set 3.1):

$$3x_1 + 2x_2 \leq 12,$$
$$x_1 + x_2 \leq 5,$$
$$x_1, x_2 \geq 0.$$

4. Maximize $f(x_1, x_2) = 2x_1 + x_2$, subject to the constraints (Exercise 10, Set 3.1):

$$3x_1 + 5x_2 \leq 15,$$
$$3x_1 + 2x_2 \leq 12,$$
$$x_1, x_2 \geq 0.$$

5. Maximize $f(x_1, x_2) = 3x_1 + 4x_2$, subject to the constraints (Exercise 11, Set 3.1):

$$x_1 + 2x_2 \leq 14,$$
$$4x_1 + 3x_2 \leq 26,$$
$$2x_1 + x_2 \leq 12,$$
$$x_1, x_2 \geq 0.$$

6. Maximize $f(x_1, x_2) = 5x_1 + 4x_2$, subject to the constraints (Exercise 12, Set 3.1):

$$x_1 + 3x_2 \leq 18,$$
$$3x_1 + 2x_2 \leq 19,$$
$$2x_1 + x_2 \leq 12,$$
$$x_1, x_2 \geq 0.$$

7. Minimize $f(y_1, y_2) = 3y_1 + 4y_2$, subject to the constraints (Exercise 13, Set 3.1):

$$3y_1 + y_2 \geq 9,$$
$$y_1 + y_2 \geq 7,$$
$$y_1 + 2y_2 \geq 8,$$
$$y_1, y_2 \geq 0.$$

8. Minimize $f(y_1, y_2) = 3y_1 + 4y_2$, subject to the constraints (Exercise 14, Set 3.1):

$$2y_1 + y_2 \geq 8,$$
$$4y_1 + 3y_2 \geq 22,$$
$$2y_1 + 5y_2 \geq 18,$$
$$y_1, y_2 \geq 0.$$

9. Minimize $f(y_1, y_2) = 2y_1 + 5y_2$, subject to the constraints (Exercise 15, Set 3.1):

$$2y_1 + y_2 \geq 9,$$
$$4y_1 + 3y_2 \geq 23,$$
$$y_1 + 3y_2 \geq 8,$$
$$y_1, y_2 \geq 0.$$

10. Minimize $f(y_1, y_2) = 9y_1 + 7y_2$, subject to the constraints (Exercise 16, Set 3.1):

$$5y_1 + 3y_2 \geq 30,$$
$$2y_1 + 3y_2 \geq 21,$$
$$3y_1 + 6y_2 \geq 36,$$
$$y_1, y_2 \geq 0.$$

11. Minimize $f(y_1, y_2) = 3y_1 + 5y_2$, subject to the constraints (Exercise 17, Set 3.1):

$$4y_1 + y_2 \geq 9,$$
$$3y_1 + 2y_2 \geq 13,$$
$$2y_1 + 5y_2 \geq 16,$$
$$y_1, y_2 \geq 0.$$

12. Minimize $f(y_1, y_2) = 2y_1 + 3y_2$, subject to the constraints (Exercise 18, Set 3.1):

$$4y_1 + y_2 \geq 7,$$
$$y_1 + y_2 \geq 4,$$
$$2y_1 + 5y_2 \geq 14,$$
$$y_1, y_2 \geq 0.$$

Find the optimum (maximum and minimum) values for each objective function and the points at which they are obtained subject to the constraints given.

13. $f(x_1, x_2) = x_1 - x_2$, subject to the constraints (Exercise 19, Set 3.1):

$$x_1 + 2x_2 \leq 6,$$
$$0 \leq x_1 \leq 5,$$
$$x_2 \geq -2.$$

14. $f(x_1, x_2) = 2x_1 + 3x_2$, subject to the constraints of Exercise 13.

15. $f(x_1, x_2) = x_1 - x_2$, subject to the constraints (Exercise 21, Set 3.1):

$$x_1 \geqslant -3,$$
$$x_1 - 2x_2 \leqslant 4,$$
$$x_2 - 3x_1 \leqslant 9,$$
$$3x_1 + x_2 \leqslant 10.$$

16. $f(x_1, x_2) = x_2 - 2x_1$, subject to the constraints of Exercise 15.

17. $f(x_1, x_2) = 3x_1 - x_2$, subject to the constraints (Exercise 22, Set 3.1):

$$3x_1 \geqslant x_2,$$
$$3x_2 \geqslant x_1,$$
$$x_1 + x_2 \geqslant 5,$$
$$2x_1 + 3x_2 \leqslant 24.$$

18. $f(x_1, x_2) = 3x_2 - x_1$, subject to the constraints of Exercise 17.

▶──

19. $f(x_1, x_2) = x_1 + x_2$, subject to the constraints (Exercise 23, Set 3.1):

$$-3x_1 + 2x_2 \geqslant 6,$$
$$2x_1 + x_2 \leqslant -2,$$
$$x_1 + x_2 \geqslant 4,$$
$$2x_1 + 7x_2 \leqslant 21.$$

20. $f(x_1, x_2) = x_1 + x_2$, subject to the constraints (Exercise 24, Set 3.1):

$$x_1 + 4x_2 \geqslant -4,$$
$$2x_1 + x_2 \leqslant 2,$$
$$x_2 \geqslant 0,$$
$$x_1 \leqslant 5.$$

21. $f(x_1, x_2) = 2x_1 + 3x_2$, subject to the constraints (Exercise 29, Set 3.1):

$$x_1 \geqslant 0,$$
$$x_2 \geqslant 0,$$
$$5x_2 - 3x_1 \leqslant 15,$$
$$x_1 \leqslant 4x_2,$$
$$2x_1 - 5x_2 \leqslant 10.$$

22. $f(x_1, x_2) = 3x_1 + 4x_2$, subject to the constraints (Exercise 28, Set 3.1):

$$3x_1 + 2x_2 \geqslant 6,$$
$$x_1 \geqslant 1,$$
$$0 \leqslant x_2 \leqslant 6,$$
$$2x_1 + 3x_2 \leqslant 24,$$
$$3x_1 + x_2 \leqslant 15.$$

23. $f(x_1, x_2) = 4x_1 + 3x_2$, subject to the constraints (Exercise 27, Set 3.1):

$$x_1 + x_2 \leqslant 0,$$
$$2x_1 - 3x_2 \leqslant 15,$$
$$x_2 \leqslant 5,$$
$$x_1 \geqslant 0,$$
$$2x_1 + x_2 \geqslant 3.$$

24. $f(x_1, x_2) = 2x_1 + 5x_2$, subject to the constraints (Exercise 30, Set 3.1):

$$x_1 \geqslant 0,$$
$$x_2 \geqslant 0,$$
$$-7x_1 + x_2 \leqslant 7,$$
$$4x_2 - 5x_1 \geqslant 20,$$
$$x_1 + x_2 \leqslant 10,$$
$$x_2 \leqslant 3.$$

3.3 FORMULATING MAXIMUM-TYPE LINEAR PROGRAMS AND SOLVING GRAPHICALLY

Example 1 Formulate (model) this problem.

A California vintner has available 660 lbs of Cabernet Sauvignon (CS) grapes, 1860 lbs of Pinot Noir (PN) grapes, and 2100 lbs of Barbera (B) grapes. The vintner makes a Pinot Noir (PN) wine which contains 20% CS, 60% PN, and 20% B grapes and sells for $3 a bottle and a Barbera (B) wine which contains 10% CS, 20% PN, and 70% B grapes and sells for $2 a bottle. Assuming that each bottle of wine requires 3 lbs of grapes, determine how many bottles of each type of wine the vintner should produce to maximize his income.

Solution This problem can be formulated using a table. The following steps show how to do this.

1. First, define the variables. Let

x_1 = the number of bottles of Pinot Noir wine to be produced,

x_2 = the number of bottles of Barbera wine to be produced, and

f = the income ($) obtained from the sale of all the wine.

Note that the variables x_1 and x_2 are defined in terms of *bottle* units while the grape supply is given in terms of *lb* units. Since one bottle of wine requires three lbs of grapes, the available supply of grapes in lb units must be divided by three to yield the available supply of grapes in bottle units, so that consistent units are used throughout the problem. Defining the variables carefully helps prevent formulation errors.

2. Set up a table with the following general headings, which actually vary depending on the specific problem. Across the top list the

	Composition			Number of units of supply available
	Product 1	Product 2	$\cdots$	
Number of units	x_1	x_2	$\cdots$	
Ingredient 1 Ingredient 2 . . .				
Unit value				Objective

OBJECTIVES

You should be able to:

a) Formulate (model) a given problem as a maximum-type linear program.

b) Express the results of (a) in matrix form.

c) Solve maximum-type linear programs graphically.

products being manufactured and at the right write "Number of units of supply available." In the next row indicate the independent variables.

Look ahead to see how this is done for this problem.

3. In the first column list the ingredients used in making each product and at the bottom write "Unit value."

Look at the table to see how this is done for this problem.

4. Enter into the table columnwise the data describing the composition of each product and in the bottom row its unit value.

Note that the percents must be converted to fractions or decimals. The value, here price, could be expressed in either dollars or cents, but be consistent. See how this is done for this problem.

5. Enter into the column headed "supply available" the appropriate data for each ingredient. Indicate at the bottom of this column the "objective" of the problem.

Entering the data for this problem, we obtain the following table:

	Composition (bottles)		Number of bottles available
	PN Wine	B Wine	
Number of bottles	x_1	x_2	
CS Grapes	0.20	0.10	220
PN Grapes	0.60	0.20	620
B Grapes	0.20	0.70	700
$ Price per bottle	3	2	Maximize income

6. For each ingredient the corresponding constraint can be read rowwise from the table. Note the similarity to the echelon tableau.

Noting that the supply available cannot be exceeded, we obtain

$$\text{CS:} \quad 0.20x_1 + 0.10x_2 \leq 220,$$
$$\text{PN:} \quad 0.60x_1 + 0.20x_2 \leq 620,$$
$$\text{B:} \quad 0.20x_1 + 0.70x_2 \leq 700.$$

7. The next two constraints are not stated *explicitly* in the problem. Rather they are *implied* by the reality of the situation. Since one cannot produce a *negative* amount of wine, we constrain the amount of wine produced to be nonnegative; that is,

$$x_1 \geq 0 \quad \text{and} \quad x_2 \geq 0.$$

The constraint that a physical quantity be realistic and hence non-negative is called the *nonnegativity constraint*.

8. Read the objective function and problem objective from the bottom row.

Here the objective is to maximize income, so that we obtain

$$\text{Maximize } f, \quad \text{where } f = 3x_1 + 2x_2.$$

The formulation can be summarized as follows:

i) $0.20x_1 + 0.10x_2 \leqslant 220,$
ii) $0.60x_1 + 0.20x_2 \leqslant 620,$
iii) $0.20x_1 + 0.70x_2 \leqslant 700,$
iv, v) $x_1, x_2 \geqslant 0,$
    vi) $\max f\colon f = 3 x_1 + 2 x_2.$

In summary, a *linear program* can be formulated, or modeled, as a system of linear inequalities, called *constraints* (both explicit and implicit), together with some linear function to be optimized—in this case maximized. That is, here we want to find the largest value of the objective function, subject to the given constraints, and the numerical values that yield it.

DO EXERCISE 13.

We can use matrices for clarification. The maximum-type linear program can be expressed in the form:

$$AX \leqslant B,$$
$$X \geqslant O,$$
$$\max f\colon f = CX,$$

where in Example 1,

$$A = \begin{bmatrix} 0.20 & 0.10 \\ 0.60 & 0.20 \\ 0.20 & 0.70 \end{bmatrix}, \qquad X = \begin{bmatrix} x_1 \\ x_2 \end{bmatrix}, \qquad B = \begin{bmatrix} 220 \\ 620 \\ 700 \end{bmatrix}, \qquad C = [3 \ \ 2].$$

The matrix A is made up of the coefficients of the constraint system. The notation $X \geqslant O$ implies that *each* component of X is nonnegative. Similarly, the notation $AX \leqslant B$ implies that each component of AX is less than or equal to the corresponding component of B.

Thus, a maximum-type linear program is composed of three parts. First, there are the linear constraints, $AX \leqslant B$, particular to each problem, and second, the nonnegativity constraints, $X \geqslant O$, which are common to most problems. Third, there is a linear objective function $f = CX$, which is to be maximized.

13. Formulate (model) the problem below using a table. Save your results.

A furniture manufacturer produces chairs and sofas. The chairs require 20 feet of wood, 1 lb of foam rubber, and 2 square yards of material. The sofas require 100 feet of wood, 50 lbs of foam rubber, and 20 square yards of material. The manufacturer has in stock 1900 feet of wood, 500 lbs of foam rubber, and 240 square yards of material. If the chairs can be sold for $20 each and the sofas for $300 each, how many of each should be produced to maximize the income?

14. Go back to Exercise 13 and express the results in matrix form.

DO EXERCISE 14.

The linear program just formulated can be solved graphically using the techniques of the preceding sections.

A general procedure for solving linear programs in two variables is:

1) **Graph the constraints;**
2) **Determine the solution set (shade this region);**
3) **Label the corner points and determine their coordinates algebraically (with the echelon method);**
4) **Evaluate the objective function at all corner points; and**
5) **Determine by inspection the optimum value (maximum or minimum) and where it is attained.**

Example 2 (Maximum-type) Solve graphically the linear program of Example 1.

Solution The formulation of this linear program was found to be:

$$\begin{aligned}
\text{i(CS):} & \quad 0.20x_1 + 0.10x_2 \leqslant 220, \\
\text{ii(PN):} & \quad 0.60x_1 + 0.20x_2 \leqslant 620, \\
\text{iii(B):} & \quad 0.20x_1 + 0.70x_2 \leqslant 700, \\
\text{vi, iv, v: } & \max f \colon f = \quad 3x_1 + \quad 2x_2; \quad x_1, x_2 \geqslant 0.
\end{aligned}$$

The constraints and solution set are shown in Fig. 3.13. Note that the nonnegativity constraints (iv, v) restrict consideration to the first quadrant.

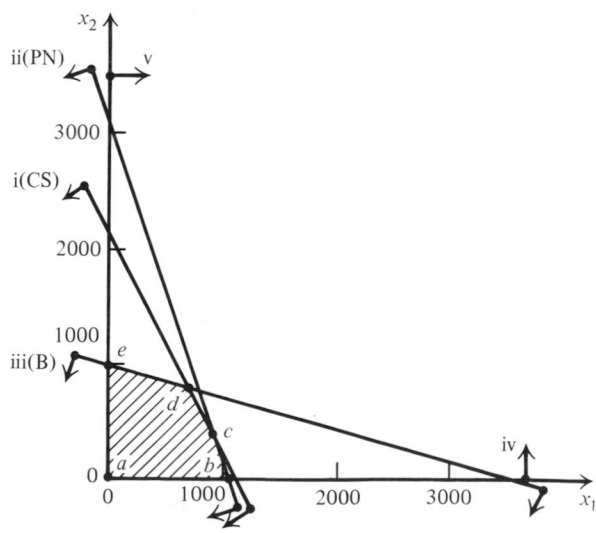

Figure 3.13

The coordinates of the corner points follow:

$$\text{iv and v yield } (a_1, a_2) = (0, 0),$$
$$\text{ii and v yield } (b_1, b_2) = \left(\frac{3100}{3}, 0\right),$$
$$\text{i and ii yield } (c_1, c_2) = (900, 400),$$
$$\text{i and iii yield } (d_1, d_2) = (700, 800),$$
$$\text{iii and iv yield } (e_1, e_2) = (0, 1000).$$

Evaluating the objective function at these points, we obtain

$$f(x_1, x_2) = 3x_1 \qquad + 2x_2,$$
$$f(0, 0) = 3 \cdot 0 \qquad + 2 \cdot 0 \qquad = 0,$$
$$f\left(\frac{3100}{3}, 0\right) = 3 \cdot \left(\frac{3100}{3}\right) + 2 \cdot 0 \qquad = 3100,$$
$$f(900, 400) = 3 \cdot 900 \qquad + 2 \cdot 400 = 3500,$$
$$f(700, 800) = 3 \cdot 700 \qquad + 2 \cdot 800 = 3700 \qquad \text{(maximum)},$$
$$f(0, 1000) = 3 \cdot 0 \qquad + 2 \cdot 1000 = 2000.$$

Thus, the maximum *income* (not necessarily profit) that the vintner can obtain is \$3700 ($f = 3700$) by producing 700 bottles of PN wine ($x_1 = 700$) and 800 bottles of B wine ($x_2 = 800$). This maximum occurs at point d, so that all the CS and B grapes are used but there is an excess of PN grapes which can be put to some other use.

DO EXERCISE 15.

15. Solve graphically the linear program of Margin Exercise 13, whose formulation was

$$20x_1 + 100x_2 \leqslant 1900,$$
$$x_1 + 50x_2 \leqslant 500,$$
$$2x_1 + 20x_2 \leqslant 240,$$
$$x_1, x_2 \geqslant 0,$$

$$\max f\colon f = 20x_1 + 300x_2.$$

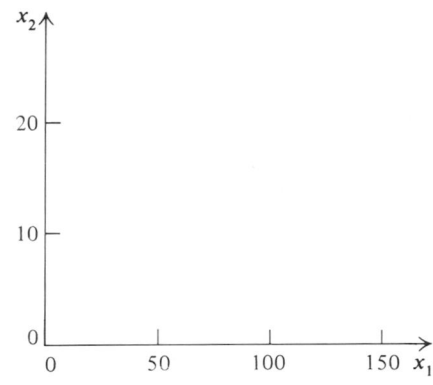

EXERCISE SET 3.3

In the following problems:

a) Formulate the model; b) Express the results in matrix form; c) Solve graphically.

1. *Business.* A clothier makes suits and dresses. Each suit requires 1 yd of polyester and 4 yds of wool while each dress requires 2 yds of polyester and 3 yds of wool. He has in stock 60 yds of polyester and 120 yds of wool. If a suit sells for \$120 and a dress sells for \$75, how many of each should he make to maximize his income?

2. *Business.* A manufacturer of hi-fi speakers makes two speaker assemblies. The inexpensive speaker assembly consists of one midrange speaker and one tweeter. It sells for \$25. The expensive speaker assembly consists of one woofer, one midrange speaker, and two tweeters. This one sells for \$150. The stock consists of 22 12" woofers, 30 5" midrange speakers, and 45 $1\frac{1}{2}$" tweeters. How many of each type of speaker assembly should be made to maximize income?

3. *Business.* A nut dealer has 1800 lbs of peanuts, 1500 lbs of cashews, and 750 lbs of almonds, from which he wishes to make two mixtures. Mixture I sells for \$0.75 per lb and contains 60% peanuts, 30% cashews, and 10% almonds. Mixture II sells for \$2.00 per lb and contains 20% peanuts, 50% cashews, and 30% almonds. How many lbs of each should he mix to maximize his income?

4. *Business.* A tea merchant has 400 lbs cut black tea, 300 lbs of pekoe, and 240 lbs of orange pekoe tea, from which he blends two mixtures. Mixture A, selling for \$1.50 per lb contains 50% cut black tea, 30% pekoe, and 20% orange pekoe. Mixture B, selling for \$4.00 per lb, contains 50% pekoe and 50% orange pekoe. How many lbs of each mixture should he blend to maximize his income?

5. *Ecology.* A certain area of forest is populated by two species of animals (A1 and A2) and the forest supplies three kinds of food (F1, F2, and F3). Species A1 requires 1 unit of food F1, 2 units of food F2, and 2 units of food F3 while species A2 requires 1.2 units of food F1, 1.8 units of food F2, and 0.6 units of food F3. If the forest can normally supply a maximum of 600 units of food F1, 960 units of food F2, and 720 units of food F3, what is the maximum total numbers of these animals that the forest can support?

7. *Ecology.* In reference to Exercise 5, if there is a wet spring, so that the maximum available supply of food becomes 720 units of food F1, 960 units of food F2, and 600 units of food F3, what maximum number of animals can now be supported? What would happen to species A1?

6. *Ecology.* In reference to Exercise 5, if species A1 is valued at $150 and species A2 is valued at $120, how many animals of each species will maximize the value of the animal stock?

OBJECTIVE

You should be able to:

a) Formulate (model) a given problem as a minimum-type linear program.

b) Express results of (a) in matrix form.

c) Solve minimum-type linear programs graphically.

3.4 FORMULATING MINIMUM-TYPE LINEAR PROGRAMS AND SOLVING GRAPHICALLY

Example 1   Formulate (model) this problem.
Ecology, Nutrition, Business. A feed supplier mixes feed for a particular animal which requires at least 160 lbs of nutrient A, at least 24 lbs of nutrient B, and at least 28 lbs of nutrient C. The supplier has available soybean meal* which costs $15 per 100-lb sack and contains 40 lbs of nutrient A, 4 lbs of nutrient B, and 4 lbs of nutrient C and triticale† which costs $10 per 100-lb sack and contains 20 lbs of nutrient A, 4 lbs of nutrient B, and 6 lbs of nutrient C. How many sacks of each ingredient should be used to mix animal feed which satisfies the minimum requirements at minimum cost?

Solution   This problem can be formulated as follows:

1. First, define the variables. Let

$$y_1 = \text{the number of 100-lb sacks of soybean meal,}$$
$$y_2 = \text{the number of 100-lb sacks of triticale, and}$$
$$f = \text{the \$ cost of the animal feed.}$$

2. As for maximum-type programs, set up a table with the following general headings. Across the top list the *sources* of materials and at

---

* Soybean meal is made from soybeans by extracting the oil, which is then used commercially for other purposes.
† *Tri'ti-cale:* a hybrid of wheat and rye.

the right write "Amount of component required." In the next row indicate the independent variables.

Look ahead to see how this is done for the present example.

	Composition			Amount of component required
	Source 1	Source 2	...	
Number of units	y_1	y_2	...	
Component 1 Component 2 . . .				
Unit cost				Objective

3. In the first column list the *components* of the sources of materials and at the bottom write "Unit cost."

Look at the table and see how this is done for this example.

4. Enter into the table columnwise the data describing the composition of each source material and its unit cost in the bottom row.

See how this is done for the present problem.

5. Enter into the column headed "Amount of component required" the appropriate data for each component. Indicate at the bottom of this column the "objective" of the problem.

Entering the data for this example, we obtain the following table:

	Composition (lbs per 100-lb sack)		Pounds required
	Soybean meal	Triticale	
Number of 100-lb sacks	y_1	y_2	
Nutrient A	40	20	160
Nutrient B	4	4	24
Nutrient C	4	6	28
$ Cost per 100-lb sack	15	10	Minimize cost

6. For each component the corresponding constraint can be read rowwise from the table by noting that the amount of the component must be at least satisfied:

$$\begin{aligned} \text{A: } 40y_1 + 20y_2 &\geqslant 160, \\ \text{B: } 4y_1 + 4y_2 &\geqslant 24, \\ \text{C: } 4y_1 + 6y_2 &\geqslant 28. \end{aligned}$$

7. As for the maximum-type programs, we have the implied non-negativity constraints:

$$y_1 \geqslant 0 \quad \text{and} \quad y_2 \geqslant 0.$$

In this example there is another implied constraint. Soybean meal and triticale are sold in 100-lb sacks. Hence, their number must be *integer*. Problems with this integer constraint (which is *not* linear) are called *integer programs*. (Their solutions can be found by considering the various integer solutions neighboring the linear program solution, but this involves much extra work.) We shall *ignore* this integer constraint and accept whatever numerical values are obtained from the solution to the *linear* program.

8. Read the objective function and problem objective from the bottom row.

Here the objective is to minimize the cost, so that we obtain

$$\text{Minimize } f, \quad \text{where } f = 15y_1 + 10y_2.$$

The formulation can be summarized as follows:

$$\begin{aligned} \text{i(A): } && 40y_1 + 20y_2 &\geqslant 160, \\ \text{ii(B): } && 4y_1 + 4y_2 &\geqslant 24, \\ \text{iii(C): } && 4y_1 + 6y_2 &\geqslant 28, \\ \text{iv, v: } && y_1, y_2 &\geqslant 0, \\ \text{vi: } && \min f\text{: } f = 15y_1 + 10y_2.& \end{aligned}$$

In matrix notation, a minimum-type linear program can be written in the form

$$AY \geqslant B,$$
$$Y \geqslant O,$$
$$\min f\text{: } f = CY,$$

where, in the preceding example,

$$A = \begin{bmatrix} 40 & 20 \\ 4 & 4 \\ 4 & 6 \end{bmatrix}, \quad Y = \begin{bmatrix} y_1 \\ y_2 \end{bmatrix}, \quad B = \begin{bmatrix} 160 \\ 24 \\ 28 \end{bmatrix}, \quad C = [15 \quad 10].$$

Thus, a minimum-type linear program is also composed of three parts. First, there are the linear constraints, $AY \geqslant B$, particular to each problem, and second, the nonnegativity constraints, $Y \geqslant O$, which are common to most problems. Third, there is a linear objective function $f = CY$, which is to be minimized.

For convenience, we use the word "optimum" to refer to either a "maximum" or a "minimum," whichever is appropriate. Thus, in general, a linear program consists of a system of particular (linear) constraints, plus nonnegativity constraints and a linear objective function to be optimized.

DO EXERCISE 16.

Example 2   Solve graphically the linear program of Example 1.

Solution   The formulation of this linear program is:

i(A): $40y_1 + 20y_2 \geqslant 160,$
ii(B): $4y_1 +\ \ 4y_2 \geqslant\ \ 24,$
iii(C): $4y_1 +\ \ 6y_2 \geqslant\ \ 28,$
vi, iv, v: $\min f : f = 15y_1 + 10y_2; \quad y_1, y_2 \geqslant 0.$

The constraints and solution set are shown in Fig. 3.14. Again, the nonnegativity constraints (iv, v) restrict consideration to the first quadrant.

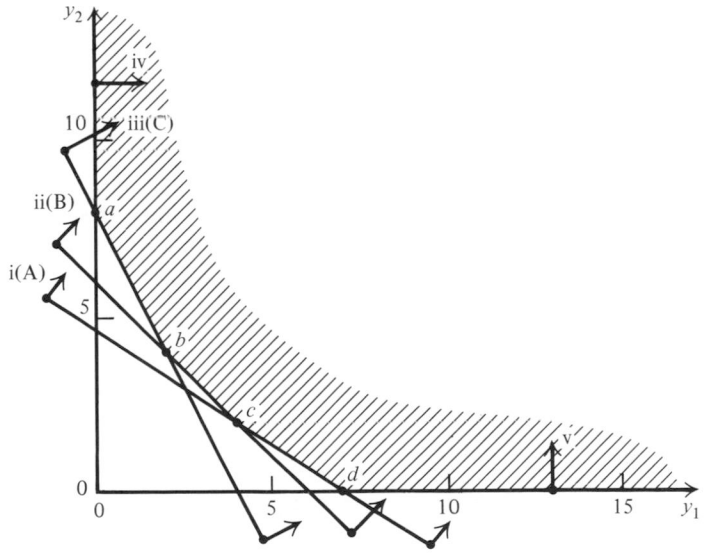

Figure 3.14

16. (a) Formulate (model) this problem using a table and (b) express the result in matrix form. Save your result.

An ore refining company has orders for 200 tons of iron, 500 tons of aluminum, and 100 tons of copper. They have available two kinds of ore. Type A contains 10% iron and 2% copper, and costs $10 per ton. Type B contains 20% aluminum and 1% copper, and costs $15 per ton. How many tons of each should be bought to minimize the cost?

17. Solve graphically the linear program of Margin Exercise 16, whose formulation was:

$$0.1y_1 + \quad 0y_2 \geqslant 200,$$
$$0y_1 + \quad 0.2y_2 \geqslant 500,$$
$$0.02y_1 + 0.01y_2 \geqslant 100,$$
$$y_1, y_2 \geqslant \quad 0,$$

$$\min f\colon f = 10y_1 + 15y_2.$$

The four corners a, b, c, and d of the solution set have coordinates as follows:

$$\text{(iii) and (iv) yield } (a_1, a_2) = (0, 8),$$
$$\text{(ii) and (iii) yield } (b_1, b_2) = (2, 4),$$
$$\text{(i) and (ii) yield } (c_1, c_2) = (4, 2),$$
$$\text{(i) and (v) yield } (d_1, d_2) = (7, 0).$$

Evaluating the objective function $f = 15y_1 + 10y_2$ at these points, we obtain:

$$f(y_1, y_2) = 15y_1 + 10y_2$$

$$f(0, 8) = 15(0) + 10(8) = \quad 80,$$
$$f(2, 4) = 15(2) + 10(4) = \quad 70 \quad \text{(minimum)},$$
$$f(4, 2) = 15(4) + 10(2) = \quad 80,$$
$$f(7, 0) = 15(7) + 10(0) = 105.$$

Thus, point b yields the minimum value of the objective function. The supplier should buy 2 sacks of soybean meal for each 4 sacks of triticale, to minimize his cost.

DO EXERCISE 17.

EXERCISE SET 3.4

a) Formulate (model) the following problems.

b) Express the results in matrix form.

c) Solve graphically.

1. *Nutrition, Business.* An animal feed to be mixed from soybean meal* and oats must contain at least 120 lbs of protein, 24 lbs of fat, and 10 lbs of mineral ash. Each 100-lb sack of soybean meal costs $15 and contains 50 lbs of protein, 8 lbs of fat, and 5 lbs of mineral ash. Each 100-lb sack of oats costs $5 and contains 15 lbs of protein, 5 lbs of fat, and 1 lb of mineral ash. How many sacks of each should be used to satisfy the minimum requirements at minimum cost?

2. *Nutrition, Business.* Suppose the oats in the preceding problem were replaced by alfalfa which costs $8 per 100 lbs and contains 20 lbs of protein, 6 lbs of fat, and 8 lbs of mineral ash. How much of each is now required to minimize the cost?

3. *Nutrition, Business.* How is the formulation of Exercise 2 changed if the mineral requirement is doubled?

4. *Nutrition, Business.* How is the formulation of Exercise 2 changed if the cost of alfalfa is increased to $10?

---

* See footnote on page 142.

5. *Business, Transportation.* An airline with two types of airplanes, P1 and P2, has contracted with a tour group to provide accommodations for a minimum of each of 2000 first-class. 1500 tourist, and 2400 economy-class passengers. Airplane P1 costs $12 thousand per mile to operate and can accommodate 40 first-class, 40 tourist, and 120 economy-class passengers, while airplane P2 costs $10 thousand per mile to operate and can accommodate 80 first-class, 30 tourist, and 40 economy-class passengers. How many of each type of airplane should be used to minimize the operating cost?

7. *Business, Transportation.* If, instead of replacing P1 by P3, P2 is replaced by P3, how many of P1 and P3 should be used, to minimize the operating cost?

9. *Business, Transportation.* If the contract requirements are changed to a minimum of 1600 first-class, 2100 tourist, and 2400 economy-class passengers, how many of airplanes P1 and P2 are now required to minimize the operating cost?

11. *Business, Transportation.* As in Exercise 9 but with airplanes P1 and P3 . . . ?

6. *Business, Transportation.* A new airplane P3 becomes available, having an operating cost of $15 thousand per mile and accommodating 40 first-class, 80 tourist, and 80 economy-class passengers. If airplane P1 of Exercise 5 were replaced by airplane P3, how many of P2 and P3 would be needed to minimize the operating cost?

8. *Business, Transportation.* If all three planes P1, P2, and P3 are used, how many of each should be used to minimize the cost? Formulate and put in matrix form only.

10. *Business, Transportation.* As in Exercise 9, but with airplanes P2 and P3 . . . ?

12. *Business, Transportation.* As in Exercise 9, but using all three-airplanes . . . ? Formulate and put in matrix form only.

The graphical method for solving linear programs, considered in Sections **3.3** and **3.4,** can be used for problems with *two* variables and, with some difficulty, *three*. However, for problems where the number of variables might run into hundreds or thousands, algebraic techniques must be used. These have the advantage that they can be adapted to high-speed computers. The *simplex algorithm*[*] is the basic technique that we will consider in this chapter.

3.5 MAXIMUM-TYPE LINEAR PROGRAMS—THE SIMPLEX ALGORITHM

Standard Form of a Linear Program

In this section we will consider the algebraic solution of maximum-type linear programs which are expressed in the following *standard form*.

OBJECTIVE

You should be able to set up the initial simplex tableau for a maximum-type linear program, solve, and read off the solution.

---

[*] An *algorithm* is a special procedure. Here the simplex algorithm is used for solving linear programs.

a) $AX \leqslant B$ (This means that all constraints are in the form

$$a_1x_1 + a_2x_2 + \cdots + a_nx_n \leqslant b.)$$

b) $X \geqslant O$ (This means that all the variables $x_1, x_2, \ldots, x_n$ are nonnegative. The given variables are called *structural* variables.)

c) $\max x_0: x_0 = CX$ (We are finding the maximum value of an objective function $x_0 = c_1x_1 + c_2x_2 + \cdots + c_nx_n.$)

(Here the objective function, previously denoted f, is denoted by the variable with the subscript "0", x_0)

We place one additional provision on the standard form, and this is the positivity constraint $B > O$, that is, *all* components of B are positive. This is assumed to avoid complications.

A set of values X which satisfies the constraints $AX \leqslant B$ together with the corresponding value of the objective function $x_0 = CX$ is called a *solution* and can be written $X; x_0$.

A solution which also satisfies the *nonnegativity constraint*, $X \geqslant O$, is called a *feasible solution*.

A feasible solution which is also a *corner point* of the region defined by $AX \leqslant B$ and $X \geqslant O$ is called a *basic feasible solution*.

A basic feasible solution which also *optimizes* the value of the objective function $x_0 = CX$ is called an *optimum basic feasible solution* or simply an *optimum feasible solution*.

Briefly, solution of a maximum-type linear program by the simplex algorithm involves the following steps:

1. Adding *slack variables* to convert the inequalities into equations,

2. Setting up the *initial simplex tableau* which is similar to the echelon tableau,

3. Finding an *initial feasible solution*, since the simplex algorithm proceeds from one feasible solution to a better one until the optimum is reached,

4. Introducing *basic and nonbasic variables* as the natural way to express *basic* feasible solutions,

5. *Choosing the proper pivot* element to advance the solution and maintain the nonnegativity of all variables,

6. *Pivoting*, which is done columnwise in exactly the same way as in the echelon method, and

7. Continuing and recognizing when the *algorithm terminates*.

We now consider these steps in detail.

Slack Variables

The first step of the simplex algorithm is to convert the inequalities of the problem into equations.

Example 1   Convert the formulation of Example 1 of Section 3.3 from a system of inequalities to a system of equations.

Solution   The problem in question had the formulation:

$$\text{i)} \qquad\qquad 0.20x_1 + 0.10x_2 \leqslant 220,$$
$$\text{ii)} \qquad\qquad 0.60x_1 + 0.20x_2 \leqslant 620,$$
$$\text{iii)} \qquad\qquad 0.20x_1 + 0.70x_2 \leqslant 700,$$
$$\text{iv)} \qquad \max x_0\colon x_0 = 3.00x_1 + 2.00x_2,$$
$$\text{v)} \qquad\qquad\qquad x_1, x_2 \geqslant 0.$$

The first of these inequalities (i) can be made into an equation by adding a variable y_1 to the lefthand side. This produces the equation

$$0.20x_1 + 0.10x_2 + y_1 = 220.$$

The quantity y_1 is called a *slack variable* since it "takes up the slack" in the equation. It follows that y_1 is nonnegative, as are x_1 and x_2. To see this in an easy way, the inequality $4 + 1 \leqslant 7$ can be made into the equation $4 + 1 + 2 = 7$ where the slack is the quantity 2.

For *each* constraint (i) through (iii), we add a *different* slack variable to the lefthand side, obtaining:

$$\text{i')} \qquad 0.20x_1 + 0.10x_2 + y_1 \qquad\qquad = 220,$$
$$\text{ii')} \qquad 0.60x_1 + 0.20x_2 \quad + y_2 \qquad\quad = 620,$$
$$\text{iii')} \qquad 0.20x_1 + 0.70x_2 \qquad\quad + y_3 \quad = 700,$$
$$\text{iv')} \qquad \max x_0\colon -3x_1 - \quad 2x_2 \qquad\qquad + x_0 = 0,$$
$$\text{v')} \qquad\qquad x_1, x_2;\quad y_1,\quad y_2,\quad y_3 \qquad \geqslant 0.$$

18. *Adding slack variables.* Convert the following formulation (Margin Exercise 13, Section 3.3) into a system of equations, and put in matrix form.

$$20x_1 + 100x_2 \leqslant 1900,$$
$$x_1 + 50x_2 \leqslant 500,$$
$$2x_1 + 20x_2 \leqslant 240,$$
$$x_1, x_2 \geqslant 0,$$
$$\max x_0 \colon x_0 = 20x_1 + 300x_2.$$

Note that the objective function

iv) $$x_0 = 3x_1 + 2x_2$$

has been rewritten

iv′) $$-3x_1 - 2x_2 + x_0 = 0.$$

We add a different slack variable each time, because what must be added to $0.20x_1 + 0.10x_2$ to get 220 may be different from what must be added to $0.60x_1 + 0.20x_2$ to get 620. Each *slack* variable must be nonnegative, as must the structural variables x_1 and x_2.

Note that the constraints (i) through (iii) can be recovered from the equations (i′) through (iii′) by setting each of the slack variables y_i, in turn, equal to 0 and replacing each = sign by $\leqslant$.

Suppose we wanted to graph the constraints. The equations related to the constraints are obtained by setting the slack variables in the equations (i′) through (iii′) equal to 0 and maintaining the equal signs. This is shown in Fig. **3.15**.

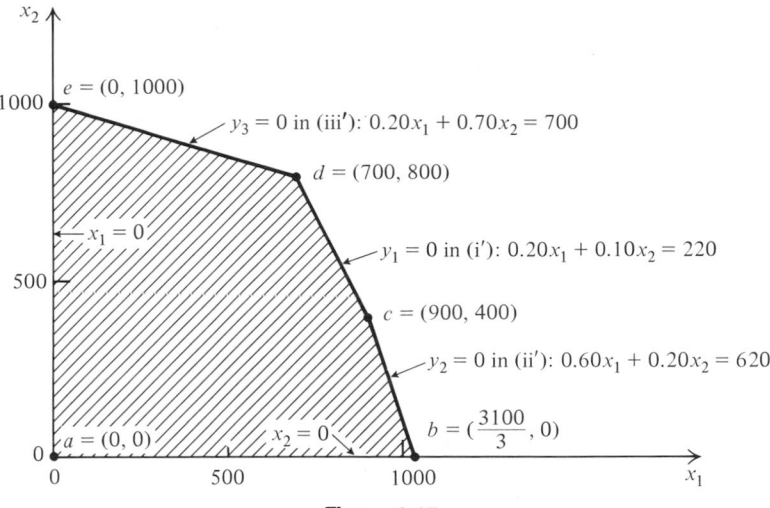

Figure 3.15

DO EXERCISE 18.

Initial Simplex Tableau

Now that we have converted the linear program from one using constraints to one using equations, we can prepare to solve it using the echelon tableau (see Sections 3.1 and 3.2). With a slight modification, the initial echelon tableau is the initial simplex tableau. See Example 2.

Example 2 Set up the initial *echelon* tableau and the initial *simplex* tableau for the problem of Example 1.

Solution The initial *echelon* tableau for the system of equations of Example 1 is

x_1	x_2	y_1	y_2	y_3	x_0	1
0.20	0.10	1	0	0	0	220
0.60	0.20	0	1	0	0	620
0.20	0.70	0	0	1	0	700
-3	-2	0	0	0	1	0

Since the x_0 column does not change in the course of the simplex algorithm, this column is usually omitted in the simplex tableau; however, the value of x_0 will always be in the lower righthand corner in the column headed "1".

Furthermore, the objective function in the *bottom row* plays a special role in the simplex algorithm and hence is set off from the constraints in the other rows with a horizontal line.* Thus, we obtain the *initial simplex* tableau:

x_1	x_2	y_1	y_2	y_3	1
0.20	0.10	1	0	0	220
0.60	0.20	0	1	0	620
0.20	0.70	0	0	1	700
-3	-2	0	0	0	0

---

* Alternately, the objective function is sometimes written in the top row. In either case it is distinguished from the constraints in the other rows by being set off with a horizontal line.

19. Set up the initial simplex tableau for the problem of Margin Exercise 18.

Matrix Formulation

Formally, in matrix notation, a problem with constraints of the form

$$AX \leq B, \qquad X \geq O,$$

$$\max x_0 : x_0 = CX,$$

or

$$-CX + x_0 = 0,$$

is converted, by addition of slack variables Y, into one of the form*

$$[A \quad I]\begin{bmatrix} X \\ Y \end{bmatrix} = B, \qquad \begin{bmatrix} X \\ Y \end{bmatrix} \geq O,$$

$$\max x_0 : x_0 = [C \quad O]\begin{bmatrix} X \\ Y \end{bmatrix}, \qquad \text{or} \qquad -[C \quad O]\begin{bmatrix} X \\ Y \end{bmatrix} + x_0 = 0.$$

In general this can be put into an initial simplex tableau of the form:

X^T	Y^T	1
A	I	B
$-C$	O	0

Note the negative sign in the bottom row.

Also the given equations can be recovered from the tableau in a straightforward manner.

DO EXERCISE 19.

Initial Feasible Solution

Recall that a feasible solution satisfies both $AX \leq B$ and $X \geq O$.

The simplex algorithm starts with one feasible solution and then generates a better one, if possible. The *initial feasible solution* is obtained in the following manner.

Example 3   Find an initial feasible solution from the initial simplex tableau of Example 2.

Solution   In Section 3.3 we solved this linear program graphically. See Fig. 3.16.

---

* You may need to review augmented matrices, Section 2.4.

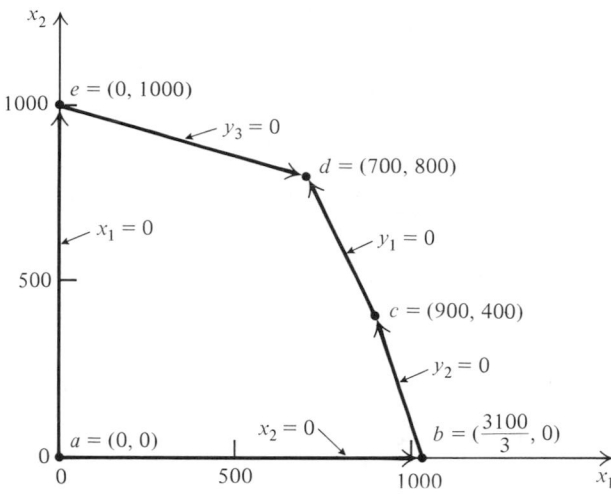

Figure 3.16

We showed previously (see Section 3.2) that the *optimum* feasible solution must occur at a *corner point*. Since we can use *any* feasible solution to start with, we choose that corner point which is simplest to find. This is the *origin* (point a), which becomes our *initial* feasible solution.

Formally, we note that our constraint equations can be written

$$[A \quad I]\begin{bmatrix} X \\ Y \end{bmatrix} = AX + IY = B.$$

Since $B > O$, for maximum-type problems we can always obtain an *initial* feasible solution by setting X (the structural variables) equal to zero and solving for Y (the slack variables):

$$X = O \quad \text{and} \quad Y = B.$$

Setting the structural variables (x_1, x_2) equal to zero in Eqs. (i′) through (iv′) of Example 1, we can determine the corresponding values of the slack variables (y_1, y_2, y_3) and objective function (x_0):

Initial feasible solution		
Structural variables	Slack variables	Objective function
$x_1 = 0$ $x_2 = 0$	$y_1 = 220$ $y_2 = 620$ $y_3 = 700$	$x_0 = 0$

20. Find an initial feasible solution from the initial simplex tableau of Margin Exercise 19.

This is a *feasible solution* since *all* variables (slack and structural) are nonnegative and satisfy the given constraints. We shall take this as our *initial* feasible solution.

DO EXERCISE 20.

The simplex algorithm starts with the initial feasible solution and proceeds to other feasible solutions. In particular, it proceeds from one **corner point** to an **adjacent** corner point. Thus, on Fig. 3.16 the solution will proceed from point a (the *initial* feasible solution) to point d (the *optimum* feasible solution) along one of two paths: $abcd$ or aed.

Basic Feasible Solutions

Consider the path $abcd$. In particular note that at

$$\text{point } a: \quad x_1 = x_2 = 0, \quad y_1 = 200, \quad y_2 = 620, \quad y_3 = 700;$$

$$\text{point } b: \quad x_2 = y_2 = 0, \quad x_1 = \frac{3100}{3}, \quad y_1 = \frac{40}{3}, \quad y_3 = \frac{1480}{3};$$

$$\text{point } c: \quad y_2 = y_1 = 0, \quad x_1 = 900, \quad x_2 = 400, \quad y_3 = 240;$$

$$\text{point } d: \quad y_1 = y_3 = 0, \quad x_1 = 700, \quad x_2 = 800, \quad y_2 = 40.$$

Also for the path aed, we have, at

$$\text{point } a: \quad x_1 = x_2 = 0, \quad y_1 = 220, \quad y_2 = 620, \quad y_3 = 700;$$

$$\text{point } e: \quad x_1 = y_3 = 0, \quad x_2 = 100, \quad y_1 = 120, \quad y_2 = 420;$$

$$\text{point } d: \quad y_3 = y_1 = 0, \quad x_1 = 700, \quad x_2 = 800, \quad y_2 = 40.$$

Note now that there are two structural variables, x_1 and x_2, and that at the origin (point a) they are both zero. As we progress from one corner to an adjacent corner, one of these variables becomes *nonzero* and a different variable becomes *zero*, so that in the present example there are always two variables that are zero in the solution at any stage.

This happens *automatically* provided we proceed along a path through adjacent corner points. Such *corner-point* solutions are important in the simplex algorithm and are called *basic* feasible solutions.

Basic and Nonbasic Variables

In general, we note that in the preceding initial feasible solution, some variables were *set* to *zero* and some variables were *computed*.

The variables which are *computed* are called *basic* variables and the variables *set to zero* are called *nonbasic*.

Basic variables can be identified from a simplex tableau as those heading a column of all zeroes except for one 1.

Nonbasic variables are the rest.

The *values* of the basic variables can be read from the *simplex* tableau simply by setting the nonbasic variables to zero and reading the tableau as in the *echelon* method. Recalling that the x_0-column is omitted from the simplex tableau, we find that the value of x_0 is the bottom element in the righthand column.

Initially, the slack variables y_1, y_2, y_3 are basic and the structural variables x_1, x_2 are nonbasic.

A basic feasible solution is written using basic and nonbasic variables. If we start with an *initial* feasible solution as in Example 3, it will be a *basic* feasible solution. Furthermore, each successive step of the simplex algorithm corresponds to a basic feasible solution.*

Choosing the Pivot

The essential difference between the echelon method and the simplex algorithm is in the way one chooses the pivot element. In the simplex algorithm we choose the pivot element to increase x_0 and to maintain feasibility.

Example 4   Find a pivot element for the initial simplex tableau of Example 2.

Solution   From Example 3, we see that the objective function can be written

$$x_0 = 0 + 3x_1 + 2x_2.$$

---

*The use of the terms *basic* and *nonbasic* variables simplifies the description of the simplex algorithm since each step is a *pivoting* operation (as in the echelon method) in which a *basic* variable is made *nonbasic* and a *nonbasic* variable is made *basic*. Thus, from one tableau to the next there is always a fixed number of basic variables and a fixed number of nonbasic variables. This interchanging the roles of the variables from basic to nonbasic and vice versa is done *automatically* as the pivoting is carried out and does *not* require any special consideration.

For each constraint there is a slack variable.

For each slack variable there is a basic variable.

Basic variables are *computed* and, with few exceptions, are greater than zero.

For each structural variable there is a nonbasic variable.

Nonbasic variables are *set to zero*.

Initially, $x_1 = x_2 = 0$ (nonbasic variables), so $x_0 = 0$. If either of these variables, x_1 or x_2, is increased, that is, made basic (nonzero), then x_0 will be increased. Therefore, either x_1 or x_2 could be chosen as the pivot column. When we choose either x_1 or x_2 as a pivot column and pivot, we increase the variable corresponding to that column.

Another way to determine the pivot column is to examine the bottom row of the simplex tableau. The equation

$$x_0 = 0 + 3x_1 + 2x_2$$

appears in the bottom row as

$$-3x_1 - 2x_2 + x_0 = 0.$$

Note the sign changes.

The bottom row entries to the left of the double vertical line are called *indicators.*

Initially, the indicators are $-3, -2, 0, 0, 0$.

We can pick our pivot column to be any column with a *negative* *indicator.* **When we do this, we are picking some nonbasic variable to become basic. If there is no obvious reason to do otherwise, we shall choose that column with the** *most negative* **indicator as our pivot column.**

Having chosen the pivot column (in this case the first column), we must now determine the pivot row. The intersection of the pivot column and pivot row determines the pivot element.

All nonbasic variables except the one involved in pivoting (the one heading the pivot column) remain nonbasic, so that x_2 is zero initially and remains zero.

Consider again the set of equations (i′) through (iv′) with $x_2 = 0$:

$$\left. \begin{array}{l} 0.20x_1 + y_1 \qquad\qquad\qquad = 220, \\ 0.60x_1 \qquad + y_2 \qquad\qquad = 620, \\ 0.20x_1 \qquad\qquad + y_3 \qquad = 700, \\ -3.00x_1 \qquad\qquad\qquad + x_0 = 0. \end{array} \right\} (E)$$

Setting *one* of the current basic variables to zero makes it nonbasic and permits us to determine the value of the other basic variables and the objective function.

The only question remaining is *which* basic variable to set to zero (that is, to make nonbasic).

If in Equations E we set *each* basic variable to zero in turn and solve for x_1, we obtain:

$$\text{for } y_1 = 0, \quad x_1 = \frac{220}{0.20} = 1100;$$

$$\text{for } y_2 = 0, \quad x_1 = \frac{620}{0.60} = 1033\tfrac{1}{3};$$

$$\text{for } y_3 = 0, \quad x_1 = \frac{700}{0.20} = 3500.$$

The simplex tableau is usually augmented with this *quotient* column headed "q."

x_1	x_2	y_1	y_2	y_3	1	q	
0.20	0.10	1	0	0	220	$1100 \left(= \dfrac{220}{0.20}\right)$	
0.60*	0.20	0	1	0	620	$1033\tfrac{1}{3} \left(= \dfrac{620}{0.60}\right)$	(Minimum)
0.20	0.70	0	0	1	700	$3500 \left(= \dfrac{700}{0.20}\right)$	
−3	−2	0	0	0	0		

The pivot row is the row with *minimum* quotient. Here the pivot row is the second row.

The intersection of the pivot column (here the first column) and the pivot row (here the second row) yields the pivot element which is *starred* (*).

To summarize:

To determine the pivot row: Divide each element in the righthand column (above the bottom row) by the rowwise corresponding entry in the pivot column. The pivot row is the row with *minimum positive* quotient.

To determine the pivot element: Find the intersection of the pivot column and the pivot row. Star (*) this element.

DO EXERCISE 21.

Pivoting

The pivoting operation for one column is exactly the same in the simplex method as in the echelon method (Sections 3.1 and 3.2). The

21. Find a pivot for the initial simplex tableau of Margin Exercise 19.

results are written in a *new* tableau with the *same* headings:

i) Divide *all* elements in the pivot row by the pivot element. Thus, the entry replacing the pivot element will be 1 and the star (*) on this element is dropped.

ii) Add some multiple of the pivot row to each other row including the bottom row (usually a different multiple for each row) to create zeroes for *all* entries of the pivot column other than the pivot element, which is now a 1. Thus, a zero is also created in the *bottom* row in the pivot column.

Example 5 Starting with the initial simplex tableau of Example 2, pivot until the simplex algorithm terminates.

Solution Pivoting, we obtain the second tableau:

x_1	x_2	y_1	y_2	y_3	1	q
0	$\frac{1}{30}$*	1	$-\frac{1}{3}$	0	$\frac{40}{3}$	400 (Minimum)
1	$\frac{1}{3}$	0	$\frac{5}{3}$	0	$\frac{3100}{3}$	3100
0	$\frac{19}{30}$	0	$-\frac{1}{3}$	1	$\frac{1480}{3}$	$\frac{14800}{19} = 778\frac{18}{19}$
0	-1	0	5	0	3100	

Here the quotient column to the right of the tableau should be ignored for the present.

In the initial tableau read horizontally from the pivot element to a column with a *1* entry in the pivot row. The basic variable heading this column becomes nonbasic (zero) through pivoting.

In this example y_2 was basic (= 620) and becomes nonbasic (= 0) as can be seen from the second tableau.

If we *interchange* the x_1 and y_2 columns, then we obtain:

y_2	x_2	y_1	x_1	y_3	1
$-\frac{1}{3}$	$\frac{1}{30}$*	1	0	0	$\frac{40}{3}$
$\frac{5}{3}$	$\frac{1}{3}$	0	1	0	$\frac{3100}{3}$
$-\frac{1}{3}$	$\frac{19}{30}$	0	0	1	$\frac{1480}{3}$
5	-1	0	0	0	3100

This tableau has the same *form* as our initial tableau provided we *reorder* the variables. However, it is *not* necessary to *physically* interchange the two columns. Solutions can be read from the tableaux in either form. Thus, it is less work simply to leave the columns in their original order and note that the *roles* of the variables have been interchanged; that is, x_1 was nonbasic and became basic, while y_2 was basic and became nonbasic. Thus, we shall speak of "interchanging the roles of x_1 and y_2." This should not be confused with physically interchanging the x_1 and y_2 columns which remain in their original location. The solution at this point is

Structural variables	Slack variables	Objective function
$x_1 = \frac{3100}{3}$, $x_2 = 0$	$y_1 = \frac{40}{3}$, $y_2 = 0$, $y_3 = \frac{1480}{3}$	$x_0 = 3100$.

or

Basic variables	Nonbasic variables	Objective function
$x_1 = \frac{3100}{3}$, $y_1 = \frac{40}{3}$, $y_3 = \frac{1480}{3}$	$x_2 = 0$, $y_2 = 0$	$x_0 = 3100$.

The solution is feasible (that is, it satisfies the given constraints including nonnegativity) and the objective function has been increased from a value of 0 to 3100.

Wrong Pivot

Let us examine what happens if the *wrong* pivot element is used. If, for example, we pivot using the first column and the *first* row, then we obtain

x_1	x_2	y_1	y_2	y_3	1
1	0.5	5	0	0	1100
0	−0.1	−3	1	0	−40
0	0.6	−1	0	1	480
0	−0.5	15	0	0	3300

The *negative* number in the righthand column is a reminder that the *wrong* pivot row has been used. In this case, we obtained $y_2 = -40$ which violates the nonnegativity constraint, which must be maintained.

Selecting the pivot row that corresponds to the *minimum* positive quotient guarantees that the solution does not violate the non-negativity constraint.

Termination

Returning our attention to the correct second tableau, we find that only the second column has a negative indicator, -1. Thus, the second column will be the next pivot column and we obtain a new set of quotients, as shown augmenting the second tableau. The minimum quotient is obtained for the first row. The pivot element is starred and the 1 in the pivot row corresponds to y_1. Thus the pivoting will interchange the roles of x_2 and y_1, that is, will make x_2 basic and y_1 nonbasic ($y_1 = 0$). Performing the pivoting, we obtain

x_1	x_2	y_1	y_2	y_3	1	q
0	1	30	-10	0	400	—
1	0	-10	5	0	900	180
0	0	-19	6^*	1	240	40 (Minimum)
0	0	30	-5	0	3500	

Again the quotient column here will be used to obtain the *next* tableau. The solution at this point is

$$x_1 = 900, \qquad y_1 = 0, \qquad x_0 = 3500.$$
$$x_2 = 400, \qquad y_2 = 0,$$
$$y_3 - 240,$$

The fourth column now has a negative indicator, -5, and thus becomes the next pivot column. The corresponding quotients are shown to the right of the tableau above. Note that the *negative* elements in the pivot column are disregarded in computing quotients. Such a pivot would make a variable *negative* and also would *decrease* the value of the objective function.

Pivoting to interchange the roles of y_2 and y_3, we obtain

x_1	x_2	y_1	y_2	y_3	1
0	1	$-\frac{5}{3}$	0	$\frac{5}{3}$	800
1	0	$\frac{35}{6}$	0	$-\frac{5}{6}$	700
0	0	$-\frac{19}{6}$	1	$\frac{1}{6}$	40
0	0	$\frac{85}{6}$	0	$\frac{5}{6}$	3700

The solution at this point is

$$x_1 = 700, \qquad y_1 = 0, \qquad x_0 = 3700.$$
$$x_2 = 800 \qquad y_2 = 40,$$
$$y_3 = 0$$

Since all indicators are nonnegative, there are no more possible pivot columns. Thus, the algorithm *terminates* and the current solution is *maximal.*

The path taken by the simplex algorithm can be followed in Fig. 3.16 as *abcd*. This particular path is a consequence of the choice of the initial pivot column. The alternate choice would have yielded the path *aed.*

Picking the most negative indicator to determine the pivot column is a good simple rule to minimize the number of steps required to achieve optimality but as can be seen in the present example, this is not always the best choice. Since there is no way to know beforehand the best possible pivot column, we use the "most negative" indicator as a good simple rule.

Summary

Let us summarize the steps in the simplex algorithm.

  I. Convert the inequalities of the problem into equations by adding slack variables.

  II. Set up the initial simplex tableau.

III. Carry out the pivoting operations as follows:

    1. **To find the pivot column, pick the column with the most negative indicator. (Actually any column with a negative indicator will do.) If there is none, go to step IV.**

    2. **To find the pivot row, find the quotients of the entries in the righthand column by the rowwise corresponding entries in the pivot column. The pivot row is the row with the minimum positive quotient.**

    3. **Star the pivot element and carry out the pivoting operation, as in the echelon method, for this column.**

    4. **To find the feasible solution at any point in the procedure, translate the tableau to the corresponding system of equations, and let the nonbasic variables be 0. The basic variables can be identified by columns with a 1 for some element and all the other elements 0. The other columns are nonbasic.**

IV. Look at the bottom row of the tableau. If all the indicators are nonnegative, there are no more possible pivot columns and the

22. Starting with the initial simplex tableau of Margin Exercise 2, pivot until the algorithm terminates. To check your answer, use more than one initial choice of pivot column and check your solution at each step against the graphical solution of Margin Exercise 15, Section 4.3.

algorithm terminates. The solution which can then be found using III(4) is maximal.

To further exemplify the procedure, let us return to the previous example, but this time start by choosing the other pivot column; that is, column two. Then pivoting would proceed as follows, starting with the initial tableau:

x_1	x_2	y_1	y_2	y_3	1	q	
0.20	0.10	1	0	0	220	2200	
0.60	0.20	0	1	0	620	3100	
0.20	0.70*	0	0	1	700	1000	(Minimum)
−3	−2	0	0	0	0		

x_1	x_2	y_1	y_2	y_3	1	q	
$\frac{6}{35}$*	0	1	0	$-\frac{1}{7}$	120	700	(Minimum)
$\frac{19}{35}$	0	0	1	$-\frac{2}{7}$	420	$773\frac{13}{19}$	
$\frac{2}{7}$	1	0	0	$\frac{10}{7}$	1000	3500	
$-\frac{17}{7}$	0	0	0	$\frac{20}{7}$	2000		

x_1	x_2	y_1	y_2	y_3	1
1	0	$\frac{35}{6}$	0	$-\frac{5}{6}$	700
0	0	$-\frac{19}{6}$	1	$\frac{1}{6}$	40
0	1	$-\frac{5}{3}$	0	$\frac{5}{3}$	800
0	0	$\frac{85}{6}$	0	$\frac{5}{6}$	3700

At this point the algorithm terminates. Following the solution from tableau to tableau, we see that on Fig. 3.16 the solution proceeded from point a to point e to point d, the same optimum as before. Note that the final *tableau* just obtained is *not* identical to the final tableau we had previously obtained. However, it is *row equivalent* (there are just row interchanges) and hence has the same solution.

Usually choosing the most negative indicator will tend to yield the solution in fewer steps but, as the present example illustrates, this need not be so. Alternate choice of pivots leads to *the* optimum solution along alternate paths.

DO EXERCISE 22.

EXERCISE SET 3.5

Exercises 1 through 6 are Exercises 1 through 6 from Set 3.2 where they were to be solved graphically. Here you are asked to solve these problems using the simplex algorithm and to check this solution with your graphical solution. Note that the simplex solution yields both structural and slack variables while the graphical solution yields only the structural variables.

1. Maximize $x_0(x_1, x_2) = x_1 + 2x_2$, subject to the constraints

$$x_1 + x_2 \leq 6,$$
$$x_2 \leq 5,$$
$$x_1, x_2 \geq 0.$$

2. Maximize $x_0(x_1, x_2) = x_1 + x_2$, subject to the constraints

$$x_1 + 2x_2 \leq 8,$$
$$x_1 \leq 6,$$
$$x_1, x_2 \geq 0.$$

3. Maximize $x_0(x_1, x_2) = 5x_1 + 4x_2$, subject to the constraints

$$3x_1 + 2x_2 \leq 12,$$
$$x_1 + x_2 \leq 5,$$
$$x_1, x_2 \geq 0.$$

4. Maximize $x_0(x_1, x_2) = 2x_1 + x_2$, subject to the constraints

$$3x_1 + 5x_2 \leq 15,$$
$$3x_1 + 2x_2 \leq 12,$$
$$x_1, x_2 \geq 0.$$

5. Maximize $x_0(x_1, x_2) = 3x_1 + 4x_2$, subject to the constraints

$$x_1 + 2x_2 \leq 14,$$
$$4x_1 + 3x_2 \leq 26,$$
$$2x_1 + x_2 \leq 12,$$
$$x_1, x_2 \geq 0.$$

6. Maximize $x_0(x_1, x_2) = 5x_1 + 4x_2$, subject to the constraints

$$x_1 + 3x_2 \leq 18,$$
$$3x_1 + 2x_2 \leq 19,$$
$$2x_1 + x_2 \leq 12,$$
$$x_1, x_2 \geq 0.$$

Use the simplex algorithm to solve the following problems. Check your answer by doing each problem twice, starting with different pivots, if possible.

7.
$$x_1 + 2x_2 \leq 60,$$
$$4x_1 + 3x_2 \leq 140,$$
$$\max x_0\colon x_0 = 120x_1 + 75x_2; \quad x_1, x_2 \geq 0$$

8.
$$4x_1 + 5x_2 \leq 35,$$
$$3x_1 + x_2 \leq 18,$$
$$\max x_0\colon x_0 = 3x_1 + 2x_2; \quad x_1, x_2 \geq 0$$

9.
$$3x_1 + 6x_2 \leq 90,$$
$$5x_1 + 3x_2 \leq 160,$$
$$x_1 + x_2 \leq 44,$$
$$\max x_0\colon x_0 = 3x_1 + 2x_2; \quad x_1, x_2 \geq 0$$

10.
$$x_1 + 2x_2 \leq 16,$$
$$3x_1 + 2x_2 \leq 20,$$
$$4x_1 + x_2 \leq 20,$$
$$\max x_0\colon x_0 = 2x_1 + x_2; \quad x_1, x_2 \geq 0$$

11.
$$5x_1 + 6x_2 \leq 60,$$
$$x_1 + x_2 \leq 11,$$
$$3x_1 + x_2 \leq 27,$$
$$\max x_0\colon x_0 = 2x_1 + x_2; \quad x_1, x_2 \geq 0$$

12.
$$-x_1 + 2x_2 \leq 10,$$
$$3x_1 + 2x_2 \leq 18,$$
$$3x_1 + x_2 \leq 15,$$
$$\max x_0\colon x_0 = x_1 + 2x_2; \quad x_1, x_2 \geq 0$$

13.
$$3x_1 + 4x_2 \le 48,$$
$$x_1 + x_2 \le 13,$$
$$2x_1 + x_2 \le 22,$$
$$\max x_0\colon x_0 = 7x_1 + 4x_2; \quad x_1, x_2 \ge 0$$

14.
$$x_1 + 3x_2 \le 18,$$
$$2x_1 + x_2 \le 11,$$
$$3x_1 + x_2 \le 15,$$
$$\max x_0\colon x_0 = 3x_1 + 4x_2; \quad x_1, x_2 \ge 0$$

15.
$$3x_1 - 2x_2 + x_3 \le 8,$$
$$-4x_1 + 3x_2 + 2x_3 \le 4,$$
$$3x_1 + x_2 - 6x_3 \le 6,$$
$$\max x_0\colon x_0 = -2x_2 + 5x_3; \quad x_1, x_2, x_3 \ge 0$$

16.
$$x_1 - 4x_2 + 3x_3 \le 12,$$
$$x_1 - 2x_2 + 6x_3 \le 4,$$
$$2x_1 + 7x_2 + x_3 \le 17,$$
$$\max x_0\colon x_0 = 5x_1 + x_2 + 2x_3; \quad x_1, x_2, x_3 \ge 0$$

17.
$$2x_1 + x_2 + 4x_3 \le 24,$$
$$x_1 + x_2 + x_3 \le 7,$$
$$2x_1 - x_2 + 3x_3 \le 12,$$
$$\max x_0\colon x_0 = x_1 + 2x_2 + 3x_3; \quad x_1, x_2, x_3 \ge 0.$$

18.
$$4x_1 + 3x_2 - 2x_3 \le 5,$$
$$5x_1 - 2x_2 + 7x_3 \le 11,$$
$$3x_1 - x_2 + 2x_3 \le 3,$$
$$\max x_0\colon x_0 = 9x_1 - 2x_2 + 11x_3; \quad x_1, x_2, x_3 \ge 0.$$

19. *Business.* A carpentry shop makes bookcases, desks, and tables. Each bookcase requires 5 man-hours of woodworking, 4 hours of finishing, 30 board feet of hardwood, and 15 board feet of inexpensive wood, and sells for $60. Each desk requires 10 man-hours of woodworking, 3 hours of finishing, 20 board feet of hardwood, and 20 board feet of inexpensive wood, and sells for $100. Each table requires 7 hours of woodworking, 2 hours of finishing, and 24 board feet of hardwood, and sells for $80. The available supply is 575 man-hours for woodworking, 220 man-hours for finishing, 1800 board feet of hardwood, and 1000 board feet of inexpensive wood. How many of each product should be made to maximize sales?

20. *Business.* A coffee merchant has a supply of 600 lbs of Mocha coffee, 2400 lbs of Columbian coffee, and 4800 lbs of Brazilian coffee. He sells 100% Mocha coffee as Turkish at $4.00 per lb. He sells a Mocha blend consisting of 25% Mocha and 75% Columbian coffee at $2.50 per lb. He sells a Columbian blend consisting of 10% Mocha, 60% Columbian, and 30% Brazilian coffee at $1.75 per lb. He sells a Brazilian blend consisting of 20% Columbian and 80% Brazilian coffee at $1.25 per lb. How many lbs of each blend should he prepare to maximize his sales?

OBJECTIVES

a) Given a formulation of a linear program you should be able to formulate its dual.

b) You should be able to set up a minimum-type linear program in a simplex tableau, solve, and read off the solution.

c) You should be able to check your solution.

3.6 DUALITY AND MINIMUM-TYPE LINEAR PROGRAMS

Minimum-Type Linear Programs

In Section 3.5 we considered maximum-type linear programs which were expressed in the standard form

$$AX \le B, \qquad X \ge O,$$
$$\max x_0\colon x_0 = CX$$

with the positivity constraint $B > O$. Now suppose we wanted to solve a minimum-type linear program, which is expressed in the form

$$AY \geqslant B, \qquad Y \geqslant O,$$
$$\min y_0 \colon y_0 = CY.$$

Duality

THE DUALITY RELATIONSHIP. Given a *maximum*-type linear program[*]

$$AX \leqslant B, \qquad X \geqslant O,$$
$$\max x_0 \colon x_0 = CX,$$

its *dual*, a ***minimum-type*** linear program is

$$A^T Y \geqslant C^T, \qquad Y \geqslant O$$
$$\min y_0 \colon y_0 = B^T Y$$

where, as in Section 2.3, the superscript T refers to matrix transpose. Note that the dual is expressed using *new* variables Y, called dual variables.

Similarly, starting with a *minimum*-type linear program

$$AY \geqslant B, \qquad Y \geqslant O, \qquad \min y_0 \colon y_0 = CY,$$

we can *derive* its *dual*, a *maximum*-type linear program

$$A^T X \leqslant C^T, \qquad X \geqslant O, \qquad \max x_0 \colon x_0 = B^T X.$$

The original physical linear program, whether maximum or minimum, is often called the *primal* and the corresponding *derived* program its *dual*. The dual frequently has significance in an applied problem.

THE DUALITY THEOREM. For a pair of primal and dual linear programs, the objective function x_0 attains its maximum if and only if the objective function y_0 attains its minimum and, furthermore,

$$\max x_0 = \min y_0.$$

---

[*] In general, we do not need the restriction $B > O$. Furthermore, any equation of the form

$$a_1 x_1 + a_2 x_2 + \cdots = b$$

can be written as two inequalities

$$a_1 x_1 + a_2 x_2 + \cdots \leqslant b$$

and

$$a_1 x_1 + a_2 x_2 + \cdots \geqslant b \quad \text{or} \quad -a_1 x_1 - a_2 x_2 - \cdots = \ \leqslant -b.$$

This way of treating equations is done to avoid later complications.

23. Formulate the linear program dual to the following program (Margin Exercise 15, Section 3.3):

$$20x_1 + 100x_2 \leqslant 1900,$$
$$x_1 + 50x_2 \leqslant 500,$$
$$2x_1 + 20x_2 \leqslant 240,$$
$$\max x_0: x_0 = 20x_1 + 300x_2; \quad x_1, x_2 \geqslant 0.$$

Thus, to solve a minimum-type linear program, we first obtain its dual, a maximum-type linear program. Then, as we shall illustrate, we can obtain the solutions of the pair of dual programs at the same time.

When working with the simplex tableau, it is customary to call the *maximum*-type linear program the *primal* and the *minimum*-type linear program the *dual* regardless of the type of physical problem. Which meaning of primal or dual is intended can be determined from the context of its use.

Example 1 Given the maximum-type linear program of Example 1 of Section 3.3, with formulation

$$0.20x_1 + 0.10x_2 \leqslant 220,$$
$$0.60x_1 + 0.20x_2 \leqslant 620,$$
$$0.20x_1 + 0.70x_2 \leqslant 700,$$
$$\max x_0: x_0 = 3x_1 + 2x_2; \quad x_1, x_2 \geqslant 0,$$

find its dual.

Solution The minimum-type linear program dual to this is

$$0.20y_1 + 0.60y_2 + 0.20y_3 \geqslant 3,$$
$$0.10y_1 + 0.20y_2 + 0.70y_3 \geqslant 2,$$
$$\min y_0: y_0 = 220y_1 + 620y_2 + 700y_3; \quad y_1, y_2, y_3 \geqslant 0.$$

DO EXERCISE 23.

24. Formulate the program dual to the following program (Margin Exercise 17 Section 3.3):

$$0.1y_1 + 0y_2 \geqslant 200,$$
$$0y_1 + 0.2y_2 \geqslant 500,$$
$$0.02y_1 + 0.01y_2 \geqslant 100,$$
$$\min y_0: y_0 = 10y_1 + 15y_2; \quad y_1, y_2 \geqslant 0.$$

Example 2 Given the minimum-type linear program (Example 1, Section 3.4) with the formulation

$$40y_1 + 20y_2 \geqslant 160,$$
$$4y_1 + 4y_2 \geqslant 24,$$
$$4y_1 + 6y_2 \geqslant 28,$$
$$\min y_0: y_0 = 15y_1 + 10y_2; \quad y_1, y_2 \geqslant 0,$$

find its dual.

Solution The maximum-type linear program dual to this is

$$40x_1 + 4x_2 + 4x_3 \leqslant 15,$$
$$20x_1 + 4x_2 + 6x_3 \leqslant 10,$$
$$\max x_0: x_0 = 160x_1 + 24x_2 + 28x_3; \quad x_1, x_2, x_3 \geqslant 0.$$

DO EXERCISE 24.

Example 3 Express the constraint set of Example 1 as a set of equations, set up the initial simplex tableau, and read off *both* the initial *primal* and *dual* solutions.

Solution Adding slack variables, we obtain

$$
\begin{aligned}
0.20x_1 + 0.10x_2 + y_1 &= 220, \\
0.60x_1 + 0.20x_2 \quad + y_2 &= 620, \\
0.20x_1 + 0.70x_2 \quad\quad + y_3 &= 700, \\
\max x_0: \quad -3x_1 \quad -2x_2 \quad\quad + x_0 &= 0, \\
x_1, x_2; \quad y_1, \quad y_2, \quad y_3 &\geqslant 0.
\end{aligned}
$$

Note that we have used the same notation for the *slack* variables as we did for the *dual* variables.

The initial simplex tableau is, as before,

x_1	x_2	y_1	y_2	y_3	1
0.20	0.10	1	0	0	220
0.60	0.20	0	1	0	620
0.20	0.70	0	0	1	700
−3	−2	0	0	0	0

Recall that the x_0-column has been suppressed. As before, the initial *primal solution* is:

$$
\begin{aligned}
x_1 &= 0, \quad & y_1 &= 220, \quad & x_0 &= 0. \\
x_2 &= 0, \quad & y_2 &= 620, \\
& & y_3 &= 700,
\end{aligned}
$$

Similarly, the dual (minimum) program can be expressed as a set of equations by *subtracting* slack variables:

$$
\begin{aligned}
0.20y_1 + 0.60y_2 + 0.20y_3 - x_1 &= 3, \\
0.10y_1 + 0.20y_2 + 0.70y_3 \quad - x_2 &= 2, \\
\min y_0: \quad 220y_1 + 620y_2 + 700y_3 \quad\quad - y_0 &= 0, \\
y_1, y_2, y_3; \quad x_1, x_2 &\geqslant 0,
\end{aligned}
$$

where we have used the same notation for the dual slack variables as for the primal structural variables.

As with the primal, setting to zero the nonbasic dual variables $y_1 = y_2 = y_3 = 0$, we obtain for the basic dual variables

$$
\begin{aligned}
x_1 &= -3, \quad \text{and} \quad & y_0 &= 0. \\
x_2 &= -2,
\end{aligned}
$$

25. For the linear program of Margin Exercise 23, set up the initial simplex tableau and read off both the initial primal and dual solutions.

Note that while the primal solution was nonnegative, the dual solution usually violates the nonnegativity constraint.

Alternately, the *dual solution* is obtained directly from the *bottom row* of the simplex tableau, so that initially

$$x_1 = -3,$$
$$x_2 = -2,$$
$$y_1 = y_2 = y_3 = 0,$$
$$y_0 = 0.$$

Note that $y_0 = x_0$ and that the x_0 column has been suppressed.

Note, further, that both primal and dual solutions use the same *notation* but are read off differently from the tableau.

DO EXERCISE 25.

Consider now setting up the initial simplex tableau for a minimum-type linear program.

Example 4   Given the minimum-type linear program of Example 2, find its dual. Then express this constraint set as a set of equations, set up the initial simplex tableau, and read off *both* the initial *primal* and *dual* solutions.

Solution   Starting now with the *primal*, that is, the maximum problem, we add slack variables and obtain:

$$40x_1 + 4x_2 + 4x_3 + y_1 \qquad\qquad = 15,$$
$$20x_1 + 4x_2 + 6x_3 \qquad + y_2 \qquad = 10,$$
$$\max x_0: -160x_1 - 24x_2 - 28x_3 \qquad\qquad + x_0 = 0,$$
$$x_1, x_2, x_3; \quad y_1, y_2 \geqslant 0.$$

Again, we have purposely used the same *notation* for the *slack* variables as for the *dual* variables (in this case, the original variables).

Putting this into a simplex tableau, we obtain:

x_1	x_2	x_3	y_1	y_2	1
40	4	4	1	0	15
20	4	6	0	1	10
−160	−24	−28	0	0	0

26. Express the constraint set of Margin Exercise 24 as a set of equations, set up the initial simplex tableau, and read off *both* the initial *primal* and *dual* solutions.

The primal solution can be read off as before:

$$x_1 = x_2 = x_3 = 0, \qquad y_1 = 15,$$
$$x_0 = 0, \qquad y_2 = 10.$$

Now, again the dual minimum program can be expressed as a set of equations by *subtracting* slack variables, obtaining;

$$40y_1 + 20y_2 - x_1 \qquad\qquad\qquad = 160,$$
$$4y_1 + 4y_2 \qquad - x_2 \qquad\qquad = 24,$$
$$4y_1 - 6y_2 \qquad\qquad - x_3 \qquad = 28,$$
$$\min y_0:\ 15y_1 + 10y_2 \qquad\qquad\qquad - y_0 = 0,$$
$$y_1, y_2;\ \ x_1, x_2, x_3 \geqslant 0,$$

where we have used the same *notation* for the dual slack variables as for the primal variables.

As with the primal, setting the nonbasic dual variables $y_1 = y_2 = 0$, we obtain, for the basic dual variables,

$$x_1 = -160, \qquad \text{and} \qquad y_0 = 0.$$
$$x_2 = -24,$$
$$x_3 = -28,$$

Alternately, we can read the dual solution directly from the bottom row of the simplex tableau, obtaining;

$$x_1 = -160, \qquad y_1 = y_2 = 0,$$
$$x_2 = -24, \qquad y_0 = 0.$$
$$x_3 = -28.$$

DO EXERCISE 26.

Thus, we see that the initial solution to the primal is

$$X = O, \qquad Y = B, \qquad x_0 = 0,$$

which is nonnegative, since we have constrained $B > O$. Similarly, the initial solution to the dual is

$$X = -C^T, \qquad Y = O, \qquad y_0 = 0,$$

which is generally not nonnegative.

The solution of the primal program has been considered in the preceding section. Consider now the solution to the dual.

Example 5 Solve the minimum-type problem of Example 2. The initial simplex tableau is given in Example 4.

Solution We proceed exactly the same way as for a maximum-type problem.

The most negative number in the bottom row is -160, so that x_1 is to become basic. Using x_1, we obtain the quotients to the right of the tableau.

x_1	x_2	x_3	y_1	y_2	1	q
40*	4	4	1	0	15	$\frac{3}{8}$ (Minimum)
20	4	6	0	1	10	$\frac{1}{2}$
-160	-24	-28	0	0	0	

The minimum quotient of $\frac{3}{8}$ implies that y_1 is to become nonbasic, or that the roles of x_1 and y_1 are to be interchanged. The pivot element is starred (*).

Pivoting to interchange the roles of x_1 and y_1, we obtain:

x_1	x_2	x_3	y_1	y_2	1	q
1	$\frac{1}{10}$	$\frac{1}{10}$	$\frac{1}{40}$	0	$\frac{3}{8}$	$\frac{15}{4}$
0	2	4*	$-\frac{1}{2}$	1	$\frac{5}{2}$	$\frac{5}{8}$ (Minimum)
0	-8	-12	4	0	60	

The *dual* solution at this point can be read from the bottom row and is

$$y_1 = 4, \qquad x_1 = 0, \qquad y_0 = 60.$$
$$y_2 = 0, \qquad x_2 = -8,$$
$$x_3 = -12,$$

Picking x_3 (most negative indicator) to become basic leads to y_2 (minimum-quotient) to become nonbasic.

Pivoting to interchange the roles of x_3 and y_2, we obtain:

x_1	x_2	x_3	y_1	y_2	1	q
1	$\frac{1}{20}$	0	$\frac{3}{80}$	$-\frac{1}{40}$	$\frac{5}{16}$	$\frac{25}{4}$
0	$\frac{1}{2}*$	1	$-\frac{1}{8}$	$\frac{1}{4}$	$\frac{5}{8}$	$\frac{5}{4}$ (Minimum)
0	-2	0	$\frac{5}{2}$	3	$\frac{135}{2}$	

The *dual* solution at this point is

$$y_1 = \tfrac{5}{2}, \quad x_1 = 0, \quad y_0 = \tfrac{135}{2}.$$
$$y_2 = 3, \quad x_2 = -2,$$
$$x_3 = 0,$$

Pivoting now to interchange the roles of x_2 and x_3, we obtain:

x_1	x_2	x_3	y_1	y_2	1
1	0	$-\frac{1}{10}$	$\frac{1}{20}$	$-\frac{1}{20}$	$\frac{1}{4}$
0	1	2	$-\frac{1}{4}$	$\frac{1}{2}$	$\frac{5}{4}$
0	0	4	2	4	70

The *dual* solution at this point is

$$y_1 = 2, \quad x_1 = 0, \quad y_0 = 70.$$
$$y_2 = 4, \quad x_2 = 0,$$
$$x_3 = 4,$$

This solution is nonnegative. It is therefore also minimal and the algorithm terminates.

Note that the nonnegativity of the dual solution corresponds to optimality of the primal.

DO EXERCISE 27.

Checking Solutions

To check either a maximum- or a minimum-type linear program, we substitute *both* primal and dual solutions into their respective programs written in *equation* form.*

---

* Actually, it is sufficient that if X is a primal feasible solution ($AX \le B, X \ge O$) and Y is a dual feasible solution ($A^T Y \ge C^T, Y \ge O$), then the condition $CX = B^T Y$ insures that both X and Y are optimal. Effecting the check in *equation* form (rather than *constraint* form as in this footnote) checks slack variables as well as structural variables and hence is useful in *locating* any errors that may exist.

27. Solve the minimum-type problem whose initial simplex tableau is given in Margin Exercise 26.

28. Check your solution to Margin Exercise 24 using primal and dual solutions. (See Margin Exercise 27 for solutions.)

Example 6 Check the solution to the linear program of Example 4.

Solution First, let us restate the programs in equation form and their solutions which can be obtained from the final tableau of Example 5.

Primal program:

$$40x_1 + 4x_2 + 4x_3 + y_1 = 15,$$
$$20x_1 + 4x_2 + 6x_3 + y_2 = 10,$$
$$\max x_0: 160x_1 + 24x_2 + 28x_3 = x_0,$$
$$x_1, x_2, x_3;\quad y_1, y_2 \geqslant 0.$$

Primal solution:

$$x_1 = \tfrac{1}{4},\quad x_2 = \tfrac{5}{4},\quad x_3 = 0,$$
$$y_1 = y_2 = 0,$$
$$x_0 = 70.$$

Dual program:

$$40y_1 + 20y_2 - x_1 = 160,$$
$$4y_1 + 4y_2 - x_2 = 24,$$
$$4y_1 + 6y_2 - x_3 = 28,$$
$$\min y_0: 15y_1 + 10y_2 = y_0,$$
$$y_1, y_2;\quad x_1, x_2, x_3 \geqslant 0.$$

Dual solution:

$$y_1 = 2,\quad y_2 = 4,$$
$$x_1 = x_2 = 0,\quad x_3 = 4,$$
$$y_0 = 70.$$

To check, we first note that all variables (structural and slack) are nonnegative. Next we evaluate the objective functions:

$$x_0 = 160(\tfrac{1}{4}) + 24(\tfrac{5}{4}) + 28(0) = 40 + 30 + 0 = 70, \quad \text{OK.}$$
$$y_0 = 15(2) + 10(4) = 30 + 40 = 70, \quad \text{OK.}$$

Next we substitute the primal solution into the primal equations:

$$40(\tfrac{1}{4}) + 4(\tfrac{5}{4}) + 4(0) + 0 = 10 + 5 + 0 + 0 = 15, \quad \text{OK.}$$
$$20(\tfrac{1}{4}) + 4(\tfrac{5}{4}) + 6(0) + 0 = 5 + 5 + 0 + 0 = 10, \quad \text{OK.}$$

Then we substitute the dual solution into the dual equations:

$$40(2) + 20(4) - 0 = 80 + 80 - 0 = 160, \quad \text{OK.}$$
$$4(2) + 4(4) - 0 = 8 + 16 - 0 = 24, \quad \text{OK.}$$
$$4(2) + 6(4) - 4 = 8 + 24 - 4 = 28. \quad \text{OK.}$$

All equations check. Hence the solutions are correct.

DO EXERCISE 28.

*(Optional) Economic Interpretation of Duality

Example 7   Formulate the linear program *dual* to that of Example 1 of Section 3.4.

Business, Nutrition. A feed supplier mixes feed for a particular animal which requires at least 160 lbs of nutrient A, at least 24 lbs of nutrient B, and at least 28 lbs of nutrient C. The supplier has available soybean meal which costs $15 per 100-lb sack and contains 40 lbs of nutrient A, 4 lbs of nutrient B, and 4 lbs of nutrient C and triticale which costs $10 per 100-lb sack and contains 20 lbs of nutrient A, 4 lbs of nutrient B, and 6 lbs of nutrient C. How many sacks of each ingredient should be used to mix animal feed which satisfies the minimum requirements at minimum cost?

Solution   The data from that problem was put into the following table:

	Composition (lbs per 100-lb sack)		Pounds required
	Soybean meal	Triticale	
Number of 100-lb sacks	y_1	y_2	
Nutrient A	40	20	160
Nutrient B	4	4	24
Nutrient C	4	6	28
$ Cost per 100-lb sack	15	10	Minimize cost

In formulating the dual we reinterpret the data. To start, we define dual variables:

$x_1 = \$$ price per lb of nutrient A
$x_2 = \$$ price per lb of nutrient B $\Big\}$ (if available in pure form),
$x_3 = \$$ price per lb of nutrient C

where

$$x_1, x_2, x_3 \geq 0.$$

If we consider selling these nutrients separately (but still for animal feed), then we cannot obtain more for them than their cost combined as soybean meal or triticale; so that

$$40x_1 + 4x_2 + 4x_3 \leq 15 \qquad \text{and} \qquad 20x_1 + 4x_2 + 6x_3 \leq 10.$$

We now seek to determine the prices x_1, x_2, x_3 such that we maximize our income and still satisfy the animal's requirements (at no more than the cost of the food sources):

$$\max x_0: x_0 = 160x_1 + 24x_2 + 28x_3.$$

It can be seen now that the program just formulated *is* the dual to the given program.

The optimal values of these dual variables x_1, x_2, x_3 are called *shadow prices* and are used in economic analyses.

Many other linear programs also have duals which have physical significance.

EXERCISE SET 3.6

Exercises 1 through 6 are Exercises 7 through 12 from Set **3.2** where they were solved graphically. Here you are asked to solve the problems using duality and the simplex algorithm and to check this solution with your graphical solution. Note that only the final solution checks and that intermediate solutions from the simplex tableaux have no corresponding points on the graph.

1. Exercise 7, Set **3.2** and Exercise 1, Set **3.5**:

Minimize $y_0(y_1, y_2) = 3y_1 + 4y_2$, subject to the constraints

$$3y_1 + y_2 \geqslant 9,$$
$$y_1 + y_2 \geqslant 7,$$
$$y_1 + 2y_2 \geqslant 8,$$
$$y_1, y_2 \geqslant 0.$$

2. Exercise 8, Set 3.2 and Exercise 2, Set 3.5:

Minimize $y_0(y_1, y_2) = 3y_1 + 4y_2$, subject to the constraints

$$2y_1 + y_2 \geqslant 8,$$
$$4y_1 + 3y_2 \geqslant 22,$$
$$2y_1 + 5y_2 \geqslant 18,$$
$$y_1, y_2 \geqslant 0.$$

3. Exercise 9, Set 3.2 and Exercise 3, Set 3.5:

Minimize $y_0(y_1, y_2) = 2y_1 + 5y_2$, subject to the constraints

$$2y_1 + y_2 \geqslant 9,$$
$$4y_1 + 3y_2 \geqslant 23,$$
$$y_1 + 3y_2 \geqslant 8,$$
$$y_1, y_2 \geqslant 0.$$

4. Exercise 10, Set 3.2 and Exercise 4, Set 3.5:

Minimize $y_0(y_1, y_2) = 9y_1 + 7y_2$, subject to the constraints

$$5y_1 + 3y_2 \geqslant 30,$$
$$2y_1 + 3y_2 \geqslant 21,$$
$$3y_1 + 6y_2 \geqslant 36,$$
$$y_1, y_2 \geqslant 0.$$

5. Exercise 11, Set 3.2 and Exercise 5, Set 3.5:

Minimize $y_0(y_1, y_2) = 3y_1 + 5y_2$, subject to the constraints

$$4y_1 + y_2 \geqslant 9,$$
$$3y_1 + 2y_2 \geqslant 13,$$
$$2y_1 + 5y_2 \geqslant 16,$$
$$y_1, y_2 \geqslant 0.$$

6. Exercise 12, Set 3.2 and Exercise 6, Set 3.5:

Minimize $y_0(y_1, y_2) = 2y_1 + 3y_2$, subject to the constraints

$$4y_1 + y_2 \geqslant 7,$$
$$y_1 + y_2 \geqslant 4,$$
$$2y_1 + 5y_2 \geqslant 14,$$
$$y_1, y_2 \geqslant 0.$$

In Exercise Set 3.4, various minimum-type linear programs were given to be formulated and solved graphically. Here you are asked to solve these problems using duality and the simplex algorithm and to check your answer with your graphical solution.

7. Exercise 1, Set 3.4 and Exercise 7, Set 3.5:

$$50y_1 + 15y_2 \geq 120,$$
$$8y_1 + 5y_2 \geq 24,$$
$$5y_1 + y_2 \geq 10,$$
$$\min y_0 \colon y_0 = 15y_1 + 5y_2; \quad y_1, y_2 \geq 0.$$

8. Exercise 2, Set 3.4 and Exercise 8, Set 3.5:

9. Exercise 3, Set 3.4 and Exercise 9, Set 3.5:

$$50y_1 + 20y_3 \geq 120,$$
$$8y_1 + 6y_3 \geq 24,$$
$$5y_1 + 8y_3 \geq 20,$$
$$\min y_0 \colon y_0 = 15y_1 + 8y_3; \quad y_1, y_3 \geq 0.$$

10. Exercise 4, Set 3.4 and Exercise 10, Set 3.5:

11. Exercise 5, Set 3.4 and Exercise 11, Set 3.5:

$$40y_1 + 80y_2 \geq 2000,$$
$$40y_1 + 30y_2 \geq 1500,$$
$$120y_1 + 40y_2 \geq 2400,$$
$$\min y_0 \colon y_0 = 12y_1 + 10y_2; \quad y_1, y_2 \geq 0.$$

12. Exercise 6, Set 3.4 and Exercise 12, Set 3.5:

13. Exercise 7, Set 3.4 and Exercise 13, Set 3.5:

$$40y_1 + 40y_3 \geq 2000,$$
$$40y_1 + 80y_3 \geq 1500,$$
$$120y_1 + 80y_3 \geq 2400,$$
$$\min y_0 \colon y_0 = 12y_1 + 15y_3; \quad y_1, y_3 \geq 0.$$

14. Exercise 8, Set 3.4 and Exercise 14, Set 3.5:

15. Exercise 9, Set 3.4 and Exercise 15, Set 3.5:

$$40y_1 + 80y_2 \geq 1600,$$
$$40y_1 + 30y_2 \geq 2100,$$
$$120y_1 + 40y_2 \geq 2400,$$
$$\min y_0 \colon y_0 = 12y_1 + 10y_2; \quad y_1, y_2 \geq 0.$$

16. Exercise 10, Set 3.4 and Exercise 16, Set 3.5:

17. Exercise 11, Set 3.4 and Exercise 17, Set 3.5:

$$40y_1 + 40y_3 \geq 1600,$$
$$40y_1 + 80y_3 \geq 2100,$$
$$120y_1 + 80y_3 \geq 2400,$$
$$\min y_0 \colon y_0 = 12y_1 + 15y_3; \quad y_1, y_3 \geq 0.$$

18. Exercise 12, Set 3.4 and Exercise 18, Set 3.5. (You were not asked to prepare a graph in Exercise 12, Set 3.4.) Check your solution using an alternate pivot.

Use duality and the simplex algorithm to solve Exercises 19 through 26. Check your answer using primal and dual solutions.

19.
$$3y_1 + y_2 \geq 14,$$
$$4y_1 + 3y_2 \geq 34,$$
$$3y_1 + 4y_2 \geq 36,$$
$$\min y_0: y_0 = 7y_1 + 8y_2; \quad y_1, y_2 \geq 0.$$

20.
$$2y_1 + y_2 \geq 11,$$
$$y_1 + y_2 \geq 9,$$
$$y_1 + 2y_2 \geq 13,$$
$$\min y_0: y_0 = 3y_1 + 2y_2; \quad y_1, y_2 \geq 0.$$

21.
$$3y_1 + 2y_2 \geq 29,$$
$$4y_1 + 5y_2 \geq 55,$$
$$y_1 + 2y_2 \geq 18,$$
$$\min y_0: y_0 = 5y_1 + 4y_2; \quad y_1, y_2 \geq 0.$$

22.
$$5y_1 + y_2 \geq 15,$$
$$4y_1 + 2y_2 \geq 24,$$
$$5y_1 + 5y_2 \geq 35,$$
$$\min y_0: y_0 = 13y_1 + 5y_2; \quad y_1, y_2 \geq 0.$$

▶

23.
$$2x_1 + 9x_2 + 4x_3 \geq 6,$$
$$3x_1 + 6x_2 - 2x_3 \geq 8,$$
$$x_1 + x_2 + x_3 \geq 3,$$
$$\min x_0: x_0 = 5x_1 + 2x_2 + x_3; \quad x_1, x_2, x_3 \geq 0.$$

24.
$$2x_1 + x_2 + 5x_3 \geq 5,$$
$$2x_1 - x_2 + 2x_3 \geq 3,$$
$$5x_1 + 3x_2 + x_3 \geq 7,$$
$$\min x_0: x_0 = 6x_1 + 2x_2 + 7x_3; \quad x_1, x_2, x_3 \geq 0.$$

25.
$$2y_1 + 3y_2 + y_3 \geq 2,$$
$$5y_1 + 2y_2 - 3y_3 \geq 4,$$
$$7y_1 + 6y_2 + 4y_3 \geq 5,$$
$$\min y_0: y_0 = 8y_1 + 9y_2 + 5y_3; \quad y_1, y_2, y_3 \geq 0.$$

26.
$$6y_1 + 3y_2 + 4y_3 \geq 2,$$
$$y_1 - 2y_2 - 5y_3 \geq 3,$$
$$2y_1 + 9y_2 + y_3 \geq 8,$$
$$\min y_0: y_0 = 12y_1 + 9y_2 + 8y_3; \quad y_1, y_2, y_3 \geq 0.$$

27. *Nutrition.* The calcium, iron, protein, and cost of various foods (per 100 g) is given in the accompanying table.

27. (*Cont.*) Using eggs, beef, cheese, and soy beans, what is the minimum-cost diet that will satisfy the minimum daily requirements?

Food (100 g)	Calcium (mg)	Iron (mg)	Protein (g)	Cost (¢)
Eggs	54	2.7	12.8	22
Beef	7	6.6	18.6	36
Chicken	14	1.5	20.2	16
Bluefish	23	0.6	20.5	24
Whole-wheat bread	96	2.2	9.3	10
Cheddar cheese	570	0.7	20.5	32
Soy beans	260	10.0	34.9	10
Sunflower seeds	57	6.0	28.0	20

Food (100 g)	Calcium (mg)	Iron (mg)	Protein (g)	Cost (¢)
Sesame seeds	72	7.7	23.4	18
Almonds	234	4.7	18.6	48
Cashews	38	3.8	17.2	48
Filberts	209	3.4	12.6	40
Millet	20	6.8	9.9	12
Minimum daily requirement	750	10	50*	

*This is an average value. Some nutritionists recommend more, some less.

28. Using data from the table of Exercise 27, select various combinations of 3 or 4 foods (other than soy beans) and determine the minimum-cost diet that will satisfy the minimum daily requirements. Answers may vary.

29. Using data from the table of Exercise 27, for eggs, beef, and cheese, what is the minimum-cost diet that will satisfy the minimum daily requirements?

CHAPTER 3 TEST

1. Formulate the following problem, giving all the constraints, and write in matrix form.

A health food store manager is preparing two mixtures of breakfast cereal out of a supply of 100 lbs rolled oats, 10 lbs chopped almonds, 5 lbs chopped dried apples, 25 lbs chopped sunflower seeds, and 15 lbs monukka raisins. The first mixture contains 80% oats, 1% almonds, no dried apple, 12% sunflower seeds, and 7% raisins, and sells for $0.95 per lb. The second mixture contains 60% oats, 3% almonds, 4% dried apple, 24% sunflower seeds, and 9% raisins, and sells for $1.35 per lb. How much of each mixture should be made to maximize the income?

Given the following linear program:

i) $\qquad y_1 + 8y_2 \geqslant 24,$
ii) $\qquad 7y_1 + y_2 \geqslant 14,$
iii) $\qquad 2y_1 + 3y_2 \geqslant 18,$
$\qquad \min f: f = y_1 + y_2; y_1, y_2 \geqslant 0,$

3. Find the optimum feasible solution and explain how you found it.

2. Graph the constraints and shade the feasible solution.

4. Using the simplex method, solve

$$x_1 + 2x_2 \leqslant 26,$$
$$x_1 + x_2 \leqslant 16,$$
$$5x_1 + 3x_2 \leqslant 70,$$
$$\max x_0: x_0 = 4x_1 + 3x_2; \quad x_1, x_2 \geqslant 0.$$

Show all work.

5. a) Write the linear program dual to that given in (1).
   b) Read off the dual solution from the final tableau of (1).
   c) Write the equations to be used in checking your solution.
   d) Check your solution.

CHAPTER FOUR

Sets and Counting Techniques

OBJECTIVES

You should be able to:

a) Determine whether an object is an element of a set.
b) Decide whether two sets are equal.
c) Name a set using the roster method or set-builder notation.
d) Decide whether one set is a subset of another.
e) Find the complement A^c of a given set.
f) Find the Cartesian Product of two sets.
g) Find the cardinality of a given set (finite).

4.1 SETS*

Sets form the foundation of our study of probability.

What do the following have in common?

| a *flock* of birds, | a *school* of fish, | a *crowd* of people, |
| a *herd* of animals, | a *pod* of whales, | a *host* of angels. |

In each case we are dealing with a collection of objects of a certain type. Rather than use a different word for each type of collection, it is convenient to denote them all by the one word "*set*."

A *set* is a collection of well-defined objects called *elements*.

One can talk about the set of all "employees" in a company since an "employee" is *well-defined*. On the other hand, one cannot talk about the set of all "good" employees unless one can provide an objective way to distinguish "good" employees.

We now develop the *notation* necessary to deal with sets.

The set of vowels can be written

$$V = \{a, e, i, o, u\}$$

where the *capital letter* V denotes the *set* as do the braces enclosing the *elements*. The elements are separated by commas, read "and." Elements of sets are usually denoted by lower-case letters. This way of describing a set is known as the "roster" method.

That "a is a vowel" or "a is an element of V" can be written

$$a \in V.$$

Similarly, that "b is not a vowel," or "b is not an element of V" can be written

$$b \notin V.$$

Two sets A and B are *equal*, written $A = B$, if and only if each element of either set is also an element of the other set. For example

a) $\{a, e, i, o, u, a\} = \{a, e, i, o, u\}$,   and

b) $\{u, o, i, e, a\} = \{a, e, i, o, u\}$,

where in (a) the *repetition* of an element does not change the set and in (b) the *order* in which we write the elements does not change the set.

Consider the set of all students in a given class. The class is not changed if some student's name is listed twice. Furthermore, the class is not changed whether the student's names are listed alphabetically, or by Social Security number, or any other way.

Example 1   Given $A = \{1, 2, 3, 4, 5, 6\}$, $B = \{2, 3, 4, 5, 6\}$,   and $C = \{2, 3, 4, 5, 6, 1\}$.

a) Is $1 \in A$?, $1 \in B$?
b) Is $A = B$?, $A = C$?

Solution
a) Yes, $1 \in A$; no, $1 \notin B$.
b) No, $A \neq B$, since $1 \in A$ but $1 \notin B$;
   yes, $A = C$, since order of elements is immaterial.

DO EXERCISE 1.

Other Set Notation

The set containing all the integers that are greater than 1 can be written

$$\{2, 3, 4, \ldots\}.$$

The dots are read "and so forth," and indicate that the pattern of the listed elements continues. This set can also be written using words as

"The set of all integers greater than 1."

or

"The set of all x such that x is an integer and $x > 1$."

or

$$\{x \mid x \text{ is an integer}, x > 1\}$$

In the latter notation, the first set brace "{" is read "The set of all," and the vertical line "|" is read "such that." The entire notation is read "The set of all x such that x is an integer and x is greater than 1." This is called "set builder" notation.

Example 2   Write $C = \{1, 4, 9, \ldots\}$ using set builder notation.

Solution

$$C = \{x \mid x \text{ is an integer and } x \text{ is a perfect square}\}$$

Answers may vary. We could have written this set

$$C = \{x \mid x = n^2, \quad n \text{ is an integer}, \quad n > 0\};$$

1. Given $A = \{a, b, c, d\}$,

   $B = \{a, e, i, o, u\}$, $C = \{a, b, c, d, c\}$.

a) Is $b \in A$?, $e \in C$?

b) Is $A = B$?, $A = C$?

2. a) Write

$$D = \{1, 3, 5, \ldots\}$$

using set-builder notation. Answers may vary.

b) Write

$$F = \{x \mid x = 3y + 1, \ y \in I, \ y \geq 0\}$$

using the roster method.

or, if we let I represent the set of all integers,

$$I = \{\ldots, -3, -2, -1, 0, 1, 2, 3, \ldots\}$$

then we could use more abbreviated notation for C.

$$C = \{x \mid x = n^2, \quad n \in I, \quad n > 0\}$$

Example 3 Write $E = \{x \mid x = 4n - 1, \quad n \in I, \quad n \geq 0\}$ using the roster method.

Solution

$$E = \{-1, 3, 7, 11, 15, \ldots\}$$

DO EXERCISE 2.

The Empty Set

The set without elements is known as the *empty set*, or *null set*, and is denoted by the symbol $\emptyset$, or { }. The following is an example of the empty set:

The set of all odd numbers whose squares are even.

The notation {0} is not appropriate for the empty set, since this set has one element in it, namely the number 0. Also, $\{\emptyset\}$ is not notation for the empty set. If $\emptyset$ represents an empty set (empty paper sack), then $\{\emptyset\}$ would represent an empty sack in a sack.

Subsets

Set A is said to be a *subset* of B if and only if every element of A is an element of B. For example, if $B = \{a, b, c\}$, the subsets of B are

$$\{a\}, \quad \{b\}, \quad \{c\}, \quad \{a, b\}, \quad \{a, c\}, \quad \{b, c\}, \quad \{a, b, c\}, \quad \emptyset$$

That "A is a subset of B" is symbolized $A \subset B$, and is often read "A is *contained* in B." Thus

$$\{a\} \subset \{a, b, c\}$$
$$\{b, c\} \subset \{a, b, c\}$$
$$\{a, b, c\} \subset \{a, b, c\}$$
$$\emptyset \subset \{a, b, c\}$$

Note that, for any set A, $A \subset A$; that is, any set is a subset of itself.*

---

* A *proper* subset is a subset such that $A \subset B$, but $A \neq B$. Thus $\{a, b\}$ is a proper subset of $\{a, b, c\}$, but $\{a, b, c\}$ is not a proper subset of $\{a, b, c\}$. When A is a proper subset of B, there is at least one element in B which is not in A. Some texts use $\subseteq$ to indicate a subset, and reserve the symbol $\subset$ only for a proper subset, but this distinction is not useful to us in this text.

Also, for any set A, $\emptyset \subset A$; that is, the empty set is a subset of every set. For example, let A be the set of all fish in a lake and C be the subset of A consisting of all fish caught by a fisherman. Thus, $C \subset A$. Since it is possible that the fisherman catches *nothing*, we can have $C = \emptyset$, so that $\emptyset \subset A$.

DO EXERCISE 3.

Example 4   Let $V = \{a, e, i, o, u\}$, $A = \{a, e\}$,   and $B = \{a, t, s\}$.
a) Is $A \subset V$?
b) Is $B \subset V$?

Solution
a) Yes, $A \subset V$, since each element of A is also in V.
b) No, $B \not\subset V$, since the elements t and s of B are not in V.

DO EXERCISE 4.

Cardinality of a Set

The *cardinality* of a set A is the *number of distinct elements* it contains and is written $\mathcal{N}(A)$. Thus, if

$$V = \{a, e, i, o, u\},$$

then

$$\mathcal{N}(V) = 5.$$

Also, $\mathcal{N}(\emptyset) = 0$; the cardinality of the empty set is 0.

DO EXERCISE 5.

Universal Sets and Complements

Any mathematical situation has a frame of reference called a *universal set*. For example, in elementary algebra the universal set is usually the set of real numbers. In plane geometry the universal set is the set of points in the plane. In probability the universal set might be, for example, the set of all possible outcomes for drawing a particular card from a well-shuffled deck. Usually it is clear what the universal set is, though it may have to be inferred from the context of a problem or application. We might think of the empty set as being on the low end of the frame of reference and the universal set as being on the high end. The universal set is usually denoted $\mathcal{U}$, and in any given application it must be true that for any set A, $A \subset \mathcal{U}$.

Example 5   A survey is to be made to determine the opinion of readers of the *National Observer* on a particular issue. What is the universal set?

---

3. List all the subsets of
$$C = \{a, b, c, d\}.$$

4. Let
$$A = \{1, 2, 3, 4, 5, 6, 7, 8, 9\},$$
$$B = \{2, 4, 6, 8\},$$
$$C = \{0, 1, 3, 5\}.$$

a) Is $B \subset A$?

b) Is $C \subset A$? Why?

5. Find:
a) $\mathcal{N}(\{a, b, c, d\})$

b) $\mathcal{N}$ (The set of all odd numbers whose squares are even)

6. A person enters a department store with the intention of spending $10. What is the universal set of his purchases?

Solution *Before* the survey is taken, the universal set consists of all readers of *National Observer*.

After the survey is taken, the universal set consists of those readers questioned in the survey. The set of readers surveyed is a subset of all the readers. The readers surveyed with opinions, pro, con, or otherwise, are all subsets of the readers surveyed.

DO EXERCISE 6.

The *absolute complement* or, simply, the *complement* of a set A, written A^c, is what is left in the universal set after the elements of A are removed.

7. Given $A = \{1, 3, 5, 7, 9\}$ and $\mathcal{U} = \{1, 2, 3, 4, 5, 6, 7, 8, 9\}$, find A^c.

Example 6 Given $A = \{a, e, i\}$ and $\mathcal{U} = \{a, e, i, o, u\}$, find A^c.

Solution $A^c = \{o, u\}$.

Note. $\mathcal{U}^c = \emptyset$ and $\emptyset^c = \mathcal{U}$.

DO EXERCISE 7.

Given two sets A and B, the *Cartesian product* of these sets, written $A \times B$, is the set of all ordered pairs (a, b) where $a \in A$ and $b \in B$. Formally

$$A \times B = \{(a, b) \mid a \in A, \quad b \in B\}.$$

Note that, as the name implies, the *order* of the elements in an *ordered pair* is important.

8. If $A = \{a, b, c\}$ and $B = \{1, 2, 3, 4\}$, find $A \times B$ and $\mathcal{N}(A \times B)$.

Example 7 If $A = \{2, 4\}$ and $B = \{1, 2, 3\}$, find $A \times B$, $\mathcal{N}(A)$, $\mathcal{N}(B)$, and $\mathcal{N}(A \times B)$.

Solution

$$A \times B = \{(2, 1), (4, 1), (2, 2), (4, 2), (2, 3), (4, 3)\}$$
$$\mathcal{N}(A) = 2 \quad \text{and} \quad \mathcal{N}(B) = 3$$
$$\mathcal{N}(A \times B) = 2 \cdot 3 = 6 \quad \text{(as can be verified by counting)}.$$

The cardinality of a Cartesian product of two sets can be found simply by counting to be

$$\mathcal{N}(A \times B) = \mathcal{N}(A) \cdot \mathcal{N}(B),$$

with the obvious extension to more than two sets.

DO EXERCISE 8.

EXERCISE SET 4.1

For Exercises 1 through 16, let $\mathcal{U} = \{x \mid x \in I, \ 0 \leqslant x \leqslant 12\}$, where I is the set of all integers, and

$$A = \{x \mid x \in \mathcal{U}, \ 0 \leqslant x \leqslant 10\}, \qquad D = \{x \mid x \in \mathcal{U}, \ x \in A, \ x \notin C\},$$

$$B = \{x \mid x \in \mathcal{U}, \ 0 < x < 10\}, \qquad E = \left\{x \mid x \in \mathcal{U}, \ \frac{x}{3} \in \mathcal{U}\right\},$$

$$C = \left\{x \mid x \in \mathcal{U}, \ \frac{x}{2} \in A\right\}, \qquad F = \{x \mid x \in \mathcal{U}, \ x \in E, \ x \leqslant 10\}.$$

1. Write the sets A through C using the roster method.

2. Write the sets D through F using the roster method.

3. Determine the complements of sets A through C.

4. Determine the complements of sets D through F.

State whether each of the following is true or false.

5. $5 \in A$ **6.** $7 \in C$ **7.** $A \subset B$ **8.** $F \subset B$

9. $\emptyset \subset D$ **10.** $\emptyset \subset A$ **11.** $F = \{0, 6, 3, 9\}$ **12.** $E = \{3, 3, 6, 9, 12\}$

13. $E = \{6, 3, 12, 9\}$ **14.** $F = \{0, 3, 6, 3, 9\}$ **15.** $E \subset F$ **16.** $D^c \subset \mathcal{U}$

For Exercises 17 through 34, let

$$\mathcal{U} = \{a, e, i, o, u, m, n, r, t, c\}$$

and

$$A = \{a, e, i, o, u\}, \qquad D = \{m, i, n, t\},$$
$$B = \{c, m, n, r, t\}, \qquad E = \{e, i\},$$
$$C = \{a, c, e\}, \qquad F = \{r, t\}.$$

17. Determine the complements of sets A through C.

18. Determine the complements of sets D through F.

State whether each of the following is true or false.

19. $E = \{c, e, i, o, u\}$ **20.** $D = \{t, i, m, e\}$ **21.** $\{c, e, i, o, u\} \subset \mathcal{U}$ **22.** $\{a, e, z\} \subset \mathcal{U}$

23. $C = \{c, a, s, e\}$ **24.** $F = \{t, r\}$ **25.** $\mathcal{N}(C) = 4$ **26.** $\mathcal{N}(E) = 2$

Answer the following.

27. $E \times F = ?$ **28.** $F \times E = ?$

29. Is $(E \times F) \subset (A \times B)$?   **30.** Is $(F \times E) \subset (A \times B)$?

31. $C \times E = ?$ **32.** $C \times F = ?$

33. $\mathcal{N}(C \times E) = ?$ **34.** $\mathcal{N}(C \times F) = ?$

35. a) Find $\mathcal{N}(\emptyset)$.
     b) List the subsets of $\emptyset$.
     c) How many subsets are there?

36. a) Find $\mathcal{N}(\{a\})$.
     b) List the subsets of $\{a\}$.
     c) How many subsets are there?

37. a) Find $\mathcal{N}(\{a, b\})$.
   b) List the subsets of $\{a, b\}$.
   c) How many subsets are there?

38. a) Find $\mathcal{N}(\{a, b, c\})$.
   b) List the subsets of $\{a, b, c\}$.
   c) How many subsets are there?

39. On the basis of Exercises 35 through 38, complete the following table.

Cardinality of a set	0	1	2	3	4	n
Number of subsets	1					

40. Complete, based on Exercise 39.

A set with n elements has _____ subsets.

OBJECTIVES

You should be able to:

a) Given two sets, A and B, find their union, $A \cup B$, and intersection, $A \cap B$.

b) Given two sets A and B, find their difference $A - B$.

c) Given a collection of subsets of a given set, determine whether it is a partition of that set.

d) Find several partitions of a given set.

e) Solve problems involving sets.

4.2 UNION AND INTERSECTION OF SETS AND VENN DIAGRAMS

Consider two sets A and B.

Let A be the set of all students in a Finite Mathematics class and B be the set of all students in an Economics class. The *union* of A and B, written $A \cup B$, is the set of all students taking Finite Mathematics *or* Economics *or both*.

Formally, the *union* of A and B is the set of all elements contained in *either A or B or both* and can be written

$$A \cup B = \{x \mid x \in A \quad or \quad x \in B\}.$$

$A \cup B$ is read "A union B."

Note. Throughout this book, "or" will be used in the *inclusive* sense of "either or both." If the *exclusive* "or"—meaning "either but not both"—is intended, it will be specifically stated.

The union of sets can be represented geometrically by means of *Venn* diagrams. If two sets A and B are each represented by the interior of some closed curve, then the union of A and B is represented by the shaded area of the following diagram.

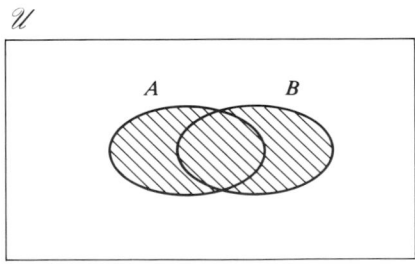

$A \cup B$ shaded

Consider again the two sets A and B, where A is the set of all students in a Finite Mathematics class, and B is the set of all students in an Economics class.

The *intersection* of A and B, written $A \cap B$, represents the set of all students taking *both* Finite Mathematics *and* Economics.

Formally, the *intersection* of A and B is the set of all elements contained in *both A and B* and can be written

$$A \cap B = \{x \mid x \in A \quad and \quad x \in B\}.$$

$A \cap B$ is read "A intersect B."

On a Venn diagram, the intersection of A and B is represented by the shaded area in the following diagram.

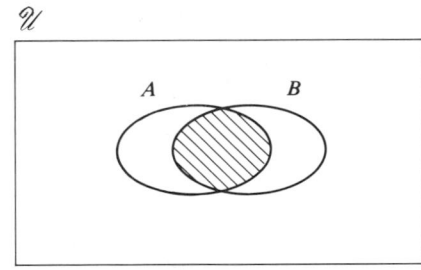

$A \cap B$ shaded

If two sets A and B have *no* elements in common, that is,

$$A \cap B = \emptyset$$

or

$$\mathcal{N}(A \cap B) = 0,$$

then A and B are said to be *disjoint*, and can be represented by the following Venn diagram:

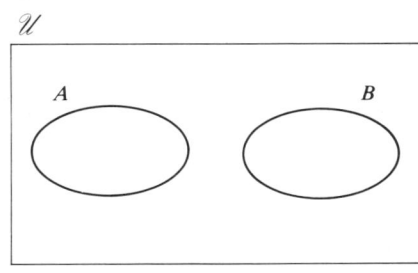

$A \cap B = \emptyset$

Note that $A \cap A^c = \emptyset$ and $A \cup A^c = \mathcal{U}$. (See the following diagram.)

9. Given $A = \{a, b, c, e, g\}$ and $B = \{b, c, e, f, s\}$. What are $A \cup B$ and $A \cap B$? Represent each of these sets by means of a Venn diagram.

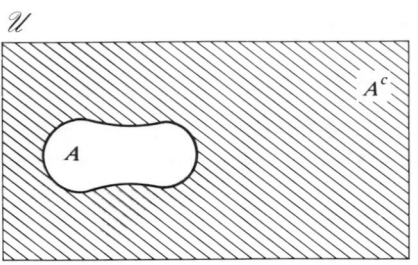

Example 1   Given $A = \{1, 2, 3, 5, 7\}$ and $B = \{0, 2, 3, 6, 9\}$. What are $A \cup B$ and $A \cap B$? Represent these sets by means of a Venn diagram.

Solution
$A \cup B = \{0, 1, 2, 3, 5, 6, 7, 9\}$
$A \cap B = \{2, 3\}$

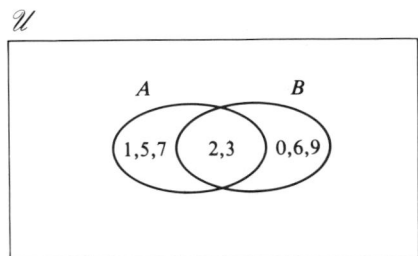

DO EXERCISE 9.

Consider the set of all students in a Finite Mathematics class. Divide the class into groups, or subsets, such that:

 i) each student in the class is in some group, and
ii) no student in the class is in more than one group.

These groups constitute a *partition* of the class.

Let A be the set of all molecules in a cookie. Break the cookie. Whatever way the cookie crumbles, the pieces are a *partition* of the cookie.

Formally, a *partition* of a set A is a division of A into subsets X_i ($i = 1, 2, \ldots, n$) such that each element of A is contained in one and only one subset. Thus, the subsets X_i are disjoint (any two different sets have no elements in common):

$$X_i \cap X_j = \emptyset \qquad \text{for all } i \neq j,$$

and their union is A:

$$X_1 \cup X_2 \cup \cdots \cup X_n = A.$$

If these conditions are satisfied, then

$$\{X_1, X_2, \ldots, X_n\}$$

expresses the statement that the X_1's are a partition of A. For example, let

A = Set of natural numbers = $\{1, 2, 3, 4, \ldots\}$,

X_1 = Set of odd natural numbers = $\{1, 3, 5, \ldots\}$

X_2 = Set of even natural numbers = $\{2, 4, 6, \ldots\}$.

Then $\{X_1, X_2\}$ is a partition of A.

This partition can be represented by the following Venn diagram:

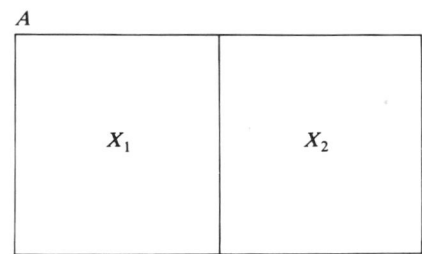

A set may be partitioned more than one way.

Example 2   Write several partitions of the set $A = \{a, e, r, s\}$ and draw the Venn diagram for each partition.

Solution

	Partitions		*Venn diagrams*
i)	$\{\{a\}, \{e\}, \{r\}, \{s\}\}$	i)	a | e | r | s
ii)	$\{\{a, e, r, s\}\}$	ii)	a, e, r, s
iii)	$\{\{a, e\}, \{r, s\}\}$	iii)	a, e | r, s
iv)	$\{\{a\}, \{e, r, s\}\}$	iv)	a | e, r, s

Note that $\{\{a, e\}, \{e, r, s\}\}$ is not a partition, because the sets are not disjoint. That is, $\{a, e\} \cap \{e, r, s\} = \{e\} \neq \emptyset$. Also, $\{\{a\}, \{r, s\}\}$ is not a partition because the union of the sets is not set A.

DO EXERCISE 10.

10. Write several partitions of the set $B = \{1, 2, 3, 4, 5\}$, and draw the Venn diagram for each. Answers may vary.

The *relative complement* of a set B with respect to set A, or, simply, the *difference* of A and B, written $A - B$, is the set of all elements contained *in A* but *not in B*. Formally,

$$A - B = \{x \mid x \in A, \quad x \notin B\}.$$

The *absolute complement* of a set A, or, simply, the *complement* of A, written A^c, can now be expressed as the relative complement of set A with respect to the universal set. Formally,

$$A^c = \mathcal{U} - A,$$

or

$$A^c = \{x \mid x \in \mathcal{U}, \quad x \notin A\}.$$

Using Venn diagrams, we can represent $A - B$, $A \cap B$, and $B - A$ by the following diagram:

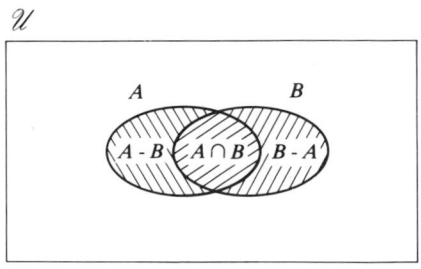

Thus, the union of two sets $A \cup B$ can always be partitioned

$$\{A - B, \quad A \cap B, \quad B - A\}.$$

Similarly, A and A^c can be represented by the Venn diagram:

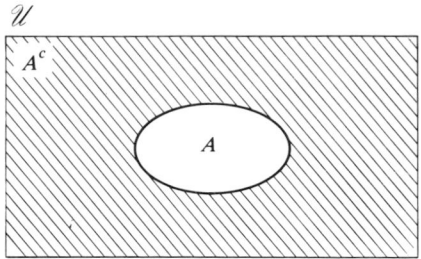

so that the universal set $\mathcal{U}$ can always be partitioned

$$\{A, A^c\}.$$

Example 3 Given $A = \{1, 2, 3, 5, 7\}$, $B = \{0, 2, 3, 6, 9\}$, and $\mathcal{U} = \{0, 1, 2, 3, 4, 5, 6, 7, 8, 9\}$. Find $A - B$, $B - A$, A^c, and $(A - B)^c$. Represent $(A - B)^c$ by a Venn diagram.

Solution

$$A - B = \{1, 5, 7\}, \quad B - A = \{0, 6, 9\}, \quad A^c = \{0, 4, 6, 8, 9\},$$
$$(A - B)^c = \{0, 2, 3, 4, 6, 8, 9\}$$

and is the shaded area in the figure.

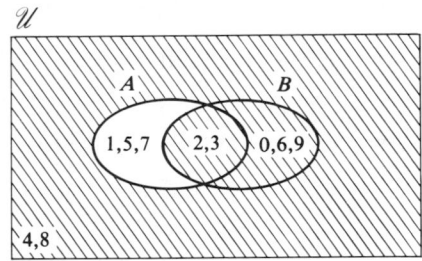

DO EXERCISE 11.

Since Venn diagrams embody set concepts, they can be used to *solve* problems involving sets. It is apparent from the following Venn diagram:

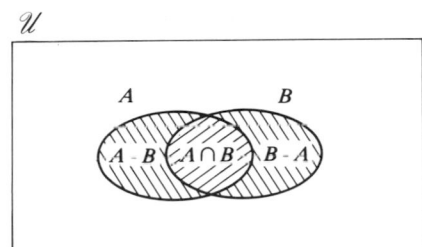

that

$$A = (A - B) \cup (A \cap B) \quad \text{and} \quad B = (B - A) \cup (A \cap B).$$

The cardinality of sets A and B is simply

$$\mathcal{N}(A) = \mathcal{N}(A - B) + \mathcal{N}(A \cap B) \tag{1}$$

and

$$\mathcal{N}(B) = \mathcal{N}(B - A) + \mathcal{N}(A \cap B). \tag{2}$$

Note that the union operation in the set relation has been replaced by the addition operation in the cardinality relation. This can be done when the components are disjoint.

Similarly, from the Venn diagram, we can write:

$$A \cup B = (A - B) \cup (A \cap B) \cup (B - A)$$

and

$$\mathcal{N}(A \cup B) = \mathcal{N}(A - B) + \mathcal{N}(A \cap B) + \mathcal{N}(B - A). \tag{3}$$

11. $A = \{a, b, c, d, e, g\},$
$B = \{b, c, d, f, h\},$
$\mathcal{U} = \{a, b, c, \dots, z\}$

Find $A - B$, $B - A$, A^c, B^c, $B^c - A^c$. Represent $B^c - A^c$ using a Venn diagram.

Solving for $\mathscr{N}(A - B)$ and $\mathscr{N}(B - A)$ in Eqs. (1) and (2) and substituting in Eq. (3), we obtain

$$\mathscr{N}(A \cup B) = \mathscr{N}(A) + \mathscr{N}(B) - \mathscr{N}(A \cap B).$$

This is a very important set property.

Thus, if $A = \{a, b, c, f\}$ and $B = \{a, b, c, d, e\}$, then

$$A \cup B = \{a, b, c, d, e, f\} \qquad \text{and} \qquad A \cap B = \{a, b, c\}.$$

Furthermore, $\mathscr{N}(A) = 4$, $\mathscr{N}(B) = 5$, $\mathscr{N}(A \cup B) = 6$, and $\mathscr{N}(A \cap B) = 3$. Substituting these quantities in

$$\mathscr{N}(A \cup B) = \mathscr{N}(A) + \mathscr{N}(B) - \mathscr{N}(A \cap B),$$

we have

$$6 = 4 + 5 - 3,$$

or

$$6 = 6,$$

which satisfies the above relation. Note that

$$\mathscr{N}(A \cup B) = \mathscr{N}(A) + \mathscr{N}(B) \qquad \text{only if } A \text{ and } B \text{ are disjoint.}$$

An example more typical of the type usually solved using Venn diagrams is the following.

Example 4   Out of a sample of people surveyed,

> 43% smoked,
> 67% drank,
> 24% smoked and drank.

What percent neither smoked nor drank?

Solution   Let

$$\mathscr{U} = \text{The set of all people surveyed,}$$
$$S = \text{The set of all smokers (surveyed),}$$

and

$$D = \text{The set of all drinkers (surveyed).}$$

Then

$$\mathscr{N}(\mathscr{U}) = 100 \quad \text{(dropping the \% sign, for simplicity),}$$
$$\mathscr{N}(S) = 43,$$
$$\mathscr{N}(D) = 67,$$

and the number of people who smoked *and* drank is given by

$$\mathscr{N}(S \cap D) = 24.$$

We think of the set symbol "∩" as corresponding to the word "and."

We want to find the number of people in the survey who neither smoked nor drank. We can express this with sets as

$$\mathcal{U} - S - D.$$

That is, we take out the smokers and drinkers and what is left are those people who neither smoke nor drink. Thus we want

$$\mathcal{N}(\mathcal{U} - S - D).$$

The appropriate Venn diagram is as follows:

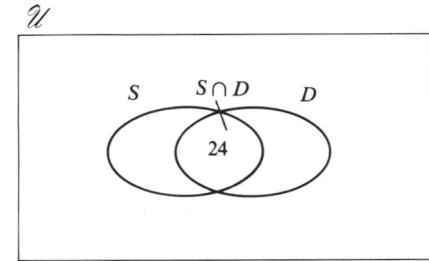

where $\mathcal{N}(S \cap D) = 24$ has been written in the area representing $S \cap D$. Now, as before, it is apparent from the Venn diagram that

$$\mathcal{N}(S - D) = \mathcal{N}(S) - \mathcal{N}(S \cap D)$$
$$= 43 - 24$$
$$= 19.$$

Similarly,

$$\mathcal{N}(D - S) = \mathcal{N}(D) - \mathcal{N}(S \cap D)$$
$$= 67 - 24$$
$$= 43.$$

Writing these two numbers, 19 and 43, into the areas representing $S - D$ and $D - S$ on the Venn diagram, we obtain:

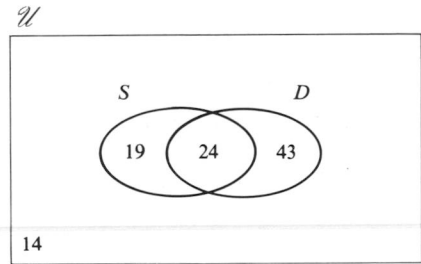

12. Out of a group of 240 students surveyed, 150 spent Friday night at a movie, 120 spent Saturday night at a basketball game, and 60 did both. How many students went to the basketball game, but not the movie?

Thus, the number of people who *either* smoked or drank is

$$\mathcal{N}(S \cup D) = \mathcal{N}(S - D) + \mathcal{N}(S \cap D) + \mathcal{N}(D - S)$$
$$= 19 + 24 + 43$$
$$= 86, \quad \text{or} \quad 86\%,$$

and the number of people who *neither* smoked nor drank is

$$\mathcal{N}(\mathcal{U} - S - D) = \mathcal{N}(\mathcal{U}) - \mathcal{N}(S \cup D)$$
$$= 100 - 86$$
$$= 14, \quad \text{or} \quad 14\%.$$

DO EXERCISE 12.

Here we used the Venn diagram both as a place to record the numbers representing the cardinality of various sets and also to indicate which differences to take to obtain new set cardinalities. The advantage of this approach may not be as apparent with the preceding example with two sets as it is in the next example with three sets.

Example 5 There are 220 students in a certain freshman class. Of these,

> 115 are taking Economics,
> 60 are taking Spanish,
> 95 are taking Mathematics,
> 20 are taking Economics and Spanish,
> 30 are taking Economics and Mathematics,
> 25 are taking Spanish and Mathematics,
> 15 are taking all three subjects.

How many students are taking only *one* of these three subjects?

Solution Formally, this problem may be posed in this manner:

Let

> $\mathcal{U}$ = set of all students in the freshman class,
> E = set of all students taking Economics,
> S = set of all students taking Spanish, and
> M = set of all students taking Mathematics.

Then

$$\mathcal{N}(\mathcal{U}) = 220,$$
$$\mathcal{N}(E) = 115,$$
$$\mathcal{N}(S) = 60,$$
$$\mathcal{N}(M) = 95,$$
$$\mathcal{N}(E \cap S) = 20 \quad \text{(those taking Economics } and \text{ Spanish),}$$
$$\mathcal{N}(E \cap M) = 30,$$
$$\mathcal{N}(S \cap M) = 25,$$
$$\mathcal{N}(E \cap S \cap M) = 15 \quad \text{(those taking Economics } and \text{ Spanish}$$
$$and \text{ Mathematics).}$$

The question is then to determine

$$\mathcal{N}(E - S - M) + \mathcal{N}(S - E - M) + \mathcal{N}(M - S - E),$$

where $\mathcal{N}(E - S - M)$ is the number of students taking Economics, but not Spanish or Mathematics, and so on.

We start with information at the bottom of the data list:

$$\mathcal{N}(E \cap S \cap M) = 15 \quad \text{or} \quad (15)_1,$$

where the number 15 is being written in the Venn diagram as $(15)_1$. Here the subscript is being used to indicate the order in which the numbers are entered on the Venn diagram, to avoid redrawing it for each step. Thus, $(15)_1$ is the first entry in this diagram.

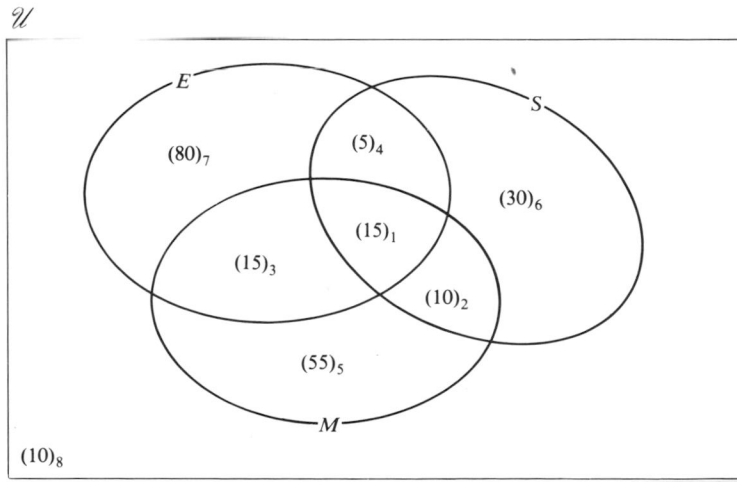

13. A survey of 195 people indicated that:

> 90 read *Time* (*T*),
> 45 read *National Review* (*R*),
> 15 read *Nation* (*N*),
> 3 read *T* and *R*,
> 5 read *T* and *N*,
> 1 read *R* and *N*,
> 1 read all three.

How many people read *none* of these three magazines? Use a Venn diagram.

Working *up* the data list, we have:

$\mathcal{N}[(S \cap M) - E]$

$\qquad = \mathcal{N}(S \cap M) - \mathcal{N}(S \cap M \cap E)$ (those taking Spanish *and* Math, *but not* Economics)

$\qquad = 25 - 15$

$\qquad = 10, \quad \text{or} \quad (10)_2;$

$\mathcal{N}[(E \cap M) - S] = 30 - 15 = 15, \quad \text{or} \quad (15)_3;$

$\mathcal{N}[(E \cap S) - M] = 20 - 15 = 5, \quad \text{or} \quad (5)_4.$

Continuing, we have

$$\mathcal{N}(M - S - E) = 95 - 15 - 15 - 10 \qquad \text{(Why?)}$$
$$= 55, \quad \text{or} \quad (55)_5.$$

That is, 40 of the elements of *M* are in *S* or *E*, leaving 55 elements for *M* − *S* − *E*. Similarly,

$$\mathcal{N}(S - E - M) = 60 - 5 - 15 - 10 = 30, \quad \text{or} \quad (30)_6;$$

and

$$\mathcal{N}(E - S - M) = 115 - 5 - 15 - 15 = 80, \quad \text{or} \quad (80)_7.$$

Thus, the number of students taking *only one* subject is

$$\mathcal{N}(E - S - M) + \mathcal{N}(S - E - M) + \mathcal{N}(M - S - E)$$
$$= 80 + 30 + 55$$
$$= 165.$$

Furthermore, the number of students taking *exactly two* subjects is

$$\mathcal{N}[(E \cap S) - M] + \mathcal{N}[(E \cap M) - S] + \mathcal{N}[(S \cap M) - E]$$
$$= 5 + 15 + 10$$
$$= 30;$$

and the number of students taking *all three* subjects is

$$\mathcal{N}(E \cap S \cap M) = 15.$$

The number of students taking *at least one* subject is the sum of these numbers:

$$165 + 30 + 15, \quad \text{or} \quad 210.$$

Since $\mathcal{N}(\mathcal{U}) = 220$, there are $220 - 210 = 10$ who are *not taking any* of the courses mentioned. At this point the Venn diagram is completely filled out and the answer to many questions can be obtained by a simple sum.

DO EXERCISE 13.

EXERCISE SET 4.2

For Exercises 1 through 5 (as in Exercises 1 through 16, Exercise Set 4.1)

$$\mathcal{U} = \{x \mid x \in I, \ \ 0 \leqslant x \leqslant 12\}, \quad \text{where } I \text{ is the set of all integers,}$$

and

$$A = \{x \mid x \in \mathcal{U}, \ \ 0 \leqslant x \leqslant 10\}, \qquad D = \{x \mid x \in \mathcal{U}, \ \ x \in A, \ \ x \notin C\}$$

$$B = \{x \mid x \in \mathcal{U}, \ \ 0 < x < 10\}, \qquad E = \left\{x \mid x \in \mathcal{U}, \ \ \frac{x}{3} \in \mathcal{U}\right\},$$

$$C = \left\{x \mid x \in \mathcal{U}, \ \ \frac{x}{2} \in A\right\}, \qquad F = \{x \mid x \in \mathcal{U}, \ \ x \in E, \ \ x \leqslant 10\}.$$

1. Determine the set indicated by each of the following:

a) $A \cup B$

b) $A \cap B$

c) $A - B$

d) $B - A$

e) $A - A^c$

f) $A - (D \cup E)$

g) $(C \cup E)^c$

h) $(A - B) \cup (C - F)$

i) $(A - B) \cap (C - F)$

j) $[(D - F)^c]^c$

2. Do any of the sets B, C, D, E, F partition set A? If yes, which ones?

3. Draw a Venn diagram illustrating the relationship among sets A, B, C, D, E, F.

For Exercises 4 through 6,

$$\mathcal{U} = \{a, e, i, o, u, c, m, n, r, t\};$$

$$A = \{a, e, i, o, u\}, \qquad D = \{m, i, n, t\},$$

$$B = \{c, m, n, r, t\}, \qquad E = \{e, i\},$$

$$C = \{a, c, e\}, \qquad F = \{r, t\}.$$

4. Answer the following:

a) Is $\{C, D, F\}$ a partition of $\mathcal{U}$?

b) Are B and C disjoint?

5. Determine the set indicated by each of the following:

a) $E \cup F$

b) $E \cap F$

c) $B - D$

d) $D - B$

e) $(A \cup B)^c$

f) $(D - E) \cup F$

g) $(D^c - E^c) \cap F$

h) $(A \cap C) - (C \cap D)$

6. Draw a Venn diagram illustrating the relationship among sets A, B, C, D, E, F.

7. Of 68 people surveyed, 33 smoked, 57 drank, and 27 did both. How many did neither?

8. *Marketing.* Of 87 people surveyed, 49 read *Time*, 21 read *Nation*, and 5 read both. How many read *only one* magazine? How many read *neither* magazine?

9. Of 73 men surveyed, 54 wore belts, 20 wore suspenders, and 3 wore neither. How many wore both?

10. Of 123 students, 79 could ride a bike, 53 could drive a car, and 15 could do both. How many could do neither?

11. *Political Science.* In a recent poll of 230 people, 50 thought only the Republicans could solve our problems, 70 thought only the Democrats could, and 25 thought neither could. How many thought one could do as well as the other (that is, both)?

12. *Medicine.* Blood can be typed as A (having Type A antigen), B (having Type B antigen), AB (having both), and O (having neither). Of 140 patients in a hospital, there were 53 with Type A blood, 47 with Type B, and 24 with Type AB. How many had Type O? *Note.* Having Type A blood implies *only* Type A antigens.

13. Of the students in a certain university,

     55% took Finite Mathematics,
     65% took English Composition,
     35% took Spanish,
     30% took Mathematics and English,
     24% took Mathematics and Spanish,
     18% took English and Spanish,

and 12% took all three.

How many students took only one of these three subjects? Only two of these three subjects? None of these courses?

14. *Marketing.* Of recent car buyers,

  65% bought automatic transmissions,
  20% bought air conditioning,
  70% bought posh seats,
  50% bought automatic transmissions and posh seats,
  10% bought automatic transmissions and air
     conditioning,
  10% bought posh seats and air conditioning,
  5% bought all three.

How many bought posh seats alone? No option? Posh seats but not air conditioning?

OBJECTIVES

You should be able to

a) Draw a tree to represent the sample space in a counting problem.

b) Count the branches of a tree to determine the number of possible outcomes in a sample space.

4.3 EXPERIMENTS, TREES, AND DISTINGUISHABILITY OF OUTCOMES

Consider an "experiment" in coin-flipping. Assuming that the coin does not land standing on its edge, there are two possible outcomes: Heads (H) or Tails (T). The set S of all possible outcomes of some particular experiment is called the *sample space* in probability theory, and is the universal set. Thus, the sample space for a single flip of the coin is:

$$S = \{H, T\}.$$

Consider flipping a coin *twice*. The first coin flip is *distinguished* from the second in time. This sample space is

$$S = \{HH, HT, TH, TT\}.$$

Each element of the set S is an ordered pair written HH rather than (H, H) for simplicity. The *first* flip is indicated by the *first* element of the ordered pair and the *second* flip is indicated by the *second* element. The sample space for flipping two coins labelled 1 and 2 (for example, a penny and a dime) is the same, with the *time ordering* replaced by the *number* (or label) *ordering*. In either case, the outcomes of each *part* of the total experiment are *considered distinguishable*.

On the other hand, we might be interested only in *how many* heads and *how many* tails occur and not in how each individual coin lands. This sample space is

$$S = \{HH, HT, TT\}.$$

Here the element TH is *considered* identical to the element HT, and is thus omitted. The labelling on the coins is ignored and they are *considered indistinguishable*.

This concept of distinguishability applies as well to more than two trials. This is one of the first concerns in probability, even though it is only implicitly, not explicitly, stated in most problems.

Example 1 An election is being held with three candidates and 101 voters. The offices of president and vice-president go to the candidates with the highest and next highest number of votes. If there is a tie for president, then there is to be a runoff election between these two candidates for president and vice-president. If there is a tie for vice-president, then there is to be a runoff election for vice-president. What is the sample space of possible outcomes?

Solution Here we are not interested in which candidate a *particular* voter votes for, so that the *voters* are indistinguishable. Furthermore, the *number* of votes each candidate gets is relevant only in a ranking sense. Thus, the president is the winner whether by one vote or 100.

Let us denote the candidates by A, B, and C. Then the sample space is:

$$S = \{AB, AC, BA, BC, CA, CB\},$$

where the first letter of each pair denotes the president and the second denotes the vice-president. Thus, AB and BA are different outcomes.

DO EXERCISE 14.

Sample spaces can be represented geometrically by *trees*. Consider again coin-flipping. For a single coin, the sample space is $S = \{H, T\}$.

14. An old sea chest contains 10 bars of silver and 5 bars of gold. If 3 bars are drawn at random, what is the sample space of possible outcomes?

The corresponding tree is:

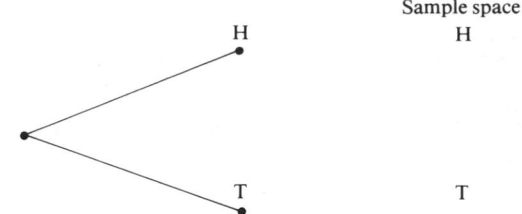

Leading out of a *vertex* (dot) representing the *flipping* of the coin, there are two *arcs* (line segments) representing the two *outcomes*, H *or* T. Thus, the tree can be *read* by saying that a coin is flipped and can land H *or* T (but *not* both H and T at the same time).

Each fork of a tree consists of a *vertex*, which represents an *experiment*, and a set of *arcs*, which represent all possible *mutually exclusive* outcomes.

Flipping a coin twice has the sample space $S = \{HH, HT, TH, TT\}$ and the tree:

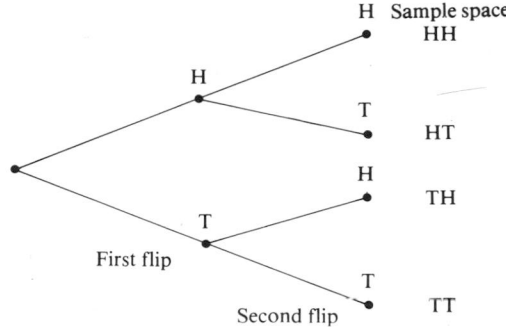

A *branch* of a tree is a sequence of connected arcs. Following along a branch of a tree from beginning to end indicates the succession of possible *individual* outcomes that constitute a single overall outcome. Thus, a tree with all its branches represents the sample space.

The upper branch of this tree indicates that the *first* coin landed H *and* the *second* coin landed H. The next lower branch indicates that the *first* coin landed H *and* the *second* coin landed T, and so forth.

The tree drawn for a given sample space may not be unique. For example, if we simultaneously flip two identical coins labelled "1" and "2," then either coin can be identified with the coin flipped first, so that two equivalent but different trees are possible.

The tree shown *distinguishes* between the two flips of the coin. If we wish to consider the coins *indistinguishable*, then we can ignore the labelling that distinguishes the coins.

Example 2   Draw a tree to represent the sample space of Example 1.

Solution

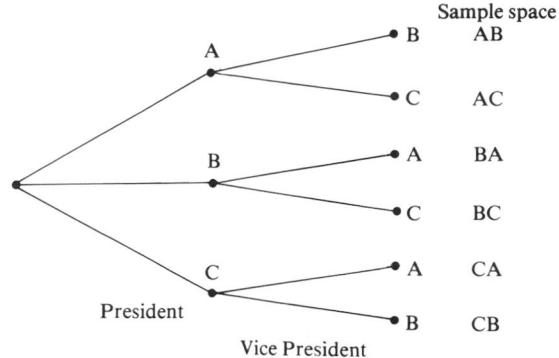

Here we are concerned only with who wins, not the actual vote. We have represented the presidential winner first on the tree, since it was represented first in the pair of the sample space.

DO EXERCISE 15.

15. Draw a tree to represent the sample space of Margin Exercise 14.

EXERCISE SET 4.3

1. A business manager has a problem that he discusses with each of his three assistants. Each comes up with a possible solution that looks equally acceptable. Hence, he resorts to his standard decision-maker: He tacks the solutions to the wall and throws darts to rank them I, II, and III. The first solution is tried. If it works, then no other solution is considered further. If the first solution doesn't work, the second solution is tried, and so forth. What is the sample space for possible outcomes? Draw a tree representing them. (Nothing has been said about what happens if III fails.)

2. Stockholders are being presented with several options on which to vote. If Option I is accepted, no further options need be considered. However, if Option I is rejected, then the voters are **asked** to vote on which option, II or III, to consider next. That is then done, with the losing option considered last. What is the sample space? Draw the tree.

3. On any particular day the weather can be classified by temperature (above normal, normal, below normal) and sky conditions (clear, cloudy, or precipitating (rain, snow, or whatever)). What is the sample space of possible weather conditions? Draw a tree.

4. *Marketing.* A survey is made to determine who buys a particular product. The people surveyed are classified by income (low, medium, or high) and education (grade school, high school, or college). What is the sample space of possible classifications? Draw a tree.

5. *Political Science.* A political survey is made and the responders are classified by sex (male or female), marital status (single or married), or age (<18 or ≥ 18). What is the sample space of the responders? Draw a tree.

6. The menu for a meal gives one the choice of (i) soup or fruit cup, (ii) fish, chicken, or meat, (iii) salad or dessert. What is the sample space of possible menus? Draw a tree.

7. Telephone lines exist between towns as indicated in the following diagram.

Draw a tree indicating all possible phone connections between *A* and *F*, assuming that no telephone *link* is used more than once.

9. *Sports.* In a single-elimination sports tournament consisting of *n* teams, a team is eliminated when it loses one game. How many games are required to complete the tournament?

8. As in Exercise 7, but with the phone lines as in the diagram below.

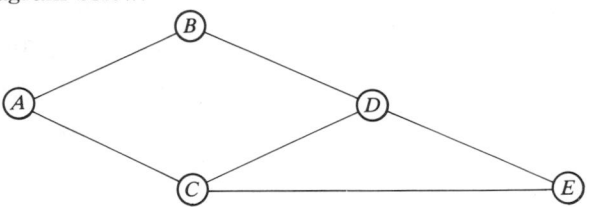

10. *Sports.* In a double-elimination softball tournament consisting of *n* teams, a team is eliminated when it loses two games. At most how many games are required to complete the tournament?

OBJECTIVES

You should be able to:

a) Express a factorial such as 6!, as a product $6 \cdot 5 \cdot 4 \cdot 3 \cdot 2 \cdot 1$, and evaluate.

b) Use the formula $P(n, k) = n(n - 1) \times (n - 2) \cdots [n - (k - 1)]$,

to find the number of permutations of *n* objects taken *k* at a time, without replacement or repetition.

c) Find the number of circular arrangements of *n* objects by evaluating $(n - 1)!$.

d) Find the number of ordered arrangements of *n* objects taken *k* at a time with replacement, or repetition, by evaluating n^k.

e) Use the Fundamental Counting Principle and the procedures of (b) through (d) to solve various types of counting problems.

4.4 COUNTING TECHNIQUES—PERMUTATIONS

Trees were used in the preceding section to show the possible outcomes of an experiment. We want to develop faster counting techniques. The following example leads up to the *Fundamental Counting Principle*.

Example 1   How many 4-letter words (not necessarily meaningful or pronounceable) can be formed using the letters P, D, Q, X *without* repetition?

Solution   Such a word would have the general form

Any of the 4 letters can be used for the first letter in the word. Once this letter has been selected, the second can be selected from the 3 remaining letters, and the third from the remaining 2 letters. The fourth letter is already determined, since only 1 possible letter remains. Thus there are

$$4 \cdot 3 \cdot 2 \cdot 1, \quad \text{or} \quad 24 \text{ words.}$$

We could, of course, have determined this by writing down all the possibilities; and even though this can be quite cumbersome in general, we list them below for reference.

PDQX   PDXQ   PQDX   PQXD   PXDQ   PXQD

DPQX   DPXQ   DQPX   DQXP   DXPQ   DXQP

QPDX   QPXD   QDPX   QDXP   QXDP   QXPD

XPDQ   XPQD   XDPQ   XDQP   XQPD   XQDP

DO EXERCISE 16.

FUNDAMENTAL COUNTING PRINCIPLE. Given a combined action, or event, in which the 1st action can be performed in n_1 ways, the 2nd action can be performed in n_2 ways, and so on, then the total number of ways the combined action can be performed is the product

$$n_1 \cdot n_2 \cdot n_3 \cdots n_k.$$

Let us demonstrate this for three actions with a tree. Assume the first action E_1 can be performed in 3 ways, the second E_2 can be performed in 4 ways, and the third E_3 in 2 ways.

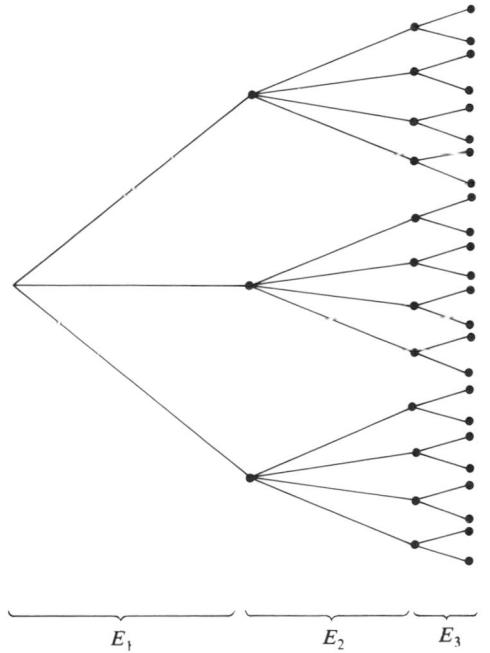

$$\underbrace{}_{E_1} \quad \underbrace{}_{E_2} \quad \underbrace{}_{E_3}$$

There are $3 \cdot 4 \cdot 2$, or 24 paths through the tree (count them), so there are 24 outcomes for all three actions.

DO EXERCISES 17 THROUGH 20.

16. How many 3-letter words can be formed using the letters P, D, Q *without* repetition?

17. How many 5-letter words can be formed using the letters P, D, Q, R, S *without* repetition?

18. How many 4-digit numbers can be formed using the digits 6, 7, 8, 9 without repetition?

19. An examination consists of ten true–false questions. How many possible different answer sheets can be turned in?

20. A woman is going out to eat. She will put on one of 5 pantsuits, one pair out of 40 pairs of shoes, and go to one of 8 restaurants. In how many ways can she dress and eat?

21. How many ways can 5 motorcycles be parked in a row?

22. How many permutations are there of a set of 5 objects? Consider a set

$$\{a, b, c, d, e\}.$$

23. Find $P(6, 6)$ and $P(5, 5)$.

Permutations

A *permutation* of a set of n objects is an ordered arrangement of all the objects. For example, consider the set of 4 objects,

$$\{P, D, Q, X\}$$

as in Example 1. There are 4 choices for the first letter, 3 choices for the second, 2 choices for the third, and 1 for the fourth. Thus, by the Fundamental Counting Principle, there are:

$$4 \cdot 3 \cdot 2 \cdot 1, \quad \text{or} \quad 24 \text{ permutations of a set of 4 objects.}$$

DO EXERCISES 21 AND 22.

In general, consider a set of n objects. There are n choices for the first object, $(n - 1)$ choices for the second, $(n - 2)$ choices for the third, and so on. The nth object is chosen in only 1 way.

The total number of permutations of a set of n objects, denoted $P(n, n)$, is given by

$$P(n, n) = \underbrace{n(n - 1)(n - 2) \cdots 3 \cdot 2 \cdot 1}_{n \text{ factors}}$$

Example 2   Find $P(7, 7)$ and $P(3, 3)$.

Solution

$$P(7, 7) = 7 \cdot 6 \cdot 5 \cdot 4 \cdot 3 \cdot 2 \cdot 1 = 5040,$$
$$P(3, 3) = 3 \cdot 2 \cdot 1 = 6$$

DO EXERCISE 23.

Factorial Notation

We will use products such as

$$6 \cdot 5 \cdot 4 \cdot 3 \cdot 2 \cdot 1 \quad \text{and} \quad 8 \cdot 7 \cdot 6 \cdot 5 \cdot 4 \cdot 3 \cdot 2 \cdot 1$$

so often that it is convenient to adopt a notation for them. We define

$$6 \cdot 5 \cdot 4 \cdot 3 \cdot 2 \cdot 1 = 6!, \quad \text{read "6 factorial".}$$

In general,

$$n! = n \text{ factorial} = \underbrace{n(n - 1)(n - 2) \cdots 3 \cdot 2 \cdot 1}_{n \text{ factors}}$$

For example,

$$n! = n \cdot (n - 1) \cdot (n - 2) \cdots 1$$

$7! = 7 \cdot 6 \cdot 5 \cdot 4 \cdot 3 \cdot 2 \cdot 1 = 5040$	
$6! = 6 \cdot 5 \cdot 4 \cdot 3 \cdot 2 \cdot 1 = 720$	
$5! = 5 \cdot 4 \cdot 3 \cdot 2 \cdot 1 = 120$	
$4! = 4 \cdot 3 \cdot 2 \cdot 1 = 24$	
$3! = 3 \cdot 2 \cdot 1 = 6$	
$2! = 2 \cdot 1 = 2$	
$1! = 1 = 1$	

DO EXERCISES 24 AND 25.

We define

$$0! = 1$$

for consistency in formulas used later.

We can now restate the formula for the total number of permutations of n objects.

$$P(n, n) = n(n - 1)(n - 2) \cdots 3 \cdot 2 \cdot 1 = n!$$

DO EXERCISE 26.

We often use other notations for factorials. Note the following.

$$7! = 7 \cdot 6 \cdot 5 \cdot 4 \cdot 3 \cdot 2 \cdot 1 = 7 \cdot (6 \cdot 5 \cdot 4 \cdot 3 \cdot 2 \cdot 1) = 7 \cdot 6!$$

$$6! = 6 \cdot 5 \cdot 4 \cdot 3 \cdot 2 \cdot 1 = 6 \cdot (5 \cdot 4 \cdot 3 \cdot 2 \cdot 1) = 6 \cdot 5!$$

In general,

$$n! = n(n - 1)!$$

DO EXERCISE 27.

Note also that

$$7! = 7 \cdot 6!$$
$$= 7 \cdot 6 \cdot 5!$$
$$= 7 \cdot 6 \cdot 5 \cdot 4!$$

In general, for any $k < n$,

$$n! = \underbrace{\underbrace{n(n - 1)(n - 2) \cdots [n - (k - 1)]}_{k \text{ factors}} \underbrace{(n - k)!}_{(n - k) \text{ factors}}}_{n \text{ factors}}$$

24. Find 8!.

25. Find 9!.

26. Use factorial notation to represent the number of permutations of 6 objects.

27. Express each in the form $n(n - 1)!$.

a) 8!.

b) 38!.

Permutations of n Objects Taken k at a Time

Consider the set

$$\{P, D, Q, X, Y\}.$$

We have 5 objects. Suppose we wanted to determine the number of permutations of 3 objects taken from the set. There would be 5 choices for the first object. Then there would remain 4 choices for the second object, and 3 for the third selection. By the Fundamental Counting Principle, there would be

$5 \cdot 4 \cdot 3$, or 60 permutations of a set of 5 objects taken 3 at a time.

In general, suppose we had a set of n objects and we wanted to determine the number of permutations of these n objects taken k at a time. There would be n choices for the first object. Then there would remain $(n - 1)$ choices for the second object, $(n - 2)$ for the third, and so on. We would make k choices in all, so there will be k factors in the product.

The total number of permutations of n objects taken k at a time, denoted $P(n, k)$, is given by

$$P(n, k) = \underbrace{n(n - 1)(n - 2) \cdots [n - (k - 1)]}_{k \text{ factors}}$$

An alternative symbol for $P(n, k)$ can be found by multiplying by 1, using $\dfrac{(n - k)!}{(n - k)!}$ as follows:

$$P(n, k) = n(n - 1)(n - 2) \cdots [n - (k - 1)] \cdot \frac{(n - k)!}{(n - k)!}$$

$$= \frac{n(n - 1)(n - 2) \cdots [n - (k - 1)](n - k)!}{(n - k)!}$$

$$= \frac{n!}{(n - k)!}$$

Thus,

The total number of permutations of n objects taken k at a time is given by

$$P(n, k) = n(n - 1)(n - 2) \cdots [n - (k - 1)] \tag{1}$$

$$P(n, k) = \frac{n!}{(n - k)!}. \tag{2}$$

Formula (1) is most useful in application, but formula (2) will be important in a development in Section **4.5**.

Example 3   Evaluate $P(7, 3)$ using both formulas.

Solution   Using formula (1), we have:

$$P(\overset{\downarrow}{7}, \overset{\downarrow}{3}) = \underbrace{\overset{\uparrow}{7} \cdot 6 \cdot \overset{\uparrow}{5}}_{} = 210$$

This number tells where to start.

This number tells how many factors.

Using formula (2) we have

$$P(7, 3) = \frac{7!}{(7 - 3)!} = \frac{7!}{4!} = \frac{7 \cdot 6 \cdot 5 \cdot 4 \cdot 3 \cdot 2 \cdot 1}{4 \cdot 3 \cdot 2 \cdot 1} = 7 \cdot 6 \cdot 5 = 210.$$

DO EXERCISES 28 AND 29.

Example 4   How many ways can the letters of "organize" be arranged

a) taking 8 at a time?   b) taking **6** at a time?
c) taking 4 at a time?   d) taking 2 at a time?

Solution

a) $P(8, 8) = 8 \cdot 7 \cdot 6 \cdot 5 \cdot 4 \cdot 3 \cdot 2 \cdot 1 = 40{,}320$
b) $P(8, \mathbf{6}) = 8 \cdot 7 \cdot 6 \cdot 5 \cdot 4 \cdot 3 \quad\quad = 20{,}160$
c) $P(8, 4) = 8 \cdot 7 \cdot 6 \cdot 5 \quad\quad\quad\quad = \quad 1680$
d) $P(8, 2) = 8 \cdot 7 \quad\quad\quad\quad\quad\quad\quad = \quad\quad 56$

DO EXERCISE 30.

*(Optional) Circular Permutations

Consider arranging the 5 letters A, B, C, D, and E in a circular permutation.

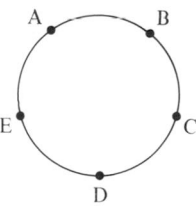

In a circle, the permutations ABCDE, BCDEA, CDEAB, DEABC, and EABCD are no longer distinguishable as they would be on a line. Therefore, for each circular permutation, there would be 5 distinguishable permutations on a line. (Think of cutting the circle open and bending it out to a line.)

Suppose we have P circular permutations. Each of these would yield $5 \cdot P$ permutations on a line, and we know there are 5! of these. Then

28. Evaluate $P(6, 4)$ using both formulas.

29. Evaluate, using formula (1):

a) $P(10, 4)$

b) $P(9, 9)$

30. How many ways can the letters of "soybean" be arranged,

a) taking 7 at a time?

b) taking 5 at a time?

c) taking 4 at a time?

d) taking 3 at a time?

31. How many ways can 6 different foods be arranged in 6 dishes *around* a Lazy Susan?

$5 \cdot P = 5!$, so

$$P = \frac{5!}{5} = 4! = 4 \cdot 3 \cdot 2 \cdot 1 = 24.$$

In general,

The number of circular permutations of n objects taken n at a time is $(n - 1)!$.

Example 5   How many ways can 10 college students sit around a campfire?

Solution

$$(10 - 1)! = 9! = 9 \cdot 8 \cdot 7 \cdot 6 \cdot 5 \cdot 4 \cdot 3 \cdot 2 \cdot 1 = 362,880.$$

DO EXERCISE 31.

Permutations "with Repetition" or "with Replacement"

32. How many 2-card permutations can be made by selecting 2 cards from a deck of 52,

a) without replacement?

Example 6   A standard deck of cards has 52 different cards. (For the exact makeup of the deck, see p. **219**.) How many 3-card permutations can be made by selecting the 3 cards without replacement? with replacement?

Solution

a) The case "without replacement" is equivalent to "without repetition," as in Example 1. This is the number of permutations of 52 things taken 3 at a time.

$$P(52, 3) = 52 \cdot 51 \cdot 50 - 132,600$$

b) with replacement?

b) The case "with replacement" is considered by first making a selection. This can be done in 52 ways. Then the card is "replaced" and we make another selection. This can still be done 52 ways. Similarly, the third selection can be made 52 ways. Thus there are

$$52 \cdot 52 \cdot 52, \quad \text{or} \quad 140,608$$

possible permutations.

The number of permutations of n objects taken k at a time, with replacement, or with repetition, is n^k.

DO EXERCISE 32.

Mixed Counting

In the following example we carry out several types of counting.

Example 7   How many ways can 3 men and 3 women be seated in a row of 6 seats

a) with no seating restrictions?
b) if men and women must alternate?
c) if a particular couple (man–woman) must sit together?
d) if a particular couple must *not* sit together?

Solution

a) There are 6 people and 6 seats. Hence there are

$$P(6,6) = 6! = 6 \cdot 5 \cdot 4 \cdot 3 \cdot 2 \cdot 1 = 720 \text{ possible arrangements.}$$

b) They can sit either MWMWMW *or* WMWMWM. These yield two possibilities. Considering the possibility MWMWMW, we think of the number of ways the *men* can be arranged within this possibility.

This is equivalent to arranging 3 objects 3 at a time, so there are

$$P(3,3) = 3! = 3 \cdot 2 \cdot 1 = 6 \text{ ways.}$$

Considering the same possibility MWMWMW, we think of the number of ways the *women* can be arranged within this possibility.

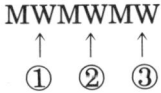

In all there are $6 \cdot 6$, or 36 ways within the possibility MWMWMW. Now consider the possibility WMWMWM. By a similar argument, there are $6 \cdot 6$, or 36 ways within the possibility WMWMWM. Then the total number of arrangements is $2 \cdot 6 \cdot 6$, or 72.

c) i) The particular couple can sit MW or WM, yielding 2! or 2 possibilities.

  ii) Considering this particular couple as *one* object and the other four men and women as 4 objects, we have a total of 5 objects to be permuted (arranged 5 at a time). Then we have

$$P(5,5) = 5! = 5 \cdot 4 \cdot 3 \cdot 2 \cdot 1 = 120 \text{ ways.}$$

 iii) The total number of arrangements, from (i) and (ii), is $2 \cdot 120 = 240$.

33. The flags of five nations are to be raised on six flagpoles arranged in a row. How many ways can the flags be raised:

a) If there are no restrictions?

b) If all the flags must be together with no gaps between them?

c) If three particular nations want their flags all together with no separation either with an empty flagpole or another flag?

d) If three particular nations do *not* want their flags together, an empty flagpole being equivalent to a flag as separation?

d) If one action can be performed in n_1 ways and another disjoint action (see p. 187) can be performed in n_2 ways, then *either* action can be performed in:

$$n_1 + n_2 \quad \text{ways.}$$

This is in contrast to the Fundamental Counting Principle where we consider *both* actions occurring and would have multiplied.

Since the couple must *either* sit together or *not* sit together, the number of arrangements with the couple sitting together (c) plus the number of arrangements with the couple not sitting together (d) must equal the number of arrangements with no seating restrictions (a). Thus, the number of arrangements with the couple not sitting together is the result of (a) minus the result of (c):

$$720 - 240 = 480.$$

DO EXERCISE 33.

EXERCISE SET 4.4

Evaluate.

1. 5!	**2.** 6!	**3.** 0!	**4.** 1!	**5.** $P(6, 6)$	**6.** $P(5, 5)$
7. $P(20, 2)$	**8.** $P(30, 2)$	**9.** $P(7, 5)$	**10.** $P(6, 4)$	**11.** $P(n, 3)$	**12.** $P(n, 2)$

13. A person can get to the airport 3 ways, fly by any of 4 airlines, and get from the airport to his final destination in 2 ways. How many different ways can a person get to his destination? [*Hint.* Use the Fundamental Counting Principle.]

14. A person driving from Boston to Los Angeles wishes to go through Washington D.C., Atlanta, New Orleans, and Denver. If there are 6 main routes between Boston and Washington, 2 between Washington and Atlanta, 4 between Atlanta and New Orleans, 5 between New Orleans and Denver, and 7 between Denver and Los Angeles, how many possible routes can he take?

15. An office manager hires 4 secretaries, one for each of his assistants. If the secretaries are assigned at random, how many different assignments are possible? Give the solution in factorial and permutation notation. (*Hint.* See Example 1 and Example 2.)

16. An ice cream store has 6 different flavors of ice cream and room under the counter for 6 cartons of ice cream. How many different ways can the ice cream cartons be arranged under the counter? Give the solution in factorial and permutation notation.

17. As in Exercise 15, 4 secretaries apply for 2 different positions. How many ways can these two positions be filled if the secretaries are hired at random? Express the solution in permutation notation. (*Hint.* See Example 3 and Example 4.)

18. As in Exercise 16, if 8 ice cream flavors are available, in how many ways can 6 flavors be selected and arranged under the counter? Express the solution in permutation notation.

19. How many words (not necessarily meaningful or pronounceable) can be formed by rearranging the letters of the word "LOVE"? Express your answer in permutation notation. (*Hint.* See Example 4.)

20. How many 3-letter words can be formed from the letters of the word "ZORCH"? Express your answer in permutation notation.

21. How many ways can 4 people be seated at a circular bridge table? Would the answer change if the table were square? (*Hint.* See Example 5.)

22. How many ways can 8 people on a committee be seated at a circular table?

23. How many 4-number license plates can be made using the digits 1, 2, 3, 4, and 5 if repetitions are permitted? not permitted? (*Hint.* See Example 6.)

24. As in Exercise 23, but the number must be even.

25. How many 3-digit numbers can be formed from the numbers 1, 2, 3, 4, and 5, if repetitions are (a) allowed, (b) not allowed?

26. As in Exercise 25, if repetitions are not allowed, how many of these are (a) larger than 300? (b) less than 500?

27. How many ways can 4 different contracts be awarded to 7 different firms? if no firm gets more than one contract?

28. How many ways can 3 people be assigned to 5 offices?

29. As in Exercise 27, but two particular contracts must be awarded to the same firm with no other restrictions.

30. As in Exercise 28, but 2 particular people want adjacent offices.

31. How many ways can 4 executives from a given company and 3 visitors from another company be seated in a row? if they must alternate host, visitor, etc.? if a member of the host company sits at each end?

32. As in Exercise 31, but (a) two particular people wish to sit together, (b) two particular people should not be seated together.

▶ ───

33. A car holds three people in the front and three people in the back. How many ways can 6 people be seated in the car? if a given couple must sit together?

34. As in Exercise 33, but two particular couples must sit together.

───

4.5 COUNTING TECHNIQUES—COMBINATIONS

We may sometimes make selections from a set without regard for order. Such selections are called *combinations*.

Example 1 How many combinations can be formed by taking elements 3 at a time from the set $\{A, B, C, D, E\}$?

Solution The combinations are

$$\{A, B, C\}, \quad \{A, B, D\}, \quad \{A, B, E\}, \quad \{A, C, D\}, \quad \{A, C, E\},$$
$$\{A, D, E\}, \quad \{B, C, D\}, \quad \{B, C, E\}, \quad \{B, D, E\}, \quad \{C, D, E\}.$$

Note that finding all the combinations of 5 objects taken 3 at a time is the same as forming all the 3-element subsets. Thus a combination is a subset. This is consistent with our earlier work regarding sets. That is, the set, or combination, $\{A, C, E\}$ is the same as the set, or combination, $\{C, A, E\}$, because *order is not considered* when describing sets.

OBJECTIVES

You should be able to

a) Evaluate $C(n, k)$ and $\binom{n}{k}$ to find the number of combinations of n objects taken k at a time.

b) Solve counting problems involving repeated and/or mixed uses of $C(n, k)$ and $P(n, k)$.

c) Find the number of distinguishable arrangements of n objects of which n_1 are of one kind, n_2 are of a second kind, and so on, where finally there are n_k of a kth kind, by evaluating

$$\frac{n!}{n_1! n_2! \cdots n_k!}$$

34. Consider the set {A, B, C, D, E}. How many combinations taken:

a) 4 at a time are there?

b) 5 at a time?

c) 2 at a time?

d) 1 at a time?

e) 0 at a time?

Query: Why should a "combination" lock really be called a "permutation" lock?

A *combination* containing k objects is a subset containing k objects.

DO EXERCISE 34.

The number of combinations of n objects taken k at a time, denoted $C(n, k)$, is the number of different subsets of k elements.

In Example 1 and Margin Exercise 34, we see that

$$C(5, 5) = 1, \qquad C(5, 2) = 10,$$
$$C(5, 4) = 5, \qquad C(5, 1) = 5,$$
$$C(5, 3) = 10, \qquad C(5, 0) = 1.$$

We can derive some general results here. First, it is always true that $C(n, n) = 1$, because a set with n objects has only 1 subset with n objects. Second, $C(n, 1) = n$ because a set with n objects has n subsets with 1 element each. Finally, $C(n, 0) = 1$, because a set with n objects has only one subset, namely the empty set $\emptyset$, with 0 elements.

We now derive a general formula for $C(n, k)$ for any $k \leq n$. Let us return to Example 1 and compare the number of combinations with the number of permutations.

Combinations		Permutations					
{A, B, C} $\longrightarrow$	ABC	BCA	CAB	CBA	BAC	ACB
{A, B, D} $\longrightarrow$	ABD	BDA	DAB	DBA	BAD	ADB
{A, B, E} $\longrightarrow$	ABE	BEA	EAB	EBA	BAE	AEB
{A, C, D} $\longrightarrow$	ACD	CDA	DAC	DCA	CAD	ADC
{A, C, E} $\longrightarrow$	ACE	CEA	EAC	ECA	CAE	AEC
{A, D, E} $\longrightarrow$	ADE	DEA	EAD	EDA	DAE	AED
{B, C, D} $\longrightarrow$	BCD	CDB	DBC	DCB	CBD	BDC
{B, C, E} $\longrightarrow$	BCE	CEB	EBC	ECB	CBE	BEC
{B, D, E} $\longrightarrow$	BDE	DEB	EBD	EDB	DBE	BED
{C, D, E} $\longrightarrow$	CDE	DEC	ECD	EDC	DCE	CED

Note that each combination of 3 objects, say {A, C, E} yields 3!, or 6, permutations, as shown above. It follows that

$$3! \cdot C(5, 3) = 60 = P(5, 3) = 5 \cdot 4 \cdot 3$$

so

$$C(5, 3) = \frac{P(5, 3)}{3!} = \frac{5 \cdot 4 \cdot 3}{3 \cdot 2 \cdot 1} = 10.$$

In general, the number of combinations of n objects taken k at a time, $C(n, k)$, times the number of permutations of these k objects, $k!$, must equal the number of permutations of n objects taken k at a time:

$$k! \cdot C(n, k) = P(n, k)$$

so

$$C(n, k) = \frac{P(n, k)}{k!} = \frac{1}{k!} \cdot P(n, k) = \frac{1}{k!} \cdot \frac{n!}{(n - k)!} = \frac{n!}{k!(n - k)!}. \quad (1)$$

We also have

$$C(n, k) = \frac{n(n - 1)(n - 2) \cdots [n - (k - 1)]}{k(k - 1)(k - 2) \cdots 3 \cdot 2 \cdot 1}. \quad (2)$$

Note that this expression for $C(n, k)$ is the quotient of two quantities, each of which has k factors.

An alternative notation, called a *Binomial Coefficient*, is also used for $C(n, k)$:

$$\binom{n}{k} = C(n, k).$$

You should be able to use either notation.

Example 2 Evaluate $\binom{7}{5}$, using expressions (1) and (2).

Solution

a) By (1),

$$\binom{7}{5} = \frac{7!}{5!2!} = \frac{7 \cdot 6 \cdot 5 \cdot 4 \cdot 3 \cdot 2 \cdot 1}{5 \cdot 4 \cdot 3 \cdot 2 \cdot 1 \cdot 2 \cdot 1} = \frac{7 \cdot 6 \cdot 5 \cdot 4 \cdot 3}{5 \cdot 4 \cdot 3 \cdot 2 \cdot 1} = \frac{7 \cdot 6}{2 \cdot 1} = 21.$$

b) By (2), This tells where to start

$$\binom{7}{5} = \frac{7 \cdot 6 \cdot 5 \cdot 4 \cdot 3}{5 \cdot 4 \cdot 3 \cdot 2 \cdot 1} = \frac{7 \cdot 6}{2 \cdot 1} = 21.$$

This tells how many factors in numerator and denominator, and where to start the denominator.

The method in (b), using formula (2), is easiest to carry out, but in some situations formula (1) does become useful.

DO EXERCISES 35 AND 36.

35. Given the set of 4 letters

$$\{A, B, C, D\},$$

a) Determine the number of permutations of this set, taking 3 letters at a time.

b) List these permutations.

c) Determine the number of combinations of this set, taking 3 letters at a time.

d) List these combinations.

36. Evaluate.

a) $\binom{10}{3}$ b) $\binom{10}{7}$

c) $C(9, 4)$ d) $C(9, 5)$

37.

a) Evaluate $\binom{8}{5}$ and $\binom{8}{3}$.

b) Which seemed easier to compute?

38. An examination consists of 10 questions. A student is required to answer any 8 of them. How many different ways can the student pick 8 questions to answer? (*Hint.* Is the *order* in which he answers the questions important, assuming the answers themselves are numbered?)

Note that

$$\binom{7}{2} = \frac{7 \cdot 6}{2 \cdot 1} = 21$$

so that

$$\binom{7}{5} = \binom{7}{2}.$$

In general,

$$\binom{n}{k} = \binom{n}{n-k}.$$

This is because every set of k elements automatically determines the *set complement* with $(n-k)$ elements. So there are the same number of k-element subsets as there are $(n-k)$-element subsets. Try this with your fingers. Hold your hands up and bend down 3 fingers. This determines not only the 3-element set you turned down, but the 7-element set still up. Knowing this may ease some computation.

DO EXERCISE 37.

Example 3 For a psychology study, 4 people are chosen at random from a group of 10 people. In how many ways can this be done?

Solution No order is implied here, nor does it seem to be important, so the number of ways the 4 people can be selected is given by

$$\binom{10}{4} = \frac{10 \cdot \overset{3}{\cancel{9}} \cdot \cancel{8} \cdot 7}{\cancel{4} \cdot \cancel{3} \cdot \cancel{2} \cdot 1} = 10 \cdot 3 \cdot 7 = 210.$$

DO EXERCISE 38.

Some problems involve repeated and/or mixed uses of combination and permutation notation.

Example 4 A university offers 5 science courses, 6 humanity courses, and 3 literature courses. How many ways can a student choose 2 science courses, 3 humanity courses, and 1 literature course?

Solution Since the *order* of choosing the subjects is irrelevant, the student can choose 2 science courses $\binom{5}{2}$ ways, 3 humanity courses $\binom{6}{3}$ ways, and 1 literature course $\binom{3}{1}$ ways. Hence, by the Fundamental Counting Principle, the total number of choices available is the product

$$\binom{5}{2} \cdot \binom{6}{3} \cdot \binom{3}{1} = \frac{5 \cdot 4}{2 \cdot 1} \cdot \frac{6 \cdot 5 \cdot 4}{3 \cdot 2 \cdot 1} \cdot \frac{3}{1} = 600.$$

DO EXERCISE 39.

Not all solutions involve products. Some require additions.

Example 5   How many ways can 2 people be chosen out of 3 men and 4 women such that *at least* one is a man?

Solution   *Either* one or two men must be chosen to satisfy the constraints of the problem. If one man is chosen, then one woman must also be chosen. This can be done $\binom{3}{1} \cdot \binom{4}{1} = 3 \cdot 4 = 12$ ways. Two men can be chosen $\binom{3}{2} = 3$ ways. Since *either* is permitted, the total number is the sum $12 + 3 = 15$.

DO EXERCISE 40.

Some problems involve partitions.

Example 6   How many ways can 9 different books be distributed among three children so that the oldest gets 4, the middle child gets 3, and the youngest gets 2?

Solution   Here the 9 books are "partitioned" into 3 groups of 4, 3, and 2. The oldest child can "choose" his books $\binom{9}{4}$ ways. The next child can "choose" his 3 books out of the remaining $9 - 4 = 5$ books $\binom{5}{3}$ ways. The youngest child can "choose" his 2 books out of the remaining $5 - 3 = 2$ books $\binom{2}{2}$ ways. The total number of "choices" for all is the product $\binom{9}{4} \cdot \binom{5}{3} \cdot \binom{2}{2}$. Making use of the fact that

$$\binom{n}{k} = \frac{n!}{k!(n-k)!},$$

we have

$$\binom{9}{4}\binom{5}{3}\binom{2}{2} = \frac{9!}{4!5!} \cdot \frac{5!}{3!2!} \cdot \frac{2!}{2!0!}$$

$$= \frac{9!}{4!3!2!} = 9 \cdot 4 \cdot 7 \cdot 5 = 1260.$$

Verify that the same solution is obtained regardless of which child "chooses" first or second.

DO EXERCISE 41.

There is a related type of problem in which not all the elements are distinguishable in an arrangement.

39. How many ways can a committee of 4 men and 3 women be chosen out of a group of 7 men and 5 women?

40. How many ways can 2 people be chosen out of 3 men and 4 women such that at least one is a woman?

41. How many ways can 7 different toys be distributed to 4 children with the oldest getting one and the others 2 each?

42. Consider the word "TENNESSEE".

a) How many letters in all are there?

b) How many T's are there?

c) How many E's?

d) How many S's?

e) How many N's?

f) Evaluate

$$\frac{9!}{1! \cdot 4! \cdot 2! \cdot 2!}.$$

(*Hint.* Write out the factorials and simplify.)

g) How many distinguishable words can be made up of all the letters of the word "TENNESSEE"?

Example 7 How many *distinguishable* words can be made up of all the letters of the word "MISSISSIPPI"?

Solution The word "MISSISSIPPI" has 11 letters. Thus, if all the letters were *distinguishable*, there would be

$$11! = 39,916,800 \text{ possible arrangements.}$$

However, we actually have

$$1 \text{ M}, \quad 4 \text{ I's}, \quad 4 \text{ S's}, \quad 2 \text{ P's.}$$

The *different* letters are distinguishable, but the 4 I's are *indistinguishable* from each other, as are the 4 S's and the 2 P's. The 4 I's, 4 S's, and 2 P's can be made distinguishable by putting *tags* on them for the moment. Thus, we obtain "eleven" letters:

$$M, \quad I_1, \quad I_2, \quad I_3, \quad I_4, \quad S_1, \quad S_2, \quad S_3, \quad S_4, \quad P_1, \quad P_2.$$

Given *any* particular word, the 4 I's can be permuted $4! = 24$ ways, the 4 S's can be permuted $4! = 24$ ways, and the 2 P's can be permuted $2! = 2$ ways. Thus the I's, S's, and P's of *each* distinguishable word can be permuted $4! \cdot 4! \cdot 2! = 1152$ ways to make words distinguishable by tags, or indistinguishable when the tags are dropped. Hence, the number of *distinguishable* words is:

$$\frac{11!}{4!4!2!} = 34,650.$$

In general,

Given a set of n objects of which n_1 are of one kind, n_2 are of a second kind, n_3 are of a third kind, and so on, where finally there are n_k of a kth kind, then the number of distinguishable arrangements is

$$\frac{n!}{n_1! \cdot n_2! \cdot n_3! \cdots n_k!}.$$

DO EXERCISE 42.

EXERCISE SET 4.5

Evaluate.

1. $C(13, 2)$

2. $C(9, 6)$

3. $\binom{13}{11}$

4. $\binom{9}{3}$

5. $C(7, 1)$

6. $C(8, 8)$

7. $C(n, 2)$

8. $C(n, 3)$

9. An office manager interviews 6 secretaries. If 4 are hired at random, how many ways can they be selected? (*Hint.* See Example 3.)

10. If ice cream comes in 8 flavors, in how many ways can a particular store select 6 flavors at random?

11. On a test, a student must answer any 7 of the first 10 questions, and any 5 of the remaining 8 questions. In how many ways can this be done? (*Hint*. See Example 4.)

13. From a group consisting of 6 men and 8 women, 5 are to be hired at random as sales representatives of a company. How many ways can this be done if:

a) It does not matter how many are men and how many are women?
b) At least 3 must be women?
c) At least 4 must be women?
d) At least 1 must be a woman?

(*Hint*. See Example 4 and Example 5.)

15. How many ways can 11 different tools be distributed among four employees if the first gets 3, the second 4, and the remaining two employees get 2 each? (*Hint*. See Example 6.)

17. How many distinguishable words can be made up of all the letters in the word "CINCINNATI"? (*Hint*. See Example 7.)

19. A psychotic professor decides to grade his 20 students on a curve without regard to test performance. There will be 2 A's, 5 B's, 8 C's, 3 D's, and 2 F's. How many ways can this be done? (*Hint*. See Example 7.)

21. In how many ways can the expression $x^3y^2z^4$ be expressed without exponents?

23. From a group of 20 employees a delegation of 3 is to be selected at random.

a) How many ways can this be done?
b) How many ways can this be done if one of the 3 is selected at random to be spokesman?

25. How many ways can a group of 3 people be selected from a group of 5 with regard to order? without regard to order?

27. A folk-dance leader has 10 records for elementary dances and 20 records for more advanced dances.

a) How many ways can 6 dances be selected at random for a program?
b) How many ways can 4 elementary dances and 2 advanced dances be selected for the program?

12. How many committees can be formed from a set of 8 senators and 5 representatives if each committee contains 4 senators and 3 representatives?

14. From a group consisting of 10 smokers and 10 nonsmokers, 4 are to be chosen for a medical study. How many ways can this be done if:

a) It does not matter how many smokers or nonsmokers there are?
b) At least 2 must be smokers?
c) At least 3 must be smokers?
d) At least 1 must be a smoker?

16. How many ways can 8 different records be distributed among three students if the first gets 2, the second gets 5, and the third gets 1?

18. How many distinguishable words can be made up of all the letters in the word "ABRACADABRA"?

20. A psychotic professor decides to grade her 24 students on a curve without regard to test performance. There will be 3 A's, 5 B's, 9 C's, 4 D's, and 3 F's. How many ways can this be done?

22. In how many ways can the expression $p^3q^5r^2$ be expressed without exponents?

24. *Psychology*. For a psychological study, a group of 3 people is selected at random from a group of 4 men and 3 women.

a) How many ways can this be done?
b) How many ways can this be done if there must be at least 1 man and at least 1 woman in the group?

26. How many ways can a group of 4 people be selected from a group of 6 with regard to order? without regard to order?

28. How many ways can a bridge team of 4 be selected at random from 6 husband–wife pairs,

a) If husband–wife pairs cannot be broken?
b) If no husband or wife is on the same team as his or her spouse, but the team still consists of 2 men and 2 women?

29. How many ways can 12 work assignments be made at random, 4 to each of 3 machinists? (*Note.* Machinists are people; people are distinguishable.)

31. As in Exercise 29, how many ways can these 12 work assignments be made on 3 work sheets *before* the work sheets are assigned to a particular machinist?

30. How many ways can 10 work assignments be made at random if one machinist gets 4 and the other two machinists 3 each?

32. As in Exercise 30, how many ways can these 10 work assignments be made on 3 work sheets *before* the work sheets are assigned to particular machinists?

▶

33. How many distinguishable words can be formed from the letters of ALGEBRA if each word is to contain 7 letters? 6 letters? 5 letters?

35.* A class consists of 10 students of whom 4 are women of whom 3 are married to 3 of the 5 married men in the class. They all spend an evening at a motel which has 2 rooms with 2 single beds each (a double) and 2 rooms with 3 single beds each (a triple). Assuming that only one sex occupies any room except for married couples who *may* share a double, how many ways can the students distribute themselves into the rooms?

37. Wendy's Hamburgers, a national firm, advertises "We Fix Hamburgers 256 Ways!" This is accomplished by various combinations of catsup, onion, mustard, pickle, mayonnaise, relish, tomato, or lettuce. Of course, one can also have a plain hamburger. Assume single portions.

a) Use combination notation to show the number of possible hamburgers.

34. How many distinguishable words can be formed from the letters of PRECEDE if each word is to contain 7 letters? 6 letters? 5 letters?

36. In Exercises 36 through 41 of Exercise Set **4**.1, we found that a set with n elements has 2^n subsets. Show that

$$2^n = \binom{n}{0} + \binom{n}{1} + \binom{n}{2} + \cdots + \binom{n}{n} = \sum_{i=0}^{n} \binom{n}{i}.$$

b) Use the result of Exercise 36 to show how the expression in (a) can be evaluated quickly.

c) Interestingly, Wendy's excludes cheese from the possibilities. This may be to avoid false advertising, because then one would have a cheeseburger. Including cheese as a possibility, how many ways does Wendy's fix hamburgers?

(Optional) Counting Numbers of POKER Hands

In answering each problem *do not* just give the answer; provide a reasoned expression as well. Read all the problems before beginning.

38. How many different 5-card hands can be dealt from a standard 52-card deck? Note that no order is considered, even though one might "order" cards after receiving them.

40. A *straight flush* consists of five cards in sequence in the same suit, but does not include royal flushes. How many are there? (Assume that an ace can be used at either the high or low end.)

39. A *royal flush* consists of a 5-card hand with A-K-Q-J-10 of the same suit. How many are there?

41. *Four of a kind* is a five-card hand where 4 of the cards are the same denomination, such as 4 jacks, 4 aces, or 4 deuces. How many are there?

---

*This problem was assigned on a test by one of the authors (JCC) and subsequently submitted by the mother of one of his students to *TIME* magazine and printed with the cover letter stating that this was an example of "nonsexist mathematics," whatever that is.

A standard deck of 52 cards is made up as follows:

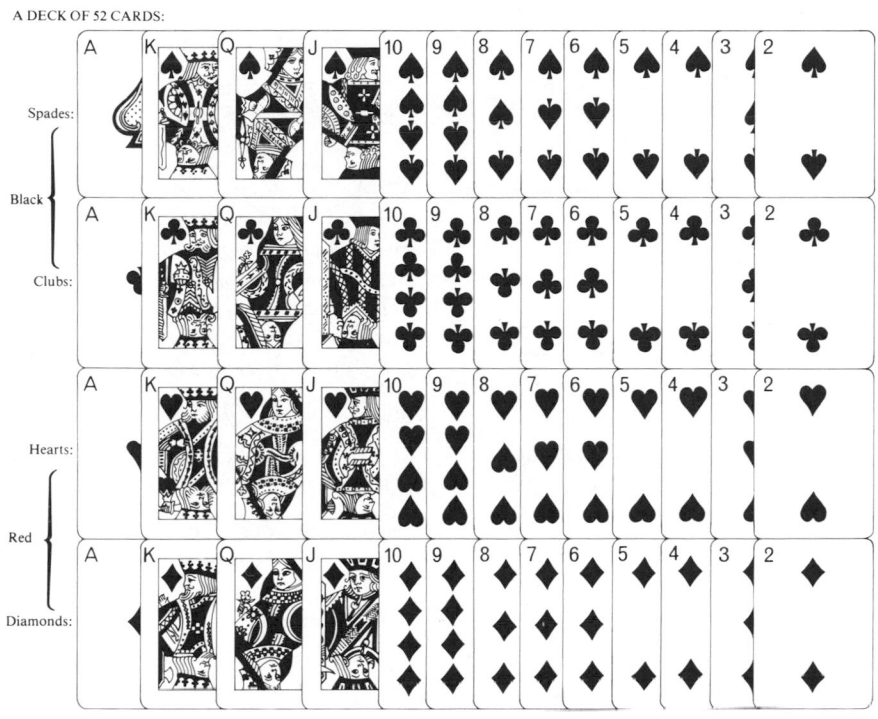

A DECK OF 52 CARDS:

42. A *full house* consists of a pair and three of a kind, such as K-K-K-7-7.

a) How many full houses are there consisting of kings and sevens?
b) How many full houses are there?

43. A *pair* is a five-card hand where just 2 of the cards are of the same denomination, such as Q-Q-8-A-3. How many are there?

44. *Three of a kind* consists of a five-card hand where 3 of the cards are of the same denomination and the other two cards are not the same denomination, such as K-K-K-10-3. How many are there?

45. A *flush* is a five-card hand where all the cards are the same suit, but not all in sequence (not a straight flush or royal flush). How many are there?

46. *Two pairs* is just what it says—a hand such as Q-Q-8-8-A. How many are there?

47. A *straight* is any five cards in sequence but not in the same suit. How many are there?

Exercises 48 through 71 provide a mixture of permutation and combination problems.

48. There are 27 people at a party. How many possible handshakes are there?

49. How many ways can 6 novels be assigned to 10 students if each gets at most one novel?

50. How many ways can 4 cars be placed in 9 garages, one car to a garage?

51. How many softball games are played in a league with 8 teams if each team plays each other team once? twice?

52. How many games are played in a league with n teams if each team plays each other team once? twice?

53. How many distinguishable words can be formed from all the letters of the word MATH? BUSINESS? PHILOSOPHICAL?

54. There are 8 points on a circle. How many triangles (inscribed) can be drawn with these points as vertices?

55. A money clip contains one each of the following bills: $1, $2, $5, $10, $20, $50, $100. How many different sums of money can be formed using the bills?

56. How many words can be formed using 4 out of 5 letters of A, B, C, D, E if the letters

a) are not repeated?
b) can be repeated?
c) are not repeated but must begin with D?
d) are not repeated but must end with DE?

57. A state forms its license plates by first listing a number which corresponds to the county the car owner dwells in (the names of the counties are alphabetized and the number is its location in the order). Then the plate lists a letter of the alphabet and this is followed by a number from 1 to 9999. How many such plates are possible if there are 80 counties?

58. How many diagonals does a hexagon have?

59. How many diagonals does an n-agon have?

60. How many distinguishable words can be formed from the letters of the word ORANGE? BIOLOGY? MATHEMATICS?

61. How many words can be formed using 5 out of 6 of the letters of G, H, I, J, K, L if the letters

a) are not repeated?
b) can be repeated?
c) are not repeated but must begin with K?
d) are not repeated but must end with IGH?

62. A set of 5 parallel lines crosses another set of 8 parallel lines. How many parallelograms are formed?

63. There are n points on a circle. How many quadrilaterals can be drawn with these points as vertices?

Solve for n.

64. $P(n, 5) = 7 \cdot P(n, 4)$

65. $C(n + 1, 3) = 2 \cdot C(n, 2)$

66. $\binom{n + 2}{4} = 6 \cdot \binom{n}{2}$

67. $P(n, 4) = 8 \cdot P(n - 1, 3)$

68. $C(n, n - 2) = 6$

69. $\binom{n}{3} = 2 \cdot \binom{n - 1}{2}$

70. $P(n, 5) = 9 \cdot P(n - 1, 4)$

71. $P(n, 4) = 8 \cdot P(n, 3)$

OBJECTIVES

You should be able to

a) Find the rth term of a binomial expansion of $(a + b)^n$.

b) Evaluate expressions like $\binom{5}{3}$ to
$$\frac{5!}{3!2!}.$$

c) Use the binomial theorem to expand expressions like $(x^2 + 3y)^9$ and 2^n.

*4.6 (OPTIONAL) THE BINOMIAL THEOREM

Consider the following expanded powers of $(a + b)^n$, where $a + b$ is any binomial. Look for patterns.

$(a + b)^0 = \qquad\qquad\qquad 1$

$(a + b)^1 = \qquad\qquad\qquad a \quad + \quad b$

$(a + b)^2 = \qquad\qquad a^2 \quad + \quad 2ab \quad + \quad b^2$

$(a + b)^3 = \qquad a^3 \quad + \quad 3ab^2b \quad + \quad 3ab^2 \quad + \quad b^3$

$(a + b)^4 = \quad a^4 \quad + \quad 4a^3b \quad + \quad 6a^2b^2 \quad + \quad 4ab^3 \quad + \quad b^4$

$(a + b)^5 = a^5 \quad + \quad 5a^4b \quad + \quad 10a^3b^2 \quad + \quad 10a^2b^3 \quad + \quad 5ab^4 \quad + \quad b^5$

There are some patterns to be noted in the expansions.

1. In each term, the sum of the exponents is n.

2. The exponents of a start with n and decrease. The last term has no factor of a. The first term has no factor of b. The exponents of b start in the second term with 1 and increase to n.

3. There is one more term than the degree of the polynomial. The expansion of $(a + b)^n$ has $n + 1$ terms.

We now find a way to determine the coefficients. Let us consider the nth power of a binomial $(a + b)$:

$$(a + b)^n = \underbrace{(a + b)(a + b)(a + b) \cdots (a + b)}_{n \text{ factors}}$$

When we multiply, we will find all possible products of a's and b's. For example, when we multiply all the first terms we will get n factors of a, or a^n. Thus the first term in the expansion is a^n, and the first coefficient is 1. Similarly, the coefficient of the last term is 1.

To get a term such as the $a^{n-r}b^r$ term, we will take a's from $n - r$ factors and b's from r factors. Thus we take n objects, $n - r$ of them a's and r of them b's. The number of ways we can do this is

$$\frac{n!}{(n - r)!r!} .$$

This is $\binom{n}{r}$. Thus the $(r + 1)$st term in the expansion is

$$\binom{n}{r}a^{n-r}b^r.$$

We now have a theorem.

THE BINOMIAL THEOREM. For any binomial $(a + b)$ and any natural number n,

$$(a + b)^n = \binom{n}{0}a^n + \binom{n}{1}a^{n-1}b + \binom{n}{2}a^{n-1}b^2 + \cdots + \binom{n}{n}b^n.$$

Sigma notation for a binomial expansion is as follows.

$$(a + b)^n = \sum_{r=0}^{n} \binom{n}{r}a^{n-r}b^r.$$

Because of the Binomial Theorem, $\binom{n}{r}$ is called a *binomial coefficient*. It can now be made apparent why 0! is defined to be 1. In the binomial expansion we want $\binom{n}{0}$ to equal 1 and we also want the

43. Find the 4th term of $(x - 3)^8$.

44. Find the 5th term of $(x - 3)^8$.

45. Find the 6th term of $(y + 2)^{10}$.

46. Find the 1st term of $(3x + 5)^5$.

definition

$$\binom{n}{r} = \frac{n!}{(n - r)!r!}$$

to hold for all whole numbers n and r. Thus we must have

$$\binom{n}{0} = \frac{n!}{(n - 0)!0!} = \frac{n!}{n!0!} = 1.$$

This will be satisfied if $0!$ is defined to be 1.

Example 1 Find the 7th term of $(4x - y^2)^9$.

Solution We let $r = 6$, $n = 9$, $a = 4x$, and $b = -y^2$ in the formula $\binom{n}{r}a^{n-r}b^r$. Then

$$\binom{9}{6}(4x)^3(-y^2)^6 = \frac{9!}{6!3!}(4x)^3(-y^2)^6$$

$$= \frac{9 \cdot 8 \cdot 7 \cdot 6!}{3! \cdot 6!} 64x^3y^{12}$$

$$= 5376x^3y^{12}.$$

DO EXERCISES 43 THROUGH 46.

Example 2 Expand $(x^2 - 2y)^5$.

Solution Note that $a = x^2$, $b = -2y$, and $n = 5$. Then, using the binomial theorem, we have

$$(x^2 - 2y)^5 = \binom{5}{0}(x^2)^5 + \binom{5}{1}(x^2)^4(-2y)$$

$$+ \binom{5}{2}(x^2)^3(-2y)^2 + \binom{5}{3}(x^2)^2(-2y)^3$$

$$+ \binom{5}{4}x^2(-2y)^4 + \binom{5}{5}(-2y)^5$$

$$= \frac{5!}{0!5!}x^{10} + \frac{5!}{1!4!}x^8(-2y)$$

$$+ \frac{5!}{2!3!}x^6(-2y)^2 + \frac{5!}{3!2!}x^4(-2y)^3$$

$$+ \frac{5!}{4!1!}x^2(-2y)^4 + \frac{5!}{5!1!}(-2y)^5$$

$$= x^{10} - 10x^8y + 40x^6y^2 - 80x^4y^3 + 80x^2y^4 - 32y^5.$$

Example 3   Expand $\left(\dfrac{2}{x} + 3\sqrt{x}\right)^4$.

Solution   Note that $a = 2/x$, $b = 3\sqrt{x}$, and $n = 4$. Then, using the binomial theorem, we have

$$\left(\frac{2}{x} + 3\sqrt{x}\right)^4 = \binom{4}{0} \cdot \left(\frac{2}{x}\right)^4 + \binom{4}{1} \cdot \left(\frac{2}{x}\right)^3 (3\sqrt{x})$$

$$+ \binom{4}{2} \cdot \left(\frac{2}{x}\right)^2 (3\sqrt{x})^2$$

$$+ \binom{4}{3}\left(\frac{2}{x}\right)(3\sqrt{x})^3 + \binom{4}{4}(3\sqrt{x})^4$$

$$= \frac{4!}{0!4!} \cdot \frac{16}{x^4} + \frac{4!}{1!3!} \cdot \frac{8}{x^3} \, 3\sqrt{x}$$

$$+ \frac{4!}{2!2!} \cdot \frac{4}{x^2} \cdot 9x$$

$$+ \frac{4!}{3!1!} \cdot \frac{2}{x} \cdot 27x^{3/2} + \frac{4!}{4!0!} \cdot 81x^2$$

$$= \frac{16}{x^4} + \frac{96}{x^{5/2}} + \frac{216}{x} + 216\sqrt{x} + 81x^2.$$

DO EXERCISES 47 THROUGH 49.

Example 3   Use $(1 + 1)^n$ for 2^n to find a binomial expansion for 2^n.

Solution

$$2^n = (1 + 1)^n = \binom{n}{0} + \binom{n}{1} + \binom{n}{2} + \cdots + \binom{n}{n}$$

The right side is an expression for the number of subsets of a set of n objects. Thus we have a proof of the following.

A set with n objects has 2^n subsets.

DO EXERCISE 50.

Pascal's Triangle

When the coefficients of the binomial expansions of $(a + b)^n$ in a triangular array are arranged as follows, we get what is known as

---

47. Expand $(x^2 - 1)^5$.

48. Expand $\left(2x + \dfrac{1}{y}\right)^4$.

49. Expand $(x - \sqrt{2})^6$.

50. Use $(2 + 1)^n$ for 3^n to find a binomial expansion of 3^n.

51. Try to write the next row of numbers in Pascal's Triangle.

Pascal's Triangle:

$$
\begin{array}{llccccccc}
(a + b)^0 & & & & & 1 & & & \\
(a + b)^1 & & & & 1 & & 1 & & \\
(a + b)^2 & & & 1 & & 2 & & 1 & \\
(a + b)^3 & & 1 & & 3 & & 3 & & 1 \\
(a + b)^4 & 1 & & 4 & & 6 & & 4 & & 1 \\
(a + b)^5 & 1 & 5 & & 10 & & 10 & & 5 & & 1 \\
\end{array}
$$

There are many patterns in the triangle. Find as many as you can.

DO EXERCISE 51.

Perhaps you have discovered a way to write the next row of numbers, given the numbers in the row above it. There are always 1's on the outside. Each remaining coefficient is found by adding the two numbers above. This is shown as follows:

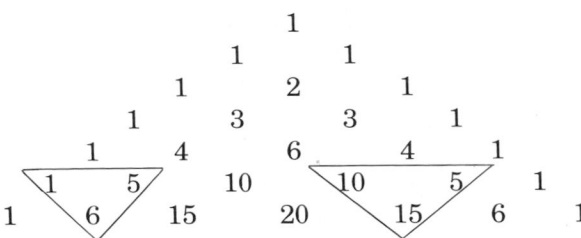

One could actually use Pascal's Triangle to find a binomial expansion. The triangle is extended until the row is obtained which corresponds to the given power. For example, using the row we just found,

$$(a + b)^6 = a^6 + 6a^5b + 15a^4b^2 + 20a^3b^3 + 15a^2b^4 + 6ab^5 + b^5.$$

EXERCISE SET 4.6

Find the indicated term of the binomial expansion.

1. 3rd, $(a + b)^6$

2. 6th, $(x + y)^7$

3. 12th, $(a - 2)^{14}$

4. 11th, $(x - 3)^{12}$

5. 5th, $(2x^3 - \sqrt{y})^8$

6. 4th, $\left(\dfrac{1}{b^2} + \dfrac{b}{3}\right)^7$

7. Middle, $(2u - 3v^2)^{10}$

8. Middle two, $(\sqrt{x} + \sqrt{3})^5$

Expand.

9. $(m + n)^5$

10. $(a - b)^4$

11. $(x^2 - 3y)^5$

12. $(3c - d)^6$

13. $(x^{-2} + x^2)^4$

14. $\left(\dfrac{1}{\sqrt{x}} - \sqrt{x}\right)^6$

15. $(1 - 1)^n$

16. $(1 + 3)^n$

17. $(\sqrt{2} + 1)^6 - (\sqrt{2} - 1)^6$

18. $(1 - \sqrt{2})^4 + (1 + \sqrt{2})^4$

19. $(\sqrt{3} - t)^4$

20. $(\sqrt{5} + t)^6$

▶

Solve for x.

21. $\displaystyle\sum_{r=0}^{8} \binom{8}{r} x^{8-r} 3^r = 0$

22. $\displaystyle\sum_{r=0}^{5} \binom{5}{r}(-1)^r x^{5-r} 3^r = 32$

23. $\displaystyle\sum_{r=0}^{4} \binom{4}{r} 5^{4-r} x^r = 64$

CHAPTER 4 TEST

Consider these sets for Questions 1 through 6:

$$A = \{a, b, c, d, e\}, \qquad B = \{a, b, c, d, e, f, g\},$$
$$C = \{a, b, c\}, \qquad D = \{d, e\}.$$

Find:

1. $A \cup B$

2. $A \cap B$

3. $C \times D$

4. $A - D$

5. Do sets C and D form a partition of A?

6. Do sets C and D form a partition of B?

7. $P(7, 4)$

8. $\binom{7}{4}$

9. $6!$

10. $1!$

11. If 3 dice of different colors are rolled simultaneously, how many ways can they land?

12. How many ways can 2 cards be drawn at random (without regard for order) from a deck of 52?

13. How many ways can 6 people be seated at a circular table?

14. How many 3-digit numbers can be formed using the digits 2, 3, 5, and 7, with repetition? without repetition?

15. How many ways can 4 men and 4 women be seated in a row if men and women occupy alternate seats?

16. How many different words can be formed from all the letters of the word JENNIFER?

17. From a group of 12 Democrats and 9 Republicans, how many committees of 3 can be formed consisting of at least 2 Democrats?

18. There are 240 students in a certain freshman class. Of these

> 120 are taking Economics,
> 65 are taking Russian,
> 84 are taking Mathematics,
> 34 are taking Economics and Russian,
> 22 are taking Economics and Mathematics,
> 18 are taking Russian and Mathematics,
> 10 are taking all three subjects.

How many are not taking any of these subjects?

19. Expand: $(x^2 + 3y)^4$.

CHAPTER FIVE

Probability

OBJECTIVES

You should be able to:

a) Compute the probability of a simple event.

b) Given $p(E)$, find $p(E^c)$ as $1 - p(E)$.

1. A die, used in craps and other games, is a cube with one of the numbers 1 through 6 on each side.

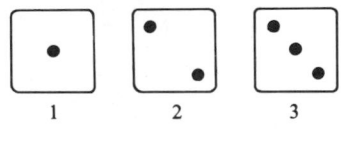

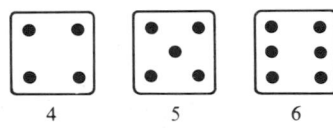

a) What would you reason to be the probability of getting a 5 on a roll of the die?

b) Roll the die 18 times and keep track of the results. What fraction of the rolls resulted in a 5?

c) Compare your answers to (a) and (b).

5.1 INTRODUCTION TO PROBABILITY

Suppose we toss a nickel 100 times and it comes up heads 53 times. We might say that the

Probability of getting a head is $\frac{53}{100}$, or 0.53.

Suppose we toss the same nickel another 100 times and it comes up heads 49 times. We might say that the

Probability of getting a head is $\frac{49}{100}$, or 0.49.

We might also reason about the probability of getting a head. There are 2 outcomes of tossing the nickel, heads and tails. If the coin is "fair," we might reason that the chances are 1 out of 2 of getting a head, so the probability of getting a head is $\frac{1}{2}$, or 0.50.

We call 0.53 and 0.49 *experimental probabilities*, and 0.50 *theoretical probability*. Which is correct? We really never know this. Such is the nature of probability. For example, to determine whether a coin is indeed fair, we may carry out an experiment: toss it a thousand times—or a million times. The information gathered may lead us to reject or accept the fairness of the coin. On the other hand, it may be quite cumbersome and time-consuming to determine probabilities experimentally, so we attempt to determine them theoretically. This will be our main objective in this chapter.

DO EXERCISE 1.

We need some terminology before we continue. Suppose we perform an *experiment* such as flipping a coin, drawing a card from a deck, or checking an item off an assembly line for quality. The results of an experiment are called the *outcomes*. The set of all possible outcomes is called the *sample space*. An *event* is a set of outcomes; that is, a subset of the sample space. For example, for the experiment "flipping a coin," an *event* is "getting a head," from the *sample space* consisting of "head, tail."

We will denote the probability that an event can occur as $p(E)$, or p_E. For example, "getting a head" may be denoted by H. Then $p(H)$, or p_H, represents the probability of getting a head. When the outcomes of an experiment all have the same probability of occurring, we say that they are *equally likely*, or *equiprobable*. To see the distinction between events that are equiprobable and those that are not, consider these dartboards (executive decision-makers).

For dart board A, the events, hitting "Yes," hitting "No," hitting "Maybe", are equally likely, but for board B they are not. A sample

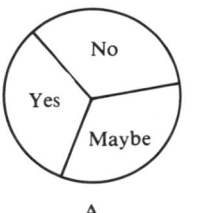
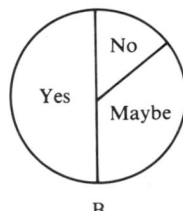

A   B

space that can be expressed as a union of equiprobable events can allow us to calculate probabilities of other events.

BASIC PROBABILITY PRINCIPLE. **If an event E can occur m ways out of n possible *equiprobable* outcomes of a sample space S, the probability of that event is given by**

$$p(E) = p_E = \frac{m}{n}.$$

That is,

$$p(E) = \frac{\mathcal{N}(E)}{\mathcal{N}(S)}.$$

We will give many examples related to a standard bridge deck of 52 cards. Such a deck is made up as follows:

A DECK OF 52 CARDS:

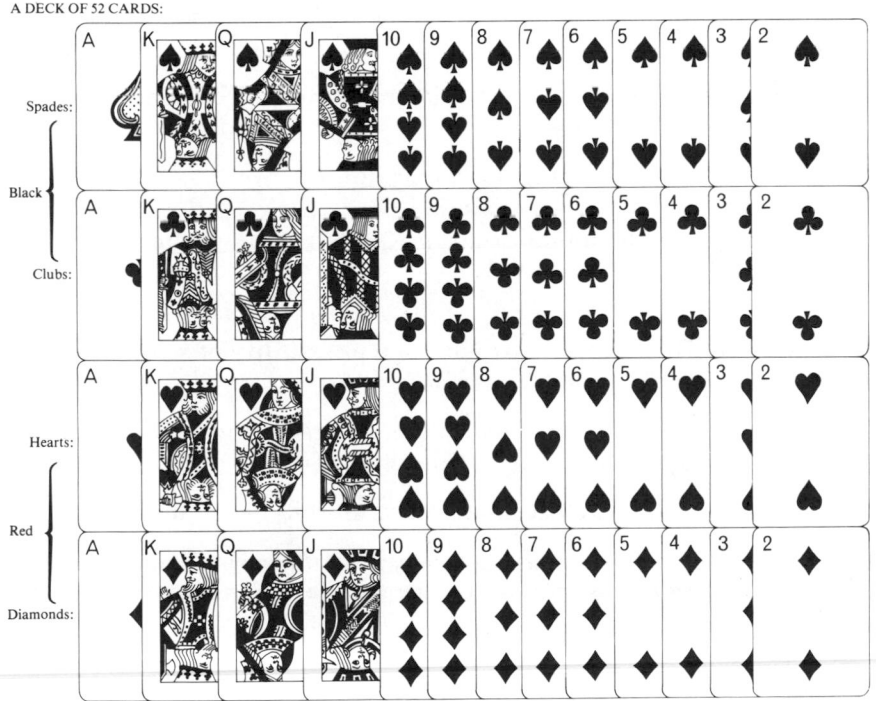

2. A card is drawn at random out of a standard deck of cards. What is the probability of drawing

a) a king?
b) a red ace?
c) a black card?
d) a face card (a king, queen, or jack)?

Example 1 What is the probability of drawing an ace at random out of a standard deck of cards?

Solution There are 52 equally likely outcomes and there are 4 ways to get an ace, so

$$p(\text{Drawing an ace}) = \tfrac{4}{52}, \quad \text{or } \tfrac{1}{13}.$$

The wording "at random" is a way of implying that each card has the same probability of being drawn as any other card; that is, is equally likely to be drawn.

DO EXERCISE 2.

3. In Example 2, what is the probability of selecting a green marble?

Example 2 Suppose we select, without looking, one marble from a sack containing 5 green marbles and 3 yellow marbles. What is the probability of selecting a yellow marble?

Solution There are 8 equally likely outcomes and 3 ways to get a yellow marble, so

$$p(\text{Selecting a yellow marble}) = \tfrac{3}{8}.$$

DO EXERCISE 3.

4. A card is drawn at random out of a *standard* deck of cards. What is the probability of getting a *blue* 18?

Example 3 In Example 2, what is the probability of selecting a red marble?

Solution There are 0 ways of selecting a red marble since there are no red marbles in the sack, so

$$p(\text{Selecting a red marble}) = \tfrac{0}{8}, \quad \text{or } 0.$$

For an event which cannot occur, $p(E) = 0$. It follows that $p(\emptyset) = 0$.

DO EXERCISE 4.

5. For flipping a fair coin, what is the probability of getting a head or a tail?

Example 4 In Example 2, what is the probability of selecting either a green marble *or* a yellow marble?

Solution Since the sack contains only green and yellow marbles, there are 8 ways of selecting either one, so

$$p(\text{Selecting a green or yellow marble}) = \tfrac{8}{8}, \quad \text{or } 1.$$

For an event which is *certain* to occur, or in the case of repeated trials, will *always* occur, $p(E) = 1$. Since the sample space contains all possible events, $p(S) = 1$.

DO EXERCISE 5.

The previous examples lead us to:

The probability that an event will occur is a number from 0 to 1; that is, for any event E,

$$0 \le p(E) \le 1.$$

Example 5 Suppose 3 cards are drawn at random from a well-shuffled deck of cards. What is the probability that all 3 are diamonds?

Solution The number of ways of drawing 3 cards out of a deck of 52 is

$$\binom{52}{3}.$$

Now 13 of the 52 cards are diamonds, so the number of ways of drawing the 3 diamonds is

$$\binom{13}{3}$$

Thus,

$$p(\text{All 3 diamonds}) = \frac{\binom{13}{3}}{\binom{52}{3}} = \frac{\frac{13 \cdot 12 \cdot 11}{3 \cdot 2 \cdot 1}}{\frac{52 \cdot 51 \cdot 50}{3 \cdot 2 \cdot 1}} = \frac{13 \cdot 12 \cdot 11}{52 \cdot 51 \cdot 50} = \frac{11}{850}.$$

DO EXERCISE 6.

Example 6 *Psychology.* For a psychology study, 2 people are selected at random from a group consisting of 8 men and 6 women. What is the probability that both are women?

Solution The number of ways of selecting 2 people from the group of 14 is $\binom{14}{2}$. The number of ways of selecting 2 women from a group of 6 is $\binom{6}{2}$. Thus,

$$p(\text{Both are women}) = \frac{\binom{6}{2}}{\binom{14}{2}} = \frac{\frac{6 \cdot 5}{2 \cdot 1}}{\frac{14 \cdot 13}{2 \cdot 1}} = \frac{6 \cdot 5}{14 \cdot 13} = \frac{3 \cdot 5}{7 \cdot 13} = \frac{15}{91}.$$

DO EXERCISE 7.

6. Suppose 2 cards are drawn at random from a well-shuffled deck of cards. What is the probability that both are clubs?

7. In Example 6, what is the probability that both are men?

8. In Example 7, what is the probability that 2 men and 1 woman are selected?

Example 7 *Psychology.* For a psychology study, 3 people are selected at random from a group consisting of 8 men and 6 women. What is the probability that 1 man and 2 women are selected?

Solution The number of ways of selecting 3 people out of a group of 14 is $\binom{14}{3}$. The number of ways of selecting 1 man out of a group of 8 is $\binom{8}{1}$. The number of ways of selecting 2 women out of a group of 6 is $\binom{6}{2}$. Then, by the Fundamental Counting Principle, we know that the number of ways of selecting 1 man and 2 women is the product

$$\binom{8}{1} \cdot \binom{6}{2}$$

Thus,

$$p(1 \text{ man and 2 women}) = \frac{\binom{8}{1} \cdot \binom{6}{2}}{\binom{14}{3}} = \frac{\dfrac{8}{1} \cdot \dfrac{6 \cdot 5}{2 \cdot 1}}{\dfrac{14 \cdot 13 \cdot 12}{3 \cdot 2 \cdot 1}} = \frac{30}{91}.$$

DO EXERCISE 8.

9. On a roll of a pair of dice, what is the probability of getting a total of

a) 8?

b) 7?

c) 11?

d) 14?

Example 8 What is the probability of getting a total of 6 on a roll of a pair of dice?

Solution We assume the dice are different, say one white and one black. There are 6 possible outcomes on one die (singular of "dice") so, by the Fundamental Counting Principle, there are $6 \cdot 6$, or 36 possible equiprobable outcomes in the sample space for rolling two dice. We show this as follows.

The pairs that total 6 are enclosed in the diagram. There are 5 such pairs, so the probability of getting a total of 6 is $\frac{5}{36}$.

DO EXERCISE 9.

Complementary Events

Drawing an ace from a deck of cards is an event. *Not* drawing an ace is also an event. If E is an event, the nonoccurrence of E is expressed by the symbol E^c, read "not E." Thus, if E is the event of drawing an ace, then E^c is the event of *not* drawing an ace. E and E^c are called *complementary* events.

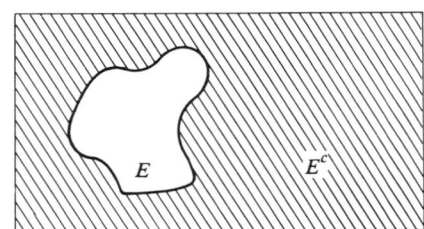

The probability that an event E will *not* occur is given by

$$p(E^c) = 1 - p(E).$$

We can demonstrate this with a Venn diagram. The total number of ways E can occur is m. Then, if there are n total outcomes, it follows that there are $(n - m)$ ways in which E does not occur, so

$$p(E^c) = \frac{n - m}{n} = \frac{n}{n} - \frac{m}{n} = 1 - p(E).$$

Example 9   Suppose $p(E) = \frac{8}{25}$. Find $p(E^c)$.

Solution $p(E^c) = 1 - \frac{8}{25} = \frac{17}{25}$.

DO EXERCISE 10.

Example 10   One card is drawn at random from a well-shuffled deck. What is the probability that it is *not* an ace?

Solution   Let $p(E)$ = probability of drawing an ace. Then $p(E^c)$ = probability that the card is not an ace. From Example 9, $p(E) = \frac{1}{13}$, so

$$p(E^c) = 1 - \frac{1}{13} = \frac{12}{13}.$$

DO EXERCISE 11.

Odds

If p is the probability for an event to occur, then, from the preceding

10. Find $p(E^c)$ if

a) $p(E) = \frac{11}{34}$

b) $p(E) = 0.63$

11. One card is drawn at random from a well-shuffled deck. What is the probability that it is

a) a red ace?

b) not a red ace?

c) a club?

d) not a club?

12. One card is drawn at random from a well-shuffled deck.

a) What are the odds for a red ace to occur?

b) What are the odds against a red ace occurring?

result, we know that $(1 - p)$ is the probability for the event not to occur. The ratio

$$p:(1 - p)$$

is the *odds for* the event to occur. The ratio

$$(1 - p):p$$

is the *odds against* the occurrence.

Example 11   One card is drawn from a well-shuffled deck.

a) What are the odds for an ace to occur?
b) What are the odds against the occurrence of an ace?

Solution   From Example 10, we let $p = p(\text{getting an ace}) = \frac{1}{13}$, and $1 - p = \frac{12}{13}$. Thus,

a) The odds are 1:12 (read "1 to 12") *for* drawing an ace.
b) The odds are 12:1 *against* drawing an ace.

DO EXERCISE 12.

Probability theory came about historically as a way to calculate odds in games of chance. Today it has ever-expanding application to fields such as business, social science, behavioral science, and physics.

EXERCISE SET 5.1

One card is drawn at random from a well-shuffled deck of 52. What is the probability of drawing

1. a queen?   **2.** a jack?   **3.** a spade?   **4.** a diamond?   **5.** an 8?

6. a 10?   **7.** a black card?   **8.** a red club?   **9.** a 7 or a jack?   **10.** an 8 or 10?

Suppose we select one billiard ball from a bag containing 6 red billiard balls and 10 white ones. What is the probability of selecting:

11. a red ball?   **12.** a white ball?   **13.** a chartreuse ball?   **14.** a red or white ball?

15. For a sociological study, a group of 4 people are chosen from a group containing 7 men and 8 women. What is the probability that 2 men and 2 women are chosen?

16. Suppose 4 pens are selected at random from a box containing 9 yellow pens and 6 blue pens. What is the probability that 2 will be yellow and 2 will be blue?

17. What is the probability of getting a total of 9 on a roll of a pair of dice?

18. What is the probability of getting a total of 10 on a roll of a pair of dice?

19. What is the probability of getting a total of 12 ("boxcars") on a roll of a pair of dice?

20. what is the probability of getting a total of 2 ("snake eyes") on a roll of a pair of dice?

21. What is the probability of getting a total of 1 on a roll of a pair of dice?

22. What is the probability of getting a total of 13 on a roll of a pair of dice?

23. From a bag containing 7 dimes, 8 nickels, and 10 quarters, 7 coins are drawn at random. What is the probability of getting 3 dimes, 2 nickels, and 2 quarters?

24. From a sack containing 7 dimes, 5 nickels, and 10 quarters, 8 coins are drawn at random. What is the probability of getting 4 dimes, 3 nickels, and 1 quarter?

Suppose 5 cards are dealt from a well-shuffled deck of 52. What is the probability of dealing:

25. 3 sevens and 2 kings? This is a type of *full house.*

26. 2 jacks and 3 aces?

27. 4 aces and 1 five?   **28.** 4 kings and 1 queen?

29. 5 aces?   **30.** 5 kings?

31. The sales force of a business consists of 10 men and 10 women. A production unit of 4 people is set up at random. What is the probability that 2 men and 2 women are chosen?

32. A union is made up of 14 women and 7 men. A bargaining unit of 3 is chosen at random. What is the probability that 2 women and 1 man are chosen?

33. At a personnel office 5 men and 3 women apply for a job. If 2 are hired at random, what is the probability that:

a) 1 is a man and 1 is a woman?
b) both are men?
c) both are women?
d) both are men *or* both are women?

34. Repeat Exercise 33, but assume that 5 men and 6 women apply for the job.

35. a) Find $p(E^c)$ if $p(E) = \frac{17}{45}$.
    b) What are the odds *for* the event E to occur?
    c) What are the odds *against* the event E occurring?

36. a) Find $p(E^c)$ if $p(E) = \frac{29}{63}$.
    b) What are the odds *for* the event E to occur?
    c) What are the odds *against* the event E occurring?

37. *Business, Advertising.* In many state-run lotteries the odds *against* winning a

1) $20 prize are 200:1,
2) $500 prize are 250,000:3, and
3) $1000 prize are 500,000:3.

What is the probability to win:

a) $20 on one lottery ticket?
b) $500 on one lottery ticket?
c) $1000 on one lottery ticket?
d) $20 on two lottery tickets?

38. *Business, Advertising.* The following is an odds chart for an actual giveway game from a national food chain.

Prize value	No. of prizes	Odds for one store visit	Odds for 13 store visits	Odds for 26 store visits
$1000	42	147,619:1	11,355:1	5678:1
$100	450	13,778:1	1,060:1	530:1
$20	895	6,927:1	533:1	267:1
$5	2,385	2,600:1	200:1	100:1
$2	7,450	832:1	64:1	32:1
$1	59,600	104:1	8:1	4:1
Total	70,822	88:1	7:1	3.5:1

a) What is the probability to win $1 in one store visit?
b) What is the probability to win $5 in 13 store visits?
c) What is the probability to win something in one store visit?
d) What is the probability to win something in 13 store visits?

You should be able to:

a) Given $p(E_1)$, $p(E_2)$, and $p(E_1 \cap E_2)$, find $p(E_1 \cup E_2)$ as

$$p(E_1) + p(E_2) - p(E_1 \cap E_2).$$

b) Given $p(E_1)$ and $p(E_2)$ and that E_1 and E_2 are *mutually exclusive*, find

$$p(E_1 \cup E_2) \quad \text{as} \quad p(E_1) + p(E_2).$$

c) Compute probabilities involving complementary and mutually exclusive events.

5.2 COMPOUND EVENTS

Let us consider an experiment where two of the outcomes are E_1 and E_2. Suppose we draw one card from a well-shuffled deck. Let

$$E_1 = \text{Event of drawing an ace}$$

and

$$E_2 = \text{Event of drawing a king.}$$

The

Event of drawing an ace or a king, or both,

is denoted

$$E_1 \cup E_2, \quad \text{or} \quad E_1 \text{ or } E_2,$$

and is called a *disjunction*. The

Event of drawing an ace and a king

is denoted

$$E_1 \cap E_2, \quad \text{or} \quad E_1 \text{ and } E_2,$$

and is called a *conjunction*. The events $E_1 \cup E_2$ and $E_1 \cap E_2$ are examples of *compound events*.

Since in *one* draw of a card it is impossible to draw an ace *and* a king at the same time, the probability of drawing an ace and a king is

$$p(E_1 \cap E_2) = p(\emptyset) = 0.$$

The probability of drawing an ace or a king is

$$p(E_1 \cup E_2) = \tfrac{8}{52} = \tfrac{2}{13}.$$

If we know the probabilities $p(E_1)$, $p(E_2)$, and $p(E_1 \cap E_2)$, we can compute $p(E_1 \cup E_2)$ using the following result.

For any events E_1 and E_2.

$$p(E_1 \cup E_2) = p(E_1) + p(E_2) - p(E_1 \cap E_2).$$

This follows from a result of Chapter 6:

$$\mathcal{N}(E_1 \cup E_2) = \mathcal{N}(E_1) + \mathcal{N}(E_2) - \mathcal{N}(E_1 \cap E_2).$$

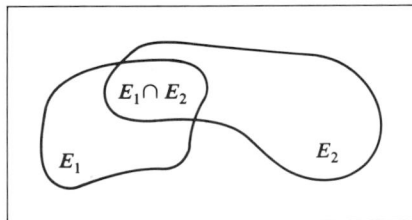

Let n = the number of elements in the sample space. Then

$$p(E_1 \cup E_2) = \frac{N(E_1 \cup E_2)}{n}$$

$$= \frac{N(E_1) + N(E_2) - N(E_1 \cap E_2)}{n}$$

$$= \frac{N(E_1)}{n} + \frac{N(E_2)}{n} - \frac{N(E_1 \cap E_2)}{n}$$

$$= p(E_1) + p(E_2) - p(E_1 \cap E_2).$$

This result is called the *addition theorem* and will be considered in more detail later in this chapter. For now we will be more interested in the following consequence of this result.

DO EXERCISE 13.

For the event E_1 = Drawing an ace in one draw of a card and the event E_2 = Drawing a king in one draw of a card, we say that they are *mutually exclusive*, meaning that they cannot both happen at the same time. That is, $p(E_1 \cap E_2) = p(\emptyset) = 0$. Then

$$p(E_1 \cup E_2) = p(E_1) + p(E_2) - p(E_1 \cap E_2)$$

$$= p(E_1) + p(E_2) - 0$$

$$= p(E_1) + p(E_2).$$

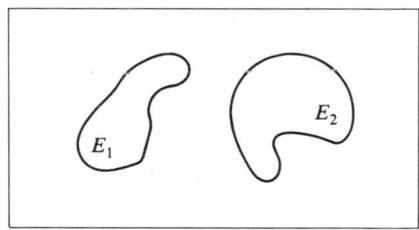

This leads us to the following.

For any events E_1 and E_2 which are mutually exclusive,

$$p(E_1 \cup E_2) = p(E_1) + p(E_2).$$

Example 1   Suppose E_1 and E_2 are mutually exclusive, and $p(E_1)$ = 0.45 and $p(E_2)$ = 0.22. Find $p(E_1 \cup E_2)$.

Solution $p(E_1 \cup E_2) = p(E_1) + p(E_2) = 0.45 + 0.22 = 0.67$

DO EXERCISE 14.

13. Suppose

$$p(E_1) = 0.34,$$
$$p(E_2) = 0.42,$$

and

$$p(E_1 \cap E_2) = 0.13.$$

Find $p(E_1 \cup E_2)$.

14. Suppose

$$p(E_1) = 0.34,$$
$$p(E_2) = 0.42,$$

and that E_1 and E_2 are mutually exclusive. Find $p(E_1 \cup E_2)$.

Two events which are mutually exclusive can be represented as *branches* of a tree.

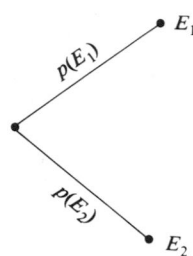

For coin-flipping, a coin cannot land both "heads up" and "tails up" at the same time, assuming the coin does not land on edge. Thus,

$$p(H \cap T) = p(\text{Heads and Tails}) = p(\emptyset) = 0,$$

so the events are mutually exclusive and

$$p(H \cup T) = p(H) + p(T) = \tfrac{1}{2} + \tfrac{1}{2} = 1.$$

Note that these events *partition* the sample space, that is, are mutually exclusive and fill the sample space.

For any events E_1 and E_2 which partition the sample space,

$$p(E_1 \cup E_2) = p(E_1) + p(E_2) = 1.$$

For any two events which partition the sample space, one event is the *complement* of the other, so

$$E_2 = E_1^c,$$

and

$$p(E_1) + p(E_2) = p(E_1) + p(E_1^c) = 1.$$

This follows from the fact that $p(E_1^c) = 1 - p(E_1)$.

Consider the example of drawing a card from a well-shuffled deck, and these 13 events:

$$E_1 = \text{drawing an ace,}$$
$$E_2 = \text{drawing a king,}$$
$$E_3 = \text{drawing a queen,}$$
$$\cdot \qquad \cdot$$
$$\cdot \qquad \cdot$$
$$\cdot \qquad \cdot$$
$$E_{12} = \text{drawing a three,}$$
$$E_{13} = \text{drawing a two.}$$

These events partition the sample space into 13 equiprobable subsets.

That is, the sample space S consisting of all possible outcomes, is given by

$$S = E_1 \cup E_2 \cup E_3 \cup \cdots \cup E_{12} \cup E_{13},$$

and any two pairs of events are mutually exclusive. We express this as:

$$E_i \cap E_j = \emptyset, \qquad \text{for all } i \neq j,$$

or

$$p(E_i \cap E_j) = 0, \qquad \text{for all } i \neq j.$$

We can represent the events as branches of a tree:

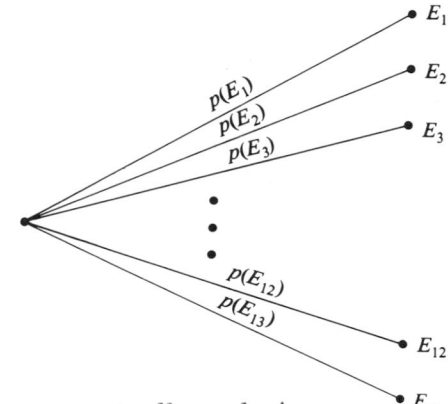

Since the events are mutually exclusive,

$$\begin{aligned} p(S) &= p(E_1 \cup E_2 \cup E_3 \cup \cdots \cup E_{13}) \\ &= p(E_1) + p(E_2) + p(E_3) + \cdots + p(E_{13}). \end{aligned}$$

The sample space S includes *all* events, so in any trial, *one* of them must occur, so that $p(S) = 1$. Thus

$$p(E_1) + p(E_2) + p(E_3) + \cdots + p(E_{13}) = 1,$$

where for each event $p(E_i)$, $0 \leq p(E_i) \leq 1$.

Let us use this partitioning idea in an example.

Example 2 *Manufacturing, Quality Control.* A box contains 20 transistors, 5 of which are defective. Three transistors are taken out at random. What is the probability that at least one is defective?

Solution Let E_i be the *event* of drawing exactly i defective transistors and $p_i = p(E_i)$. Drawing different numbers of transistors are mutually exclusive events, so that

$$p_0 + p_1 + p_2 + p_3 = 1.$$

Note that no more than 3 defective transistors can be drawn. Thus,

15. An old sea chest contains 10 bars of silver and 5 bars of gold. If 3 bars are drawn at random, what is the probability that exactly one is gold?

the probability of drawing at least one defective transistor, $p_{i \geqslant 1}$, is given by

$$p_{i \geqslant 1} = p_1 + p_2 + p_3 = 1 - p_0.$$

We have a choice of computing the three probabilities p_1, p_2, p_3, or the one probability p_0 and subtracting from 1. Since computing p_0 not only involves less work, but less work in turn involves less opportunity for error, we shall compute p_0.

Remember that this is a combination (unordered) rather than a permutation (ordered), since the order in which the transistors is drawn is not relevant. Thus, the number of ways 3 transistors can be drawn out of a total of 20 is $\binom{20}{3}$, and the number of ways 3 nondefective transistors can be drawn is $\binom{15}{3}$, since there are $20 - 5$ or 15 nondefective transistors. Thus, the probability p_0 is:

$$p_0 = \frac{\binom{15}{3}}{\binom{20}{3}} = \frac{\dfrac{15 \cdot 14 \cdot 13}{3 \cdot 2 \cdot 1}}{\dfrac{20 \cdot 19 \cdot 18}{3 \cdot 2 \cdot 1}} = \frac{15 \cdot 14 \cdot 13}{20 \cdot 19 \cdot 18} = \frac{91}{228}$$

and

$$p_{i \geqslant 1} = 1 - p_0 = 1 - \frac{91}{228} = \frac{137}{228}.$$

Even though it is more work, consider obtaining $p_{i \geqslant 1}$ from

$$p_{i \geqslant 1} = p_1 + p_2 + p_3.$$

Now, p_1 is the ratio of the number of ways one defective and two nondefective transistors can be drawn, or

$$p_1 = \frac{\binom{5}{1}\binom{15}{2}}{\binom{20}{3}} = \frac{105}{228}.$$

Similarly,

$$p_2 = \frac{\binom{5}{2}\binom{15}{1}}{\binom{20}{3}} = \frac{30}{228}, \quad \text{and} \quad p_3 = \frac{\binom{5}{3}\binom{15}{0}}{\binom{20}{3}} = \frac{2}{228}.$$

Thus,

$$p_{i \geqslant 1} = \frac{105}{228} + \frac{30}{228} + \frac{2}{228} = \frac{137}{228}, \quad \text{as before.}$$

DO EXERCISE 15.

EXERCISE SET 5.2

1. a) Suppose $p(E_1) = 0.73$, $p(E_2) = 0.24$, and $p(E_1 \cap E_2) = 0.20$. Find $p(E_1 \cup E_2)$.
   b) Suppose $p(E_1) = \frac{2}{7}$ and $p(E_2) = \frac{5}{14}$ and E_1 and E_2 are mutually exclusive. Find $p(E_1 \cup E_2)$.

2. a) Suppose $p(E_1) = 0.46$, $p(E_2) = 0.50$, and $p(E_1 \cap E_2) = 0.35$. Find $p(E_1 \cup E_2)$.
   b) Suppose $p(E_1) = \frac{3}{8}$ and $p(E_2) = \frac{5}{15}$ and E_1 and E_2 are mutually exclusive. Find $p(E_1 \cup E_2)$.

3. Two fair dice are rolled and the sum of the numbers showing is noted. What is the probability that the sum is 7? (Express this as the union of mutually exclusive events).

4. As in Exercise 3, what is the probability that the sum is even?

5. *Manufacturing, Quality Control.* A crate of 20 machine parts contains 3 defective parts. Two parts are drawn at random. What is the probability that:

a) Neither is defective?
b) One is defective (the other could be either defective or not defective)?
c) Only one is defective?
d) Both are defective?

6. *Public Health.* If 5 workers in a group of 25 have mononucleosis and 3 workers are chosen at random, what is the probability that:

a) None have mono?
b) All have mono?
c) Only one has mono? (1 has mono, 2 do not)
d) At least one has mono? (See Example 5 for a hint.)

7. There are 5 married couples in a room. If two people are chosen at random, what is the probability that:

a) One is a man and the other a woman?
b) They are of the same sex? (2 men or 2 women)
c) They are married (to each other)?

8. There are 3 married couples and 3 unmarried couples in a room. If two people are chosen at random, what is the probability that:

a) They are of the opposite sex?
b) They are of the same sex?
c) They are married?
d) They are married to each other?

9. Given a standard deck of cards. Two cards are dealt from a shuffled deck. What is the probability that:

a) Both are aces?
b) They are a pair?
c) Both are the same suit?
d) They are neither a pair nor the same suit?

10. As in Exercise 9, what is the probability that:

a) they are a pair but not aces?
b) One is an ace and the other is of the same suit?
c) Both are the same color?
d) They are a pair of different colors?

11. Two couples (four people) go to a theatre and find a row of six seats. If all are seated at random, what is the probability that a given couple will sit together?

*12. Two couples (four people) go to a restaurant and are seated at a round table with six seats. If all are seated at random, what is the probability that a given couple will sit together? [*Hint.* Look up circular permutations, Section 4.4, p. 207.]

13. If the letters of the word "hooch" are scrambled and reassembled by a chimpanzee (that is, randomly), what is the probability that they spell "hooch" correctly? [*Hint.* See Section 4.4, p. 207, and/or Section 4.5, p. 216. Compare methods.]

14. A little red wagon consists of a base with four wheels and a handle. Each wheel is held on the axle with a cotter pin. The handle is attached with a larger cotter pin. If the wheels and handle are removed, what is the probability that a chimpanzee without instructions (that is, at random) will put the wagon together correctly?

OBJECTIVES

You should be able to:

a) Given that E_1 and E_2 are independent, and $p(E_1)$ and $p(E_2)$, find $p(E_1 \cap E_2)$ as $p(E_1) \cdot p(E_2)$.

b) Compute probabilities involving independent events.

5.3 INDEPENDENT EVENTS—MULTIPLICATION THEOREM

In Section 8.2 we considered, among other things, probabilities of disjunctions

$$p(E_1 \cup E_2),$$

where the joint probability $p(E_1 \cap E_2) = 0$. Such events were *mutually exclusive*. That is, we could compute the probability by adding the respective probabilities:

$$p(E_1 \cup E_2) = p(E_1) + p(E_2).$$

Now we want to consider probabilities of conjunctions

$$p(E_1 \cap E_2).$$

We shall see that, under certain conditions, we can compute these probabilities by multiplying the respective probabilities:

$$p(E_1 \cap E_2) = p(E_1) \cdot p(E_2).$$

Let us consider some examples.

Example 1   A black die and a white die are rolled. What is the probability that a 5 is obtained on the black die and an odd number is obtained on the white die?

Solution   Let

E_1 = the event of a 5 on the black die and

E_2 = the event of an odd number on the white die.

a) A 5 is obtained on the black die in 1 out of 6 outcomes, so

$$p(5 \text{ on black}) = p(E_1) = \tfrac{1}{6}.$$

b) An odd number is obtained on the white die in 3 of the 6 outcomes, so

$$p(\text{Odd on white}) = p(E_2) = \tfrac{3}{6} = \tfrac{1}{2}.$$

c) Look at the sample space in Example 8 of Section 5.1. there are 3 outcomes where a 5 on black and an odd number on white occur, so

$$p(5 \text{ on black and odd on white}) = p(E_1 \cap E_2) = \tfrac{3}{36} = \tfrac{1}{12}.$$

Note that this same result can be obtained from the product of the individual probabilities:

$$p(5 \text{ on black } and \text{ odd on white}) = p(5 \text{ on black}) \cdot p(\text{Odd on white})$$

or

$$p(E_1 \cap E_2) = p(E_1) \cdot p(E_2)$$
$$= \tfrac{1}{6} \cdot \tfrac{1}{2} = \tfrac{1}{12}.$$

DO EXERCISE 16.

In Example 1 E_1 (the event of a 5 on the black die) is *independent* of E_2 (the event of an odd number on the white die.)

Not all events are independent, however. Consider the probability of rain on either of two particular days. Weather patterns being what they are, the weather on the second of two successive days is very much influenced by the weather on the preceding day.* Thus, the weather on the second day is *dependent* (to some extent) on the weather on the first day. On the other hand if the two days are sufficiently separated, say Easter and Christmas, then the weather on the second day is not likely to be influenced by the weather on the first day. Thus, in this case the weather on the second day is (essentially) *independent* of the weather on the first day. Dependent events are considered further later.

The independence of two events usually has to be inferred from the nature of the problem.

For two events E_1 and E_2 which do *not* affect each other, we have

MULTIPLICATION THEOREM FOR INDEPENDENT EVENTS. If neither of the events E_1 and E_2 affects the other, we say that the events are *independent*. Then

$$p(E_1 \cap E_2) = p(E_1) \cdot p(E_2).$$

The probability for both E_1 and E_2 to occur is the product of their individual probabilities.

DO EXERCISES 17 AND 18.

Example 2   A fair coin is flipped and two fair dice are rolled. What is the probability for a tail to show on the coin *and* the total, or sum, on the dice to be 7?

Solution   The flipping of the coin does not affect the sum on the dice nor does the sum on the dice affect the flipping of the coin (unless there was some unstated strange condition, such as that the dice and coins were glued together). Thus the events

$$E_1 = \text{Getting a tail}, \qquad E_2 = \text{The total is 7}$$

are independent.

a) $p(\text{Getting a tail}) = p(E_1) = \frac{1}{2}.$

---

*In fact, if you used today's weather as a prediction of tomorrow's, you would be correct about 80% of the time.

16. A black die and a white die are rolled.

a) What is the probability that a 2 is obtained on the black die?

b) What is the probability that an even number is obtained on the white die?

c) What is the probability that a 2 is obtained on the black die *and* an even number on the white die?

d) Does

$p(2 \text{ on black and even on white}) = p(2 \text{ on black}) \cdot p(\text{Even on white})$?

17. A sack contains 4 red and 4 yellow marbles. One marble is drawn and replaced, and a second is drawn. What is the probability that the

a) first is red?

b) second is yellow?

c) first is red *and* the second is yellow?

18. Given that E_1 and E_2 are independent, find $p(E_1 \cap E_2)$ where

a) $p(E_1) = \frac{3}{4}, \qquad p(E_2) = \frac{2}{9};$

b) $p(E_1) = 0.44, \qquad p(E_2) = 0.3.$

19. A fair dime and nickel are flipped and one fair die is rolled. What is the probability that the coins show a head on the dime and a tail on the nickel and the die shows a 4?

20. A die is rolled three times. What is the probability of getting a 2 on all three rolls?

21. A coin is flipped four times. What is the probability that the flips come out in the order H, H, T, H?

b) $p(\text{Total is } 7) = p(E_2) = \frac{1}{6}$. See the sample space in Example 8 of Section 8.1.

c) Thus,

$p(\text{Getting a tail } and \text{ the total is } 7) = p(\text{Getting a tail}) \cdot p(\text{Total is } 7)$

or

$$p(E_1 \cap E_2) = p(E_1) \cdot p(E_2)$$
$$= \frac{1}{2} \cdot \frac{1}{6}$$
$$= \frac{1}{12}.$$

DO EXERCISE 19

Sometimes we have to infer the "and" quality of an event from the wording of a problem.

Example 3 A die is rolled four times. What is the probability of getting a 5 on all four rolls?

Solution The event

$$E = \text{Getting a 5 on all four rolls}$$

can be reexpressed with "and" as

$E = (\text{1st roll 5}) \text{ and } (\text{2nd roll 5}) \text{ and } (\text{3rd roll 5}) \text{ and } (\text{4th roll 5}).$

Each of these four events is independent of the others, so that from the Multiplication Theorem

$$p(E) = \frac{1}{6} \cdot \frac{1}{6} \cdot \frac{1}{6} \cdot \frac{1}{6} = \frac{1}{1296}.$$

DO EXERCISE 20.

Example 4 A coin is flipped five times. What is the probability that the flips come out in the order H, T, T, H, T?

Solution The event

$$E = \text{flips come out in the order H, T, T, H, T}$$

can be reexpressed with "and" as

$$E = (\text{1st flip H}) \text{ and } (\text{2nd flip T}) \text{ and } (\text{3rd flip T}) \text{ and }$$
$$(\text{4th flip H}) \text{ and } (\text{5th flip T}).$$

Each of these five events is independent of the others, so that, from the Multiplication Theorem

$$p(E) = \frac{1}{2} \cdot \frac{1}{2} \cdot \frac{1}{2} \cdot \frac{1}{2} \cdot \frac{1}{2} = \frac{1}{32}.$$

DO EXERCISE 21.

Example 5   The probability that a man will live another 20 years is $\frac{1}{5}$ and the probability that his wife will live another 20 years is $\frac{1}{4}$. What is the probability that:

a) Both will live another 20 years?
b) Neither will live another 20 years?
c) Only one will live another 20 years?

Solution   Let

$$E_1 = \text{Event that the man lives another 20 years,}$$
$$E_2 = \text{Event that the woman lives another 20 years.}$$

a) Assuming that the longevity of each is independent (this assumption might be questioned, especially in the case of older people, where the death of one mate sometimes affects the death of the other), the joint probability that *both* will be alive in 20 years can be expressed as

$$p_{\text{Both}} = p(E_1 \text{ and } E_2) = p(E_1 \cap E_2),$$

and from the Multiplication Theorem is given by

$$p_{\text{Both}} = p(E_1 \cap E_2) = p(E_1) \cdot p(E_2) = \tfrac{1}{5} \cdot \tfrac{1}{4} = \tfrac{1}{20}.$$

b) That *neither* lives another 20 years can be expressed

(The man does not live another 20 years)

and

(The woman does not live another 20 years)

or, in set language,

$$E_1^c \cap E_2^c.$$

Assuming these events independent, we have from the Multiplication Theorem

$$\begin{aligned}
p_{\text{Neither}} = p(E_1^c \cap E_2^c) &= p(E_1^c) \cdot p(E_2^c) \\
&= [1 - p(E_1)] \cdot [1 - p(E_2)] \\
&= (1 - \tfrac{1}{5})(1 - \tfrac{1}{4}) = \tfrac{4}{5} \cdot \tfrac{3}{4} \\
&= \tfrac{3}{5}.
\end{aligned}$$

c) We look at this two ways. That only one (either) will live another 20 years can be expressed as

(The man is alive and the woman is not)

or

(The man is not alive and the woman is),

or, in set language,

$$(E_1 \cap E_2^c) \cup (E_1^c \cap E_2).$$

22. The probability that a student passes Economics is $\frac{4}{5}$ and the probability that the student passes Finite Mathematics is $\frac{2}{3}$. Assuming no dependence between passing or failing one course and passing or failing the other course, what is the probability that the student:

a) Passes both courses?
b) Passes neither course?
c) Passes one of the courses?

Do this two ways.

Since the events $E_1 \cap E_2^c$ and $E_1^c \cap E_2$ are mutually exclusive, both cannot occur at the same time. Thus,

$$p_{\text{Either}} = p(E_1 \cap E_2^c) + p(E_1^c \cap E_2).$$

Now we can assume E_1 and E_2^c to be independent, and E_1^c and E_2 to be independent, so

$$
\begin{aligned}
p_{\text{Either}} &= p(E_1 \cap E_2^c) + p(E_1^c \cap E_2) \\
&= p(E_1) \cdot p(E_2^c) + p(E_1^c) \cdot p(E_2) \\
&= \tfrac{1}{5} \cdot \tfrac{3}{4} + \tfrac{4}{5} \cdot \tfrac{1}{4} = \tfrac{3}{20} + \tfrac{4}{20} = \tfrac{7}{20}.
\end{aligned}
$$

Considering this another way, the probability that *either* will be alive in 20 years, plus the probability that *both* will be alive, plus the probability that *neither* will be alive exhausts the possibilities, so that

$$p_{\text{Either}} + p_{\text{Both}} + p_{\text{Neither}} = 1,$$

and

$$
\begin{aligned}
p_{\text{Either}} &= 1 - p_{\text{Both}} - p_{\text{Neither}} \\
&= 1 - \tfrac{1}{20} - \tfrac{3}{5} \\
&= \tfrac{7}{20}.
\end{aligned}
$$

DO EXERCISE 22.

Let us solve the problem in Example 5 using trees. Since either the man lives another 20 years, or he does not, the events E_1 and E_1^c are *mutually exclusive*, so that they can be represented along different branches of a tree, as in Section 5.2.

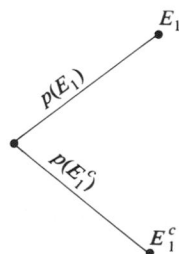

Since the events E_2 and E_2^c are also mutually exclusive whether or not the woman lives another 20 years, we can represent these events as *different branches* starting at the end of the previous branches,

that is

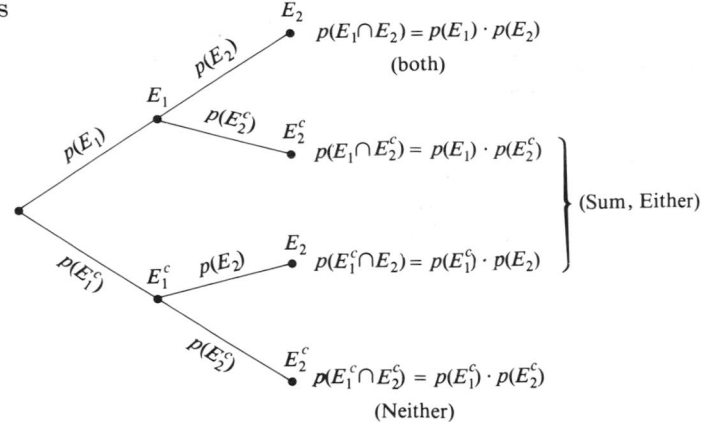

23. Solve Margin Exercise 22 using trees.

Note that since events E_1 and E_2 are independent, either event could have been written first on the tree.

From the Multiplication Theorem, the joint probability for two events (along the *same branch* of the tree) to both happen is the *product* of their individual probabilities as indicated to the right of the tree.

Using numerical values, we have:

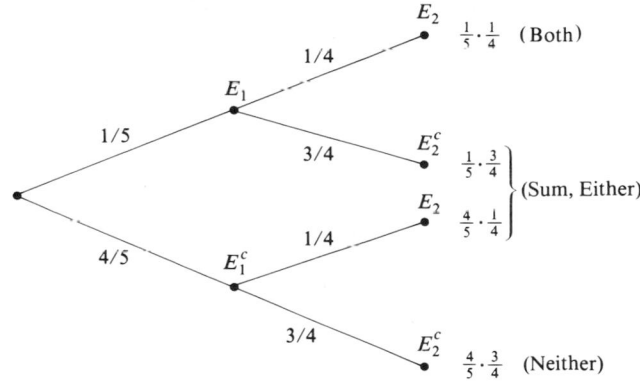

Thus, as before

$$p_{\text{Both}} = p(E_1 \cap E_2) = \tfrac{1}{5} \cdot \tfrac{1}{4} = \tfrac{1}{20},$$

$$p_{\text{Neither}} = p(E_1^c \cap E_2^c) = \tfrac{4}{5} \cdot \tfrac{3}{4} = \tfrac{3}{5},$$

and

$$p_{\text{Either}} = p(E_1 \cap E_2^c) + p(E_1^c \cap E_2) = \tfrac{1}{5} \cdot \tfrac{3}{4} + \tfrac{4}{5} \cdot \tfrac{1}{4} = \tfrac{7}{20}.$$

DO EXERCISE 23.

Example 6 *Manufacturing, Quality Control.* A box contains 20 transistors, 5 of which are defective. An inspector takes out 1 transistor at random, examines it for defects, and replaces it. After it has been replaced, another inspector does the same thing, and then so does a third inspector. What is the probability that at least one of the examiners finds a defective transistor? (Compare with Example 2, Section 5.2.)

Solution   The probability that at least one of the examiners finds a defective transistor, denoted $p_{i \geq 1}$, plus the probability that none of them finds a defective transistor, denoted p_0, sums to 1 because the events partition the sample space, so that

$$p_{i \geq 1} = 1 - p_0.$$

Suppose we use this fact and try to compute p_0 to then get $p_{i \geq 1}$. This time we must allow for *replacement*. The probability that a defective transistor is not drawn in the first trial is $\frac{15}{20}$. Since each trial is *independent* and *identical*, the probability that no defective transistors are drawn in three trials is the product

$$p_0 = \tfrac{15}{20} \cdot \tfrac{15}{20} \cdot \tfrac{15}{20} = (\tfrac{3}{4})^3 = \tfrac{27}{64},$$

so that $p_{i \geq 1} = 1 - \frac{27}{64} = \frac{37}{64}$.

This problem can also be solved using a tree. If D_1 is the event that the first transistor drawn is defective and D_1^c is the event that it is not defective, then (as in Example 5) we can draw the following tree:

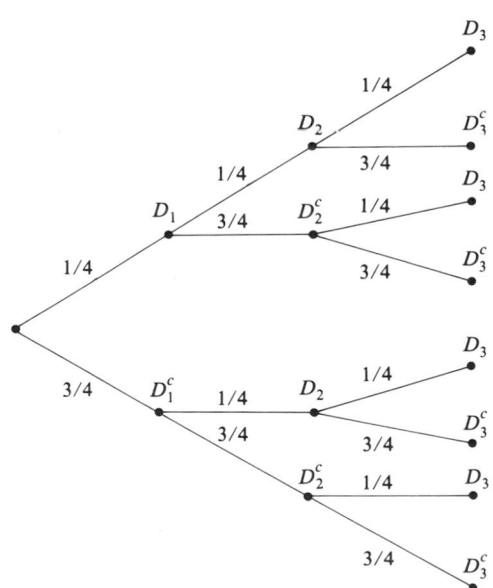

The probability that at least one defective transistor is drawn is the probability that:

(The first is defective) plus the probability that
(The first is not defective but the second is) plus the probability that
(The first and second are not defective but the third is). That is,

$$p_{i \geqslant 1} = p(D_1) + p(D_1^c \cap D_2) + p(D_1^c \cap D_2^c \cap D_3)$$

$$= \quad \tfrac{1}{4} \quad + \quad \tfrac{3}{4} \cdot \tfrac{1}{4} \quad + \quad \tfrac{3}{4} \cdot \tfrac{3}{4} \cdot \tfrac{1}{4}$$

$$= \tfrac{37}{64}, \quad \text{as before.}$$

DO EXERCISE 24.

(Optional) The Birthday Problem

THE BIRTHDAY PROBLEM. Of n people in a group, what is the probability that at least two of them have the same birthday (day and month, but not necessarily the same year)?

If $p(E)$ is the probability that at least two people have the same birthday,

then $p(E^c)$ is the probability that no two people have the same birthday and

$$p(E) = 1 - p(E^c).$$

The probability $p(E^c)$ will be evaluated two ways.

i) From Section 7.1,

$$p(E^c) = \frac{\mathcal{N}(E^c)}{\mathcal{N}(S)}.$$

One person can have a birthday on any of 365 days (ignoring Leap Year possibilities). Two people can have birthdays on 365^2 days. Thus, in general, n people can have birthdays on 365^n days, so that

$$\mathcal{N}(S) = 365^n.$$

Now if the second person has a birthday other than that of the first person, he has 364 possibilities. The third person, has then 363 possibilities. In general, the nth person has $[365 - (n - 1)]$ possibilities, so that the n people have $\mathcal{N}(E^c)$ possibilities, where

$$\mathcal{N}(E^c) = 365 \cdot 364 \cdot 363 \cdots [365 - (n - 1)].$$

Thus,

$$p(E^c) = \frac{365 \cdot 364 \cdot 363 \cdots [365 - (n - 1)]}{365^n}.$$

24. In Example 6, what is the probability that at least two of the examiners find a defective transistor?

25. From Table 5.1, determine the probability that two or more students in your class have the same birthday. If your class is large (more than 40 students), divide the class into two groups and examine the results for each half as well as for the whole class.

ii) Alternately, let E_n be the event that the nth person does not have a birthday in common with the preceding $(n - 1)$ people. Thus,

$$p(E_n) = \frac{[365 - (n - 1)]}{365}.$$

(Note that $p(E_1) = 365/365 = 1$ is the probability that the first person has no common birthday with his predecessors, of which he has none.) Since each person's birthday is independent of another person's birthday, the events E_n are also independent, so that from *this* section

$$
\begin{aligned}
p(E^c) &= p(E_1 \cap E_2 \cap \cdots \cap E_n) \\
&= p(E_1) \cdot p(E_2) \cdots p(E_n) \\
&= \frac{365}{365} \cdot \frac{364}{365} \cdots \frac{[365 - (n - 1)]}{365},
\end{aligned}
$$

which is equivalent to the first value for $p(E^c)$.

Evaluating $p(E) = 1 - p(E^c)$, we obtain Table 5.1.

Table 5.1.
Probability for two or more people to have the same birthday.

n	$p(E)$	n	$p(E)$	n	$p(E)$	n	$p(E)$	n	$p(E)$
2	0.00274	16	0.284	24	0.538	32	0.753	40	0.891
5	0.0271	17	0.315	25	0.569	33	0.775	50	0.970
10	0.117	18	0.347	26	0.598	34	0.795	60	0.9951
11	0.141	19	0.379	27	0.627	35	0.814	70	0.99916
12	0.167	20	0.411	28	0.654	36	0.832	80	0.999914
13	0.194	21	0.444	29	0.681	37	0.849	90	0.999994
14	0.223	22	0.476	30	0.706	38	0.864	100	0.9999997
15	0.253	23	0.507	31	0.732	39	0.878		

For 23 or more people, the probability that two or more people will have the same birthday is greater than $\frac{1}{2}$, that is, it is more likely to occur than not!

DO EXERCISE 25.

In situations where class sizes are large, it may be of interest to consider:

THE EXTENDED BIRTHDAY PROBLEM. **Of n people in a group,**

what is the probability that at least k of them have the same birthday (day and month, but not necessarily the same year)?

We have just solved this for $k = 2$.

Let

$p_n(k)$ = the probability that of n people in a group, there is at least one k-tuplet but no higher multiplets,

and

$p_n(\geqslant k)$ = the probability that of n people in a group, there is at least one k-tuplet.

Thus,

$p_n(1)$ = what we previously called $p(E^c)$

and

$p_n(\geqslant 2)$ = what we previously called $p(E)$;

so that

$$p_n(\geqslant 2) = 1 - p_n(1).$$

The Birthday Problem can be restated as seeking an expression for $p_n(\geqslant k)$.

We can express $p_n(\geqslant k)$ as

$$p_n(\geqslant k) = p_n[\geqslant (k-1)] - p_n(k-1).$$

Thus, for $k = 3$

$$p_n(\geqslant 3) = p_n(\geqslant 2) - p_n(2).$$

We have $p_n(\geqslant 2)$ and need only calculate $p_n(2)$ to solve the Birthday Problem for $k = 3$.

Similarly, for $k = 4$

$$p_n(\geqslant 4) = p_n(\geqslant 3) - p_n(3).$$

Once we have $p_n(\geqslant 3)$, we need calculate only $p_n(3)$ to be able to solve the Birthday Problem for $k = 4$, and so on for any value of k.

Consider now the Birthday Problem for $k = 3$.

Let

$p_n(2, m)$ = the probability that of n people in a group, there are m pairs of people with the same birthday and no more than two people with the same birthday.

The probability $p_n(2, m)$ can be obtained by seeking the probability for m pairs with the same birthday in one configuration (for example, no pairs in the first $n - m$ people and the last matches the first and the next-to-last matches the second, and so on) times the number of possible configurations (how many ways m pairs can be distributed among n people):

$$p_n(2, m) = \frac{p_{n-m}(1)}{365^m} \cdot \frac{\binom{n}{2m}}{k!(2!)^k}.$$

The probability $p_n(2)$ is then the sum over all possible values of m:

$$p_n(2) = \sum_{m=1}^{m_L} p_n(2, m),$$

where the upper limit of the sum is

$$m_L = \text{the maximum possible number of pairs}$$

$$= \text{the largest integer} \leqslant \frac{n}{2}.$$

This computation is rather awkward in the form given. However, let

$$A_n(m) = \frac{(n - 2m + 2)(n - 2m + 1)}{2m(365 + m - n)};$$

then after some algebra, we can obtain

$$p_n(2) = p_n(1)A_n(1)[1 + A_n(2)(1 + A_n(3)(1 + \cdots (1 + A_n(m_L)))) \cdots].$$

In this form the computations were performed by one of the authors (JCC) on an HP–25. This handheld calculator has 49 programmable steps and for large values of n the program ran 2 to 3 minutes.

Table 5.2, showing values of $p_n(\geqslant 3)$, was obtained.

Table 5.2.
Probability that three or more people have the same birthday.

n	$p_n(\geqslant 3)$	n	$p_n(\geqslant 3)$	n	$p_n(\geqslant 3)$
20	0.00824	88	0.51107	160	0.98216
30	0.02853	90	0.53420	170	0.99169
40	0.06689	100	0.64586	180	0.99645
50	0.12638	110	0.74553	190	0.99862
60	0.20723	120	0.82796	200	0.999512
70	0.30649	130	0.89108	210	0.999844
80	0.41817	140	0.93570	220	0.9999552
87	0.49945	150	0.96477	240	0.99999999

Thus, for 88 or more people it is more likely than not for *three* or more people in a group to have the same birthday.

In a similar manner (see Exercise 30), we can obtain $p_n(3)$ and the following table of $p_n(\geqslant 4)$:

Table 5.3.
Probability that four or more people have the same birthday.

n	$p_n(\geqslant 4)$	n	$p_n(\geqslant 4)$	n	$p_n(\geqslant 4)$
50	0.00428	140	0.21109	210	0.65703
60	0.00884	150	0.26463	220	0.71873
70	0.01621	160	0.32390	230	0.77493
80	0.02723	170	0.38777	240	0.82737
90	0.04275	180	0.45480	250	0.88904
100	0.06358	186	0.49583	260	0.94889
110	0.09041	187	0.50269	270	0.98618
120	0.12380	190	0.52327	280	0.99751
130	0.16403	200	0.59130	290	0.99966

Thus, for 187 or more people it is more likely than not for *four* or more people to have the same birthday.

Effect of Leap Years

So far we have assumed 365 days in a year. Actually, 3 out of 4 years have 365 days and 1 out of 4 years has 366 days (to a good approximation).

Let

$p_{n,365} = p_n$ based on a 365-day year (these values are the ones previously obtained),

$p_{n,366} = p_n$ based on a 366-day year (these values can be obtained from the preceding formulas by replacing 365 by 366),

and

$$p_{n,\mathrm{L}} = p_n \text{ allowing for leap years.}$$

Then, to this approximation, we have

$$p_{n,\mathrm{L}} = \frac{3}{4} p_{n,365} + \frac{1}{4} p_{n,366}.$$

Consider $k = 3$. Then

$$p_{87,365}(\geqslant 3) = 0.49945 \qquad \text{(See Table 5.2)}$$

and

$$p_{87,\text{L}}(\geqslant 3) = 0.49901.$$

Thus, the leap-year correction does not change the results appreciably nor the break-even point of 88.

Similarly, for $k = 2$ and $k = 4$ the leap-year correction does not change the results appreciably nor the break-even points of 23 and 187.

Effect of Twins

Twins are born with a frequency of one pair in approximately 89 pregnancies; thus the probability of twins per pregnancy is

$$p^{(2)} \approx \frac{1}{89}.$$

If $p^{(k)}$ is the probability of k-tuplets per pregnancy, then Hellin's law states that

$$p^{(3)} \approx [p^{(2)}]^2,$$
$$p^{(4)} \approx [p^{(2)}]^3,$$

and, in general,

$$p^{(k)} \approx [p^{(2)}]^{k-1} \quad \text{for } k \geqslant 3.$$

The twinning rate is sufficiently small that for determining the break-even point of the Birthday Problem, we need not consider multiple births other than twins and a single occurrence of twins at that.

If twins are born once in about 89 pregnancies, or

$$p^{(2)} \approx \frac{1}{89},$$

then one pair of twins will be found in about 90 people, or

$$p_{\text{T}} \approx \frac{1}{90}.$$

The probability for two or more people to have the same birthday *allowing for twins* is given by

$$\tilde{p}(\geqslant 2) = (1 - p_{\text{T}})p_n(\geqslant 2) + p_{\text{T}}$$
$$= p_n(\geqslant 2) + p_{\text{T}}[1 - p_n(\geqslant 2)].$$

Thus, we have
$$p_{22}(\geqslant 2) = 0.47570$$
and obtain
$$\tilde{p}_{22}(\geqslant 2) = 0.48153;$$

so that, even though the presence of twins increases the probability for two or more people to have the same birthday, the break-even point of 23 is not changed.

The probability for *three* or more people to have the same birthday *allowing for twins* is given by

$$\tilde{p}_n(\geqslant 3) = (1 - p_T)p_n(\geqslant 3) + p_T p_n^T(\geqslant 3)$$
$$= p_n(\geqslant 3) + p_T[p_n^T(\geqslant 3) - p_n(\geqslant 3)],$$

where the superscript "T" indicates that the n people include one pair of twins, and
$$p_n^T(\geqslant 3) = 1 - p_n^T(2).$$

Here $p_n^T(2)$ can be obtained in a manner similar to that used for $p_n(2)$. (See Exercise 31.)

Thus, we obtain
$$p_{87}(\geqslant 3) = 0.49945,$$
$$p_{87}^T(\geqslant 3) = 0.58653,$$
and
$$\tilde{p}_{87}(\geqslant 3) = 0.50042;$$

so that the break-even point of 88, for three or more people to have the same birthday, drops to 87 with allowance for twins.

The break-even point of 187, for four or more people to have the same birthday, allowing for twins is considered in Exercise 32.

EXERCISE SET 5.3

Given that E_1 and E_2 are independent, find $p(E_1 \cap E_2)$:

1. $p(E_1) = \frac{7}{9}$, $p(E_2) = \frac{11}{14}$.

2. $p(E_1) = \frac{4}{5}$, $p(E_2) = \frac{3}{8}$.

3. $p(E_1) = 0.48$, $p(E_2) = 0.33$.

4. $p(E_1) = 0.77$, $p(E_2) = 0.101$.

5. One card is drawn from a well-shuffled deck of 52 and replaced, and a second card is drawn. What is the probability that:

a) The first is a spade?
b) The second is an ace?
c) The first is a spade and the second is an ace?

6. As in Exercise 5, what is the probability that

a) The first is a face card?
b) The second is a king?
c) The first is a face card and the second is a king?

7. One card is drawn from a well-shuffled deck of 52, but not replaced, and a second card is drawn.

a) What is the probability that the first card is a spade?

b) What is the conditional probability that the second is a diamond, given that the first is a spade?

c) What is the probability that the first is a spade and the second a diamond?

9. For an unfair coin, $p(H) = \frac{2}{3}$ and $p(T) = \frac{1}{3}$. The coin is flipped five times. What is the probability that the flips come out in the order T, H, H, T, H?

11. A student entering college has a $\frac{1}{3}$ probability for getting married while in college and a 0.6 probability of graduating. What is the probability that he will graduate married?

13. *Quality Control.* Candles are molded on a production line such that 10% are defective. What is the probability that a box of six are all good? All defective?

15. A class is one-third women and two-thirds men. Also, 60% are blond and 40% have dark hair. What is the probability that a person chosen at random is a blond woman?

17. Suppose your probability of passing a test over this chapter is $\frac{3}{4}$, and the probability of your passing a psychology test the same day is $\frac{5}{8}$. Assuming the events are independent, what is the probability that you:

a) Pass both tests?

b) Pass one test, but not the other?

c) Fail both tests?

19. *Quality Control.* A box contains 24 transistors, 6 of which are defective. An inspector takes out 1 transistor at random, examines it for defects, and replaces it. After it has been replaced, another inspector does the same thing, and then so does the third inspector. What is the probability that at least one of the examiners finds a defective transistor?

*21. *On the Birthday Problem.* Check an almanac for the birthdays of all the presidents of the U.S. How many presidents have there been? Do any two have the same birthday? Does this seem reasonable based on the table?

8. As in Exercise 7,

a) What is the probability that the first card is a face card?

b) What is the conditional probability that the second card is a four, given that the first card is a face card?

c) What is the probability that the first card is a face card and the second is a four?

10. For an unfair coin, $p(H) = \frac{4}{5}$ and $p(T) = \frac{1}{5}$. the coin is flipped three times. What is the probability that the flips come out in the order T, H, T?

12. Two pilots are trying to communicate with each other during a severe thunderstorm. If the probability for a malfunction of either transceiver is 25%, what is the probability that they *can* communicate?

14. *Quality Control.* A vintner is making three separate batches of wine. Due to circumstances beyond his control, the probability for success is 0.9 for any one batch. What is the probability all three are successful?

16. In a lake there are several kinds of fish of which 20% are pike. Of each kind 10% is tagged. What is the probability that a fish caught at random will be a tagged pike?

18. Suppose your probability of passing a test over this chapter is $\frac{5}{6}$, and the probability of your passing a sociology test the same day is $\frac{2}{3}$. Assuming the events are independent, what is the probability that you:

a) Pass both tests?

b) Pass one test, but not the other?

c) Fail both tests?

20. As in Exercise 19 but the box has 45 transistors of which 9 are defective.

*22. *On the Birthday Problem.* The probabilities regarding birthdays apply also to death dates. Check an almanac for the death dates of the U.S. presidents. Do any two have the same death date (excluding year)? Does this seem reasonable based on the table?

23. Democrats and Republicans are running for Congress (both Senate and House of Representatives). If the odds for the Democrat to win the Senate seat are 3:2 and the odds for the Republican to win the House seat are 5:4, what is the probability that both Democrats win? one Democrat and one Republican? Assume that one campaign does not affect the other. (*Hint.* Convert the odds to probabilities. For example, 3:2 converts to a probability of $\frac{3}{5}$.)

25. Five cards are dealt at random from a deck of 52 cards. What is the probability for:

a) A pair?
b) Three of a kind?
c) Two pairs?
d) A full house (a pair and three of a kind)?
e) Four of a kind?

(*Hint.* See Exercises 38 through 47 of Exercise Set 4.5.)

27. There are three identical urns containing purple and yellow balls. The first has 7 purple and 2 yellow, the second, 3 purple and 3 yellow, and the third, 2 purple and 4 yellow. An urn is chosen at **random** and a ball selected at random. What is the probability it is purple? Draw a tree.

29. A coin weighted so that heads is twice as likely as tails is to be flipped three times. What is the probability for at least two heads? Solve using a tree diagram.

▶

*30. *On the Birthday Problem.* Obtain an expression for $p_n(3)$.

*32. *On the Birthday Problem.* Is the break-even point of 187 for four or more people to have the same birthday changed by allowing for twins?

24. A baseball team is playing a doubleheader. If the odds are 4:3 to win the first game and 2:3 to win the second game, what is the probability that the team will win both games? only one game?

26. As in Exercise 25 (read the whole problem for definitions), what is the probability for:

a) A royal flush (an ace-high straight flush)?
b) A straight flush (not including an ace as the high card)?
c) A flush (five cards of the same suit, not all in sequence)?
d) A straight (five cards in sequence, not all the same suit)?

(*Hint.* See Exercises 38 through 47 of Exercise Set 4.5.)

28. As in Exercise 27, a second ball is drawn at random from a random choice of one of the two urns not selected in the first choice. What is the probability the ball is yellow?

*31. *On the Birthday Problem.* Obtain an expression for $p_n^{\mathrm{T}}(2)$.

5.4. THE ADDITION THEOREM

Addition Theorem—Two Events

So far we have studied

i) *Mutually exclusive events:* Two events E_1 and E_2 which cannot both happen at the same time, that is, $p(E_1 \cap E_2) = 0$;

OBJECTIVE

You should be able to calculate probabilities using the Addition Theorem (where events are independent or mutually exclusive, or neither) and trees.

ii) *Independent events:* The probability of *both* occurring is the *product* of their individual probabilities (Multiplication Theorem); that is,

$$p(E_1 \cap E_2) = p(E_1) \cdot p(E_2).$$

For mutually exclusive events, the probability for *either* to happen is the *sum* of the individual probabilities; that is,

$$p(E_1 \cup E_2) = p(E_1) + p(E_2).$$

We also considered briefly $p(E_1 \cup E_2)$ for the general case. Let us reconsider such probabilities.

In Section 5.2 we developed the following result.

THE ADDITION THEOREM. **For any events E_1 and E_2,**

$$p(E_1 \cup E_2) = p(E_1) + p(E_2) - p(E_1 \cap E_2).$$

The probability of either E_1 or E_2 is the probability of E_1 plus the probability of E_2 minus the probability of both E_1 and E_2.

Example 1 Suppose the probability that a student will pass Finite Mathematics is 68%, or 0.68, and the probability that the student will pass an Accounting course is 70%, or 0.70. The probability that the student will pass both courses is 64%, or 0.64. What is the probability that the student will pass either Finite Mathematics or Accounting, or both?

Solution Let

$$E_1 = \text{Event of passing Finite Mathematics,}$$
$$E_2 = \text{Event of passing Accounting.}$$

Then

$$E_1 \cap E_2 = \text{Event of passing both courses,}$$

and

$$E_1 \cup E_2 = \text{Event of passing Finite Mathematics}$$
$$\textit{or} \text{ Accounting (or both.)}$$

We are seeking $p(E_1 \cup E_2)$, and we already know $p(E_1)$, $p(E_2)$, and $p(E_1 \cap E_2)$. Thus, by the Addition Theorem, we have

$$p(E_1 \cup E_2) = p(E_1) + p(E_2) - p(E_1 \cap E_2)$$
$$= 0.68 + 0.70 - 0.64$$
$$= 0.74.$$

DO EXERCISE 26.

Note in Example 1 that the events E_1 and E_2 were not independent. That is,

$$(0.68)(0.70) = 0.476 \neq 0.64,$$

or

$$p(E_1) \cdot p(E_2) \neq p(E_1 \cap E_2).$$

Since E_1 and E_2 were *not* independent, the extra information about $p(E_1 \cap E_2)$ had to be supplied.

Now let us consider a problem where events are independent.

Example 2 If the probability for rain is 0.4 on April 1, and 0.3 on November 29, what is the probability for rain on either day?

Solution Let

$$E_1 = \text{The event of rain on April 1}$$

and

$$E_2 = \text{The event of rain on November 29.}$$

Then

$$E_1 \cup E_2 = \text{The event of rain on April 1 or November 29.}$$

From the wording of the problem, we may infer that E_1 and E_2 are independent. Thus, we determine $p(E_1 \cap E_2)$ from the Multiplication Theorem and $p(E_1 \cup E_2)$ from the Addition Theorem:

$$
\begin{aligned}
p(E_1 \cup E_2) &= p(E_1) + p(E_2) - p(E_1 \cap E_2) \\
&= p(E_1) + p(E_2) - p(E_1) \cdot p(E_2) \\
&= 0.4 + 0.3 - (0.4)(0.3) \\
&= 0.58.
\end{aligned}
$$

DO EXERCISE 27.

Using Trees to Solve Probability Problems

Let us see how we can use trees to facilitate solving problems.

In Example 3, we reconsider the Addition Theorem for the case where events are mutually exclusive, but now the events are themselves compound events.

26. Suppose the probability that a woman lives to be 70 is 0.81 and that she will go bald is 0.37. The probability that she will live to be 70 and also go bald is 0.28. What is the probability that she will live to be 70 or go bald?

27. A farmer faces a season of drought with a probability of 0.1 and an insect invasion with probability of 0.3.

a) What is the probability that either or both events will occur?

b) What is the probability that neither event will occur? [*Hint*.

$$p(E_1^c \cap E_2^c) = p(E_1^c) \cdot p(E_2^c).]$$

Example 3 Experience shows that, with fatal accidents involving two cars, the probability that neither driver is drunk is $\frac{1}{4}$, and the probability that both are drunk is $\frac{1}{8}$. What is the probability that only one driver is drunk?

Solution Let

$$E_1 = \text{Event that the } \textit{first} \text{ driver is drunk,}$$

and

$$E_2 = \text{Event that the } \textit{second} \text{ driver is drunk.}$$

Now the event that only one driver is drunk translates to:

(First drunk *and* the second not drunk)

or

(First not drunk *and* the second drunk),

or, using set language, we have

$$(E_1 \cap E_2^c) \cup (E_1^c \cap E_2).$$

This is a union of mutually exclusive events $E_1 \cap E_2^c$ and $E_1^c \cap E_2$. We are seeking the probability, p_1, of this event:

$$p_1 = p[(E_1 \cap E_2^c) \cup (E_1^c \cap E_2)] = p(E_1 \cap E_2^c) + p(E_1^c \cap E_2).$$

To find this, consider a tree.

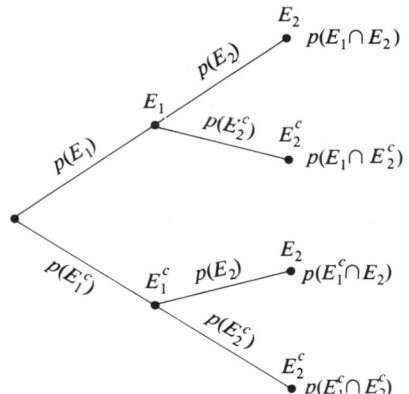

From the fact that $E_1 \cap E_2^c$ and $E_1^c \cap E_2$ represent *different* branches of the tree, we could have deduced that they are mutually exclusive events. But we also get more information from the tree, by noting that:

$$E_1 \cap E_2 = \text{The event that both are drunk,}$$
$$E_1^c \cap E_2^c = \text{The event that neither is drunk.}$$

Then since $E_1 \cap E_2$, $E_1 \cap E_2^c$, $E_1^c \cap E_2$, and $E_1^c \cap E_2^c$ represent *all* branches of the tree, they fill the sample space, so:

$$p(E_1 \cap E_2) + p(E_1 \cap E_2^c) + p(E_1^c \cap E_2) + p(E_1^c \cap E_2^c) = 1$$

or

$$p(E_1 \cap E_2) + p_1 + p(E_1^c \cap E_2^c) = 1,$$

so

$$p_1 = 1 - [p(E_1 \cap E_2) + p(E_1^c \cap E_2^c)]$$
$$= 1 - (\tfrac{1}{8} + \tfrac{1}{4})$$
$$= \tfrac{5}{8}.$$

Note that we never actually determined $p(E_1)$ or $p(E_2)$, nor can we from the data given (unless we assume that $p(E_1) = p(E_2)$).

We can also solve this problem using a Venn diagram.

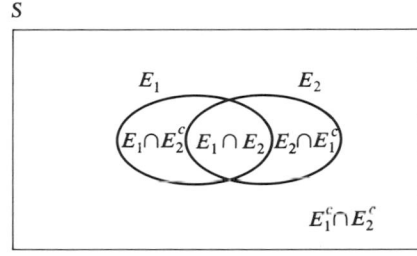

Putting in the available information we obtain:

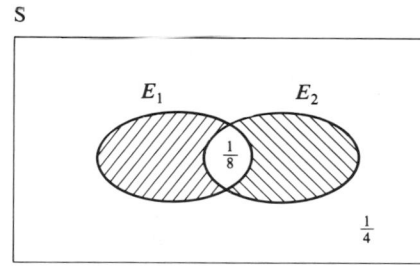

The probability that only one driver is drunk corresponds to the shaded area, and is equal to

$$p_1 = 1 - (\tfrac{1}{8} + \tfrac{1}{4})$$
$$= \tfrac{5}{8}, \quad \text{as before.}$$

DO EXERCISE 28.

28. *Public Health.* A country is afflicted with two kinds of flu during one winter. The probability that a person will get both kinds of flu is 0.43. The probability that a person will get only one kind of flu is 0.51. What is the probability that a person will get neither kind of flu?

(Optional) The Addition Theorem—Three or More Events

The Addition Theorem for two events E_1 and E_2 is:

$$p(E_1 \cup E_2) = p(E_1) + p(E_2) - p(E_1 \cap E_2).$$

This theorem can be extended to three events, E_1, E_2, and E_3, by replacing E_2 in the above relation by $(E_2 \cup E_3)$, yielding

$$p(E_1 \cup E_2 \cup E_3) = p(E_1) + p(E_2 \cup E_3) - p[E_1 \cap (E_2 \cup E_3)].$$

From a Venn diagram, we can verify the relation

$$p[E_1 \cap (E_2 \cup E_3)] = p[(E_1 \cap E_2) \cup (E_1 \cap E_3)].$$

Applying the Addition Theorem for two events, we obtain the Addition Theorem for three events:

THE ADDITION THEOREM FOR THREE EVENTS. **For any events E_1, E_2 and E_3,**

$$p(E_1 \cup E_2 \cup E_3) = p(E_1) + p(E_2) + p(E_3)$$
$$- p(E_1 \cap E_2) - p(E_1 \cap E_3) - p(E_2 \cap E_3)$$
$$+ p(E_1 \cap E_2 \cap E_3).$$

For more than three events, the Addition Theorem has the same pattern as the preceding with appropriate alternating changes of sign.

Example 4 a group of n men, each with a different type of car, would like to swap cars. So they toss their keys into a pile. Each man takes out a key at random. What is the probability that at least one man gets his own car key?

Solution Let E_i be the event that the ith man gets his own car back. Then the probability that at least one man out of n gets his own car back is

$$p_n = p(E_1 \cup E_2 \cup \cdots \cup E_n).$$

Using the Addition Theorem, this can be expanded into a series of terms representing events corresponding to one or more men getting their own cars back.

For $n = 1$, it is certain that the man will get his car back. That is, $p_1 = 1$.

Consider $n = 2$.

$$p_2 = p(E_1 \cup E_2) = p(E_1) + p(E_2) - p(E_1 \cap E_2).$$

Each man has a $\frac{1}{2}$ probability of getting his own car back; that is,

$$p(E_1) = p(E_2) = \tfrac{1}{2}.$$

But if one man gets his own car back, so must the other man, so that

$$p(E_1 \cap E_2) = \tfrac{1}{2}.$$

Thus,

$$p_2 = \tfrac{1}{2} + \tfrac{1}{2} - \tfrac{1}{2} = 1 - \tfrac{1}{2} = \tfrac{1}{2},$$

or

$$p_2 = p_1 - \tfrac{1}{2}.$$

For $n = 3$, we use the Addition Theorem for three events. The probability that any one man gets his own car back is:

$$p(E_i) = \tfrac{1}{3},$$

so that

$$p(E_1) + p(E_2) + p(E_3) = 1 = p_1.$$

Two men out of three can be chosen to get their cars back $\binom{3}{2}$ ways. A given pair can get their cars back with probability 1/3! Thus,

$$p(E_1 \cup E_2) + p(E_1 \cup E_3) + p(E_2 \cup E_3) = \binom{3}{2} \cdot \frac{1}{3!} = \frac{1}{2!}.$$

All three men can get their own cars back with probability 1/3!, so that we have

$$p_3 = 1 - \frac{1}{2!} + \frac{1}{3!} - \frac{2}{3},$$

or

$$p_3 = p_2 + \frac{1}{3!}.$$

If we keep up this process, we find that:

$$p_n = p_{n-1} + (-1)^{n-1} \cdot \frac{1}{n!},$$

or

$$p_n = 1 - \frac{1}{2!} + \frac{1}{3!} - \frac{1}{4!} + \cdots + (-1)^{n-1} \cdot \frac{1}{n!}.$$

DO EXERCISE 29.

29.

a) Evaluate p_n for several increasing values of n. (*Note*. As n becomes increasingly large, p_n approaches the quantity

$$(1 - 1/e) = 0.632121\ldots,$$

e being $2.71828\ldots$, an important number in mathematics.

b) Pair off with someone, each person having a shuffled deck of cards. Each person turns up a first card, and the two cards are compared for a match. Then each turns up a second card, and these cards are compared for a match. This is continued until both have exhausted their cards. What approximate odds should you bet for at least one match? Did you get a match?

EXERCISE SET 5.4

Find $p(E_1 \cup E_2)$, where:

1. $p(E_1) = 0.67$, $p(E_2) = 0.65$, and $p(E_1 \cap E_2) = 0.61$.

2. $p(E_1) = 0.76$, $p(E_2) = 0.57$, and $p(E_1 \cap E_2) = 0.46$.

3. *Biology.* Examination of fruitflies indicates that 30% have an eye defect, 60% have a color variation, and 10% have both. What is the probability that a random fly will have either an eye defect or a color variation? [*Hint.* 30% have an eye defect, so the probability of an eye defect is 0.3.]

4. *Agriculture.* A farmer finds that 24% of his corn crop has a blight, and 30% has received insufficient rain, and 8% has both. What part of the crop had either blight or insufficient rain?

Find $p(E_1 \cup E_2)$, assuming E_1 and E_2 are independent.

5. $p(E_1) = \frac{3}{5}$, $p(E_2) = \frac{1}{3}$.

6. $p(E_1) = \frac{5}{8}$, $p(E_2) = \frac{2}{5}$.

7. If the probability for snow is 0.7 and the probability of having a fire in your home is 0.0006, what is the probability of snow *or* a fire in your home?

8. If the probability for a flood is 0.0004 and the probability of passing math is 0.73, what is the probability of a flood or passing math?

9. *Business.* The manager of a company is faced with the prospect of a snowstorm, which would keep 40% of the employees out, and a flu epidemic, which would keep 15% out. What percent of his employees should be expected to show up for work?

10. *Business.* A manufacturer is coming out with two new products. The first has a 70% chance of being successful and the second 80%. What is the probability that *either* will be successful?

11. *Public Affairs.* A survey indicates that 40% of car drivers do not wear seat belts. That is, the probability that a driver is wearing a seat belt is 0.6. In an accident between two cars what is the probability that neither driver was wearing a seat belt? just one driver? [*Hint.* Assume events independent, and draw a tree.]

12. If the probability for divorce in a marriage is $\frac{1}{3}$, what is the probability that two given couples will both stay married or both get divorced? [*Hint.* Draw a tree, assuming events independent.]

*__13.__ (This problem can be solved using the Addition Theorem for Three Events. It can also be solved—and more simply—using a Venn diagram.) Cars can be bought with any of the following options:

A) An engine package (higher horsepower, etc.)
B) A suspension package (heavy duty suspension, etc.)
C) An appointment package (vinyl seats, etc.)

If

65% buy option A,	30% buy options A and B
40% buy option B,	20% buy options A and C
40% buy option C,	10% buy options B and C
	10% buy options A, B, and C.

What percent of the car buyers buy no options at all?

OBJECTIVES

You should be able to

a) Compute conditional probabilities using the expression

$$p(E_2 \mid E_1) = \frac{p(E_1 \cap E_2)}{p(E_1)}$$

5.5 CONDITIONAL PROBABILITY— MULTIPLICATION THEOREM—DEPENDENT AND INDEPENDENT EVENTS

In the preceding sections we considered events E_1 and E_2 as *independent*. Now let us consider events E_1 and E_2 as *dependent*. The probability that E_1 occurs *provided* that E_2 occurs is written

$$p(E_1 \text{ provided } E_2) = p(E_1 \mid E_2).$$

Similarly, the probability that E_2 occurs *provided* that E_1 occurs is written

$$p(E_2 \text{ provided } E_1) = p(E_2 \,|\, E_1).$$

Example 1   A box contains 3 red billiard balls and 2 white billiard balls. One ball is selected, but not replaced, and a second is selected. What is the probability that the second is white given that the first is red?

Solution   Let

$$E_1 = \text{The event that the first ball drawn is red,}$$

and

$$E_2 = \text{The event that the second ball drawn is white.}$$

Then drawing a tree and letting the ordered pair (R, W) represent the number of red and white balls remaining in the box, we obtain

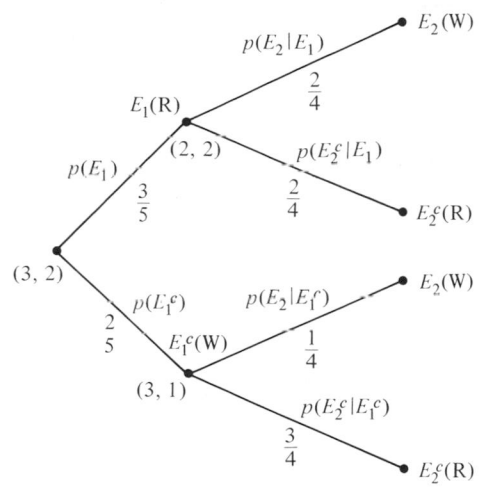

Thus, we find

$$p(\text{First red}) = p(E_1) = \tfrac{3}{5},$$

and

$$p(\text{Second white, provided first red}) = p(E_2 \,|\, E_1) = \tfrac{1}{2}.$$

Now consider the sample space as follows, where we label the red balls R_1, R_2, and R_3 and the white balls W_1, W_2. The first element in a pair represents the first ball selected and the second element, the second ball selected.

b) Given a probability problem, determine whether two events are independent.

c) Compute probabilities using the Multiplication Theorem for n events.

$$
\begin{array}{cccc}
(R_1, R_2) & (R_1, R_3) & (R_1, W_1) & (R_1, W_2) \\
(R_2, R_1) & (R_2, R_3) & (R_2, W_1) & (R_2, W_2) \\
(R_3, R_1) & (R_3, R_2) & (R_3, W_1) & (R_3, W_2)
\end{array}
$$

$$
\begin{array}{cccc}
(W_1, R_1) & (W_1, R_2) & (W_1, R_3) & (W_1, W_2) \\
(W_2, R_1) & (W_2, R_2) & (W_2, R_3) & (W_2, W_1)
\end{array}
$$

There are 20 equiprobable outcomes in all. Note that

$$p(\text{First red and second white}) = p(E_1 \cap E_2) = \tfrac{6}{20} = \tfrac{3}{10}.$$

Thus, we find that

$\qquad p(\text{First red and second white})$

$$= p(\text{First red}) \cdot p(\text{Second white, provided first red})$$

or

$$p(E_1 \cap E_2) = p(E_1) \cdot p(E_2 \,|\, E_1)$$

since

$$\tfrac{3}{10} = \tfrac{3}{5} \cdot \tfrac{1}{2}.$$

This result can be formalized as the

MULTIPLICATION THEOREM FOR ANY TWO EVENTS.   If the occurrence of event E_2 depends on the occurrence of event E_1, then the probability for their joint occurrence is given by

$$p(E_1 \cap E_2) = p(E_1) \cdot p(E_2 \,|\, E_1).$$

Actually, this equation is valid for any two events, dependent or independent.

Assuming $p(E_1) \neq 0$, we can divide and obtain the following alternative form of the Multiplication Theorem:

$$p(E_2 \,|\, E_1) = \frac{p(E_1 \cap E_2)}{p(E_1)}.$$

Suppose S is the sample space. Then the preceding expression can be expressed as

$$p(E_2 \,|\, E_1) = \frac{p(E_1 \cap E_2)}{p(E_1)} = \frac{\dfrac{\mathcal{N}(E_1 \cap E_2)}{\mathcal{N}(S)}}{\dfrac{\mathcal{N}(E_1)}{\mathcal{N}(S)}} = \frac{\mathcal{N}(E_1 \cap E_2)}{\mathcal{N}(E_1)}.$$

We can interpret this meaningfully by considering a Venn diagram.

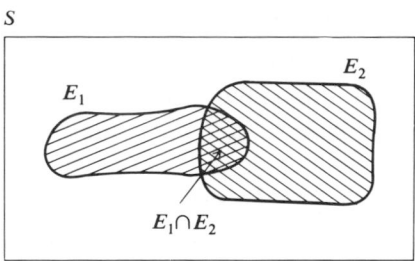

We can compute $p(E_2 \mid E_1)$ by first determining how many elements are in the *reduced sample space* E_1. This is $\mathcal{N}(E_1)$. Of those elements in E_1, how many are in E_2? This is $\mathcal{N}(E_1 \cap E_2)$. The division of $\mathcal{N}(E_1 \cap E_2)$ by $\mathcal{N}(E_1)$ give us $p(E_2 \mid E_1)$.

Returning to Example 1, we note that

$$p(\text{First red}) = p(E_1) = \tfrac{12}{20}$$

and

$$p(\text{First red and second white}) = p(E_1 \cap E_2) = \tfrac{6}{20}.$$

It then follows that the *conditional* probability $p(E_2 \mid E_1)$ can be obtained from the Multiplication Theorem as

$$p(\text{Second white, provided first red}) = p(E_2 \mid E_1)$$

$$= \frac{p(E_1 \cap E_2)}{p(E_1)}$$

$$= \frac{\frac{6}{20}}{\frac{12}{20}} = \tfrac{6}{12} = \tfrac{1}{2}.$$

Note that we can also obtain this probability by considering the reduced sample space, enclosed in the diagram. It has 12 equiprobable outcomes, of which 6 have the second ball selected as white; that is,

$$\mathcal{N}(E_1) = 12 \quad \text{and} \quad \mathcal{N}(E_1 \cap E_2) = 6.$$

Thus the conditional probability is

$$p(E_2 \mid E_1) = \frac{\mathcal{N}(E_1 \cap E_2)}{\mathcal{N}(E_1)} = \frac{6}{12} = \frac{1}{2}.$$

DO EXERCISES 30 AND 31.

30. Given

$$p(E_1 \cap E_2) = 0.125,$$

and

$$p(E_1) = 0.625,$$

find $p(E_2 \mid E_1)$.

31. In Example 2, what is the probability that the second ball is white, given that the first ball is white?

Suppose the events E_1 and E_2 are independent. Then

$$p(E_2 \mid E_1) = \frac{p(E_1 \cap E_2)}{p(E_1)} = \frac{p(E_1) \cdot p(E_2)}{p(E_1)} = p(E_2),$$

which is consistent with an earlier statement regarding independent events. That is, the occurrence of E_1 does not affect E_2.

Determining Whether Events are Independent

Let us now consider an example where we try to determine whether two events are independent. That is, we will try to decide whether either $p(E_2 \mid E_1) = p(E_2)$ or $p(E_1 \cap E_2) = p(E_1) \cdot p(E_2)$ is true.

Example 2 *Public Health.* A medical survey of 1000 people over the age of fifty-five was made to investigate the dependence of smoking on lung cancer. Of those surveyed 500 were steady smokers. Among the smokers, 200 had some form of lung cancer, while among the nonsmokers, only 120 had lung cancer.

a) What is the probability that a person smokes?
b) What is the probability that a person has lung cancer?
c) What is the probability that a person has lung cancer provided that person smokes?
d) Are smoking and lung cancer independent events?
e) What is the probability that a person has lung cancer given that person is a nonsmoker?

Solution Let

$$S = \text{Event of smoking,}$$
$$C = \text{Event of having lung cancer.}$$

To compute the various probabilities, we first organize the data into a table in terms of the number of people involved.

	C	C^c	Total
S	200	300	500
S^c	120	380	500
Total	320	680	1000

This data can also be organized, or represented, on a Venn diagram, as follows.

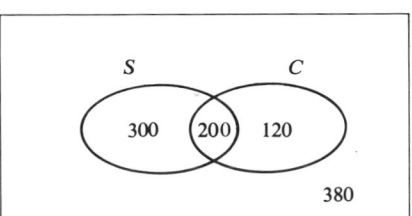

Either representation leads to the desired probabilities.
a) The probability that a person smokes is given by

$$p(S) = \frac{500}{1000} = 0.5.$$

b) The probability that a person has lung cancer (whether or not he smokes) is given by

$$p(C) = \frac{320}{1000} = 0.32.$$

c) The (conditional) probability that a person has lung cancer given that the person smokes is given by

$$p(C \mid S) = \frac{p(S \cap C)}{p(S)}.$$

From the table, we see that $p(S \cap C) = \dfrac{200}{1000} = 0.2$, so

$$p(C \mid S) = \frac{0.2}{0.5} = 0.4.$$

d) Now $p(C \mid S) = 0.4$ and $p(C) = 0.32$, and since

$$p(C \mid S) \neq p(C),$$

the events are dependent (not independent).
e) The probability that a person is a nonsmoker is given by

$$p(S^c) = \frac{500}{1000} = 0.50;$$

and the probability of the person being a nonsmoker and having lung cancer is given by

$$p(S^c \cap C) = \frac{120}{1000} = 0.12.$$

Thus, the probability that a person has lung cancer given that the

32. A sociological survey of 1000 people who were married at one time was made to investigate the dependence of age of marriage upon divorce. Of those surveyed 500 were divorced. Among those divorced, 280 were married as teenagers, while among those never divorced, only 190 were married as teenagers.

a) What is the probability that a person gets divorced?

b) What is the probability that a person was married as a teenager?

c) What is the probability that a person was married as a teenager given that the person gets divorced?

d) Are teenage marriages and divorce independent events?

e) What is the probability that a person was married as a teenager given that person does not get divorced?

33. The probability that a coed passes Economics is $\frac{4}{5}$, that she passes Mathematics is $\frac{2}{3}$, and that she passes both is $\frac{3}{5}$. Are the events of passing these two courses independent? That is, does

$$p(E_1 \cap E_2) = p(E_1) \cdot p(E_2)?$$

Compare this with Margin Exercise 22.

person is a nonsmoker is given by

$$p(C \mid S^c) = \frac{p(S^c \cap C)}{p(S^c)} = \frac{0.12}{0.50} = 0.24.$$

Note that

$$p(C \mid S^c) = 0.24 < 0.40 = p(C \mid S).$$

This confirms that (from the given data) nonsmokers will have less lung cancer than will smokers.

DO EXERCISES 32 AND 33.

Consider two *dependent* events E_1 and E_2. E_1 and E_1^c are mutually exclusive, as are E_2 and E_2^c, so that we can draw a tree:

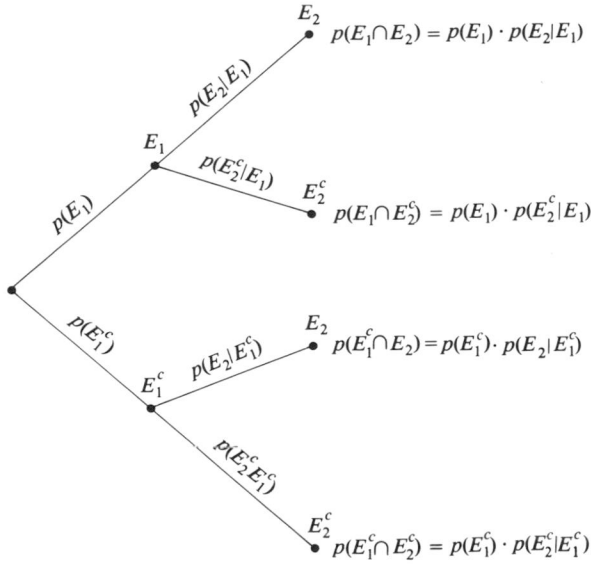

Consider, for example, the uppermost branch. The first event is E_1, and the probability that E_1 occurs is written in along the arc to E_1. If E_1 does happen, then E_2 may or may not happen. For the uppermost branch, E_2 happens, and the quantity written along the second arc represents the probability that E_2 happens provided E_1 happened, that is $p(E_2 \mid E_1)$. From the Multiplication Theorem, the joint probability for both E_1 and E_2 to happen is the product of the two probabilities written along the arcs of the branch. Thus, trees can be used for dependent events (as here), as well as for independent events (as in the preceding section).

Trees need not be limited to two branches or two arcs. Each *branch* of a tree can consist of any number of connected arcs. If some event E_1

precedes another event E_2, then the arc corresponding to E_1 must be closer to the vertex of the tree than is that of E_2. The *joint* probability for all events represented by the connected arcs of a branch of a tree is given by the *product* of the *conditional* probabilities written in along arcs. Thus, if $E_1, E_2, E_3, \ldots, E_n$ are events corresponding to the connected arcs of a branch of a tree, then

$$p(E_1 \cap E_2 \cap E_3 \cap \cdots \cap E_n) = p(E_1) \cdot p(E_2 \mid E_1) \cdot p(E_3 \mid E_1 \cap E_2)$$
$$\cdots p(E_n \mid E_1 \cap E_2 \cap \cdots \cap E_{n-1}).$$

This is the *Multiplication Theorem* for *n* events.

Example 3 A box contains 20 transistors, 5 of which are defective. Three transistors are taken out at random. What is the probability that at least one is defective? (This is Example 2 of Section 5.2 reconsidered, using the concepts of conditional probability.)

Solution Let D_i be the event that the *i*th transistor is defective. Then we obtain the following tree, where the ordered pair (d, g) represents the numbers of defective and good transistors, respectively, remaining in the box.

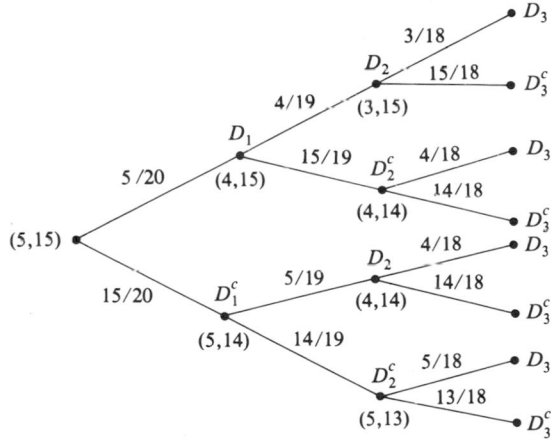

The probability for drawing at least one defective transistor is

$$p_{i \geqslant 1} = 1 - p_0,$$

where p_0 is the probability that none of the three transistors is defective, $p(D_i^c)$ or

$$p_0 = p(D_1^c \cap D_2^c \cap D_3^c).$$

Using the Multiplication Theorem, we obtain

$$p_0 = p(D_1^c) \cdot p(D_2^c \mid D_1^c) \cdot p(D_3^c \mid D_1^c \cap D_2^c).$$

34. An old sea chest contains 10 bars of silver and 5 bars of gold. If 3 bars are drawn at random, what is the probability that exactly one is gold? Solve using a tree and conditional-probability concepts. (This is also Margin Exercise 15.)

Since there are 5 defective transistors out of a total of 20, there must be 15 nondefective transistors, so that the probability that the first transistor is not defective is

$$p(D_1^c) = \tfrac{15}{20}.$$

If the first transistor is nondefective, then there are still 5 defective transistors remaining out of a reduced total of 19. Thus, there are 14 nondefective transistors and the probability that the second transistor is not defective *provided* the first transistor is not defective is

$$p(D_2^c \mid D_1^c) = \tfrac{14}{19}.$$

If both the first and second transistors are not defective, then there are 13 remaining nondefective transistors out of 18, so that the conditional probability that the third is not defective, given that the first and second were not defective, is

$$p(D_3^c \mid D_1^c \cap D_2^c) = \tfrac{13}{18}.$$

These three probabilities

$$p(D_1^c), \qquad p(D_2^c \mid D_1^c), \qquad p(D_3^c \mid D_1^c \cap D_2^c)$$

are written in along successive arcs of the lowest branch of the tree. Their *product* is their *joint* probability,

$$p_0 = \tfrac{15}{20} \cdot \tfrac{14}{19} \cdot \tfrac{13}{18} = \tfrac{91}{228}.$$

Then

$$p_{i \,\geqslant\, 1} = 1 - p_0 = \tfrac{137}{228}, \quad \text{as before.}$$

From the tree, it can be seen that the actual calculations involved here are quite simple, although the notation required to describe them can become quite lengthy.

DO EXERCISE 34.

Each *fork* of a tree represents a partition of the possible outcomes. That is, the arcs out of that vertex represent mutually exclusive events which fill that sample space. Thus, all *branches* represent mutually exclusive events, so that the probability for the occurrence of *either* of several events is the *sum* of the relevant probabilities.

Example 4 Using the data and tree from Example 3 and conditional-probability concepts, what is the probability for drawing:

a) Exactly one defective transistor?
b) Either one or two defective transistors?

Solution

a) The probability for drawing one defective transistor is

$$p_1 = p(D_1 \cap D_2^c \cap D_3^c) + p(D_1^c \cap D_2 \cap D_3^c) + p(D_1^c \cap D_2^c \cap D_3).$$

Each of these *joint* probabilities is, from the Multiplication Theorem, a *product* of the conditional probabilities written in along the arcs of appropriate *branches* of the tree. Since each joint event is mutually exclusive of the other, the probability for any *one* of the alternatives to occur is the *sum* of the various alternatives, as indicated by the appropriate branches,

$$p_1 = (\tfrac{5}{20} \cdot \tfrac{15}{19} \cdot \tfrac{14}{18}) + (\tfrac{15}{20} \cdot \tfrac{5}{19} \cdot \tfrac{14}{18}) + (\tfrac{15}{20} \cdot \tfrac{14}{19} \cdot \tfrac{5}{18}) = \tfrac{105}{228}, \quad \text{as before.}$$

b) Similarly, the probability for drawing two defective transistors is the sum

$$p_2 = p(D_1 \cap D_2 \cap D_3^c) + p(D_1 \cap D_2^c \cap D_3) + p(D_1^c \cap D_2 \cap D_3).$$

Evaluating, using the Multiplication Theorem, we obtain:

$$p_2 = (\tfrac{5}{20} \cdot \tfrac{4}{19} \cdot \tfrac{15}{18}) + (\tfrac{5}{20} \cdot \tfrac{15}{19} \cdot \tfrac{4}{18}) + (\tfrac{15}{20} \cdot \tfrac{5}{19} \cdot \tfrac{4}{18}) = \tfrac{30}{228}, \quad \text{as before.}$$

If E_i is the probability of drawing i defective transistors, then E_1 and E_2 are mutually exclusive events. The probability for *either* is the *sum*

$$p(E_1 \cup E_2) = p_1 + p_2 = \tfrac{105}{228} + \tfrac{30}{228} = \tfrac{135}{228},$$

and represents the sum of the probabilities corresponding to the branches of the tree representing one or two defective transistors.

DO EXERCISE 35.

---

35. Using the data and tree from Margin Exercise 34 and conditional-probability concepts, what is the probability for drawing 3 silver bars *or* 3 gold bars?

EXERCISE SET 5.5

1. Given $p(E_1 \cap E_2) = 0.0625$ and $p(E_1) = 0.125$, find $p(E_2 \mid E_1)$.

2. Given $p(E_1 \cap E_2) = 0.16$ and $p(E_1) = 0.64$, find $p(E_2 \mid E_1)$.

3. Given $p(E_1 \cap E_2) = \tfrac{12}{45}$ and $p(E_1) = \tfrac{13}{45}$, find $p(E_2 \mid E_1)$.

4. Given $p(E_1 \cap E_2) = \tfrac{16}{35}$ and $p(E_1) = \tfrac{19}{35}$, find $p(E_2 \mid E_1)$.

5. One card is drawn from a well-shuffled deck. What is the probability that it is an ace given that it is a red card?

6. One card is drawn from a well-shuffled deck. What is the probability that it is a king given that it is a black card?

7. A sack contains 5 blue marbles and 3 yellow marbles. One marble is selected, but not replaced, and a second is selected. What is the probability that

a) The second is yellow given that the first was blue?
b) The second is blue given that the first was blue?

8. Repeat Exercise 7, where the sack contains 5 blue marbles and 2 yellow marbles.

9. *Political Science.* The probability that Democrat A wins in the coming election is $\frac{3}{5}$; that of Democrat B winning the election for a separate seat is $\frac{2}{3}$; that for both is $\frac{1}{2}$. What is the probability that Democrat A wins provided Democrat B wins? Are their elections independent?

10. *Business.* A manufacturer is introducing two new products. From a survey, the first has 70% chance of success, the second 80%, and both 65%. What is the probability that the first product is successful given that the second is? Is the success of one product dependent on that of the other?

11. *Business.* The manager of a company is faced with a snowstorm, which usually keeps 40% of his employees out, and a flu epidemic, which usually keeps 15% out. If 45% of his employees show up, is the absentee rate what should be expected if the events were independent? Or is it possible that the employees were taking advantage of the situations? If 50% show up?

12. *Business, Sports.* The manager of a sports arena knows that bad weather will reduce attendance by 30% and another game on TV will reduce attendance by 25%. Both events occur (bad weather and another game on TV) and the attendance is 60% below normal. What drop in attendance would be expected if the events were independent? Or is it possible people are using bad weather as an excuse to stay home and watch TV?

13. *Business.* A restaurant buyer orders some food. Due to a failure of refrigeration, 50% of the food was spoiled, and due to a fuel shortage, only 80% of the food could be transported. The buyer says she has received only 40% of her order. Is this reasonable if the two events are independent?

14. *Business.* The manager of a shop has ordered some stock. Due to a severe rainstorm, 60% of the order was ruined; and due to a bridge washout, only 70% of the order arrived. The foreman claims that only 10% is usable. Is this reasonable if the rain damage and bridge washout are independent events? If only 30% is usable?

15. A box contains 10 candles, of which 7 are green and 3 purple. Three candles are taken out at random, one at a time. What is the probability that they alternate in color?

16. A class contains 6 men and 12 women. Four people are selected at random, one at a time. What is the probability that the first two are of the same sex and opposite to that of the last two?

17. Statistics indicate that 4% of men are colorblind and 0.3% of women are colorblind. Assuming that a population is half male and half female, what is the probability that a person selected at random is colorblind? Draw a tree.

18. The employees of a company are one-third women and two-thirds men. If men are twice as likely as women to have a car and a third of the women have cars, what is the probability that an employee selected at random has a car? Draw a tree.

19. *Business, Public Health.* A stockyard gets cattle from three ranches. The first ranch supplies 300 cattle, the second 500, and the third 200. Due to an outbreak of hoof-and-mouth disease, 10%, 15%, and 20% of the cattle, respectively, have the disease. What percent of the combined stock from the first and third are infected? Overall? [*Hint.* Draw two tree diagrams.]

20. *Business.* Three manufacturers, A, B, and C, supply respectively, 5, 5, and 6 cases of lightbulbs. Each case contains 24 lightbulbs. Manufacturer A makes lightbulbs which are 1% defective, B 2% defective, and C 5% defective. What percent of the combined order of A and B should be defective? Of the total order? [*Hint.* Draw two tree diagrams.]

21. A person holds a two-tailed coin in one hand and a fair coin in the other. If a hand is chosen at random and that coin is flipped, what is the probability that *tails* shows? Draw a tree.

22. In an urn are three coins, of which two are fair and one is two-tailed. A person reaches in and takes out at random one in his right hand and one in his left hand. What is the probability that the two-tailed coin is in either hand? Draw a tree.

23. In an urn are three coins of which two are fair and one two-tailed. A person reaches in and takes one out in each hand at random. (See Exercise 22.) Someone selects a hand at random. The coin in the hand selected is flipped. What is the probability that tails shows? Draw a tree.

25. One box contains 3 brass washers and 2 steel washers. Another box contains 4 brass washers and 3 steel ones. A box is chosen at random and a washer is taken out at random. The washer is then put into the other box (the one not chosen). A box is again chosen at random and a washer taken out at random. What is the probability it is brass? Draw a tree. What is the probability the same washer is picked both times?

27. A multiple-choice exam is being given. If a student knows the answer, he gets it right. If the student doesn't know the answer, he picks at random any of the four possible answers. It is also possible he "isn't sure" but has it narrowed down to one of two answers. If a student knows 80% of the answers and does not know 10% at all, what is the probability that he will get an arbitrary question correct? Draw a tree.

29. As in Exercise 28: If a voter is selected at random, what is the probability he or she will switch-vote (that is, a Republican voting Democratic or a Democrat voting Republican)?

24. Two urns contain 3 and 4 fair coins, respectively. A two-tailed coin is dropped at random into one of these urns. An urn is then selected at random and a coin is taken out at random and flipped. What is the probability for tails to show? [*Hint.* How many stages does the tree have?]

26. *Business, Banking.* A savings-and-loan institution classifies borrowers as AAA, AA, or A risks. AAA risks constitute 10% of the borrowers and default 5% of the time. AA risks constitute 25% and default 10% of the time. A risks constitute 65% and default 20% of the time. If a borrower is selected at random, what is the probability he will default? Draw a tree.

28. *Political Science.* In a certain city, registered voters are 40% Republican, 35% Democrat, and 25% independent. A Republican and a Democratic candidate are running for office. From a survey, 70% of the Republicans and 80% of the Democrats will vote for their party's candidate, while 75% of the independents will vote Democrat and the rest Republican. Which candidate has the better chance of winning, and by what odds? Draw a tree.

5.6 CONDITIONAL PROBABILITY—BAYES' THEOREM

In order to illustrate Bayes' theorem, consider first:

Example 1 Three manufacturers, I, II, and III, supply all the calculators to a particular store. I supplies 50, with 4% defective, II supplies 60 with 1% defective, and III supplies 30 with 2% defective. If a calculator is purchased at random, what is the probability that it is defective?

Solution Let

D = event that the calculator is defective,
E_1 = event that it came from I,
E_2 = event that it came from II,
E_3 = event that it came from III.

OBJECTIVE

You should be able to use Bayes' Theorem to determine conditional probabilities.

We first draw a tree noting that the events E_1, E_2, and E_3 are mutually exclusive. Here we use the Multiplication Theorem (Section 5.5) to rewrite the joint probabilities.

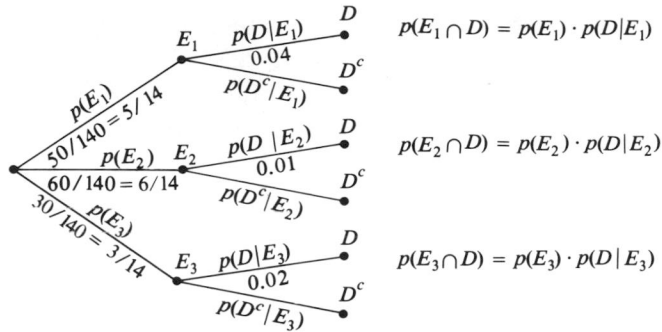

$$p(E_1 \cap D) = p(E_1) \cdot p(D|E_1)$$

$$p(E_2 \cap D) = p(E_2) \cdot p(D|E_2)$$

$$p(E_3 \cap D) = p(E_3) \cdot p(D|E_3)$$

Since the outcome can be either D or D^c, the set of outcomes D is actually a *reduced sample space* which can be expressed as a union of mutually exclusive events:

$$D = (E_1 \cap D) \cup (E_2 \cap D) \cup (E_3 \cap D)$$

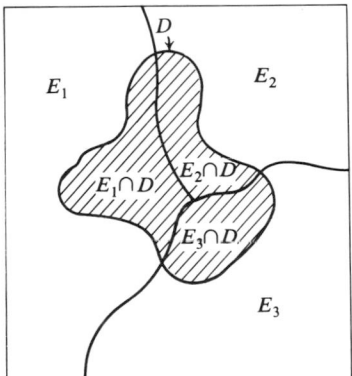

Then the probability that the calculator is defective, $p(D)$, is the sum:

$$
\begin{aligned}
p(D) &= p(E_1 \cap D) + p(E_2 \cap D) + p(E_3 \cap D) \\
&= p(E_1) \cdot p(D\,|\,E_1) + p(E_2) \cdot p(D\,|\,E_2) + p(E_3) \cdot p(D\,|\,E_3) \\
&= \tfrac{5}{14}(0.04) + \tfrac{6}{14}(0.01) + \tfrac{3}{14}(0.02) \\
&= \tfrac{5}{14} \cdot \tfrac{4}{100} + \tfrac{6}{14} \cdot \tfrac{1}{100} + \tfrac{3}{14} \cdot \tfrac{2}{100} \\
&= \tfrac{32}{1400} = \tfrac{4}{175}.
\end{aligned}
$$

Now with Example 1 in mind, consider:

Example 2   Given the data of Example 1 with a calculator purchased at random. What is the probability that, if it is defective, it came from manufacturer I?, II?, III?

Solution   The probability that if a calculator is defective it came from manufacturer I is $p(E_1 \mid D)$; from II, $p(E_2 \mid D)$; from III, $p(E_3 \mid D)$.

To determine any of these conditional probabilities, we use the reduced sample space D with probability $p(D)$. Of these, $E_1 \cap D$ are defective and came from I. From the Multiplication Theorem (Section 8.5), we find $p(E_1 \mid D)$:

$$p(E_1 \mid D) = \frac{p(E_1 \cap D)}{p(D)} = \frac{p(E_1) \cdot p(D \mid E_1)}{p(D)} .$$

Using probabilities already computed in Example 1, we obtain:

$$p(E_1 \mid D) = \frac{\frac{5}{14} \cdot \frac{4}{100}}{\frac{32}{1400}} = \frac{20}{32} = \frac{5}{8} .$$

Similarly, the probability that if the calculator is defective it came from manufacturer II, $p(E_2 \mid D)$, is given by:

$$p(E_2 \mid D) = \frac{p(E_2) \cdot p(D \mid E_2)}{p(D)} = \frac{\frac{6}{14} \cdot \frac{1}{100}}{\frac{32}{1400}} = \frac{6}{32} = \frac{3}{16} .$$

And the probability that if the calculator is defective it came from manufacturer III, $p(E_3 \mid D)$, is given by

$$p(E_3 \mid D) = \frac{p(E_3) \cdot p(D \mid E_3)}{p(D)} = \frac{\frac{3}{14} \cdot \frac{2}{100}}{\frac{32}{1400}} = \frac{6}{32} = \frac{3}{16} .$$

Alternately, using the expression for $p(D)$ from Example 1, we can write:

$$p(E_1 \mid D) = \frac{p(E_1 \cap D)}{p(E_1 \cap D) + p(E_2 \cap D) + p(E_3 \cap D)} .$$

Each of these joint probabilities can be obtained from the Multiplication Theorem, so that we can write

$$p(E_1 \mid D) = \frac{p(E_1) \cdot p(D \mid E_1)}{p(E_1) \cdot p(D \mid E_1) + p(E_2) \cdot p(D \mid E_2) + p(E_3) \cdot p(D \mid E_3)} .$$

This is Bayes' Theorem for three events.

Note that here we first calculated $p(D)$ and then $p(E_1 \mid D)$ using the concept of reduced sample space. On the other hand Bayes' Theorem incorporates this concept implicitly so that one calculates $p(E_1 \mid D)$

36. Three manufacturers A, B, and C supply all the fire alarms to a group of residences. A supplies 200 with 3% defective, B supplies 150 with 4% defective, and C supplies 100 with 5% defective. If an alarm is selected at random, what is the probability

a) It is defective?

b) If it is defective, it came from A? from B? from C?

directly. Either way the result is the same but the first method may be easier to remember.

Note that the probabilities

$$p(D \mid E_1) = 0.04, \qquad p(D \mid E_2) = 0.01, \qquad \text{and} \qquad p(D \mid E_3) = 0.02$$

represent probabilities "before" the calculator is purchased. These are sometimes called *a priori* probabilities. The probabilities, given that a calculator has been purchased,

$$p(E_1 \mid D) = \tfrac{5}{8}, \qquad p(E_2 \mid D) = \tfrac{3}{16}, \qquad p(E_3 \mid D) = \tfrac{3}{16},$$

can be thought of as "after," or *a posteriori*, probabilities. That is, information known beforehand allows one to compute probabilities of what will later occur.

DO EXERCISE 36

The general Bayes' Theorem for n events as follows.

BAYES' THEOREM. For any events $E_1, E_2, \ldots, E_n$ which partition a sample space, if the probability of each event is greater than 0 and if the events are conditional on some event C with $p(C) > 0$, then for each value of i $(i = 1, 2, \ldots, n)$,

$$p(E_i \mid C) = \frac{p(E_i \cap C)}{p(E_1 \cap C) + p(E_2 \cap C) + \cdots + p(E_n \cap C)}$$

$$= \frac{p(E_i) \cdot p(C \mid E_i)}{p(E_1) \cdot p(C \mid E_1) + p(E_2) \cdot p(C \mid E_2) + \cdots + p(E_n) \cdot p(C \mid E_n)}.$$

Example 3 *Political Science.* The public relations agent for a small political party wants to convince people how many people support his party even though his party received only 10% of the vote for the past three elections. The agent chooses for a survey an issue which is supported by 75% of his party and opposed by 75% of the other parties. If a voter is selected at random, what is the probability that:

a) The voter supports the issue?

b) If the voter supports the issue, the voter also supports the party?

Solution Let

P represent voters of the party of the PR agent,

P^c represent voters of the opposition,

S represent the voters who support the issue,

S^c represent the voters who oppose the issue.

Then we can draw a tree to represent this situation and incorporate the data:

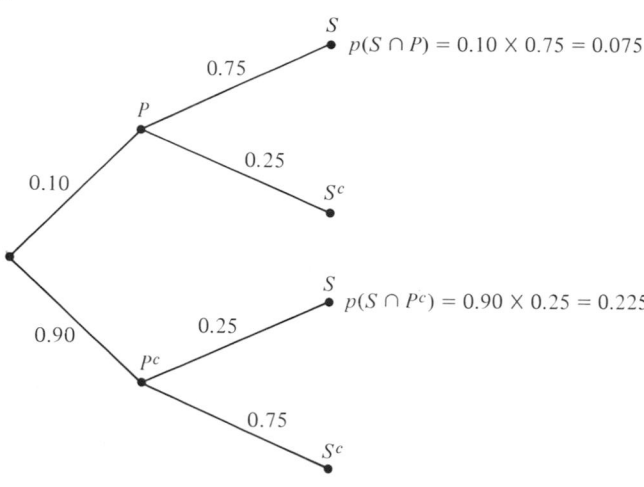

$p(S \cap P) = 0.10 \times 0.75 = 0.075$

$p(S \cap P^c) = 0.90 \times 0.25 = 0.225$

a) The probability that a voter supports the issue is

$$p(S) = p(S \cap P) + p(S \cap P^c)$$
$$= 0.075 + 0.225 = 0.30.$$

b) The probability that, if a voter supports the issue, he also supports the party is

$$p(P \mid S) = \frac{p(S \cap P)}{p(S)} = \frac{0.075}{0.30} = \frac{1}{4} = 0.25.$$

Note how different this probability is from the probability that a voter supports the party, $p(P) = 0.10$. The results of this calculation can be used quite effectively (albeit deceptively) to make support for the party seem greater than it is.

EXERCISE SET 5.6

1. *Public Health.* Statistics indicate that 4% of men are colorblind and 0.3% of women are colorblind. Assume that a population is half male and half female. If a person selected at random is colorblind, what is the probability the person is a woman? (See Exercise 17, Exercise Set 5.5.)

2. The employees of a company are one-third women and two-thirds men. Men are twice as likely as women to have a car and a third of the women have cars. If an employee selected at random has a car, what is the probability that the employee is a woman? (See Exercise 18, Exercise Set 5.5.)

3. *Business, Public Health.* A stockyard gets cattle from three ranches. The first supplies 300 cattle, the second 500, and the third 200. Due to an outbreak of hoof-and-mouth-disease 10%, 15%, and 20% of the cattle, respectively, have the disease. If an animal selected at random from those shipped by the first and third ranches is infected, what is the probability it came from the third ranch? If an animal selected at random from all three ranches is infected, what is the probability it came from the third ranch? (See Exercise 19, Exercise Set 5.5.)

5. A person holds a two-tailed coin in one hand and a fair coin in the other. A hand is chosen at random and the coin is flipped. If tails shows, what is the probability that the coin is two-tailed? (See Exercise 21, Exercise Set 5.5.)

7. Two urns contain 3 and 4 fair coins, respectively. A two-tailed coin is dropped at random into one of these urns. An urn is then selected at random and a coin is taken out at random and flipped. If tails shows, what is the probability that the coin came from the urn with the two-tailed coin? (See Exercise 24, Exercise Set 5.5.)

9. *Business, Banking.* A savings-and-loan institution classifies borrowers as AAA, AA, or A risks. AAA risks constitute 10% of their borrowers and default 5% of the time. AA risks constitute 25% and default 10%. A risks constitute 65% and default 20%. If a borrower defaults, what is the probability he was a AAA risk? (See Exercise 26, Exercise Set 5.5.)

11. *Business.* In drilling for oil, it is known that terrain with geological characteristics A yields oil once in 10 strikes and constitutes 50% of available land. Terrain with characteristics B yields oil once in 5 strikes and constitutes 30% of the available land. Terrain with characteristics C yields oil once in 4 strikes and constitutes 20% of the available land. If a site is selected at random, what is the probability for an oil strike? If oil is struck, what is the probability it came from land with characteristics A?

4. *Business.* Three manufacturers A, B, and C, supply, respectively, 5, 5, and 6 cases of lightbulbs. Each case contains 24 lightbulbs. Manufacturer A makes lightbulbs that are 1% defective, B's are 2% defective, and C's are 5% defective. If a lightbulb selected at random from those provided by manufacturers A and B is defective, what is the probability it came from B? selected from all three? (See Exercise 20, Exercise Set 5.5.)

6. In an urn are three coins of which two are fair and one is two-tailed. A person reaches in and takes out at random one coin in each hand. A hand is then selected at random and the coin in that hand is flipped. If tails shows, what is the probability the coin is fair? (See Exercise 22, Exercise Set 5.5.)

8. One box contains 3 brass washers and 2 steel washers. Another box contains 4 brass washers and 3 steel ones. A box is chosen at random and a washer is taken out at random and put into the other box (the one not chosen). A box is again chosen at random and a washer taken out at random. If the washer selected is brass, what is the probability that it came from the box from which a washer had been removed? (See Exercise 25, Exercise Set 5.5.)

10. A multiple-choice exam is being given. If a student knows the answer, he gets it right. If he doesn't know the answer, he picks at random any of the four given choices. It is also possible he "isn't sure" but has it narrowed down to one of two answers. A student knows 80% of the answers and does not know 10% at all. If he gets a question correct what is the probability he knew the answer? (See Exercise 27, Exercise Set 5.5.)

12. *Public Health, Biomedicine.* Patients entering a clinic have one of three (mutually exclusive) diseases A, B, or C. They also have one or more of the symptoms of fever, sore throat, or faintness. From the records of the clinic, 25% of the entering patients have disease A with symptoms of fever and sore throat, 35% disease B with symptoms of fever and faintness, and 40% disease C with all three symptoms. What is the probability that an undiagnosed patient has disease C given that he has a sore throat?

13. *Public Health, Biomedicine.* A test for mononucleosis administered to students among whom 10% have the disease is 90% accurate; that is, 90% of those with the disease will have a positive reaction and vice versa. If a student has a positive reaction, what is the probability he does not have mononucleosis?

14. As in Exercise 13, how accurate must the test be such that if a student has a positive reaction, the probability he does not have mononucleosis is 0.1?

15. *Pyschology, Biology.* In a laboratory there are 3 boxes (numbered I, II, and III). In the first are 6 white mice, in the second 3 white and 3 black, and in the third 6 black mice. While cleaning up, an assistant bumps into one box (at random) and a mouse (at random) escapes. This mouse is then caught but the assistant doesn't know which box to put the mouse back in. Therefore, he decides to take a mouse out of each box in turn (I, II, then III). If the two mice match in color, he puts both into that box; otherwise, he goes on to the next box to try for a match. What is the probability that the mouse is put back into the box from which it escaped? Draw a tree. If the mouse is put into the third box, what is the probability it came from that box?

16. *Psychology, Biology.* As in Exercise 15, but two mice (instead of one) escape from the same box. If they are both white, he puts them in the first box. If they are one of each color, he puts them in the second box. If they are both black, he puts them in the third box. What is the probability that he puts them back in the correct box? Draw a tree. If he puts them in the third box, what is the probability that they came from that box?

17. *Business, Quality Control.* A box of 6 clocks contains 3 defective ones. If 3 are chosen at random what is the probability that more than one is defective? In an effort to minimize the apparent number of defective clocks, if *one* or *no* clock chosen is defective, the sample is displayed. However, if more than one clock is defective, the defective clocks are replaced and that many are again drawn at random. Draw the tree. What is the probability that only one clock in the sample displayed is defective? If only one clock in the sample displayed is defective, what is the probability that some clocks had been redrawn?

18. As in Exercise 17, but the new clocks are redrawn *before* the old ones are replaced?

19. *Sports.* The winner of the National League playoffs is the first team to win three out of five games. The two teams A and B are evenly matched. What is the probability team A will win if:

a) Team A loses the first game?
b) Team A loses one of the first two games?
c) Team A loses the first two games?
d) Check a baseball almanac to compare these theoretical probabilities with the experimental probabilities you can determine from the almanac.

CHAPTER 5 TEST

1. a) Find $p(E^c)$ if $p(E) = \frac{23}{39}$.

b) What are the odds *for* the event E?

c) What are the odds *against* the event E?

3. Find $p(E_1 \cap E_2)$, where E_1 and E_2 are independent, $p(E_1) = \frac{11}{17}$ and $p(E_2) = \frac{34}{35}$.

5. a) Find $p(E_2 \mid E_1)$, where

$$p(E_1 \cap E_2) = 0.0043$$

and $p(E_1) = 0.125$.

b) Are the events E_1 and E_2 of (a) independent given that $p(E_2) = 0.1$?

7. A shipment of 125 stereos contains 5 defective stereos. Two stereos are selected at random. What is the probability that:

a) Both are defective?

b) Exactly one is defective?

c) Neither is defective?

9. A city is afflicted with two kinds of disease during one summer. The probability that a person gets both diseases is 0.28. the probability that a person gets neither of the diseases is 0.41. What is the probability that a person gets exactly one of the diseases?

11. Three manufacturers supply all the stereos to a particular music store. A supplies 20, with 5% defective, B supplies 70 with 3% defective, and C supplies 10 with 8% defective. If a stereo is purchased at random, what is the probability that:

a) It is defective?

b) If it is defective, it came from A? from B? from C?

2. Find $p(E_1 \cup E_2)$, where E_1 and E_2 are mutually exclusive, $p(E_1) = 0.34$ and $p(E_2) = 0.56$.

4. Find $p(E_1 \cup E_2)$, where E_1 and E_2 are independent, $p(E_1) = 0.11$ and $p(E_2) = 0.42$.

6. What is the probability of getting a 2 or 3 on a single roll of a die?

8. For an unfair coin, $p(H) = \frac{3}{5}$ and $p(T) = \frac{2}{5}$. The coin is flipped five times. What is the probability that the flips come out in the order H, T, T, T, H?

10. A manufacturer is introducing two new products. From a marketing survey, the first has an 80% chance of success, the second 60%, and both 48%. What is the probability that the first product is successful, given that the second is? Is the success of one product dependent on that of the other?

CHAPTER SIX

Random Variables—Statistics

OBJECTIVES

Given a probability problem, you should be able to determine what the random variable is, its values, and the corresponding probability function.

6.1 RANDOM VARIABLES AND PROBABILITY FUNCTIONS

It is convenient in the study of probability and statistics to introduce the concepts of "random variable" and "probability function." We shall define them formally after illustrating them.

Example 1 Two fair dice are rolled. The *sum* of the numbers showing is noted. What is the random variable? What is the probability function for this random variable?

Solution A fair die has 6 sides each of which has the same probability to show (face up). The following is a list of the $6 \cdot 6$ or 36 possible equiprobable outcomes, the sample space, as developed in Example 8 of Section 5.1:

$$
\begin{array}{cccccc}
(1,1) & (1,2) & (1,3) & (1,4) & (1,5) & (1,6) \\
(2,1) & (2,2) & (2,3) & (2,4) & (2,5) & (2,6) \\
(3,1) & (3,2) & (3,3) & (3,4) & (3,5) & (3,6) \\
(4,1) & (4,2) & (4,3) & (4,4) & (4,5) & (4,6) \\
(5,1) & (5,2) & (5,3) & (5,4) & (5,5) & (5,6) \\
(6,1) & (6,2) & (6,3) & (6,4) & (6,5) & (6,6)
\end{array}
$$

Now let us form a table listing those outcomes (a, b) whose sum $a + b$ is a particular value x:

x	2	3	4	5	6	7	8	9	10	11	12
(a, b)	$(1,1)$	$(1,2)$ $(2,1)$	$(1,3)$ $(2,2)$ $(3,1)$	$(1,4)$ $(2,3)$ $(3,2)$ $(4,1)$	$(1,5)$ $(2,4)$ $(3,3)$ $(4,2)$ $(5,1)$	$(1,6)$ $(2,5)$ $(3,4)$ $(4,3)$ $(5,2)$ $(6,1)$	$(2,6)$ $(3,5)$ $(4,4)$ $(5,3)$ $(6,2)$	$(3,6)$ $(4,5)$ $(5,4)$ $(6,3)$	$(4,6)$ $(5,5)$ $(6,4)$	$(5,6)$ $(6,5)$	$(6,6)$

Counting the number of times a given sum is obtained, we obtain a table of frequencies f:

x	2	3	4	5	6	7	8	9	10	11	12
f	1	2	3	4	5	6	5	4	3	2	1

Converting these frequencies into probabilities p as in Section 5.1, we obtain the table:

x	2	3	4	5	6	7	8	9	10	11	12
p	$\frac{1}{36}$	$\frac{2}{36}$	$\frac{3}{36}$	$\frac{4}{36}$	$\frac{5}{36}$	$\frac{6}{36}$	$\frac{5}{36}$	$\frac{4}{36}$	$\frac{3}{36}$	$\frac{2}{36}$	$\frac{1}{36}$

Here we have taken the *random variable X* (capital letter) to be the "sum of the numbers showing" and *x* (small letter) to be the *value* of the random variable. A *particular* value of the random variable is denoted by x_i.

Note that $p_1 + p_2 + \cdots + p_n = 1$, or simply,

$$\sum_{i=1}^{n} p_i = 1,$$

for probability functions.

In general, consider an experiment with *n* outcomes.

A *random variable X* is a *rule* (function) which assigns a numerical value x_i to each outcome.

As in the preceding example, the outcomes of an experiment need not be equiprobable.

To each outcome we assign not only a number x_i but also a probability p_i. This set of all ordered pairs $\{(x_i, p_i) \mid i = 1, 2, \ldots, n\}$, for example, as displayed in the preceding table, is called the *probability frequency function* of the random variable X, or simply, the *probability function*.

DO EXERCISE 1.

Many of the problems in the preceding chapter involved random drawings

i) of objects restricted to *two* kinds,
ii) simultaneously (in sequence and *without* replacement).

Specifically, consider the following example.

Example 2 *Business, Quality Control.* Given a box of 20 transistors of which 5 are defective. Three are drawn at random and the number defective is noted. What is the random variable? What is the probability function for this random variable?

Solution The random variable *X* here is the number of defective transistors drawn. It takes on the values $x = 0, 1, 2, 3$. The probability of drawing a particular number x_i of defective transistors is $p(x_i)$ or p_{x_i}.

Thus, to obtain p_0 we calculate the probability of drawing 0 defective transistors out of 5 available and 3 nondefective transistors out of 15

1. Two fair dice are rolled and the *difference* (in magnitude) between the numbers showing is noted. What is the probability function?

2. A sea chest contains 10 silver bars and 5 gold ones. Three are drawn at random and the number of gold bars is noted. What is the probability function?

available; that is,

$$p_0 = \frac{\binom{5}{0}\binom{15}{3}}{\binom{20}{3}} = \frac{91}{228}.$$

Similarly, we obtain:

$$p_1 = \frac{\binom{5}{1}\binom{15}{2}}{\binom{20}{3}} = \frac{105}{228},$$

$$p_2 = \frac{\binom{5}{2}\binom{15}{1}}{\binom{20}{3}} = \frac{30}{228},$$

$$p_3 = \frac{\binom{5}{3}\binom{15}{0}}{\binom{20}{3}} = \frac{2}{228}.$$

Thus, we obtain the probability function given in the following table:

x	0	1	2	3
p	$\dfrac{91}{228}$	$\dfrac{105}{228}$	$\dfrac{30}{228}$	$\dfrac{2}{228}$

DO EXERCISE 2.

Example 2 and Margin Exercise 2 just considered have certain characteristics in common. Specifically, such problems have

s objects,
m of which have a particular characteristic and
$s - m$ do not have this characteristic;
n objects are drawn at random and
r of those drawn have this characteristic.

If we take as the random variable R the number of those drawn that have this characteristic, then the values that this random variable can assume are $r = 0, 1, 2, \ldots, n$ and the probability that r have this characteristic is:

$$p_r = H(s, m; n, r) = \frac{\binom{m}{r}\binom{s-m}{n-r}}{\binom{s}{n}},$$

where

$$0 \leq r \leq n \leq s \quad \text{and} \quad 0 \leq r \leq m \leq s.$$

We call p_r the *hypergeometric* probability. We have already used it in Chapter 5. All we have done here is give it a name.

EXERCISE SET 6.1

1. The sales force of a business consists of 20 people, half of whom are men and the other half women. Four people are chosen at random. What is the probability function for the number of women chosen? (See Exercise 31, Set 5.1.)

2. A union has 21 members, 14 of whom are women and the other 7 are men. Three people are chosen at random. What is the probability function for the number of women chosen? men? (See Exercise 32, Set 5.1.)

3. Eight people apply for a job, 5 men and 3 women. Four are hired at random. What is the probability function for the number of women hired?

4. Eight people apply for a job, 4 men and 4 women. Four are hired at random. What is the probability function for the number of women hired?

5. *Business, Quality Control.* A crate of 20 machine parts contains 6 defective parts. Five parts are drawn at random. What is the probability function for the number of defective parts drawn?

6. *Public Health.* Five workers in a group of 25 have mononucleosis. Three workers are chosen at random. What is the probability distribution for the number who have mono? (See Exercise 6, Set 5.2.)

7. *Political Science.* If a party fields 3 candidates for office, each opposed by an equiprobable candidate, what is the probability function for the number of winners for the party in an election?

8. As in Exercise 7, but the first candidate has a probability of $\frac{1}{3}$ to win, the second $\frac{1}{2}$, and the third $\frac{2}{3}$?

9. *Business, Quality Control.* A wine rack contains 7 bottles of red wine and 2 of white wine. If 3 bottles are taken out at random, what is the probability function for the number of bottles of white wine? red wine?

10. *Business, Quality Control.* A case of 12 bottles of wine contains 3 which have spoiled. If 3 bottles are taken out at random, what is the distribution function for the number of bottles of spoiled wine among those drawn? If the case contains 4 spoiled bottles?

11. *Psychology, Biology.* One cage contains 3 white mice and 2 black ones and another cage contains 2 white mice and 3 black ones. A cage is chosen at random and 3 mice are taken out at random. What is the probability function for the number of white mice in the sample?

12. *Psychology, Biology.* There are 2 cages of mice as in Exercise 11. A cage is chosen at random. A mouse of unidentified color escapes from the chosen cage. Then 3 mice are taken out at random. What is the probability function for the number of white mice in the sample?

13. *Psychology, Biology.* There are 2 cages of mice as in Exercise 11. A white mouse escapes from an unidentified cage. Then a cage is chosen at random and 3 mice are taken out at random. What is the probability function for the number of white mice in the sample?

14. *Psychology, Biology.* There are 2 cages of mice as in Exercise 11. All the mice escape and are put back at random, 5 to each cage. A cage is then selected at random and 3 mice are taken out at random. What is the probability function for the number of white mice in the sample?

15. *Sports.* The winner in a World Series is the first team to win 4 out of seven games. What is the probability function for the number of games in the series if the two teams are evenly matched?

16. *Sports.* As in Exercise 15, but the winning team must win by two games. After 7 games, one extra game is to be played if one team is winning by only one game but the series ends if there is a 4-4 tie?

17. Two gamblers toss fair coins. One wins the toss if they match, the other if they don't. The winner of the game is the first to win two tosses in a row. What is the probability function for the number of tosses required for someone to win the game?

18. Three gamblers each toss a fair coin. The winner of a toss is the odd man, if there is one. The winner of the game is the first one to win two tosses in a row. What is the probability function for the number of tosses required for someone to win the game?

OBJECTIVE

You should be able to calculate the expected value of a random variable.

6.2 AVERAGE AND EXPECTED VALUE

Frequently we have some data and would like some way of determining a "center" point. This is usually done by computing the *mean* or *average* value.

Example 1 ▦ The test scores for a particular class are 76, 72, 88, 90, 74, 83, 52, 79, 81, 84, and 69. What is the average score?

Solution The mean or average value $\bar{x}$ is simply the sum of the various test scores divided by the number of test scores; that is,

$$\bar{x} = \tfrac{1}{11}(76 + 72 + 88 + 90 + 74 + 83 + 52 + 79 + 81 + 84 + 69)$$
$$= \tfrac{1}{11}(848)$$
$$= 77.09.$$

3. ▦ If the test scores for a class are 69, 72, 83, 74, 89, 67, 77, 82, 84, 93, 68, and 79, what is their average value?

In general, if there are n data points $x_1, x_2, \ldots, x_n$, then their *average* value is:

$$\bar{x} = \frac{1}{n}(x_1 + x_2 \cdots x_n),$$

or, using summation notation,

$$\bar{x} = \frac{1}{n}\sum_{i=1}^{n} x_i.$$

DO EXERCISE 3.

Sometimes a given data point is present more than once.

Example 2 ▦ *Sports.* On the fourteenth hole of the 1976 Andy Williams San Diego Open Golf Tournament, scores were obtained as given in the following table. What was the average score on this hole?

Score	Number with score
3 (Eagle)	1
4 (Birdie)	29
5 (Par)	176
6 (Bogie)	30
7 (Double bogie)	2

Solution Here each score x_i occurs with frequency f_i (number with a particular score). The total number of scores is:

$$N = 1 + 29 + 176 + 30 + 2 = 238,$$

so that the average score is

$$\bar{x} = \tfrac{1}{238}(1 \cdot 3 + 29 \cdot 4 + 176 \cdot 5 + 30 \cdot 6 + 2 \cdot 7)$$
$$= \tfrac{1193}{238}$$
$$= 5.0126 \qquad \text{(to four decimal places)}.$$

Note that the average score is quite close to par (5).

In general, if each data point x_i occurs with frequency f_i, then the total number of data points is:

$$N = f_1 + f_2 + \cdots + f_n = \sum_{i=1}^{n} f_i,$$

and the average value is given by:

$$\bar{x} = \frac{1}{N}\sum_{i=1}^{n} f_i x_i.$$

DO EXERCISE 4.

Example 3 ▦ Two fair dice are rolled. The sum of the numbers showing is noted. The random variable X corresponds to the sum.

a) Suppose we roll the dice 144 times and obtain the following frequency table.

x	2	3	4	5	6	7	8	9	10	11	12
f	4	9	12	17	21	23	19	16	13	7	3

4. ▦ *Sports.* On the sixteenth hole of the 1976 Andy Williams San Diego Open Golf Tournament, scores were obtained as given in the following table.

Score	Number with score
1 (Hole-in-one)	3
2 (Birdie)	164
3 (Par)	61
4 (Bogies)	3

a) What was the average score on this hole?

b) Based on your answer to part (a), was this an easy or difficult hole?

What is the average value?

b) Suppose we roll the dice 1440 times and obtain the following frequency table.

x	2	3	4	5	6	7	8	9	10	11	12
f	38	84	119	163	207	239	193	159	122	75	41

What is the average value?

Solution

a) The average value of x is given by:

$$\bar{x} = \tfrac{1}{144} \sum_{i=1}^{11} x_i f_i = \tfrac{1}{144}(994) = 6.90278.$$

b) The average value of x is given by:

$$\bar{x} = \tfrac{1}{1440} \sum_{i=1}^{11} x_i f_i = \tfrac{1}{1440}(10{,}046) = 6.97639.$$

We might call the average values calculated in Example 3 *experimental estimates* of the "center" point. Suppose we had no data, or did not want to bother to obtain any and we wanted to determine a theoretical "center" point. Then we start with the probability distribution function, if such is available. It is. We considered it in Example 1 of the preceding section:

x	2	3	4	5	6	7	8	9	10	11	12
p	$\frac{1}{36}$	$\frac{2}{36}$	$\frac{3}{36}$	$\frac{4}{36}$	$\frac{5}{36}$	$\frac{6}{36}$	$\frac{5}{36}$	$\frac{4}{36}$	$\frac{3}{36}$	$\frac{2}{36}$	$\frac{1}{36}$

Note that the average value can be expressed as

$$\bar{x} = \frac{1}{N} \sum_{i=1}^{n} x_i f_i = \sum_{i=1}^{n} \frac{f_i}{N} x_i.$$

In the long run, after many rolls of the dice, we would expect that the probability p_i for a given outcome would be quite close to the quotient of the frequency f_i for that outcome and the total number of trials N:

$$p_i \approx \frac{f_i}{N}.$$

Thus,

$$\sum_{i=1}^{n} \frac{f_i}{N} x_i \approx \sum_{i=1}^{n} p_i x_i.$$

Then, rather than compute an *average value* from the data, we can compute an *expected value* from the probability function. This is given by:

$$E(X) = \mu = \sum_{i=1}^{n} p_i x_i,$$

where μ is the Greek letter "mu."

Here x_i are now the values of the random variable. Note that we use $\bar{x}$ to denote the average value computed from experimental data and μ to denote the expected value computed theoretically from the probability function.

The expected value of the sum of the dice is given by

$$E(X) = 2 \cdot \tfrac{1}{36} + 3 \cdot \tfrac{2}{36} + 4 \cdot \tfrac{3}{36} + 5 \cdot \tfrac{4}{36} + 6 \cdot \tfrac{5}{36}$$
$$+ 7 \cdot \tfrac{6}{36} + 8 \cdot \tfrac{5}{36} + 9 \cdot \tfrac{4}{36} + 10 \cdot \tfrac{3}{36} + 11 \cdot \tfrac{2}{36} + 12 \cdot \tfrac{1}{36},$$
$$E(X) = 7.$$

In the long run, the more we roll the dice, the closer we "expect" the average values to be to the expected value. Note in Example 3 that the average values 6.90278 and 6.97639 are getting closer to the expected value 7.

DO EXERCISE 5.

For the *hypergeometric frequency function* (see preceding section), it can be shown that the expected value is given by

$$E(X) = n \cdot p$$

where

$$p = \frac{m}{s}.$$

As before, n is the number of trials to draw a random number of objects of which m out of s have a particular characteristic, so that p represents the probability of drawing one with this characteristic *initially*.

Example 4 *Business, Quality Control.* Given a box of 20 transistors, of which 5 are defective. Three are drawn at random and the number defective is noted. What is the expected number of defective transistors in the sample? Use both the general formula and the special one for hypergeometric probabilities. [See Example 2, Section 6.1.]

5. Two fair dice are rolled and the *difference* (in magnitude) between the numbers showing is noted. What is the expected value of this difference? See Margin Exercise 1.

6. A sea chest contains 10 silver bars and 5 gold ones. Three are drawn at random and the number of gold bars is noted. Determine the expected number of gold bars to be drawn using both the general formula and that for hypergeometric frequency functions. See Margin Exercise 2.

Solution   From Example 2, Section 6.1, we have the probability function

x	0	1	2	3
p	$\frac{91}{228}$	$\frac{105}{228}$	$\frac{30}{228}$	$\frac{2}{228}$

Using the general formula for expected value and the probability function previously obtained, we have

$$E(X) = 0 \cdot \tfrac{91}{228} + 1 \cdot \tfrac{105}{228} + 2 \cdot \tfrac{30}{228} + 3 \cdot \tfrac{2}{228} = \tfrac{171}{228} = \tfrac{3}{4}.$$

Using the special formula, we have

$$E(X) = 3 \cdot \tfrac{5}{20} = \tfrac{3}{4}, \qquad \text{as above.}$$

Note that the expected value $E(X)$ need not be a possible value of the random variable; that is, no value of the random variable is $\frac{3}{4}$.

DO EXERCISE 6.

In a game of chance, the game is said to be *favorable* or *unfavorable* to the player as the expected value is positive or negative. The game is considered *fair* if the expected value is zero.

Example 5   Consider a lottery in which 10,000 tickets are sold at $1 each. Five tickets are drawn at random. The first-place winner gets a $5000 car, the second-place winner gets a $700 stereo, and the next three winners get $100 each. What is the expected value of a ticket? Is the game fair?

Solution   We let the amount of winnings per ticket be the random variable. The probability function is as follows:

x	0	100	700	5000
p	$\frac{9995}{10000}$	$\frac{3}{10000}$	$\frac{1}{10000}$	$\frac{1}{10000}$

The expected value of a *ticket* is then

$$E_{\mathrm{T}} = 0 \cdot \tfrac{9995}{10000} + 100 \cdot \tfrac{3}{10000} + 700 \cdot \tfrac{1}{10000} + 5000 \cdot \tfrac{1}{10000} = \tfrac{60}{100}, \quad \text{or } \$0.60.$$

Since one is paying $1 for a ticket worth $0.60, one would be suspicious that the game is not fair. The expected value of the *game* E_{G} can be obtained by identifying a new random variable X_i' with the *net* winnings; that is, total winnings minus the cost. Thus, $x_i' = $

$x_i - 1$, so that

$$E_G = E_T - 1 = 0.60 - 1.00 = -0.40.$$

Since $E_G \neq 0$, the game is not fair. Now let c be the cost of a ticket for a fair game. The expected value of a *ticket* E_T is still $0.60, but the expected value of the *game* is now

$$E_G = E_T - c = 0.60 - c,$$

so that $E_G = 0$ for $c = 0.60$ or $0.60. Thus the game is fair when the cost of a ticket equals the expected value.

DO EXERCISE 7.

Example 6   You have a choice between buying one chance for $1 in the lottery of Example 5 or 4 chances for $0.25 each in the following lottery. There are 1000 chances being sold for $0.25 each with a first prize of a $100 TV set. Assuming that the prize can always be exchanged for some other article of equal value, which lottery is the better buy?

Solution   The probability function for this lottery is:

x	0	100
p	$\frac{999}{1000}$	$\frac{1}{1000}$

where the random variable is the winnings per ticket. Thus, the expected value per ticket is

$$E_T = 0 \cdot \frac{999}{1000} + 100 \cdot \frac{1}{1000} = 0.1, \quad \text{or } \$0.10.$$

The expected value of 4 tickets is

$$E_{4T} = 4 \cdot E'_T = 4(0.1) = 0.4, \quad \text{or } \$0.40.$$

The same answer would have been obtained had we taken the random variable as the winnings for 4 tickets. In that case:

x	0	100
p	$\frac{996}{1000}$	$\frac{4}{1000}$

and

$$E_{4T} = 0 \cdot \frac{996}{1000} + 100 \cdot \frac{4}{1000} = 0.4.$$

Thus $1 would buy four tickets in this lottery with an expected value of $0.40 compared to one ticket in the lottery of Example 5 with an

7. A raffle is being held to raise money for a charity. There are 1000 tickets to be sold for $10 each, with a first prize of a three-week vacation in Europe worth $1500, a second prize of one week in lovely downtown Burbank worth $250, and 5 third prizes of $2 tickets to a movie travelogue. What is the expected value of a ticket? How much of the cost of each ticket goes to charity if the printing expenses are $40 and all other labor is volunteer?

8. Consider a comparison between the lotteries of Examples 5 and 6. If the random variable is taken as winning something (rather than a given amount), which lottery is the better buy

expected value of $0.60, so that the better buy is from the lottery of Example 5.

DO EXERCISE 8.

*(Optional) Craps

Example 7 *Business, Casinos.* Craps is a dice game with many variations of the basic rules. The rules used in casinos are the following: A shooter rolls two dice. If the sum of the numbers showing totals 7 or 11, he wins; if the sum totals 2, 3, or 12, he loses. If the sum is anything else (that is, 4, 5, 6, 7, 8, 9, or 10), this becomes his "point." To win, he must roll his point before he rolls a 7, in which case he loses. What is the probability of winning? If even money is bet, what is the expected value?

Solution First let us draw a tree and indicate for each branch the *conditional* probability of occurrence using the probabilities determined in Example 1 of Section 6.1.

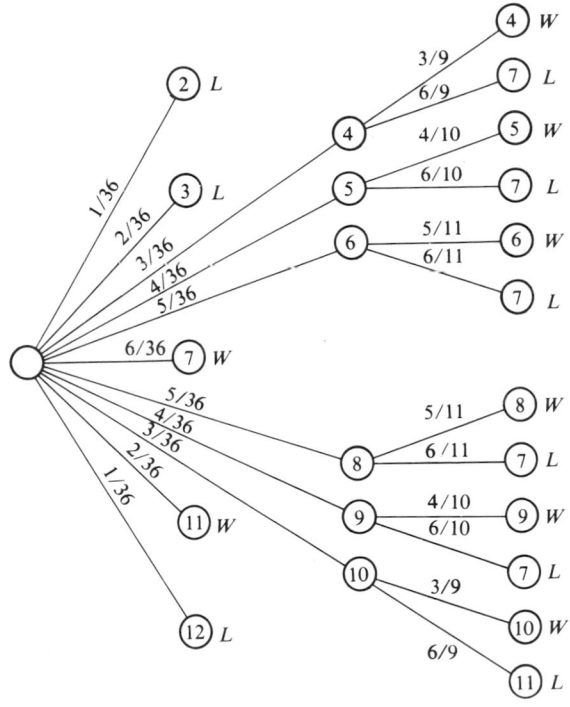

The tree has been simplified in the following respect. If, for example, the "point" is 4, then the branch of the tree for this point would be

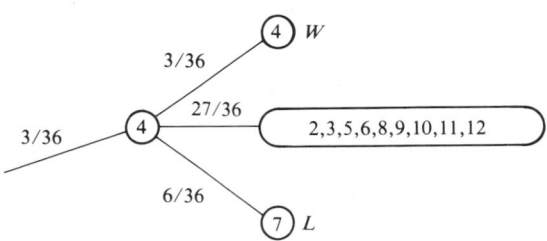

If the point 4 is obtained, the player wins and the game ends. If a 7 is obtained, the player loses and the game ends. If anything but a 4 or a 7 is obtained, the game continues until a 4 or 7 is obtained.

Using Bayes' Theorem, we can determine the probability of obtaining the point 4 provided the game ends, that is, the player gets a 4 or 7:

$$p(4 \mid 4 \text{ or } 7) = \frac{\frac{3}{36}}{\frac{3}{36} + \frac{6}{36}} = \frac{3}{9}.$$

Similarly, we can determine the probability of obtaining a 7 provided the game ends:

$$p(7 \mid 4 \text{ or } 7) = \frac{\frac{6}{36}}{\frac{3}{36} + \frac{6}{36}} = \frac{6}{9}.$$

Using only these two options, we can simplify this branch of the tree to

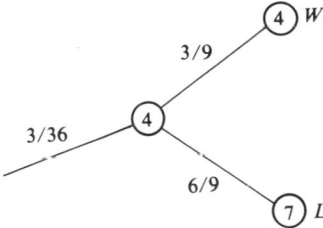

"Points" other than 4 are considered in the same manner. Thus, the probability to win p_W is:

$$p_\text{W} = \frac{3}{36} \cdot \frac{3}{9} + \frac{4}{36} \cdot \frac{4}{10} + \frac{5}{36} \cdot \frac{5}{11} + \frac{6}{36} + \frac{5}{36} \cdot \frac{5}{11}$$

$$+ \frac{4}{36} \cdot \frac{4}{10} + \frac{3}{36} \cdot \frac{3}{9} + \frac{2}{36} = \frac{244}{495}.$$

The probability to lose, p_L, is

$$p_\text{L} = 1 - p_\text{W} = \frac{251}{495}.$$

Thus the odds to win are 244:251. The expected value is:

$$E(X) = 1 \cdot \frac{244}{495} + (-1) \cdot \frac{251}{495} = -\frac{7}{495},$$

indicating that, in the long run, the *shooter* will lose.

DO EXERCISE 9.

9. *Business, Casinos.* From Example 7, it is apparent that one can win in the long run by betting *against* the shooter. Since this is the traditional role of the "house," in the long run a casino would lose if it accepted such bets. Since gambling casinos cannot afford to lose, they modify the rules as follows: If one bets *with* the shooter, then the same rules prevail. However, if one bets *against* the shooter, then a standoff feature is added; namely, if the initial roll is 2, the game ends with no win and no loss. What is the expectation of winning by betting against the shooter? Is it better to bet for or against the shooter?

EXERCISE SET 6.2

1. ▦ *Business, Racing.* The lap speeds in the speed trials of a stock-car race were

90.791, 89.237, 89.108, 87.926, 86.401, 85.858, 83.271, and 79.669 mph.

What is the average lap speed?

2. ▦ *Business, Racing.* The lap speeds for a set of speed trials were

91.101, 90.973, 89.257, 86.118, 85.879, 82.438, 81.962, 78.113, and 78.104 mph.

What is the average lap speed?

3. ▦ *Business.* In the course of an evening, a restaurant served 23 lobster dinners at $7.95, 47 steak dinners at $6.95, 53 roast beef dinners at $6.45, 33 shrimp dinners at $5.95, 29 Salisbury steak dinners at $4.95, 37 fried chicken dinners at $4.45, and 23 fish dinners at $3.75. What was the average price of a dinner?

4. ▦ *Business.* A theater has 320 seats for which tickets cost $6.90, 480 seats at $5.90, 624 seats at $4.70, and 484 seats at $3.60. What is the average cost of a ticket?

5. A union has 21 members, 14 of whom are women and the other 7 are men. Three people are chosen at random. What is the expected number of women chosen? men? Use both formulas. [See Exercise 2, Set 6.1.]

6. The sales force of a business consists of 20 people, half of whom are men and the other half women. Four people are chosen at random. What is the expected number of women chosen? Use both formulas. [See Exercise 1, Set 6.1.]

7. Eight people apply for a job, 4 men and 4 women. Four are hired at random. What is the expected number of men hired? Use both formulas. [See Exercise 4, Set 6.1.]

8. Eight people apply for a job, 5 men and 3 women. Four are hired at random. What is the expected number of women hired? Use both formulas. [See Exercise 3, Set 6.1.]

9. *Public Health.* Five workers in a group of 25 have mononucleosis. Three workers are chosen at random. What is the expected number of those chosen who have mono? [See Exercise 6, Set 6.1.]

10. *Business.* A crate of 20 machine parts contains 6 defective parts. Five parts are taken out at random. What is the expected number of defective parts taken out? [See Exercise 5, Set 6.1.]

11. *Political Science.* If a party fields 3 candidates for office, each opposed by an equiprobable candidate, what is the expected number of winners for the party? [See Exercise 7, Set 6.1.]

12. *Political Science.* As in Exercise 11, but the first candidate has a probability of $\frac{1}{3}$ to win, the second $\frac{1}{2}$, and the third $\frac{2}{3}$? [See Exercise 8, Set 6.1.]

13. *Business.* A case of 12 bottles of wine contains 3 which have spoiled. Three bottles are taken out at random. What is the expected number of bottles of spoiled wine among those drawn? if the case contains 4 spoiled bottles? [See Exercise 10, Set 6.1.]

14. *Business.* A wine rack contains 7 bottles of red wine and 2 of white wine. If 3 bottles are taken out at random, what is the expected number of bottles of white wine drawn? red wine? [See Exercise 9, Set 6.1.]

▶ ───────────────────────────────

15. *Psychology, Biology.* One cage contains 3 white mice and 2 black ones and another cage contains 2 white mice and 3 black ones. A cage is chosen at random and 3 mice are taken out at random. What is the expected number of white mice in the sample? [See Exercise 11, Set 6.1.]

16. *Psychology, Biology.* There are two cages of mice, as in Exercise 15. A cage is chosen at random. A mouse of unidentified color escapes from the chosen cage. Then 3 mice are taken out at random. What is the expected number of white mice in the sample? [See Exercise 12, Set 6.1.]

17. *Psychology, Biology.* There are 2 cages of mice, as in Exercise 15. A white mouse escapes from an unidentified cage. Then a cage is chosen at random and 3 mice are taken out at random. What is the expected number of white mice in the sample? [See Exercise 13, Set 6.1.]

18. *Psychology, Biology.* There are 2 cages of mice, as in Exercise 11. All the mice escape and are put back at random, 5 to each cage. A cage is then selected at random and 3 mice are taken out at random. What is the expected number of white mice in the sample? [See Exercise 14, Set 6.1.]

19. *Sports.* The winner in a World Series is the first team to win 4 out of 7 games. If the two teams are evenly matched, what is the expected number of games required for one team to win the series? [See Exercise 15, Set 6.1.]

20. *Sports.* As in Exercise 19, but the winning team must win by two games. After 7 games one extra game is to be played if one team is winning by only one game but the series ends if there is a 4-4 tie [See Exercise 16, Set 6.1.]

6.3 VARIANCE AND STANDARD DEVIATION

In addition to wanting to know the mean or average value of the data or the expected value of a probability-distribution function for some random variable, we may want to know about the "spread" of the data. To do this, we use the *variance* and the *standard deviation*.

Let us consider the test scores from Example 1 of Section 6.2:

76, 72, 88, 90, 74, 83, 52, 79, 81, 84, 69.

In that example we found that the average value was

$$\bar{x} = 77.09.$$

We want to consider how the data *varies* from the average value. How can we do this? One way might be to consider the differences between a score and the average value

$$x_i - \bar{x}.$$

For example, $76 - 77.09 = -1.09$ represents the deviation of the score 76 from the mean. If we add all of these deviations and average them, we would have

$$\frac{1}{n} \sum_{i=1}^{n} (x_i - \bar{x}),$$

which is called the *average deviation from the mean.* Unfortunately, this quantity always adds to 0. Thus, it would not yield any results which would vary between different sets of data to yield information about the spread of the data. (The reader should verify that for the above set of data, the average deviation from the mean is 0.) To avoid this difficulty, we consider the *deviation squared:*

$$(x_i - \bar{x})^2.$$

OBJECTIVES

You should be able to calculate the variance and standard deviation of:

a) Given data
b) A random variable, given its probability function.

10. ▦ Determine the variance of the test scores of Margin Exercise 3: 69, 72, 83, 74, 89, 67, 77, 82, 84, 93, 68, and 79.

11. ▦ A sea chest contains 10 silver bars and 5 gold ones. Three are drawn at random. What is the variance of the number of gold bars drawn? (See Margin Exercise 2, Section 6.1, and Margin Exercise 6, Section 6.2.)

The average of the deviation squared, called the *sample variance*, is given by *

$$s^2 = \frac{1}{n} \sum_{i=1}^{n} (x_i - \bar{x})^2.$$

Now calculating the variance, we obtain

$$s^2 = \tfrac{1}{11}[(76 - 77.09)^2 + (72 - 77.09)^2 + \cdots + (69 - 77.09)^2]$$

or

$$s^2 = 101.72.$$

DO EXERCISE 10.

When a probability function is known, we can find a "theoretical" variance defined as the *expected value* of the *deviation squared*:

$$\sigma^2 = E[(X - \mu)^2] = \sum_{i=1}^{n} p_i (x_i - \mu)^2.$$

Example 1 ▦ Given a box of 20 transistors, of which 5 are defective. What is the variance of the number of defective transistors?

Solution   From Example 2, Section 6.1, we have the probability function:

x	0	1	2	3
p	$\frac{91}{228}$	$\frac{105}{228}$	$\frac{30}{228}$	$\frac{2}{228}$

From Example 4, Section 6.2, we have the expected value

$$\mu = E(X) = \tfrac{3}{4}.$$

The variance is given by:

$$\sigma^2 = \tfrac{91}{228}(0 - \tfrac{3}{4})^2 + \tfrac{105}{228}(1 - \tfrac{3}{4})^2 + \tfrac{30}{228}(2 - \tfrac{3}{4})^2 + \tfrac{2}{228}(3 - \tfrac{3}{4})^2$$

$$= \tfrac{91}{228} \cdot \tfrac{9}{16} + \tfrac{105}{228} \cdot \tfrac{1}{16} + \tfrac{30}{228} \cdot \tfrac{25}{16} + \tfrac{2}{228} \cdot \tfrac{81}{16}$$

$$= 0.503.$$

DO EXERCISE 11.

Note that we started with certain units. For example, in Example 1 the test scores were measured in percentile "points." So is the average

---

* Here we consider *all* the data. If we were using only a *sample* of the data, then the n in the denominator should be replaced by $(n - 1)$, for statistical reasons which we cannot go into in this text. We need the definition of variance using n for later use.

value. On the other hand, the variance is measured in the *square* of these units, or "points-squared." We can obtain a measure of the "spread" of the data in the same units as the data by using the *standard deviation* which is the square root of the variance.

Example 2 ▦ Obtain the standard deviation of the test scores of Example 1, Section 6.2.

Solution The standard deviation is the positive square root of the variance:

$$s = +\sqrt{s^2} = \sqrt{\frac{1}{n}\sum_{i=1}^{n}(x_i - \bar{x})^2}.$$

From earlier work in this section, we know that $s^2 = 101.72$, so that $s = \sqrt{101.72} = 10.09$. (Use a square-root table or hand calculator.)

DO EXERCISE 12.

In general, a small standard deviation relative to the mean indicates that the data has little spread about the mean, while a large standard deviation relative to the mean indicates a large spread about the mean.

Example 3 ▦ Find the standard deviation for the number of defective transistors of Example 1.

Solution The standard deviation is

$$\sigma = \sqrt{\sigma^2} = \sqrt{\sum_{i=1}^{n} p_i(x_i - \mu)^2}.$$

In Example 1, we found that $\sigma^2 = 0.503$, so that

$$\sigma = \sqrt{0.503} = 0.709.$$

DO EXERCISE 13.

12. ▦ Find the standard deviation of the test scores of Margin Exercise 10.

13. ▦ Find the standard deviation of the number of gold bars of Margin Exercise 11.

EXERCISE SET 6.3

1. ▦ *Business, Racing.* Find the variance and standard deviation for the lap speeds of Exercise 1, Set 6.2:

    90.791,   89.237,   89.108,   87.926,   86.401,

    85.858,   83.271,   and   79.669 mph.

2. ▦ *Business, Racing.* Find the variance and standard deviation for the lap speeds of Exercise 2, Set 6.2:

    91.101,   90.973,   89.257,   86.118,   85.879,

    82.438,   81.962,   78.113,   and   78.104 mph.

3. ▦ *Business.* Find the variance and standard deviation for the dinners of Exercise 3, Set 6.2:

23 at \$7.95, 47 at \$6.95, 53 at \$6.45, 33 at \$5.95, 29 at \$4.95, 37 at \$4.45, and 23 at \$3.75.

5. A union has 21 members, of whom 14 are women and 7 are men. Three people are chosen at random. What is the variance and standard deviation of the number of women chosen? [See Exercise 5, Set 6.2.]

7. *Public Health.* Five workers in a group of 25 have mononucleosis. Three workers are chosen at random. What is the variance and standard deviation of the number with mono? [See Exercise 9, Set 6.2.]

4. ▦ *Business.* Find the variance and standard deviation for the tickets of Exercise 4, Set 6.2:

320 at \$6.90, 480 at \$5.90, 624 at \$4.70, and 484 at \$3.60.

6. The sales force of a business consists of 20 people, half of whom are men and the other half women. Four people are chosen at random. What is the variance and standard deviation of the number of women chosen? [See Exercise 6, Set 6.2.]

8. *Business.* A crate of 20 machine parts contains 6 defective parts. Five parts are taken out at random. What is the variance and standard deviation of the number of defective parts taken out? [See Exercise 10, Set 6.2.]

OBJECTIVES

You should be able to:

a) Calculate the expected value of a random variable (two ways) in a binomial-probability problem.

b) Solve binomial-probability problems.

6.4 BERNOULLI TRIALS—BINOMIAL PROBABILITY

Many experiments have outcomes which fall naturally into two disjoint sets designated simply "success" or "failure." For example,

i) The flipping of a coin and its landing "heads" or "tails" (barring its standing on edge);

ii) The winning or losing of an election (allowing for the ultimate resolution of a tie);

iii) The passing or failing of a manufactured article to a particular tolerance for quality control.

Experiments or trials with two possible outcomes are called *Bernoulli trials.*

Most of the problems considered so far involved repeated trials in which each trial changed the conditions of subsequent trials. In particular for hypergeometric probabilities, an object was drawn at random from a group of objects and *not* replaced before the next trial.

Now we consider trials such that whatever is removed in one trial is *replaced* before the next trial. Each trial is the same and hence *independent* of the others. Thus, we consider *repeated independent Bernoulli trials.*

Example 1 A fair coin is flipped repeatedly. What is the probability for heads to show 3 times out of 5 flips?

Solution Here a "success" can be identified with a coin showing "heads." Since the coin is fair, the probability for either heads or tails to show is $\frac{1}{2}$. One way for heads to show 3 times out of 5 is for the first 3 flips to show heads and the last two to show tails, that is

HHHTT.

Each particular outcome (H or T) occurs with probability $\frac{1}{2}$. In the outcome HHHTT, each individual flip is independent of the others. Thus the combined outcome HHHTT will happen with a probability that is the product of the individual probabilities (Multiplication Theorem):

$$\left(\tfrac{1}{2}\right)^3\left(\tfrac{1}{2}\right)^2 = \tfrac{1}{32}.$$

There are 10 configurations in which we can get 3 heads out of 5 flips. They are

HHHTT	TTHHH
HHTHT	THTHH
HTHHT	THHTH
HHTTH	HTTHH
HTHTH	THHHT.

Each of these configurations is equally probable, since it is calculated exactly the same way as the previous calculation for HHHTT.

Using techniques from Chapter 5, we can compute directly the number of ways 3 heads can show in 5 flips:

$$C(5, 3) = \binom{5}{3} = \frac{5 \cdot 4 \cdot 3}{1 \cdot 2 \cdot 3} = 10.$$

The probability for heads to show 3 times out of five is the product of the probability for heads to show 3 times out of five in a particular configuration times the number of different ways heads can show 3 times out of five, that is

$$p_{3H} = \tfrac{1}{32} \cdot 10 = \tfrac{10}{32}.$$

Note that the answer is the same whether *one* coin is flipped *5* times in a row or *5* coins are flipped *once* at the same time. In either case each trial is identical.

DO EXERCISE 14.

14. What is the probability for tails to show twice in 6 flips of a fair coin? List the configurations.

Problems of this type have a *binomial* probability. This means that for a *binomial* random variable,

  i) There are two outcomes (Bernoulli trials);
  ii) Each trial is independent of preceding trials;
  iii) Each trial is identical to preceding trials.

Binomial probability is similar to *hypergeometric* probability in that objects in random drawings are restricted to *two* kinds, but differs in that *binomial* probability assumes *replacement* or identical trials while *hypergeometric* probability assumes *no replacement* or that successive trials differ.

Example 2 *Business, Quality Control.* Electrical switches are man-ufactured with 10% being defective. Five switches are drawn at random and without replacement. What is the probability that two of these are defective?

Solution If this problem had specified that 5 switches were to be drawn from some *fixed* number of switches, then the probability would be hypergeometric. However, a fixed number is *not* specified. Rather, switches are being manufactured, as on a production line, and continually fed into some container from which the sample of 5 is taken. Thus, within the limits of the information available to us, the trials are independent and hence the probability is binomial.

Let the *event* be the drawing of a *defective* switch. This event has a probability $p = 0.10$. The probability for drawing a nondefective switch is $q = 1 - p = 0.90$.

The probability for getting the *first two* switches defective and the *last three* switches nondefective is

$$(0.10)^2(0.90)^3.$$

The number of ways two defective switches can be drawn out of a sample of five is $\binom{5}{2}$.

Thus, the probability that there will be two defective switches in a sample of five is

$$\binom{5}{2}(0.10)^2(0.90)^3 = 10(0.10)(0.729) = 0.0729.$$

In general, if the probability is p that some event *will* happen in one trial and $q = 1 - p$ that it will *not*, then the binomial probability p_k for the event to happen k times out of n trials is given by

$$p_k = B(n, k, p) = \binom{n}{k}p^k q^{n-k}.$$

In Example 2, we have

$$p = 0.10, \quad q = 0.90, \quad n = 5, \quad k = 2,$$

so that

$$p_2 = \binom{5}{2}(0.10)^2(0.90)^3 = 0.0729, \quad \text{as before.}$$

Alternately, we may take the *event* as the drawing of a *nondefective* switch. Then we seek the probability of drawing *three nondefective* switches out of a sample of five. Thus

$$p = 0.90, \quad q = 0.10, \quad n = 5, \quad k = 3,$$

and

$$p_3 = \binom{5}{3}(0.90)^3(0.10)^2 = 0.0729.$$

This is the same answer as before but with different notation. Recall that

$$\binom{n}{k} = \binom{n}{n-k}.$$

DO EXERCISE 15.

Example 3 What is the probability function for the problem of Example 2?

Solution The probability in this case is binomial. Thus, taking the random variable K to be the number of defective switches observed in the sample, we have

k	p_k
0	$\binom{5}{0}(0.1)^0(0.9)^5 = 0.59049$
1	$\binom{5}{1}(0.1)^1(0.9)^4 = 0.32805$
2	$\binom{5}{2}(0.1)^2(0.9)^3 = 0.07290$
3	$\binom{5}{3}(0.1)^3(0.9)^2 = 0.00810$
4	$\binom{5}{4}(0.1)^4(0.9)^1 = 0.00045$
5	$\binom{5}{5}(0.1)^5(0.9)^0 = 0.00001$

15. *Public Health.* Treatment for a certain disease is effective 80% of the time. If six treated patients are surveyed, what is the probability that four of them will be cured?

16. What is the probability function for the problem of Margin Exercise 15?

17. Determine the expected value for the problem of Margin Exercise 15 using both formulas.

DO EXERCISE 16.

In Section 6.2 the expected value of a random variable X was defined by

$$\mu = E(X) = \sum_{i=1}^{n} x_i p_i.$$

For binomial probability, the random variable is K, the number of successes in a series of trials. It takes on the values $k = 0, 1, \ldots, n$. Thus, we write

$$\mu = E(K) = \sum_{k=1}^{n} k p_k.$$

It can be shown that the expected value for the random variable with binomial probability is given by

$$\mu = E(K) = n \cdot p,$$

where n is the number of trials and p is the probability for the event to occur in one trial.

Note that the expected value for the *hypergeometric* probability distribution is the same as that for the *binomial* probability *provided* that, in the former case, p is taken as the probability of an *initial* success; that is, $p = m/s$, as in Section 6.2.

Example 4 Determine the expected value for the problem of Example 2 using both formulas.

Solution Using

$$E(K) = \sum_{k=1}^{n} k p_k,$$

we have

$$E(K) = 0(0.59049) + 1(0.32805) + 2(0.07290) + 3(0.00810)$$
$$+ 4(0.00045) + 5(0.00001) = 0.50000.$$

Using $E(K) = n \cdot p$, we have $n = 5$ and $p = 0.1$, so that

$$E(K) = 5(0.1) = 0.5, \quad \text{as above.}$$

DO EXERCISE 17.

We can also determine the variance and standard deviation for a binomial probability distribution.

Example 5 ▦ Find the variance and standard deviation for the problem of Example 2.

Solution   The expected value for this problem was found in Example 4 to be
$$\mu = 0.5.$$

Thus, the variance is

$$\sigma^2 = 0.59049(0 - 0.5)^2 + 0.32805(1 - 0.5)^2 + 0.07290(2 - 0.5)^2$$
$$+ 0.00810(3 - 0.5)^2 + 0.00045(4 - 0.5)^2 + 0.00001(5 - 0.5)^2$$

or

$$\sigma^2 = 0.45000,$$

so that the standard deviation is

$$\sigma = \sqrt{0.45000} = 0.67082.$$

The variance and standard deviation for a binomial probability-distribution function can be obtained simply from the formula

$$\sigma^2 = np(1 - p).$$

In this case, $n = 5$ and $p = 0.1$ (see Example 2), so that

$$\sigma^2 = 5 \cdot 0.1(1 - 0.1), \quad \text{or} \quad \sigma^2 = 0.45, \quad \text{as before.}$$

DO EXERCISE 18.

Example 6   *Business, Quality Control.* Given a box of 20 transistors of which 5 are defective. Three transistors are drawn at random, the number defective is noted, and the transistors are replaced. This is repeated 5 times. What is the probability that at least one is defective in at least 4 trials?

Solution   The probability p that at least one transistor is defective is the *hypergeometric* probability given by the quantity $p_{i \geq 1}$ in Example 5 of Section 5.2; that is,

$$p = \tfrac{137}{228} \quad \text{and} \quad q = 1 - p = \tfrac{91}{228}.$$

This corresponds to a "success" (that is, the event happening) in the second part of the problem. The probability for at least 4 successes is

$$p_{i \geq 4} = p_4 + p_5$$

where p_4 and p_5 are the *binomial* probabilities

$$p_4 = \tbinom{5}{4}\left(\tfrac{137}{228}\right)^4\left(\tfrac{91}{228}\right)^1 \quad \text{and} \quad p_5 = \tbinom{5}{5}\left(\tfrac{137}{228}\right)^5\left(\tfrac{91}{228}\right)^0.$$

Note that while in Example 2 the probability was given, in this Example it had to be computed. Furthermore, one problem may involve more than one type of probability, in this case both hypergeometric and binomial.

DO EXERCISE 19.

18. ▦ Determine the variance and standard deviation for the problem of Margin Exercise 15 using both formulas.

19. An old sea chest contains 10 bars of silver and 5 bars of gold. Three are drawn out at random, the number of gold bars is noted, and the bars are replaced. This is repeated four times. What is the probability for drawing two gold bars three times? *Caution.* Interpret subscripts carefully.

EXERCISE SET 6.4

1. Five fair coins are tossed. What is the probability function for the number of tails showing? What is the expected number of tails? What is the variance and standard deviation? Use both formulas.

3. *Political Science.* Half the people in a community favor a certain political stand and half oppose it. Of 6 people selected at random, what is the expected number to favor the stand? to oppose the stand? What is the probability that of these 6 people 4 or more will either favor the stand or oppose it?

5. *Business.* An impostor applies for a job as a wine taster. As a test he is given 5 wines to taste to determine whether they are *vin ordinaire* or a great vintage wine. What is the probability that he gets at least 4 out of 5 correct by guessing? What is the expected number of correct evaluations? What is the variance and standard deviation?

7. *Science.* A complex experiment consists of 6 components each with a reliability of 0.9 (that is, the probability for the component to work is 0.9). If the experiment is so constructed that it can be run if no more than one out of the 6 components fails to function properly, what is the probability that the experiment can be run? What is the expected number of failures? What is the variance and standard deviation?

9. *Public Health.* Treatment for a certain disease is effective 80% of the time. If 5 patients are sampled, what is the probability function for the number of effective treatments? What is the expected value? What is the probability that the treatment is ineffective for at least one patient?

11. *Business, Quality Control.* Sparkplugs are manufactured and pass along a conveyor belt for inspection. A sample of 5 is taken at random. What is the probability function for the number of defective plugs if the defective rate is 10%? if the defective rate is 20%? if the defective rate is 30%? Which defective rate is most probable if *no* defective plugs are found in the sample? if one defective plug is found in the sample? if two are found? if three are found?

2. *Public Health.* If the birth rate for boys and girls were equal, what would be the distribution of girls in a four-child family? What is the expected number of girls? What is the variance and standard deviation? Use both formulas.

4. *Political Science.* One-third of the people in a community favor a certain political stand and two thirds oppose it. Of 6 people selected at random, what is the expected number to favor the stand? to oppose the stand? What is the probability that of these 6 people at least half will oppose the issue?

6. *Business.* As in Exercise 5, but the applicant is genuine and can distinguish the two wines 4 times out of 5. What is the probability that he fails the test (that is, fails to get at least 4 out of 5 correct)? What is the expected number of correct evaluations? What is the variance and standard deviation?

8. *Science.* A successful flight of an exploratory space rocket requires that no more than two of the 10 components fail to function properly (due to use of interlocking failsafe circuits). If each component has a reliability of 0.98, what is the probability for a successful flight? What is the expected number of component failures? What is the variance and standard deviation?

10. *Demographics.* If 30% of marriages end in divorce by the fifth year of marriage, what is the probability function for the number of couples out of a sample of 6 who have been divorced after no more than 5 years of marriage? What is the probability that half or more of the sample has been divorced? What is the expected number of divorced couples? the variance? the standard deviation?

12. *Business, Quality Control.* Machine A makes ballpoint pens with a deficiency rate of 10% and machine B makes pens with a deficiency rate of 30%. Two pens are taken from each machine. What is the probability that two of the four are defective? If the pens from the two machines are mixed half and half (that is, with a deficiency rate of 20%) before a sample of 4 is taken, what is the probability that there will be two defective pens in the sample?

13. *Business.* A company manufactures a type of mousetrap which is 50% effective. They would like to claim that it is at least 80% effective. A sample of 5 traps is tested for effectiveness. What is the probability that the sample is at least 80% effective? If the testing of a sample of 5 traps is repeated 5 times, what is the expected number of trials for which the samples tested are at least 80% effective?

15. As in Exercise 14, what is the probability that at least two thermometers are defective? If the test is repeated 5 times, what is the probability that at least two defective thermometers are found in at least 2 trials out of the five?

14. *Business.* A company manufactures thermometers with a deficiency rate of 20%. A sample of 5 thermometers is tested. What is the probability that at least one thermometer is defective? If this test is repeated 5 times, what is the probability that there is at least one defective thermometer in each trial? in at least 4 out of 5 trials?

CHAPTER 6 TEST

1. A multiple-choice test consists of 5 questions, each with a choice of 4 answers. If a student guesses answers at random, what is the probability function for the number of correct answers?

3. In question 2, what is the number of questions the student should expect to get correct?

2. A multiple-choice test consists of 5 questions each with a choice of 4 answers. If the probability is $\frac{3}{4}$ that a student knows an answer to a question, what is the probability that he gets at least 4 correct?

4. Given the following probability function, find the expected value.

x	0	1	2	3	4
p	$\frac{1}{9}$	$\frac{2}{9}$	$\frac{3}{9}$	$\frac{2}{9}$	$\frac{1}{9}$

5. Using the data of question 4, find the variance.

6. Using the data of question 5, find the standard deviation.

Decision Theory—
Markov Chains
and Games

OBJECTIVES

You should be able to:

a) Given a problem involving a Markov chain, draw the transition diagram, and find the transition matrix;
b) Given a matrix, decide whether it qualifies as a transition matrix;
c) Given an initial probability vector P_0 and a transition matrix T, find P_n either as

$$P_n = P_{n-1}T,$$

or

$$P_n = P_0T^n.$$

7.1 TRANSITION MATRICES AND PROBABILITY VECTORS

A *Markov chain* is a sequence of experiments with certain features which we shall illustrate before presenting a formal definition.

Example 1   *Business, Marketing Surveys.* A child, looking back over the many ice cream cones he has eaten through the years, recalls that:

a) After he had eaten a vanilla cone, the probability was:

   i) 0 that he would pick vanilla next time,
   ii) $\frac{1}{2}$ that he would pick chocolate next time,
   iii) $\frac{1}{2}$ that he would pick strawberry next time;

b) After he had eaten a chocolate cone, the probability was:

   i) $\frac{1}{5}$ that he would pick vanilla next time,
   ii) $\frac{2}{5}$ that he would pick chocolate next time,
   iii) $\frac{2}{5}$ that he would pick strawberry next time;

c) After he had eaten a strawberry cone, the probability was:

   i) $\frac{1}{3}$ that he would pick vanilla next time,
   ii) 0 that he would pick chocolate next time,
   iii) $\frac{2}{3}$ that he would pick strawberry next time.

Assuming that the child's first ice cream cone is vanilla, draw the tree describing possible outcomes through his third ice cream cone.

Solution   The tree can be drawn in a straightforward manner.

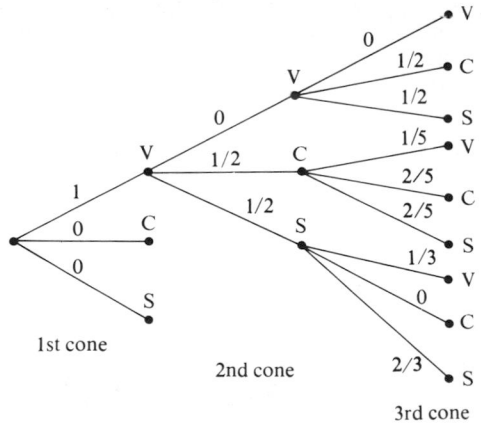

In view of the *repetitive* nature of the example, continuing the tree through further cycles (that is, the fourth and following ice cream cones) becomes increasingly awkward. Hence, it is useful to adopt an alternate way of representing these trials—that is, by means of a *transition diagram*.

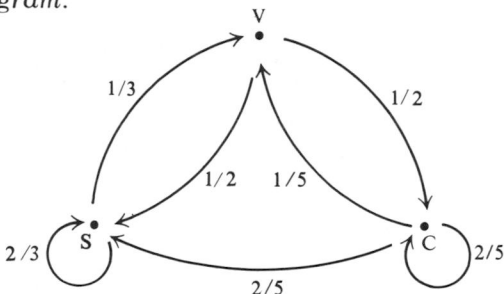

Here the directed line represents a *transition* from one *state* to another (here the state corresponds to a flavor). The number along the line corresponds to the probability that, if one starts in one state, next time he will be in the other state; that is, the probability of transition from one state to the other. The "transition" from one state at one stage to the *same* state at the next stage is represented by a "self" loop. The line is omitted where the transition probability is zero.

This transition diagram illustrates the general features of a *Markov chain*:

1. **The outcome of each experiment (or process, or choice) is one of a *set of discrete states* (a state is another name for an outcome).**
2. **The probability for transition from one state to another depends only on the present state (that is, the state one is in and is leaving).**

In the present example the states correspond to flavors. And since the next choice of flavor depends only on the previous choice, the whole process is a Markov chain.

DO EXERCISE 1.

Transition *diagrams* provide a *graphical* way of representing Markov chains. However, for computational purposes, *transition matrices* are more convenient.

We can define a transition matrix *T* by

$$T = [t_{ij}]_{n \times n},$$

a square matrix, where there are *n* states and t_{ij} represents the transition probability from state *i* to state *j*, so that $0 \le t_{ij} \le 1$. Note that the order of the indices *is* important. Since the object in question must be in one of the *n* states,

$$\sum_{j=1}^{n} t_{ij} = 1 \qquad \text{for all } i = 1, \ldots, n;$$

1. *Political Science.* Of voters sampled, 60% of the Democrats (that is, who voted Democrat in the last election) will vote Democrat in the next election, 20% will vote Republican, and 20% will vote Independent.

Of the Republicans, 40% will vote Democrat and 60% will vote Republican. Of the Independents, 40% will vote Democrat, 20% will vote Republican, and 40% will vote Independent.

a) Assuming that voters are split evenly among the three parties, draw a tree indicating voting patterns through one election.

b) Draw the transition diagram.

2. What is the transition matrix for the problem of Margin Exercise 1?

that is, **the sum of the *row* entries must be 1. There is no corresponding restriction for *column* entries.**

Example 2   What is the transition matrix for the problem of Example 1?

Solution   Let vanilla be state 1, chocolate be state 2, and strawberry be state 3. Then

$$T = \begin{array}{c} \\ \\ \end{array} \begin{array}{ccc} V & C & S \\ \begin{bmatrix} 0 & \frac{1}{2} & \frac{1}{2} \\ \frac{1}{5} & \frac{2}{5} & \frac{2}{5} \\ \frac{1}{3} & 0 & \frac{2}{3} \end{bmatrix} & \begin{array}{c} V \\ C \\ S \end{array} \end{array}$$

Note that the *row* elements do sum to 1.

DO EXERCISE 2.

Example 3   Given the following matrix, determine whether it qualifies as a transition matrix. If it does, draw the corresponding transition diagram.

$$T = \begin{array}{c} \\ \\ \end{array} \begin{array}{cccc} 1 & 2 & 3 & 4 \\ \begin{bmatrix} \frac{1}{2} & \frac{1}{2} & 0 & 0 \\ \frac{1}{2} & 0 & \frac{1}{2} & 0 \\ 0 & \frac{1}{3} & \frac{1}{3} & \frac{1}{3} \\ \frac{1}{2} & 0 & 0 & \frac{1}{2} \end{bmatrix} & \begin{array}{c} 1 \\ 2 \\ 3 \\ 4 \end{array} \end{array}$$

Solution

i) Since the elements are all nonnegative and the row elements all sum to one, we have a transition matrix and can proceed.
ii) Labelling the states 1, 2, 3, and 4, we have:

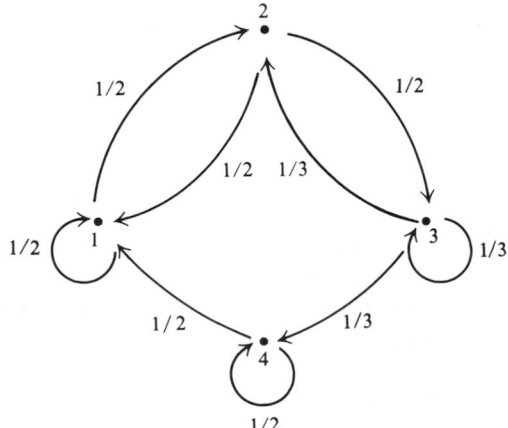

Note that the various states can be located where convenient or advantageous, so that pictorially the transition diagram may not be unique.

DO EXERCISE 3.

Transition matrices are useful in determining the probability of being in various states at later stages.

Example 4   For Example 1, determine the probability that the child will choose each of the different flavors for the second cone; for the third cone.

Solution   Let us solve this problem first by using the tree diagram, then using the transition matrix.

Method 1   It is convenient to describe the initial state of the system by a probability *vector* (here a row matrix),

$$P_0 = [p_1 \quad p_2 \cdots p_n],$$

where p_i $(i = 1, \ldots, n)$ is the probability of being in state i at that stage. Thus, $0 \leqslant p_i \leqslant 1$, because p_i is a probability and $\sum_{i=1}^{n} p_i = 1$, since the n states exhaust the possibilities. In the present case, the initial probability vector is

$$P_0 = [1 \quad 0 \quad 0],$$

since the child's first ice cream cone was vanilla (the first state) and there were three flavors. If the child were equally likely to have

3. For each matrix, determine whether it qualifies as a transition matrix. If it does, draw the corresponding transition diagram.

a) $T = \begin{bmatrix} \frac{1}{3} & \frac{1}{3} & \frac{1}{3} & 0 \\ \frac{1}{2} & \frac{1}{2} & 0 & 0 \\ \frac{1}{3} & 0 & \frac{1}{3} & \frac{1}{3} \\ 0 & 0 & \frac{1}{2} & \frac{1}{2} \end{bmatrix}$

b) $T = \begin{bmatrix} \frac{1}{8} & \frac{2}{8} & 0 & \frac{5}{8} \\ \frac{1}{9} & \frac{2}{9} & 0 & \frac{5}{9} \\ \frac{1}{2} & 0 & 0 & \frac{1}{2} \\ \frac{1}{4} & \frac{3}{4} & 0 & 0 \end{bmatrix}$

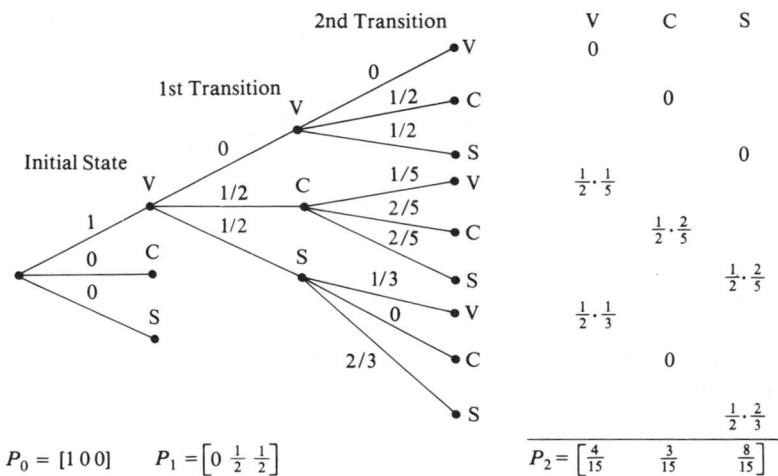

$P_0 = [1\,0\,0] \qquad P_1 = \begin{bmatrix} 0 & \frac{1}{2} & \frac{1}{2} \end{bmatrix} \qquad\qquad P_2 = \begin{bmatrix} \frac{4}{15} & \frac{3}{15} & \frac{8}{15} \end{bmatrix}$

picked any flavor, then the initial probability vector would have been

$$P_0 = [\tfrac{1}{3} \quad \tfrac{1}{3} \quad \tfrac{1}{3}].$$

Multiplying along branches of the tree and adding among the branches, we obtain the results of the *first* choice as the probability vector

$$P_1 = [0 \quad \tfrac{1}{2} \quad \tfrac{1}{2}]. \qquad \text{(See figure on page 313.)}$$

The components represent the probability that vanilla, chocolate, or strawberry will be chosen for the *second* cone, given that the *first* cone was vanilla.

Continuing along the branches of the tree for the *second* choice (that is, the third cone), we obtain

$$P_2 = [\tfrac{4}{15} \quad \tfrac{3}{15} \quad \tfrac{8}{15}].$$

The components represent the probability that vanilla, chocolate, or strawberry will be chosen for the *third* cone, given that the *first* cone was vanilla.

Method 2 As can be seen from the transition *diagram*, the same results can be obtained from the transition *matrix* in the following manner:

$$P_1 = P_0 T,$$
$$P_2 = P_1 T,$$
$$\begin{matrix} \cdot & \cdot \\ \cdot & \cdot \\ \cdot & \cdot \end{matrix}$$

Here,

$$P_1 = [1 \quad 0 \quad 0] \cdot \begin{bmatrix} 0 & \tfrac{1}{2} & \tfrac{1}{2} \\ \tfrac{1}{5} & \tfrac{2}{5} & \tfrac{2}{5} \\ \tfrac{1}{3} & 0 & \tfrac{2}{3} \end{bmatrix} = [0 \quad \tfrac{1}{2} \quad \tfrac{1}{2}],$$

$$P_2 = [0 \quad \tfrac{1}{2} \quad \tfrac{1}{2}] \cdot \begin{bmatrix} 0 & \tfrac{1}{2} & \tfrac{1}{2} \\ \tfrac{1}{5} & \tfrac{2}{5} & \tfrac{2}{5} \\ \tfrac{1}{3} & 0 & \tfrac{2}{3} \end{bmatrix} = [\tfrac{4}{15} \quad \tfrac{3}{15} \quad \tfrac{8}{15}],$$

$$\begin{matrix} \cdot \\ \cdot \\ \cdot \end{matrix}$$

The numerical computation is the same in both cases, but the transition matrix facilitates the computation.

Alternately, we could substitute the expression for P_1 into the expression for P_2, obtaining:

$$P_2 = (P_0 T)T = P_0 T^2.$$

Thus, here we have

$$P_2 = [1 \quad 0 \quad 0] \cdot \begin{bmatrix} 0 & \frac{1}{2} & \frac{1}{2} \\ \frac{1}{5} & \frac{2}{5} & \frac{2}{5} \\ \frac{1}{3} & 0 & \frac{2}{3} \end{bmatrix}^2 = [1 \quad 0 \quad 0] \cdot \begin{bmatrix} \frac{4}{15} & \frac{3}{15} & \frac{8}{15} \\ \frac{32}{150} & \frac{39}{150} & \frac{79}{150} \\ \frac{4}{18} & \frac{3}{18} & \frac{11}{18} \end{bmatrix}$$

or

$$P_2 = [\tfrac{4}{15} \quad \tfrac{3}{15} \quad \tfrac{8}{15}], \quad \text{as before.}$$

In general, we can obtain each probability vector from its predecessor,

$$P_n = P_{n-1}T;$$

or directly from the initial probability vector,

$$P_n = P_0 T^n.$$

DO EXERCISE 4.

4. For Margin Exercise 1, determine the probability vector for the first election using both a tree diagram and the transition matrix. Assuming the transition matrix does not change for the subsequent election, determine the probability vector for the second election from the transition matrix, in two ways.

EXERCISE SET 7.1

1. *Business.* A taxi company in a certain town has set up three zones. Taxis picking up a passenger in the first zone have 50% probability of delivering the passenger to that zone and are twice as likely to deliver a passenger to the second zone as to the third zone. A passenger picked up in the second zone will be let off there with a probability equal to that for being delivered to either other zone. A passenger picked up in the third zone is twice as likely to go to the first zone as either to go to the second zone or to stay in the third zone. Draw the transition diagram and find the transition matrix.

2. *Business, Marketing Surveys.* Of car owners surveyed, 60% of the VW owners would buy a VW for their next car while 20% each would buy a Ford or Chevy. Of the Ford owners, 30% would buy a Ford next time, 30% would buy VW, and 40% a Chevy. Of the Chevy owners, 40% would buy a Chevy, 40% a VW, and 20% a Ford. Draw the transition diagram and find the transition matrix.

For each matrix of Exercises 3 through 14, determine whether it qualifies as a transition matrix. If not, state why. If so, draw the transition diagram.

3. $\begin{bmatrix} \frac{1}{2} & -\frac{1}{8} & \frac{5}{8} \\ \frac{1}{3} & \frac{1}{3} & \frac{1}{3} \\ \frac{1}{5} & \frac{2}{5} & \frac{2}{5} \end{bmatrix}$

4. $\begin{bmatrix} \frac{2}{5} & \frac{2}{5} & \frac{1}{5} \\ 0 & 1 & 0 \\ 1 & 0 & 0 \end{bmatrix}$

5. $\begin{bmatrix} \frac{1}{2} & \frac{1}{2} & 0 & 0 \\ \frac{1}{2} & \frac{1}{2} & 0 & 0 \\ 0 & 0 & \frac{2}{3} & \frac{1}{3} \\ 0 & 0 & \frac{1}{3} & \frac{2}{3} \end{bmatrix}$

6. $\begin{bmatrix} \frac{1}{3} & \frac{2}{3} & 0 \\ \frac{1}{2} & \frac{3}{8} & \frac{3}{8} \\ 0 & \frac{2}{3} & \frac{1}{3} \end{bmatrix}$

7. $\begin{bmatrix} 1 & 0 \\ 0 & 1 \end{bmatrix}$

8. $\begin{bmatrix} 1 & 0 \\ 0 & 1 \end{bmatrix}$

9. $\begin{bmatrix} 1 & 0 & 0 \\ 0 & 1 & 0 \\ \frac{1}{2} & \frac{1}{2} & 0 \end{bmatrix}$

10. $\begin{bmatrix} 0 & 1 & 0 \\ 0 & 0 & 1 \\ 1 & 0 & 0 \end{bmatrix}$

11. $\begin{bmatrix} 0 & 1 & 0 \\ 0 & 0 & 1 \\ 0 & 0 & 1 \end{bmatrix}$

12. $\begin{bmatrix} \frac{1}{2} & \frac{1}{2} & 0 & 0 \\ 0 & \frac{1}{2} & \frac{1}{2} & 0 \\ 0 & 0 & \frac{1}{2} & \frac{1}{2} \\ 0 & 0 & \frac{1}{2} & \frac{1}{2} \end{bmatrix}$

13. $\begin{bmatrix} \frac{1}{2} & \frac{1}{2} & 0 & 0 \\ 0 & 1 & 0 & 0 \\ 0 & \frac{1}{3} & \frac{1}{3} & \frac{1}{3} \\ 0 & 0 & 0 & 1 \end{bmatrix}$

14. $\begin{bmatrix} 0 & 1 & 0 & 0 \\ \frac{1}{3} & \frac{1}{3} & \frac{1}{3} & 0 \\ 0 & \frac{1}{3} & \frac{1}{3} & \frac{1}{3} \\ 0 & 0 & 1 & 0 \end{bmatrix}$

15. *Business.* As in Exercise 1, a taxi starts in the second zone. Using a tree, determine its probable location after discharging its second passenger. What is the initial probability vector? Determine the probability vector after the second passenger, two ways.

16. *Business, Marketing Surveys.* As in Exercise 2, assume that initially car ownership is equally divided among VW, Ford, and Chevy. Using a tree, determine the probable ownership distribution for the second car; for the third car. What is the initial probability vector? Determine the probability vector for the second car, and for the third car, two ways.

17. As in Exercise 15, but the taxi starts in the *third* zone. Which method is less work?

18. As in Exercise 16, but the initial distribution is all VW's; is all Fords; is all Chevies.

In each of Exercises 19 through 35, given P_0 and T, determine P_n.

19. $P_0 = [1 \quad 0 \quad 0]$, T from Exercise 4, $P_2 = ?$

20. $P_0 = [1 \quad 0 \quad 0]$, T from Exercise 4, $P_3 = ?$

21. $P_0 = [\frac{1}{4} \quad \frac{1}{4} \quad \frac{1}{4} \quad \frac{1}{4}]$, T from Exercise 5, $P_2 = ?$

22. $P_0 = [\frac{1}{5} \quad \frac{4}{5}]$, T from Exercise 7, $P_5 = ?$

23. $P_0 = [\frac{1}{2} \quad \frac{1}{2}]$, T from Exercise 8, $P_1 = ?$

24. $P_0 = [1 \quad 0]$, T from Exercise 8, $P_1 = ?$

25. $P_0 = [1 \quad 0]$, T from Exercise 8, $P_2 = ?$

26. $P_0 = [0 \quad 0 \quad 1]$, T from Exercise 9, $P_1 = ?$

27. $P_0 = [0 \quad 0 \quad 1]$, T from Exercise 9, $P_2 = ?$

28. $P_0 = [\frac{1}{3} \quad \frac{1}{3} \quad \frac{1}{3}]$, T from Exercise 11, $P_2 = ?$

29. $P_0 = [\frac{1}{3} \quad \frac{1}{3} \quad \frac{1}{3}]$, T from Exercise 11, $P_3 = ?$

30. $P_0 = [\frac{1}{4} \quad \frac{1}{4} \quad \frac{1}{4} \quad \frac{1}{4}]$, T from Exercise 12, $P_1 = ?$

31. $P_0 = [\frac{1}{4} \quad \frac{1}{4} \quad \frac{1}{4} \quad \frac{1}{4}]$, T from Exercise 12, $P_2 = ?$

32. $P_0 = [\frac{1}{4} \quad \frac{1}{4} \quad \frac{1}{4} \quad \frac{1}{4}]$, T from Exercise 13, $P_1 = ?$

33. $P_0 = [\frac{1}{4} \quad \frac{1}{4} \quad \frac{1}{4} \quad \frac{1}{4}]$, T from Exercise 13, $P_2 = ?$

34. $P_0 = [\frac{1}{4} \quad \frac{1}{4} \quad \frac{1}{4} \quad \frac{1}{4}]$, T from Exercise 14, $P_1 = ?$

35. $P_0 = [\frac{1}{4} \quad \frac{1}{4} \quad \frac{1}{4} \quad \frac{1}{4}]$, T from Exercise 14, $P_2 = ?$

OBJECTIVES

You should be able to

a) Determine whether a transition matrix is regular.
b) Determine whether an ergodic transition matrix is regular.

7.2 REGULAR AND IRREGULAR MARKOV CHAINS

In order to determine the long-range characteristics of a Markov chain, we shall be interested in whether or not its transition matrix is *regular*.

A Markov chain is *regular* if its transition matrix is regular. A transition matrix

$$T = [t_{ij}]_{n \times n}$$

is regular if some power k of T has all positive elements; that is,

$$T^k = [t_{ij}^{(k)}]_{n \times n}$$

where

$$t_{ij}^{(k)} > 0 \quad \text{for all } i, j.$$

Example 1   Determine whether the following transition matrix is regular and draw its transition diagram:

$$T = \begin{bmatrix} \frac{1}{3} & \frac{2}{3} & 0 \\ \frac{1}{2} & 0 & \frac{1}{2} \\ 0 & \frac{1}{3} & \frac{2}{3} \end{bmatrix}$$

Solution   The transition diagram is

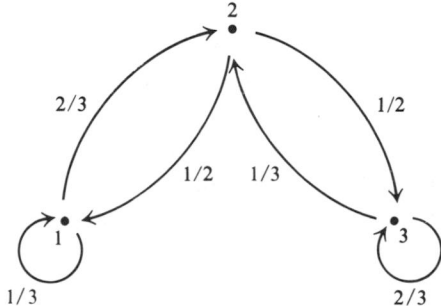

Since the first power of T contains zero elements, we proceed to square T and obtain:

$$T^2 = \begin{bmatrix} \frac{8}{18} & \frac{4}{18} & \frac{6}{18} \\ \frac{3}{18} & \frac{9}{18} & \frac{6}{18} \\ \frac{3}{18} & \frac{4}{18} & \frac{11}{18} \end{bmatrix}.$$

The elements of T^2 are all positive. Thus T is regular. If regularity had not been determined with T^2, then it is usually faster to obtain T^4, T^8, and so on, rather than T^3, T^4, and so on.

DO EXERCISE 5.

A Markov chain is *ergodic* if it is possible to go from each state to each other state. This might mean going to a state by first going to an intermediate state. If it is *not* possible to go from each state to each other state, then the Markov chain and its transition diagram are *not* ergodic and *not* regular. This frequently permits *irregularity* to be determined by inspection of the transition diagram.

Example 2   Determine from its transition diagram whether or not the following transition matrix is regular:

$$T = \begin{bmatrix} 1 & 0 & 0 \\ 0 & 1 & 0 \\ \frac{1}{3} & \frac{1}{3} & \frac{1}{3} \end{bmatrix}.$$

5. Determine whether the following transition matrix is regular and draw its transition diagram:

$$T = \begin{bmatrix} 0 & 1 & 0 \\ \frac{1}{3} & \frac{1}{3} & \frac{1}{3} \\ 0 & 1 & 0 \end{bmatrix}$$

6. Determine whether or not the following transition matrix is regular, using both the transition matrix and its transition diagram:

$$T = \begin{bmatrix} \frac{2}{5} & \frac{2}{5} & \frac{1}{5} \\ 0 & 1 & 0 \\ \frac{1}{4} & \frac{1}{4} & \frac{1}{2} \end{bmatrix}$$

Solution The transition diagram is

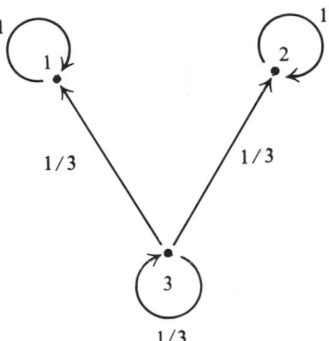

Since there is no arc leading either out of state 1 or out of state 2, it is not possible to leave either of these states once they are entered. Hence the chain is not ergodic and hence not regular.

Successive powers of T are:

$$T^2 = \begin{bmatrix} 1 & 0 & 0 \\ 0 & 1 & 0 \\ \frac{4}{9} & \frac{4}{9} & \frac{1}{9} \end{bmatrix}, \qquad T^4 = \begin{bmatrix} 1 & 0 & 0 \\ 0 & 1 & 0 \\ \frac{40}{81} & \frac{40}{81} & \frac{1}{81} \end{bmatrix}, \qquad \ldots,$$

and further squaring will not eliminate the zeros, so T is not regular, as already determined from the transition diagram.

States that cannot be exited once entered are called *absorbing states*, and can be detected by a "1" in the corresponding location in the main diagonal (upper left to lower right) of the transition matrix.

In Example 2 there are 1's as the first two elements in the main diagonal of the transition matrix. Thus, states 1 and 2 are absorbing. Markov chains with absorbing states are not regular (since they are not ergodic).

DO EXERCISE 6.

If a Markov chain is ergodic (that is, it is possible to go from each state to each other state), then it is regular if there is at least one nonzero element on the main diagonal of the transition matrix.

Example 3 Determine whether the transition matrix of Example 1 is regular without using the power test.

Solution From the transition diagram it can be seen that this Markov chain is ergodic. Since there is at least one (in this case, two) nonzero element(s) on the main diagonal of the transition matrix, the Markov chain is regular.

DO EXERCISE 7.

If a Markov chain is ergodic, then it is either regular or *periodic* (cyclic). In a periodic Markov chain, some sequence of transitions repeats regularly, as do some powers of the transition matrix. Thus, zeroes cannot be eliminated, so that periodic Markov chains are *not* regular. The distinction between regular and periodic Markov chains can be determined using the power test given at the beginning of this section.

If the main diagonal of an ergodic Markov chain contains *at least one* nonzero element, then it is regular. However, if the main diagonal of an ergodic Markov chain contains *no* zero element, it *may* still be regular. To determine whether it is regular or periodic, consider successive powers or squarings of the transition matrix. If this yields a *one* on the main diagonal, the transition matrix is periodic but if there are any nonzero elements less than one on the main diagonal, the transition matrix is regular.

Example 4 Determine whether the following transition matrix is regular and draw its transition diagram:

$$T = \begin{bmatrix} 0 & 1 & 0 \\ \frac{1}{2} & 0 & \frac{1}{2} \\ 0 & 1 & 0 \end{bmatrix}.$$

Solution The transition diagram is:

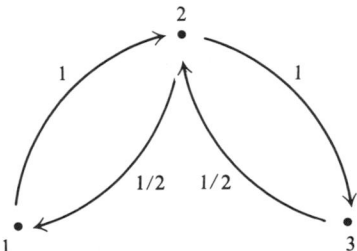

Here it is possible to go from each state to each other state, so that the transition matrix is ergodic. However, is *not* apparent whether or

7. Determine whether the transition matrix of Margin Exercise 5 is regular without using the power test.

8. Determine whether the following transition matrix is regular and draw its transition diagram:

$$T = \begin{bmatrix} 0 & \frac{1}{2} & \frac{1}{2} \\ \frac{1}{2} & 0 & \frac{1}{2} \\ \frac{1}{2} & \frac{1}{2} & 0 \end{bmatrix}.$$

not it is periodic. Hence we must resort to the power test. Squaring T, we obtain

$$T^2 = \begin{bmatrix} \frac{1}{2} & 0 & \frac{1}{2} \\ 0 & 1 & 0 \\ \frac{1}{2} & 0 & \frac{1}{2} \end{bmatrix}.$$

Further squaring yields

$$T^4 = \begin{bmatrix} \frac{1}{2} & 0 & \frac{1}{2} \\ 0 & 1 & 0 \\ \frac{1}{2} & 0 & \frac{1}{2} \end{bmatrix} = T^2.$$

Thus, the zeros cannot be eliminated and T is *not* regular. Also the presence of a one on the main diagonal of T^2 indicates that T is periodic rather than regular.

DO EXERCISE 8.

EXERCISE SET 7.2

For each of the transition matrices of Exercises 1 through 10, determine which are regular using, if possible, information along the main diagonal and the transition diagrams.

1. $\begin{bmatrix} \frac{2}{5} & \frac{2}{5} & \frac{1}{5} \\ 0 & 1 & 0 \\ 1 & 0 & 0 \end{bmatrix}$
2. $\begin{bmatrix} \frac{1}{2} & \frac{1}{2} & 0 & 0 \\ \frac{1}{2} & \frac{1}{2} & 0 & 0 \\ 0 & 0 & \frac{2}{3} & \frac{1}{3} \\ 0 & 0 & \frac{1}{3} & \frac{2}{3} \end{bmatrix}$
3. $\begin{bmatrix} 1 & 0 \\ 0 & 1 \end{bmatrix}$
4. $\begin{bmatrix} 0 & 1 \\ 1 & 0 \end{bmatrix}$
5. $\begin{bmatrix} 1 & 0 & 0 \\ 0 & 1 & 0 \\ \frac{1}{2} & \frac{1}{2} & 0 \end{bmatrix}$

6. $\begin{bmatrix} 0 & 1 & 0 \\ 0 & 0 & 1 \\ 1 & 0 & 0 \end{bmatrix}$
7. $\begin{bmatrix} 0 & 1 & 0 \\ 0 & 0 & 1 \\ 0 & 0 & 1 \end{bmatrix}$
8. $\begin{bmatrix} \frac{1}{2} & \frac{1}{2} & 0 & 0 \\ 0 & \frac{1}{2} & \frac{1}{2} & 0 \\ 0 & 0 & \frac{1}{2} & \frac{1}{2} \\ 0 & 0 & \frac{1}{2} & \frac{1}{2} \end{bmatrix}$
9. $\begin{bmatrix} \frac{1}{2} & \frac{1}{2} & 0 & 0 \\ 0 & 1 & 0 & 0 \\ 0 & \frac{1}{3} & \frac{1}{3} & \frac{1}{3} \\ 0 & 0 & 0 & 1 \end{bmatrix}$
10. $\begin{bmatrix} 0 & 1 & 0 & 0 \\ \frac{1}{3} & \frac{1}{3} & \frac{1}{3} & 0 \\ 0 & \frac{1}{3} & \frac{1}{3} & \frac{1}{3} \\ 0 & 0 & 1 & 0 \end{bmatrix}$

Determine which of the following ergodic transition matrices are regular using the power test if necessary.

11. $\begin{bmatrix} 0 & 1 \\ \frac{1}{2} & \frac{1}{2} \end{bmatrix}$
12. $\begin{bmatrix} 0 & 1 & 0 \\ \frac{2}{3} & 0 & \frac{1}{3} \\ \frac{1}{3} & \frac{2}{3} & 0 \end{bmatrix}$
13. $T = \begin{bmatrix} 0 & 1 & 0 \\ 0 & 0 & 1 \\ \frac{2}{5} & \frac{3}{5} & 0 \end{bmatrix}$
14. $T = \begin{bmatrix} 0 & 1 & 0 \\ 0 & \frac{2}{5} & \frac{3}{5} \\ 1 & 0 & 0 \end{bmatrix}$

15. $T = \begin{bmatrix} 0 & 0 & 1 \\ 0 & 0 & 1 \\ \frac{2}{5} & \frac{3}{5} & 0 \end{bmatrix}$
16. $T = \begin{bmatrix} \frac{1}{5} & \frac{2}{5} & \frac{2}{5} \\ \frac{1}{2} & 0 & \frac{1}{2} \\ \frac{1}{4} & \frac{1}{4} & \frac{1}{2} \end{bmatrix}$
17.* $T = \begin{bmatrix} 0 & \frac{2}{5} & 0 & \frac{3}{5} \\ 0 & 0 & 0 & 1 \\ 1 & 0 & 0 & 0 \\ 0 & 0 & 1 & 0 \end{bmatrix}$
18. $T = \begin{bmatrix} 0 & \frac{1}{3} & \frac{2}{3} & 0 \\ 0 & 0 & 1 & 0 \\ 0 & 0 & 0 & 1 \\ \frac{2}{3} & \frac{1}{3} & 0 & 0 \end{bmatrix}$

---

* This problem is of special interest because there are four states and only five nonzero elements.

19. $T = \begin{bmatrix} 0 & 0 & 0 & 1 \\ \frac{1}{2} & 0 & \frac{1}{2} & 0 \\ 0 & \frac{1}{2} & 0 & \frac{1}{2} \\ 0 & 0 & 1 & 0 \end{bmatrix}$

20. $\begin{bmatrix} 0 & 0 & \frac{3}{4} & \frac{1}{4} \\ \frac{1}{2} & 0 & 0 & \frac{1}{2} \\ \frac{1}{4} & \frac{3}{4} & 0 & 0 \\ 0 & \frac{1}{2} & \frac{1}{2} & 0 \end{bmatrix}$

7.3 FIXED POINTS AND "THE LONG RUN"

In Section 7.2, we sought to determine the regularity of transition matrices. Here we make use of their regularity to determine what happens in the long run.

Example 1 ▦ *Business, Marketing Surveys.* A travel agency recommends three vacation resorts: Acapulco, Bermuda, and the Caribbean. The probability that a vacationer who has been to one place will go to the same or another place is given by the following transition matrix (Example 1, Section 7.2):

$$T = \begin{array}{c} \\ \\ \end{array} \begin{array}{ccc} A & B & C \\ \end{array}$$

$$T = \begin{bmatrix} \frac{1}{3} & \frac{2}{3} & 0 \\ \frac{1}{2} & 0 & \frac{1}{2} \\ 0 & \frac{1}{3} & \frac{2}{3} \end{bmatrix} \begin{array}{c} A \\ B \\ C \end{array}$$

Assuming that one starts by going to Acapulco, what is the probability that one will be in any one of the three places after *many* visits?

Solution If one starts in Acapulco, then the initial probability vector is

$$P_0 = [1 \quad 0 \quad 0].$$

Successive probability vectors can be determined as in Section 7.1:

$$P_1 = P_0 T = [\tfrac{1}{3} \quad \tfrac{2}{3} \quad 0],$$
$$P_2 = P_1 T = [\tfrac{4}{9} \quad \tfrac{2}{9} \quad \tfrac{3}{9}],$$
$$P_3 = P_2 T = [\tfrac{7}{27} \quad \tfrac{11}{27} \quad \tfrac{9}{27}],$$
$$P_4 = P_3 T = [\tfrac{47}{162} \quad \tfrac{46}{162} \quad \tfrac{69}{162}], \quad \text{and so forth.}$$

Although it may not be apparent, this sequence of vectors is approaching some limit vector, although rather slowly. To speed up the calculation, let us consider *successive squarings* of T.

From Section 7.1,

$$P_n = P_0 T^n.$$

OBJECTIVES

You should be able to

a) Determine the fixed probability vector of a transition matrix;
b) Solve probability problems involving Markov Chains and fixed probability vectors.

Thus,

$$T^2 = \begin{bmatrix} \frac{8}{18} & \frac{4}{18} & \frac{6}{18} \\ \frac{3}{18} & \frac{9}{18} & \frac{6}{18} \\ \frac{3}{18} & \frac{4}{18} & \frac{11}{18} \end{bmatrix},$$

and

$$P_2 = P_0 T^2 = [\tfrac{4}{9} \quad \tfrac{2}{9} \quad \tfrac{3}{9}], \quad \text{as before.}$$

Continuing,

$$T^4 = \frac{1}{324} \begin{bmatrix} 94 & 92 & 138 \\ 69 & 117 & 138 \\ 69 & 92 & 163 \end{bmatrix},$$

so that $P_4 = P_0 T^4 = [\tfrac{47}{162} \quad \tfrac{46}{162} \quad \tfrac{69}{162}]$, as before; and

$$T^8 = \frac{1}{104976} \begin{bmatrix} 24706 & 32108 & 48162 \\ 24081 & 32733 & 48162 \\ 24081 & 32108 & 48787 \end{bmatrix},$$

so that $P_8 = [24706 \quad 32108 \quad 48162]/104976$, and so forth.

This calculation is quite tedious, so let us seek an alternate method.

Note that, if we started with the initial probability vector

$$P_0 = [0 \quad 1 \quad 0],$$

then we would have obtained

$$P_8 = [24081 \quad 32733 \quad 48162]/104976.$$

Furthermore, if we had started with

$$P_0 = [0 \quad 0 \quad 1],$$

then

$$P_8 = [24081 \quad 32108 \quad 48787]/104976.$$

By comparing these three values of P_8, we find that, regardless where we start, we seem to be approaching the same probability vector.[*]

If these sequences tend to approach some limit vector, than after a while there should be little change from one transition to the next. This leads us to ask what some probability vector $\bar{P}$ must be such that it is the same after a transition as before. That is, we want to find $\bar{P}$ such that

$$\bar{P} = \bar{P}T.$$

---

[*] This is also true for any convex combination of these states.

Such a probability vector is called a *fixed probability vector*, or *fixed point*.

The equation $\bar{P} = \bar{P}T$ can be written $O = \bar{P}T - \bar{P}$, or, since matrix multiplication is distributive,

$$\bar{P}(T - I) = O$$

where I is the identity matrix of appropriate order.

This system of equations, $\bar{P}(T - I) = O$, is homogeneous (which we cannot pursue here in detail, but see Section 2.2) and hence does not have a unique (nontrivial) solution. The solution can be made unique, however, since the probability vector

$$\bar{P} = [\bar{p}_1 \quad \bar{p}_2 \quad \cdots \quad \bar{p}_n]$$

must have components which sum to 1; that is,

$$\sum_{i=1}^{n} \bar{p}_i = 1.$$

The system can now be solved using the echelon method of Chapter 2.

For Example 1, we have

$$\sum_{i=1}^{3} \bar{p}_i = \bar{p}_1 + \bar{p}_2 + \bar{p}_3 = 1,$$

and

$$\bar{P}(T - I) = O$$

or

$$[\bar{p}_1 \quad \bar{p}_2 \quad \bar{p}_3]\left(\begin{bmatrix} \frac{1}{3} & \frac{2}{3} & 0 \\ \frac{1}{2} & 0 & \frac{1}{2} \\ 0 & \frac{1}{3} & \frac{2}{3} \end{bmatrix} - \begin{bmatrix} 1 & 0 & 0 \\ 0 & 1 & 0 \\ 0 & 0 & 1 \end{bmatrix}\right) = [0 \quad 0 \quad 0],$$

so that

$$[\bar{p}_1 \quad \bar{p}_2 \quad \bar{p}_3]\begin{bmatrix} -\frac{2}{3} & \frac{2}{3} & 0 \\ \frac{1}{2} & -1 & \frac{1}{2} \\ 0 & \frac{1}{3} & -\frac{1}{3} \end{bmatrix} = [0 \quad 0 \quad 0].$$

Thus, the initial echelon tableau is

$\bar{p}_1$	$\bar{p}_2$	$\bar{p}_3$	1
1	1	1	1
$-\frac{2}{3}$	$\frac{1}{2}$	0	0
$\frac{2}{3}$	-1	$\frac{1}{3}$	0
0	$\frac{1}{2}$	$-\frac{1}{3}$	0

9. Convert P_8 and $\bar{P}$ to decimals, and compare their numerical values.

Note that the *columns* of $(T - I)$ become the *rows* of the echelon tableau. Solving, we obtain the final tableau.

$\bar{p}_1$	$\bar{p}_2$	$\bar{p}_3$	1
1	0	0	$\frac{3}{13}$
0	1	0	$\frac{4}{13}$
0	0	1	$\frac{6}{13}$
0	0	0	0

The last row of zeros is evidence of the linear dependence of the system $P(T - I) = O$. The fixed probability vector is:

$$\bar{P} = [\tfrac{3}{13} \quad \tfrac{4}{13} \quad \tfrac{6}{13}],$$

which does satisfy the fixed-point equation. Note that $\bar{P}$ is fairly close to P_8 previously obtained. (See Margin Exercise 9.)

DO EXERCISE 9.

10. Determine the fixed probability vector and long-range properties of the transition matrix (Margin Exercise 5 Section 7.2):

$$T = \begin{bmatrix} 0 & 1 & 0 \\ \frac{1}{3} & \frac{1}{3} & \frac{1}{3} \\ 0 & 1 & 0 \end{bmatrix}.$$

The calculation of the fixed probability vector $\bar{P}$ does not depend on the initial state. Thus, regardless of the initial probability vector, in the long run, after sufficient transitions of a *regular* matrix chain have occurred, the probability vector will approach a unique fixed probability vector.

Thus, the significance of *regularity* is that where one ends up in the long run does *not* depend upon where one started.

DO EXERCISE 10.

Example 2 Determine the fixed probability vector and long-range properties of the transition matrix (Example 4, Section 7.2):

$$T = \begin{bmatrix} 0 & 1 & 0 \\ \frac{1}{2} & 0 & \frac{1}{2} \\ 0 & 1 & 0 \end{bmatrix}.$$

Solution This transition matrix is ergodic and hence has a unique fixed probability vector, which we find to be

$$\bar{P} = [\tfrac{1}{4} \quad \tfrac{1}{2} \quad \tfrac{1}{4}].$$

Consider now successive powers of the transition matrix T:

$$T = T^3 = T^5 = \cdots = \begin{bmatrix} 0 & 1 & 0 \\ \frac{1}{2} & 0 & \frac{1}{2} \\ 0 & 1 & 0 \end{bmatrix}$$

and

$$T^2 = T^4 = T^6 = \cdots = \begin{bmatrix} \frac{1}{2} & 0 & \frac{1}{2} \\ 0 & 1 & 0 \\ \frac{1}{2} & 0 & \frac{1}{2} \end{bmatrix}$$

This cyclic transition matrix does not exhibit the same long-range characteristics as does the *regular* transition matrix of Example 1.

Since an ergodic transition matrix which is not regular must be *periodic*, there is no "*long-run*" probability vector to end up with since the transitions keep on cycling even though a *fixed* probability vector exists.

DO EXERCISE 11.

Example 3 Determine the fixed probability vector and long-range properties of the transition matrix (Example 2, Section 7.2):

$$T = \begin{bmatrix} 1 & 0 & 0 \\ 0 & 1 & 0 \\ \frac{1}{3} & \frac{1}{3} & \frac{1}{3} \end{bmatrix}.$$

Solution This transition matrix is neither regular nor ergodic. Since states 1 and 2 are absorbing states, once one enters either of these states, one cannot exit. Thus, by inspection,

$$\bar{P}_1 = [1 \quad 0 \quad 0] \quad \text{and} \quad \bar{P}_2 = [0 \quad 1 \quad 0]$$

must each be fixed probability vectors, as can be verified by evaluating the product $\bar{P}_1 T$ obtaining $\bar{P}_1$ and by evaluating the product $\bar{P}_2 T$ obtaining $\bar{P}_2$.*

For nonergodic transition matrixes, the fixed probability vector may *not* be *unique* and hence where one ends up in the "long run" *may* depend on where one started.

Regularity of the transition matrix ensures that the "long run" *will* correspond to the fixed probability vector.

11. Attempt to determine the fixed probability vector and long range properties of the (ergodic) transition matrix

$$T = \begin{bmatrix} 0 & \frac{2}{3} & \frac{1}{3} \\ 1 & 0 & 0 \\ 1 & 0 & 0 \end{bmatrix}$$

Why can no "long-run" probability vector be found?

EXERCISE SET 7.3

Determine the fixed probability vector for each transition matrix of Exercises 1 through 10 (Exercises 11 through 20, Set 7.2).

1. $T = \begin{bmatrix} 0 & 1 \\ \frac{1}{2} & \frac{1}{2} \end{bmatrix}$ **2.** $T = \begin{bmatrix} 0 & 1 & 0 \\ \frac{2}{3} & 0 & \frac{1}{3} \\ \frac{1}{3} & \frac{2}{3} & 0 \end{bmatrix}$ **3.** $T = \begin{bmatrix} 0 & 1 & 0 \\ 0 & 0 & 1 \\ \frac{2}{5} & \frac{3}{5} & 0 \end{bmatrix}$ **4.** $T = \begin{bmatrix} 0 & 1 & 0 \\ 0 & \frac{2}{3} & \frac{3}{5} \\ 1 & 0 & 0 \end{bmatrix}$ **5.** $T = \begin{bmatrix} 0 & 0 & 1 \\ 0 & 0 & 1 \\ \frac{2}{5} & \frac{3}{5} & 0 \end{bmatrix}$

*Furthermore, any convex combination of $\bar{P}_1$ and $\bar{P}_2$ is also a fixed probability vector.

6. $T = \begin{bmatrix} \frac{1}{5} & \frac{2}{5} & \frac{2}{5} \\ \frac{1}{2} & 0 & \frac{1}{2} \\ \frac{1}{4} & \frac{1}{4} & \frac{1}{2} \end{bmatrix}$ **7.** $T = \begin{bmatrix} 0 & \frac{2}{5} & 0 & \frac{3}{5} \\ 0 & 0 & 0 & 1 \\ 1 & 0 & 0 & 0 \\ 0 & 0 & 1 & 0 \end{bmatrix}$ **8.** $T = \begin{bmatrix} 0 & \frac{1}{3} & \frac{2}{3} & 0 \\ 0 & 0 & 1 & 0 \\ 0 & 0 & 0 & 1 \\ \frac{2}{3} & \frac{1}{3} & 0 & 0 \end{bmatrix}$ **9.** $T = \begin{bmatrix} 0 & 0 & 0 & 1 \\ \frac{1}{2} & 0 & \frac{1}{2} & 0 \\ 0 & \frac{1}{2} & 0 & \frac{1}{2} \\ 0 & 0 & 1 & 0 \end{bmatrix}$ **10.** $T = \begin{bmatrix} 0 & 0 & \frac{3}{4} & \frac{1}{4} \\ \frac{1}{2} & 0 & 0 & \frac{1}{2} \\ \frac{1}{4} & \frac{3}{4} & 0 & 0 \\ 0 & \frac{1}{2} & \frac{1}{2} & 0 \end{bmatrix}$

11. Using the data from Exercise 1, Set 7.1, determine the "long-run" distribution (fixed-point probability vector) of taxi location. By successive squaring, show that each row of T^m is approaching the fixed point.

12. Using the data from Exercise 2, Set 7.1, determine the "long-run" distribution of car ownership. By successive squaring, show that each row of T^m is approaching the fixed point.

---

13. Consider the "main" street of a town with cross streets numbered 1, 2, 3, 4, and 5. A drunk is standing at the intersection of 3rd and Main. He flips a fair coin. If heads shows, he walks one block toward his home at 5th and Main, while if tails shows, he walks one block toward the bar at 1st and Main. At each intersection he repeats this procedure. If he reaches either his home or the bar he stays there.

Find the transition matrix and the transition diagram. By inspection of these, determine *two* fixed points of the transition matrix. What is the initial probability vector? Determine successive values of the probability vector and find how this is related to the fixed points. What is the probability that the drunk eventually (that is, in the long run) reaches home?

15. As in Exercise 14, but the bar is closed, so that, after reaching the bar, the next step is *always* to walk a block toward home (this corresponds to a *reflecting* barrier).

14. As in Exercise 13, but:

i) if the drunk gets home, he stays there, as before (this corresponds to an *absorbing* barrier);
ii) but if the drunk gets to the bar, he again flips a coin: Heads, he walks one block toward home, and tails, he stays a while and has a drink (this corresponds to a *retaining* barrier).

Find the transition matrix and transition diagram. By inspection of these, determine the fixed point. Determine successive values of the probability vector and find how this is related to the fixed point. What is the probability that the drunk gets home?

16. Two gamblers, one with $2 and the other with $3, flip a fair coin. If tails shows, the first pays the second $1 and if heads shows, the second pays the first $1. The game ends when either becomes broke.

Set up the transition matrix and draw the transition diagram. Is the transition matrix regular? Explain. How much should the first gambler expect to have after 3 flips of the coin? after 4 flips? Note that the probability is 60% that the first gambler will go broke and 40% that the second one will go broke.

17. *Psychology.* A mouse is placed in the maze in the accompanying figure. There is an equal probability that it will pass through each door to an adjoining room. Find the transition matrix and the fixed point. What happens in the long run?

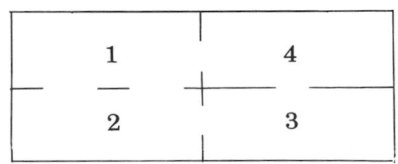

7.4 PURE STRATEGIES AND MATRIX GAMES

Many situations involving two or more people can be represented as *games* whose solutions aid in *decision*-making.

Formulation of Games as Matrices

Example 1   *The Prisoner's Dilemma.* (*Psychology*) A district attorney has two prisoners whom he suspects of a big robbery. Lacking evidence, he puts each in a separate isolation cell and makes each the following proposition: "If neither of you confess, I will see that you both get 1 year on trumped up charges. If you both confess, you will both get 10 years. However, if just one of you confesses, that one will get off with just probation (0 years) while the other one will get 20 years. So you better confess before he does." Draw the tree for this game and convert to matrix form.

Before discussing this problem,

DO EXERCISE 12.

Solution   Let us first represent this situation using a tree diagram. Starting at node A, Prisoner I can decide to confess (C) or not to confess (C^c):

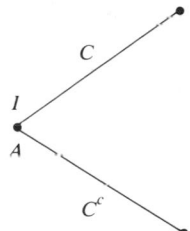

Prisoner II has the same choices. Thus, we write:

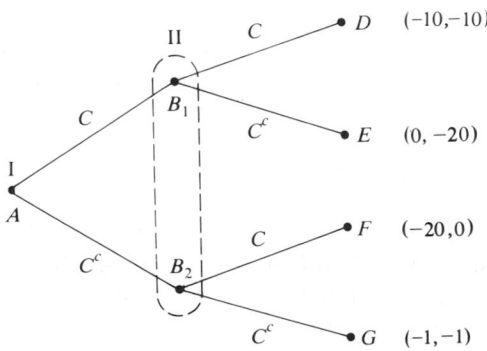

OBJECTIVE

You should be able to find the pure-strategy solution of a matrix game.

12. Partition your class into student pairs with each student of a pair representing a prisoner. Then, *without any discussion*, let each student write his decision on a slip of paper. The instructor can then compile the results.

Here the dashed line about nodes B_1 and B_2 indicates that Prisoner II does *not* know Prisoner I's decision; that is, he doesn't know whether he is at node B_1 or B_2.

The nodes that a player cannot distinguish belong to the same *information set*.

The choices available to Prisoner II at either node B_1 or node B_2 (referred to simply as B) are to confess or not to confess. Prisoner I has the same two choices at node A. All of I's choices are represented by information set A, $\mathscr{I}_A = \{C, C^c\}$, and all of II's choices are represented by information set B, $\mathscr{I}_B = \{C, C^c\}$. The *pure strategies* for I are $S_I = \mathscr{I}_A$ and the *pure strategies* for II are $S_{II} = \mathscr{I}_B$.

In general, the pure strategies for I can be represented by $S_I = \{\alpha_1, \alpha_2, \ldots, \alpha_m\}$ and for II by $S_{II} = \{\beta_1, \beta_2, \ldots, \beta_n\}$. Corresponding to each pair of pure strategies $\alpha_i \beta_j$, there is a resultant payoff to each, denoted r_{ij}.

A *game* is a situation involving two or more players with rules leading to choices at various information sets and a payoff for each possible outcome.

So far we have represented a game by a tree. Now, in order to facilitate the solution, we represent the game by a matrix using pure strategies. Let I's pure strategy α_i correspond to the ith row of a matrix and II's pure strategy β_j correspond to the jth column. The payoff r_{ij} then corresponds to the ij-element of the payoff matrix R.

Thus, for the Prisoner's Dilemma example, we have

$$
\begin{array}{c}
\quad\quad\quad\quad\quad\quad \text{II} \\
\quad\quad\quad\quad \beta_1 = C \quad\quad \beta_2 = C^c \\
\text{I} \quad
\begin{array}{c}
\alpha_1 = C \\
\alpha_2 = C^c
\end{array}
\left[
\begin{array}{cc}
(-10, -10) & (0, -20) \\
(-20, 0) & (-1, -1)
\end{array}
\right]
\end{array}
$$

Here the resultant payoff is written as an ordered pair, each component (coordinate) being the payoff to a particular player. For example, $r_{11} = r_D = (-10, -10)$.

We will solve this game later in Example 4.

Pure strategies play a very important role in game theory. The Prisoner's Dilemma problem illustrates the determination of the pure strategies and their use in obtaining the matrix form of a game when each player has only *one* information set.

We will limit our consideration in this chapter mainly to *zero-sum* games. Then what one player wins, the other loses, and vice versa, so that the sum of their combined winnings is zero. Thus, for a zero-sum game, the payoff is simply the amount Player I *wins* from Player II. A negative payoff indicates the amount Player I *loses* to Player II. Note that the games in Examples 1 and 2 are *not* zero-sum.

Solution of Matrix Games

The basis for solution of zero-sum games is the principle of *individual rationality*. This states that each player will try to *maximize* the payoff to *himself*.

Example 3   Solve the following zero-sum game.

$$
\begin{array}{c}
 & & \text{II} \\
 & & \begin{array}{cccc} \beta_1 & \beta_2 & \beta_3 & \beta_4 \end{array} \\
\text{I} \begin{array}{c} \alpha_1 \\ \alpha_2 \\ \alpha_3 \end{array} & & \left[\begin{array}{cccc} 3 & 2 & 4 & 1 \\ 1 & 2 & 2 & 3 \\ 7 & 3 & 3 & 5 \end{array} \right]
\end{array}
$$

Solution   The first step is to try to identify those pure strategies that are unprofitable to a player. Positive payoffs represent amounts that player II must pay player I. Thus, player I wants to maximize what he wins and player II wants to minimize what he loses.

After a little searching, it is apparent that player II will *always* choose pure strategy β_2 in preference to pure strategy β_3, since whatever player I chooses (α_1, α_2, or α_3), player II's loss is less or no greater. That is, if player II chooses β_2 rather than β_3, then,

  i) if player I chooses α_1, then player II loses 2 rather than 4;
 ii) if player I chooses α_2, then player II loses 2 rather than 2;
iii) if player I chooses α_3, then player II loses 3 rather than 3.

Thus, the column pure strategy β_2 *dominates* the column pure strategy β_3.

A *column* strategy β_j dominates another *column* strategy β_i if:

a) Each element of the jth column is less than or equal to the rowwise corresponding element of the ith column and
b) At least one such element is less (rather than less than *or equal*).

Since β_2 dominates β_3, we can eliminate the β_3 column from consideration and obtain:

$$
\begin{array}{c}
 & \text{II} \\
\begin{array}{cc}
 & \\
I & \begin{array}{c}\alpha_1\\ \alpha_2 \\ \alpha_3\end{array}
\end{array}
\begin{array}{c}
\begin{array}{cccc}\beta_1 & \beta_2 & \beta_3 & \beta_4\end{array} \\
\left[\begin{array}{cccc}
3 & 2 & 4 & 1 \\
1 & 2 & 2 & 3 \\
7 & 3 & 3 & 5
\end{array}\right]
\end{array}
\end{array}
$$

Having done this, we now see that player I will *always* choose pure strategy α_3 in preference to pure strategy α_1, since whatever player II chooses (β_1, β_2, or β_4), player I's winnings are equal or greater. That is, if player I chooses α_3 rather than α_1, then,

  i) if player II chooses β_1, then player I wins 7 rather than 3,
 ii) if player II chooses β_2, then player I wins 3 rather than 2,
iii) if player II chooses β_4, then player I wins 5 rather than 1.

Thus, the row pure strategy α_3 *dominates* the row pure strategy α_1.

A *row* strategy α_j dominates another *row* strategy α_i if:

a) Each element of the *j*th row is greater than or equal to the columnwise corresponding element of the *i*th row and
b) At least one such element is greater (rather than greater than *or equal*).

Since α_3 dominates α_1, we can eliminate the α_1 row from consideration and obtain:

$$
\begin{array}{c}
\text{II} \\
\begin{array}{cccc}\beta_1 & \beta_2 & \beta_3 & \beta_4\end{array}
\end{array}
$$

In a similar manner, we find that α_3 dominates α_2, so that we obtain:

$$
\begin{array}{c}
\text{II} \\
\begin{array}{cccc}\beta_1 & \beta_2 & \beta_3 & \beta_4\end{array}
\end{array}
$$

Player I is now reduced to the single pure strategy α_3. Knowing this, player II can now see that pure strategy β_2 dominates β_1 and β_4.

Thus, we obtain:

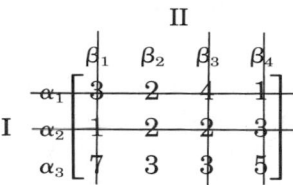

The game is now reduced to a single pure strategy α_3 for player I and a single pure strategy β_2 for player II.

The payoff that player I obtains from player II with this resulting optimum pair of pure strategies is called the *value* of the game, denoted by v; that is, $v = 3$ here.

The solution to this game is thus:

a) the optimum pure strategies, $\alpha_3\beta_2$;
b) the value of the game, $v = 3$.

DO EXERCISE 13.

The concept of dominance can sometimes be used to solve problems which are *not* zero sum.

Example 4 Solve the Prisoner's Dilemma problem (Example 1) which in matrix form is:

$$
\begin{array}{c}
 & \text{II} \\
\begin{array}{cc}
\beta_1 = C & \beta_2 = C^c
\end{array} \\
\text{I} \quad
\begin{array}{c}
\alpha = C \\
\alpha_2 = C^c
\end{array}
\left[
\begin{array}{cc}
(-10, -10) & (0, -20) \\
(-20, 0) & (-1, -1)
\end{array}
\right]
\end{array}
$$

Solution If the two prisoners could talk to each other, then it would be to their *mutual* advantage for *neither* to confess: $\alpha_2\beta_2 = (C^c, C^c)$. However, the prisoners are being held in separate isolation cells and *cannot* confer.

Writing the payoffs to each prisoner as a separate matrix, we have:

$$
\begin{array}{c}
 & \text{II} \\
\begin{array}{cc}
\beta_1 = C & \beta_2 = C^c
\end{array} \\
\text{I} \quad
\begin{array}{c}
\alpha_1 = C \\
\alpha_2 = C^c
\end{array}
\left[
\begin{array}{cc}
-10 & 0 \\
-20 & -1
\end{array}
\right] \\
\text{Payoff to I}
\end{array}
\qquad
\begin{array}{c}
 & \text{II} \\
\begin{array}{cc}
\beta_1 = C & \beta_2 = C^c
\end{array} \\
\begin{array}{c}
\alpha_1 = C \\
\alpha_2 = C^c
\end{array}
\left[
\begin{array}{cc}
-10 & -20 \\
0 & -1
\end{array}
\right] \\
\text{Payoff to II}
\end{array}
$$

Looking at the matrix of the payoff to Prisoner I, we see that strategy α_1 dominates strategy α_2. Similarly, looking at the matrix of the

13. Solve the following game.

$$
\begin{array}{c}
 & \text{II} \\
\begin{array}{ccc}
\beta_1 & \beta_2 & \beta_3
\end{array} \\
\text{I} \quad
\begin{array}{c}
\alpha_1 \\
\alpha_2 \\
\alpha_3 \\
\alpha_4
\end{array}
\left[
\begin{array}{ccc}
4 & 2 & 8 \\
1 & 2 & 4 \\
5 & 4 & 5 \\
2 & 4 & 7
\end{array}
\right]
\end{array}
$$

14. Compare the results of Margin Exercise 1 with the solution obtained in Example 4.

payoff to Prisoner II, we see that strategy β_1 dominates strategy β_2. Thus, the solution is given by the strategy pair $\alpha_1\beta_1 = (C, C)$; that is, *both* confess.

If the prisoners could confer, it would be to their *mutual* advantage for them both *not* to confess. On the other hand, if they cannot confer, then it is to their *individual* advantage for them both to confess. The dilemma of this problem arises from the difference between these two solutions and psychological arguments about the logical basis for making decisions. While game theory says that both should confess, would this be psychologically satisfying to you as a prisoner knowing you would go to jail for 10 years?

DO EXERCISE 14.

Not all games can be reduced to a pair of pure strategies using the concept of dominance.

Example 5 Solve the following game.

$$
\begin{array}{c}
 & \text{II} \\
 & \begin{array}{cccc} \beta_1 & \beta_2 & \beta_3 & \beta_4 \end{array} \\
\text{I} \begin{array}{c} \alpha_1 \\ \alpha_2 \\ \alpha_3 \end{array} & \left[\begin{array}{cccc} 7 & 2 & 1 & 4 \\ 4 & 3 & 4 & 5 \\ 1 & 2 & 8 & 3 \end{array}\right]
\end{array}
$$

Solution Here we can use dominance to eliminate pure strategy β_4, but that is as far as we can go toward the solution without using other techniques.

Now if player I chooses α_1, then the *minimum* he can win is 1 if player II chooses β_3. We write this number to the right of the payoff matrix in the row corresponding to α_1. Thus,

	II				Row minimum	Maximum minimum
	β_1	β_2	β_3	β_4		
α_1	7	2	1	4	1	
I α_2	4	3	4	5	3	3
α_3	1	2	8	3	1	
Column maximum	7	3	8	5		
Minimum maximum		3				

Continuing, if player I chooses α_2, then the *minimum* he can win is 3 if player II chooses β_2. If player I chooses α_3, then the *minimum* he can win is 1 if player II chooses β_1.

If player I now chooses the *maximum* of these *minima*, called the *maximin*, which is equal to 3, then player I can guarantee that, regardless of what player II does, he can win *at least* this maximum value of 3. This is player I's *security level*.

Now consider player II. Since the entries in the payoff matrix represent amounts that player II pays player I, player II wants to minimize his loss. Thus, if player II chooses β_1, then the *maximum* he can lose is 7 if player I chooses α_1. We write this number below the payoff matrix in the column corresponding to β_1.

Continuing, if player II chooses β_2, then the *maximum* he can lose is 3 if player I chooses α_2. If player II chooses β_3, then the *maximum* he can lose is 8 if player I chooses α_3. Also, if player II chooses β_4, then the *maximum* he can lose is 5 if player I chooses α_2. Here we have included β_4 even though it is a dominated strategy.

If player II now chooses the *minimum* of these *maxima*, called the *minimax*, which is equal to 3, then player II can *guarantee* that regardless of what player I does, he can lose *no more* than this minimax value of 3. This is player II's *security level*.

If the values of the maximin and minimax are *equal*, as they are here, then this is the value of the game and the corresponding pure strategies are optimum. Such strategies are said to be in *equilibrium* since neither player can gain by a *unilateral* change from them—that is, by one player changing his strategy while the other players do not. Thus, the solution to this game is

a) The optimum pure strategies, $\alpha_2\beta_2$;
b) The value of the game, $v = 3$.

A game with *zero* value v is called a fair game.

It is quite possible that there may be more than one equilibrium point for a given matrix game, although each will necessarily have the same game value. This causes no particular problem.

If the values of the maximin and minimax are *not* equal, then we must seek the solution in terms of mixed strategies, as in the next section.

Whether or not we had used dominance to eliminate β_4, we would have found the *same* equilibrium point. While dominance is not needed to find equilibrium points, it is useful in reducing the number of strategies that need be considered in the solution of a matrix game and will be useful in the next section.

DO EXERCISE 15.

15. Solve the following game by seeking the equilibrium solution.

$$
I\quad
\begin{array}{c}
 \\ \alpha_1 \\ \alpha_2 \\ \alpha_3 \\ \alpha_4
\end{array}
\begin{array}{c}
\overset{\text{II}}{\begin{array}{ccc} \beta_1 & \beta_2 & \beta_3 \end{array}} \\
\left[\begin{array}{ccc}
7 & 2 & 1 \\
3 & 4 & 4 \\
3 & 1 & 8 \\
6 & 5 & 7
\end{array}\right]
\end{array}
$$

EXERCISE SET 7.4

Solve the following zero-sum matrix games for the optimum pure strategies and the value of the game. Player I chooses the row strategy α_i and player II chooses the column strategy β_j.

1.
$$\begin{bmatrix} 3 & 4 & 7 \\ 2 & 1 & 3 \\ 2 & 5 & 3 \end{bmatrix}$$

2.
$$\begin{bmatrix} 2 & 4 & -1 & 5 \\ 3 & 7 & 2 & 4 \\ 5 & 6 & 8 & 9 \end{bmatrix}$$

3.
$$\begin{bmatrix} 1 & -3 & 4 \\ 2 & 7 & 8 \\ -3 & 2 & 5 \\ 1 & 7 & 1 \end{bmatrix}$$

4.
$$\begin{bmatrix} 7 & -8 & 1 & 2 \\ 2 & 5 & 5 & 4 \\ 9 & 7 & 8 & 5 \end{bmatrix}$$

5.
$$\begin{bmatrix} 7 & 2 & 3 \\ 5 & 4 & 6 \\ 2 & 4 & 5 \\ 6 & 1 & 4 \end{bmatrix}$$

6.
$$\begin{bmatrix} 5 & 3 & 3 & 5 \\ 3 & 2 & 7 & 1 \\ 2 & 3 & 8 & 0 \\ 6 & 1 & 0 & 5 \end{bmatrix}$$

7.
$$\begin{bmatrix} 11 & 8 & 2 & 3 \\ 10 & 7 & 1 & 2 \\ 3 & 4 & 2 & 1 \\ 9 & 8 & 7 & 3 \end{bmatrix}$$

8.
$$\begin{bmatrix} 3 & 4 & 11 & 9 & 7 \\ 6 & 4 & 8 & 3 & 2 \\ 8 & 5 & 7 & 10 & 6 \\ 9 & 2 & 7 & 3 & 4 \end{bmatrix}$$

OBJECTIVE

You should be able to find the mixed strategies for a game that can be reduced to 2×2.

7.5 MIXED STRATEGIES: $m \times n$ AND 2×2 GAMES

Not all matrix games can be solved for an optimum strategy in terms of a pair of *pure* strategies.

Example 1 Solve the following matrix game.

$$\text{I} \begin{array}{cc} & \text{II} \\ & \begin{array}{cc} \beta_1 & \beta_2 \end{array} \\ \begin{array}{c} \alpha_1 \\ \alpha_2 \end{array} & \begin{bmatrix} 1 & 5 \\ 3 & 2 \end{bmatrix} \end{array}$$

Solution It can be seen here that the number of strategies cannot be reduced using dominance. Furthermore, the maximin value is 2 while the minimax value is 3. Since these two values are *not equal*, we must seek a solution in terms of *mixed strategies*. A *mixed strategy* for a player is a *probability distribution* over his *pure* strategies.

Let

$$X = [x_1 \ x_2]$$

be a probability distribution over the pure strategies of player I:

$$S_I = \{\alpha_1, \alpha_2\}.$$

Here x_i is the probability of choosing pure strategy α_i.

In general, $X = [x_1 \ x_2 \cdots x_i \cdots x_m]$ is the probability distribution over the pure strategies

$$S_I = \{\alpha_1, \alpha_2, \ldots, \alpha_i, \ldots, \alpha_m\}.$$

Similarly, let

$$Y = [y_1 \ y_2]$$

be a probability distribution over the pure strategies of player II:

$$S_{\text{II}} = \{\beta_1, \beta_2\}.$$

Here y_j is the probability of choosing pure strategy β_j. In general, $Y = [y_1 \ y_2 \cdots y_j \cdots y_n]$ is the probability distribution over the pure strategies

$$S_{\text{II}} = \{\beta_1, \beta_2, \ldots, \beta_j, \ldots, \beta_n\}.$$

For simplicity, we replace α_i and β_j by x_i and y_j and write:

$$
\begin{array}{c}
\text{II} \\
\begin{array}{cc}
y_1 & y_2
\end{array} \\
\text{I} \ \begin{array}{c} x_1 \\ x_2 \end{array} \begin{bmatrix} 1 & 5 \\ 3 & 2 \end{bmatrix}
\end{array}
$$

Player I seeks to determine X such that, if player II chooses β_1, that is, $Y = [1 \ \ 0]$, then he (player I) gains a payoff of at least v_{I}; that is,

$$1x_1 + 3x_2 \geq v_{\text{I}}.$$

Also, if player II chooses β_2, that is $Y = [0 \ \ 1]$, then he (player I) again seeks to gain a payoff of at least v_{I}; that is,

$$5x_1 + 2x_2 \geq v_{\text{I}}.$$

For any *convex combination** of player II's pure strategies β_1 and β_2, that is, $Y = [y_1 \ y_2]$,

$$(x_1 \cdot 1 + x_2 \cdot 3)y_1 + (x_1 \cdot 5 + x_2 \cdot 2)y_2 \geq v_{\text{I}}y_1 + v_{\text{I}}y_2 = v_{\text{I}}(y_1 + y_2) = v_{\text{I}}.$$

Thus, player I's minimum gain is still v_{I}. Note that since X and Y are probability distributions, we must have

$$x_1 + x_2 = y_1 + y_2 = 1.$$

The quantity v_{I} is the *minimum* amount that player I can win and he wants to *maximize* this value, so it is called the *maximin*.

Similarly, player II seeks to determine Y such that if player I chooses α_1, that is, $X = [1 \ \ 0]$, then he (player II) loses no more than v_{II}; that is,

$$1y_1 + 5y_2 \leq v_{\text{II}}.$$

---

* Recall p. 128 for the definition of convex combination.

Also, if player I chooses α_2, that is, $X = [0 \quad 1]$, then he (player II) also loses no more than v_{II}; that is,

$$3y_1 + 2y_2 \leq v_{II}.$$

As before, for any convex combination of player I's pure strategies α_1 and α_2, that is, $X = [x_1\ x_2]$,

$$x_1(1y_1 + 5y_2) + x_2(3y_1 + 2y_2) \leq v_{II}x_1 + v_{II}x_2 = v_{II}(x_1 + x_2) = v_{II}.$$

Thus, player II's maximum loss is v_{II}.

The quantity v_{II} is the *maximum* amount that player II can lose and he wants to *minimize* this value, so it is called the *minimax*.

The *von Neumann Minimax Theorem* states that† the inequality signs can be replaced by equal signs, so that $v_I = v_{II}$.

From a reasoning point of view, if the *strict* inequality held for some pure strategy of his opponent, then it would not be in the interest of his opponent to choose that pure strategy. Thus, the inequality signs can be replaced by equal signs. In either case, we obtain:

$$x_1 + 3x_2 = v, \tag{1}$$

and

$$5x_1 + 2x_2 = v, \tag{2}$$

$$y_1 + 5y_2 = v, \tag{3}$$

$$3y_1 + 2y_2 = v. \tag{4}$$

The value v of the game can be eliminated from each pair of equations by subtraction, so that we obtain the following systems of equations:

$x_1 + x_2 = 1,$ (The sum of probability components must be 1.)

$4x_1 - x_2 = 0,$ (Subtracting Eq. (1) from Eq. (2).)

and

$y_1 + y_2 = 1,$ (The sum of probability components must be 1.)

$2y_1 - 3y_2 = 0.$ (Subtracting Eq. (3) from Eq. (4).)

Solving each of these systems, we obtain the optimum strategies:

$$X = [x_1 \quad x_2] = [\tfrac{1}{5} \quad \tfrac{4}{5}]$$

and

$$Y = [y_1 \quad y_2] = [\tfrac{3}{5} \quad \tfrac{2}{5}].$$

To obtain v, we substitute these values of x_1 and x_2 into Eq. (1) or into Eq. (2) or these values of y_1 and y_2 into Eq. (3) or into Eq. (4) and obtain $v = \tfrac{13}{5}$.

---

† There are many alternate formulations of the Minimax Theorem including an equivalence with the Duality Theorem of linear programming.

Thus, if player I adopts the strategy X, then he will win v independent of player II's strategy. If player II adopts the strategy Y, then he will lose v independent of player I's strategy. Only if *both* deviate from the optimum strategies can player I win more and player II lose more.

The optimum *mixed* strategy for player I, $X = [\frac{1}{5} \ \frac{4}{5}]$, means that $\frac{1}{5}$ of the time he will choose *pure* strategy α_1 and $\frac{4}{5}$ of the time he will choose *pure* strategy α_2. However, if player II knows when player I is going to choose a particular pure strategy (regardless of the probability distribution), then player II may hold player I's winnings to less than v. Thus, in order for player I to win at least v, he must keep his opponent from knowing his choice. One way of doing this is for him to randomize his choices by using a spinning arrow pinned to a disk as illustrated below:

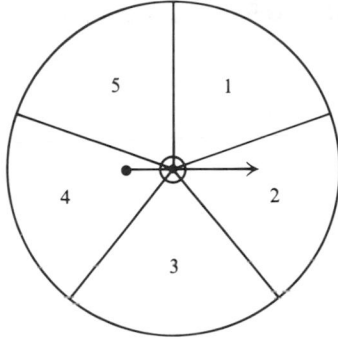

Here the circumference has been divided into 5 equal arcs. One of these is selected to represent x_1 (and α_1) and the other 4 to represent x_2 (and α_2). The arrow is spun by flipping with the finger. Where the pointer land determines which strategy is to be used. Why is this valid?

Similarly, a spinner can be used by player II, in this case the same spinner. Player II need only designate any 3 regions to represent y_1 (and β_1) and the remaining 2 to represent y_2 (and β_2).

How can a sweep second hand on a watch be used to randomize the pure strategies?

DO EXERCISE 16.

As we have solved 2×2 matrix games, so can we solve larger square-matrix games except that the amount of work involved is increased. Note that if the size of a matrix game can be reduced by eliminating dominated strategies, this should be done *before* setting up the equations for the optimum mixed strategies.

16. Solve the following matrix game for the optimum strategy for each player and the value of the game.

$$
\begin{array}{c c}
 & \text{II} \\
 & \begin{array}{cc} \beta_1 & \ \ \beta_2 \end{array} \\
\text{I} \ \begin{array}{c} \alpha_1 \\ \alpha_2 \end{array} & \begin{bmatrix} 7 & 2 \\ 1 & 4 \end{bmatrix}
\end{array}
$$

EXERCISE SET 7.5

Solve the following matrix games for the optimum strategy for each player and the value of the game. Player I chooses the row strategy α_i and player II chooses the column strategy β_j. (*Hint*. It may be necessary to use dominance to reduce these games to 2×2.)

1. $\begin{bmatrix} 1 & 3 \\ 4 & 2 \end{bmatrix}$
2. $\begin{bmatrix} 5 & 2 \\ 3 & 4 \end{bmatrix}$
3. $\begin{bmatrix} 2 & 4 & -1 & 5 \\ 3 & 7 & 2 & 4 \\ 2 & 3 & 8 & 9 \end{bmatrix}$
4. $\begin{bmatrix} 5 & 1 & 3 & 5 \\ 3 & 2 & 7 & -1 \\ 4 & 4 & 8 & 0 \\ 6 & 1 & 0 & 5 \end{bmatrix}$
5. $\begin{bmatrix} 1 & 7 & 2 & 9 \\ 1 & 8 & 4 & 7 \\ 2 & 4 & 7 & 3 \\ 6 & 6 & 1 & 7 \end{bmatrix}$
6. $\begin{bmatrix} 7 & -7 & 4 & -3 \\ 4 & 3 & -1 & 8 \\ 9 & -7 & 5 & -3 \\ 5 & 4 & -1 & 6 \end{bmatrix}$

7. *Business.* A new industrial plant is trying to minimize its costs by dumping its refuse rather than recycling it. Thus, they either truck the refuse to an isolated place in the country or dump it into the local stream. The government inspector can inspect only one site at a time and must choose between them. If the inspector catches them dumping far out in the country, there is a small fine; however, if he catches them dumping in the local stream, they will be put out of business. Representing this as a zero-sum game and assigning numerical values to various outcomes, we obtain:

		Inspector's visit	
		Far out	Local stream
Plant's	Far out	$\begin{bmatrix} -1 \end{bmatrix}$	2
disposal	Local stream	2	$-25 \end{bmatrix}$

What should the strategy of each be?

8. Two players individually and simultaneously show one or two fingers. If the numbers of fingers match, then player I wins from player II an amount equal to the *sum* of the number of fingers shown by each. Otherwise, player II wins from player I this sum. What should their strategies be? Is the game fair?

9. As in Exercise 7, what should the payoff be for *not* getting caught in order for the gain to be just worth the risk, that is, for the game to be fair ($v = 0$)?

10. As in Exercise 8, if *each* player chooses a $[\frac{1}{2} \ \frac{1}{2}]$ strategy, what is the value of the game?

11. There are two poker players. At a certain point in the game the first player has a nothing hand and a choice of bluffing (that he holds good cards) or not bluffing. The other player can either call or not call. The payoffs can be represented by

$$\begin{array}{cc} & \text{II} \\ & \begin{array}{cc} C & C^c \end{array} \\ \text{I} \begin{array}{c} B \\ B^c \end{array} & \begin{bmatrix} -5 & 5 \\ 1 & 0 \end{bmatrix} \end{array}$$

What should each player do and what is the value of the game?

12. As in Exercise 8, if one player chooses the optimum strategy and the other player chooses a $[\frac{1}{2} \ \frac{1}{2}]$ strategy, what is the value of the game?

7.6 VOTING COALITIONS AND CHARACTERISTIC FORM

For games with more than 2 players, especially those for which we are more concerned with the formation of coalitions than with strategies, *characteristic* form is more convenient.

Let $N = \{1, 2, \ldots, n\}$ be the set of n players and $S = \{1, 2, \ldots, s\} \subset N$ be any coalition of s of these players. Then the best that S can guarantee itself is obtained by considering its worst opposition, namely, the coalition $N - S$ of the remaining players. We now have a 2-person game of the coalition S versus the coalition $(N - S)$ with $v(S)$ the security level of the coalition S and $v(N - S)$ the security level of the coalition $(N - S)$. If the game is zero-sum, then $v(S) + v(N - S) = 0$, but this need not be the case.

How the quantity $v(S)$ is to be divided among the members of the coalition S is for *them* to decide and is not our concern here.

Since a coalition of *no* players cannot win anything, we can write

$$v(\emptyset) = 0.$$

Furthermore, we shall assume that the game is *proper;* that is, two disjoint cooperating coalitions can win at least as much together as separately. Then we may write:

$$v(S \cup T) \geq v(S) + v(T),$$

where $S \subset N$, $T \subset N$, and $S \cap T = \emptyset$.

If $v(S)$ is obtainable for each possible coalition $S \subset N$, then these $v(S)$ comprise the *characteristic function* form of a game.

The general solution of games in characteristic form is beyond the scope of this text. Instead, we consider two solutions for an "expected" value: the *Shapley* value and the *Banzhaf* value.

To derive the Shapley value, consider all possible $n!$ *arrangements* (permutations) of the n players. If each arrangement is equally probable, then each can be assigned a probability $1/n!$.

Suppose we let the players enter into coalitions randomly and focus our attention on those for which a particular player i is the *last* to

OBJECTIVE

You should be able to find the Shapley and Banzhaf values for the strength of a coalition.

form coalition S; that is, N is partitioned as follows:

$$N = \{\underbrace{j_1, j_2, \ldots, j_{s-1}}, \quad i, \quad \underbrace{j_{s+1}, \ldots, j_n}\}$$

$(s - 1)$ players in Coalition $(S - \{i\})$ $\qquad$ $(n - s)$ players

Coalition S $\qquad$ Coalition $(N - S)$

The first $(s - 1)$ players can enter the coalition S in $(s - 1)!$ ways. Similarly, the last $(n - s)$ players can enter the coalition $(N - S)$ in $(n - s)!$ ways. Since the total number of possible orderings is $n!$, the probability that player i is the *last* member to enter the coalition S is:

$$p_n(s) = \frac{(s - 1)!(n - s)!}{n!}.$$

The contribution of player i to the coalition S is

$$v(S) - v(S - \{i\}).$$

Thus, the *expected* value of player i's contribution to coalition S as the last player to enter this coalition is:

$$\phi_i = \sum p_n(s)[v(S) - v(S - \{i\})] \qquad \text{for all } S \subset N \quad \text{and} \quad \text{for all } i \in S.$$

While this is a *formal* expression for ϕ_i, we will soon present a simple way to obtain ϕ_i. This is the *Shapley value* for player i. The Shapley value for all players is written

$$\phi = [\phi_1, \phi_2, \ldots, \phi_n],$$

where the components satisfy

$$\sum_{i=1}^{n} \phi_i = v(N).$$

Example 1 *Business, Political Science.* Four stockholders have, respectively, 5, 5, 3, and 2 shares of stock. Each share of stock carries with it one vote. Each stockholder votes his shares as a bloc. A proposition requires a simple majority (8 votes) to pass. Find the Shapley value for each stockholder.

Solution Games of this sort are called *voting* or *quota* games and are represented by

$$[q; w_1, w_2, \ldots, w_n].$$

Here there are n voters with a voting weight w_i (nonnegative) for each voter i. A coalition W is winning if the sum of the voting weights w_i of all voters i in the coalition W *is* $\geq q$. Here the quota q must be such that

$$q > \tfrac{1}{2}(\text{Sum of } all \text{ the voting weights } w_i) = \tfrac{1}{2}\sum_i w_i.$$

If a coalition S is winning,

$$v(S) = 1;$$

otherwise, it is losing and

$$v(S) = 0.$$

Under these conditions, player i contributes to a coalition only when

$$v(S) - v(S - \{i\}) = 1,$$

so that his Shapley value ϕ_i is the sum of $p_n(s)$ for all coalitions in which voter i is pivotal; that is, for each coalition in which voter i changes coalition S from winning to losing.

This example can be written as the quota game

$$[8; 5, 5, 3, 2].$$

If we let the various players enter into coalitions in all 24 possible arrangements and indicate with an asterisk when a player is *pivotal* in changing a winning coalition to a losing coalition, we have:

$$
\begin{array}{cccc}
12^*34 & 21^*34 & 31^*24 & 412^*3 \\
12^*43 & 21^*43 & 31^*42 & 413^*2 \\
13^*24 & 23^*14 & 32^*14 & 421^*3 \\
13^*42 & 23^*41 & 32^*41 & 423^*1 \\
142^*3 & 241^*3 & 341^*2 & 431^*2 \\
143^*2 & 243^*1 & 342^*1 & 432^*1
\end{array}
$$

Thus, of the 24 possible arrangements, player 1 is pivotal in establishing the quota in 8, player 2 in 8, player 3 in 8, and player 4 in none. That is,

$$\phi_1 = \phi_2 = \phi_3 = \tfrac{8}{24} = \tfrac{1}{3} \quad \text{and} \quad \phi_4 = 0,$$

so that

$$\phi = [\tfrac{1}{3}, \tfrac{1}{3}, \tfrac{1}{3}, 0].$$

Note that player 3 with 3 votes is pivotal in the same number of coalitions as players 1 and 2 and thus has the same Shapley value.

17. Find the Shapley value for the quota game:

$$[9; 5, 5, 3, 2].$$

Player 4 with zero Shapley value contributes nothing to *any* coalition. Such a player is called a *dummy*.

Alternately, we can obtain the Shapley value from the formula:

$$\phi_i = \sum p_n(s) \quad \text{for all coalitions in which voter } i \text{ is pivotal.}$$

For $s = 1$ or 4, there are *no* winning coalitions with player 1 pivotal. For $s = 2$, there are 2 winning coalitions ($S = \{2, 1\}$ and $S = \{3, 1\}$) with player 1 pivotal and for $s = 3$, there are also 2 winning coalitions ($S = \{2, 4, 1\}$ and $S = \{3, 4, 1\}$) with player 1 pivotal. Thus,

$$\phi_1 = 0\left(\frac{0!3!}{4!}\right) + 2\left(\frac{1!2!}{4!}\right) + 2\left(\frac{2!1!}{4!}\right) + 0\left(\frac{3!0!}{4!}\right)$$
$$= 0 + 2 \cdot \tfrac{1}{12} + 2 \cdot \tfrac{1}{12} + 0$$
$$= \tfrac{1}{3}, \quad \text{as before.}$$

The other ϕ_i can be obtained similarly.

Note that if the quota is raised from 8 to 9, then the quota game

$$[9; 5, 5, 3, 2]$$

has no dummies. (See Margin Exercise 17 following.)

DO EXERCISE 17.

The Shapley value has been derived various ways by different mathematicians. We have used the foregoing derivation to facilitate comparison between the Shapley value and the Banzhaf value. Both value concepts measure voting power although they sometimes yield somewhat different *numerical* values. This difference does not seem to be *qualitatively* significant. However, Banzhaf is a lawyer and his value has been used in various legal arguments.[*]

While Shapley considered the *ordered arrangement* of players entering a coalition, Banzhaf considered *unordered arrangements*.

Each voter can vote *yea* or *nay* on a particular issue, so that n voters can vote 2^n possible ways, each assumed equally probable and assuming no abstentions. Since we are interested in *how* a voter votes regardless of the *order* in which he casts his vote, we are dealing with combinations rather than permutations. A voter is considered *critical* if in a given combination, *his vote can reverse the outcome* of the total vote. The Banzhaf value for voter i, β_i, is the fraction of votes for

[*] See Lucas, W. F., "Measuring Power in Weighted Voting Systems," Technical Report No. 227, Department of Operations Research, Cornell University, Ithaca, N.Y., September 1974.

which voter i is critical. For all voters, we write

$$\beta = [\beta_1, \beta_2, \ldots, \beta_n],$$

where

$$\sum_{i=1}^{n} \beta_i = 1.$$

Example 2 *Business, Political Science.* Determine the Banzhaf value for the quota game

$$[8; 7, 4, 2, 1],$$

and compare the results with the Shapley value.

Solution To solve for the Banzhaf value, we construct a table which lists the n players across the top and, below, each possible voting combination for which the issue would *pass* (since for each *passing* combination a reversal of *all* votes would yield a *losing* combination). Then we place an asterisk on each critical vote. Note that when there is more than one asterisk in a row, each is to be interpreted independently. Thus, we obtain:

Player	1	2	3	4	Total
Number of votes	7	4	2	1	Quota = 8
Vote: Y or N	Y*	Y	Y	Y	1
	Y*	Y	Y	N	1
	Y*	Y	N	Y	1
	Y*	N	Y	Y	1
	Y*	Y*	N	N	2
	Y*	N	Y*	N	2
	Y*	N	N	Y*	2
Number critical	7	1	1	1	10

Thus,

$$\beta = [7, 1, 1, 1]/10 \quad \text{or} \quad [\tfrac{7}{10}, \tfrac{1}{10}, \tfrac{1}{10}, \tfrac{1}{10}].$$

Solving for the Shapley value, using the table as before, we have:

12*34	21*34	31*24	41*23
12*43	21*43	31*42	41*32
13*24	231*4	321*4	421*3
13*42	2341*	3241*	4231*
14*23	241*3	341*2	431*2
14*32	2431*	3421*	4321*

18. Show that for the quota game of Example 1, the Banzhaf value is the *same* as the Shapley value.

so that

$$\phi = [9, 1, 1, 1]/12 \quad \text{or} \quad [\tfrac{9}{12}, \tfrac{1}{12}, \tfrac{1}{12}, \tfrac{1}{12}].$$

Alternately, we can use the formula and obtain:

$$\phi_1 = 0 + 3 \cdot \tfrac{2}{24} + 3 \cdot \tfrac{2}{24} + 1 \cdot \tfrac{6}{24} = \tfrac{18}{24},$$

$$\phi_2 = 0 + 1 \cdot \tfrac{2}{24} + 0 + 0 = \tfrac{2}{24},$$

$$\phi_3 = 0 + 1 \cdot \tfrac{2}{24} + 0 + 0 = \tfrac{2}{24},$$

$$\phi_4 = 0 + 1 \cdot \tfrac{2}{24} + 0 + 0 = \tfrac{2}{24},$$

so that

$$\phi = [9, 1, 1, 1]/12, \quad \text{as before.}$$

DO EXERCISE 18.

EXERCISE SET 7.6

1. Find the Shapley and Banzhaf values for the quota game $[7; 6, 4, 2, 1]$. Show that they are identical in this case.

2. Find the Shapley and Banzhaf values for the quota game $[6; 4, 3, 2, 1]$. Show that they are identical in this case.

3. Find the Shapley and Banzhaf values for the voting game $[7; 7, 3, 2, 1]$. Player 1 is called a *dictator*. Why?

4. Find the Shapley and Banzhaf values for the voting game $[3; 1, 1, 1, 1]$ in which there are no blocs of more than one vote per voter.

5. *Political Science.* Assume the Democrats have 49 votes, the Republicans have 48, the Independents have 3, and each party votes as a bloc. If a simple majority (51) is required for passage of a bill, what are the voting strengths (Shapley and Banzhaf values) for each party? Write as a quota game. Note the strength of the small third party.

6. *Political Science.* Find the Shapley and Banzhaf values for the quota game $[7; 6, 3, 2, 1]$. Note that no coalition can win without player 1. Such a player is said to have *veto* power. Compare with Exercise 3 where player 1 was a *dictator*.

▶

7. *Political Science.* For the quota game of Example 1, player 4 is a dummy. What is the minimum quota required to eliminate any dummies from the game? Find the voting strength (Shapley and Banzhaf values) for each player under the revised conditions.

8. *Political Science.* A committee consists of 4 members, each with one vote. A simple majority (3) is required for a decision; however, the chairman can break ties. What is his voting strength? (Both values.)

9. *Political Science.* If the chairman of the committee has veto power instead of just the power to break ties, what is his voting strength?

10. *Political Science.* As in Exercise 9, if two of the members agree to pool their votes, how do they change the voting strengths? three of the members?

CHAPTER 7 TEST

1. In a certain city a review of electoral records indicated that in an election in which the incumbent was:

a) Democratic, the probability was:

   i) 0.4 that the next mayor would be Democratic,
   ii) 0.3 that the next mayor would be Republican,
   iii) 0.3 that the next mayor would be Independent;

b) Republican, the probability was:

   i) 0.4 that the next mayor would be Democratic,
   ii) 0.6 that the next mayor would be Republican,
   iii)   0 that the next mayor would be Independent;

c) Independent, the probability was:

   i) 0.3 that the next mayor would be Democratic,
   ii)   0 that the next mayor would be Republican,
   iii) 0.7 that the next mayor would be Independent.

Assuming that the mayor is originally a Republican, draw the tree describing the possible outcomes. Draw the transition diagram and set up the transition matrix.

Given the matrix game $\begin{bmatrix} 1 & 7 & 5 & 0 \\ -7 & 4 & 1 & -3 \\ 2 & 8 & 2 & 7 \\ 3 & 3 & 2 & 8 \end{bmatrix}.$

4. Find the pure-strategy solution if it exists. Show your work.

6. Illustrate the solution graphically.

Given the quota game [10; 7, 5, 3, 1].

8. Find the Shapley value.

2. Determine which of the following transition matrices are regular. Explain why.

a) $\begin{bmatrix} 0 & 0 & 1 \\ 0 & 1 & 0 \\ 1 & 0 & 0 \end{bmatrix}$

b) $\begin{bmatrix} 0 & 1 & 0 & 0 \\ 0 & 0 & 1 & 0 \\ 0 & 0 & 0 & 1 \\ 1 & 0 & 0 & 0 \end{bmatrix}$

c) $\begin{bmatrix} \frac{1}{2} & \frac{1}{2} & 0 & 0 \\ \frac{1}{3} & \frac{2}{3} & 0 & 0 \\ 0 & 0 & \frac{2}{5} & \frac{3}{5} \\ 0 & 0 & \frac{1}{4} & \frac{3}{4} \end{bmatrix}$

d) $\begin{bmatrix} 0 & 1 & 0 \\ 0 & 0 & 1 \\ \frac{1}{2} & \frac{1}{2} & 0 \end{bmatrix}$

3. Given

$$P_0 = [1 \quad 0 \quad 0] \quad \text{and} \quad T = \begin{bmatrix} \frac{1}{2} & \frac{1}{2} & 0 \\ 0 & \frac{2}{3} & \frac{1}{3} \\ \frac{1}{4} & 0 & \frac{3}{4} \end{bmatrix},$$

find P_2 and the fixed probability vector.

5. Indicate all dominating strategies.

7. Solve the game for the strategies for each player, and the value of the game.

9. Find the Banzhaf value.

CHAPTER EIGHT

Mathematics of Finance

OBJECTIVES

You should be able to

a) Find certain terms of a sequence given the nth term.

b) Identify the first term and the common difference of an arithmetic sequence.

c) Find a specified term of an arithmetic sequence.

d) Find the sum of the first n terms of an arithmetic sequence.

1. A sequence is given by

$$a_n = n^2 + 3.$$

Find the first 4 terms and the 15th term.

8.1 ARITHMETIC SEQUENCES*

Sequences

A *sequence* is an ordered set of numbers. Here is an example

$$3, 5, 7, 9, 11, \ldots$$

The three dots mean that there are more and more numbers in the sequence. A sequence that does not end is called *infinite*. A sequence that does end is called *finite*.

Each number is called a *term* of the sequence. The first term is 3, the second term is 5, the third term is 7, and so on. We can describe the terms as follows:

$$a_1 = 3,$$
$$a_2 = 5,$$
$$a_3 = 7,$$
$$a_4 = 9,$$

and so on, where the nth term is $a_n = 2n + 1$. That is, a sequence is a function whose domain is a set of consecutive natural numbers.

Instead of using $a(n)$ for the nth term, we are using a_n. We also call a_n the *general term*.

Example 1 A sequence is given by

$$a_n = 2^n$$

Find the first 4 terms and the 17th term.

Solution
$$a_1 = 2^1 = 2,$$
$$a_2 = 2^2 = 4,$$
$$a_3 = 2^3 = 8,$$
$$a_4 = 2^4 = 16,$$
$$\vdots$$
$$a_{17} = 2^{17} = 131{,}072.$$

DO EXERCISE 1.

---

* This chapter is intended principally for business and economics students. It can be omitted without loss of continuity although it provides a nice lead-in to the study of calculus. A hand calculator with an $\boxed{x^y}$ key will be quite helpful.

Arithmetic Sequences

Consider the sequence

$$3, 5, 7, 9, 11, \ldots$$

Note that the number 2 can be added to each term to obtain the next term. Sequences in which a certain number can be added to any term to get the next term are called *arithmetic sequences* (or *arithmetic progressions*). The number d which we add to one term to get the next is called the *common difference*. This is because we can subtract any term from the one that follows it and get d.

$$a_{k+1} - a_k = d \quad \text{for any } k \geqslant 1.$$

Examples The following are arithmetic sequences. Identify the first term and the common difference.

Solution

Sequence	First term	Common difference
2. $3, 5, 7, 9, 11, \ldots$	3	2
3. $34, 27, 20, 13, 6, -1, -8, \ldots$	34	-7
4. $\$5200, \$4687.50, \$4175, \ldots$	$\$5200$	$\$512.50$

DO EXERCISES 2 THROUGH 5.

For an arithmetic sequence

        The 1st term is a_1,

        the 2nd term is $a_2 = a_1 + d,$

        the 3rd term is $a_3 = (a_1 + d) + d = a_1 + 2d,$

        the 4th term is $a_4 = [(a_1 + d) + d] + d = a_1 + 3d,$

and so on. Generalizing, we obtain the following.

The nth term of an arithmetic sequence is given by

$$a_n = a_1 + (n - 1)d, \quad \text{for any } n \geqslant 1.$$

Example 5 Find the 15th term of the sequence $4, 7, 10, 13, \ldots$

Solution First note that

$$a_1 = 4, \quad d = 3, \quad \text{and} \quad n = 15.$$

The following are arithmetic sequences. Identify the first term and the common difference.

2. $2, 5, 8, 11, 14, \ldots$

3. $19, 14, 9, 4, -1, -6, \ldots$

4. $\$6300, \$5953.25, \$5606.50, \ldots$

5. $2, 2\frac{1}{2}, 3, 3\frac{1}{2}, 4, 4\frac{1}{2}, \ldots$

6. Find the 18th term of the sequence

$$2, 6, 10, 14, \ldots$$

Then using the formula

$$a_n = a_1 + (n - 1)d,$$

we have

$$a_{15} = 4 + (15 - 1)3 = 4 + 14 \cdot 3 = 4 + 42 = 46.$$

We could check this by writing out 15 terms of the sequence.

DO EXERCISES 6 AND 7.

Sum of the First n Terms of an Arithmetic Sequence

Suppose we add the first 4 terms of the sequence

$$3, 5, 7, 9, 11, \ldots$$

We get

$$3 + 5 + 7 + 9, \quad \text{or} \quad 24.$$

7. Find the 11th term of the sequence

$$\$6300, \$5953.25, \$5606.50, \ldots$$

The sum of the first n terms of a sequence is denoted S_n. Thus, for the preceding sequence, $S_4 = 24$. We want to find a formula for S_n when the sequence is arithmetic. We can denote an arithmetic sequence as

$$a_1, (a_1 + d), (a_1 + 2d), \ldots, (a_n - 2d), (a_n - d), a_n.$$

Then S_n is given by

$$S_n = a_1 + (a_1 + d) + (a_1 + 2d) + \cdots + (a_n - 2d) + (a_n - d) + a_n. \tag{1}$$

If we reverse the order of addition we get

$$S_n = a_n + (a_n - d) + (a_n - 2d) + \cdots + (a_1 + 2d) + (a_1 + d) + a_1. \tag{2}$$

Suppose we add corresponding terms of each side of Eqs. (1) and (2). Then we get

$$2S_n = [a_1 + a_n] + [(a_1 + d) + (a_n - d)]$$
$$+ [(a_1 + 2d) + (a_n - 2d)] + \cdots + [(a_n - 2d) + (a_1 + 2d)]$$
$$+ [(a_n - d) + (a_1 + d)] + [a_n + a_1].$$

This simplifies to

$$2S_n = (a_1 + a_n) + (a_1 + a_n) + (a_1 + a_n) + \cdots + (a_1 + a_n).$$

Since there are n binomials $(a_1 + a_n)$ being added, it follows that $2S_n = n(a_1 + a_n)$, from which we get the following formula.

The sum of the first n terms of an arithmetic sequence is given by

$$S_n = \frac{n}{2}(a_1 + a_n).$$

Example 6   Find the sum of the first 100 natural numbers.

Solution   The sum is

$$1 + 2 + 3 + \cdots + 99 + 100$$

This is the sum of the first 100 terms of the arithmetic sequence for which

$$a_1 = 1, \qquad a_n = 100, \qquad \text{and} \qquad n = 100.$$

Then substituting in the formula

$$S_n = \frac{n}{2}(a_1 + a_n),$$

we get

$$S_{100} = \tfrac{100}{2}(1 + 100) = 50(101) = 5050.$$

DO EXERCISE 8.

The preceding formula is useful when we know a_1 and a_n, the first and last terms, but it often happens that a_n is not known. We thus need a formula in terms of a_1, n, and d.

Substituting $a_1 + (n-1)d$ for a_n in the formula $S_n = \frac{n}{2}(a_1 + a_n)$, we get

$$S_n = \frac{n}{2}(a_1 + [a_1 + (n-1)d]),$$

from which we get the following formula.

The sum of the first n terms of an arithmetic sequence is given by

$$S_n = \frac{n}{2}[2a_1 + (n-1)d].$$

Example 7   Find the sum of the first 15 terms of the arithmetic sequence 4, 7, 10, 13, ...

Solution   Note that

$$a_1 = 4, \qquad d = 3, \qquad \text{and} \qquad n = 15.$$

8. Find the sum of the first 200 natural numbers.

9. Find the sum of the first 16 terms of the sequence

$$1, 3, 5, 7, 9, \ldots$$

Then, substituting in the formula

$$S_n = \frac{n}{2}[2a_1 + (n - 1)d],$$

we get

$$S_{15} = \tfrac{15}{2}[2 \cdot 4 + (15 - 1)3] = \tfrac{15}{2}[8 + 14 \cdot 3] = \tfrac{15}{2}[8 + 42]$$
$$= \tfrac{15}{2}[50] = 375.$$

DO EXERCISE 9.

10. A family saves money in an arithmetic sequence. They save $500 the 1st year, $700 the 2nd, $900 the 3rd year, and so on, for 13 years. How much do they save in all?

Example 8 A family saves money in an arithmetic sequence. They save $600 the first year, $700 the second, and so on, for 20 years. How much do they save in all (disregarding interest)?

Solution The amount saved is the sum

$$\$600 + \$700 + \$800 + \cdots$$

Here the dots mean that the pattern continues, even though this is not an infinite sequence. In short, we need not bother to determine the last term. We can find the sum by noting that

$$a_1 = \$600, \quad d = \$100, \quad \text{and} \quad n = 20.$$

Then, substituting in the formula

$$S_n = \frac{n}{2}[2a_1 + (n - 1)d],$$

we get

$$S_{20} = \tfrac{20}{2}[2 \cdot \$600 + (20 - 1)\$100] = 10[\$1200 + 19 \cdot \$100]$$
$$= 10[\$1200 + \$1900] = 10[\$3100] = \$31,000.$$

DO EXERCISE 10.

EXERCISE SET 8.1

In each of the following sequences, the nth term is given. Find the first 4 terms, and the 15th term.

1. $a_n = \dfrac{n}{n + 1}$ **2.** $a_n = n + \dfrac{1}{n}$ **3.** $a_n = \dfrac{n^2 - 1}{n^3 + 1}$ **4.** $a_n = (-\tfrac{1}{2})^n$

The following are arithmetic sequences. Identify the first term and the common difference.

5. $2, 7, 12, 17, \ldots$ **6.** $7, 3, -1, -5, \ldots$ **7.** $\$1.06, \$1.12, \$1.18, \$1.24, \ldots$

8. $\$214, \$211, \$208, \$205, \ldots$ **9.** $5, 4\tfrac{1}{3}, 3\tfrac{2}{3}, 3, 2\tfrac{1}{3}, \ldots$ **10.** $\tfrac{3}{2}, \tfrac{9}{4}, 3, \tfrac{15}{4}, \ldots$

11. Find the 12th term of the arithmetic sequence

$$3, 7, 11, \ldots$$

12. Find the 11th term of the arithmetic sequence

$$\$0.08, \$0.13, \$0.18, \ldots$$

13. Find the 13th term of the arithmetic sequence

$$\$1200, \$964.32, \$728.64, \ldots$$

14. Find the 10th term of the arithmetic sequence

$$\$200, \$198.32, \$196.64, \ldots$$

15. Find the sum of the first 300 natural numbers.

16. Find the sum of the first 400 natural numbers.

17. Find the sum of the first 20 terms of the sequence

$$6, 9, 12, 15, \ldots$$

18. Find the sum of the first 14 terms of the sequence

$$12, 8, 4, \ldots$$

19. Find a formula for the sum of the first n natural numbers:

$$1 + 2 + 3 + \cdots + n$$

20. Find a formula for the sum of the first n consecutive odd natural numbers starting with 1:

$$1 + 3 + 5 + \cdots + (2n - 1)$$

21. If a student saves 1¢ on October 1, 2¢ on October 2, 3¢ on October 3, etc., how much would be saved in October? (October has 31 days.)

22. If a student saves \$40 on September 1, \$60 on September 2, \$80 on September 3, how much would be saved in September? (September has 30 days.)

23. Find the sum of the first 8 terms of the arithmetic sequence

$$\$512.50, \$1025.00, \$1537.50, \ldots$$

24. Find the sum of the first 10 terms of the arithmetic sequence

$$\$78.90, \$157.80, \$236.70, \ldots$$

8.2 GEOMETRIC SEQUENCES

Geometric Sequences

Consider the sequence

$$2, 6, 18, 54, 162, \ldots$$

If we multiply each term by 3 we get the next term. Sequences in which each term can be multiplied by a certain number to get the next term are called *geometric*. We usually denote this number r. We refer to it as the *common ratio* because we can get r by dividing any term by the preceding term.

$$\frac{a_{k+1}}{a_k} = r, \quad \text{or} \quad a_{k+1} - a_k = d \quad \text{for any } k \geqslant 1.$$

Examples The following are geometric sequences. Identify the common ratio.

OBJECTIVES

You should be able to
a) Identify the common ratio of a geometric sequence.
b) Find the nth term of a geometric sequence.
c) Find the sum of the first n terms of a geometric sequence.
d) Find the sum of an infinite geometric series, if it exists.

The following are geometric sequences. Identify the common ratio.

11. $1, 5, 25, 125, \ldots$

12. $3, -9, 27, -81, \ldots$

13. $6000, $5100, $4335, $3684.75, \ldots$

14. $100, $109, $118.81, \ldots$

15. $1, \frac{1}{2}, \frac{1}{4}, \frac{1}{8}, \ldots$

16. Find the 9th term of the geometric sequence

$$2, 4, 8, 16, \ldots$$

Solution

Sequence	Common ratio
1. $3, 6, 12, 24, 48, 96, \ldots$	2
2. $3, -6, 12, -24, 48, -96, \ldots$	-2
3. $5200, $3900, $2925, $2193.75, \ldots$	0.75
4. $1000, $1080, $1166.40, \ldots$	1.08

DO EXERCISES 11–15.

If we let a_1 be the 1st term and r be the common ratio, then $a_1 r$ is the 2nd term, $a_1 r^2$ is the 3rd term, and so on. Generalizing, we obtain the following.

The nth term of a geometric sequence is given by

$$a_n = a_1 r^{n-1}, \quad \text{for any } n \geqslant 1.$$

Note that the exponent is 1 less than the number of the term.

Example 5 Find the 7th term of the geometric sequence 4, 20, 100, . . .

Solution First note that

$$a_1 = 4, \quad n = 7, \quad \text{and} \quad r = \tfrac{20}{4}, \text{ or } 5.$$

Then, using the formula

$$a_n = a_1 r^{n-1},$$

we have

$$a_7 = 4 \cdot 5^{7-1} = 4 \cdot 5^6 = 4 \cdot 15{,}625 = 62{,}500.$$

DO EXERCISE 16.

Example 6 Find the 10th term of the geometric sequence

$$64, 32, 16, 8, \ldots$$

Solution First note that

$$a_1 = 64, \quad n = 10, \quad \text{and} \quad r = \tfrac{32}{64}, \text{ or } \tfrac{1}{2}.$$

Then using the formula

$$a_n = a_1 r^{n-1},$$

we have

$$a_{10} = 64 \cdot \left(\frac{1}{2}\right)^{10-1} = 64 \cdot \left(\frac{1}{2}\right)^9 = 2^6 \cdot \frac{1}{2^9} = \frac{1}{2^3} = \frac{1}{8}.$$

DO EXERCISE 17.

Sum of the First n Terms of a Geometric Sequence

We want to find a formula for the sum S_n of the first n terms of a geometric sequence

$$a_1, a_1 r, a_1 r^2, a_1 r^3, \ldots, a_1 r^{n-1}, \ldots$$

The sum S_n is given by

$$S_n = a_1 + a_1 r + a_1 r^2 + \cdots + a_1 r^{n-2} + a_1 r^{n-1}.$$

We want to develop a formula that allows us to find this sum without a great amount of adding. If we multiply both sides of the preceding equation by r, we have

$$r S_n = a_1 r + a_1 r^2 + a_1 r^3 + \cdots + a_1 r^{n-1} + a_1 r^n.$$

When we multiply S_n by -1, we get

$$-S_n = -a_1 - a_1 r - a_1 r^2 - \cdots - a_1 r^{n-2} - a_1 r^{n-1}.$$

Then, when we add $r S_n$ and $-S_n$, we get

$$r S_n - S_n = a_1 r^n - a_1$$

or

$$(r - 1)S_n = a_1(r^n - 1),$$

from which we get the following formula.

The sum of the first n terms of a geometric sequence is given by

$$S_n = \frac{a_1(r^n - 1)}{r - 1}, \qquad \text{for any } r \neq 1.$$

Example 7   Find the sum of the first 7 terms of the geometric sequence $3, 15, 75, 375, \ldots$

Solution   First note that

$$a_1 = 3, \qquad n = 7, \qquad \text{and} \qquad r = \tfrac{15}{3}, \text{or } 5.$$

Then, using the formula

$$S_n = \frac{a_1(r^n - 1)}{r - 1},$$

we have

$$S_7 = \frac{3(5^7 - 1)}{5 - 1} = \frac{3(78{,}125 - 1)}{4} = \frac{3(78{,}124)}{4} = 58{,}593.$$

17. Find the 6th term of the geometric sequence

$$3, 1, \tfrac{1}{3}, \tfrac{1}{9}, \ldots$$

18. Find the sum of the first 8 terms of the geometric sequence

$$2, 4, 8, 16, \ldots$$

19. Under the conditions of Example 8, how much would you make in October, which has 31 days?

DO EXERCISE 18.

Doubling Your Salary

Example 8 Suppose someone offered you a job during the month of September (30 days) under the following conditions. You will be paid $0.01 for the first day, $0.02 for the second, $0.04 for the third, and so on, doubling your previous day's salary each day. How much would you earn? (Would you take the job? Make a decision before reading further.)

Solution The amount earned is the sum

$$\$0.01 + \$0.01(2) + \$0.01(2^2) + \$0.01(2^3) + \cdots + \$0.01(2^{29}),$$

where

$$a_1 = \$0.01, \quad n = 30, \quad \text{and} \quad r = 2.$$

Then, using the formula

$$S_n = \frac{a_1(r^n - 1)}{r - 1},$$

we have

$$S_{30} = \frac{\$0.01(2^{30} - 1)}{2 - 1}$$

$$\approx \$0.01(1{,}074{,}000{,}000 - 1) \qquad \text{(Use a calculator to approximate } 2^{30})$$

$$\approx \$0.01(1{,}074{,}000{,}000)$$

$$\approx \$10{,}740{,}000.$$

Now would you take the job?

DO EXERCISE 19.

Note. One could find 2^{30} in various ways. It can be found directly on a calculator with an $\boxed{x^y}$ key or by expressing the power as, say, $2^{10} \cdot 2^{10} \cdot 2^{10}$. Then one could find 2^{10} and multiply that number by itself 3 times. Another way would be $2^5 \cdot 2^5 \cdot 2^5 \cdot 2^5 \cdot 2^5 \cdot 2^5$.

Infinite Geometric Series

Suppose we consider the sum of the terms of an infinite geometric sequence, such as 2, 4, 8, 16, 32, ... We get what is called an *infinite geometric series*

$$2 + 4 + 8 + 16 + 32 + \cdots$$

As n grows larger and larger, the sum of the first n terms, S_n, becomes larger and larger without bound. There are infinite series that get closer and closer to some specific number. Here is an example:

$$\frac{1}{2} + \frac{1}{4} + \frac{1}{8} + \frac{1}{16} + \cdots + \frac{1}{2^n} + \cdots$$

Let's consider S_n for some values of n.

$$S_1 = \tfrac{1}{2} \qquad\qquad\qquad\qquad = \tfrac{1}{2} = 0.5$$
$$S_2 = \tfrac{1}{2} + \tfrac{1}{4} \qquad\qquad\qquad = \tfrac{3}{4} = 0.75$$
$$S_3 = \tfrac{1}{2} + \tfrac{1}{4} + \tfrac{1}{8} \qquad\qquad = \tfrac{7}{8} = 0.875$$
$$S_4 = \tfrac{1}{2} + \tfrac{1}{4} + \tfrac{1}{8} + \tfrac{1}{16} \qquad = \tfrac{15}{16} = 0.9375$$
$$S_5 = \tfrac{1}{2} + \tfrac{1}{4} + \tfrac{1}{8} + \tfrac{1}{16} + \tfrac{1}{32} = \tfrac{31}{32} = 0.96875$$

Perhaps you have noticed that we can describe S_n as follows:

$$S_n = \frac{2^n - 1}{2^n}.$$

Note that the numerator is less than the denominator for all values of n, but as n gets larger and larger, the values of S_n get closer and closer to 1. We say that 1 is the *limit* of S_n and that 1 is the *sum* of the *infinite geometric series*. The sum of an infinite series, if it exists, is denoted S_∞. It can be shown (but we will not do it here) that the sum of the terms of a geometric series exists if and only if $|r| < 1$ (that is, the absolute value of the common ratio is less than 1).

We want to find a formula for the sum of an infinite geometric series. We first consider the sum of the first n terms:

$$S_n = \frac{a_1(r^n - 1)}{r - 1} = \frac{a_1 - a_1 r^n}{1 - r}.$$

For $|r| < 1$, it follows that values of r^n get closer and closer to 0 as n gets large. (Pick a number between -1 and 1 and check this by finding larger and larger powers on your calculator). As r^n gets closer and closer to 0, so does $a_1 r^n$, so S_n gets closer and closer to $a_1/(1 - r)$.

When $|r| < 1$, the sum of an infinite geometric series is given by

$$S_\infty = \frac{a_1}{1 - r}.$$

Example 9 Determine whether this infinite geometric series has a sum. If so, find it:

$$1 + 3 + 9 + 27 + \cdots$$

Determine whether each infinite geometric series has a sum. If so, find it.

20. $1 + 7 + 49 + 343 + \cdots$

21. $1 + (-1) + 1 + (-1) + \cdots$

22. $\frac{1}{2} + \frac{1}{4} + \frac{1}{8} + \frac{1}{16} + \frac{1}{32} + \cdots$

23. $625 + 250 + 100 + 40 + \cdots$

24. Rework Example 11 when the proportion is 95%.

Solution $|r| = |3| = 3$, and since $|r| \not< 1$ the series does *not* have a sum.

Example 10 Determine whether this infinite geometric series has a sum. If so, find it:

$$1 - \frac{1}{2} + \frac{1}{4} - \frac{1}{8} + \frac{1}{16} - \cdots$$

Solution

a) $|r| = |-\frac{1}{2}| = \frac{1}{2}$, and since $|r| < 1$ the series does have a sum.
b) The sum is given by

$$S_\infty = \frac{1}{1 - (-\frac{1}{2})} = \frac{2}{3}.$$

DO EXERCISES 20 THROUGH 23.

Example 11 *Economic multiplier.* The United States banking laws require most banks to maintain a reserve equivalent to a certain proportion of their outstanding deposits. This enables such banks, when they wish and when they can find borrowers, to loan out a certain proportion of the funds that have been deposited in them. Let us assume that this proportion is 0.90 (or 90%). Now suppose a corporation deposits $1000 in a bank which, subsequently, is able to loan the maximum legally possible amount, and this loan is redeposited elsewhere, and so on. What is the total effect of the $1000 on the economy?

Solution The total effect can be modeled as the sum of the infinite geometric series

$$\$1000 + \$1000(0.90) + \$1000(0.90)^2 + \$1000(0.90)^3 + \cdots$$

which is given by

$$S_\infty = \frac{\$1000}{1 - 0.90} = \$10,000.$$

The sum $10,000 is the result of what is referred to in economics as the *multiplier effect.*

DO EXERCISE 24.

EXERCISE SET 8.2

The following are geometric sequences. Identify the common ratio.

1. $7, 14, 28, 56, \ldots$

2. $5, -15, 45, -135, \ldots$

3. $12, -4, \frac{4}{3}, -\frac{4}{9}, \ldots$

4. $4, 2, 1, \frac{1}{2}, \frac{1}{4}, \ldots$

5. $\$5600, \$5320, \$5054, \$4801.30, \ldots$

6. $\$780, \$858, \$943.80, \$1038.18, \ldots$

7. Find the 8th term of the geometric **sequence**

$$1, 3, 9, \ldots$$

8. Find the 10th term of the geometric sequence

$$7, 35, 175, 875, \ldots$$

9. Find the 9th term of the geometric sequence

$$25, 5, 1, \frac{1}{5}, \frac{1}{25}, \ldots$$

10. Find the 10th term of the geometric sequence

$$64, 16, 4, 1, \frac{1}{4}, \frac{1}{16}, \ldots$$

11. Find the 12th term of the geometric sequence

$$\$1000, \$1080, \$1166.40, \ldots$$

Round to the nearest cent.

12. Find the 9th term of the geometric sequence

$$\$1000, \$1070, \$1144.90, \ldots$$

Round to the nearest cent.

13. Find the sum of the first 7 terms of the geometric sequence

$$8, 16, 32, \ldots$$

14. Find the sum of the first 8 terms of the geometric sequence

$$24, -48, 96, \ldots$$

15. Find the sum of the first 5 terms of the geometric sequence

$$\$1000, \$1000(1.08), \$1000(1.08)^2, \ldots$$

Round to the nearest cent.

16. Find the sum of the first 6 terms of the geometric sequence

$$\$200, \$200(1.06), \$200(1.06)^2, \ldots$$

Round to the nearest cent.

17. Suppose someone offered you a job during the month of February (28 days) under the following conditions. You will be paid $0.01 the 1st day, $0.02 the 2nd, $0.04 the 3rd, and so on, doubling your previous day's salary each day. How much would you earn?

18. In Exercise 17, how much would you earn during a February in a leap year (29 days)?

Determine whether each of the following infinite geometric series has a sum. If so, find it.

19. $4 + 20 + 100 + 500 + \cdots$

20. $-6 + 18 - 54 + 162 - \cdots$

21. $10 + 2 + \frac{2}{5} + \frac{2}{25} + \frac{2}{125} + \cdots$

22. $14 + 2 + \frac{2}{7} + \frac{2}{49} + \frac{2}{343} + \cdots$

23. $162 + 108 + 72 + 48 + \cdots$

24. $128 + 96 + 72 + 54 + \cdots$

25. $\$1000(1.08)^{-1} + \$1000(1.08)^{-2} + \$1000(1.08)^{-3} + \cdots$

26. $\$500(1.02)^{-1} + \$500(1.02)^{-2} + \$500(1.02)^{-3} + \cdots$

27. *Economics.* The government makes an $8,000,000,000 expenditure for a new type of aircraft. If 75% of this gets spent again, and 75% of that gets spent, and so on, what is the total effect on the economy?

28. Repeat Exercise 29 for $9,400,000,000 and 99%.

29. *Advertising effect.* A company is marketing a new product in a city of 5,000,000 people. They plan an advertising campaign which they think will induce 40% of the people to buy the product. They then estimate that if those people like the product, they will induce 40% (of the 40% of 5,000,000) more to buy the product, and those will induce 40% more to buy the product, and so on. In all, how many people will buy the product as a result of the advertising campaign? What percentage is this of the population?

30. Repeat Exercise 31 for 6,000,000 people and 45%.

OBJECTIVES

You should be able to

a) Use the straight-line method and prepare a depreciation schedule for a situation. Also, find a formula for the book values V_n, and find the common difference.

b) Use the double-declining balance method and prepare a depreciation schedule for a situation. Also, find a formula for the book values V_n.

c) Use the sum of the year's digits method, find the depreciation fractions for a situation, and prepare a depreciation schedule.

8.3 *(OPTIONAL) DEPRECIATION

A company buys an office machine for $5200 on January 1 of a given year. It is expected to last for 8 years, at which time its *trade-in*, or *salvage*, *value* will be $1100.

Over its lifetime it declines or *depreciates* $5200 − $1100, or $4100. The decline in value from $5200 to $1100 can occur in many ways, as shown in the table below.

Method (1) is called the *straight-line method*, Method (2) the *double-declining balance method*, and Method (3) the *sum of the year's digits method*. We shall consider each of these.

0 yrs.	1	2	3	4	5	6	7	8 yrs.	
$5200	$4687.50	$4175.00	$3662.50	$3150.00	$2637.50	$2125.00	$1612.50	$1100	(1)
$5200	$3900.00	$2925.00	$2193.75	$1645.31	$1233.98	$1100.00	$1100.00	$1100	(2)
$5200	$4288.89	$3491.67	$2808.34	$2238.90	$1783.34	$1441.67	$1213.89	$1100	(3)

25. For the following situation, find the total depreciation, annual depreciation, and rate of depreciation.

Item: Automobile.

Cost = $8700,
Expected life = 5 years,
Salvage value = $1600.

Straight-Line Depreciation

Suppose, for the machine above, the company figures the decline in value to be the *same* each year, that is $\frac{1}{8}$, or 12.5%, of $4100, which is $512.50. After 1 year the *book value*, or simply *value*, is

$$\$5200 - \$512.50, \text{ or } \$4687.50.$$

After 2 years it is

$$\$4687.50 - \$512.50, \text{ or } \$4175.00.$$

After 3 years it is

$$\$4175.00 - \$512.50, \text{ or } \$3662.50,$$

and so on.

For straight-line depreciation,

1. The total depreciation = Cost − Salvage value.

2. The annual depreciation = $\dfrac{\text{Cost} - \text{Salvage value}}{\text{Expected life}}$.

3. The rate of depreciation = $\dfrac{\text{Annual depreciation}}{\text{Total depreciation}}$.

DO EXERCISE 25.

A depreciation schedule gives a complete listing of the book values and total depreciation throughout the life of an item.

Example 1   Prepare a depreciation schedule for the following situation.

Item: Office machine.

$$\text{Cost} = \$5200,$$
$$\text{Expected life} = 8 \text{ years},$$
$$\text{Salvage value} = \$1100.$$

Solution

Year	Rate of depreciation	Annual depreciation	Book value	Total depreciation
0			$5200	
1	$\frac{1}{8}$ or 12.5%	$512.50	4687.50	$ 512.50
2	12.5%	512.50	4175.00	1025.00
3	12.5%	512.50	3662.50	1537.50
4	12.5%	512.50	3150.00	2050.00
5	12.5%	512.50	2637.50	2562.50
6	12.5%	512.50	2125.00	3075.00
7	12.5%	512.50	1612.50	3587.50
8	12.5%	512.50	1100.00	4100.00

The rate of depreciation is the same each year.

The annual depreciation is the same each year.

We find the book values by starting with the initial cost, $5200, and successively subtracting $512.50.

We find the total depreciations by starting with $512.50 after the first year and successively adding $512.50.

DO EXERCISE 26.

Why do we call this *straight-line depreciation*? If we make a graph of book values versus time, the values lie on a straight line.

26. Prepare a depreciation schedule for the situation in Margin Exercise 25.

Year	Rate of depreciation	Annual depreciation	Book value	Total depreciation
0				
1				
2				
3				
4				
5				

27. For the situation in Margin Exercise 25, find

a) A formula for the book values V_n.
b) The common difference.

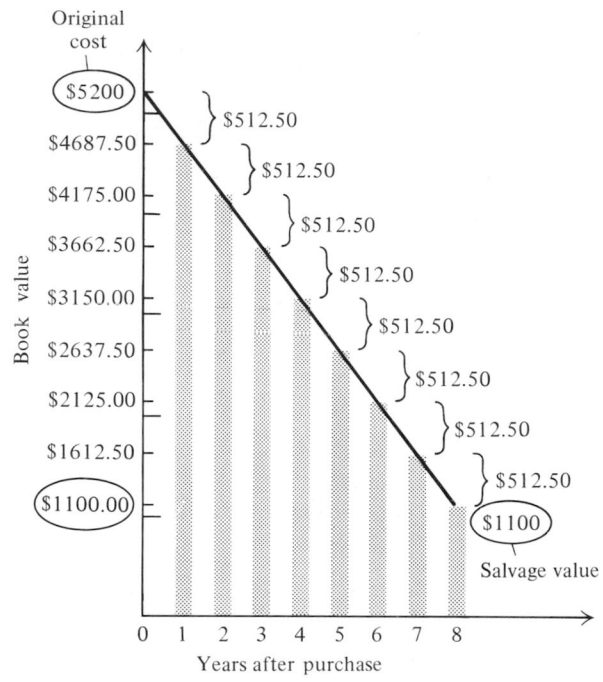

The book values V_n of an item n years after purchase form an arithmetic sequence for which

$$V_n = C - n\left(\frac{C - S}{N}\right),$$

where

$$C = \text{original cost of an item,}$$
$$N = \text{years of expected life,}$$
$$S = \text{salvage value.}$$

For the machine in Example 1,

$$V_n = \$5200 - n\left(\frac{\$5200 - \$1100}{8}\right) = \$5200 - (\$512.50)n,$$

and the common difference is $-\$512.50$.

DO EXERCISE 27.

Declining Balance Depreciation

A company buys a machine for $5200. It is expected to last for 8 years, at which time its salvage value will be $1100. The straight-line rate of depreciation would be $\frac{1}{8}$, or 12.5%. Depreciation can be deducted as a business expense when a business computes its taxes.

When a business is starting out it has many expenses and less income and therefore needs all the tax advantages it can get. For this and other reasons, the Internal Revenue Service allows certain assets to be depreciated at a rate which is larger than the straight-line rate, but *no more* than twice the straight-line rate. (Such a rate could be, for example, $1\frac{1}{4}$, $1\frac{1}{2}$, or 2 times the straight-line rate.) Suppose for the above, the rate is $2 \cdot \frac{1}{8}$, or 25%. This is called the *double-declining balance method*. Then the book value after 1 year is:

$5200 − (25% × $5200)   (We subtract 25% of the initial book value.)

= $5200 − (0.25 × $5200)

= $5200 − $1300

= $3900.

After 2 years it is

$3900 − (0.25 × $3900)   (We subtract 25% of preceding book value.)

= $3900 − $975

= $2925.

After 3 years it is

$2925 − (0.25 × $2925)

= $2925 − $731.25

= $2193.75.

After 4 years it is

$2193.75 − (0.25 × $2193.75)

= $2193.75 − $548.44   (Rounded to the nearest cent)

= $1645.31,

and so on.

Example 2   Prepare a depreciation schedule for the situation below. Use the double-declining balance method.

Item: Office machine.

$$\text{Cost} = \$5200,$$
$$\text{Expected life} = 8 \text{ years},$$
$$\text{Salvage value} = \$1100.$$

28. Prepare a depreciation schedule for the situation below. Use the double-declining balance method.

Item: Automobile.

$$\text{Cost} = \$8700,$$
$$\text{Expected life} = 5 \text{ years},$$
$$\text{Salvage value} = \$1600.$$

Year	Rate of depreciation	Annual depreciation	Book value	Total depreciation
0				
1				
2				
3				
4				
5				

Solution

Year	Rate of depreciation	Annual depreciation	Book value	Total depreciation
0			$5200	
1	$\frac{2}{8}$ or 25%	$1300.00	3900.00	$1300
2	25%	975.00	2925.00	2275
3	25%	731.25	2193.75	3006.25
4	25%	548.44	1645.31	3554.69
5	25%	411.33	1233.98	3966.02
6		133.98	1100.00	4100.00
7		0 *	1100.00 *	4100.00 *
8		0	1100.00	4100.00

> The rate of depreciation is the same each year, twice the straight-line rate.

> We find the book values by starting with the initial cost, $5200, and successively subtracting 0.25 times the book value. For example, $5200 − (0.25 × $5200) = $3900. Then, $3900 − (0.25 × $3900) = $2925, and so on.

> We find the annual depreciations when we multiply each successive book value by 0.25. For example, 0.25 × $5200 = $1300, and 0.25 × $3900 = $975.

DO EXERCISE 28.

The book values V_n of an item n years after purchase form a geometric sequence

$$V_n = C\left(1 - \frac{m}{N}\right)^n, \qquad 0 < m \leqslant 2,$$

where

$$C = \text{original cost of an item.}$$

This holds until V_n drops below the salvage value S.

---

* Note that

$$\$1233.98 - (0.25 \times \$1233.98) = \$1233.98 - \$308.50$$
$$= \$925.48,$$

but the book value cannot drop below the salvage value. Thus, after $1233.98 the next book value becomes $1100.00, and the annual depreciation for that year is $1233.98 − $1100.00, or $133.98.

DO EXERCISE 29.

29. For the situation in Margin Exercise 28, find a formula for the book values V_n.

Sum of the Year's Digits Depreciation

Another method of depreciation which allows larger amounts of depreciation in early years and smaller amounts in later years is the *sum of the year's digits method*. Each year a different rate of depreciation is used, which is a fraction.

Example 3   For the situation below,

a) Find the depreciation fractions;
b) Find the depreciation and book values after 1 year, 2 years.

Item: Office machine.

$$\text{Cost} = \$5200,$$
$$\text{Expected life} = 8 \text{ years},$$
$$\text{Salvage value} = \$1100.$$

Solution

a) To find the depreciation we first find the sum of the year's digits:

$$8 + 7 + 6 + 5 + 4 + 3 + 2 + 1 = \boxed{36}.*$$

The number 36 will be the denominator of each fraction. We then find the depreciation fractions (rates) by dividing each number in the sum by 36:

$$\frac{8}{36}, \ \frac{7}{36}, \ \frac{6}{36}, \ \frac{5}{36}, \ \frac{4}{36}, \ \frac{3}{36}, \ \frac{2}{36}, \ \frac{1}{36}.$$

b) The total depreciation is $\$5200 - \$1100$, or $\$4100$. First year depreciation is

$$\frac{8}{36} \times \$4100 = \frac{8 \times \$4100}{36} = \frac{\$32{,}800}{36} = \$911.11 \qquad \text{(Rounded to the nearest cent.)}$$

---

* The sum

$$1 + 2 + 3 + \cdots + n$$

is the sum of the terms of an arithmetic sequence where $a_1 = 1$, and $a_n = n$. The sum is given by

$$S_n = \frac{n}{2}(a_1 + a_n) = \frac{n}{2}(1 + n)$$
$$= \frac{n(n + 1)}{2}.$$

Thus,

$$8 + 7 + 6 + 5 + 4 + 3 + 2 + 1 = 1 + 2 + 3 + 4 + 5 + 6 + 7 + 8$$
$$= \frac{8(8 + 1)}{2} = \frac{8(9)}{2} = \frac{72}{2} = 36.$$

30. For the situation below,

a) Find the depreciation fractions.
b) Find the depreciation and book values after 1 year, 2 years, and 3 years.

Item: Automobile.

$$\text{Cost} = \$8700,$$
$$\text{Expected life} = 5 \text{ years},$$
$$\text{Salvage value} = \$1600.$$

31. Prepare a depreciation schedule for the situation below. Use the sum of the year's digits method.

$$\text{Cost} = \$8700,$$
$$\text{Expected life} = 5 \text{ years},$$
$$\text{Salvage value} = \$1600.$$

Year	Rate of depreciation	Annual depreciation	Book value	Total depreciation
0				
1				
2				
3				
4				
5				

The book value after 1 year is

$$\$5200 - \$911.11, \text{ or } \$4288.89.$$

Second-year depreciation is

$$\frac{7}{36} \times \$4100 = \frac{7 \times \$4100}{36} = \frac{\$28,700}{36} = \$797.22.$$

The book value after 2 years is

$$\$4288.89 - \$797.22, \text{ or } \$3491.67.$$

DO EXERCISE 30.

Example 4 Prepare a depreciation schedule for the situation below. Use the sum of the year's digits method.

$$\text{Cost} = \$5200,$$
$$\text{Expected life} = 8 \text{ years},$$
$$\text{Salvage value} = \$1100.$$

Solution

Year	Rate of depreciation	Annual depreciation	Book value	Total depreciation
0			$5200	
1	$\frac{8}{36}$ or 22.2%	$911.11	4288.89	$ 911.11
2	$\frac{7}{36}$ or 19.4%	797.22	3491.67	1708.33
3	$\frac{6}{36}$ or 16.7%	683.33	2808.34	2391.66
4	$\frac{5}{36}$ or 13.9%	569.44	2238.90	2961.10
5	$\frac{4}{36}$ or 11.1%	455.56	1783.34	3416.66
6	$\frac{3}{36}$ or 8.3%	341.67	1441.67	3758.33
7	$\frac{2}{36}$ or 5.6%	227.78	1213.89	3986.11
8	$\frac{1}{36}$ or 2.8%	113.89	1100.00	4100.00

The rate of depreciation gets lower each year.

We find the annual depreciations first. To do this we multiply the total depreciation by each fraction. For example, $\frac{8}{36} \times \$4100 = \$911.11$, $\frac{7}{36} \times \$4100 = \$797.22$, and so on.

We find the book values by subtracting each annual depreciation in succession. For example, $\$5200 - \$911.11 = \$4288.89$, $\$4288.89 - \$797.22 = \$3491.67$, and so on.

DO EXERCISE 31.

EXERCISE SET 8.3

Use the straight-line method.

a) Prepare a depreciation schedule.
b) Find a formula for the book values V_n.
c) Find the common difference.

1. *Item:* Automobile.
   Cost = $8000,
   Expected life = 4 years,
   Salvage value = $2000.

2. *Item:* Automobile.
   Cost = $12,000,
   Expected life = 3 years,
   Salvage value = $4800.

3. *Item:* Postage machine.
   Cost = $450,
   Expected life = 8 years,
   Salvage value = $0.

4. *Item:* Typewriter.
   Cost = $2500,
   Expected life = 6 years,
   Salvage value = $0.

In Exercises 5 through 8, use the double-declining balance method.

a) Prepare a depreciation schedule.
b) Find a formula for the book values V_n.

5. (See Exercise 1.)   6. (See Exercise 2.)   7. (See Exercise 3.)   8. (See Exercise 4.)

In Exercises 9 through 12, use the sum of the year's digits method.

a) Find the depreciation fractions.
b) Prepare a depreciation schedule.

9. (See Exercise 1.)   10. (See Exercise 2.)   11. (See Exercise 3.)   12. (See Exercise 4.)

8.4 SIMPLE AND COMPOUND INTEREST

Simple Interest

You put $100 in a savings account for 1 year. This is called *principal*.

The *interest rate* is 8%. This means you get back 8% of the principal,

$$8\% \text{ of } \$100,$$

or

$$8\% \times \$100,$$

or

$$\$8.00,$$

in addition to the principal. The $8.00 is called *interest*.

OBJECTIVES

You should be able to

a) Given a principal P, an interest rate i, and a time t, find the amount to which P grows at simple interest.
b) Given a principal P, an interest rate i, and a time t, find the amount to which P grows at interest compounded n times a year.

32. Suppose $1000 is invested in a savings account at 6% simple interest. What is the amount in the account at the end of 3 months?

Hint. 3 months = $\frac{3}{12}$ yr. = $\frac{1}{4}$ yr.

The *amount* you get back is

$$\text{(Principal)} + \text{(Interest)}, \quad \text{or } \$100 + \$8, \quad \text{or } \$108.$$

To find interest for a fraction t of a year (or for any time t), we compute the interest for 1 year and multiply by t. Thus, $100 principal invested at an interest rate of 8% for $\frac{1}{4}$ of a year, yields interest of

$$(8\% \times \$100) \times \tfrac{1}{4}, \quad \text{or } \$2.00.$$

We have the following formulas.

SIMPLE INTEREST. **Principal P invested at simple interest rate i for time t, in years, yields interest I given by**

$$I = P \cdot i \cdot t.$$

AMOUNT. **The amount A to which principal P will grow at simple interest rate i, for t years, is given by**

$$A = P + Pit = P(1 + it).$$

Note that $I = Pit$ usually is written as $I = Prt$, but we are reserving the letter r for later use.

DO EXERCISE 32.

33. Under the conditions of Example 1, how much is due on $1700 left unpaid for 1 month?

Example 1 A loan charges 18% simple interest. How much is due on $1000 left unpaid for 1 month?

Solution $P = \$1000$, $i = 18\%$, or 0.18, and $t = \frac{1}{12}$ yr. Then the amount due is

$$A = P(1 + it) = \$1000(1 + 0.18 \times \tfrac{1}{12}) = \$1000(1 + 0.015)$$
$$= \$1000(1.015) = \$1015.$$

DO EXERCISE 33.

Compound Interest

Suppose you invested $1000 at an interest rate of 8%, compounded annually. The amount A_1 in the account at the end of 1 year is given by

$$A_1 = \$1000(1 + 0.08) = \$1000(1.08) = \$1080.$$

Going into the second year you have a new principal of $1080, so by the end of 2 years you would have the amount A_2 given by

$$A_2 = \$1080(1 + 0.08) = \$1080(1.08) = \$1166.40.$$

Going into the third year you have a new principal of $1166.40, so by the end of 3 years you would have the amount A_3 given by

$$A_3 = \$1166.40(1 + 0.08) = \$1166.40(1.08) \approx \$1259.71.$$

Note the following:

$$A_1 = \$1000(1.08)^1,$$
$$A_2 = \$1000(1.08)^2,$$
$$A_3 = \$1000(1.08)^3.$$

The amounts A_n form a geometric sequence with common ratio 1.08. In general, suppose you invest a principal of P dollars at interest rate i, compounded annually. The amount A_1 in the account at the end of 1 year is given by

$$A_1 = P(1 + i) = Pr,$$

where, for convenience, $r = 1 + i$.

Going into the second year you would have a new principal of Pr dollars, so by the end of 2 years you would have the amount A_2 given by

$$A_2 = A_1 \cdot r = (Pr)r = Pr^2.$$

Going into the third year you have a new principal of Pr^2, so by the end of 3 years you would have the amount A_3 given by

$$A_3 = A_2 \cdot r = (Pr^2)r = Pr^3.$$

The amounts A_n form a geometric sequence with common ratio r, which is $1 + i$.

INTEREST COMPOUNDED ANNUALLY. **If principal P is invested at interest rate i, compounded annually, in t years it will grow to the amount A given by**

$$A = P(1 + i)^t.$$

Example 2   Suppose $1000 is invested at 5% compounded annually. How much is in the account at the end of 3 years?

Solution   We substitute $1000 for P, 0.05 for i, and 3 for t in the equation $A = P(1 + i)^t$, and get

$$A = \$1000(1 + 0.05)^3 = \$1000(1.05)^3 = \$1000(1.157625)$$
$$= \$1157.625 \approx \$1157.63.$$

DO EXERCISE 34.

34. Suppose $2000 is invested at 6% compounded annually. How much is in the account at the end of 4 years?

If interest is compounded quarterly, we can find a formula like the one above as follows:

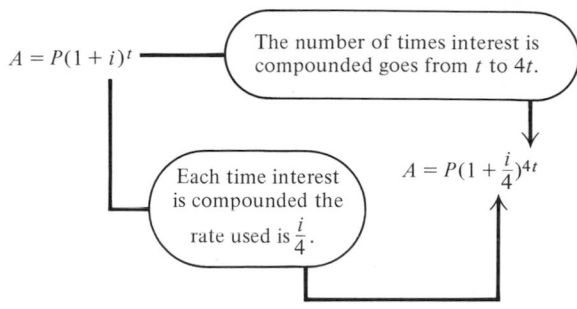

$A = P(1 + i)^t$ — The number of times interest is compounded goes from t to $4t$.

Each time interest is compounded the rate used is $\frac{i}{4}$.

$A = P(1 + \frac{i}{4})^{4t}$

In general,

INTEREST COMPOUNDED n TIMES PER YEAR. **If principal P is invested at interest rate i, compounded n times per year, in t years it will grow to an amount A given by**

$$A = P\left(1 + \frac{i}{n}\right)^{nt}.$$

The number nt is the total number of payment periods.

Example 3 Suppose $1000 is invested at 8%. How much is in the account at the end of 3 years if interest is

a) simple?   b) compounded annually?
c) compounded semiannually?   d) compounded quarterly?
e) compounded daily?   f) hourly?

Solution

a) $A = P(1 + it) = \$1000(1 + 0.08 \times 3) = \$1000(1 + 0.24)$

$= \$1000(1.24) = \$1240.00.$

b) $A = P(1 + i)^t = \$1000(1 + 0.08)^3 = \$1000(1.08)^3$

$= \$1000(1.259712) \approx \$1259.71.$

c) $A = P\left(1 + \frac{i}{n}\right)^{nt} = \$1000\left(1 + \frac{0.08}{2}\right)^{2 \times 3} = \$1000(1 + 0.04)^6$

$= \$1000(1.04)^6 = \$1000(1.265319)$

$\approx \$1265.32.$

d) $A = P\left(1 + \dfrac{i}{n}\right)^{nt} = \$1000\left(1 + \dfrac{0.08}{4}\right)^{4 \times 3}$

$= \$1000(1 + 0.02)^{12}$

$= \$1000(1.02)^{12} = \$1000(1.268242)$

$\approx \$1268.24$

e) $A = P\left(1 + \dfrac{i}{n}\right)^{nt} = \$1000\left(1 + \dfrac{0.08}{365}\right)^{365 \times 3}$

$= \$1000(1 + 0.000219)^{1095}$

$= \$1000(1.000219)^{1095} = \$1000(1.270967)$

$\approx \$1270.97.$

f) $A = P\left(1 + \dfrac{i}{n}\right)^{nt} = \$1000\left(1 + \dfrac{0.08}{8760}\right)^{8760 \times 3}$

$= \$1000(1 + 0.00000913)^{26,280}$

$= \$1000(1.00000913)^{26,280} = \$1000(1.271168)$

$\approx \$1271.17.$

CALCULATOR NOTE. One can find these powers on a calculator with an x^y key, by a compound interest table, or by the method described earlier where the larger power is broken down to a product of smaller powers. The number of places on the calculator may affect the accuracy of the answer. Thus, you may occasionally find that your answers do not agree with those in the answer section which have been found on a calculator with a ten-digit readout. In general, if you are using a calculator, do all your computations and round only at the end.

Compare the amounts found in Example 3:

$\$1240, \quad \$1259.71, \quad \$1265.32, \quad \$1268.24, \quad \$1270.97, \quad \$1271.17.$

Note that as the number of periods of compounding increases within a fixed time, the greater the *amount* becomes, but the *increase* gets less and less. If we keep using more compounding periods, the amount gets closer and closer to an amount found by *continuous compounding*, $\$1271.25$. We will study continuous compounding in a later chapter.

DO EXERCISE 35.

Present Value

A representative of a financial institution is often asked to solve a problem like the following.

35. Suppose $1000 is invested at 6%. How much is in the account at the end of 2 years if interest is

a) simple
b) compounded annually?
c) compounded semiannually?
d) compounded quarterly?
e) compounded daily?

Example 4 Following the birth of a child, a parent wants to make an initial investment P which will grow to $10,000 by the child's twentieth birthday. Interest is compounded semiannually at 8%. What should the initial investment be?

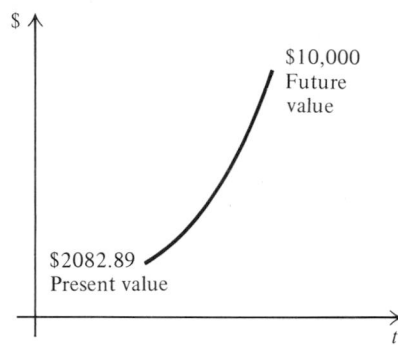

Solution Using the formula

$$A = P\left(1 + \frac{i}{n}\right)^{nt},$$

we find P such that

$$\$10{,}000 = P\left(1 + \frac{0.08}{2}\right)^{2 \times 20}$$

or

$$\$10{,}000 = P(1.04)^{40}.$$

Then

$$\$10{,}000 \approx P(4.801021),$$

and

$$P \approx \frac{\$10{,}000}{4.801021} \approx \$2082.89.$$

Thus a principal of $2082.89 would have to be invested at 8%, compounded semiannually, to grow to $10,000 in 20 years. The amount $2082.89 is called the *present value* of $10,000 for 20 years at 8% compounded semiannually. We can also say that the *future value* of $2082.89 is $10,000.

By solving $A = P(1 + i)^t$ and $A = P(1 + (i/n))^{nt}$ for P, we get general formulas for present value.

PRESENT VALUE. The present value P of an amount A at interest rate i, compounded annually, for t years is given by

$$P = A(1 + i)^{-t}.$$

For interest compounded n times per year, the present value P is given by

$$P = A\left(1 + \frac{i}{n}\right)^{-nt}.$$

DO EXERCISE 36.

36. Find the present value of $4000 for 5 years at 6%, compounded quarterly.

EXERCISE SET 8.4

Suppose $2000 is invested at the given simple interest rate and for the time indicated. What is the amount in the account?

1. 9%, 4 months   **2.** 10%, 8 months   **3.** 14%, 2 years   **4.** 11%, 3 years

5. Suppose $2000 is invested at 7%. How much is in the account at the end of 2 years if interest is

a) simple?
b) compounded annually?
c) compounded semiannually?
d) compounded quarterly?
e) compounded daily?

6. Suppose $1500 is invested at 10%. How much is in the account at the end of 3 years if interest is

a) simple?
b) compounded annually?
c) compounded semiannually?
d) compounded every 2 months?
e) compounded daily?

Find the present value of:

7. $1000 at 8% compounded annually for 3 years.

8. $1000 at 9% compounded annually for 4 years.

9. $1000 at 8% compounded quarterly for 3 years.

10. $1000 at 9% compounded semiannually for 4 years.

11. $10,000 at 6% compounded semiannually for 18 years.

12. $15,000 at 7% compounded semiannually for 18 years.

13. *Personal debt.* On the average every person in this country has a debt of $1000. How much will this debt be in 2 years at 7%, compounded annually?

14. *Personal debt.* In Exercise 13, how much will be due in 2 years at 12%, compounded annually?

15. *Inflation.* Inflation is based on what a person could buy in 1967 for $1. In 1980 what was bought for $1 in 1967 will cost $1(1 + 0.07)^{13}$, assuming a rate of inflation of 7%. How much is that cost?

16. *Inflation.* In 1984 what was bought for $1 in 1967 will cost $1(1 + 0.07)^{17}$, assuming a rate of inflation of 7%. How much is that cost?

17. *Finding the interest rate.* $2560 is invested at interest rate i, compounded annually. In 2 years it grows to $2890. What is the interest rate?

18. *Finding the interest rate.* $1000 is invested at interest rate i, compounded annually. In 2 years it grows to $1210. What is the interest rate?

OBJECTIVES

You should be able to

a) Find the effective annual yield of an amount invested at interest rate i compounded n times per year.

b) Find the annual percentage rate on a loan at a given add-on interest rate.

37. Find the effective annual yield, when the nominal interest rate is 9%, compounded semiannually.

8.5 ANNUAL PERCENTAGE RATE

Effective Yield

Suppose $1000 is invested at 8%, compounded quarterly for 1 year. We know that this will grow to an amount

$$\$1000\left(1 + \frac{0.08}{4}\right)^4, \text{ or } \$1082.43,$$

which is an increase of 8.243%. This is the same as if $1000 were invested at 8.243% compounded once a year (simple interest). The 8.243% is called the *effective annual yield* or *annual percentage rate*, and the 8% is called the *nominal rate*. In general, if P is invested at interest rate i, compounded n times per year, then the effective annual yield is that number E satisfying

$$P(1 + E) = P\left(1 + \frac{i}{n}\right)^n.$$

Then

$$1 + E = \left(1 + \frac{i}{n}\right)^n$$

and

$$E = \left(1 + \frac{i}{n}\right)^n - 1.$$

Example 1 Find the effective annual yield, when the nominal interest rate is 7%, compounded semiannually.

Solution

$$E = \left(1 + \frac{i}{n}\right)^n - 1$$

$$= \left(1 + \frac{0.07}{2}\right)^2 - 1 = (1.035)^2 - 1$$

$$= 1.071225 - 1 = 0.071225 \approx 7.123\%.$$

DO EXERCISE 37.

Add-on Interest

Consider a car loan.

Situation: Car loan of $1000 at 7% for 1 year

Question: Couldn't the borrower put the $1000 in a savings account at 7.5% and make money?

Car loans are examples of what lending institutions call *add-on-interest*. The nominal, or stated, interest rate is 7%. This is *not* the true rate, the *annual percentage rate*, APR. Lenders use the simple interest formula, $I = Prt$, and figure that the loan will earn interest of $1000 \times 0.07 \times 1$, or $70. They "add on" the $70, so you have to pay back $1070. For simplicity, suppose you pay back the loan in 4 payments. Each payment is $1070 \div 4$, or $267.50. Your loan decreases as follows:

$$\$1070, \quad \$802.50, \quad \$535, \quad \$267.50.$$

What's the catch? The lending institution *does not* allow you the full use of the $1070 for the year. The average principal you have is

$$\frac{\$1070 + \$802.50 + \$535 + \$267.50}{4} = \$668.75.$$

How do we find APR? It is defined to be that interest rate such that

$$\begin{pmatrix} \text{Interest} \\ \text{for first} \\ \text{3 months} \end{pmatrix} + \begin{pmatrix} \text{Interest} \\ \text{for second} \\ \text{3 months} \end{pmatrix} + \begin{pmatrix} \text{Interest} \\ \text{for third} \\ \text{3 months} \end{pmatrix} + \begin{pmatrix} \text{Interest} \\ \text{for fourth} \\ \text{3 months} \end{pmatrix}$$

$$= \text{Total interest,}$$

or

$$(\$1070 \times \text{APR} \times \tfrac{1}{4}) + (\$802.50 \times \text{APR} \times \tfrac{1}{4})$$
$$+ (\$535 \times \text{APR} \times \tfrac{1}{4}) + (\$267.50 \times \text{APR} \times \tfrac{1}{4}) = \$70.$$

Factoring out APR, we get

$$[(\$1070 \times \tfrac{1}{4}) + (\$802.50 \times \tfrac{1}{4}) + (\$535 \times \tfrac{1}{4}) + (\$267.50 \times \tfrac{1}{4})] \cdot \text{APR}$$
$$= \$70$$

or

$$\left[\frac{\$1070 + \$802.50 + \$535 + \$267.50}{4}\right] \cdot \text{APR} = \$70.$$

Then

$$\$668.75 \cdot \text{APR} = \$70$$

$$\text{APR} = \frac{\$70}{\$668.75} \approx 10.5\%$$

In general,

APR = (Total interest) ÷ (Average principal).

For 12 payments in the above situation, the APR would have been 12.1%. In either case the true interest rate, or APR, is almost double

38. Find the APR. Assume a car loan at the given add-on interest rate for 1 year and 12 payments.

$$\text{Loan} = \$4000,$$
$$\text{Add-on rate} = 9\%.$$

the stated rate. You would not save money by putting the money in the bank. The Truth In Lending Law *requires* lenders to inform you of the APR.

DO EXERCISE 38.

EXERCISE SET 8.5

Find the effective yield.

1. 8%, compounded semiannually

2. 10%, compounded semiannually

3. 9%, compounded quarterly

4. 12%, compounded quarterly

5. 8%, compounded 6 times per year

6. 10%, compounded every 2 months

7. 8%, compounded daily

8. 10%, compounded daily

9. 8%, compounded hourly

10. 10%, compounded hourly

Find the *APR*. Assume that these are car loans at the given add-on interest rate for 1 year and 12 payments.

11. Loan = $1000,
   Add-on rate = 8%.

12. Loan = $1000,
   Add-on rate = 6%.

13. Loan = $2000,
   Add-on rate = 10%.

14. Loan = $5000,
   Add-on rate = 9%.

OBJECTIVES

You should be able to

a) Find the amount of an annuity where P dollars is being invested n times per year at interest rate i, compounded n times per year for N years.

b) Find what payment P will have to be made n times a year for N years at interest rate i, compounded n times per year, so that V dollars will have accumulated in N years.

8.6 ANNUITIES AND SINKING FUNDS

Annuities

An *annuity* is a series of equal payments made at equal time intervals. Rent payments are an example of an annuity. Fixed deposits in a savings account can also be an annuity. For example, suppose someone makes a sequence of deposits of $1000 each in a savings account on which interest is compounded annually at 8%. The total amount in the account, including interest, is called the *amount of the annuity*, or the *future value of the annuity*. Let us find the amount of the given annuity for a period of 5 years. The following time diagram can help. Note that we do not make a deposit until the end of the first year.

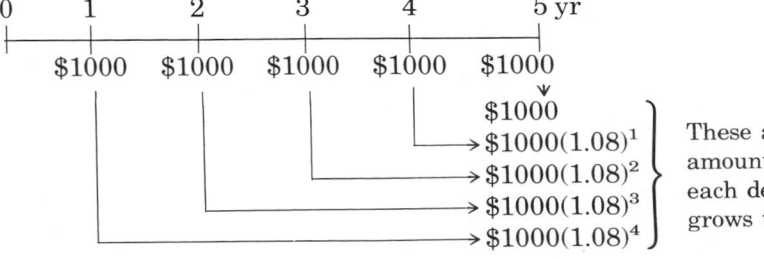

The amount of the annuity is the sum

$$\$1000 + \$1000(1.08)^1 + \$1000(1.08)^2 + \$1000(1.08)^3 + \$1000(1.08)^4.$$

This is the sum of the terms of a geometric sequence where

$$a_1 = \$1000, \qquad n = 5, \qquad \text{and} \qquad r = 1.08.$$

We can find this sum using the formula

$$S_n = \frac{a_1(r^n - 1)}{r - 1}.$$

We have

$$S_5 = \frac{\$1000(1.08^5 - 1)}{1.08 - 1} \approx \frac{\$1000(1.469328 - 1)}{0.08} \approx \$5866.60.$$

When equal deposits are made at equal time intervals which are the same as the periods of compounding and the first deposit is not made until the end of the first year, we have what is called an *ordinary annuity*. We shall consider only ordinary annuities.

In general, suppose we have an ordinary annuity in which P dollars are deposited each year for N years, at interest rate i, compounded annually. The first payment P, invested at the end of the first year, will be invested for $N - 1$ years and will grow to

$$P(1 + i)^{N-1}.$$

The second payment will be invested for $N - 2$ years and grow to

$$P(1 + i)^{N-2},$$

and so on; the next-to-last payment will be invested for 1 year and grow to

$$P(1 + i)^1,$$

and the last payment will not have time to grow, and will be

$$P.$$

Then the amount of the annuity will be the sum V given by

$$V = P + P(1 + i)^1 + P(1 + i)^2 + \cdots + P(1 + i)^{N-2} + P(1 + i)^{N-1}.$$

This is the sum of the terms of a geometric sequence where

$$a_1 = P, \qquad n = N, \qquad \text{and} \qquad r = 1 + i.$$

We can find this sum using the formula

$$S_n = \frac{a_1(r^n - 1)}{r - 1}.$$

39. Find the amount of an annuity where $200 per year is being invested at 6% compounded annually, for 16 years.

We have

$$V = \frac{P[(1 + i)^N - 1]}{(1 + i) - 1},$$

from which we get the following formula:

THE AMOUNT OF AN ANNUITY. The amount of an annuity, V, where P dollars are invested at the end of each of N years at interest rate i, compounded annually, is given by

$$V = \frac{P[(1 + i)^N - 1]}{i}$$

If interest is compounded n times per year and deposits are being made every compounding period, the formula for V is found by replacing i by $\dfrac{i}{n}$ and N by nN.

Example 1   Find the amount of an annuity where $1000 per year is being invested at 7%, compounded annually, for 15 years.

Solution

$$V = \frac{\$1000[(1 + 0.07)^{15} - 1]}{0.07} = \frac{\$1000[(1.07)^{15} - 1]}{0.07}$$

$$= \frac{\$1000[2.759032 - 1]}{0.07} \approx \$25{,}129.03.$$

DO EXERCISE 39.

40. Find the amount of an annuity where $200 every 3 months is being invested at 6%, compounded quarterly, for 16 years.

Example 2   Find the amount of an annuity where $1000 every 3 months is being invested at 7%, compounded quarterly, for 15 years.

Solution

$$V = \frac{\$1000\left[\left(1 + \dfrac{0.07}{4}\right)^{60} - 1\right]}{\dfrac{0.07}{4}} = \frac{\$1000[(1.0175)^{60} - 1]}{0.0175}$$

$$= \frac{\$1000[2.831816 - 1]}{0.0175} \approx \$104{,}675.20.$$

Note that much more money is being deposited in this annuity than in Example 1.

DO EXERCISE 40.

Sinking Funds

The following is an adaptation of a problem considered before.

Example 3 Following the birth of a child, a parent wants to make a deposit on each of the child's subsequent birthdays so that $10,000 will have accumulated by the child's 20th birthday. Interest will be compounded annually at 8%. What should each deposit be?

Solution We can use the formula

$$V = \frac{P[(1 + i)^N - 1]}{i}.$$

We know that $V = \$10,000$, $i = 0.08$, and $N = 20$. Then we substitute,

$$\$10,000 = \frac{P[(1 + 0.08)^{20} - 1]}{0.08},$$

and solve for P:

$$\$10,000(0.08) = P[(1.08)^{20} - 1],$$

$$\$800 = P[4.660957 - 1] = P[3.660957],$$

$$P = \frac{\$800}{3.660957} \approx \$218.52.$$

Each birthday after the child's birth, the parent will need to deposit $218.52.

DO EXERCISE 41.

The situation in Example 3 illustrates what is called a *sinking fund*. Any financial arrangement in which periodic payments are made for the purpose of growing to a specific future amount is called a *sinking fund*. The word "sinking" is somewhat of a misnomer in that one is making deposits to get to a future amount. The word probably comes from "sinking" the future amount back to now to consider what deposits need to be made.

41. Following the birth of a child, a parent wants to make a deposit on each of the child's subsequent birthdays so that $10,000 will have accumulated by the child's 10th birthday. Interest will be compounded annually at 7%. What should each deposit be?

EXERCISE SET 8.6

Find the amount of an annuity where;

1. $1000 is being invested each year at 7%, compounded annually for 4 years.

2. $2000 is being invested each year at 9%, compounded annually for 5 years.

3. $1000 is being invested each year at 7%, compounded annually for 10 years.

4. $3000 is being invested each year at 9%, compounded annually for 10 years.

5. $2000 is being invested every 3 months at 8%, compounded quarterly for 5 years.

6. $300 is being invested every 3 months at 6%, compounded quarterly for 8 years.

7. $10 is being invested each month at 6%, compounded monthly for 8 years.

8. $20 is being invested each month at 7%, compounded monthly for 10 years.

9. A person decides to save money for retirement. $1000 is invested each year at 7.5%, compounded annually. How much will be in the retirement fund at the end of 30 years?

10. A person decides to save money for retirement, investing $500 each year at 8.5%, compounded annually. How much will be in the retirement fund at the end of 40 years?

11. A family expects to buy a new car 5 years from now. They decide to put away $50 a month. At 8% interest, compounded monthly, how much will they have at the end of 5 years?

12. A company decides to put away money for future expansion of their business. They save $1000 a month. At 9% interest, compounded monthly, how much will they have at the end of 6 years?

13. Due to increased business a company expects to have to buy a $10,000 machine 8 years from now. They decide to make deposits each year at 6.5%, compounded annually. What should each deposit be so that the company will accumulate the $10,000?

14. A young couple wants to have a down payment of $8000 to buy a new home in 9 years. They decide to make deposits each year at 5%, compounded annually. What should each deposit be so that they will accumulate the $8000?

15. A family expects to pay $7000 for a car 5 years from now. They decide to make a deposit each month. Interest at 6% will be compounded monthly. What should each deposit be so that they will have accumulated the $7000?

16. A family expects to pay $800 for a freezer 3 years from now. They decide to make a deposit each month. Interest at 5.5% will be compounded monthly. What should each deposit be so that they will have accumulated the $800?

▶

Solve each formula for P.

17. $V = \dfrac{P[(1 + i)^N - 1]}{i}$

18. $V = \dfrac{P\left[\left(1 + \dfrac{i}{n}\right)^{nN} - 1\right]}{\dfrac{i}{n}}$

OBJECTIVES

You should be able to

a) Find what lump sum would have to be deposited now at interest rate k, compounded n times a year, so that P dollars can be withdrawn n times a year for N years.

b) Find the equal payment P required to pay off a loan of S dollars at interest rate k, compounded n times per year, in N years.

8.7 PRESENT VALUE OF AN ANNUITY AND AMORTIZATION

Present Value of an Annuity

The *present value of an annuity* is the sum of the present values of each payment of the annuity. Celebrities such as movie stars or athletes sometimes have years when they make lots of money, and then their incomes decline. Suppose a person wants to make a deposit in a lump sum right now, so that for each of the following 5 years $1000 can be drawn from the account. Interest is to be compounded annually at 8%. In effect, we can think of the deposit consisting of five different parts, the first being that amount which should be deposited now so that there will be $1000 one year from now, plus the second being that amount which should be deposited now so that there will be another $1000 two years from now, plus the third being that amount which should be deposited now so that there will be another $1000 three years from now, and so on. Each of these is a present value of $1000 a certain number of years from now. This can be shown in the following time diagram.

These are the
present values
of each withdrawal.

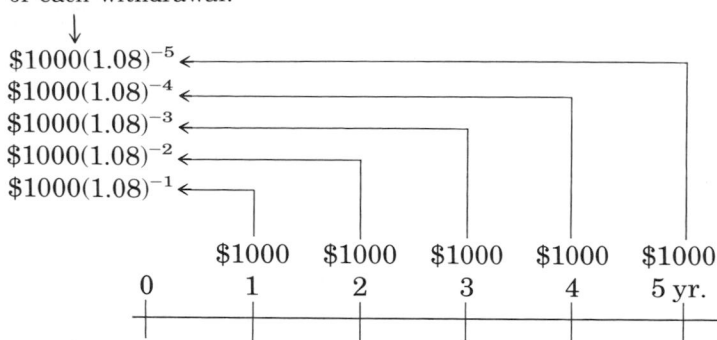

The present value of the annuity is the sum

$$\$1000(1.08)^{-1} + \$1000(1.08)^{-2} + \$1000(1.08)^{-3}$$
$$+ \$1000(1.08)^{-4} + \$1000(1.08)^{-5}.$$

This is the sum of a geometric sequence where

$$a_1 = \$1000(1.08)^{-1}, \quad n = 5, \quad \text{and} \quad r = 1.08^{-1}.$$

We can find this sum using the formula

$$S_n = \frac{a_1(r^n - 1)}{r - 1}.$$

We have

$$S_5 = \frac{\$1000(1.08)^{-1}[(1.08^{-1})^5 - 1]}{1.08^{-1} - 1} = \frac{\$1000(1.08)^{-1}[(1.08)^{-5} - 1]}{1.08^{-1} - 1}.$$

Multiplying S_5 by 1 using $\dfrac{1.08}{1.08}$ will ease our calculations.

$$S_5 = \frac{1.08}{1.08} \times \frac{\$1000(1.08)^{-1}[(1.08)^{-5} - 1]}{1.08^{-1} - 1}$$

$$= \frac{\$1000[(1.08)^{-5} - 1]}{1 - 1.08} = \frac{\$1000[1 - (1.08)^{-5}]}{1.08 - 1}$$

$$= \frac{\$1000[1 - 0.680583]}{0.08} \approx \$3992.71.$$

In general, suppose a lump sum S is to be deposited now at interest
rate i, compounded annually, so that P dollars can be withdrawn for
each of the next N years; then S is the present value of the annuity
and is given by the sum

$$P(1 + i)^{-1} + P(1 + i)^{-2} + \cdots + P(1 + i)^{-(N-1)} + P(1 + i)^{-N}.$$

42. What lump sum would have to be deposited now at 8%, compounded annually, so that withdrawals of $2000 can be made every year for 20 years?

We can find this sum using the formula

$$S_n = \frac{a_1(r^n - 1)}{r - 1},$$

where $a_1 = P(1 + i)^{-1}$, $n = N$, and $r = (1 + i)^{-1}$. We have

$$S = \frac{P(1 + i)^{-1}[(1 + i)^{-N} - 1]}{(1 + i)^{-1} - 1},$$

from which we get the following formula:

THE PRESENT VALUE OF AN ANNUITY

$$S = \frac{P[1 - (1 + i)^{-N}]}{i}.$$

If interest is compounded n times per year and withdrawals are to be made every compounding period, the formula for S is found by replacing i by $\dfrac{i}{n}$ and N by nN.

Example 1   What lump sum would have to be deposited now at 7%, compounded annually, so that withdrawals of $1000 can be made every year for 15 years?

Solution

$$S = \frac{\$1000[1 - (1.07)^{-15}]}{0.07} \approx \frac{\$1000[1 - 0.362446]}{0.07} \approx \$9107.91.$$

DO EXERCISE 42.

43. What lump sum would have to be deposited now at 8%, compounded quarterly, so that withdrawals of $500 can be made every 3 months for 10 years?

Example 2   What lump sum would have to be deposited now at 6%, compounded quarterly, so that withdrawals of $500 can be made every 3 months for 10 years?

Solution

$$S = \frac{\$500\left[1 - \left(1 + \dfrac{0.06}{4}\right)^{-40}\right]}{\dfrac{0.06}{4}} = \frac{\$500[1 - (1.015)^{-40}]}{0.015}$$

$$\approx \frac{\$500[1 - 0.551262]}{0.015} \approx \$14{,}957.93.$$

DO EXERCISE 43.

Amortization

In the formula for the present value of an annuity, what application is there for the situation where we know S, i, and N, and want to compute P?

Example 3   A person borrows $6000 at an interest rate of 14%, compounded monthly. The loan is to be paid off by 36 equal monthly payments over the next 3 years. How much is each payment?

Solution   We can use the formula

$$S = \frac{P\left[1 - \left(1 + \frac{i}{n}\right)^{-nN}\right]}{\frac{i}{n}}.$$

We know that $S = \$6000$, $i = 0.14$, $n = 12$, and $N = 3$. Then we substitute

$$\$6000 = \frac{P\left[1 - \left(1 + \frac{0.14}{12}\right)^{-36}\right]}{\frac{0.14}{12}},$$

and solve for P:

$$\$6000\left(\frac{0.14}{12}\right) = P[1 - 0.658646],$$

$$\$70.00 = P[0.341354],$$

$$P = \frac{\$70.00}{0.341354} \approx \$205.07.$$

DO EXERCISE 44.

One might ask why we cannot use the sinking-fund formula for this problem. In a sinking fund a payment is made and then allowed to grow at a certain interest rate. The payee gets the money back at the end. When one *amortizes* a loan, as in Example 3, the money is received at the outset. In Example 3, the borrower gets the $6000 at the outset.

When financial institutions loan money, they start computing interest right away. After 1 month a payment of $205.08 was made. In effect a part of the $6000, a present value, has grown to $205.08 during the first month and consists of principal and interest. Each monthly payment is such a payment of interest and principal; the amounts of each vary, but they always total $205.08. When we amortize a loan, we know the present value of an annuity and want to find the amount of each equal payment.

44. A family buys a house for $52,000, makes a down payment of $10,000 and borrows the remaining $42,000 at 10.75% interest, compounded monthly. The loan is to be paid off by 300 equal monthly payments over the next 25 years. How much is each payment?

EXERCISE SET 8.7

What lump sum would have to be deposited now at

1. 7%, compounded annually, so that withdrawals of $1000 can be made every year for 4 years?

2. 9%, compounded annually, so that withdrawals of $2000 can be made every year for 5 years?

3. 7%, compounded annually, so that withdrawals of $1000 can be made every year for 10 years?

4. 9%, compounded annually, so that withdrawals of $3000 can be made every year for 10 years?

5. 8%, compounded quarterly, so that withdrawals of $2000 can be made every 3 months for 5 years?

6. 6%, compounded semiannually, so that withdrawals of $300 can be made every 6 months for 8 years?

7. 6%, compounded monthly, so that withdrawals of $10 can be made every month for 8 years?

8. 7%, compounded monthly, so that withdrawals of $20 can be made every month for 10 years?

9. A family pays $4800 for a remodeling job. A down payment of $600 is made, and $4200 is borrowed at 12%, compounded monthly. The loan is to be paid off by 36 equal payments over the next 3 years. How much is each payment?

10. A family pays $10,000 for an addition to their home. A down payment of $1000 is made, and $9000 is borrowed at 10.5%, compounded monthly. The loan is to be paid off by 24 equal payments over the next 2 years. How much is each payment?

11. A family buys a house for $75,000. A down payment of $15,000 is made, and $60,000 is borrowed at 10.5%, compounded monthly. The loan is to be paid off by 360 equal payments over the next 30 years. How much is each payment?

12. A family buys a condominium for $40,000. A down payment of $10,000 is made, and $30,000 is borrowed at 11%, compounded monthly. The loan is to be paid off by 300 equal payments over the next 25 years. How much is each payment?

▶──

Solve each formula for P.

13. $S = \dfrac{P[1 - (1 + i)^{-N}]}{i}$.

14. $S = \dfrac{P\left[1 - \left(1 + \dfrac{i}{n}\right)^{-nN}\right]}{\dfrac{i}{n}}$.

A *perpetuity* is an annuity in which payments will be made forever.

15. What lump sum would have to be deposited now at 8%, compounded annually so that withdrawals of $1000 can be made every year forever? *Hint.* Use the formula for the sum of an infinite geometric series. What other method can be used?

16. A family wishes to set up an endowed professorship at a college. It will pay a salary of $12,000 each year forever. What lump sum would have to be deposited now at 7.5%, compounded annually, to provide this salary?

CHAPTER 8 TEST

1. The following is an arithmetic sequence. Identify the first term and the common difference.

$$5, \quad 8, \quad 11, \quad \ldots$$

2. Find the 20th term of the sequence in Question 1.

3. Find the sum of the first 20 terms of this arithmetic sequence.

$$\$1.00, \quad \$1.06, \quad \$1.12, \quad \ldots$$

4. The following is a geometric sequence. Identify the common ratio.

$$\$100, \quad \$105, \quad \$110.25, \quad \ldots$$

5. Find the 10th term of the geometric sequence in Question 4. Round to the nearest cent.

6. Find the sum of the first 10 terms of the sequence in Question 4. Round to the nearest cent.

7. Determine whether this infinite geometric series has a sum. If so, find it.

$$\$1000, \quad \$80, \quad \$6.40, \quad \ldots$$

Consider this situation for Questions 8 through 10.

Item. Automobile.

$$\text{Cost} = \$8500,$$
$$\text{Expected life} = 4 \text{ years,}$$
$$\text{Salvage value} = \$2550.$$

8. Use the straight-line method,

a) Prepare a depreciation schedule.
b) Find a formula for the book values V_n.
c) Find the common difference.

9. Use the double-declining balance method,

a) Prepare a depreciation schedule.
b) Find a formula for the book values V_n.

10. Use the sum of the year's digits method,

a) Find the depreciation fractions.
b) Prepare a depreciation schedule.

11. Suppose $1000 is invested at 6%. How much is in the account at the end of 3 years if interest is

a) simple?
b) compounded annually?
c) compounded semiannually?
d) compounded quarterly?

12. Find the effective annual yield.

$$9\%, \text{compounded quarterly}$$

13. Find the APR. Assume a car loan at the given add-on interest rate for 1 year and 12 payments.

$$\text{Loan} = \$5000, \qquad \text{Add-on rate} = 11\%.$$

14. Find the amount of an annuity where $2000 is being invested semiannually at 8.4%, compounded semiannually, for 7 years.

15. Due to increasing business, a company expects to have to pay $8000 for a machine 6 years from now. They decide to make deposits each year at 8%, compounded annually. What should each payment be so that the company will accumulate the $8000?

16. What lump sum would have to be deposited now at 5%, compounded annually, so that withdrawals of $10,000 can be made every year for 15 years?

17. A family buys a condominium for $100,000. A down payment of $25,000 is made, and $75,000 is borrowed at 10%, compounded monthly. The loan is to be paid off by 240 equal payments over the next 20 years. How much is each payment?

CHAPTER NINE

Which peak appears to have maximum height? (*French Government Tourist Office*).

Differential Calculus

OBJECTIVES

You should be able to

a) Decide if a graph is of a continuous function.
b) Decide if a function is continuous at a given point a.
c) Find

$$\lim_{x \to a} f(x),$$

if such a limit exists.
d) Find a limit like

$$\lim_{h \to 0} (3x^2 + 3xh + h^2).$$

9.1 CONTINUITY AND LIMITS

In this section we give an intuitive (meaning "based on prior and present experience") treatment of two important concepts: continuity and limits.

Continuity

The following are graphs of functions that are *continuous* over the whole real line $(-\infty, \infty)$.

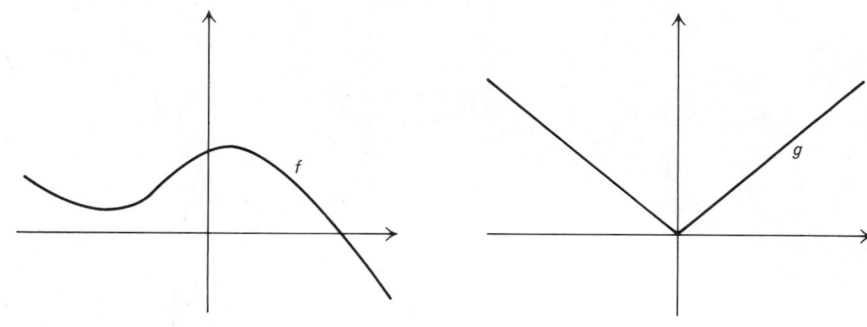

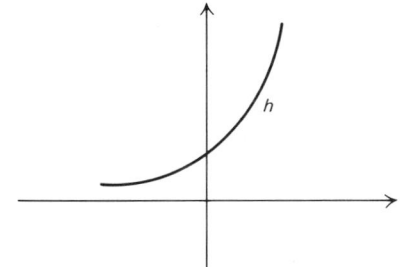

Note that there are no "jumps" or holes in the graphs. For now we will use a somewhat intuitive definition of continuity, which we will refine later. We say that a function is *continuous* over, or on, some interval of the real line if its graph can be traced without lifting a pencil from the paper. The following are graphs of functions that are *not* continuous over the whole real line.

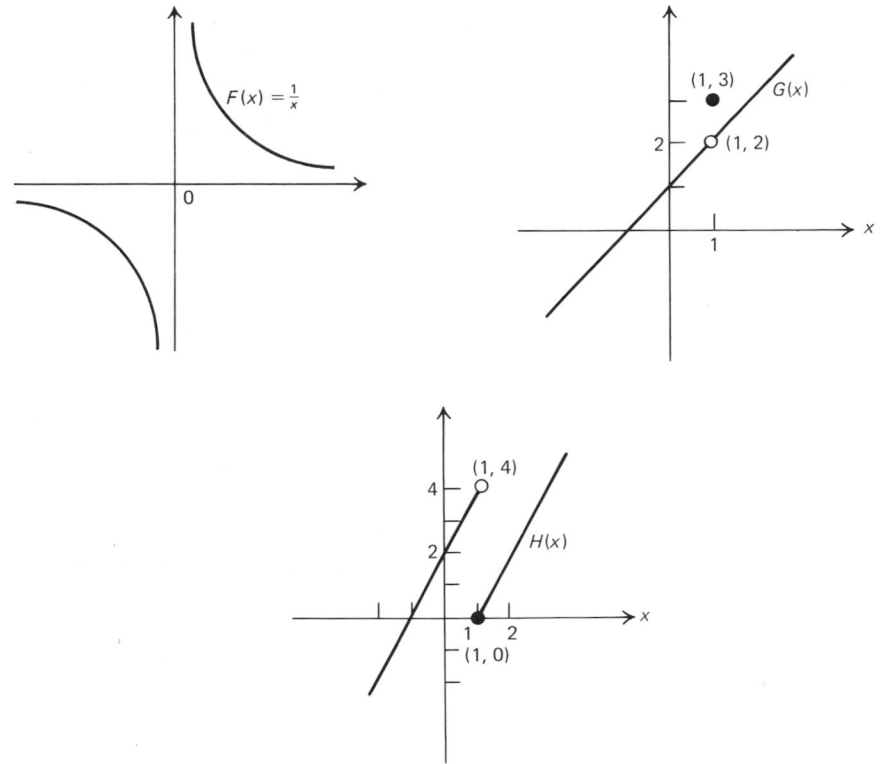

For *G* and *H*, the open circle indicates that the circled point is not part of the graph.

In each case the graph *cannot* be traced without lifting the pencil from the paper. However, each case represents a different situation. Let us discuss why each case fails to be continuous over the whole real line.

The function F fails to be continuous over the *whole* real line $(-\infty, \infty)$. Since F is not defined at $x = 0$, the point $x = 0$ is not part of the domain, so $f(0)$ does not exist and there is no point $(0, f(0))$ on the graph. Thus there is no point to trace at $x = 0$. However, F is continuous on the intervals $(-\infty, 0)$ and $(0, \infty)$.

The function G is not continuous over the whole real line since it is not continuous at $x = 1$. Let us trace the graph of G to the left of $x = 1$. As x approaches 1, $G(x)$ seems to approach 2; but at $x = 1$, $G(x)$ *jumps* up to 3, while to the right of $x = 1$, $G(x)$ *jumps* back to some value close to 2. Thus, G is discontinuous at $x = 1$.

1. Which functions are continuous?

a)

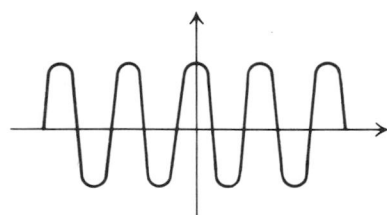

b)

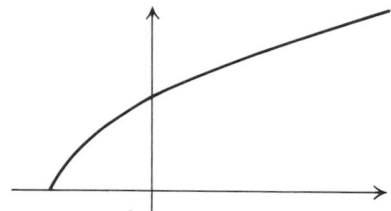

c)

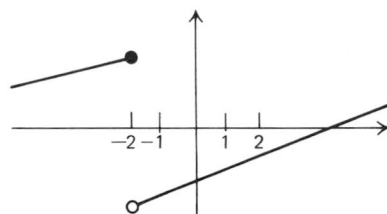

d)

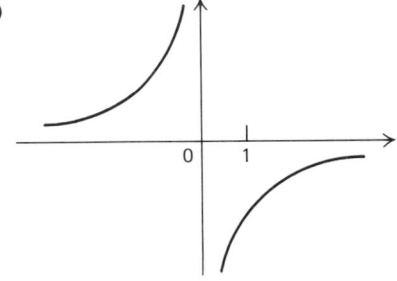

2. a) Decide whether the function in Margin Exercise 1(c) is continuous at −2, at 1.

b) Decide whether the function in Margin Exercise 1(d) is continuous at 1, at 0.

The function H is not continuous over the whole real line since it is not continuous at $x = 1$. Let us trace the graph of H starting to the left of $x = 1$. As x approaches 1, $H(x)$ seems to approach 4; but at $x = 1$, $H(x)$ jumps down to 0, while just to the right of $x = 1$, $H(x)$ is close to 0. Thus $H(x)$ is discontinuous at $x = 1$.

DO EXERCISES 1 AND 2.

Limits

These notions about *continuity* can be formalized by introducing the concept of *limits*. The study of limits has application not just to continuity but throughout calculus.

Consider the function f given by

$$f(x) = 2x + 3.$$

Suppose we select input numbers x closer and closer to the number 4, and look at the output numbers $2x + 3$. Study the following tables.

R			**L**	
Inputs x	Outputs $2x + 3$		Inputs x	Outputs $2x + 3$
5	13		2	7
4.8	12.6		3.6	10.2
4.3	11.6		3.8	10.6
4.1	11.2		3.9	10.8
4.01	11.02		3.99	10.98
4.001	11.002		3.999	10.998

In Table R, the input numbers approach 4 from numbers greater than 4 or, on a graph, from numbers to the right of 4. In Table L, the input numbers approach 4 from numbers less than 4 or, on a graph, from numbers to the left of 4. In both cases the outputs approach 11. Thus we say,

As x *approaches* 4, $2x + 3$ *approaches* 11.

An arrow, →, is often used for the word "approaches." Thus the above can be written

As $x \rightarrow 4$, $2x + 3 \rightarrow 11$.

The number 11 is said to be the *limit* of $2x + 3$ as x approaches 4. We can abbreviate this statement as follows:

$$\lim_{x \to 4} (2x + 3) = 11.$$

This is read, "The limit, as x approaches 4, of $2x + 3$ is 11."

DEFINITION. A function, f, has the *limit*, L, as x approaches a, written

$$\lim_{x \to a} f(x) = L,$$

if we can get $f(x)$ as close to L as we wish by restricting x to a sufficiently small interval about a but excluding a.

DO EXERCISE 3.

The limit at a point a does not depend on the function value at a, $f(a)$, should that exist. To see this more clearly, let us consider again the function G whose graph we showed as an example of a discontinuous function. G can be defined as follows:

$$G(x) = \frac{x^2 - 1}{x - 1}, \qquad x \neq 1$$

$$\left(\text{If an input } x \text{ is not 1, the output is } \frac{x^2 - 1}{x - 1}. \right)$$

$$G(x) = 3, \qquad x = 1$$

(If an input is 1, the output is 3.)

Let us set up input–output tables for $G(x)$ as x approaches 1 from the left and from the right.

R

Inputs x	Outputs $\dfrac{x^2 - 1}{x - 1}$
2	3
1.6	2.6
1.2	2.2
1.1	2.1
1.01	2.01
1.001	2.001

L

Inputs x	Outputs $\dfrac{x^2 - 1}{x - 1}$
0	1
0.7	1.7
0.8	1.8
0.9	1.9
0.99	1.99
0.999	1.999

3. Consider

$$f(x) = 3x - 1.$$

a) Complete each table. (▦ helpful, though not essential.)

R

Inputs x	Outputs $3x - 1$
7	
6.9	
6.4	
6.1	
6.01	
6.001	

L

Inputs x	Outputs $3x - 1$
5	
5.7	
5.8	
5.9	
5.99	
5.999	

b) Complete.

As $x \to 6,\ 3x - 1 \to \square$

c) Complete.

$$\lim_{x \to 6} (3x - 1) = \square$$

4. Consider

$$f(x) = \frac{x^2 - 9}{x - 3}.$$

a) Complete each table. (▦ helpful, though not essential.)

R

Inputs x	Outputs $\dfrac{x^2 - 9}{x - 3}$
4	
3.6	
3.2	
3.1	
3.01	
3.001	

L

Inputs x	Outputs $\dfrac{x^2 - 9}{x - 3}$
1	
2	
2.4	
2.9	
2.99	
2.999	

b) Find

$$\lim_{x \to 3} \frac{x^2 - 9}{x - 3}.$$

Note that as the inputs x approach 1 from either the left or right, the outputs approach 2. Thus,

$$\lim_{x \to 1} G(x) = 2,$$

while the function value at $x = 1$ is

$$G(1) = 3.$$

Since these two values are not equal at $x = 1$, G is discontinuous at $x = 1$.

DEFINITION. A function f is *discontinuous* at $x = a$, if

$$\lim_{x \to a} f(x) \neq f(a).$$

Note that

$$\frac{x^2 - 1}{x - 1}$$

does not have a value at $x = 1$, but that it does have a limit, which we found using the input–output tables. Specifically, the function

$$\frac{x^2 - 1}{x - 1},$$

when factored, equals

$$\frac{(x - 1)(x + 1)}{x - 1},$$

which equals $x + 1$, provided $x \neq 1$. Accordingly, the limit as $x \to 1$ could have been found by substituting approaching inputs directly into $x + 1$.

DO EXERCISE 4

Consider again the function H whose graph we showed as an example of a discontinuous function. H can be defined as follows.

$$H(x) = 2x + 2, \qquad x < 1$$

(If an input x is less than 1, the output is $2x + 2$.)

$$H(x) = 2x - 2, \qquad 1 \leqslant x$$

(If an input x is greater than or equal to 1, the output is $2x - 2$.)

DO EXERCISE 5

We see that as x approaches 1 from the left, $H(x)$ approaches 4:

$$x \to 1, \qquad H(x) \to 4.$$
(from the left)

As x approaches 1 from the right, $H(x)$ approaches 0:

$$x \to 1, \qquad H(x) \to 0.$$
(from the right)

Thus the limiting values of $H(x)$ as x approaches 1 from the left and from the right are not equal: $4 \neq 0$. The limit, therefore, does *not* exist, and the function is discontinuous at this point.

DEFINITION. A function is *discontinuous* at a point, a, if the limit of the function as x approaches a from the left is not equal to the limit of the function as x approaches a from the right.

In such a case, we can also say that the function is discontinuous because it does not have a limit as x approaches a.

Summarizing ideas illustrated in the foregoing examples, we obtain the following definition of continuity.

**DEFINITION. A function f is *continuous* at $x = a$ if
1. $f(a)$ exists,
2. $\lim_{x \to a} f(x)$ exists, and
3. $\lim_{x \to a} f(x) = f(a)$.
A function is *continuous over an interval, I,* if it is continuous at each point in I.**

We previously considered the function f given by

$$f(x) = 2x + 3.$$

This function is continuous at 4 because
1. $f(4)$ exists, $f(4) = 11$,
2. $\lim_{x \to 4} f(x)$ exists, $\quad \lim_{x \to 4} f(x) = 11$ (as shown earlier),
3. $\lim_{x \to 4} f(x) = 11 = f(4)$.

In fact, $f(x) = 2x + 3$ is continuous at any point on the real line.

DO EXERCISE 6.

CONTINUITY PRINCIPLES. The following continuity principles, which we will not prove, allow us to build up continuous functions.
 i. **Any constant function is continuous (such a function never varies).**
 ii. **For any positive integer n, x^n and $\sqrt[n]{x}$ are continuous. When n is even, the inputs of $\sqrt[n]{x}$ are restricted to $[0, \infty)$.**
 iii. **If $f(x)$ and $g(x)$ are continuous, then so are $f(x) + g(x)$, $f(x) - g(x)$, and $f(x) \cdot g(x)$.**
 iv. **If $f(x)$ is continuous, so is $1/f(x)$, as long as the inputs x are not such that the outputs $f(x) = 0$.**

5. Complete these input–output tables as x approaches 1 from the left and right, and determine the limiting values.

L

Inputs x	Outputs $2x + 2$
0.5	
0.7	
0.9	
0.99	
0.999	

R

Inputs x	Outputs $2x - 2$
1.8	
1.6	
1.1	
1.01	
1.001	

6. Consider

$$f(x) = 3x - 1$$

(See Margin Exercise 3.)
a) Does $f(6)$ exist? If so, what is it?
b) Does $\lim_{x \to 6} f(x)$ exist? If so, what is it?
c) Does

$$\lim_{x \to 6} f(x) = f(6)?$$

d) Is f continuous at 6?

7. Provide an argument to show that the function given by

$$f(x) = \frac{\sqrt[3]{x} - 7x^2}{x - 2}$$

is continuous as long as $x \neq 2$.

Let us convince ourselves that $x^2 - 3x + 2$ is continuous. Now x^2 is continuous, by (ii). The constant function 3 is continuous by (i), the function x is continuous by (ii), so the product $3x$ is continuous by (iii). Thus $x^2 - 3x$ is continuous by (iii), and since the constant 2 is continuous, we can apply (iii) again to show that $x^2 - 3x + 2$ is continuous. In similar fashion, any polynomial such as

$$f(x) = x^4 - 5x^3 + x^2 - 7$$

is continuous. A rational function is a quotient of two polynomials

$$r(x) = \frac{f(x)}{q(x)}.$$

Thus by (iv), a rational function is continuous as long as the inputs x are not such that $q(x) = 0$.

DO EXERCISE 7.

Find each limit, if it exists.

8. $\lim\limits_{x \to -2} (x^4 - 5x^3 + x^2 - 7)$

If a function is continuous at a, we can substitute to find the limit.

Example 1 Find $\lim\limits_{x \to 2} (x^4 - 5x^3 + x^2 - 7)$.

Solution. It follows from the continuity principles that $x^4 - 5x^3 + x^2 - 7$ is continuous. Thus the limit can be found by substitution:

$$\lim_{x \to 2} (x^4 - 5x^3 + x^2 - 7) = 2^4 - 5 \cdot 2^3 + 2^2 - 7$$

$$= 16 - 40 + 4 - 7 = -27.$$

9. $\lim\limits_{x \to 1} \sqrt{x^2 + 3x + 4}$

Example 2 Find $\lim\limits_{x \to 0} \sqrt{x^2 - 3x + 2}$.

Solution By using the continuity principles, we have shown that $x^2 - 3x + 2$ is continuous; and as long as x is restricted to values for which $x^2 - 3x + 2$ is nonnegative, it follows from principle (ii) that $\sqrt{x^2 - 3x + 2}$ is continuous. Thus, we can substitute to find the limit:

$$\lim_{x \to 0} \sqrt{x^2 - 3x + 2} = \sqrt{0^2 - 3 \cdot 0 + 2} = \sqrt{2}.$$

DO EXERCISES 8 AND 9.

More on Limits

Let us consider another limit that does not exist. Consider

$$f(x) = \frac{1}{x - 1}.$$

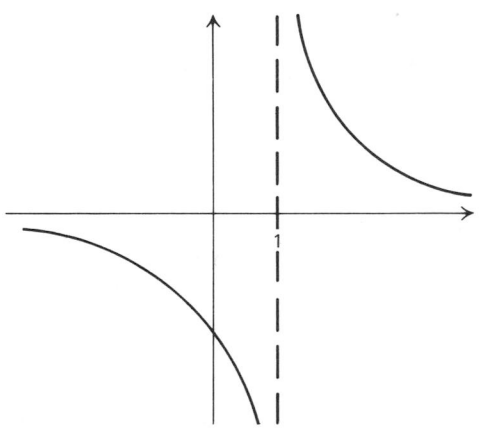

Let us try to determine

$$\lim_{x \to 1} \frac{1}{x - 1}.$$

Note that we cannot substitute 1 for x, since it would result in division by 0; that is, the function is not continuous at 1. Let us investigate the behavior of the function as x approaches 1.

R		**L**	
Inputs x	Outputs $\frac{1}{x-1}$	Inputs x	Outputs $\frac{1}{x-1}$
2	1	0	−1
1.6	1.667	0.7	−3.333
1.2	5.0	0.8	−5
1.1	10	0.9	−10
1.01	100	0.99	−100
1.001	1000	0.999	−1000

As x approaches 1 from the right, the outputs get larger and larger. These numbers do not approach any real number, though it might be said that "the limit from the right is ∞ (infinity)." As x approaches 1 from the left, the outputs get smaller and smaller. These numbers do not approach any real number, though it might be said that "the limit from the left is $-\infty$ (negative infinity)." Thus

$$\lim_{x \to 1} \frac{1}{x - 1}$$

does not exist.

DO EXERCISES 10 AND 11.

10. *Earned-run average.* A pitcher's earned-run average (the average number of runs given up every 9 innings or 1 game) is given by

$$A = 9 \cdot \frac{n}{i},$$

where n = the number of earned runs allowed, and i = the number of innings pitched. Suppose we fix the number of earned runs allowed at 4 and let i vary. We get a function given by

$$A(i) = 9 \cdot \frac{4}{i}.$$

a) Complete the following table, rounding to two decimal places.

Innings pitched (i)	Earned-run average (A)
9	
8	
7	
6	
5	
4	
3	
2	
1	
$\frac{2}{3}$ (2 outs)	
$\frac{1}{3}$ (1 out)	

b) Find

$$\lim_{i \to 0} A(i).$$

c) On the basis of (a) and (b), what might a pitcher's earned run average be if 4 runs were allowed and there were 0 outs?

$$f(x) = \frac{1}{x - 3}.$$

a) ▦ Complete each table.

Inputs x	Outputs $\dfrac{1}{x - 3}$
0	
−0.6	
−0.8	
−0.9	
−0.99	
−0.999	

Inputs x	Outputs $\dfrac{1}{x - 3}$
−2	
−1.5	
−1.4	
−1.1	
−1.01	
−1.001	

b) Find

$$\lim_{x \to -1} \frac{1}{x - 3},$$

if it exists.

Find each limit, if it exists.

12. $\lim\limits_{x \to -4} \dfrac{x^2 - 16}{x + 4}$

13. $\lim\limits_{x \to 3} \dfrac{x - 3}{x^2 - 9}$

Limit Principles

There are limit principles that correspond to the continuity principles.

If $\lim\limits_{x \to a} f(x) = L$ and $\lim\limits_{x \to a} g(x) = M$, then

L1. $\lim\limits_{x \to a} c = c$. (The limit of a constant is the constant.)

L2. $\lim\limits_{x \to a} x^n = a^n$, $\lim\limits_{x \to a} \sqrt[n]{x} = \sqrt[n]{a}$, for any positive integer n. (When n is even the inputs of $\sqrt[n]{x}$ must be restricted to $[0, \infty)$.)

L3. $\lim\limits_{x \to a} [f(x) \pm g(x)] = \lim\limits_{x \to a} f(x) \pm \lim\limits_{x \to a} g(x) = L \pm M$. (The limit of a sum or difference is the sum or difference of the limits.)

$\lim\limits_{x \to a} [f(x) \cdot g(x)] = [\lim\limits_{x \to a} f(x)] \cdot [\lim\limits_{x \to a} g(x)] = L \cdot M$. (The limit of a product is the product of the limits.)

L4. $\lim\limits_{x \to a} \dfrac{1}{f(x)} = \dfrac{1}{\lim\limits_{x \to a} f(x)} = \dfrac{1}{L}$, provided $L \neq 0$. (The limit of a reciprocal is the reciprocal of the limit.)

Example 3 Find $\lim\limits_{x \to -3} \dfrac{x^2 - 9}{x + 3}$.

Solution The function $(x^2 - 9)/(x + 3)$ is not continuous at $x = -3$. We use some algebraic simplification and then some limit principles.

$$\lim_{x \to -3} \frac{x^2 - 9}{x + 3} = \lim_{x \to -3} \frac{(x + 3)(x - 3)}{x + 3}$$

$$= \lim_{x \to -3} (x - 3), \quad \text{assuming } x \neq -3$$

$$= \lim_{x \to -3} x - \lim_{x \to -3} 3, \quad \text{(by L3)}$$

$$= -3 - 3 = -6$$

DO EXERCISES 12 AND 13.

In the next section we encounter expressions with two variables, x and h; our interest is in limits where x is fixed as a constant and $h \to 0$.

Example 4 Find $\lim\limits_{h\to 0} (3x^2 + 3xh + h^2)$

Solution If we treat x as a constant, using the limit principles, it follows that

$$\lim_{h\to 0} (3x^2 + 3xh + h^2) = 3x^2 + 3x0 + 0^2 = 3x^2.$$

The reader can check any limit about which there is uncertainty by using an input–output table. Below is a table for this limit.

h	$3x^2 + 3xh + h^2$
1	$3x^2 + 3x \cdot 1 + 1^2$, or $\qquad 3x^2 + 3x + 1$
0.8	$3x^2 + 3x(0.8) + (0.8)^2$, or $\qquad 3x^2 + 2.4x + 0.64$
0.5	$3x^2 + 3x(0.5) + (0.5)^2$, or $\qquad 3x^2 + 1.5x + 0.25$
0.1	$3x^2 + 3x(0.1) + (0.1)^2$, or $\qquad 3x^2 + 0.3x + 0.01$
0.01	$3x^2 + 3x(0.01) + (0.01)^2$, or $\quad 3x^2 + 0.03x + 0.0001$
0.001	$3x^2 + 3x(0.001) + (0.001)^2$, or $3x^2 + 0.003x + 0.000001$

From the pattern in the table it appears that

$$\lim_{h\to 0} (3x^2 + 3xh + h^2) = 3x^2.$$

DO EXERCISE 14.

Limits at Infinity

Sometimes we need to determine limits when the inputs get larger and larger. For example, consider

$$f(x) = 3 - \frac{1}{x}.$$

Look at the input–output table to the right.

Inputs x	Outputs $3 - \dfrac{1}{x}$
1	2.0
10	2.9
50	2.98
100	2.99
2000	2.9995

Note that as the inputs get larger and larger, the outputs get closer to 3. We say "the limit as x goes to infinity of $3 - (1/x)$ is 3." We can abbreviate this:

$$\lim_{x\to\infty}\left(3 - \frac{1}{x}\right) = 3.$$

DO EXERCISE 15.

14. a) Complete the table.

h	$2x + h$
1	
0.7	
0.4	
0.1	
0.01	
0.001	

b) Find

$$\lim_{h\to 0} (2x + h).$$

15. Consider

$$f(x) - \frac{2x + 5}{x}.$$

a) Complete this table. (A hand calculator would be helpful.)

Inputs x	Outputs $\dfrac{2x + 5}{x}$
4	
20	
80	
200	
1000	
10,000	

b) Find

$$\lim_{x\to\infty}\left(\frac{2x + 5}{x}\right).$$

EXERCISE SET 9.1

Which functions are continuous?

1.   **2.**   **3.**   **4.**

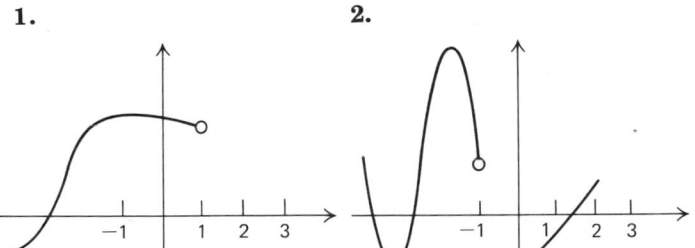 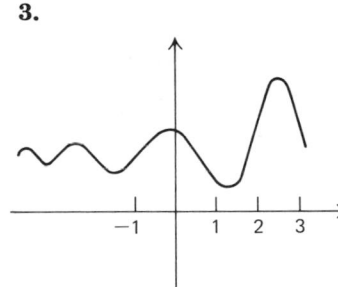 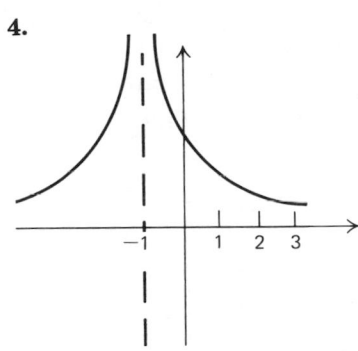

5–8. For each of Exercises 1–4, decide (a) whether the function is continuous at −1, and (b) whether the function is continuous at 1.

The Postage Function. Postal rates established in 1978 were as follows: 15¢ for the 1st ounce and 13¢ for each additional ounce or fraction thereof. Formally, if x is the weight of a letter in ounces, then $p(x)$ is the cost of mailing the letter, where

$$p(x) = 15¢, \quad \text{if } 0 < x \leqslant 1,$$
$$p(x) = 28¢, \quad \text{if } 1 < x \leqslant 2,$$
$$p(x) = 41¢, \quad \text{if } 2 < x \leqslant 3,$$

and so on, up to 12 ounces (at which point postal cost also depends on distance). The graph of p is shown at the right.

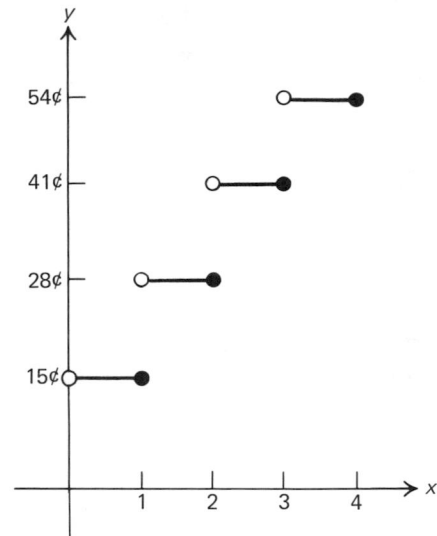

9. Is p continuous at 1? $1\frac{1}{2}$? 2? 2.53?

10. Is p continuous at 3? $3\frac{1}{4}$? 4? 3.98?

Using the graph, find each limit.

11. $\lim\limits_{x \to 1} p(x)$ **12.** $\lim\limits_{x \to 1/2} p(x)$ **13.** $\lim\limits_{x \to 2.3} p(x)$ **14.** $\lim\limits_{x \to 2} p(x)$

15. a) Complete.

Inputs x	Outputs $x^2 - 3$
2	
1.5	
1.2	
1.1	
1.01	
1.001	

b) Make up and complete a table for the limit from the left.

c) Find $\lim\limits_{x \to 1} (x^2 - 3)$.

16. a) Complete.

Inputs x	Outputs $x^2 + 4$
2	
1.4	
1.2	
1.1	
1.01	
1.001	

b) Make up and complete a table for the limit from the left.

c) Find $\lim\limits_{x \to 1} (x^2 + 4)$.

Find each limit, if it exists. If you have trouble, make up your own input–output table.

17. $\lim\limits_{x \to -5} \dfrac{x^2 - 25}{x + 5}$

18. $\lim\limits_{x \to -4} \dfrac{x^2 - 16}{x + 4}$

19. $\lim\limits_{x \to 0} \dfrac{1}{x}$

20. $\lim\limits_{x \to 0} \dfrac{3}{x}$

21. $\lim\limits_{x \to -2} \dfrac{3}{x}$

22. $\lim\limits_{x \to -4} \dfrac{1}{x}$

23. $\lim\limits_{x \to 2} \dfrac{x^3 - 8}{x - 2}$

24. $\lim\limits_{x \to 1} \dfrac{x^3 - 1}{x - 1}$

25. $\lim\limits_{x \to 1} \dfrac{x^2 - 1}{x^3 - 1}$

26. $\lim\limits_{x \to 1} \dfrac{x^3 - 1}{x^2 - 1}$

27. (▦ with $\sqrt{x}$ key.) $\lim\limits_{x \to 4} \dfrac{2 - \sqrt{x}}{4 - x}$

28. (▦ with $\sqrt{x}$ key.) $\lim\limits_{x \to 1} \dfrac{2 - \sqrt{x + 3}}{x - 1}$

29. (▦ with $\sqrt{x}$ key.) $\lim\limits_{x \to 1} \dfrac{2 - \sqrt{x + 3}}{x - 1}$

30. $\lim\limits_{h \to 0} \dfrac{-1}{x(x + h)}$

31. $\lim\limits_{h \to 0} (2x + h + 1)$

32. a) Complete.

x	$\dfrac{2x - 4}{5x}$
8	
60	
100	
400	
6000	
20,000	

b) Find

$\lim\limits_{x \to \infty} \left(\dfrac{2x - 4}{5x} \right)$.

33. a) Complete.

x	$\dfrac{3x + 1}{4x}$
10	
80	
200	
500	
8000	
40,000	

b) Find

$\lim\limits_{x \to \infty} \left(\dfrac{3x + 1}{4x} \right)$.

Find each limit, if it exists. If you have trouble, make up your own input–output tables.

34. $\lim\limits_{x \to \infty} \dfrac{1}{x}$ **35.** $\lim\limits_{x \to \infty} \dfrac{2}{x}$ **36.** $\lim\limits_{x \to \infty} \left(2 + \dfrac{1}{x}\right)$ **37.** $\lim\limits_{x \to \infty} \left(5 - \dfrac{1}{x}\right)$ **38.** $\lim\limits_{x \to \infty} \dfrac{2x^2 - 5}{3x^2 - x + 7}$

39. *Business—Depreciation.* A new conveyor system costs $10,000. In any year it depreciates 8% of its value at the beginning of that year.
a) What is the annual depreciation in each of the first five years?
b) What is the total depreciation at the end of 10 years?
c) What is the limit of the sum of the annual depreciation costs?

40. *Business—Depreciation.* A new car costs $6000. In any year it depreciates 30% of its value at the beginning of that year.
a) What is the annual depreciation in each of the first five years?
b) What is the total depreciation at the end of 10 years?
c) What is the limit of the sum of the annual depreciation costs?

41. Inside its own 5-yd line, a defensive football team is penalized half the distance to the goal. Suppose a defensive team keeps getting penalized. What is the limit of the distance of the offensive team from the goal? Can the offensive team ever score a touchdown in this manner?

OBJECTIVES

You should be able to
a) Compute an average rate of one variable with respect to another.
b) Find a simplified difference quotient.

9.2 AVERAGE RATES OF CHANGE

The graph below shows the total production of suits by Raggs, Ltd. during one morning of work. Industrial psychologists have found curves like this typical of the production of factory workers.

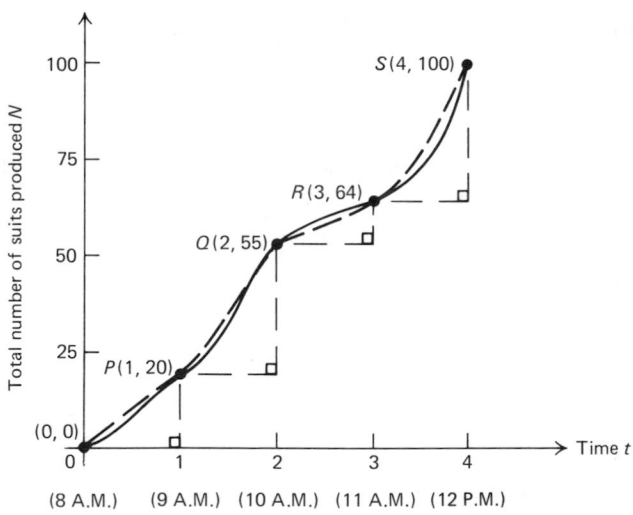

Example 1 What is the number of suits produced from 9 A.M. to 10 A.M.?

Solution At 10 A.M., 55 suits had been produced. At 9 A.M., 20 suits had been produced. In the hour from 9 A.M. to 10 A.M. the number of suits produced was

$$55 \text{ suits} - 20 \text{ suits, or } 35 \text{ suits.}$$

Note that this is the slope of the line from P to Q.

Example 2 What was the average number of suits produced per hour from 9 A.M. to 11 A.M.?

Solution $\dfrac{64 \text{ suits} - 20 \text{ suits}}{11 \text{ A.M.} - 9 \text{ A.M.}} = \dfrac{44 \text{ suits}}{2 \text{ hr}} = 22 \dfrac{\text{suits}}{\text{hr}}$ (suits per hour)

This is the slope of the line from P to R. It is not shown in the graph.

DO EXERCISE 16.

Let us consider a function $y = f(x)$ and two inputs x_1 and x_2. The *change in input* or the *change in x* is

$$x_2 - x_1.$$

The *change in output* or the *change in y* is

$$y_2 - y_1.$$

The *average rate of change of y with respect to x*, as x changes from x_1 to x_2, is the ratio of the change in output to the change in input.

$$\frac{y_2 - y_1}{x_2 - x_1}.$$

If we look at a graph of the function, we see that

$$\frac{y_2 - y_1}{x_2 - x_1} = \frac{f(x_2) - f(x_1)}{x_2 - x_1}$$

and that this is the slope of the line from $P(x_1, y_1)$ to $Q(x_2, y_2)$. The line $\overleftrightarrow{PQ}$ is called a *secant* line.

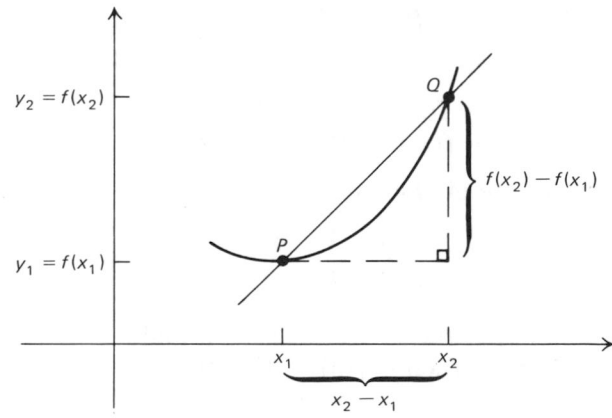

16. Referring to the graph of suits produced by Raggs Ltd.:

a) Find the number of suits produced per hour from

8 A.M. to 9 A.M.,
9 A.M. to 10 A.M.,
10 A.M. to 11 A.M.,
11 A.M. to 12 P.M.

b) Which interval in (a) had the highest number?

c) Why do you think this happened?

d) Which interval in (a) had the lowest number?

e) Why do you think this happened?

f) What was the average number of suits produced per hour from 8 A.M. to 12 P.M.?

17. For

$$f(x) = x^3$$

find the average rates of change and
sketch the secant lines as
a) x changes from 1 to 4,
b) x changes from 1 to 2,
c) x changes from 2 to 4,
d) x changes from -1 to -4.

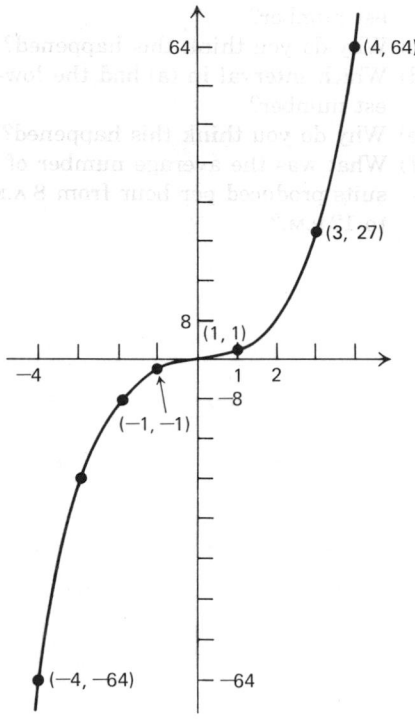

Example 3 For $y = f(x) = x^2$, find the average rates of change as
a) x changes from 1 to 3,
b) x changes from 1 to 2,
c) x changes from 2 to 3.

Solution The following graph is not necessary to the computations,
but gives us a look at the secant lines whose slopes are being
computed.

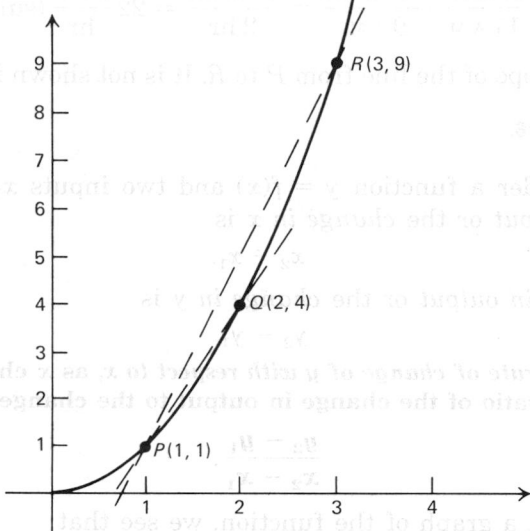

a) When $x_1 = 1$, $y_1 = f(x_1) = f(1) = 1^2 = 1$, and
when $x_2 = 3$, $y_2 = f(x_2) = f(3) = 3^2 = 9$.

The average rate of change is

$$\frac{y_2 - y_1}{x_2 - x_1} = \frac{f(x_2) - f(x_1)}{x_2 - x_1} = \frac{9 - 1}{3 - 1} = \frac{8}{2} = 4.$$

b) When $x_1 = 1$, $y_1 = f(x_1) = f(1) = 1^2 = 1$, and
when $x_2 = 2$, $y_2 = f(x_2) = f(2) = 2^2 = 4$.

The average rate of change is

$$\frac{4 - 1}{2 - 1} = \frac{3}{1} = 3.$$

c) When $x_1 = 2$, $y_1 = f(x_1) = f(2) = 2^2 = 4$, and
when $x_2 = 3$, $y_2 = f(x_2) = f(3) = 3^2 = 9$.

The average rate of change is

$$\frac{9 - 4}{3 - 2} = \frac{5}{1} = 5.$$

DO EXERCISES 17 AND 18.

For a linear function average rates of change are the same, for any choice of x_1 and x_2, being equal to the slope m of the line. As seen in Example 3 and in Margin Exercise 17, a function that is not linear has average rates of change that vary with the choice of x_1 and x_2.

Difference Quotients

Let us now simplify our notation a bit, by doing away with subscripts. Instead of x_1, we will simply write x.

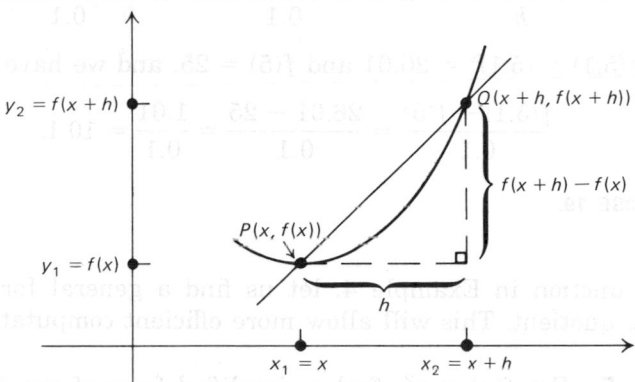

To get from x_1, or x, to x_2 we move a distance h. Thus $x_2 = x + h$. Then the average rate of change, also called a *difference quotient*, is given by

$$\frac{y_2 - y_1}{x_2 - x_1} = \frac{f(x_2) - f(x_1)}{x_2 - x_1} = \frac{f(x + h) - f(x)}{(x + h) - x} = \frac{f(x + h) - f(x)}{h}.$$

We shall be using the expression on the right.

The average rate of change of f with respect to x is also called the *difference quotient*. It is given by

$$\frac{f(x + h) - f(x)}{h}.$$

The difference quotient is equal to the slope of the line from $P(x, f(x))$ to $Q(x + h, f(x + h))$.

18. For

$$f(x) = \tfrac{1}{2}x + 1$$

find the average rates of change and sketch the secant lines as
a) x changes from 2 to 4,
b) x changes from 2 to 3,
c) x changes from -1 to 4.

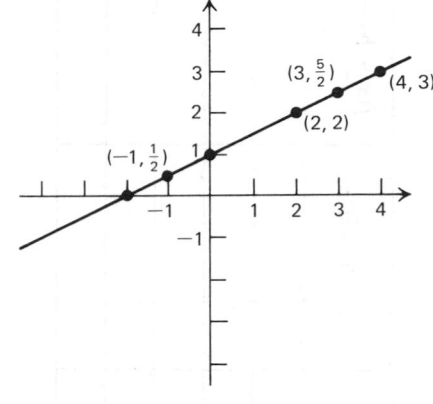

19. For

$$f(x) = 4x^2,$$

complete the following table to find the difference quotients.

x	3	3	3	3	3
h	2	1	0.1	0.01	0.001
$x + h$					
$f(x)$					
$f(x + h)$					
$f(x + h) - f(x)$					
$\dfrac{f(x + h) - f(x)}{h}$					

Example 4 For $f(x) = x^2$, find the difference quotient when
a) $x = 5$ and $h = 3$,
b) $x = 5$ and $h = 0.1$.

Solutions
a) We substitute $x = 5$ and $h = 3$ into the formula,

$$\frac{f(x + h) - f(x)}{h} = \frac{f(5 + 3) - f(5)}{3} = \frac{f(8) - f(5)}{3}.$$

Now $f(8) = 8^2 = 64$, and $f(5) = 5^2 = 25$, and we have

$$\frac{f(8) - f(5)}{3} = \frac{64 - 25}{3} = \frac{39}{3} = 13.$$

b) We substitute $x = 5$ and $h = 0.1$ into the formula,

$$\frac{f(x + h) - f(x)}{h} = \frac{f(5 + 0.1) - f(5)}{0.1} = \frac{f(5.1) - f(5)}{0.1}.$$

Now $f(5.1) = (5.1)^2 = 26.01$ and $f(5) = 25$, and we have

$$\frac{f(5.1) - f(5)}{0.1} = \frac{26.01 - 25}{0.1} = \frac{1.01}{0.1} = 10.1.$$

DO EXERCISE 19.

For the function in Example 4, let us find a general form of the difference quotient. This will allow more efficient computations.

Example 5 For $f(x) = x^2$, find a simplified form of the difference quotient. Then find the value of the difference quotient when $x = 5$ and $h = 0.1$.

Solution

$$f(x) = x^2,$$

so

$$f(x + h) = (x + h)^2 = x^2 + 2xh + h^2.$$

Then

$$f(x + h) - f(x) = (x^2 + 2xh + h^2) - x^2 = 2xh + h^2.$$

So

$$\frac{f(x + h) - f(x)}{h} = \frac{2xh + h^2}{h} = \frac{h(2x + h)}{h} = 2x + h.$$

It is important to note that a difference quotient is defined *only* when $h \neq 0$. The simplification above is valid only for nonzero values of h.

When $x = 5$ and $h = 0.1$,

$$\frac{f(x + h) - f(x)}{h} = 2x + h = 2 \cdot 5 + 0.1 = 10 + 0.1 = 10.1.$$

DO EXERCISE 20.

Example 6 For $f(x) = x^3$ find a simplified form of the difference quotient.

Solution Now $f(x) = x^3$, so

$$f(x + h) = (x + h)^3 = x^3 + 3x^2 h + 3xh^2 + h^3.$$

This is shown on p. 9. Then

$$f(x + h) - f(x) = (x^3 + 3x^2 h + 3xh^2 + h^3) - x^3$$
$$= 3x^2 h + 3xh^2 + h^3.$$

So

$$\frac{f(x + h) - f(x)}{h} = \frac{3x^2 h + 3xh^2 + h^3}{h} = \frac{h(3x^2 + 3xh + h^2)}{h}$$
$$= 3x^2 + 3xh + h^2.$$

Again, this is true *only* for $h \neq 0$.

DO EXERCISE 21.

Example 7 For $f(x) = \dfrac{3}{x}$ find a simplified form of the difference quotient.

Solution Now

$$f(x) = \frac{3}{x}, \quad \text{so} \quad f(x + h) = \frac{3}{x + h}.$$

Then

$$f(x + h) - f(x) = \frac{3}{x + h} - \frac{3}{x}$$

$$= \frac{3}{x + h} \cdot \frac{x}{x} - \frac{3}{x} \cdot \frac{x + h}{x + h} \quad \text{(Here we are multiplying by 1 to get a common denominator.)}$$

$$= \frac{3x - 3(x + h)}{x(x + h)}$$

$$= \frac{3x - 3x - 3h}{x(x + h)} = \frac{-3h}{x(x + h)}.$$

20. For

$$f(x) = 4x^2$$

find a simplified form of the difference quotient by completing steps (a) through (c). Then complete the table in (d) using the simplified form.
a) Find $f(x + h)$.
b) Find $f(x + h) - f(x)$.
c) Find $\dfrac{f(x + h) - f(x)}{h}$ and simplify.

d) Complete.

x	h	$\dfrac{f(x + h) - f(x)}{h}$
6	-3	
6	-2	
6	-1	
6	-0.1	
6	-0.01	
6	-0.001	

21. a) For

$$f(x) = 4x^3$$

find a simplified difference quotient.
b) Complete.

x	h	$\dfrac{f(x + h) - f(x)}{h}$
-2	1	
-2	0.1	
-2	0.01	
-2	0.001	

22. a) For
$$f(x) = \frac{1}{x},$$

find a simplified difference quotient.

b) Complete.

x	h	$\dfrac{f(x + h) - f(x)}{h}$
2	3	
2	1	
2	0.1	
2	0.01	
2	0.001	

So

$$\frac{f(x + h) - f(x)}{h} = \frac{\dfrac{-3h}{x(x + h)}}{h} = \frac{-3h}{x(x + h)} \cdot \frac{1}{h} = \frac{-3}{x(x + h)}.$$

This is true only for $h \neq 0$.

DO EXERCISE 22.

EXERCISE SET 9.2

1. *Economics—Utility.* Utility is a type of function that arises in economics. When a consumer receives x units of a certain product, a certain amount of pleasure, or utility, U, is derived from them. Below is a typical graph of a utility function.

a) Find the average rate of change of U as x changes from 0 to 1, 1 to 2, 2 to 3, 3 to 4.

b) Why do you think the average rates of change are decreasing?

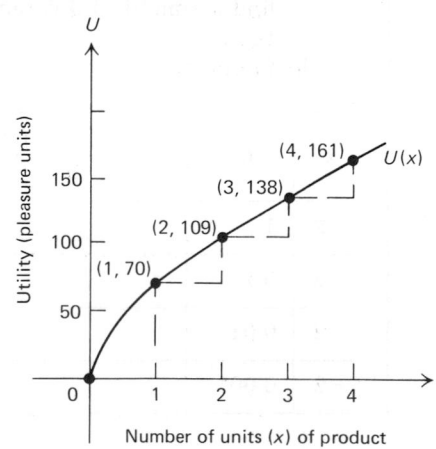

2. *Business—Advertising results.* The graph below shows a typical response to advertising. After an amount a is spent on advertising, the company sells $N(a)$ units of a product.

a) Find the average rate of change of N, as a changes from 0 to 1, 1 to 2, 2 to 3, 3 to 4.

b) Why do you think the average rates of change are decreasing?

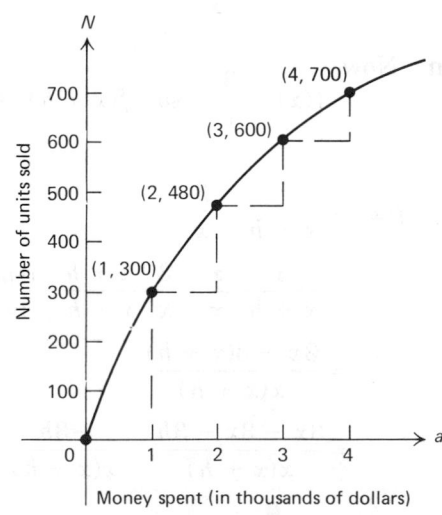

3. *Psychology—Memory.* The total number of words, $M(t)$, that a person can memorize in time t, in minutes, is shown in the graph below.
a) Find the average rate of change of M as t changes from 0 to 8, 8 to 16, 16 to 24, 24 to 32, 32 to 36.
b) Why do the average rates of change become 0 after 24 minutes?

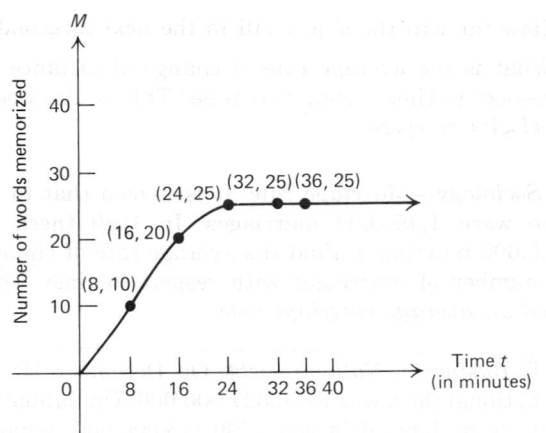

5. *Sociology—Population growth.* The two curves at the right describe the number of people in each of two countries, A and B, at time t, in years.
a) Find the average rate of change of each population (number of people in the population) with respect to time t, as t changes from 0 to 4. This is often called an *average growth rate.*
b) If the calculation in (a) were the only one made, would we detect the fact that the populations were growing differently?
c) Find the average rates of change of each population as t changes from 0 to 1, 1 to 2, 2 to 3, and 3 to 4.
d) For which population does the statement "the population grew 125 million each year" convey the least information about what really took place?

6. *Business—Cost.* A firm determines that the total cost, C, of producing x units of a certain product is given by
$$C(x) = -0.05x^2 + 50x,$$
where $C(x)$ is in dollars.
a) Find $C(301)$.
b) Find $C(300)$.
c) Find $C(301) - C(300)$.
d) Find $\dfrac{C(301) - C(300)}{301 - 300}$.

4. *Biomedical—Temperature during an illness.* The °F temperature T, of a patient during an illness is given by the graph below, where t = time, in days.

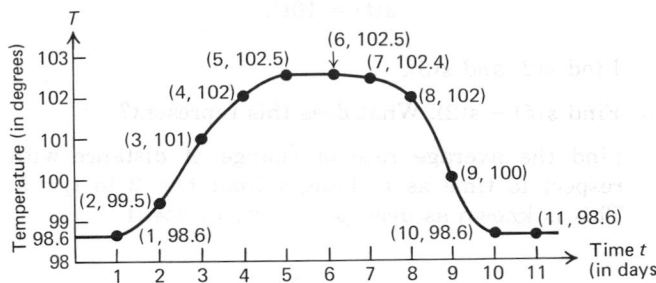

a) Find the average rate of change of T as t changes from 1 to 10. Using this rate of change, would you know that the person was sick?
b) Find the average rate of change of T with respect to t, as t changes from 1 to 2, 2 to 3, 3 to 4, 4 to 5, 5 to 6, 6 to 7, 7 to 8, 8 to 9, 9 to 10, 10 to 11.
c) When do you think the temperature began to rise?
d) When do you think the temperature reached its peak?
e) When do you think the temperature began to subside?
f) When was the temperature back to normal?

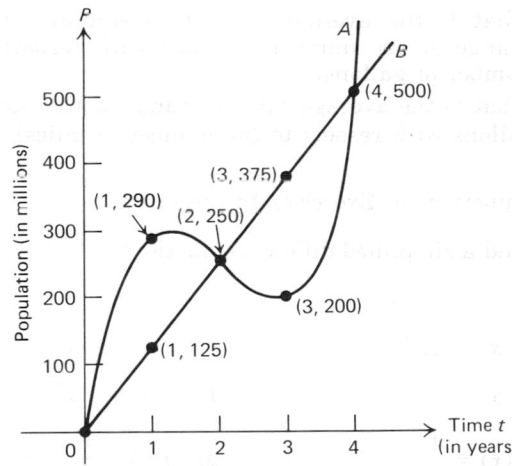

7. *Business—Revenue.* A firm determines that the total revenue (money coming in) from the sale of x units of a certain product is given by
$$R(x) = -0.01x^2 + 1000,$$
where $R(x)$ is in dollars.
a) Find $R(301)$.
b) Find $R(300)$.
c) Find $R(301) - R(300)$.
d) Find $\dfrac{R(301) - R(300)}{301 - 300}$.

8. *Average velocity.* A car is at a distance s (in miles) from its starting point in t hours, given by

$$s(t) = 10t^2.$$

a) Find $s(2)$ and $s(5)$.

b) Find $s(5) - s(2)$. What does this represent?

c) Find the average rate of change of distance with respect to time as t changes from $t_1 = 2$ to $t_2 = 5$. This is known as *average velocity* or *speed*.

9. *Average velocity.* An object is dropped from a certain height. It is known that it will fall a distance s (in feet) in t seconds, given by

$$s(t) = 16t^2.$$

a) How far will the object fall in 3 seconds?

b) How far will the object fall in the next 2 seconds?

c) What is the average rate of change of distance with respect to time during this time? This is also *average velocity* or *speed*.

10. *Sociology—Divorce rate.* It is known that in 1960 there were 400,000 divorces. In 1980 there were 990,000 divorces. Find the average rate of change in the number of divorces with respect to time. This is called an *average divorce rate*.

11. *Sociology—Marriage rate.* It is known that in 1960 there were 1,450,000 marriages. In 1980 there were 2,900,000 marriages. Find the average rate of change of the number of marriages with respect to time. This is called an *average marriage rate*.

12. At the beginning of a trip, the odometer on a car reads 30,680 and it has a full tank of gas. At the end of the trip the odometer reads 30,970. It takes 20 gallons of gas to fill the tank again.

a) What is the average rate of consumption (rate of change of the number of miles with respect to the number of gallons)?

b) What is the average rate of change of the number of gallons with respect to the number of miles?

13. ▦ *Business—National debt.* On December 12, 1978 the national debt was $776,327,000,000. On January 31, 1979 the national debt was $790,116,000,000. What was the average rate of change in the national debt with respect to time?

For functions in Exercises 14 through 25,

a) Find a simplified difference quotient,

b) Complete the table to the right,

x	h	$\dfrac{f(x + h) - f(x)}{h}$
4	2	
4	1	
4	0.1	
4	0.01	

14. $f(x) = 5x^2$

15. $f(x) = 7x^2$

16. $f(x) = -5x^2$

17. $f(x) = -7x^2$

18. $f(x) = 5x^3$

19. $f(x) = 7x^3$

20. $f(x) = \dfrac{4}{x}$

21. $f(x) = \dfrac{5}{x}$

22. $f(x) = 2x + 3$

23. $f(x) = -2x + 5$

24. $f(x) = x^2 + x$

25. $f(x) = x^2 - x$

▶

Find the simplified difference quotient.

26. $f(x) = mx + b$

27. $f(x) = ax^2 + bx + c$

28. $f(x) = ax^3 + bx^2$

29. $f(x) = \sqrt{x}$

9.3 DIFFERENTIATION USING LIMITS

Tangent Lines

A line tangent to a circle is a line that touches the circle exactly once.

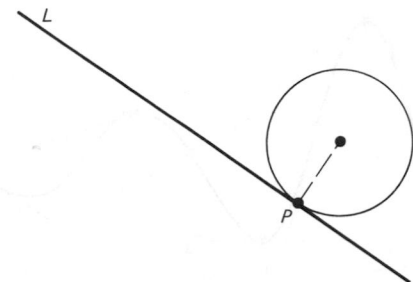

OBJECTIVES

Given a formula of a function, you should be able to find a formula for its derivative, then find various values of the derivative.

This definition becomes unworkable with other curves. For example, consider the following curve. Line L touches the curve at point P but meets the curve at other places. It will be considered a tangent line, but "touching at one point" cannot be its definition.

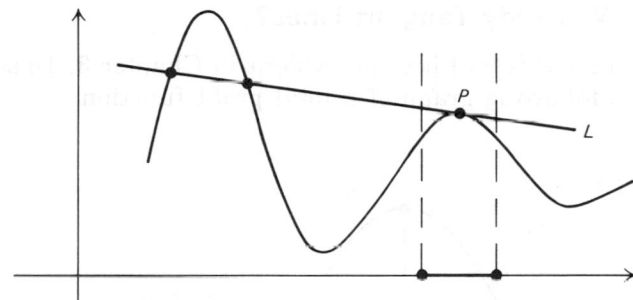

Note in the preceding figure that, over a suitably small interval containing P, line L does touch the curve exactly once. This is still not a suitable definition of *tangent line* because it allows a line like M in the following figure to be a tangent, which we will not accept.

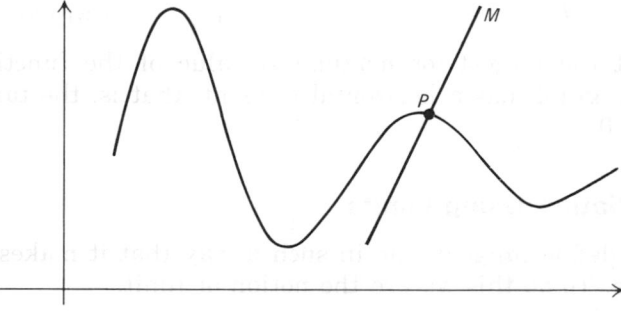

23. a) Which appear to be tangent lines?

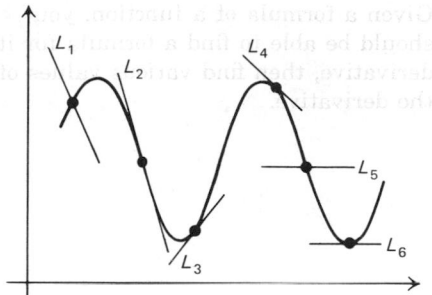

b) Below is a graph of $y = x^2$. Tangent lines are drawn at various points on the graph. Let $m(x) = $ slope at the point $(x, f(x))$.

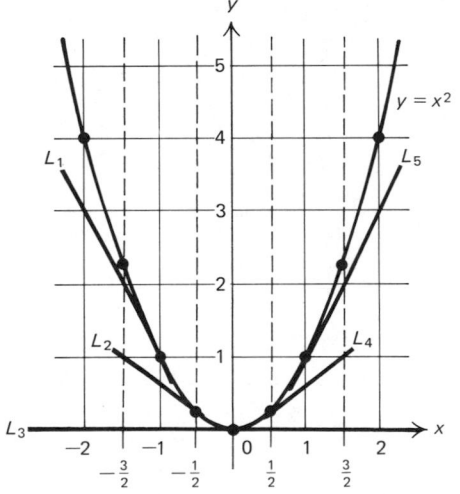

Estimate the slope of each line and complete this table.

Lines	x	$m(x)$
L_1	-1	
L_2	$-\frac{1}{2}$	
L_3	0	
L_4	$\frac{1}{2}$	
L_5	1	

c) Derive a formula for $m(x)$.

Later we will give a definition of a tangent line, but for now we will rely on intuition (experience). In the figure L_1 and L_2 are not tangents. All the others are tangent lines.

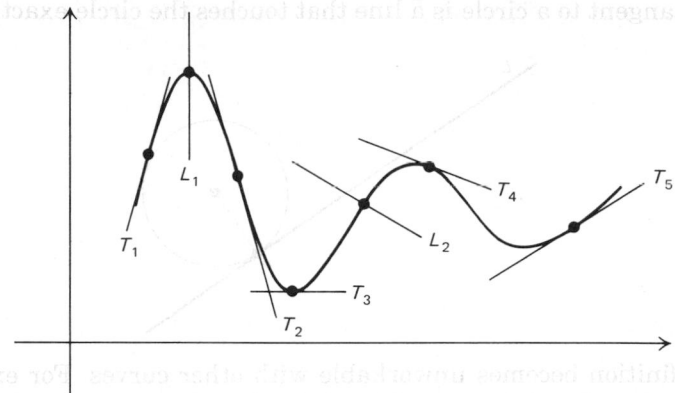

DO EXERCISE 23.

Why Do We Study Tangent Lines?

The reason for this will become evident in Chapter 3. To see briefly, look at the following graph of a total profit function.

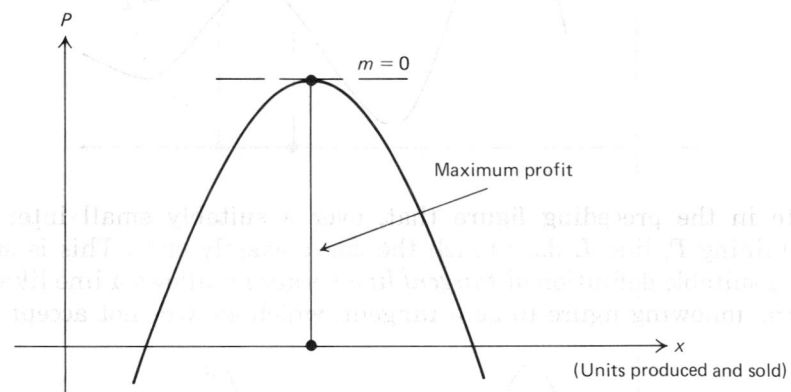

Note that the largest (or maximum) value of the function occurs where the graph has a horizontal tangent; that is, the tangent line has slope 0.

Differentiation Using Limits

We shall define *tangent* line in such a way that it makes sense for *any* curve. To do this we use the notion of limit.

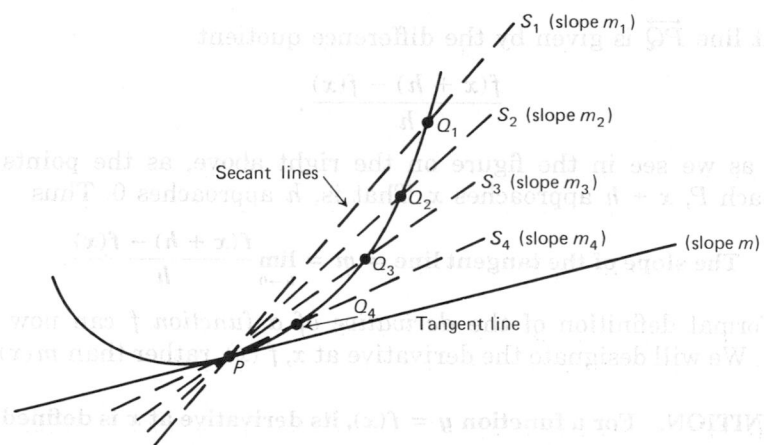

In this figure, we obtain the line tangent to the curve at point P by considering secant lines through P and neighboring points Q_1, Q_2, and so on. As the points Q approach P, the secant lines approach the tangent line. Each secant has a slope. The slopes of the secant lines approach the slope of the tangent line. In fact, we *define* the *tangent line* to be the line that contains the point P and has slope m, where m is the limit of the slopes of the secant lines as the points Q approach P.

How might we calculate the limit m?

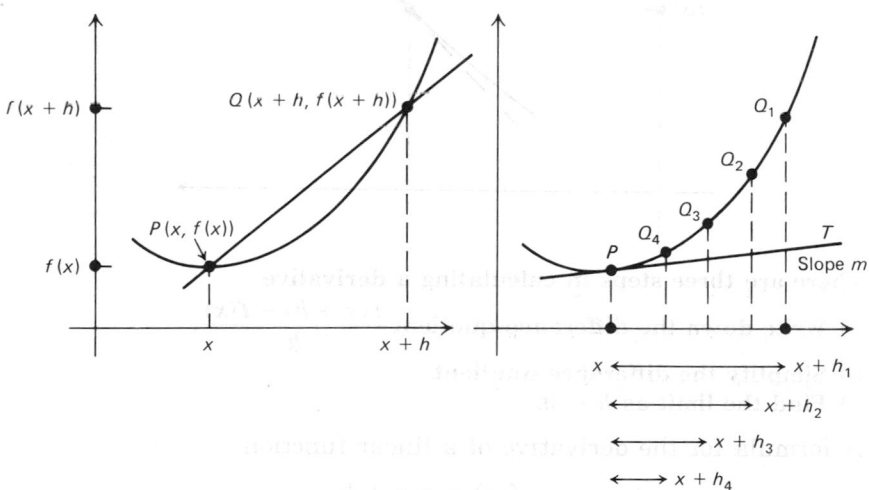

Suppose P has coordinates $(x, f(x))$. Then the first coordinate of Q is x plus some number h, or $x + h$. The coordinates of Q are $(x + h, f(x + h))$. Now from Section 9.2, we know that the slope of the secant line $\overleftrightarrow{PQ}$ is given by the difference quotient

$$\frac{f(x + h) - f(x)}{h}.$$

Now, as we see in the figure on the right above, as the points Q approach P, $x + h$ approaches x. That is, h approaches 0. Thus

$$\text{The slope of the tangent line} = m = \lim_{h \to 0} \frac{f(x + h) - f(x)}{h}.$$

The formal definition of the *derivative of a function f* can now be given. We will designate the derivative at x, $f'(x)$, rather than $m(x)$.

DEFINITION.   For a function $y = f(x)$, its derivative at x is defined as follows:

$$f'(x) = \lim_{h \to 0} \frac{f(x + h) - f(x)}{h}.$$

This is the basic definition of *differential calculus*.

Let us now calculate some formulas for derivatives. That is, given a formula for a function f, we will be trying to find a formula for f'.

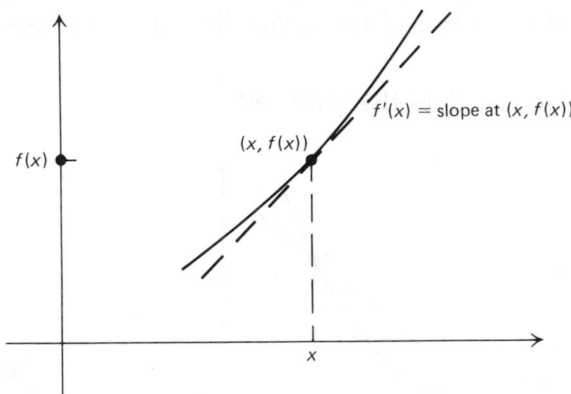

$f'(x) = $ slope at $(x, f(x))$

$(x, f(x))$

$f(x)$

x

There are three steps in calculating a derivative.
a) **Write down the difference quotient** $\dfrac{f(x + h) - f(x)}{h}.$
b) **Simplify the difference quotient.**
c) **Find the limit as $h \to 0$.**

A formula for the derivative of a linear function

$$f(x) = mx + b$$

is

$$f'(x) = m.$$

Let us verify it using the definition.

"*Nothing in this world is so powerful as an idea whose time has come.*"

VICTOR HUGO

Example 1 For $f(x) = mx + b$, find $f'(x)$.

Solution

a) $\dfrac{f(x + h) - f(x)}{h} = \dfrac{[m(x + h) + b] - (mx + b)}{h}$

b) $\dfrac{f(x + h) - f(x)}{h} = \dfrac{mx + mh + b - mx - b}{h} = \dfrac{mh}{h} = m$

c) $\displaystyle\lim_{h \to 0} \dfrac{f(x + h) - f(x)}{h} = \lim_{h \to 0} m = m,$

since m does not involve h. Thus

$$f'(x) = m.$$

In Margin Exercise 23 you may have conjectured that the function

$$f(x) = x^2$$

has derivative

$$f'(x) = 2x.$$

This would mean that the tangent line at $x = 4$ has slope $f'(4) = 8$. Let us verify this particular case and then the general formula.

Example 2 For $f(x) = x^2$, find $f'(4)$.

Solution

a) $\dfrac{f(4 + h) - f(4)}{h} = \dfrac{(4 + h)^2 - 4^2}{h}$

b) $\dfrac{f(4 + h) - f(4)}{h} = \dfrac{16 + 8h + h^2 - 16}{h} = \dfrac{8h + h^2}{h}$

$$= \dfrac{h(8 + h)}{h} = 8 + h$$

c) $\displaystyle\lim_{h \to 0} \dfrac{f(4 + h) - f(4)}{h} = \lim_{h \to 0} (8 + h) = 8.$ Thus $f'(4) = 8.$

DO EXERCISE 24.

Example 3 For $f(x) = x^2$, find (the general formula) $f'(x)$.

Solution

a) $\dfrac{f(x + h) - f(x)}{h} = \dfrac{(x + h)^2 - x^2}{h}$

24. For $f(x) = x^2$, find $f'(5)$ using the definition of a derivative.

25. For $f(x) = 4x^2$, find $f'(x)$. Then find $f'(5)$ and interpret the meaning.

b) In Example 5 of Section 9.2, p. 404, we showed how this difference quotient can be simplified as follows:

$$\frac{f(x + h) - f(x)}{h} = 2x + h.$$

c) We want to find

$$\lim_{h \to 0} \frac{f(x + h) - f(x)}{h} = \lim_{h \to 0} (2x + h).$$

As $h \to 0$, we see that $2x + h \to 2x$. Thus

$$\lim_{h \to 0} (2x + h) = 2x,$$

and we have

$$f'(x) = 2x,$$

which tells us, for example, that at $x = -3$, the curve has a tangent line whose slope is

$$f'(-3) = 2(-3), \quad \text{or} \quad -6.$$

We may say, simply, "The curve has slope -6."

DO EXERCISE 25.

26. For $f(x) = 4x^3$, find $f'(x)$. Then find $f'(-5)$ and $f'(0)$.

Example 4 For $f(x) = x^3$, find $f'(x)$. Then find $f'(-1)$ and $f'(10)$.

Solution

a) $\dfrac{f(x + h) - f(x)}{h} = \dfrac{(x + h)^3 - x^3}{h}$

b) In Example 6 of Section 9.2, p. 405, we showed how this difference quotient can be simplified as follows:

$$\frac{f(x + h) - f(x)}{h} = 3x^2 + 3xh + h^2.$$

c) $\displaystyle\lim_{h \to 0} \frac{f(x + h) - f(x)}{h} = \lim_{h \to 0} (3x^2 + 3xh + h^2) = 3x^2$

An input–output table for this is shown on p. 397 of Section 9.1. Thus for $f(x) = x^3$, we have $f'(x) = 3x^2$. Then

$$f'(-1) = 3(-1)^2 = 3, \quad \text{and} \quad f'(10) = 3(10)^2 = 300.$$

DO EXERCISE 26.

Example 5   For $f(x) = \dfrac{3}{x}$, find $f'(x)$. Then find $f'(1)$ and $f'(2)$.

Solution

a) $\dfrac{f(x + h) - f(x)}{h} = \dfrac{[3/(x + h)] - (3/x)}{h}$

b) In Example 7 of Section 9.2, p. 405, we showed that this difference quotient can be simplified as follows:

$$\frac{f(x + h) - f(x)}{h} = \frac{-3}{x(x + h)}.$$

c) We want to find

$$\lim_{h \to 0} \frac{f(x + h) - f(x)}{h} = \lim_{h \to 0} \frac{-3}{x(x + h)}.$$

As $h \to 0$, $x + h \to x$, so we have

$$f'(x) = \lim_{h \to 0} \frac{-3}{x(x + h)} = \frac{-3}{x^2}.$$

Then

$$f'(1) = \frac{-3}{1^2} = -3, \quad \text{and} \quad f'(2) = \frac{-3}{2^2} = -\frac{3}{4}.$$

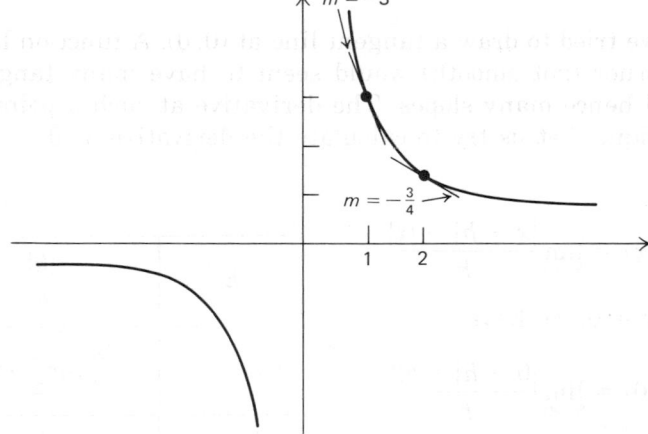

Note that $f'(0)$ does not exist because $f(0)$ does not exist. We say, "f is not differentiable at 0." When a function is not defined at a point, it is not differentiable at that point. In fact, if a function is discontinuous at a point, it is not differentiable at that point.

DO EXERCISE 27.

27. For

$$f(x) = \frac{1}{x},$$

find $f'(x)$. Then find

$$f'(-10) \quad \text{and} \quad f'(-2).$$

It can happen that a function f is defined and continuous at a point but that its derivative f' is not. The function f given by

$$f(x) = |x|$$

is an example. Note that

$$f(0) = |0| = 0,$$

so the function is defined at 0.

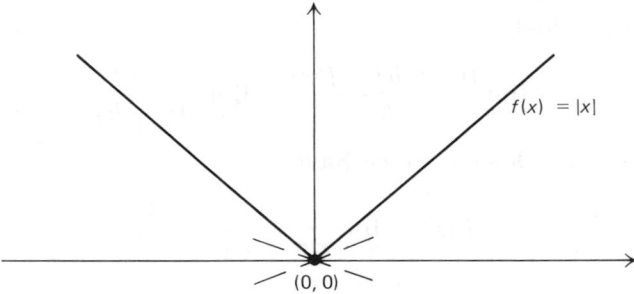

Suppose we tried to draw a tangent line at $(0, 0)$. A function like this with a corner (not smooth) would seem to have many tangents at $(0, 0)$, and hence many slopes. The derivative at such a point would not be unique. Let us try to calculate the derivative at 0.

Now

$$f'(x) = \lim_{h \to 0} \frac{|x + h| - |x|}{h}.$$

Thus at $x = 0$, we have

$$f'(0) = \lim_{h \to 0} \frac{|0 + h| - |0|}{h}$$

$$= \lim_{h \to 0} \frac{|h|}{h}.$$

h	$\dfrac{|h|}{h}$
2	$\dfrac{|2|}{2}$, or $\dfrac{2}{2}$, or 1
1	1
0.1	1
0.01	1
0.001	1

Look at the input–output tables. Note that as h approaches 0 from the right, $|h|/h$ approaches 1, but as h approaches 0 from the left, $|h|/h$ approaches -1. Thus

$$\lim_{h \to 0} \frac{|h|}{h} \text{ does not exist,}$$

so

$$f'(0) \text{ does not exist.}$$

h	$\dfrac{\lvert h \rvert}{h}$
-2	$\dfrac{\lvert -2 \rvert}{-2}$, or $\dfrac{2}{-2}$, or -1
-1	-1
-0.1	-1
-0.01	-1
-0.001	-1

If a function has a "sharp point" or "corner," it will not have a derivative at that point.

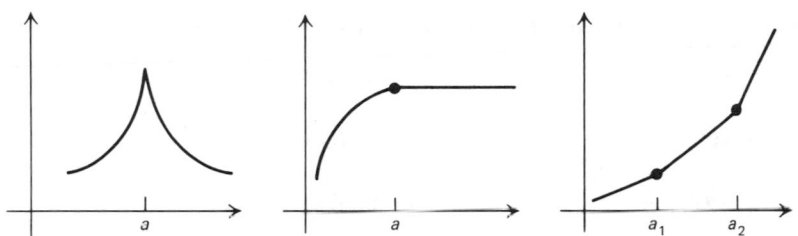

DO EXERCISE 28.

A function may also fail to be differentiable at a point by having a vertical tangent at that point. The following function has a vertical tangent at point a.

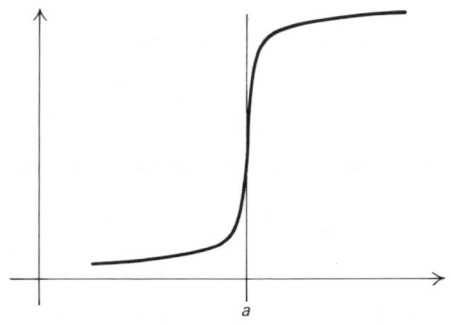

28. List the points at which the function is not differentiable.

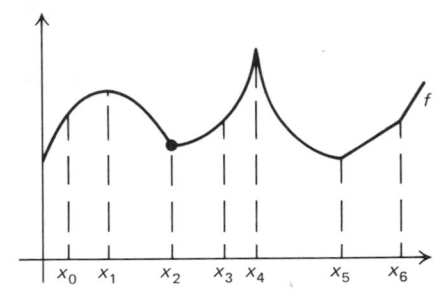

Recall that vertical lines have no slope, and hence there is no derivative at such a point.

Each of the preceding examples, including $f(x) = |x|$, is continuous at each point in an interval I but not differentiable at each point in I. That is, continuity does not imply differentiability. On the other hand, *if* we know that a function is differentiable at each point in an interval I, *then* it is continuous over I. The function $f(x) = x^2$ is an example. Also, *if* a function is discontinuous at some point a, *then* it is not differentiable at a. Thus when we know a function is differentiable over an interval, it is *smooth* in the sense that there are no "sharp points," "corners," or "breaks" in the graph.

EXERCISE SET 9.3

For each function, find $f'(x)$. Then find $f'(-2)$, $f'(-1)$, $f'(0)$, $f'(1)$, and $f'(2)$, if they exist.

1. $f(x) = 5x^2$

2. $f(x) = 7x^2$

3. $f(x) = -5x^2$

4. $f(x) = -7x^2$

5. $f(x) = 5x^3$

6. $f(x) = 7x^3$

7. $f(x) = 2x + 3$

8. $f(x) = -2x + 5$

9. $f(x) = -4x$

10. $f(x) = \frac{1}{2}x$

11. $f(x) = x^2 + x$

12. $f(x) = x^2 - x$

13. $f(x) = \dfrac{4}{x}$

14. $f(x) = \dfrac{5}{x}$

15. $f(x) = mx$

16. $f(x) = ax^2 + bx + c$

17. List the points in the graph below at which the function is not differentiable.

18. *The postage function.* Consider the postage function defined in Exercise Set 9.1. At what values is the function not differentiable?

19. Consider the function f given by

$$f(x) = \frac{x^2 - 9}{x + 3}.$$

For what values is this function not differentiable?

9.4 DIFFERENTIATION TECHNIQUES: POWER RULE AND SUM–DIFFERENCE RULE

Leibniz's Notation

When y is a function of x, we will also designate the derivative, $f'(x)$, as follows

$$\frac{dy}{dx},$$

which is read "the derivative of y with respect to x." This notation was invented by the German mathematician Leibniz. It does *not* mean dy divided by dx! (That is, we cannot interpret dy/dx as a quotient until meanings are given to dy and dx, which we will not do here.) For example, if $y = x^2$, then

$$\frac{dy}{dx} = 2x.$$

We may also write

$$\frac{d}{dx}f(x)$$

to denote the derivative of f with respect to x. For example,

$$\frac{d}{dx}x^2 = 2x.$$

The value of $\frac{dy}{dx}$ when $x = 5$ can be denoted by

$$\left.\frac{dy}{dx}\right|_{x=5}.$$

Thus for $\frac{dy}{dx} = 2x$,

$$\left.\frac{dy}{dx}\right|_{x=5} = 2 \cdot 5, \text{ or } 10.$$

In general, for $y = f(x)$,

$$\left.\frac{dy}{dx}\right|_{x=a} = f'(a).$$

DO EXERCISE 29.

The Power Rule

In the remainder of this section we will develop rules and techniques for efficient differentiating.

OBJECTIVES

You should be able to

a) Differentiate using the Power Rule, the Sum–Difference Rule, or the rule for differentiating a constant or a constant times a function.

b) Find the points on the graph of a function where the tangent line has a given slope.

29. For

$$y = x^3,$$

use the results of previous work and find

a) $\dfrac{dy}{dx}$

b) $\dfrac{d}{dx}x^3$

c) $\left.\dfrac{dy}{dx}\right|_{x=4}$

The German mathematician and philosopher Gottfried Wilhelm von Leibniz (1646–1716) and the English mathematician, philosopher, and physicist Sir Isaac Newton (1642–1727) are both credited with the invention of the calculus, though each made the invention independent of the other. Newton used the dot notation $\dot{y}$ for dy/dt, where y is a function of time, and this notation is still used, though it is not as prevalent as Leibniz's notation.

Find $\dfrac{dy}{dx}$ (differentiate).

30. $y = x^6$

31. $y = x^{-7}$

32. $y = \sqrt[3]{x}$

33. $y = x^{-1/4}$

This table contains functions and derivatives that we have found in previous work. Look for a pattern.

Function	Derivative
x^2	$2x^1$
x^3	$3x^2$
x^{-1}, or $\dfrac{1}{x}$	$-1 \cdot x^{-2}$, or $-\dfrac{1}{x^2}$

Perhaps you have discovered the following.

POWER RULE.   For any real number a,

$$\frac{d}{dx} x^a = a \cdot x^{a-1}.$$

Note that this rule holds no matter what the exponent. That is, to differentiate x^a, write down the exponent a, followed by x with an exponent 1 less than a.

① Bring down the exponent as a factor.

② Subtract 1 from the exponent.

Example 1 $\dfrac{d}{dx} x^5 = 5x^4$

Example 2 $\dfrac{d}{dx} x = 1 \cdot x^{1-1} = 1 \cdot x^0 = 1$

Example 3 $\dfrac{d}{dx} x^{-4} = -4 \cdot x^{-4-1} = -4x^{-5}$, or $-4 \cdot \dfrac{1}{x^5}$, or $-\dfrac{4}{x^5}$

The Power Rule allows us to differentiate $\sqrt{x}$.

Example 4 $\dfrac{d}{dx} \sqrt{x} = \dfrac{d}{dx} x^{1/2} = \dfrac{1}{2} \cdot x^{1/2-1} = \dfrac{1}{2} x^{-1/2}$,   or $\dfrac{1}{2} \cdot \dfrac{1}{x^{1/2}}$,   or $\dfrac{1}{2} \cdot \dfrac{1}{\sqrt{x}}$, or $\dfrac{1}{2\sqrt{x}}$

Example 5 $\dfrac{d}{dx} x^{-2/3} = -\dfrac{2}{3} x^{(-2/3)-1} = -\dfrac{2}{3} x^{-5/3}$, or $-\dfrac{2}{3} \dfrac{1}{x^{5/3}}$, or $-\dfrac{2}{3\sqrt[3]{x^5}}$

DO EXERCISES 30 THROUGH 33.

The Derivative of a Constant Times a Function

Look at the graph of the constant function $f(x) = c$. What is the slope at each point P on the graph? It follows that

the derivative of a constant function is 0.

Examples $\dfrac{d}{dx} 3 = 0,$ $\dfrac{d}{dx}\left(-\dfrac{1}{4}\right) = 0$

34. Find $g'(x)$, if
$$g(x) = -14.$$

$f(x) = c$ P

DO EXERCISE 34.

Now let us consider differentiating functions like

$$f(x) = 5x^2 \qquad \text{and} \qquad g(x) = -7x^4.$$

Note that we already know how to differentiate x^2 and x^4. Let us again look for a pattern in the results of Section 9.3.

Function	Derivative
$5x^2$	$10x$
$-4x$	-4
$-7x^2$	$-14x$
$5x^3$	$15x^2$

Perhaps you have discovered the following:

The derivative of a constant times a function is the constant times the derivative of the function. Using derivative notation this can be written

$$\frac{d}{dx}[c \cdot f(x)] = c \cdot f'(x).$$

Combining this rule with the Power Rule allows us to find many derivatives.

Example 6 $\dfrac{d}{dx} 5x^4 = 5 \dfrac{d}{dx} x^4 = 5 \cdot 4 \cdot x^{4-1} = 20x^3$

Example 7 $\dfrac{d}{dx} - 9x = -9 \dfrac{d}{dx} x = -9 \cdot 1 = -9$

With practice you will be able to differentiate many such functions in one step.

Find $\dfrac{dy}{dx}$.

35. $y = 5x^{20}$

36. $y = -\dfrac{3}{x}$

37. $y = -8\sqrt{x}$

38. $y = 0.16x^{6.25}$

Example 8 $\dfrac{d}{dx}\dfrac{-4}{x^2} = \dfrac{d}{dx} - 4x^{-2} = -4 \cdot \dfrac{d}{dx} x^{-2} = -4(-2)x^{-2-1}$

$$= 8x^{-3}, \text{ or } \dfrac{8}{x^3}.$$

Example 9 $\dfrac{d}{dx} - x^{0.7} = -1 \cdot \dfrac{d}{dx} x^{0.7} = -1 \cdot 0.7 \cdot x^{0.7-1} = -0.7x^{-0.3}$

DO EXERCISES 35 THROUGH 38.

The Derivative of a Sum or Difference

In Exercise 11 of Exercise Set 9.3 you found that for
$$f(x) = x^2 + x$$
the derivative is
$$f'(x) = 2x + 1.$$

Note that the derivative of x^2 is $2x$, and the derivative of x is 1; and the sum of these derivatives is $f'(x)$. This illustrates the following:

SUM–DIFFERENCE RULE

a) **The derivative of a sum is the sum of the derivatives:**
$$\text{if } t(x) = f(x) + g(x), \quad \text{then } t'(x) = f'(x) + g'(x).$$

b) **The derivative of a difference is the difference of the derivatives:**
$$\text{if } t(x) = f(x) - g(x), \quad \text{then } t'(x) = f'(x) - g'(x).$$

Any function that is a sum or difference of several terms can be differentiated term by term.

Example 10 $\dfrac{d}{dx}(3x + 7) = \dfrac{d}{dx}(3x) + \dfrac{d}{dx}(7)$

$$= 3\dfrac{d}{dx}x + 0 = 3 \cdot 1 = 3.$$

Example 11 $\dfrac{d}{dx}(5x^3 - 3x^2) = \dfrac{d}{dx}(5x^3) - \dfrac{d}{dx}(3x^2)$

$$= 5\dfrac{d}{dx}x^3 - 3\dfrac{d}{dx}x^2$$

$$= 5 \cdot 3x^2 - 3 \cdot 2x = 15x^2 - 6x.$$

Example 12 $\dfrac{d}{dx}\left(24x - \sqrt{x} + \dfrac{2}{x}\right) = \dfrac{d}{dx}(24x) - \dfrac{d}{dx}(\sqrt{x}) + \dfrac{d}{dx}\left(\dfrac{2}{x}\right)$

$$= 24 \cdot \dfrac{d}{dx}x - \dfrac{d}{dx}x^{1/2} + 2 \cdot \dfrac{d}{dx}x^{-1}$$

$$= 24 \cdot 1 - \dfrac{1}{2}x^{(1/2)-1} + 2(-1)x^{-1-1}$$

$$= 24 - \dfrac{1}{2}x^{-1/2} - 2x^{-2}$$

$$= 24 - \dfrac{1}{2\sqrt{x}} - \dfrac{2}{x^2}.$$

DO EXERCISES 39 THROUGH 41.

A word of caution! The derivative of

$$f(x) + c,$$

a function plus a constant, is just the derivative of the function

$$f'(x).$$

The derivative of

$$c \cdot f(x),$$

a function times a constant, is the constant times the derivative

$$c \cdot f'(x).$$

That is, for a product the constant is retained, but for a sum it is not.

It is important to be able to determine points at which the tangent line to a curve has a certain slope—that is, points at which the derivative attains a certain value.

Example 13   Find the points on the graph of $y = -x^3 + 6x^2$ at which the tangent line is horizontal.

Solution.   A horizontal tangent has slope 0. Thus we seek the values of x for which $dy/dx = 0$. That is, we want to find x such that

$$-3x^2 + 12x = 0.$$

We factor and solve

$$x(-3x + 12) = 0.$$

$$x = 0 \quad \text{or} \quad -3x + 12 = 0$$
$$x = 0 \quad \text{or} \quad -3x = -12$$
$$x = 0 \quad \text{or} \quad x = 4$$

Find $\dfrac{dy}{dx}$ (differentiate).

39. $y = -\tfrac{1}{4}x - 9$

40. $y = 7x^4 + 6x^2$

41. $y = 15x^2 + \dfrac{4}{x} + \sqrt{x}$

42. Find the points on the graph of

$$y = \tfrac{1}{3}x^3 - 2x^2 + 4x$$

at which the tangent line is horizontal.

We are to find the points *on the graph*, so we have to determine the second coordinates from the original equation $y = -x^3 + 6x^2$.

$$\text{For } x = 0, \qquad y = -0^3 + 6 \cdot 0^2 = 0.$$
$$\text{For } x = 4, \qquad y = -(4)^3 + 6 \cdot 4^2 = -64 + 96 = 32.$$

Thus the points we are seeking are $(0, 0)$ and $(4, 32)$.

DO EXERCISE 42.

Example 14 Find the points on the graph of $y = -x^3 + 6x^2$ at which the tangent has slope 6.

Solution We want to find values of x for which $dy/dx = 6$. That is, we want to find x such that

$$-3x^2 + 12x = 6.$$

To solve, we add -6 and get

$$-3x^2 + 12x - 6 = 0.$$

We can simplify this equation by multiplying by $-\tfrac{1}{3}$, since each term has a common factor of -3. We get

$$x^2 - 4x + 2 = 0.$$

This is a quadratic equation, not readily factorable, so we use the quadratic formula, where $a = 1$, $b = -4$, and $c = 2$:

$$x = \frac{-b \pm \sqrt{b^2 - 4ac}}{2a} = \frac{-(-4) \pm \sqrt{(-4)^2 - 4 \cdot 1 \cdot 2}}{2 \cdot 1} = \frac{4 \pm \sqrt{8}}{2}$$

$$= \frac{2 \cdot 2 \pm 2\sqrt{2}}{2 \cdot 1} = \frac{2}{2} \cdot \frac{2 \pm \sqrt{2}}{1} = 2 \pm \sqrt{2}.$$

The solutions are $2 + \sqrt{2}$ and $2 - \sqrt{2}$. We determine the second coordinates from the original equation. For $x = 2 + \sqrt{2}$,

$$\begin{aligned}
y &= -(2 + \sqrt{2})^3 + 6(2 + \sqrt{2})^2 \\
&= -[(2 + \sqrt{2})^2(2 + \sqrt{2})] + 6(4 + 4\sqrt{2} + 2) \\
&= -[(6 + 4\sqrt{2})(2 + \sqrt{2})] + 6(6 + 4\sqrt{2}) \\
&= -[12 + 6\sqrt{2} + 8\sqrt{2} + 8] + 36 + 24\sqrt{2} \\
&= -[20 + 14\sqrt{2}] + 36 + 24\sqrt{2} \\
&= -20 - 14\sqrt{2} + 36 + 24\sqrt{2} = 16 + 10\sqrt{2}.
\end{aligned}$$

Similarly, for $x = 2 - \sqrt{2}$,

$$y = 16 - 10\sqrt{2}.$$

Thus the points we are seeking are $(2 + \sqrt{2}, 16 + 10\sqrt{2})$ and $(2 - \sqrt{2}, 16 - 10\sqrt{2})$.

DO EXERCISE 43.

We illustrate the results of Examples 13 and 14 in the following graph. You will not be asked to sketch such graphs at this time.

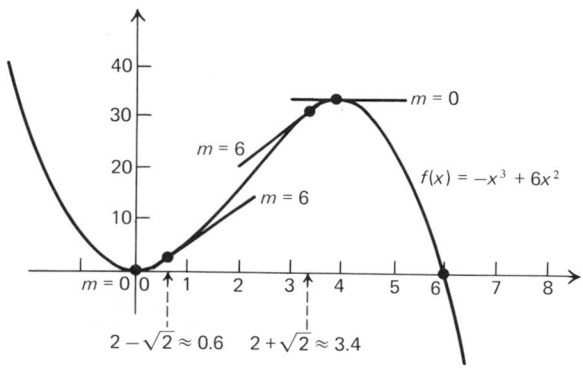

43. Find the points on the graph of

$$y = \tfrac{1}{3}x^3 - 2x^2 + 4x$$

at which the tangent line has slope 3.

EXERCISE SET 9.4

Find $\dfrac{dy}{dx}$.

1. $y = x^7$

2. $y = x^8$

3. $y = 15$

4. $y = 78$

5. $y = 4x^{150}$

6. $y = 7x^{200}$

7. $y = x^3 + 3x^2$

8. $y = x^4 - 7x$

9. $y = 8\sqrt{x}$

10. $y = 4\sqrt{x}$

11. $y = x^{0.07}$

12. $y = x^{0.78}$

13. $y = \tfrac{1}{2}x^{4/5}$

14. $y = -4.8x^{1/3}$

15. $y = x^{-3}$

16. $y = x^{-4}$

17. $y = 3x^2 - 8x + 7$

18. $y = 4x^2 - 7x + 5$

19. $y = \sqrt[4]{x} - \dfrac{1}{x}$

20. $y = \sqrt[5]{x} - \dfrac{2}{x}$

Find $f'(x)$.

21. $f(x) = 0.64x^{2.5}$

22. $f(x) = 0.32x^{12.5}$

23. $f(x) = \dfrac{5}{x} - x$

24. $f(x) = \dfrac{4}{x} - x$

25. $f(x) = 4x - 7$

26. $f(x) = 7x + 11$

27. $f(x) = 4x + 9$

28. $f(x) = 7x - 14$

29. $f(x) = \dfrac{x^4}{4}$

30. $f(x) = \dfrac{x^3}{3}$

31. $f(x) = -0.01x^2 - 0.5x + 70$

32. $f(x) = -0.01x^2 + 0.4x + 50$

33. $f(x) = 3x^{-2/3} + x^{3/4} + x^{6/5} + \dfrac{8}{x^3}$

34. $f(x) = x^{-3/4} - 3x^{2/3} + x^{5/4} + \dfrac{2}{x^4}$

For each function, find the points on the graph at which the tangent line is horizontal.

35. $y = x^2$

36. $y = -x^2$

37. $y = -x^3$

38. $y = x^3$

39. $y = 3x^2 - 5x + 4$

40. $y = 5x^2 - 3x + 8$

41. $y = -0.01x^2 - 0.5x + 70$

42. $y = -0.01x^2 + 0.4x + 50$

43. $y = 2x + 4$

44. $y = -2x + 5$

45. $y = 4$

46. $y = -3$

47. $y = -x^3 + x^2 + 5x - 1$

48. $y = -\frac{1}{3}x^3 + 6x^2 - 11x - 50$

49. $y = \frac{1}{3}x^3 - 3x + 2$

50. $y = x^3 - 6x + 1$

For each function, find the points on the graph at which the tangent line has slope 1.

51. $y = 20x - x^2$

52. $y = 6x - x^2$

53. $y = -0.025x^2 + 4x$

54. $y = -0.01x^2 + 2x$

55. $y = \frac{1}{3}x^3 + 2x^2 + 2x$

56. $y = \frac{1}{3}x^3 - x^2 - 4x + 1$

▶

57. Find the points on the graph of

$$y = x^4 - \frac{4}{3}x^2 - 4$$

at which the tangent line is horizontal.

58. Find the points on the graph of

$$y = 2x^6 - x^4 - 2$$

at which the tangent line is horizontal.

Find dy/dx. Each of the following can be differentiated using the rules developed in this section, but some algebra may be required beforehand.

59. $y = x(x - 1)$

60. $y = (x - 1)(x + 1)$

61. $y = (x - 2)(x + 3)$

62. $y = \dfrac{5x^2 - 8x + 3}{8}$

63. $y = \dfrac{x^5 + x}{x^2}$

64. $y = (5x)^2$

65. $y = (-4x)^3$

66. $y = \sqrt{7x}$

67. $y = \sqrt[3]{8x}$

68. $y = (x - 3)^2$

69. $y = (x + 1)^3$

70. $y = (x - 2)^3(x + 1)$

OBJECTIVES

You should be able to,

a) Given a distance function $s(t)$, find a formula for the velocity $v(t)$ and the acceleration $a(t)$, and evaluate $s(t)$, $v(t)$, and $a(t)$ for given values of t.

b) Given y as a function of x, find the rate of change of y with respect to x, and evaluate this rate of change for values of x.

9.5 APPLICATIONS AND RATES OF CHANGE

Instantaneous Rate of Change

A car travels 100 miles in 2 hours. Its *average* speed (or velocity) is 100 mi/2 hr, or 50 mi/hr. This is the *average rate of change* of distance with respect to time. At various times during the trip the speedometer did not read 50, however. Thus we say that 50 is the *average*. A snapshot of the speedometer taken at any instant would indicate *instantaneous* speed, or rate of change.

Average rates of change are given by difference quotients. If distance s is a function of time t, then average velocity is given by

$$\text{Average velocity} = \frac{s(t + h) - s(t)}{h}.$$

Instantaneous rates of change are found by letting $h \to 0$. Thus

$$\text{Instantaneous velocity} = \lim_{h \to 0} \frac{s(t + h) - s(t)}{h} = s'(t)$$

Example 1   An object travels in such a way that distance, s, (in miles) from the starting point is a function of time, t, (in hours) as follows:

$$s(t) = 10t^2.$$

a) Find the average velocity between the times $t = 2$ and $t = 5$.
b) Find the (instantaneous) velocity when $t = 4$.

Solution

a) From $t = 2$ to $t = 5$, $h = 3$, so

$$\frac{s(t + h) - s(t)}{h} = \frac{s(2 + 3) - s(2)}{3} = \frac{s(5) - s(2)}{3} = \frac{10 \cdot 5^2 - 10 \cdot 2^2}{3}$$

$$= \frac{250 - 40}{3} = \frac{210}{3} = 70\frac{mi}{hr}$$

b) Instantaneous velocity $= s'(t) = 20t$.

Then

$$s'(4) = 20 \cdot 4 = 80\frac{mi}{hr}$$

DO EXERCISE 44.

We usually use the letter v for velocity. Then

$$v(t) = \lim_{h \to 0} \frac{s(t + h) - s(t)}{h} = s'(t).$$

The rate of change of velocity is called *acceleration*. We usually use the letter a for acceleration,

$$\text{Acceleration} = a(t) = v'(t).$$

Example 2   For $s(t) = 10t^2$, find $v(t)$ and $a(t)$.

Solution
$$v(t) = s'(t) = 20t,$$
$$a(t) = v'(t) = 20.$$

For an automobile we may give the velocity in *miles per hour*, and the acceleration, which is the change in velocity per unit time, in (*miles per hour*) *per hour*. We abbreviate this mi/hr². Thus in Example 2, the acceleration is a constant, 20 mi/hr².

DO EXERCISES 45 AND 46.

44. For the function in Example 1,
a) find the average velocity between the times $t = 1$ and $t = 6$.
b) find the (instantaneous) velocity when $t = 5$.

45. An object is dropped from a certain height. It will fall downward a distance of s feet in t seconds as given by

$$s(t) = 16t^2.$$

a) Find the velocity $v(t)$.
b) Find its velocity 2 sec after it has been dropped.
c) Find its velocity 10 sec after it has been dropped.

46. In reference to Margin Exercise 45, find $a(t)$. In what units should it be expressed?

47. The volume V of a cubical carton with a side of length s, in feet, is given by

$$V = s^3.$$

a) Find the rate of change of the volume V with respect to the length s of a side.
b) Find the rate of change of volume when $s = 10$ ft.

In general, derivatives give instantaneous rates of change.

RATE OF CHANGE. If y is a function of x, then the (instantaneous) *rate of change of y with respect to x is given by the derivative*

$$\frac{dy}{dx}, \quad \text{or } f'(x).$$

Example 3 The spherical volume V of a cancer tumor is given by

$$V = \frac{4}{3}\pi r^3,$$

where r is the radius of the tumor, in centimeters.

a) Find the rate of change of the volume with respect to the radius.
b) Find the rate of change of volume at $r = 1.2$ cm.

Solution

a) $\dfrac{dV}{dr} = V'(r) = \dfrac{4}{3} \cdot 3 \cdot \pi r^2 = 4\pi r^2.$

b) $V'(1.2) = 4\pi(1.2)^2 = 5.76\pi \approx 18 \dfrac{\text{cm}^3}{\text{cm}} = 18 \text{ cm}^2.$

DO EXERCISE 47.

48. The initial population of a bacteria colony is 10,000. After t hours the colony grows to a number $P(t)$ given by

$$P(t) = 10{,}000(1 + 0.97t + t^2).$$

a) Find the growth rate of the population.
b) Find the number of bacteria present (the population) when $t = 5$ hr. Find the growth rate when $t = 5$ hr.
c) Find the number of bacteria present when $t = 6$ hr. Find the growth rate when $t = 6$.

Example 4 The initial population in a bacteria colony is 10,000. After t hours the colony grows to a number $P(t)$ given by

$$P(t) = 10{,}000(1 + 0.86t + t^2)$$

a) Find the rate of change of the population P with respect to time t. This is also known as the *growth rate*.
b) Find the number of bacteria present after 5 hours. Also find the growth rate when $t = 5$.

Solution

a) Note $P(t) = 10{,}000 + 8600t + 10{,}000t^2$. Then

$$P'(t) = 8600 + 20{,}000t.$$

b) The number of bacteria present when $t = 5$ hr is given by

$$P(5) = 10{,}000 + 8600 \cdot 5 + 10{,}000 \cdot 5^2 = 303{,}000.$$

The growth rate when $t = 5$ is given by

$$P'(5) = 8600 + 20{,}000 \cdot 5 = 108{,}600 \frac{\text{bacteria}}{\text{hr}}.$$

Thus at $t = 5$, there are 303,000 bacteria present, and the colony is growing at the rate of 108,600 bacteria per hour.

DO EXERCISE 48.

Rates of Change in Economics

In the study of economics we are frequently interested in how such quantities as cost, revenue, and profit change with an increase in product quantity. In particular, we are interested in what is called *marginal** cost or profit (or whatever). This term is used to signify *rate of change with respect to quantity*. Thus, if

$\qquad C(x)$ = the total cost of producing x units of a product
$\qquad\qquad$ (usually considered in some time period),

then

$\qquad C'(x)$ = the marginal cost
$\qquad\qquad$ = the rate of change of the total cost with respect to
$\qquad\qquad$ the number of units, x, produced.

Let us think about these interpretations. The total cost of producing 5 units of a product is $C(5)$. The rate of change $C'(5)$ is the cost per unit at that stage in the production process. That this cost per unit does not include fixed costs is seen in this example.

$$C(x) = \underbrace{(x^2 + 4x)}_{\text{Variable costs}} + \underbrace{\$10{,}000}_{\text{Fixed costs (constant)}}$$

Then

$$C'(x) = 2x + 4.$$

This is because the derivative of a constant is 0. This verifies an economic principle that says the fixed costs of a company have no effect on marginal cost.

Following are some other marginal functions. Recall that

$\qquad R(x)$ – the total revenue from the sale of x units.

Then

$\qquad R'(x)$ = the marginal revenue
$\qquad\qquad$ = the rate of change of the total revenue with respect
$\qquad\qquad$ to the number x of units sold.

$\qquad P(x)$ = the total profit from the production and sale
$\qquad\qquad$ of x units of a product,
$\qquad\qquad$ = $R(x) - C(x)$.

$\qquad P'(x)$ = the marginal profit
$\qquad\qquad$ = the rate of change of the total profit with respect
$\qquad\qquad$ to the number of units x produced and sold
$\qquad\qquad$ = $R'(x) - C'(x)$.

---

* The word "marginal" comes from the Marginalist School of Economic Thought, which originated in Austria for the purpose of applying mathematics and statistics to the study of economics.

49. Given

$$R(x) = 50x - 0.5x^2,$$
$$C(x) = 10x + 3,$$

find
a) $P(x)$
b) $R(40), C(40), P(40)$
c) $R'(x), C'(x), P'(x)$
d) $R'(40), C'(40), P'(40)$
e) Is the marginal revenue constant?

Example 5   Given

$$R(x) = 50x,$$
$$C(x) = 2x^3 - 12x^2 + 40x + 10,$$

find
a) $P(x)$
b) $R(2), C(2), P(2)$
c) $R'(x), C'(x), P'(x)$
d) $R'(2), C'(2), P'(2)$

Solution

a) $P(x) = R(x) - C(x) = 50x - (2x^3 - 12x^2 + 40x + 10)$
$$= -2x^3 + 12x^2 + 10x - 10.$$

b) $R(2) = 50 \cdot 2 = \$100$ (the total revenue from the sale of the first 2 units)

$C(2) = 2 \cdot 2^3 - 12 \cdot 2^2 + 40 \cdot 2 + 10 = \$58$ (the total cost of producing the first 2 units)

$P(2) = R(2) - C(2) = \$100 - \$58 = \$42$ (the total profit from the production and sale of the first 2 units)

c) $R'(x) = 50.$
$C'(x) = 6x^2 - 24x + 40,$
$P'(x) = R'(x) - C'(x) = 50 - (6x^2 - 24x + 40) = -6x^2 + 24x + 10$

d) $R'(2) = \$50$ per unit
$C'(2) = 6 \cdot 2^2 - 24 \cdot 2 + 40 = \$16$ per unit
$P'(2) = \$50 - \$16 = \$34$ per unit

Note that marginal revenue is constant. No matter how much is produced and sold, the revenue per unit stays the same. This may not always be the case. Also note that $C'(2)$, or $16 per unit, is not the average cost per unit, which is given by

$$\frac{\text{Total cost of producing 2 units}}{2 \text{ units}} = \frac{\$58}{2} = \$29 \text{ per unit.}$$

In general,

$$A(x) = \text{average cost of producing } x \text{ units} = \frac{C(x)}{x}.$$

DO EXERCISE 49.

Let us look at a typical marginal cost function, C', and its associated total cost function C.

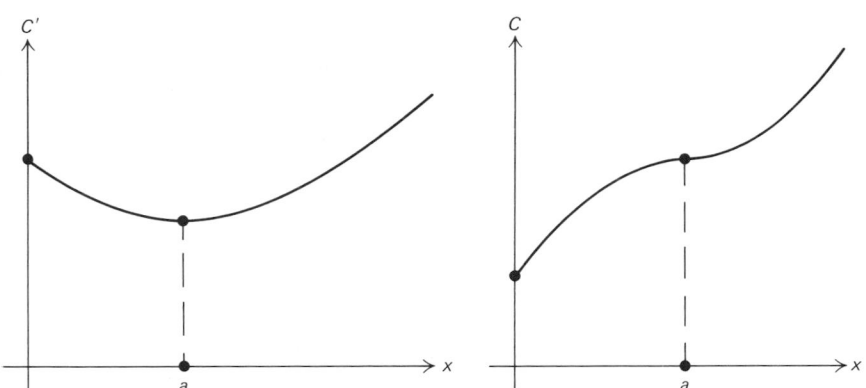

Marginal cost normally decreases as more units are produced until it reaches some minimum value at a, and then it increases. This is probably due to something like having to pay overtime or buying more machinery. Since $C'(x)$ represents slope of $C(x)$ and is positive and decreasing up to a, the graph turns downward as x goes from 0 to a. Then past a it turns upward.

EXERCISE SET 9.5

1. Given

$$s(t) = t^3 + t,$$

where s is measured in feet, and t in seconds, find
a) $v(t)$,
b) $a(t)$,
c) the velocity and acceleration when $t = 4$ sec.

3. *View to the horizon.* The view, or distance, in miles, which one can see to the horizon from a height h, in feet, is given by
$$V = 1.22\sqrt{h}$$
a) Find the rate of change of V with respect to h.
b) How far can one see to the horizon out an airplane window from a height of 40,000 ft?
c) Find the rate of change at $h = 40,000$.

5. *Sociology—Percentage of the population in college.* The percentage of the population in college is given by a linear function
$$P(t) = 1.25t + 15,$$
where $P(t) = $ percentage in college the t'th year after 1940. Find the rate of change of the percentage P with respect to time t.

2. Given
$$s(t) = 3t + 10,$$
where s is measured in miles, and t in hours, find
a) $v(t)$,
b) $a(t)$,
c) the velocity and acceleration when $t = 2$ hr. When the distance function is given by a linear function, we have what is called *uniform motion.*

4. *Stopping distance on glare ice.* The stopping distance (at some fixed speed) of regular tires is given by a linear function of the air temperature F:

$$D(F) = 2F + 115,$$

where $D(F) = $ stopping distance, in feet, when the air temperature is F, in degrees Fahrenheit. Find the rate of change of the stopping distance D with respect to the air temperature F.

6. *Biomedical—Healing wound.* The circular area A, in square centimeters, of a healing wound is given by

$$A = \pi r^2,$$

where r is the radius, in centimeters. Find the rate of change of the area with respect to the radius.

7. *Biomedical—Healing wound.* The circumference C, in centimeters, of a healing wound is given by

$$C = 2\pi r,$$

where r is the radius, in centimeters. Find the rate of change of the circumference with respect to the radius.

9. *Biomedical—Fever.* The temperature, T, of a person during an illness is given by

$$T(t) = -0.1t^2 + 1.2t + 98.6,$$

where T is the temperature (°F) at time t, measured in days.
a) Find the rate of change of the temperature with respect to time.
b) Find the temperature at $t = 1.5$ days.
c) Find the rate of change at $t = 1.5$ days.

11. *Biomedical—Blood pressure.* For a certain dosage of x cc (cubic centimeters) of a drug, the resultant blood pressure B is given by

$$B(x) = 0.05x^2 - 0.3x^3.$$

Find the rate of change of the blood pressure with respect to the dosage.

12. *Ecology—Home range.* The home range, H, of an animal is defined as the region to which it confines its movements. The area of that region is related to its body weight by

$$H = W^{1.41}$$

Find $\dfrac{dH}{dW}$.

14. Given

$$R(x) = 50x - 0.5x^2,$$
$$C(x) = 4x + 10,$$

find
a) $P(x)$,
b) $R(20)$, $C(20)$, $P(20)$,
c) $R'(x)$, $C'(x)$, $P'(x)$,
d) $R'(20)$, $C'(20)$, $P'(20)$.

8. *Sociology—Population growth rate.* The population of a city grows from an initial size of 100,000 to an amount P given by

$$P = 100{,}000 + 2000t^2,$$

where t is measured in years.
a) Find the growth rate.
b) Find the number of people in the city after 10 years (at $t = 10$ yr).
c) Find the growth rate at $t = 10$ yr.

10. *Business—Advertising.* A firm estimates that it will sell N units of a product after spending a dollars on advertising, where

$$N(a) = -a^2 + 300a + 6,$$

and a is measured in thousands of dollars.
a) What is the rate of change of the number of units sold with respect to the amount spent on advertising?
b) How many units will be sold after spending $10 thousand dollars on advertising?
c) What is the rate of change at $a = 10$?

These lions may be determining territory area. (*Ian Cleghorn: Photo Researchers, Inc.*)

13. *Ecology—Territory area.* The territory area, T, of an animal is defined to be its defended, or exclusive, region. The area T of that region is related to its body weight by

$$T = W^{1.31}$$

Find $\dfrac{dT}{dW}$.

15. Given

$$R(x) = 5x,$$
$$C(x) = 0.001x^2 + 1.2x + 60,$$

find
a) $P(x)$,
b) $R(100)$, $C(100)$, $P(100)$,
c) $R'(x)$, $C'(x)$, $P'(x)$,
d) $R'(100)$, $C'(100)$, $P'(100)$.

9.6 DIFFERENTIATION TECHNIQUES: PRODUCT AND QUOTIENT RULES

The derivative of a sum is the sum of the derivatives, but the derivative of a product is *not* the product of the derivatives. To see this, consider x^2 and x^5. The product is x^7, and the derivative of this product is $7x^6$. The individual derivatives are $2x$ and $5x^4$, and the product of these derivatives is $10x^5$, which is not $7x^6$.

The following is the rule for finding the derivative of a product.

PRODUCT RULE.   If $p(x) = f(x) \cdot g(x)$, then,

$$p'(x) = f(x) \cdot g'(x) + f'(x) \cdot g(x).$$

The derivative of a product is the first factor times the derivative of the second factor, plus the derivative of the first factor times the second factor.

Let us check this for $x^2 \cdot x^5$. There are four steps.

1. Write down the first factor.

2. Multiply it by the derivative of the second factor.

3. Write the derivative of the first factor.

4. Multiply it by the second factor.

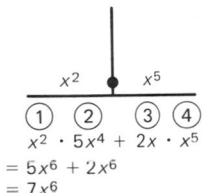

Example 1

$$\frac{d}{dx}(x^4 - 2x^3 - 7)(3x^2 - 5x) = (x^4 - 2x^3 - 7)(6x - 5)$$

$$+ (4x^3 - 6x^2)(3x^2 - 5x)$$

Note that we could have multiplied the polynomials and then differentiated, avoiding the Product Rule, but this would have been more work.

DO EXERCISES 50 AND 51.

The derivative of a quotient is *not* the quotient of the derivatives. To see why, consider x^5 and x^2. The quotient x^5/x^2 is x^3, and the derivative of this quotient is $3x^2$. The individual derivatives are $5x^4$ and $2x$, and the quotient of these derivatives $5x^4/2x$ is $(5/2)x^3$, which is not $3x^2$. The rule for differentiating quotients is as follows:

OBJECTIVES

You should be able to differentiate using the Product and Quotient Rules.

Use the Product Rule to find $f'(x)$.

50. $f(x) = 3x^8 \cdot x^{10}$

51. $f(x) = (9x^3 + 4x^2 + 10)(-7x^2 + x^4)$

52. For
$$f(x) = \frac{x^9}{x^5},$$

find $f'(x)$ using the Quotient Rule.

QUOTIENT RULE. If $q(x) = \dfrac{f(x)}{g(x)}$, then

$$q'(x) = \frac{g(x) \cdot f'(x) - g'(x) \cdot f(x)}{[g(x)]^2}.$$

The derivative of a quotient is the denominator times the derivative of the numerator, minus the derivative of the denominator times the numerator, all divided by the square of the denominator.

Another way to remember this is shown below. It starts with squaring the denominator. The denominator is also used as the first factor of the first term above.

1. Square the denominator.
2. Write down the denominator.
3. Multiply the denominator by the derivative of the numerator.
4. Write a minus sign.
5. Find the derivative of the denominator.
6. Multiply it by the numerator.

Example 2 For $q(x) = x^5/x^3$, find $q'(x)$.

Solution $q'(x) = \dfrac{x^3 \cdot 5x^4 - 3x^2 \cdot x^5}{[x^3]^2} = \dfrac{5x^7 - 3x^7}{x^6} = \dfrac{2x^7}{x^6} = 2x$

DO EXERCISE 52.

Example 3 Differentiate $\dfrac{1 + x^2}{x^3}$.

This means find

$$\frac{d}{dx} f(x)$$

where $f(x)$ is given by

$$f(x) = \frac{1 + x^2}{x^3}.$$

Solution

$$\frac{d}{dx}\left(\frac{1 + x^2}{x^3}\right) = \frac{x^3 \cdot 2x - 3x^2(1 + x^2)}{(x^3)^2} = \frac{2x^4 - 3x^2 - 3x^4}{x^6} = \frac{-x^4 - 3x^2}{x^6}$$

$$= \frac{-1 \cdot x^2 \cdot x^2 - 3x^2}{x^6} = \frac{x^2(-x^2 - 3)}{x^6} = \frac{-x^2 - 3}{x^4}$$

Example 4   Differentiate $\dfrac{x^2 - 3x}{x - 1}$.

Solution

$$\frac{d}{dx}\left(\frac{x^2 - 3x}{x - 1}\right) = \frac{(x - 1)(2x - 3) - 1(x^2 - 3x)}{(x - 1)^2}$$

$$= \frac{2x^2 - 5x + 3 - x^2 + 3x}{(x - 1)^2}$$

$$= \frac{x^2 - 2x + 3}{(x - 1)^2}.$$

It is not necessary to multiply out $(x - 1)^2$.

DO EXERCISES 53 AND 54.

An Application

We discussed earlier that it is more typical for a total revenue function to vary depending on the number x of units sold. Let us see what can determine this. Recall the consumer's demand function $p = D(x)$, discussed on p. 64. It is the price p a seller must charge in order to sell exactly x units of a product. This is typically a decreasing function.

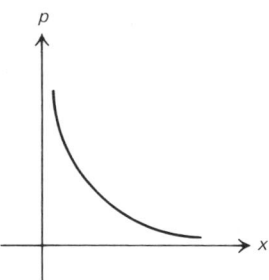

The total revenue from the sale of x units is then given by

$$R(x) = (\text{number of units sold}) \cdot (\text{price charged to sell the units}),$$

or

$$R(x) = x \cdot p = xD(x).$$

Differentiate. That is, find

$$\frac{d}{dx} f(x)$$

where $f(x)$ is as follows:

53. $\dfrac{1 - x^2}{x^5}$

54. $\dfrac{x^2 - 1}{x^3 + 1}$

55. A company determines that the demand function for a certain product is given by

$$p = D(x) = 200 - x.$$

a) Find an expression for total revenue $R(x)$.
b) Find the marginal revenue $R'(x)$.

A typical graph of a revenue function is shown below.

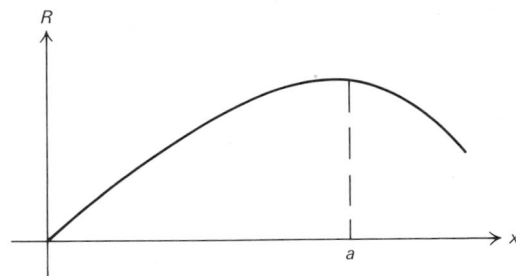

To sell more units, $D(x)$ decreases. Because we have a product $x \cdot D(x)$, the revenue typically rises for a while as x increases, but tapers off as $D(x)$ gets smaller and smaller.

Using the Product Rule, one can obtain an expression for the marginal revenue $R'(x)$ in terms of x and $D'(x)$ as follows:

$$R(x) = xD(x),$$

so

$$R'(x) = 1 \cdot D(x) + x \cdot D'(x) = D(x) + xD'(x).$$

You need not memorize this. One can merely repeat the Product Rule where necessary.

DO EXERCISE 55.

EXERCISE SET 9.6

Differentiate. That is, find $\dfrac{d}{dx} f(x)$ where $f(x)$ is as follows.

1. $x^3 \cdot x^8$; two ways

2. $x^4 \cdot x^9$; two ways

3. $\dfrac{-1}{x}$; two ways

4. $\dfrac{1}{x}$; two ways

5. $\dfrac{x^8}{x^5}$; two ways

6. $\dfrac{x^9}{x^5}$; two ways

7. $(8x^5 - 3x^2 + 20)(8x^4 - 3\sqrt{x})$

8. $(7x^6 + 4x^3 - 50)(9x^{10} - 7\sqrt{x})$

9. $x(300 - x)$

10. $x(400 - x)$

11. $\dfrac{x}{300 - x}$

12. $\dfrac{x}{400 - x}$

13. $\dfrac{3x - 1}{2x + 5}$

14. $\dfrac{2x + 3}{x - 5}$

15. $\dfrac{x^2 + 1}{x^3 - 1}$

16. $\dfrac{x^3 - 1}{x^2 + 1}$

17. $\dfrac{x}{1 - x}$

18. $\dfrac{x}{3 - x}$

19. $\dfrac{x - 1}{x + 1}$

20. $\dfrac{x + 2}{x - 2}$

21. $\dfrac{1}{x - 3}$

22. $\dfrac{1}{x + 2}$

23. $\dfrac{3x^2 + 2x}{x^2 + 1}$

24. $\dfrac{3x^2 - 5x}{x^2 - 1}$

25. $\dfrac{3x^2 - 5x}{x^8}$

26. $\dfrac{3x^2 + 2x}{x^5}$

In each of Exercises 27–30, a demand function $p = D(x)$ is given. Find
a) total revenue $R(x)$,
b) marginal revenue $R'(x)$.

27. $D(x) = 400 - x$

28. $D(x) = 500 - x$

29. $D(x) = \dfrac{4000}{x} + 3$

30. $D(x) = \dfrac{3000}{x} + 5$

31. In Section 9.5, we defined the average cost of producing x units of a product in terms of the total cost $C(x)$ by

$$A(x) = \frac{C(x)}{x}.$$

Use the Quotient Rule to find a general expression for *marginal average cost* $A'(x)$.

32. In this section we determined that

$$R(x) = xD(x).$$

Then

$$D(x) = \frac{R(x)}{x} = \text{average revenue from the sale of } x \text{ units.}$$

Use the Quotient Rule to find a general expression for *marginal average revenue* $D'(x)$.

▶──

Differentiate each function.

33. $f(x) = \dfrac{x^3}{\sqrt{x} - 5}$

34. $g(t) = \dfrac{1 + \sqrt{t}}{t^5 + 3}$

35. $f(v) = \dfrac{3}{1 + v + v^2}$

36. $g(z) = \dfrac{1 + z + z^2}{1 - z + z^2}$

37. $p(t) = \dfrac{t}{1 - t + t^2 - t^3}$

38. $f(x) = \dfrac{\dfrac{2}{3x} - 1}{\dfrac{3}{x^2} + 5}$

39. $h(x) = \dfrac{x^3 + 5x^2 - 2}{\sqrt{x}}$

40. $y(t) = 5t(t - 1)(2t + 3)$

41. $f(x) = x(3x^3 + 6x - 2)(3x^4 + 7)$

42. $g(x) = (x^3 - 8) \cdot \dfrac{x^2 + 1}{x^2 - 1}$

43. $f(t) = (t^5 + 3) \cdot \dfrac{t^3 - 1}{t^3 + 1}$

44. $f(x) = \dfrac{(x^2 + 3x)(x^5 - 7x^2 - 3)}{x^4 - 3x^3 - 5}$

45. $f(x) = \dfrac{(2x^2 + 3)(4x^3 - 7x + 2)}{x^7 - 2x^6 + 9}$

46. $s(t) = \dfrac{5t^8 - 2t^3}{(t^5 - 3)(t^4 + 7)}$

──

9.7 THE EXTENDED POWER RULE/THE CHAIN RULE

The Extended Power Rule

How do we differentiate more complicated functions such as

$$y = (1 + x^2)^3, \quad y = (1 + x^2)^{89}, \quad \text{or} \quad y = (1 + x^2)^{1/3}?$$

For $(1 + x^2)^3$ we can expand and then differentiate, but while this could be done for $(1 + x^2)^{89}$, it would certainly be time-consuming, and such an expansion of the Power Rule would not work for $(1 + x^2)^{1/3}$. Not knowing this, we might conjecture that the derivative of the function $y = (1 + x^2)^3$ is

$$3(1 + x^2)^2. \tag{1}$$

OBJECTIVES

You should be able to differentiate using the Extended Power Rule.

Differentiate.

56. $(1 + x^2)^{10}$

57. $(1 - x^2)^{1/2}$

To check this, we expand $(1 + x^2)^3$ and then differentiate. We know $(a + h)^3 = a^3 + 3a^2h + 3ah^2 + h^3$, so

$$(1 + x^2)^3 = 1^3 + 3 \cdot 1^2 \cdot (x^2)^1 + 3 \cdot 1 \cdot (x^2)^2 + (x^2)^3$$
$$= 1 + 3x^2 + 3x^4 + x^6$$

(We could also have done this by finding $(1 + x^2)^2$ and then multiplying again by $1 + x^2$.) It follows that

$$\frac{dy}{dx} = 6x + 12x^3 + 6x^5 = (1 + 2x^2 + x^4)6x$$
$$= 3(1 + x^2)^2 \cdot 2x \qquad (2)$$

Comparing this with Eq. (1), we see that the Power Rule is not sufficient for such a differentiation. Note that the factor $2x$ in the actual derivative Eq. (2) is the derivative of the "inside" function, $1 + x^2$. This is consistent with the following new rule.

THE EXTENDED POWER RULE. Suppose ▭ is some function of x. Then

$$\frac{d}{dx} \boxed{}^a = a \boxed{}^{a-1} \cdot \frac{d}{dx} \boxed{}.$$

More formally, if $g(x)$ is a function of x, then

$$\frac{d}{dx}[g(x)]^a = a[g(x)]^{a-1} \cdot \frac{d}{dx} g(x).$$

Let us differentiate $(1 + x^3)^5$. There are four steps to carry out.

$\boxed{}^5$ **1.** Mentally block out the "inside" function $1 + x^3$.

$5\boxed{}^4$ **2.** Differentiate the "outside" function $\boxed{}^5$.

$5(1 + x^3)^4$ **3.** Write in the "inside" function.

$5(1 + x^3)^4 \cdot 3x^2$
$= 15x^2(1 + x^3)^4$ **4.** Multiply by the derivative of the "inside" function.

Step 4 is most commonly overlooked. Try not to forget it!

Example 1 $\dfrac{d}{dx}(1 + x^3)^{1/2} = \frac{1}{2}(\boxed{1 + x^3})^{1/2-1} \cdot 3x^2 = \frac{1}{2}(1 + x^3)^{-1/2} \cdot 3x^2$

$$= \frac{3x^2}{2\sqrt{1+x^3}}$$

DO EXERCISES 56 AND 57.

Example 2   Differentiate $(1 - x^2)^3 - (1 - x^2)^2$.

Solution   Here we combine the Difference Rule and the Extended Power Rule.

$\dfrac{d}{dx}[(1 - x^2)^3 - (1 - x^2)^2]$

$= 3(1 - x^2)^2(-2x) - 2(1 - x^2)(-2x)$ (We differentiate each term using the Extended Power Rule.)

$= -6x(1 - x^2)^2 + 4x(1 - x^2)$

$= x(1 - x^2)[-6(1 - x^2) + 4]$ (Here we factor out $x(1 - x^2)$.)

$= x(1 - x^2)[-6 + 6x^2 + 4]$

$= x(1 - x^2)(6x^2 - 2) = 2x(1 - x^2)(3x^2 - 1)$.

DO EXERCISE 58.

Example 3   Differentiate $(x - 5)^4(7 - x)^{10}$.

Solution   Here we combine the Product Rule and the Extended Power Rule.

$\dfrac{d}{dx}(x - 5)^4(7 - x)^{10}$

$= (x - 5)^4 10(7 - x)^9(-1) + 4(x - 5)^3(7 - x)^{10}$

$= -10(x - 5)^4(7 - x)^9 + 4(x - 5)^3(7 - x)^{10}$

$= (x - 5)^3(7 - x)^9[-10(x - 5) + 4(7 - x)]$ (We factored out $(x - 5)^3(7 - x)^9$.)

$= (x - 5)^3(7 - x)^9[-10x + 50 + 28 - 4x]$

$= (x - 5)^3(7 - x)^9(78 - 14x)$

$= 2(x - 5)^3(7 - x)^9(39 - 7x)$.

DO EXERCISE 59.

Example 4   Differentiate $\sqrt[4]{\dfrac{x + 3}{x - 1}}$.

58. Differentiate.

$$(1 + x^2)^2 - (1 + x^2)^3$$

59. Differentiate.

$$(x - 4)^5(6 - x)^3$$

60. Differentiate.

$$\sqrt[3]{\frac{x + 5}{x - 4}}$$

Solution We have to use the Quotient Rule to differentiate the inside function $(x + 3)/(x - 1)$.

$$\frac{d}{dx} \sqrt[4]{\frac{x + 3}{x - 1}} = \frac{d}{dx}\left(\frac{x + 3}{x - 1}\right)^{1/4} = \frac{1}{4}\left(\frac{x + 3}{x - 1}\right)^{1/4 - 1}\left[\frac{(x - 1)1 - 1(x + 3)}{(x - 1)^2}\right]$$

$$= \frac{1}{4}\left(\frac{x + 3}{x - 1}\right)^{-3/4}\left[\frac{x - 1 - x - 3}{(x - 1)^2}\right]$$

$$= \frac{1}{4}\left(\frac{x + 3}{x - 1}\right)^{-3/4} \cdot \frac{-4}{(x - 1)^2}$$

$$= \left(\frac{x + 3}{x - 1}\right)^{-3/4} \cdot \frac{-1}{(x - 1)^2}$$

DO EXERCISE 60.

The Chain Rule

The Extended Power Rule is a special case of a more general rule called the *Chain Rule*. Before discussing it, we shall define the *composition* of functions. Consider the following, for example.

$$f(x) = x^3 \qquad \text{(This function cubes each input.)}$$

and

$$g(x) = 1 + x^2 \qquad \text{(This function adds 1 to the square of each input.)}$$

We define a new function that first does what g does (adds 1 to the square) and then does what f does (cubes). The new function is called the *composition* of f and g and is symbolized $f(g(x))$. We can visualize the composition of functions as follows.

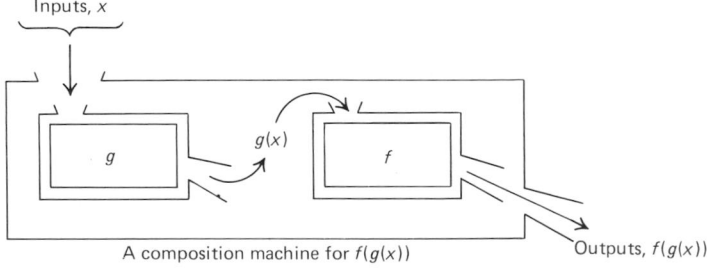

Inputs, x

$g(x)$

g

f

A composition machine for $f(g(x))$

Outputs, $f(g(x))$

Example 5 Given $f(x) = x^3$ and $g(x) = 1 + x^2$, find $f(g(x))$ and $g(f(x))$.

Solution   We find $f(g(x))$ by substituting $g(x)$ for x.

$$f(g(x)) = f(1 + x^2) \qquad \text{(Substituting } 1 + x^2 \text{ for } x\text{)}$$
$$= (1 + x^2)^3 = 1 + 3x^2 + 3x^4 + x^6$$

We find $g(f(x))$ by substituting $f(x)$ for x.

$$g(f(x)) = g(x^3) \qquad \text{(Substituting } x^3 \text{ for } x\text{)}$$
$$= 1 + (x^3)^2$$
$$= 1 + x^6$$

DO EXERCISE 61.

Example 6   Given $f(x) = \sqrt{x}$ and $g(x) = x - 1$, find $f(g(x))$ and $g(f(x))$.

Solution

$$f(g(x)) = f(x - 1) = \sqrt{x - 1}, \qquad g(f(x)) = g(\sqrt{x}) = \sqrt{x} - 1$$

DO EXERCISE 62.

THE CHAIN RULE. **Suppose** ▭ **is some function of** x**. Then**

$$\frac{d}{dx} f(\boxed{}) = f'(\boxed{}) \cdot \frac{d}{dx} \boxed{}.$$

More formally, the derivative of the composition $f(g(x))$ is given by

$$\frac{d}{dx} f(g(x)) - f'(g(x)) \cdot \frac{d}{dx} g(x).$$

Note how the Extended Power Rule is a special case.

$$\frac{d}{dx} [g(x)]^a = a[g(x)]^{a-1} \cdot \frac{d}{dx} g(x).$$

The Chain Rule often appears in another form. Suppose $y = f(u)$ and $u = g(x)$. Then

$$\frac{dy}{dx} = \frac{dy}{du} \cdot \frac{du}{dx}.$$

For example, if $y = 2 + \sqrt{u}$ and $u = x^3 + 1$, then

$$\frac{dy}{du} = \frac{1}{2} u^{-1/2} \qquad \text{and} \qquad \frac{du}{dx} = 3x^2,$$

61. Given $f(x) = 3x$ and $g(x) = x^2 - 1$, find $f(g(x))$ and $g(f(x))$.

62. Given $f(x) = 4x + 5$ and $g(x) = \sqrt[3]{x}$, find $f(g(x))$ and $g(f(x))$.

so

$$\frac{dy}{dx} = \frac{dy}{du} \cdot \frac{du}{dx} = \frac{1}{2\sqrt{u}} \cdot 3x^2$$

$$= \frac{3x^2}{2\sqrt{x^3 + 1}} \qquad \text{(Substituting } x^3 + 1 \text{ for } u\text{)}$$

EXERCISE SET 9.7

Differentiate.

1. $(1 - x)^{55}$ 　　　**2.** $(1 - x)^{100}$ 　　　**3.** $\sqrt{1 + 8x}$ 　　　**4.** $\sqrt{1 - x}$ 　　　**5.** $\sqrt{3x^2 - 4}$

6. $\sqrt{4x^2 + 1}$ 　　　**7.** $(3x^2 - 6)^{-40}$ 　　　**8.** $(4x^2 + 1)^{-50}$ 　　　**9.** $x\sqrt{2x + 3}$ 　　　**10.** $x\sqrt{4x - 7}$

11. $x^2\sqrt{x - 1}$ 　　　　　**12.** $x^3\sqrt{x + 1}$ 　　　　　**13.** $\dfrac{1}{(3x + 8)^2}$ 　　　　**14.** $\dfrac{1}{(4x + 5)^2}$

15. $(1 + x^3)^3 - (1 + x^3)^4$ 　　　　　　　　　**16.** $(1 + x^3)^5 - (1 + x^3)^4$

17. $x^2 + (200 - x)^2$ 　　　**18.** $x^2 + (100 - x)^2$ 　　　**19.** $(x + 6)^{10}(x - 5)^4$ 　　　**20.** $(x - 4)^8(x + 3)^9$

21. $(x - 4)^8(3 - x)^4$ 　　　**22.** $(x + 6)^{10}(5 - x)^9$ 　　　**23.** $-4x(2x - 3)^3$ 　　　**24.** $-5x(3x + 5)^6$

25. $\sqrt{\dfrac{1 - x}{1 + x}}$ 　　　　　**26.** $\sqrt{\dfrac{3 + x}{2 - x}}$

27. Consider

$$f(x) = \frac{x^2}{(1 + x)^5}.$$

a) Find $f'(x)$ using the Quotient Rule and the Extended Power Rule.

b) Note that $f(x) = x^2(1 + x)^{-5}$. Find $f'(x)$ using the Product Rule and the Extended Power Rule.

c) Compare answers to (a) and (b).

28. Consider

$$g(x) = (x^3 + 5x)^2.$$

a) Find $g'(x)$ using the Extended Power Rule.

b) Note that $g(x) = x^6 + 10x^4 + 25x^2$. Find $g'(x)$.

c) Compare answers to (a) and (b).

29. A total cost function is given by

$$C(x) = 1000\sqrt{x^3 + 2}.$$

Find the marginal cost $C'(x)$.

30. A total revenue function is given by

$$R(x) = 2000\sqrt{x^2 + 3}.$$

Find the marginal revenue $R'(x)$.

▶

Differentiate the following functions.

31. $y = \sqrt[3]{x^3 - 6x + 1}$ 　　　**32.** $s = \sqrt[4]{t^4 + 3t^2 + 8}$ 　　　**33.** $y = \dfrac{x}{\sqrt{x - 1}}$ 　　　**34.** $y = \dfrac{(x + 1)^2}{(x^2 + 1)^3}$

35. $u = \dfrac{(1 + 2v)^4}{v^4}$ 　　　**36.** $y = x\sqrt{1 + x^2}$ 　　　**37.** $y = \dfrac{\sqrt{1 - x^2}}{1 - x}$ 　　　**38.** $w = \dfrac{u}{\sqrt{1 + u^2}}$

39. $y = \left(\dfrac{x^2 - x - 1}{x^2 + 1}\right)^3$ 　　　**40.** $y = \sqrt{1 + \sqrt{x}}$ 　　　**41.** $s = \dfrac{\sqrt{t-1}}{\sqrt{t+1}}$ 　　　**42.** $y = x^{2/3} \cdot \sqrt[3]{1 + x^2}$

9.8 HIGHER DERIVATIVES

Consider the function given by

$$f(x) = x^5 - 3x^4 + x.$$

Its derivative f' is given by

$$f'(x) = 5x^4 - 12x^3 + 1.$$

This function f' can be differentiated. We use the notation f'' for the derivative $(f')'$. We call f'' the *second derivative* of f. It is given by

$$f''(x) = 20x^3 - 36x^2.$$

Continuing in this manner, we have

$$f'''(x) = 60x^2 - 72x, \quad \text{(The third derivative of } f.\text{)}$$
$$f''''(x) = 120x - 72, \quad \text{(The fourth derivative of } f.\text{)}$$
$$f'''''(x) = 120. \quad \text{(The fifth derivative of } f.\text{)}$$

When notation, like $f'''''(x)$, gets lengthy we can abbreviate it using a numeral in parentheses. Thus

$$f^{(4)}(x) = 120x - 72, \quad f^{(5)}(x) = 120, \quad \text{and} \quad f^{(6)}(x) = 0.$$

DO EXERCISE 63.

Leibniz's notation for the second derivative of a function given by $y = f(x)$ is

$$\frac{d^2y}{dx^2} \quad \text{or} \quad \frac{d}{dx}\left(\frac{dy}{dx}\right),$$

read "the second derivative of y with respect to x." The 2s in this notation are *not* exponents. If $y = x^5 - 3x^4 + x$, then

$$\frac{d^2y}{dx^2} = 20x^3 - 36x^2.$$

Leibniz's notation for the third derivative is d^3y/dx^3, for the fourth derivative d^4y/dx^4, and so on:

$$\frac{d^3y}{dx^3} = 60x^2 - 72x, \quad \frac{d^4y}{dx^4} = 120x - 72, \quad \frac{d^5y}{dx^5} = 120.$$

DO EXERCISE 64.

Example   For $y = \dfrac{1}{x}$, find $\dfrac{d^2y}{dx^2}$.

OBJECTIVES

You should be able to find a higher-order derivative.

63. Find the first six derivatives.

$$f(x) = 2x^6 - x^5 + 10.$$

64. For

$$y = x^7 - x^3,$$

find

a) $\dfrac{dy}{dx}$,

b) $\dfrac{d^2y}{dx^2}$,

c) $\dfrac{d^3y}{dx^3}$,

d) $\dfrac{d^4y}{dx^4}$.

65. For $y = \dfrac{2}{x}$, find $\dfrac{d^2y}{dx^2}$.

Solution $y = x^{-1}$, so

$$\frac{dy}{dx} = -1 \cdot x^{-1-1} = -x^{-2}, \quad \text{or} \quad -\frac{1}{x^2}.$$

Then

$$\frac{d^2y}{dx^2} = (-2)(-1)x^{-2-1} = 2x^{-3}, \quad \text{or} \quad \frac{2}{x^3}.$$

DO EXERCISE 65.

Acceleration can be thought of as a second derivative. As an object moves, its distance from a fixed point after time t is some function of the time, say $s(t)$. Then

$$v(t) = s'(t) = \text{velocity at time } t,$$

and

$$a(t) = v'(t) = s''(t) = \text{acceleration at time } t.$$

66. For $s(t) = 3t + t^4$, find the acceleration $a(t)$.

Whenever a quantity is a function of time, the first derivative gives the rate of change with respect to time and the second derivative gives the acceleration. For example, if $y = P(t)$ gives the number of people in a population at time t, then $P'(t)$ gives how fast the size of the population is changing and $P''(t)$ gives the acceleration in the size of the population.

DO EXERCISE 66.

EXERCISE SET 9.8

Find $\dfrac{d^2y}{dx^2}$.

1. $y = 3x + 5$ **2.** $y = -4x + 7$ **3.** $y = -\dfrac{1}{x}$ **4.** $y = -\dfrac{3}{x}$ **5.** $y = x^{1/4}$

6. $y = \sqrt{x}$ **7.** $y = x^4 + \dfrac{4}{x}$ **8.** $y = x^3 - \dfrac{3}{x}$ **9.** $y = x^{-3}$ **10.** $y = x^{-4}$

11. $y = x^n$ **12.** $y = x^{-n}$ **13.** $y = x^4 - x^2$ **14.** $y = x^4 + x^3$ **15.** $y = \sqrt{x - 1}$

16. $y = \sqrt{x + 1}$ **17.** $y = ax^2 + bx + c$ **18.** $y = mx + b$

19. For $y = x^4$, find $\dfrac{d^4y}{dx^4}$.   **20.** For $y = x^5$, find $\dfrac{d^4y}{dx^4}$.   **21.** For $y = x^6 - x^3 + 2x$, find $\dfrac{d^5y}{dx^5}$.

22. For $y = x^7 - 8x^2 + 2$, find $\dfrac{d^6y}{dx^6}$.   **23.** For $y = x^n$, find $\dfrac{d^6y}{dx^6}$.   **24.** For $y = x^k$, find $\dfrac{d^5y}{dx^5}$.

25. If s is a distance given by $s(t) = t^3 + t^2 + 2t$, find the acceleration.

26. If s is a distance given by $s(t) = t^4 + t^2 + 3t$, find the acceleration.

27. A population grows from an initial size of 100,000 to an amount $P(t)$ given by

$$P(t) = 100{,}000(1 + 0.6t + t^2).$$

What is the acceleration in the size of the population?

28. A population grows from an initial size of 100,000 to an amount $P(t)$ given by

$$P(t) = 100{,}000(1 + 0.4t + t^2).$$

What is the acceleration in the size of the population?

▶ ──

Find y', y'', and y'''.

29. $y = x^{-1} + x^{-2}$

30. $y = \dfrac{1}{1 - x}$

31. $y = x\sqrt{1 + x^2}$

32. $y = 3x^5 + 8\sqrt{x}$

33. $y = \dfrac{3x - 1}{2x + 3}$

34. $y = \dfrac{1}{\sqrt{x - 1}}$

35. $y = \dfrac{x}{\sqrt{x - 1}}$

36. $y = \dfrac{\sqrt{x} - 1}{\sqrt{x} + 1}$

Differentiate implicitly to find $\dfrac{dy}{dx}$ and $\dfrac{d^2y}{dx^2}$.

37. $xy + x - 2y = 4$

38. $y^2 - xy + x^2 = 5$

39. $x^2 - y^2 = 5$

40. $x^3 - y^3 = 8$

───

9.9 THE SHAPE OF A GRAPH: FINDING MAXIMUM AND MINIMUM VALUES

First and second derivatives give us information about the shape of a graph that may be relevant in finding maximum and minimum values of functions. Throughout this section we will assume that the functions are continuous.

Increasing and Decreasing Functions

We have seen how the slope of a linear function determines whether it is increasing or decreasing (or neither). For a general function, the derivative yields similar information. Let us investigate how this happens in the margin exercise.

DO EXERCISE 67.

The following is how we can use derivatives to determine whether a function is increasing or decreasing.

If $f'(x) > 0$, for all x in an interval I, then f is increasing over I.

If $f'(x) < 0$, for all x in an interval I, then f is decreasing over I.

OBJECTIVES

You should be able to find maximum and minimum values of functions.

67. Consider the following graph.

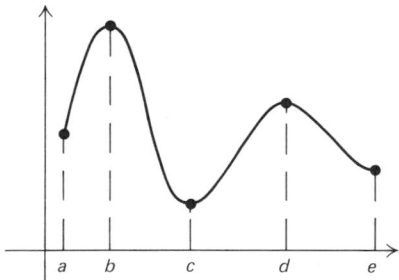

a) Over what intervals is the function increasing?

b) Over what intervals is the derivative positive? To determine this, locate a straightedge at points on the graph and decide whether slopes of tangent lines are positive.

c) Over what intervals is the function decreasing?

d) Over what intervals is the derivative negative?

Look for a pattern.

68. Consider the following graph.

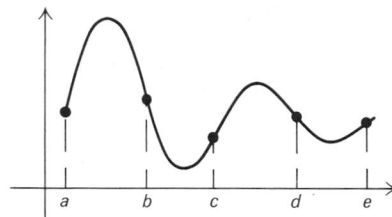

a) Over what intervals is the *derivative* increasing?

b) Over what intervals is the *derivative* decreasing?

69. Consider the following graph.

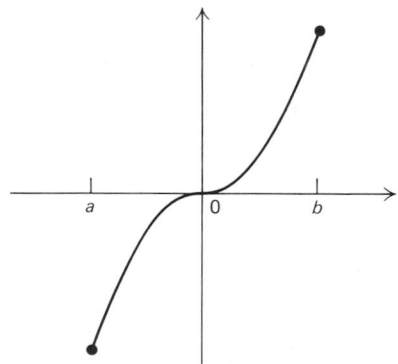

a) Over what intervals is the graph concave up?

b) Over what intervals is the graph concave down?

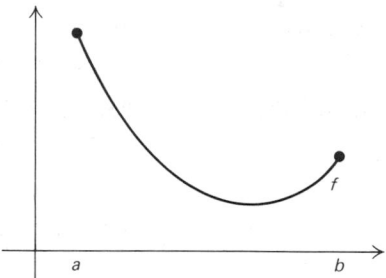

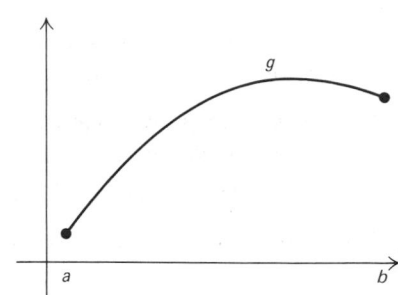

Concavity: Increasing and Decreasing Derivatives

The following are two functions. The graph on the left is turning upward and the other is turning downward. Let's see if we can relate this to their derivatives.

Consider the graph of *f*. Take a ruler, or straightedge, and move along the curve from left to right. What happens to the slopes of the tangent lines? Do the same for the graph of *g*. Look for a pattern.

DO EXERCISE 68.

We have the following.

1. If $f''(x) > 0$ on an interval, *I*, then *f* is turning upward on *I* (since *f'* is increasing on *I*). Such a graph is said to be *concave up* over *I*.

2. If $f''(x) < 0$ on an interval, *I*, then *f* is turning downward on *I* (since *f'* is decreasing on *I*). Such a graph is said to be *concave down* over *I*.

The following is a helpful memory device.

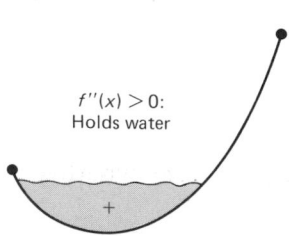

$f''(x) > 0$:
Holds water

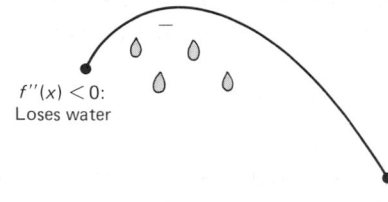

$f''(x) < 0$:
Loses water

DO EXERCISE 69.

A *point of inflection*, or an *inflection point*, is a point across which the direction of concavity changes. For example, point P is an inflection point of the graph on the left. Points P, Q, R, and S are inflection points of the graph on the right. In Margin Exercise 69, the point $(0, 0)$ is an inflection point.

70. Which are points of inflection?

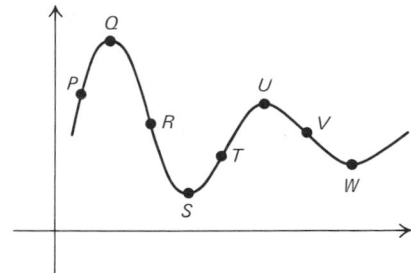

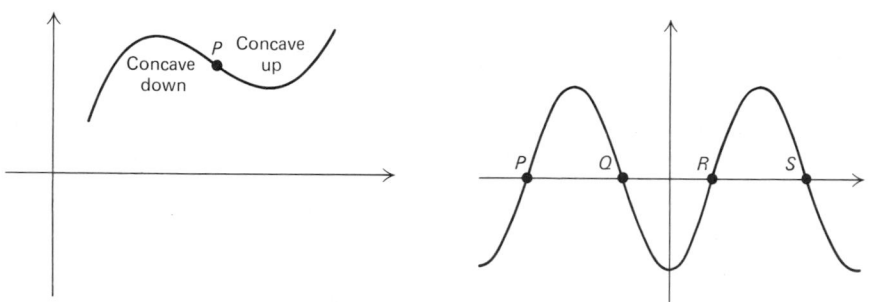

DO EXERCISE 70.

Just knowing the values of f' and f'' at some specific point x_0 can yield a lot of information about the shape of the graph over some (possibly small) interval containing x_0 as an interior point (assuming f'' exists and is continuous over the interval).

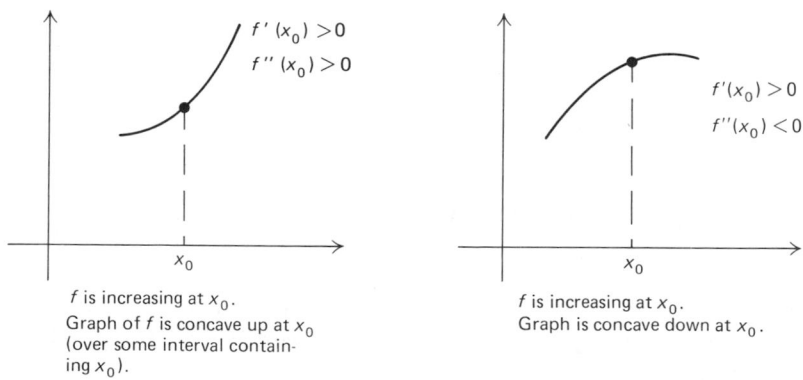

f is increasing at x_0.
Graph of f is concave up at x_0 (over some interval containing x_0).

f is increasing at x_0.
Graph is concave down at x_0.

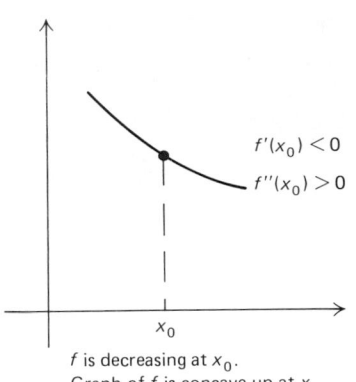

$f'(x_0) < 0$
$f''(x_0) > 0$

f is decreasing at x_0.
Graph of f is concave up at x_0.

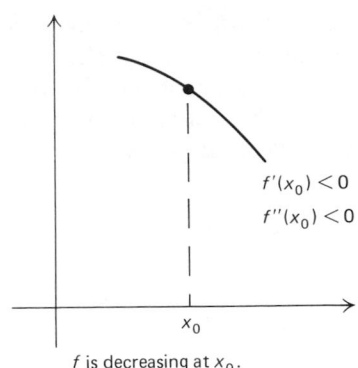

$f'(x_0) < 0$
$f''(x_0) < 0$

f is decreasing at x_0.
Graph is concave down at x_0.

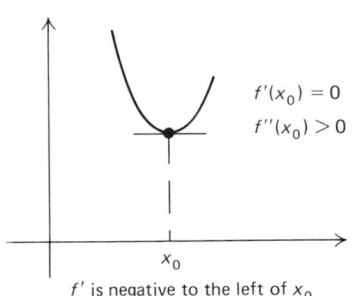

$f'(x_0) = 0$
$f''(x_0) > 0$

f' is negative to the left of x_0
and positive to the right.
Graph is concave up at x_0.

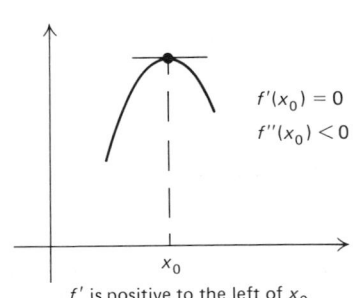

$f'(x_0) = 0$
$f''(x_0) < 0$

f' is positive to the left of x_0
and negative to the right.
Graph is concave down at x_0.

Example 1. Determine the shape of the graph of $f(x) = x^3 - x^2$ at $x = -1$.

Solution For $f(x) = x^3 - x^2$,

$$f'(x) = 3x^2 - 2x \quad \text{and} \quad f''(x) = 6x - 2.$$

Then

$$f'(-1) = 3(-1)^2 - 2(-1) \quad \text{and} \quad f''(-1) = 6(-1) - 2$$
$$= 3 + 2 = 5 \qquad\qquad\qquad = -6 - 2 = -8.$$

Thus the function is increasing at $x = -1$ since $f'(-1) > 0$, and also concave down since $f''(-1) < 0$. This is shown below, where $f(-1) = (-1)^3 - (-1)^2 = -2$.

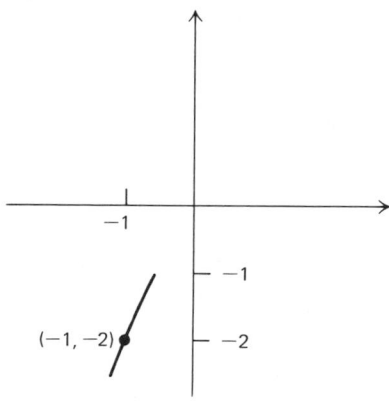

71. Determine the shape of the graph of $f(x) = x^4 - 4x$, at

a) $x = 3$,

b) $x = 1$.

DO EXERCISE 71.

Let us take a more global look at two functions from the standpoint of the concepts we have considered. For example, consider

$$f(x) = x^2.$$

Now

$$f'(x) = 2x.$$

Note that $f'(0) = 0$. The graph has a horizontal tangent at 0. Also, when $x < 0$, $2x < 0$, so $f'(x) < 0$. Thus the function is decreasing on the interval $(-\infty, 0)$. When $x > 0$, $2x > 0$, so $f'(x) > 0$. This tells us that the function is increasing on the interval $(0, \infty)$. Check these facts on the graph.

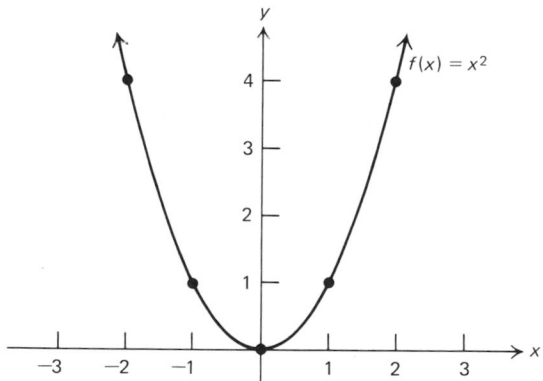

Let us look at the second derivative

$$f''(x) = 2.$$

The second derivative is positive for all values of x, since it is the constant 2; thus the graph is concave up over the entire real line. Check this on the graph.

As another example, consider

$$f(x) = x^3.$$

Now

$$f'(x) = 3x^2.$$

Note that $f'(0) = 0$. The graph has a horizontal tangent at 0. Also, when $x < 0$, $x^2 > 0$, so $3x^2 > 0$ and $f'(x) > 0$. Thus the function is increasing over the interval $(-\infty, 0)$. When $x > 0$, $x^2 > 0$, so $3x^2 > 0$ and $f'(x) > 0$. Thus the function is increasing over the interval $(0, \infty)$. In fact it is increasing over the entire real line. Check this on the graph.

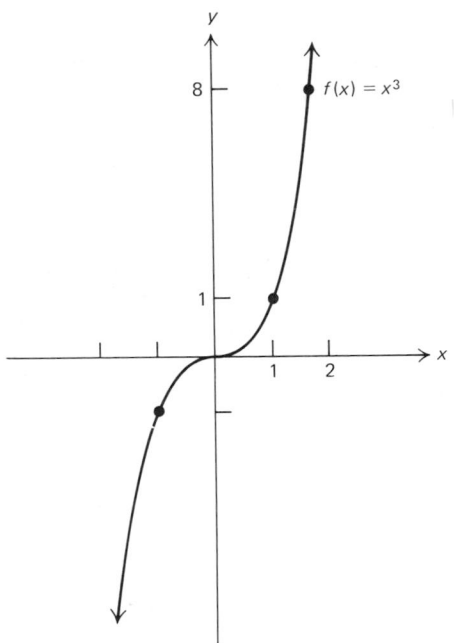

Let us look at the second derivative,

$$f''(x) = 6x.$$

When $x < 0$, $6x < 0$, so $f''(x) < 0$. Thus the graph is concave down over the interval $(-\infty, 0)$. When $x > 0$, $6x > 0$, so $f''(x) > 0$. Thus the graph is concave up over the interval $(0, \infty)$. Check this on the graph, noting also that $f''(0) = 0$ and that the graph has an inflection point $(0, 0)$.

Critical Points

A *critical point* of a function is an interior point c of its domain at which the function has a horizontal tangent, or at which the derivative does not exist. That is, c is a critical point if

$$f'(c) = 0 \quad \text{or} \quad f'(c) \text{ does not exist.}$$

Consider the following graph.

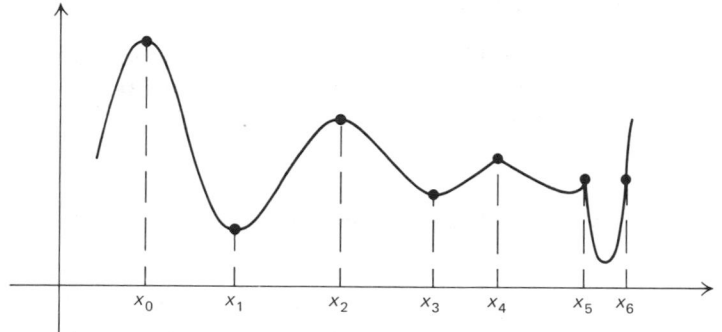

The points x_0, x_1, x_2, and x_3 are all critical points because the derivative is 0 at each of these points. The points x_4, x_5, and x_6 are all critical points because the derivative does not exist at these points.

DO EXERCISE 72.

The Shape of a Graph
Between Critical Points and Endpoints

Suppose we have a continuous function defined over an interval $[a, b]$.

DO EXERCISES 73 AND 74.

Consider the following graph.

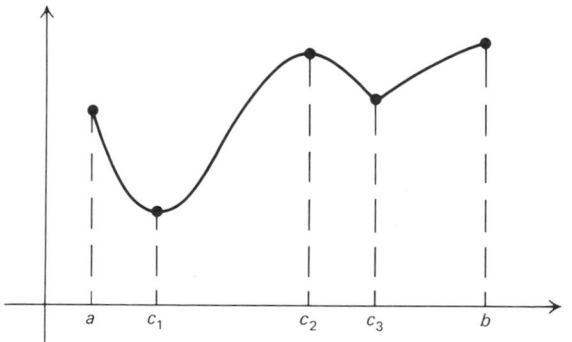

72. Consider this graph.

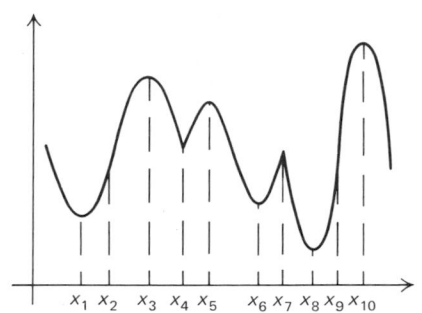

a) At which points are there horizontal tangents?

b) At which points does the derivative not exist?

c) Which are critical points?

73. Try to draw a graph of a continuous function from P to Q that increases on part or parts of $[a, b]$, and decreases on part or parts of $[a, b]$.

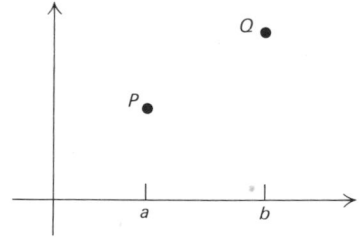

Does the function you drew have any critical points between a and b?

74. Now try to draw a graph of a continuous function from P to Q that increases on part or parts of $[a, b]$, and decreases on part or parts of $[a, b]$, but in such a way that no critical points occur between a and b.

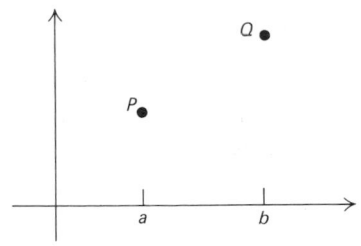

75. In each graph find where maximum and minimum values occur.

a)

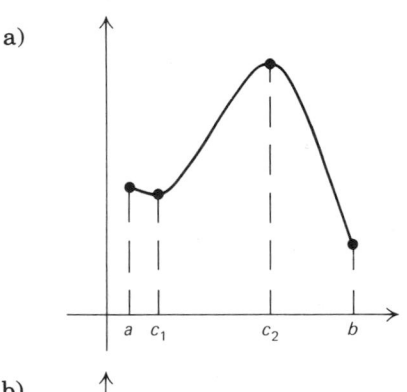

b)

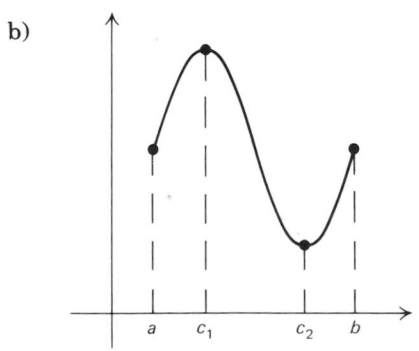

We have three critical points—c_1, c_2, and c_3. These, together with the endpoints, will be referred to as *key* points. That is, the key points are

$$a, \quad b, \quad c_1, \quad c_2, \quad c_3.$$

Note, in the foregoing graph, that between any two consecutive key points the function is either increasing or decreasing.

This graph and the experience with Margin Exercises 73 and 74 lead us to the following principle.

SHAPE PRINCIPLE. Suppose f is a continuous function over an interval $[a, b]$. Then between any two consecutive key points (a, b, plus critical points $c_1, c_2, c_3, \ldots c_n$) the function is increasing, or it is decreasing.

Finding Maximum and Minimum Values

Consider the function f whose graph over the interval $[a, b]$ is as follows. The function value $f(c_1)$ is called a *minimum* value of the function, and $f(b)$ is called a *maximum value* of the function.

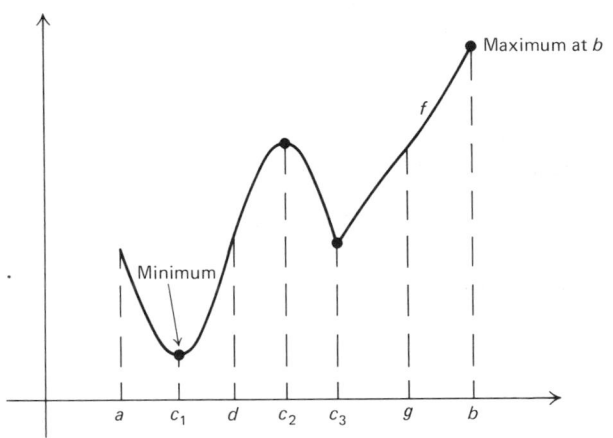

A function f on an interval $[a, b]$ has a *maximum* at x_0 if

$$f(x_0) \geq f(x) \quad \text{for all } x \text{ in } [a, b].$$

A function f on an interval $[a, b]$ has a *minimum* at x_0 if

$$f(x_0) \leq f(x) \quad \text{for all } x \text{ in } [a, b].$$

DO EXERCISE 75.

Look again at the preceding graph, but consider only the smaller interval $[d, g]$. Over that interval $f(c_2)$ is a maximum, and $f(c_3)$ is a minimum. In relation to the larger interval $[a, b]$, we sometimes call $f(c_2)$ a relative maximum and $f(c_3)$ a relative minimum, because there is a smaller interval, namely $[d, g]$, over which they are indeed maximum and minimum values. We shall restrict our attention to finding *absolute*, or overall, *maximum* and *minimum* values.

Two results follow from Margin Exercise 75. The first is:

A continuous function f defined on a closed interval $[a, b]$ must have a maximum and minimum value at points in $[a, b]$.

The second is a modification of the Shape Principle.

MAXIMUM–MINIMUM PRINCIPLE 1. Suppose f is a continuous function over an interval $[a, b]$, with critical points $c_1, c_2, \ldots c_n$. The key points are $a, b, c_1, c_2, \ldots c_n$. Consider function values of the key points:

$$f(a), f(b), f(c_1), f(c_2), \ldots f(c_n).$$

The largest of these is the *maximum* of f on the interval $[a, b]$.
The smallest of these is the *minimum* of f on the interval $[a, b]$.

This follows from the Shape Principle because between two key points the function is either increasing or decreasing. Thus whatever the maximum and minimum values are, they occur among function values of the key points.

Example 2 Find the maximum and minimum values of $f(x) = 3x^2 - x^3$ on the interval $[-\frac{1}{2}, 5]$.

Solution
a) First find $f'(x)$.

$$f'(x) = 6x - 3x^2$$

b) Determine the critical points. The derivative exists for all real numbers. Thus, the only candidates for critical points are those x's such that $f'(x) = 0$. Setting $f'(x)$ equal to 0 and solving, we get

$$f'(x) = 6x - 3x^2 = 0,$$
$$3x(2 - x) = 0,$$

$$3x = 0 \quad \text{or} \quad 2 - x = 0,$$
$$x = 0 \quad \text{or} \quad -x = -2,$$
$$x = 2.$$

76. Find the maximum and minimum values of

$$f(x) = x^3 - x^2 - x + 2$$

on the interval $[-1, 2]$.

The critical points are 0 and 2. The key points are $-\frac{1}{2}$, 5, 0, and 2.

c) We compute the *function* values at the key points.

$$f(-\tfrac{1}{2}) = 3(-\tfrac{1}{2})^2 - (-\tfrac{1}{2})^3 = 3 \cdot \tfrac{1}{4} + \tfrac{1}{8} \qquad\qquad = \tfrac{7}{8}$$

$$f(5) = 3 \cdot 5^2 - 5^3 = 3 \cdot 25 - 125 = 75 - 125 = -50 \qquad \text{Minimum}$$

$$f(0) = 3 \cdot 0^2 - 0^3 = 0 - 0 \qquad\qquad\qquad\qquad = 0$$

$$f(2) = 3 \cdot 2^2 - 2^3 = 3 \cdot 4 - 8 \qquad\qquad\qquad\quad = 4 \qquad \text{Maximum}$$

Thus

$$\text{Maximum} = 4 \text{ at } x = 2, \quad \text{and} \quad \text{Minimum} = -50 \text{ at } x = 5.$$

DO EXERCISE 76.

Example 3 Find the maximum and minimum values of $f(x) = 3x^2 - x^3$ on the interval $[7, 10]$.

Solution   As in Example 2 the derivative is 0 at 0 and 2. But neither 0 nor 2 is in the interval $[7, 10]$, so there are no critical points in this interval. Thus the maximum and minimum values occur at the endpoints.

$$f(7) = 3 \cdot 7^2 - 7^3 = 3 \cdot 49 - 343 \qquad\qquad = -196 \qquad \text{Maximum}$$

$$f(10) = 3 \cdot 10^2 - 10^3 = 3 \cdot 100 - 1000 = 300 - 1000 = -700 \qquad \text{Minimum}$$

Note that a maximum can be a negative number.

DO EXERCISE 77.

77. Find the maximum and minimum values of

$$f(x) = x^3 - x^2 - x + 2$$

on the interval $[5, 6]$.

When there is only *one* critical point c_0 in I, it can work out that we do not need to check the endpoint values. Consider these cases.

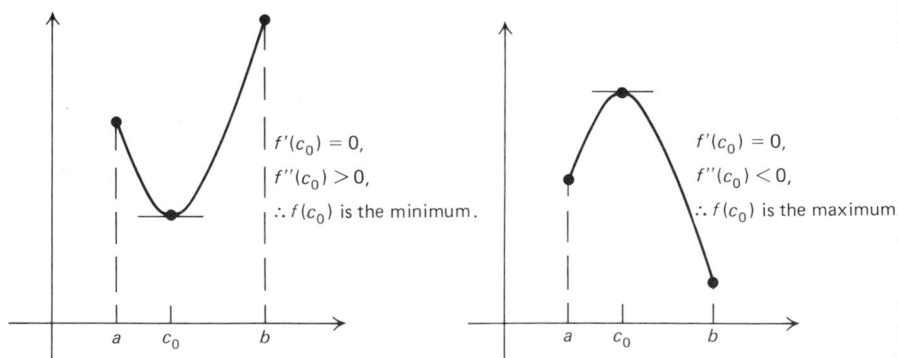

When $f'(c_0) = 0$ and $f''(c_0) > 0$, $f'(x)$ changes from negative to positive as x goes from the left of c_0 to the right. That is, the function f is decreasing to the left of c_0 and increasing to the right of c_0. It follows that $f(c_0)$ is the minimum value of f on I. Similarly, if $f'(c_0) = 0$ and $f''(c_0) < 0$, $f'(x)$ changes from positive to negative as x goes from the left of c_0 to the right. That is, the function f is increasing to the left of c_0 and decreasing to the right of c_0. It follows that $f(c_0)$ is the maximum value of f on I. The above turns out to hold no matter what the interval I, whether it is open, closed, or extends to infinity.

MAXIMUM–MINIMUM PRINCIPLE 2. **Suppose f is a function such that $f'(x)$ exists for every x in an interval I, and that there is *exactly one* (critical) point c_0, interior to I, for which $f'(c_0) = 0$. Then**

$$f(c_0) \text{ is the maximum value on } I \text{ if } f''(c_0) < 0$$

or

$$f(c_0) \text{ is the minimum value on } I \text{ if } f''(c_0) > 0.$$

If $f''(c_0) = 0$, we would have to use Maximum–Minimum Principle 1, or we would have to know more about the behavior of the function on the given interval.

Example 4 Find the maximum and minimum values of $f(x) = 4x - x^2$.

Solution When no interval is specified, we consider the entire domain of the function. In this case the domain is the set of all real numbers.

a) Find $f'(x)$.

$$f'(x) = 4 - 2x.$$

b) Determine the critical points. The derivative exists for all real numbers. Thus we merely solve $f'(x) = 0$.

$$4 - 2x = 0$$
$$-2x = -4$$
$$x = 2.$$

Since there is only one critical point, we can use the second derivative

$$f''(x) = -2.$$

Now the second derivative is constant, so $f''(2) = -2$, and since this is negative, we have the

$$\text{Maximum} = f(2) = 4 \cdot 2 - 2^2 = 8 - 4 = 4 \quad \text{at} \quad x = 2.$$

The function has no minimum, as the graph at the top of the next page indicates.

78. Find the maximum and minimum values of

$$f(x) = x^2 - 4x.$$

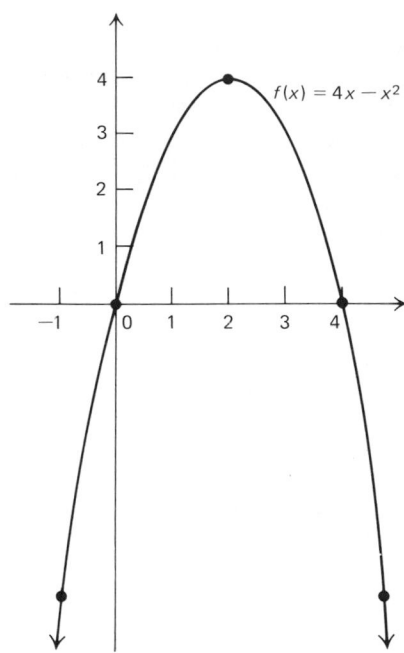

DO EXERCISE 78.

Example 5 Find the maximum and minimum values of $f(x) = 4x - x^2$ on the interval $[0, 4]$.

Solution By the reasoning in Example 4 we know that the maximum value is $f(2)$, or 4. We know this here also, without checking the endpoints. This time we have to check for the minimum:

$$f(0) = 4 \cdot 0 - 0^2 = 0 \quad \text{and} \quad f(4) = 4 \cdot 4 - 4^2 = 0.$$

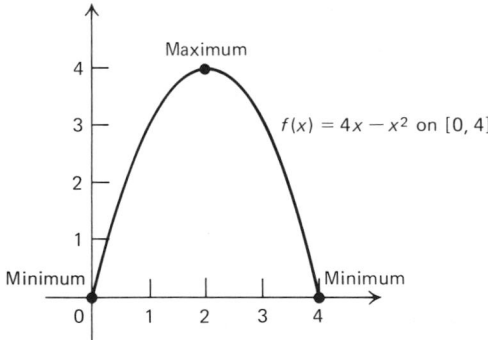

Thus the minimum is 0. It occurs twice at $x = 0$ and $x = 4$. Thus, the

$$\text{Maximum} = 4 \quad \text{at} \quad x = 2,$$

and the

$$\text{Minimum} = 0 \quad \text{at} \quad x = 0 \quad \text{and} \quad x = 4.$$

DO EXERCISE 79.

79. Find the maximum and minimum values of

$$f(x) = x^2 - 4x$$

on the interval $[0, 4]$.

Example 6 Find the maximum and minimum values of $f(x) = x^3$.

Solution

a) Find $f'(x)$.
$$f'(x) = 3x^2.$$

b) Find the critical points.
$$3x^2 = 0$$
$$x^2 = 0$$
$$x = 0.$$

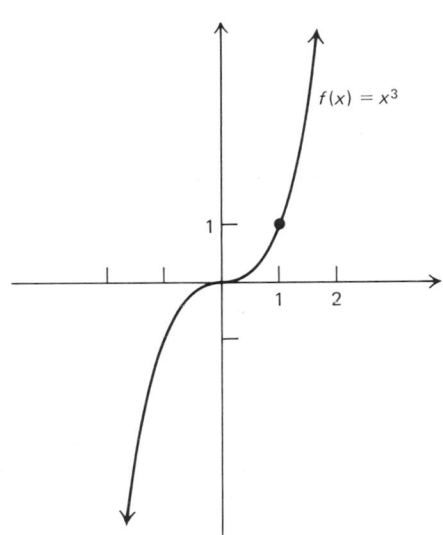

$f(x) = x^3$

Since there is only one critical point we can use the second derivative
$$f''(x) = 6x.$$

Now $f''(0) = 6 \cdot 0 = 0$, so Maximum–Minimum Principle 2 fails.

We cannot use Maximum–Minimum Principle 1 because there are no endpoints. But note that $f'(x) = 3x^2$ is never negative. Thus it is increasing everywhere but at $x = 0$, so there is no maximum or minimum.

DO EXERCISE 80.

80. a) Find the maximum and minimum values of
$$f(x) = x^3$$
on the interval $[-2, 2]$. [*Hint:* This function must have maximum and minimum values because it is restricted to a closed interval.] What are the only numbers at which these can occur?

b) Find the maximum and minimum values of
$$f(x) = x^3 - 4.$$

81. Find the maximum and minimum values of

$$f(x) = 10x + \frac{1}{x}$$

on $(0, \infty)$.

Example 7 Find the maximum and minimum values of $f(x) = 5x + (35/x)$ on the interval $(0, \infty)$.

Solution

a) Find $f'(x)$. We first express $f(x)$ as
$$f(x) = 5x + 35x^{-1}.$$
Then
$$f'(x) = 5 - 35x^{-2} = 5 - \frac{35}{x^2}.$$

b) Now $f'(x)$ exists for all values of x in $(0, \infty)$. Thus the only critical points are those for which $f'(x) = 0$.

$$5 - \frac{35}{x^2} = 0$$

$$5 = \frac{35}{x^2}$$

$$5x^2 = 35 \quad \text{(Multiplying by } x^2 \text{, since } x \neq 0\text{)}$$

$$x^2 = 7$$

$$x = \pm\sqrt{7}.$$

The only critical point in $(0, \infty)$ is $\sqrt{7}$. Thus we can use the second derivative

$$f''(x) = 70x^{-3} = \frac{70}{x^3}$$

to determine whether we have a maximum or minimum. Now $f''(x)$ is positive for all values of x in $(0, \infty)$, so $f''(\sqrt{7}) > 0$, and the

$$\text{Minimum} = f(\sqrt{7}) = 5 \cdot \sqrt{7} + \frac{35}{\sqrt{7}} \quad \text{at} \quad x = \sqrt{7}.$$

The function has no maximum value.

In general,

Suppose a function has only one critical point c in an interval that does not have endpoints, or does not contain its endpoints, such as $(-\infty, \infty)$, $(0, \infty)$, or (a, b). Then, if the function has a maximum, it will have no minimum; and if it has a minimum, it will have no maximum.

See Example 4 and Example 7.

DO EXERCISE 81.

EXERCISE SET 9.9

The curves on the graph at the right show the gasoline mileage obtained when traveling at a constant speed, for an average-size car and a compact car.

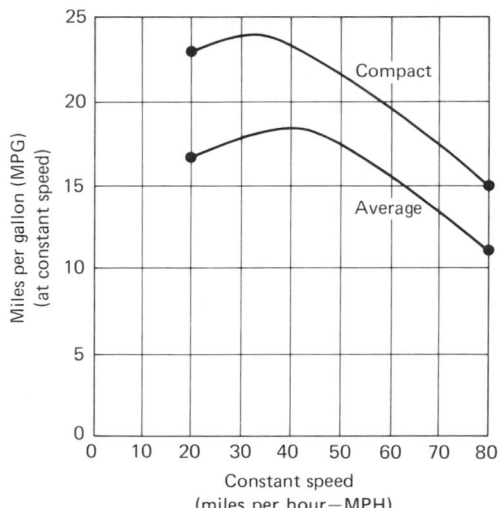

1. Consider the graph for the average-size car over the interval $[20, 80]$.

a) Estimate the speed at which the maximum gasoline mileage is obtained.

b) Estimate the speed at which the minimum gasoline mileage is obtained.

c) What is the mileage obtained at 70 mph?

d) What is the mileage obtained at 55 mph?

e) What percent increase in mileage is there by traveling at 55 mph rather than at 70 mph?

2. Answer the same questions as in Exercise 1 for the compact car.

For the following functions, find the maximum and minimum values, if they exist, over the indicated interval. When no interval is specified, use the real line $(-\infty, \infty)$.

3. $f(x) = 5 + x - x^2$; $[0, 2]$ **4.** $f(x) = 4 + x - x^2$; $[0, 2]$

5. $f(x) = x^3 - x^2 - x + 2$; $[0, 2]$ **6.** $f(x) = x^3 + \frac{1}{2}x^2 - 2x + 5$; $[0, 1]$

7. $f(x) = x^3 - x^2 - x + 2$; $[-1, 0]$ **8.** $f(x) = x^3 + \frac{1}{2}x^2 - 2x + 5$; $[-2, 0]$

9. $f(x) = 3x - 2$; $[-1, 1]$ **10.** $f(x) = 2x + 4$; $[-1, 1]$ **11.** $f(x) = 3x - 2$

12. $f(x) = 2x + 4$ **13.** $f(x) = x(70 - x)$ **14.** $f(x) = x(50 - x)$

15. $f(x) = 2x^2 - 40x + 400$ **16.** $f(x) = 2x^2 - 20x + 100$ **17.** $f(x) = x - \frac{4}{3}x^3$; $(0, \infty)$

18. $f(x) = 16x - \frac{4}{3}x^3$; $(0, \infty)$ **19.** $f(x) = 17x - x^2$ **20.** $f(x) = 27x - x^2$

21. $f(x) = \frac{1}{3}x^3 - 3x$; $[-2, 2]$ **22.** $f(x) = \frac{1}{3}x^3 - 5x$; $[-3, 3]$

23. $f(x) = -0.001x^2 + 4.8x - 60$ **24.** $f(x) = -0.01x^2 + 1.4x - 30$

25. $f(x) = -\frac{1}{3}x^3 + 6x^2 - 11x - 50$; $(0, 3)$ **26.** $f(x) = -x^3 + x^2 + 5x - 1$; $(0, \infty)$

27. $f(x) = 15x^2 - \frac{1}{2}x^3$; $[0, 30]$ **28.** $f(x) = 4x^2 - \frac{1}{2}x^3$; $[0, 8]$ **29.** $f(x) = 2x + \dfrac{72}{x}$; $(0, \infty)$

30. $f(x) = x + \dfrac{3600}{x}$; $(0, \infty)$ **31.** $f(x) = x^2 + \dfrac{432}{x}$; $(0, \infty)$ **32.** $f(x) = x^2 + \dfrac{250}{x}$; $(0, \infty)$

33. $f(x) = 2x^4 - x$; $[-1, 1]$ **34.** $f(x) = 2x^4 + x$; $[-1, 1]$ **35.** $f(x) = \sqrt[3]{x}$; $[0, 8]$

36. $f(x) = \sqrt{x}$; $[0, 4]$ **37.** $f(x) = (x + 1)^3$ **38.** $f(x) = (x - 1)^3$

39. See Exercise 10 in Exercise Set 9.6. What is the maximum number of units sold? What must be spent on advertising to sell that number of units?

40. See Exercise 9 in Exercise Set 9.6. What is the maximum temperature during the illness and on what day does it occur?

41. At travel speed (constant velocity) x there are y accidents in daytime for every 100 million miles of travel, where y is given by

$$y = x^2 - 122.5x + 3775.$$

At what travel speed do the fewest accidents occur?

42. At travel speed (constant velocity) x, the cost y, in cents per mile, of operating a car is given by

$$y = 0.02x^2 - 1.3x + 30.$$

At what travel speed is the cost of operating a car a minimum?

▶ ───

Find the maximum and minimum values, if they exist, over the indicated interval. When no interval is specified, use the real line $(-\infty, \infty)$.

43. $g(x) = x\sqrt{x + 3};\ [-3, 3]$

44. $h(x) = x\sqrt{1 - x};\ [0, 1]$

45. $f(x) = x^{2/3};\ [-1, 1]$

46. $g(x) = x^{2/3}$

47. $f(x) = \frac{1}{3}x^3 - x + \frac{2}{3}$

48. $f(x) = \frac{1}{3}x^3 - \frac{1}{2}x^2 - 2x + 1$

49. $f(x) = \frac{1}{3}x^3 - 2x^2 + x;\ [0, 4]$

50. $g(x) = \frac{1}{3}x^3 + 2x^2 + x;\ [-4, 0]$

51. $t(x) = x^4 - 2x^2$

52. $f(x) = 2x^4 - 4x^2 + 2$

53. *Business.* Several costs in a business environment can be separated into two components: those that increase with volume and those that decrease with volume. Quality of customer service, although more expensive as it is increased, has part of its increased cost offset by customer goodwill. A firm has determined that its cost of service is the following function of "quality units,"

$$C(x) = (2x + 4) + \left(\frac{2}{x - 6}\right), \qquad x > 6.$$

Find the number of "quality units" the firm should use to minimize its total cost of service.

───

OBJECTIVE

You should be able to solve maximum–minimum problems.

9.10 MAXIMUM–MINIMUM PROBLEMS

One very important application of the differential calculus is the solving of maximum–minimum problems, that is, finding the maximum or minimum value of some varying quantity Q and the point where that maximum or minimum occurs.

Example 1 A hobby store has 20 ft of fencing to fence off a rectangular electric train area in one corner of its display room. What dimensions of the rectangle will maximize the area? What is the maximum area?

Exploratory Solution   Intuitively, one might think that it does not matter what dimensions one uses; they will all yield the same area. To show that this is not true, as well as to conjecture a possible solution, consider the exploratory exercises in Margin Exercise 82. But, before doing those exercises let us express the area in terms of one variable. If we let x = the length of one side, and y = the length of the other, then since the sum of the lengths must be 20 ft,

$$x + y = 20, \quad \text{and} \quad y = 20 - x.$$

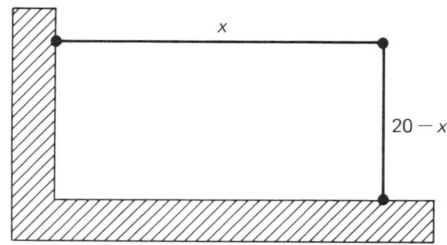

Then the area is given by

$$A = xy$$
$$A = x(20 - x) - 20x - x^2.$$

DO EXERCISE 82.

Calculus Solution   We are trying to find the maximum value of

$$A = 20x - x^2 \text{ on the interval } (0, 20).$$

We consider the interval $(0, 20)$ because x is the length of one side and cannot be negative. Since there is only 20 ft of fencing, x cannot be greater than 20. Also, x cannot be 20 because the length of y would be 0.

a) We first find $A'(x)$, where $A(x) = 20x - x^2$.

$$A'(x) = 20 - 2x.$$

b) This derivative exists for all values of x in $(0, 20)$. Thus the only critical points are where

$$A'(x) = 20 - 2x = 0,$$
$$-2x = -20,$$
$$x = 10.$$

Since there is only one critical point in the interval, we can use the

82. *Exploratory exercises.*
a) Complete this table.

x	y $20 - x$	A $x(20 - x)$
0		
4		
6.5		
8		
10		
12		
13.2		
20		

b) Make a graph of x versus A, that is, of points (x, A) from the table; and connect them with a smooth curve.

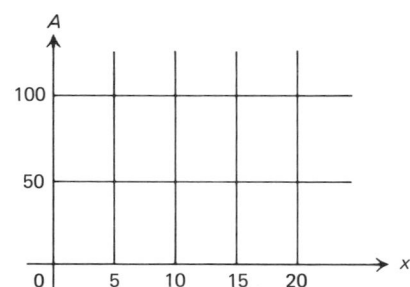

c) Does it matter what dimensions we use?

d) Make a conjecture about what the maximum might be and where it would occur.

83. A rancher has 50 ft of fencing to fence off a rectangular animal pen in the corner of a barn. What dimensions of the rectangle will yield the maximum area? What is the maximum area?

second derivative to determine whether we have a maximum. Note that

$$A''(x) = -2,$$

which is a constant. Thus $A''(10)$ is negative, so $A(10)$ is a maximum. Now

$$A(10) = 10(20 - 10) = 10 \cdot 10 = 100.$$

Thus the maximum area of 100 sq. ft is obtained using 10 ft for the length of one side, and $20 - 10$, or 10 ft, for the other. Note that while you may have conjectured this in Margin Exercise 82, the tools of calculus allowed us to prove it.

DO EXERCISE 83.

Example 2 A stereo manufacturer determines that in order to sell x units of a new stereo its price per unit must be

$$p = D(x) = 1000 - x.$$

It also determines that the total cost of producing x units is given by

$$C(x) = 3000 + 20x.$$

a) Find the total revenue $R(x)$.

b) Find the total profit $P(x)$.

c) How many units must the company produce and sell to maximize profit?

d) What is the maximum profit?

e) What price per unit must be charged to make this maximum profit?

Solution

a) $R(x) = $ Total revenue $= $ (number of units) $\cdot$ (price per unit)

$$= \quad x \qquad\qquad p$$

$$= x(1000 - x) = 1000x - x^2.$$

b) $P(x) = R(x) - C(x) = (1000x - x^2) - (3000 + 20x)$

$$= -x^2 + 980x - 3000.$$

c) To find the maximum value of $P(x)$ we first find $P'(x)$.

$$P'(x) = -2x + 980.$$

This is defined for all real numbers (actually we are interested in numbers x in $[0, \infty)$ only, since we cannot produce a negative number of stereos). Thus we solve

$$P'(x) = -2x + 980 = 0$$

$$-2x = -980$$

$$x = 490.$$

Since there is only one critical point, we can try to use the second derivative to determine whether we have a maximum. Note that

$$P''(x) = -2, \text{ a constant.}$$

Thus $P''(490)$ is negative, so $P(490)$ is a maximum.

d) The maximum profit is given by

$$P(490) = -(490)^2 + 980 \cdot 490 - 3000 = \$237{,}100.$$

Thus the stereo manufacturer makes a maximum profit of $237,100 by producing and selling 490 stereos.

e) The price per unit to make the maximum profit is $p = 1000 - 490 = \$510$.

DO EXERCISE 84.

Marginal Analysis

Let us take a general look at the total profit function and its related functions.

In the first graph we have the total cost and total revenue functions. We can estimate what the maximum profit might be by looking for the widest gap between $R(x)$ and $C(x)$. Points B_0 and B_2 are "break-even" points.

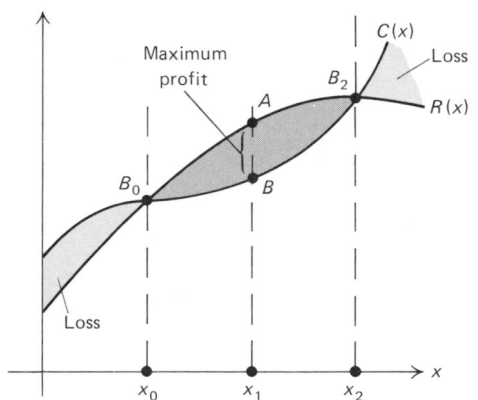

84. A company determines that in order to sell x units of a certain product its price per unit must be

$$p = D(x) = 200 - x.$$

It also determines that its total cost of producing x units is given by

$$C(x) = 5000 + 8x.$$

a) Find the total revenue $R(x)$.

b) Find the total profit $P(x)$.

c) How many units must the company produce and sell in order to maximize profit?

d) What is the maximum profit?

e) What price per unit must be charged to make this maximum profit?

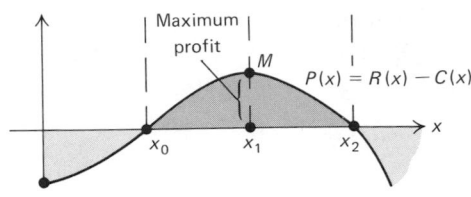

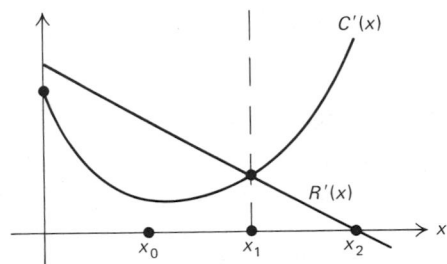

In the second graph we have the total profit function. Note that when production is too low ($<x_0$) there is a loss because of high fixed or initial costs and low revenue. When production is too high ($>x_2$), there is also a loss due to high marginal costs and low marginal revenues, as seen in the third graph.

The business operates at a profit everywhere between x_0 and x_2. Note that maximum profit occurs at a critical point x_1 of $P(x)$. If we assume that $P'(x)$ exists for all x in some interval, usually $[0, \infty)$, this critical point occurs at some number x such that

$$P'(x) = 0.$$

Since $P(x) = R(x) - C(x)$, it follows that

$$P'(x) = R'(x) - C'(x).$$

Thus the maximum profit occurs at some number x such that

$$R'(x) - C'(x) = 0,$$

or

$$R'(x) = C'(x).$$

In summary,

Maximum profit is achieved when marginal revenue equals marginal cost:

$$R'(x) = C'(x).$$

Here is a general strategy for solving maximum–minimum problems. While it may not guarantee success, it should certainly enhance one's chances.

1. **Read the problem carefully. If relevant, draw a picture.**
2. **Label the picture with appropriate variables and constants, noting what varies and what stays fixed.**
3. **Translate the problem to an equation, involving a quantity Q to be maximized or minimized.**
4. **Try to express Q as a function of *one* variable. Use the procedures developed in Section 3.2 to determine the maximum or minimum values and the points where they occur.**

Example 3 From a thin piece of cardboard 8 in. by 8 in., square corners are cut out so that the sides can be folded up to make a box. What dimensions will yield a box of maximum volume? What is the maximum volume?

Exploratory Solution One might again think that it does not matter what the dimensions are, but our experience with Example 1 should lead us to think otherwise. We make a drawing as shown below.

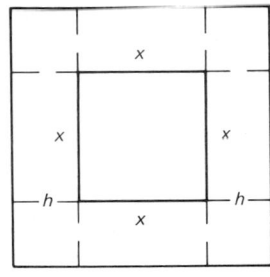   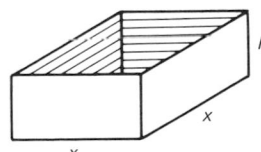

When squares of length h on a side are cut out of the corners we are left with a square base of length x. The volume of the resulting box is

$$V = lwh = x \cdot x \cdot h.$$

We want to express V in terms of one variable. Note that the overall length of a side of the cardboard is 8 in. We see from the drawing that

$$h + x + h = 8,$$

or

$$x + 2h = 8.$$

85. *Exploratory exercises.*

a) Complete this table.

x	h $\frac{1}{2}(8-x)$	V $x \cdot x \cdot \frac{1}{2}(8-x)$
0		
1		
2		
3		
4		
4.6		
5		
6		
6.8		
7		
8		

b) Make a graph of x versus V.

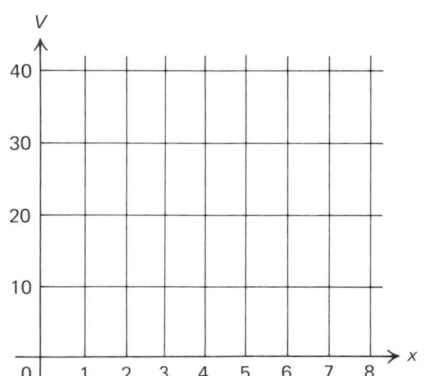

c) Make a conjecture about what the maximum might be and where it would occur.

Solving for h we get

$$2h = 8 - x$$
$$h = \tfrac{1}{2}(8 - x) = \tfrac{1}{2} \cdot 8 - \tfrac{1}{2}x = 4 - \tfrac{1}{2}x.$$

Thus

$$V = x \cdot x \cdot (4 - \tfrac{1}{2}x) = x^2(4 - \tfrac{1}{2}x) = 4x^2 - \tfrac{1}{2}x^3.$$

In Margin Exercise 85 you will compute some values of V.

DO EXERCISE 85.

Calculus Solution You probably noted in Margin Exercise 85 that it was a bit more difficult than in Example 1 to conjecture where the maximum occurs. At the least it seems reasonable that it occurs for some x between 5 and 6. Let us find out for certain, using calculus. We are trying to find the maximum value of

$$V(x) = 4x^2 - \tfrac{1}{2}x^3 \quad \text{on the interval } (0, 8).$$

We first find $V'(x)$.

$$V'(x) = 8x - \tfrac{3}{2}x^2.$$

Now $V'(x)$ exists for all x in the interval $(0, 8)$ so we set it equal to 0 to find the critical values.

$$V'(x) = 8x - \tfrac{3}{2}x^2 = 0,$$
$$x(8 - \tfrac{3}{2}x) = 0,$$
$$x = 0 \quad \text{or} \quad 8 - \tfrac{3}{2}x = 0,$$
$$x = 0 \quad \text{or} \quad -\tfrac{3}{2}x = -8,$$
$$x = 0 \quad \text{or} \quad x = -\tfrac{2}{3}(-8) = \tfrac{16}{3}.$$

The only critical point in $(0, 8)$ is $\tfrac{16}{3}$. Thus we can use second derivative

$$V''(x) = 8 - 3x$$

to determine whether we have a maximum. Since

$$V''(\tfrac{16}{3}) = 8 - 3 \cdot \tfrac{16}{3} = -8,$$

$V''(\tfrac{16}{3})$ is negative, so $V(\tfrac{16}{3})$ is a maximum, and

$$V(\tfrac{16}{3}) = 4 \cdot (\tfrac{16}{3})^2 - \tfrac{1}{2}(\tfrac{16}{3})^3 = \tfrac{1024}{27} = 37\tfrac{25}{27}.$$

The maximum volume is $37\tfrac{25}{27}$ cu. in. The dimensions that yield this maximum volume are

$$x = \tfrac{16}{3} = 5\tfrac{1}{3}\text{ in.,} \quad \text{by } x = 5\tfrac{1}{3}\text{ in.,} \quad \text{by } h = 4 - \tfrac{1}{2}(\tfrac{16}{3}) = 1\tfrac{1}{3}\text{ in.}$$

It would surely have been difficult to guess this from Margin Exercise 85.

DO EXERCISE 86.

In the following problem, an open-top container of fixed volume is to be constructed. We want to determine the dimensions that will allow it to be built with the least amount of material. Such a problem could be important from an ecological standpoint.

Example 4 A container firm is designing an open-top rectangular box, with a square base, which will hold 108 cubic centimeters (cc). What dimensions yield the minimum surface area? What is the minimum surface area?

Solution The surface area of the box is

$$S = x^2 + 4xy.$$

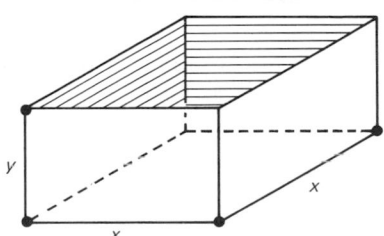

The volume must be 108 cc, and is given by

$$V = x^2y = 108.$$

To express S in terms of one variable, we solve $x^2y = 108$ for y:

$$y = \frac{108}{x^2}.$$

Then

$$S = x^2 + 4x\left(\frac{108}{x^2}\right) = x^2 + \frac{432}{x}.$$

Now S is defined only for positive numbers, and the problem dictates that the length x be positive, so we are minimizing S on the interval $(0, \infty)$. We first find dS/dx.

$$\frac{dS}{dx} = 2x - \frac{432}{x^2}.$$

86. Repeat Example 3, but for a piece of cardboard that is 10 in. by 10 in.

87. Repeat Example 4, but for a fixed volume of 500 cc.

Since dS/dx exists for all x in $(0, \infty)$, the only critical points are where $dS/dx = 0$. Thus, we solve the following equation:

$$2x - \frac{432}{x^2} = 0$$

$$x^2\left(2x - \frac{432}{x^2}\right) = x^2 \cdot 0 \quad \text{(We multiply by } x^2 \text{ to clear of fractions.)}$$

$$2x^3 - 432 = 0$$

$$2x^3 = 432$$

$$x^3 = 216$$

$$x = 6.$$

This is the only critical point, so we can use the second derivative to determine whether we have a minimum.

$$\frac{d^2S}{dx^2} = 2 + \frac{864}{x^3}.$$

Note that this is positive for all positive values of x. Thus we have a minimum at $x = 6$. When $x = 6$, it follows that $y = 3$:

$$y = \frac{108}{6^2} = \frac{108}{36} = 3.$$

Thus the surface area is minimized when $x = 6$ cm (centimeters) and $y = 3$ cm. The minimum surface area is

$$S = 6^2 + 4 \cdot 6 \cdot 3 = 108 \text{ cm}^2.$$

This, by coincidence, is the same number as the fixed volume.

DO EXERCISE 87.

Example 5 *Determining a ticket price.* Fight promoters ride a thin line between profit and loss, especially in determining the price to charge for admission to closed-circuit television showings in local theaters. By keeping records, a theater determines that, if the admission price is $20, it averages 1000 people in attendance. But, for every increase of $1, it loses 100 customers from the average. Every customer spends an average of $0.80 on concessions. What admission price should the theater charge to maximize total revenue?

Solution   Let x = the amount by which the price of $20 should be increased (if x is negative the price would be decreased). We first express total revenue R as a function of x. Note that

$R(x)$ = (Revenue from tickets) + (Revenue from concessions)

$\quad$ = (Number of people) · (Ticket price) + \$0.80(Number of people)

$\quad$ = $(1000 - 100x)(20 + x) + 0.80(1000 - 100x)$

$\quad$ = $20{,}000 - 2000x + 1000x - 100x^2 + 800 - 80x$

$R(x)$ = $-100x^2 - 1080x + 20{,}800.$

We are trying to find the maximum value of R over the set of all real numbers. To find x such that $R(x)$ is a maximum we first find $R'(x)$.

$$R'(x) = -200x - 1080.$$

This derivative exists for all real numbers x; thus the only critical points are where $R'(x) = 0$, so we solve that equation.

$$-200x - 1080 = 0$$
$$-200x = 1080$$
$$x = -5.4 = -\$5.40$$

Since this is the only critical point, we can use the second derivative,

$$R''(x) = -200,$$

to determine whether we have a maximum. Since $R''(-5.4)$ is negative, $R(-5.4)$ is a maximum. Thus to maximize revenue the theater should charge

$$\$20 + (-\$5.40) \quad \text{or} \quad \$14.60 \text{ per ticket.}$$

That is, this reduced ticket price will get more people into the theater $(1000 - 100(-5.4)$, or 1540), and will result in maximum revenue.

DO EXERCISE 88.

88. Transit companies also ride a thin line between profit and loss. A company determines that, at a fare of 30¢, it will average 10,000 fares a day. For every increase of 10¢, it loses 2000 customers. What fare should be charged to maximize revenue?

[*Hint:* Let x = the number of 10¢ fare increases (if x is negative the fare would be decreased). Then the new fare would be $30 + 10x$.]

EXERCISE SET 9.10

1. Of all the numbers whose sum is 50, find the two that have the maximum product. That is, maximize $Q = xy$, where $x + y = 50$.

2. Of all the numbers whose sum is 70, find the two that have the maximum product. That is, maximize $Q = xy$, where $x + y = 70$.

3. In Exercise 1, can there be a minimum product? Explain.

4. In Exercise 2, can there be a minimum product? Explain.

5. Of all numbers whose difference is 4, find the two that have the minimum product.

6. Of all numbers whose difference is 6, find the two that have the minimum product.

7. Maximize $Q = xy^2$, where x and y are positive numbers, such that $x + y^2 = 1$.

8. Maximize $Q = xy^2$, where x and y are positive numbers such that $x + y^2 = 4$.

9. Minimize $Q = x^2 + y^2$, where $x + y = 20$.

10. Minimize $Q = x^2 + y^2$, where $x + y = 10$.

11. Maximize $Q = xy$, where x and y are positive numbers such that $(4/3)x^2 + y = 16$.

12. Maximize $Q = xy$, where x and y are positive numbers such that $x + (4/3)y^2 = 1$.

13. A rancher wants to build a rectangular fence next to a river, using 120 yd of fencing. What dimensions of the rectangle will maximize the area? What is the maximum area? Note that the rancher does not have to fence in the side next to the river.

14. A rancher wants to enclose two rectangular areas near a river, one for sheep and one for cattle. There are 240 yd of fencing available. What is the largest total area that can be enclosed?

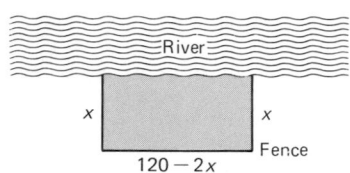

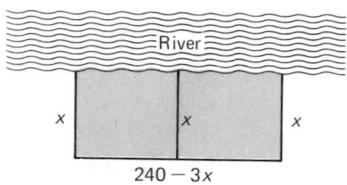

15. A carpenter is building a room with a fixed perimeter of 54 ft. What are the dimensions of the largest room that can be built? What is its area?

16. Of all rectangles that have a perimeter of 34 ft, find the dimensions of the one with the largest area. What is its area?

Business—Maximizing profit. Find the maximum profit and the number of units that must be produced and sold to yield the maximum profit.

17. $R(x) = 50x - 0.5x^2$, $C(x) = 4x + 10$

18. $R(x) = 50x - 0.5x^2$, $C(x) = 10x + 3$

19. $R(x) = 2x$, $C(x) = 0.01x^2 + 0.6x + 30$

20. $R(x) = 5x$, $C(x) = 0.001x^2 + 1.2x + 60$

21. $R(x) = 9x - 2x^2$, $C(x) = x^3 - 3x^2 + 4x + 1$; $R(x)$ and $C(x)$ are in thousands of dollars, and x is in thousands of units.

22. $R(x) = 100x - x^2$, $C(x) = \frac{1}{3}x^3 - 6x^2 + 89x + 100$; $R(x)$ and $C(x)$ are in thousands of dollars, x is in thousands of units.

23. Raggs, Ltd., a clothing firm, determines that to sell x suits its price per suit must be

$$p = D(x) = 150 - 0.5x.$$

It also determines that its total cost of producing x suits is given by

$$C(x) = 4000 + 0.25x^2.$$

a) Find the total revenue $R(x)$.
b) Find the total profit $P(x)$.
c) How many suits must the company produce and sell to maximize profit?
d) What is the maximum profit?
e) What price per suit must be charged to make this maximum profit?

24. An appliance firm is marketing a new refrigerator. It determines that to sell x refrigerators its price per refrigerator must be

$$p = D(x) = 280 - 0.4x.$$

It also determines that its total cost of producing x refrigerators is given by

$$C(x) = 5000 + 0.6x^2.$$

a) Find the total revenue $R(x)$.
b) Find the total profit $P(x)$.
c) How many refrigerators must the company produce and sell to maximize profit?
d) What is the maximum profit?
e) What price per refrigerator must be charged to make this maximum profit?

25. From a thin piece of cardboard 30 in. by 30 in., square corners are cut out so the sides can be folded up to make a box. What dimensions will yield a box of maximum volume? What is the maximum volume?

26. From a thin piece of cardboard 20 in. by 20 in. square corners are cut out so the sides can be folded up to make a box. What dimensions will yield a box of maximum volume? What is the maximum volume?

27. A container company is designing an open-top, square-based, rectangular box that will have a volume of 62.5 cubic inches. What dimensions yield the minimum surface area? What is the minimum surface area?

28. A soup company is constructing an open-top, rectangular, metal tank with a square base, that will have a volume of 32 cubic feet. What dimensions yield the minimum surface area? What is the minimum surface area?

29. A university is trying to determine what price to charge for football tickets. At a price of $6 per ticket it averages 70,000 per game. For every increase of $1 it loses 10,000 people from the average. Every person at the game spends an average of $1.50 on concessions. What price per ticket should be charged to maximize revenue? How many people will attend at that price?

30. Suppose you are the owner of a 30-unit motel. All units are occupied when you charge $20 a day per unit. For every increase of x dollars in the daily rate, there are x units vacant. Each occupied room costs $2 per day to service and maintain. What should you charge per unit to maximize profit?

31. An apple farm yields an average of 30 bushels of apples per tree when 20 trees are planted on an acre of ground. Each time 1 more tree is planted per acre, the yield decreases 1 bu. per tree due to the extra congestion. How many trees should be planted to get the highest yield?

32. When a theater owner charges $3 for admission there is an average attendance of 100 people. For every $0.10 increase in admission, there is a loss of 1 customer from the average. What admission should be charged to maximize revenue?

33. The postal service places a limit of 84 inches on the combined length and girth (distance around) of a package to be sent parcel post. What dimensions of a rectangular box with square cross section will contain the largest volume that can be mailed? [*Hint:* There are two different girths.]

34. A rectangular play area is to be laid out in a person's back lot, and is to contain 48 square yards. The neighbor agrees to pay half the cost of the side of the play area that lines the lot. What dimensions will minimize the cost of the fence?

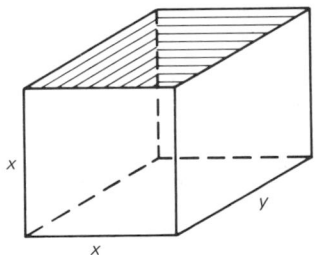

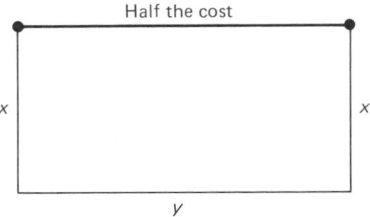

35. For what positive number is the sum of its reciprocal and five times its square a minimum?

36. For what positive number is the sum of its reciprocal and four times its square a minimum?

37. A rectangular box with a volume of 320 cubic feet is to be constructed with a square base and top. The cost per square foot for the bottom is 15¢, for the top is 10¢, and for the sides is 2.5¢. What dimensions will minimize the cost?

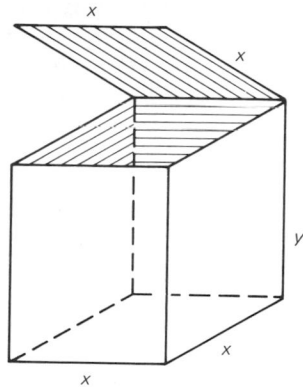

38. A merchant who was purchasing a display sign from a salesclerk said, "I want a sign 10 ft by 10 ft." The salesman responded, "That's just what we'll give you; only to make it more aesthetic, why don't we change it to 7 ft by 13 ft?" Comment.

39. A Norman window is a rectangle with a semicircle on top. Suppose the perimeter of a particular Norman window is to be 24 ft. What should be its dimensions so the maximum amount of light will be allowed to enter through the window?

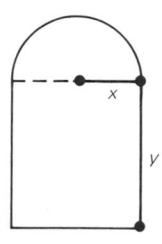

40. Solve Exercise 39, but this time the semicircle is to be stained glass, which transmits only half as much light as the semicircle in Exercise 39.

41. The amount of money deposited in a financial institution in savings accounts is directly proportional to the interest rate the financial institution pays on the money. Suppose a financial institution can loan *all* the money it takes in on its savings accounts at an interest rate of 18%. What interest rate should it pay on its savings accounts to maximize profits?

42. ▦ A page in this book is 73.125 square inches. On the average there is a 0.75-in. margin at the top and at the bottom of each page, and a 0.5-in. margin on each of the sides. What should the outside dimensions of each page be so the printed area is a maximum? Measure the outside dimensions to see whether the actual dimensions maximize the printed area.

A Norman window. (*Owen Franken: Stock, Boston*)

43. A 24-inch piece of string is cut in two pieces. One piece is used to form a circle and the other to form a square. How should the string be cut so the sum of the areas is a minimum? maximum?

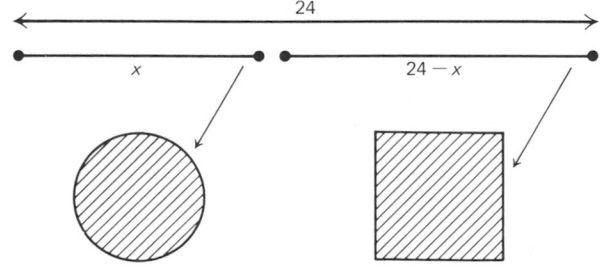

44. A power line is to be constructed from a power station at point A (see the figure) to an island at point C, which is directly 1 mile out in the water from a point B on the shore. Point B is 4 miles down-shore from the power station at A. It costs \$5000 per mile to lay the power line under water and \$3000 per mile to lay the line under ground. At what point S down-shore from A should the line come to the shore to minimize cost? Note that S could very well be B or A. [*Hint:* The length of $\overline{CS}$ is $\sqrt{1 + x^2}$.]

45. ▦ *Biology—Flights of homing pigeons.* It is known that homing pigeons tend to avoid flying over water in the daytime, perhaps because the downdrafts of air over water make flying difficult. Suppose a homing pigeon is released on an island at point C, which is directly 3 miles out in the water from a point B on shore. Point B is 8 miles down-shore from the pigeon's home loft at point A. Assume a pigeon requires 1.28 times the rate of energy over land, to fly over water. Toward what point S down-shore from A should the pigeon fly to minimize the total energy required to get to home loft A? Assume (Total energy) = (Energy rate over water) · (Distance over water) + (Energy rate over land) · (Distance over land).

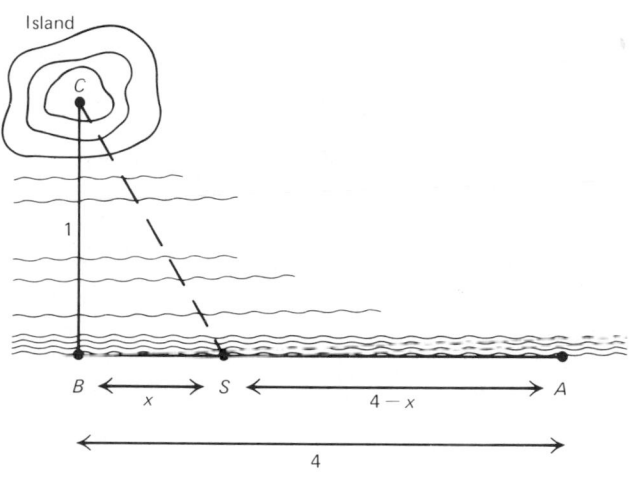

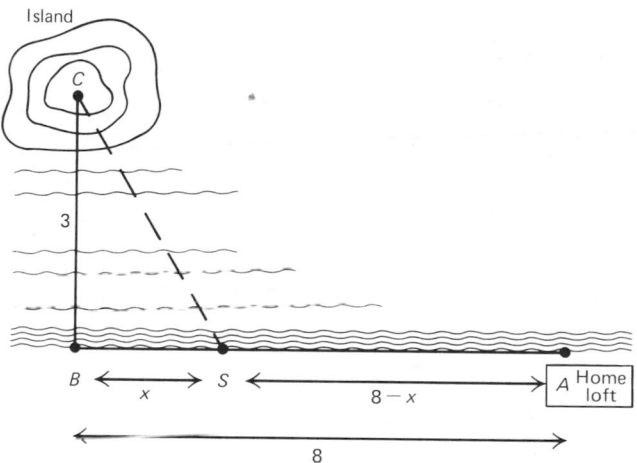

46. A road is to be built between two cities, C_1 and C_2, on opposite sides of a river of uniform width r. Because of the river, a bridge must be built. C_1 is a units from the river, and C_2 is b units from the river; $a \leqslant b$. Where should the bridge be located to minimize the total distance between the cities? Give a general solution using the constants a, b, p, and r in the drawing to the right.

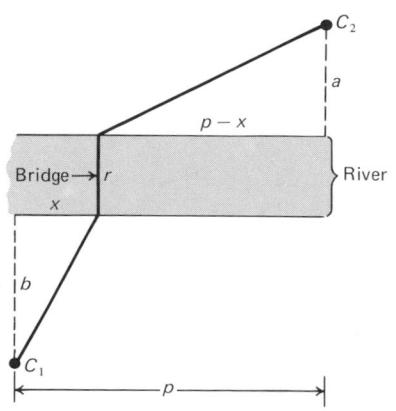

47. The total cost function for producing x units of a certain product is given by

$$C(x) = 8x + 20 + \frac{x^3}{100}.$$

a) Find the marginal cost $C'(x)$.
b) Find the average cost $A(x) = C(x)/x$.
c) Find the *marginal average cost* $A'(x)$.
d) Find the minimum of $A(x)$ and the value x_0 at which it occurs. Find the marginal cost at x_0.
e) Compare $A(x_0)$ and $C'(x_0)$.

49. Minimize $Q = x^3 + 2y^3$, where x and y are positive numbers such that $x + y = 1$.

48. Consider $A(x) = C(x)/x$.
a) Find $A'(x)$ in terms of $C'(x)$ and $C(x)$.
b) Show that $A(x)$ has a minimum at that value of x_0 such that

$$C'(x_0) = A(x_0) = \frac{C(x_0)}{x_0}.$$

This shows that when marginal cost and average cost are the same, a product is being produced at the least average cost.

50. Minimize $Q = 3x + y^3$, where $x^2 + y^2 = 2$.

How can inventory costs be minimized? (*Jan Lukas: Photo Researchers, Inc.*).

OBJECTIVES

You should be able, given certain inventory costs, to find how many times a year a store should reorder a product, and in what lot size, to minimize total inventory costs.

9.11 BUSINESS APPLICATIONS: MINIMIZING INVENTORY COSTS

A retail outlet of a business is usually concerned about inventory costs. Suppose, for example, an appliance store sells 2500 tv sets per year. One way it could operate is to order all the tv sets at once. But then the owners would face the carrying costs (insurance, building space, and so on) of storing all those tv's. Thus they might make several smaller orders, say 5, so that the largest number they would ever have to store is 500. On the other hand, each time they reorder there are certain reorder costs such as paperwork, delivery charges, manpower, and so on. It would, therefore, seem that there is some balance between carrying costs and reorder costs. We will see how calculus can help to determine what that balance might be. We will be trying to minimize the following function:

$$\text{Total inventory costs} = \begin{pmatrix} \text{Yearly carrying} \\ \text{costs} \end{pmatrix} + \begin{pmatrix} \text{Yearly reorder} \\ \text{costs} \end{pmatrix}.$$

The *lot size x* refers to the largest amount ordered each reordering period. Note the following graphs. Thus if the lot size is x, then $x/2$ represents the average amount held in stock over the course of the year.

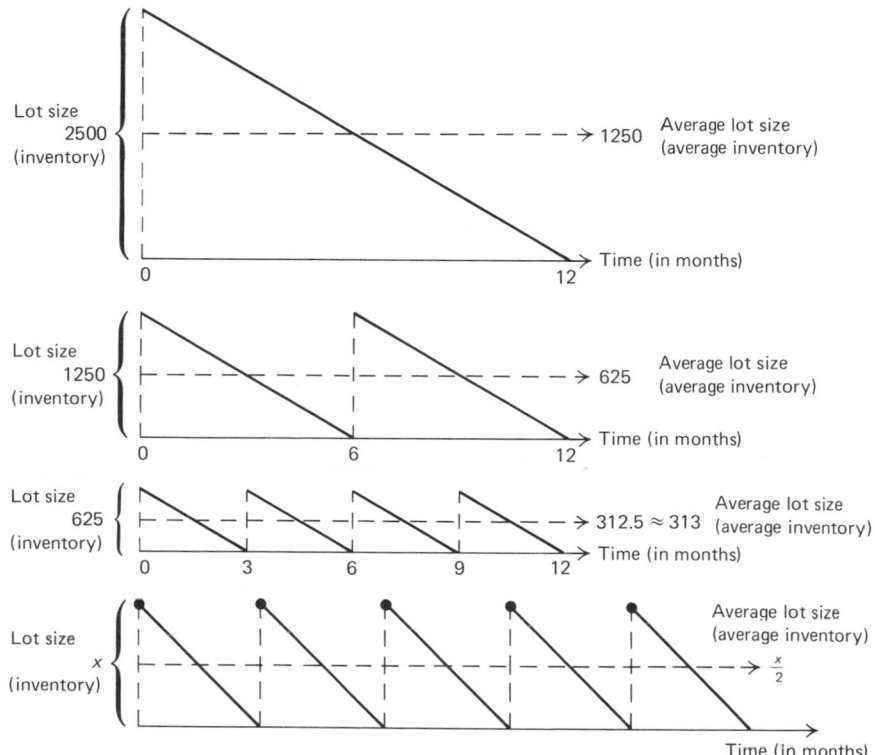

Example 1 A retail appliance store sells 2500 tv's per year. It costs $10 to store one tv for a year. To reorder tv's there is a fixed cost of $20 plus $9 for each tv. How many times per year should the store reorder tv's, and in what lot size, so that inventory costs are minimized?

Solution Let x = the lot size. Now inventory costs are given by

$$C(x) = \text{(Yearly carrying costs)} + \text{(Yearly reorder costs)}.$$

We consider each separately:

a) *Yearly carrying costs.* The average amount held in stock is $x/2$, and it costs $10 per tv for storage. Thus

$$\text{Yearly carrying costs} = \begin{pmatrix} \text{Yearly cost} \\ \text{per item} \end{pmatrix} \begin{pmatrix} \text{Average number} \\ \text{of items} \end{pmatrix} = 10 \cdot \frac{x}{2}.$$

89. ▦ Without a knowledge of calculus one might make a rough estimate of the lot size that will minimize total inventory costs by completing a table like the following. Complete the table and make such an estimate.

Lot size x	Number of reorders $\dfrac{2500}{x}$	Average inventory $\dfrac{x}{2}$	Carrying costs $10 \cdot \dfrac{x}{2}$	Cost of each order $20 + 9x$	Reorder costs $(20 + 9x)\dfrac{2500}{x}$	Total inventory costs $C(x)$ $10 \cdot \dfrac{x}{2} + (20 + 9x)\dfrac{2500}{x}$
2500	1	1250	$12,500	$22,520	$22,520	$35,020
1250	2	625	$6,250	$11,270	$22,540	
500	5	250	$2,500	$4,520		
250	10	125				
167	15	84				
125	20					
100	25					
90	28					
50	50					

b) *Yearly reorder costs.* Now $x =$ lot size, and suppose there are N reorders each year. Then $Nx = 2500$, and $N = 2500/x$. Thus

$$\text{Yearly reorder costs} = \begin{pmatrix}\text{Cost of each} \\ \text{order}\end{pmatrix}\begin{pmatrix}\text{Number of} \\ \text{reorders}\end{pmatrix}$$

$$= (20 + 9x)\frac{2500}{x}.$$

c) Hence

$$C(x) = 10 \cdot \frac{x}{2} + (20 + 9x)\frac{2500}{x},$$

$$C(x) = 5x + \frac{50,000}{x} + 22,500.$$

DO EXERCISE 89.

d) We want to find a minimum value of C on the interval $[1, 2500]$. We first find $C'(x)$:

$$C'(x) = 5 - \frac{50,000}{x^2}.$$

e) Now $C'(x)$ exists for all x in $[1, 2500]$, so the only critical points are those x such that $C'(x) = 0$. We solve $C'(x) = 0$.

$$5 - \frac{50,000}{x^2} = 0$$

$$5 = \frac{50,000}{x^2}$$

$$5x^2 = 50,000$$

$$x^2 = 10,000$$

$$x = \pm 100$$

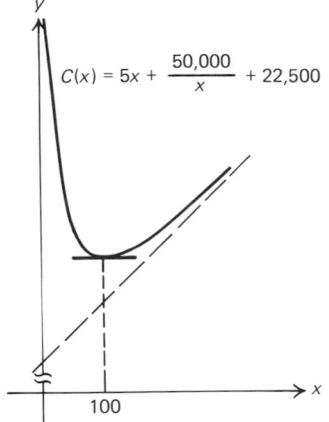

Now there is only one critical point in the interval $[1, 2500]$, $x = 100$, so we can use the second derivative to see if we have a maximum or minimum:

$$C''(x) = \frac{100,000}{x^3}.$$

Now $C''(x)$ is positive for all x in $[1, 2500]$, so we do have a minimum at $x = 100$. Thus, to minimize inventory costs, the store should order tv's (2500/100) or 25 times per year. The lot size is 100.

DO EXERCISE 90.

What happens in such problems when the answer is not a whole number? For functions of this type, we consider the two whole numbers closest to the answer, and substitute them into $C(x)$. The value that yields the smallest $C(x)$ is the lot size.

Example 2   Repeat Example 1, using all the data given, but change the $10 storage cost to $20. How many times per year should the store reorder tv's and in what lot size, so that inventory costs are minimized?

Solution   Comparing this with Example 1, we find that the inventory cost function becomes

$$C(x) = 20 \cdot \frac{x}{2} + (20 + 9x)\frac{2500}{x} = 10x + \frac{50,000}{x} + 22,500.$$

Then we find $C'(x)$, set it equal to 0, and solve for x.

$$C'(x) = 10 - \frac{50,000}{x^2} = 0$$

$$10 = \frac{50,000}{x^2}$$

$$10x^2 = 50,000$$

$$x^2 = 5000$$

$$x = \sqrt{5000} \approx 70.7 \qquad (\text{▦ or Table 1}).$$

Since it does not make sense to reorder 70.7 tv's each time, we consider the two numbers closest to 70.7, which are 70 and 71. Now

$$C(70) \approx \$23,914.29 \qquad \text{and} \qquad C(71) \approx \$23,204.23.$$

It follows that the lot size that will minimize cost is 71. [*Note:* Such a procedure will not work for all types of functions, but will work for the type we are considering here. The number of times an order should be placed is 2500/71 $\approx$ 35, so there is still some estimating involved.]

DO EXERCISE 91.

90. An appliance store sells 600 refrigerators per year. It costs $30 to store one refrigerator for one year. To reorder refrigerators there is a fixed cost of $40 plus $11 for each refrigerator. How many times per year should the store order refrigerators, and in what lot size, to minimize inventory costs?

91. Repeat Margin Exercise 90, using all the data given, but change the $30 storage cost to $50.

The value of the lot size that minimizes total inventory costs is often referred to as the *economic ordering quantity*. There are three assumptions made in using the foregoing method to determine the economic ordering quantity. The first is that the demand for the product is the same throughout the year. For television sets this may be reasonable, but for seasonal items such as clothing or skis, this assumption may not be reasonable. The second assumption is that the time between the placing of an order and the time of its receipt should be consistent throughout the year. The third assumption is that the various costs involved, such as storage, shipping charges, and so on, do not vary. This may not be reasonable in a time of inflation, although one may account for them by anticipating what they might be and using average costs. Nevertheless, the model described above can be useful, and it allows us to analyze a seemingly difficult problem using the calculus.

EXERCISE SET 9.11

1. A sporting goods store sells 100 pool tables per year. It costs $20 to store one pool table for one year. To reorder pool tables there is a fixed cost of $40, plus $16 for each pool table. How many times per year should the store order pool tables, and in what lot size, to minimize inventory costs?

2. A pro shop in a bowling alley sells 200 bowling balls per year. It costs $4 to store one bowling ball for one year. To reorder bowling balls there is a fixed cost of $1, plus $0.50 for each bowling ball. How many times per year should the shop order bowling balls, and in what lot size, to minimize inventory costs?

3. A retail outlet for Boxowitz Calculators sells 360 calculators per year. It costs $8 to store one calculator for one year. To reorder calculators, there is a fixed cost of $10, plus $8 for each calculator. How many times per year should the store order calculators, and in what lot size, to minimize inventory costs?

4. A sporting goods store in southern California sells 720 surfboards per year. It costs $2 to store one surfboard for one year. To reorder surfboards there is a fixed cost of $5, plus $2.50 for each surfboard. How many times per year should the store order surfboards, and in what lot size, to minimize inventory costs?

5. Repeat Exercise 3, using all the data given, but change the $8 storage charge to $9.

6. Repeat Exercise 4, using all the data given, but change the $5 fixed cost to $4.

7. *Minimizing inventory costs—A general solution.* A store sells Q units of a product per year. It costs a dollars to store one unit for one year. To reorder units, there is a fixed cost of b dollars, plus c dollars for each unit. In what lot size should the store reorder to minimize inventory costs?

8. Use the general solution found in Exercise 7 to find how many times per year a store should reorder, and in what lot size, when $Q = 2500$, $a = \$10$, $b = \$20$, and $c = \$9$.

CHAPTER 9 TEST

Which functions are continuous?

1.

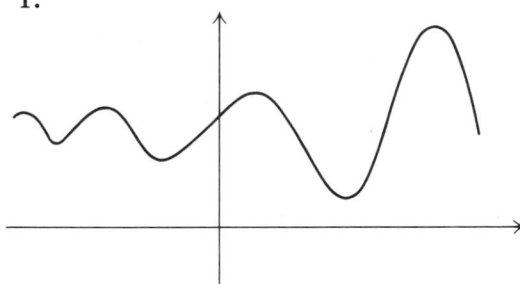

2.

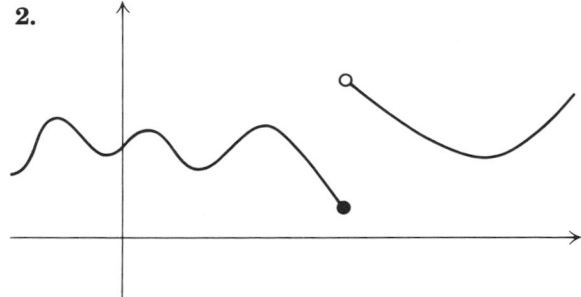

3. a) Complete.

Inputs x	Outputs $\dfrac{x^3 - 8}{x - 2}$
3	
2.5	
2.1	
2.01	
2.1	
2.001	

b) Find

$$\lim_{x \to 2} \frac{x^3 - 8}{x - 2}.$$

4. Find a simplified difference quotient for

$$f(x) = 3x^2 + 1.$$

5. a) Complete.

x	$\dfrac{4x - 3}{x}$
5	
80	
200	
10,000	

b) Find

$$\lim_{x \to \infty} \left(\frac{4x - 3}{x} \right).$$

6. Find the points on the graph of $y = x^3 - 3x^2$ at which the tangent is horizontal.

Find $\dfrac{dy}{dx}$.

7. $y = x^{84}$

8. $y = 10\sqrt{x}$

9. $y = \dfrac{-10}{x}$

10. $y = x^{5/4}$

11. $y = -0.5x^2 + 0.61x + 90$

Differentiate.

12. $\frac{1}{3}x^3 - x^2 + 2x + 4$

13. $\dfrac{2x - 5}{x^4}$

14. $\dfrac{x}{5 - x}$

15. $(x + 3)^4(7 - x)^5$

16. $(x^5 - 4x^3 + x)^{-5}$

17. $x\sqrt{x^2 + 5}$

18. Given $R(x) = 50x$ and $C(x) = 0.001x^2 + 1.2x + 60$, find
a) $P(x)$.
b) $R(10)$, $C(10)$, $P(10)$.
c) $R'(x)$, $C'(x)$, $P'(x)$.
d) $R'(10)$, $C'(10)$, $P'(10)$.

19. In a certain memory experiment a person is able to memorize M words after t minutes, where

$$M = -0.001t^3 + 0.1t^2.$$

a) Find the rate of change of the number of words memorized with respect to time.
b) How many words are memorized the first 10 minutes (at $t = 10$)?
c) What is the memory rate at $t = 10$ minutes?

20. For $y = x^4 - 3x^2$, find $\dfrac{d^3y}{dx^3}$.

Find the maximum and minimum values, if they exist, over the indicated interval. Where no interval is specified, use the real line.

21. $f(x) = x(6 - x)$

22. $f(x) = x^3 + x^2 - x + 1$; $[-2, \frac{1}{2}]$

23. $f(x) = -x^2 + 8.6x + 10$

24. $f(x) = -2x + 5$; $[-1, 1]$

25. $f(x) = -2x + 5$

26. $f(x) = 3x^2 - x - 1$

27. $f(x) = x^2 + \dfrac{128}{x}$; $(0, \infty)$

28. Of all numbers whose difference is 8, find the two that have the minimum product.

29. Minimize $Q = x^2 + y^2$, where $x - y = 10$.

30. Find the maximum profit and the number of units that must be produced and sold to yield the maximum profit.

$$R(x) = x^2 + 110x + 60,$$
$$C(x) = 1.1x^2 + 10x + 80$$

31. From a piece of cardboard 60 in. by 60 in., square corners are cut out so the sides can be folded up to make a box. What dimensions will yield a box of maximum volume? What is the maximum volume?

32. A sporting-goods store sells 1225 tennis rackets per year. It costs $2 to store one tennis racket for one year. To reorder tennis rackets, there is a fixed cost of $1, plus $0.50 for each tennis racket. How many times per year should the sporting-goods store order tennis rackets, and in what lot size, to minimize inventory costs?

Exponential growth. (*Marshall Henrichs*)

Exponential and Logarithmic Functions

OBJECTIVES

You should be able to

a) Graph equations like $y = 2^x$ and $y = \log_2 x$.

b) Given an exponential equation, write an equivalent logarithmic equation.

c) Given a logarithmic equation, write an equivalent exponential equation.

d) Given $\log_a 3 = 1.099$ and $\log_a 5 = 1.609$, find logarithms like $\log_a 15$ and $\log_a 5a$.

e) Use Table 2 to find logarithms such as $\log 546$ and $\log 0.0546$.

f) Given an equation like $y = a \cdot b^x$, find $\log y$.

g) Solve an equation like $e^t = 40$, for t.

h) Solve problems involving applications of logarithms.

1. Consider $y = 3^x$.

a) Complete this table of function values.

x	0	$\frac{1}{2}$	1	2	-1	-2
3^x						

b) Graph $y = 3^x$.

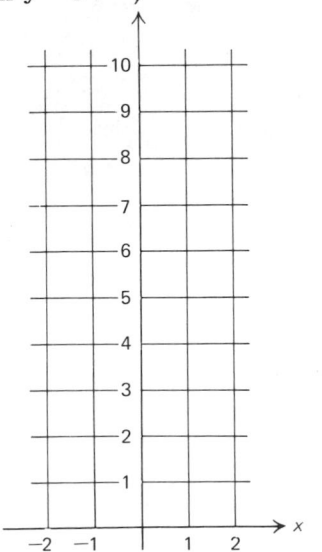

10.1 EXPONENTIAL AND LOGARITHMIC FUNCTIONS

Exponential Functions

The following are examples of exponential functions:

$$y = 2^x, \qquad y = (\tfrac{1}{2})^x, \qquad y = (0.4)^x.$$

Note, in contrast to power functions like $y = x^2$ or $y = x^3$, that the variable in an exponential function is in the exponent. Exponential functions have extensive application. Let us consider their graphs.

Example 1   Graph $y = 2^x$.

Solution

a) First we find some function values.

x	0	$\frac{1}{2}$	1	2	3	-1	-2
y (or 2^x)	1	1.4	2	4	8	$\frac{1}{2}$	$\frac{1}{4}$

Note: For

$x = 0, y = 2^0 = 1$

$x = \frac{1}{2}, y = 2^{1/2} = \sqrt{2} \approx 1.4$

$x = 1, y = 2^1 = 2$

$x = 2, y = 2^2 = 4$

$x = 3, y = 2^3 = 8$

$x = -1, y = 2^{-1} = \frac{1}{2}$

$x = -2, y = 2^{-2} = \dfrac{1}{2^2} = \dfrac{1}{4}$

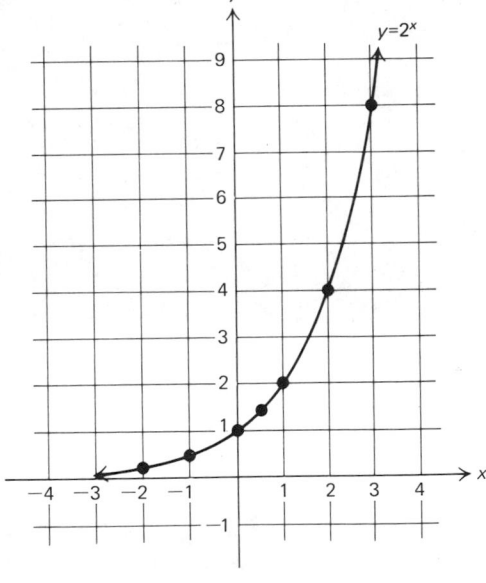

b) Next, we plot the points and connect them with a smooth curve as shown in the figure.

DO EXERCISE 1.

Example 2   Graph $y = (\frac{1}{2})^x$.

Solution

a) We first find some function values. Before we do this, note that

$$y = (\tfrac{1}{2})^x = (2^{-1})^x = 2^{-x}.$$

This will ease our work.

x	0	$\frac{1}{2}$	1	2	-1	-2	-3
y	1	0.7	$\frac{1}{2}$	$\frac{1}{4}$	2	4	8

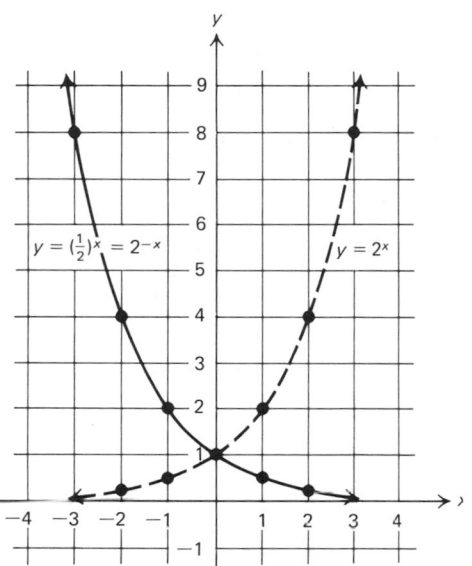

Note: For,

$x = 0, \quad y = 2^{-0} = 1$

$x = \dfrac{1}{2}, \quad y = 2^{-1/2}$

$\qquad = \dfrac{1}{\sqrt{2}} \approx \dfrac{1}{1.4} \approx 0.7$

$x = 1, \quad y = 2^{-1} = \frac{1}{2}$

$x = 2, \quad y = 2^{-2} = \frac{1}{4}$

$x = -1, \quad y = 2^{-(-1)} = 2$

$x = -2, \quad y = 2^{-(-2)} = 4$

$x = -3, \quad y = 2^{-(-3)} = 8$

b) We plot these points and connect them with a smooth curve as shown by the solid curve in the figure. The dashed curve shows $y = 2^x$ for comparison.

DO EXERCISE 2.

Logarithmic Functions

The definition of logarithms is as follows:

$$\text{``}y = \log_a x\text{''} \quad \text{means} \quad \text{``}x = a^y\text{''}$$

The number a is called the *logarithmic base*. Thus, for logarithms base 10, $\log_{10} x$ is that number y such that $x = 10^y$. A logarithm can thus be thought of as an exponent. We can convert from a logarithmic

2. Consider $y = (\frac{1}{3})^x$.

a) Complete this table of function values.

x	0	$\frac{1}{2}$	1	2	-1	-2
y						

b) Graph $y = (\frac{1}{3})^x$.

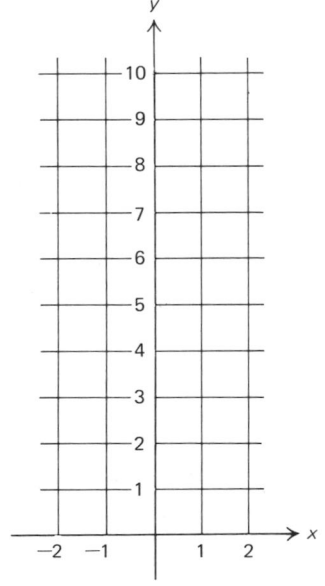

3. Write equivalent exponential equations.

a) $\log_b P = T$
b) $\log_9 3 = \frac{1}{2}$
c) $\log_{10} 1000 = 3$
d) $\log_{10} 0.1 = -1$

equation to an exponential equation, and conversely, as follows.

Logarithmic equation	*Exponential equation*
$\log_a M = N$	$a^N = M$
$\log_{10} 100 = 2$	$10^2 = 100$
$\log_{10} 0.01 = -2$	$10^{-2} = 0.01$
$\log_{49} 7 = \frac{1}{2}$	$49^{1/2} = 7$

DO EXERCISES 3 AND 4.

To graph a logarithmic equation, we can graph its equivalent exponential equation.

Example 3 Graph $y = \log_2 x$.

Solution We first write the equivalent exponential equation

$$x = 2^y.$$

We select values for y and find the corresponding values of 2^y.

4. Write equivalent logarithmic equations.

a) $e^k = T$
b) $16^{1/4} = 2$
c) $10^4 = 10{,}000$
d) $10^{-3} = 0.001$

x (or 2^y)	1	2	4	8	$\frac{1}{2}$	$\frac{1}{4}$
y	0	1	2	3	-1	-2

Next, we plot points, remembering that x is still the first coordinate.

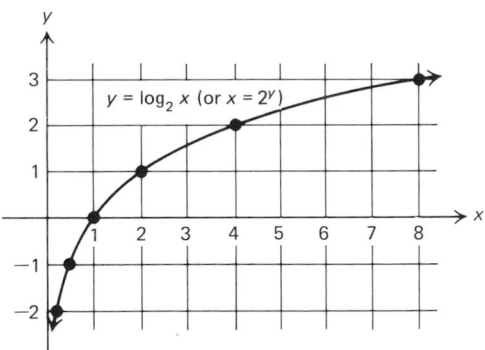

DO EXERCISE 5.

Basic Properties of Logarithms

The following are some basic properties of logarithms. The proofs are optional, but follow from properties of exponents.

5. Graph $y = \log_3 x$.

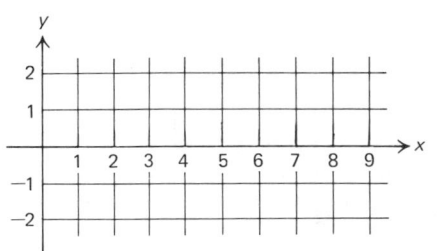

PROPERTY 1. $\log_a MN = \log_a M + \log_a N$

PROPERTY 2. $\log_a \dfrac{M}{N} = \log_a M - \log_a N$

PROPERTY 3. $\log_a M^k = k \cdot \log_a M$

PROPERTY 4. $\log_a a = 1$

PROPERTY 5. $\log_a a^k = k$

PROPERTY 6. $\log_a 1 = 0$

Proof of 1 and 2 Let $X = \log_a M$ and $Y = \log_a N$. Then, writing the equivalent exponential equations, we have

$$M = a^X \quad \text{and} \quad N = a^Y.$$

Then by properties of exponents (see Section 1.1),

$$MN = a^X \cdot a^Y = a^{X+Y}, \quad \text{so} \quad \log_a MN = X + Y$$
$$= \log_a M + \log_a N,$$

$$\frac{M}{N} = a^X \div a^Y = a^{X-Y}, \quad \text{so} \quad \log_a \frac{M}{N} = X - Y$$
$$= \log_a M - \log_a N.$$

Proof of 3 Let $X = \log_a M$. Then

$$a^X = M, \quad \text{so} \quad (a^X)^k = M^k, \quad \text{or} \quad a^{Xk} = M^k.$$

Thus

$$\log_a M^k = Xk = k \cdot \log_a M.$$

Proof of 4 $\log_a a = 1$ because $a^1 = a$.

Proof of 5 $\log_a a^k = k$ because $(a^k) = a^k$.

Proof of 6 $\log_a 1 = 0$ because $a^0 = 1$.

Let us illustrate these properties.

Examples Given

$$\log_a 2 = 0.301$$
$$\log_a 3 = 0.477,$$

find each of the following.

a) $\log_a 6$. $\quad \log_a 6 = \log_a (2 \cdot 3) = \log_a 2 + \log_a 3 \quad \quad \text{(Property 1)}$
$$= 0.301 + 0.477$$
$$= 0.778$$

Given

$$\log_a 2 = 0.301$$
$$\log_a 5 = 0.699,$$

find each of the following.

6. $\log_a 10$

7. $\log_a \frac{2}{5}$

8. $\log_a \frac{5}{2}$

9. $\log_a \frac{1}{5}$

10. $\log_a \sqrt{a^3}$

11. $\log_a 5a$

12. $\log_a 16$

b) $\log_a \frac{2}{3}$. $\log_a \frac{2}{3} = \log_a 2 - \log_a 3$ (Property 2)
$$= 0.301 - 0.477$$
$$= -0.176$$

c) $\log_a 81$. $\log_a 81 = \log_a 3^4 = 4\log_a 3$ (Property 3)
$$= 4(0.477)$$
$$= 1.908$$

d) $\log_a \frac{1}{3}$. $\log_a \frac{1}{3} = \log_a 1 - \log_a 3$ (Property 2)
(Property 6)
$$= 0 - 0.477$$
$$= -0.477$$

e) $\log_a \sqrt{a}$. $\log_a \sqrt{a} = \log_a a^{1/2} = \frac{1}{2}$ (Property 5)

f) $\log_a 2a$. $\log_a 2a = \log_a 2 + \log_a a$ (Property 1)
(Property 4)
$$= 0.301 + 1$$
$$= 1.301$$

g) $\log_a 5$. *No way to find using these properties.*
($\log_a 5 \neq \log_a 2 + \log_a 3$)

h) $\dfrac{\log_a 3}{\log_a 2}$. $\dfrac{\log_a 3}{\log_a 2} = \dfrac{0.477}{0.301} = 1.58.$

We simply divided, not using any of the properties.

DO EXERCISES 6 THROUGH 12.

Natural Logarithms

The number e, which is approximately 2.718282, has extensive application in many fields. *We will develop e thoroughly in Section* 10.2. The number $\log_e x$ is called the *natural logarithm* of x and is abbreviated $\ln x$; that is,

$$\ln x = \log_e x.$$

The following is a restatement of the basic properties of logarithms in terms of natural logarithms.

PROPERTY 1. $\ln MN = \ln M + \ln N$

PROPERTY 2. $\ln \dfrac{M}{N} = \ln M - \ln N$

PROPERTY 3. $\ln a^k = k \cdot \ln a$

PROPERTY 4.   ln $e = 1$

PROPERTY 5.   ln $e^k = k$

PROPERTY 6.   ln $1 = 0$

Let us illustrate these properties.

Examples   Given

$$\ln 2 = 0.6931$$
$$\ln 3 = 1.0986,$$

find each of the following.

a) ln 6. $\ln 6 = \ln(2 \cdot 3) = \ln 2 + \ln 3$ (Property 1)
$$= 0.6931 + 1.0986$$
$$= 1.7917$$

b) ln 81. $\ln 81 = \ln(3^4)$
$$= 4 \ln 3 \qquad\qquad \text{(Property 3)}$$
$$= 4(1.0986)$$
$$= 4.3944$$

c) ln $\frac{2}{3}$. $\ln \frac{2}{3} = \ln 2 - \ln 3$ (Property 2)
$$= 0.6931 - 1.0986$$
$$= -0.4055$$

d) ln $\frac{1}{3}$. $\ln \frac{1}{3} = \ln 1 - \ln 3$ (Property 2)
                 (Property 6)
$$= 0 - 1.0986$$
$$= -1.0986$$

e) ln 2e. $\ln 2e = \ln 2 + \ln e$ (Property 1)
                 (Property 4)
$$= 0.6931 + 1$$
$$= 1.6931$$

f) ln $\sqrt{e^3}$. $\ln \sqrt{e^3} = \ln e^{3/2}$
$$= \tfrac{3}{2} \qquad\qquad \text{(Property 5)}$$

DO EXERCISES 13 THROUGH 19.

Finding Natural Logarithms Using a Caculator   If you have a $\boxed{\ln}$ key on your calculator, you can find natural logarithms directly.

Given

$$\ln 2 = 0.6931$$
$$\ln 5 = 1.6094,$$

find each of the following.

13. ln 10

14. ln $\frac{2}{5}$

15. ln $\frac{5}{2}$

16. ln 16 [*Hint:* $16 = 2^4$]

17. ln $5e$

18. ln $\sqrt{e}$

19. ln $\frac{1}{5}$

▥ Find each logarithm. Round to six decimal places.

20. ln 2

21. ln 20

22. ln 100

23. ln 0.07432

24. ln 1.08

25. ln 0.9999

Examples Find each logarithm on your calculator. Round to six decimal places.

$$\ln 5.24 = 1.656321, \quad \ln 0.00001277 \approx -11.268412.$$

DO EXERCISES 20–25.

Finding Natural Logarithms Using a Table (Optional) If you do not have a calculator with a natural logarithm key, you can use Table 3. Part of Table 3 is shown below. It shows some values of ln x.

x	0.00	0.01	0.02	0.03	0.04	0.05	0.06	0.07	0.08	0.09
5.0	1.6094	1.6114	1.6134	1.6154	1.6174	1.6194	1.6214	1.6233	1.6253	1.6273
5.1	1.6292	1.6312	1.6332	1.6351	1.6371	1.6390	1.6409	1.6429	1.6448	1.6467
5.2	1.6487	1.6506	1.6525	1.6544	1.6563	1.6582	1.6601	1.6620	1.6639	1.6658
5.3	1.6677	1.6696	1.6715	1.6734	1.6752	1.6771	1.6790	1.6808	1.6827	1.6845
5.4	1.6864	1.6882	1.6901	1.6919	1.6938	1.6956	1.6974	1.6993	1.7011	1.7029

Example 1 Find ln 5.24. Use Table 3. To find ln 5.24, locate the row headed 5.2; then move across to the column headed 0.04. Note the shaded number in the table. Thus

$$\ln 5.24 = 1.6563$$

We find natural logarithms of numbers not in Table 3 as follows. We first express in scientific notation the number whose natural logarithm we are finding; that is, as a product $M \times 10^k$, where $1 \leq M < 10$.

Example 2 Find ln 5240. Use Table 3.

$$
\begin{aligned}
\ln 5240 &= \ln (5.24 \times 1000) \\
&= \ln (5.24 \times 10^3) && (1000 = 10^3) \\
&= \ln 5.24 + \ln 10^3 && \text{(Property 1)} \\
&= \ln 5.24 + 3 \ln 10 && \text{(Property 3)} \\
&= 1.6563 + 6.9078 && \text{(Find ln 5.24 in the body of} \\
& && \text{Table 3 and 3 ln 10 at the bottom.)} \\
&= 8.5641
\end{aligned}
$$

Example 3   Find $\ln 0.000524$. Use Table 3.

$$\ln 0.000524 = \ln (5.24 \times 0.0001)$$
$$= \ln (5.24 \times 10^{-4}) \quad (0.0001 = 10^{-4})$$
$$= \ln 5.24 + \ln 10^{-4} \quad \text{(Property 1)}$$
$$= \ln 5.24 - 4 \ln 10 \quad \text{(Property 3)}$$
$$= 1.6563 - 9.2103 \quad \text{(Find } \ln 5.24 \text{ in the body of Table 3 and } 4 \ln 10 \text{ at the bottom.)}$$
$$= -7.554$$

A number like $\ln 5.243$ can not be found in Table 3. In such cases we round and then use Table 3:

$$\ln 5.243 \approx \ln 5.24$$
$$\approx 1.6563.$$

DO EXERCISES 26–30.

Common Logarithms (Optional)

Logarithms to the base 10 are called *common logarithms*. When we write

$$\log M,$$

with no base indicated, base 10 is to be understood. Note the following comparison of common logarithms and powers of 10.

$1000 = 10^3$	The common	$\log 1000 = 3$
$100 = 10^2$	logarithms	$\log 100 = 2$
$10 = 10^1$	at the right	$\log 10 = 1$
$1 = 10^0$	follow from	$\log 1 = 0$
$0.1 = 10^{-1}$	the powers at	$\log 0.1 = -1$
$0.01 = 10^{-2}$	the left.	$\log 0.01 = -2$
$0.001 = 10^{-3}$		$\log 0.001 = -3$

DO EXERCISE 31.

Finding Common Logarithms Using a Calculator

If you have a $\boxed{\log}$ key on your calculator, you can find common logarithms directly.

Examples   Find each logarithm on your calculator. Round to four decimal places.

$$\log 4258 \approx 3.6292, \qquad \log 0.00001277 \approx -4.8938.$$

DO EXERCISE 32.

Using Table 3, find each logarithm.

26. $\ln 8.13$

27. $\ln 81{,}300$

28. $\ln 0.0813$

29. $\ln 2000$

30. $\ln 0.0001$

31. Find these logarithms.
a) $\log 10{,}000$
b) $\log 0.0001$

32. ▦ Find each logarithm. Round to four decimal places.

a) $\log 31{,}456$

b) $\log 0.9080701$

33. Find these logarithms. First find
log 7.86

a) log 7860

b) log 786

c) log 78.6

d) log 7.86

e) log 0.786

f) log 0.0786

g) log 0.00786

34. Find these logarithms.
a) log 76.4

b) log 2330

c) log 0.0087

Finding Common Logarithms Using a Table

If you do not have a calculator with a common logarithm key, you can
use Table 2. It is arranged in a manner similar to Table 3. For
example,
$$\log 5.24 = 0.7193$$

Now suppose we wanted to find log 524. We could estimate this
roughly by noting that 524 is between 100 and 1000, so log 524 is
between 2 and 3. To get a closer approximation, note that
$$\log 524 = \log(5.24 \times 100) - \log 5.24 + \log 100$$
$$= \log 5.24 + 2$$
$$= 0.7193 + 2 \qquad \text{(Table 2)}$$
$$= 2.7193$$

The number 2 is called the *characteristic*. The number 0.7193 is
called the *mantissa*. In the following, as we move the decimal point,
note how the mantissa recurs and the characteristic changes.

$$
\begin{aligned}
\log 5240 &= 3 + 0.7193 = 3.7193 \\
\log 524 &= 2 + 0.7193 = 2.7193 \\
\log 52.4 &= 1 + 0.7193 = 1.7193 \\
\log 5.24 &= 0 + 0.7193 = 0.7193 \\
\log 0.524 &= -1 + 0.7193 = -0.2807 \\
\log 0.0524 &= -2 + 0.7193 = -1.2807 \\
\log 0.00524 &= -3 + 0.7193 = -2.2807
\end{aligned}
$$

DO EXERCISES 33 AND 34.

Exponential Equations

In an equation where a variable occurs in an exponent, we call the
equation *exponential*. Logarithms can be used to manipulate or solve
exponential equations.

Example 4 Solve for t: $e^t = 40$

Solution
$$\ln e^t = \ln 40 \qquad \text{(Taking the natural logarithm on both sides)}$$
$$t = \ln 40 \qquad \text{(Property 5)}$$
$$t = 3.688879 \qquad (\text{▦ or Table 3.})$$
$$t \approx 3.7$$

It should be noted that this is an approximation for t even though an
equal sign is often used.

Example 5   Solve for t: $e^{-0.04t} = 0.05$

Solution

$$\ln e^{-0.04t} = \ln 0.05 \qquad \text{(Taking the natural logarithm on both sides.)}$$

$$-0.04t = \ln 0.05 \qquad \text{(Property 5)}$$

$$t = \frac{\ln 0.05}{-0.04}$$

$$t = \frac{-2.995732}{-0.04} \qquad (\text{▦ or Table 3.})$$

$$t \approx 75$$

CALCULATOR NOTE: For purposes of space and explanation, we have rounded the value of $\ln 0.05$ to -2.995732 in an intermediate step. On a calculator you should find

$$\frac{\ln 0.05}{-0.04},$$

obtaining

$$\frac{-2.995732274}{-0.04}.$$

Divide, and round at the end. Answers in the key are found in this manner. Remember, the number of places in a table or on a calculator may affect the accuracy of the answer. Usually, your answer should agree to at least three digits.

DO EXERCISE 35.

Application*

Exponential and logarithmic functions have many applications. One is to the *loudness* of sound.

The *loudness L*, in bels†, of a sound of intensity I is defined to be

$$L = \log \frac{I}{I_0},$$

where I_0 is the minimum intensity detectable by the human ear (the tick of a watch at 20 feet under very quiet conditions).

When one sound is 10 times as intense as another, its loudness is 1 bel louder. If one sound is 100 times as intense as another, it is louder by 2 bels, and so on. This unit of loudness called the *bel* is rather large, so in practice, a subunit 1/10th as large, called a *decibel* (dB), is used.

_____
* This application may be omitted if common logarithms were omitted.
† After Alexander Graham Bell.

35. Solve for t.

a) $e^t = 80$

b) $e^{-0.06t} = 0.07$

Noise Pollution. Sounds at 90 decibels and higher cause temporary and, eventually, permanent hearing loss due to deterioration of tiny cells that transmit sound from the ear to the brain.

---

36. Find the loudness, in decibels, of the sound of a heavy truck that has an intensity of $10^9 \cdot I_0$.

37. Find the loudness, in decibels, of the sound in a broadcasting studio for which the intensity I is $199 \cdot I_0$.

The preceding formula for L, in dB becomes $L = 10 \log \dfrac{I}{I_0}$. In summary,

$$L, \text{ in bels} = \log \frac{I}{I_0},$$

$$L, \text{ in dB} = 10 \log \frac{I}{I_0}.$$

Sounds can be interpreted as multiples of the minimum intensity I_0.

Example 6   Find the loudness, in decibels, of conversational speech, having an intensity I which is $10^6 \cdot I_0$ (1 million times as intense as I_0).

Solution

$$L = 10 \log \frac{10^6 \cdot I_0}{I_0} = 10(\log 10^6) = 10 \cdot 6 \qquad \text{(Property 5)}$$

$$= 60 \text{ decibels}$$

Example 7   Find the loudness, in decibels, of the sound in a library which is 2510 times as intense as the minimum intensity I_0.

Solution

$$L = 10 \log \frac{2510 \cdot I_0}{I_0} = 10(\log 2510) = 10(3.3997) \qquad (\blacksquare \text{ or Table 2})$$

$$\approx 34 \text{ decibels}$$

DO EXERCISES 36 AND 37.

EXERCISE SET 10.1

Graph.

1. $y = 4^x$ **2.** $y = 5^x$ **3.** $y = (0.4)^x$ **4.** $y = (0.2)^x$

5. $y = \log_4 x$ **6.** $y = \log_5 x$

Write equivalent exponential equations.

7. $\log_2 8 = 3$ **8.** $\log_3 81 = 4$ **9.** $\log_8 2 = \frac{1}{3}$ **10.** $\log_{27} 3 = \frac{1}{3}$

11. $\log_a K = J$ **12.** $\log_a J = K$ **13.** $\log_b T = v$ **14.** $\log_c Y = t$

Write equivalent logarithmic equations.

15. $e^M = b$ **16.** $e^t = p$ **17.** $10^2 = 100$ **18.** $10^3 = 1000$

19. $10^{-1} = 0.1$ **20.** $10^{-2} = 0.01$ **21.** $M^p = V$ **22.** $Q^n = T$

Given $\log_b 3 = 1.099$ and $\log_b 5 = 1.609$, find each of the following. Do not use tables.

23. $\log_b 15$ **24.** $\log_b \frac{3}{5}$ **25.** $\log_b \frac{5}{3}$ **26.** $\log_b \frac{1}{3}$

27. $\log_b \frac{1}{5}$ **28.** $\log_b \sqrt{b}$ **29.** $\log_b \sqrt{b^3}$ **30.** $\log_b 3b$

31. $\log_b 5b$ **32.** $\log_b 9$ **33.** $\log_b 25$ **34.** $\log_b 75$

Given $\ln 4 = 1.3863$ and $\ln 5 = 1.6094$, find each of the following. Do not use tables.

35. $\ln 20$ **36.** $\ln \frac{4}{5}$ **37.** $\ln \frac{5}{4}$ **38.** $\ln \frac{1}{5}$ **39.** $\ln \frac{1}{4}$ **40.** $\ln 5e$

41. $\ln 4e$ **42.** $\ln \sqrt{e^6}$ **43.** $\ln \sqrt{e^8}$ **44.** $\ln 25$ **45.** $\ln 16$ **46.** $\ln 100$

▦ Find each logarithm. Round to six decimal places.

47. $\ln 5894$ **48.** $\ln 99{,}999$ **49.** $\ln 0.0182$ **50.** $\ln 0.00087$

Using Table 3, find the following logarithms.

51. $\ln 1.88$ **52.** $\ln 18.8$ **53.** $\ln 0.0188$ **54.** $\ln 0.188$

55. $\ln 906$ **56.** $\ln 8100$ **57.** $\ln 0.011$ **58.** $\ln 0.00056$

▦ Find each logarithm. Round to four decimal places.

59. $\log 876{,}502$ **60.** $\log 677.3$ **61.** $\log 0.1112$ **62.** $\log 0.00092$

Using Table 2, find the following logarithms.

63. $\log 2.13$ **64.** $\log 213$ **65.** $\log 0.213$ **66.** $\log 0.00213$

67. $\log 4500$ **68.** $\log 702$ **69.** $\log 0.0008$ **70.** $\log 0.999$

Solve for t.

71. $e^t = 100$ **72.** $e^t = 1000$ **73.** $e^t = 60$ **74.** $e^t = 90$

75. $e^{-t} = 0.1$ **76.** $e^{-t} = 0.01$ **77.** $e^{-0.02t} = 0.06$ **78.** $e^{0.07t} = 2$

Applied Problems

79. Find the loudness, in decibels, of a dishwasher (in operation) that has an intensity of $2{,}500{,}000 \cdot I_0$.

80. Find the loudness, in decibels, of an automobile engine that has an intensity of $3{,}100{,}000 \cdot I_0$.

81. Find the loudness, in decibels, of a three-engine jet aircraft (500 ft away) which has an intensity of $10^{12} \cdot I_0$.

82. Find the loudness, in decibels, of the threshold of sound pain for which the intensity is $10^{14} \cdot I_0$.

Find the loudness of this aircraft. (*Daniel S. Brody: Stock, Boston*)

Earthquake magnitude The magnitude R (measured on the Richter scale) of an earthquake of intensity I is defined to be

$$R = \log \frac{I}{I_0},$$

where I_0 is a minimum intensity used for comparison. When one earthquake is 10 times as intense as another, its magnitude on the Richter scale is 1 higher. If one earthquake is 100 times as intense as another, its magnitude on the Richter scale is 2 higher, and so on. Thus an earthquake whose magnitude is 7 on the Richter scale is 10 times as intense as an earthquake whose magnitude is 6. Earthquakes can be interpreted as multiples of the minimum intensity I_0.

83. The Mexico City earthquake of 1978 had an intensity of $10^{7.85} \cdot I_0$. What was its magnitude on the Richter scale?

84. The Anchorage, Alaska earthquake on March 27, 1964 had an intensity of $10^{8.4} \cdot I_0$. What was its magnitude on the Richter scale?

This photograph shows part of the damage of the earthquake in Anchorage, Alaska in 1964. (*Pro Pix, from Monkmeyer*)

pH: In chemistry pH is defined as pH $= -\log [H^+]$, where $[H^+] =$ hydrogen ion concentration in moles per liter. For example, the hydrogen ion concentration in milk is $4 \cdot 10^{-7}$ moles per liter, so

$$pH = -\log (4 \cdot 10^{-7}) = -[\log 4 + (-7)] = -[0.6021 - 7] \approx 6.4.$$

85. For eggs, $[H^+] = 1.6 \cdot 10^{-8}$. Find the pH.

86. For tomatoes, $[H^+] = 6.3 \cdot 10^{-5}$. Find the pH.

Solve for t.

87. $P = P_0 e^{kt}$

88. $P = P_0 e^{-kt}$

▦ Use input–output tables. Find each limit.

89. $\lim\limits_{x \to \infty} \ln x$

90. $\lim\limits_{x \to 1} \ln x$

Verify each of the following.

91. $\ln x = \dfrac{\log x}{\log e} \approx 2.3026 \log x$

92. $\log x = \dfrac{\ln x}{\ln 10} \approx 0.4343 \ln x$

OBJECTIVES

You should be able to

a) Graph functions like
   $f(x) = 2e^x$, and $g(x) = 1 - e^{-x}$.

b) Differentiate functions involving e.

c) Solve applied problems involving exponential functions.

10.2 THE EXPONENTIAL FUNCTION, BASE e

The exponential functions

$$f(x) = ce^{kx} \quad \text{and} \quad f(x) = ce^{-kx}$$

are some of the most important ones in mathematics and in the applications of mathematics.

The General Base a

In Chapter 1 we reviewed definitions of expressions of the type a^x, where x was a rational number. For example,

$$a^{2.34}, \quad \text{or} \quad a^{234/100},$$

means "raise a to the 234th power and take the 100th root."

What about expressions with irrational exponents, such as $2^{\sqrt{2}}$, 2^{π}, or $2^{-\sqrt{3}}$? An irrational number is a number named by an infinite, nonrepeating decimal. Let us consider 2^{π}. We know π is irrational with infinite, nonrepeating decimal expansion

$$3.141592654 \cdots .$$

This means that π is approached as a limit by the rational numbers

$$3, \quad 3.1, \quad 3.14, \quad 3.141, \quad 3.1415, \cdots ;$$

so it seems reasonable that 2^{π} should be approached as a limit by the rational powers

$$2^{3}, \quad 2^{3.1}, \quad 2^{3.14}, \quad 2^{3.141}, \quad 2^{3.1415}, \cdots$$

DO EXERCISE 38.

In general, a^{x} is approximated by the values of a^{r} for rational numbers r near x; a^{x} is the limit of a^{r} as r approaches x through rational values. In summary, for $a > 0$, the definition of a^{x} for rational numbers x can be extended to arbitrary real numbers x in such a way that the usual laws of exponents, such as

$$a^{x} \cdot a^{y} - a^{x+y}, \quad a^{x} \div a^{y} = a^{x-y}, \quad (a^{x})^{y} = a^{xy}, \quad \text{and} \quad a^{-x} = \frac{1}{a^{x}},$$

still hold. Moreover, the function so obtained,

$$f(x) = a^{x},$$

is continuous.

The following are some properties of the exponential function for various bases.

1. The function $f(x) = a^{x}$, where $a > 1$, is a positive, increasing, continuous function; and as x gets smaller, a^{x} approaches 0.

38. (▦ with a $\boxed{y^{x}}$ key.) Complete this table. Round to six decimal places.

r	2^{r}
3	
3.1	
3.14	
3.141	
3.1415	

What seems to be the value of 2^{π} to two decimal places?

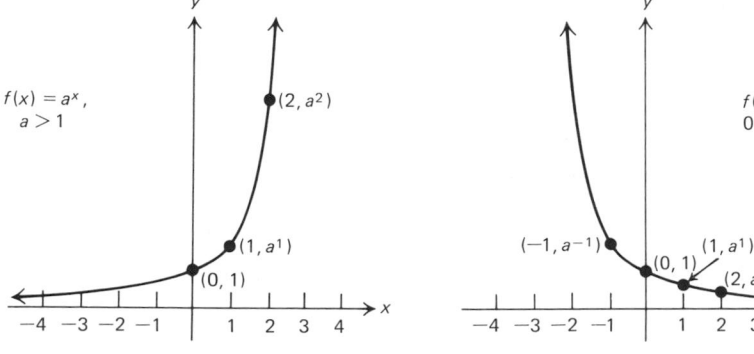

39. In order to investigate

$$\lim_{h \to 0} \frac{2^h - 1}{h}$$

we choose a sequence of numbers h, approaching 0, and compute

$$\frac{2^h - 1}{h}.$$

a) Complete this table.
   Values of 2^h can be found using a calculator with a square-root key. Just take successive square roots.

h	$\dfrac{2^h - 1}{h}$
$\frac{1}{2}$	
$\frac{1}{4}$	
$\frac{1}{8}$	
$\frac{1}{16}$	
$\frac{1}{32}$	

b) To the nearest tenth, what is the value of

$$\lim_{h \to 0} \frac{2^h - 1}{h}?$$

2. The function $f(x) = a^x$, where $0 < a < 1$, is a positive, decreasing, continuous function; and as x gets larger, a^x approaches 0. When $a = 1$, $f(x) = a^x = 1^x = 1$, and is a constant function.

Let us consider finding derivatives of exponential functions.

The Derivative of a^x, the Number e

Let us consider finding the derivative of the exponential function

$$f(x) = a^x.$$

The derivative is given by

$$f'(x) = \lim_{h \to 0} \frac{f(x + h) - f(x)}{h} \qquad \text{(Definition of the derivative.)}$$

$$= \lim_{h \to 0} \frac{a^{x+h} - a^x}{h} \qquad \text{(Substituting } a^{x+h} \text{ for } f(x + h) \text{ and } a^x \text{ for } f(x))$$

$$= \lim_{h \to 0} \frac{a^x \cdot a^h - a^x \cdot 1}{h}$$

We get

$$f'(x) = a^x \cdot \lim_{h \to 0} \frac{a^h - 1}{h}. \qquad (1)$$

In particular, for $g(x) = 2^x$,

$$g'(x) = 2^x \cdot \lim_{h \to 0} \frac{2^h - 1}{h}.$$

Note that the limit does not depend on the value of x at which we are evaluating the derivative. For $g'(x)$ to exist, we must determine if

$$\lim_{h \to 0} \frac{2^h - 1}{h} \text{ exists.}$$

Let us investigate this question.

DO EXERCISE 39.

The margin exercise suggests that $(2^h - 1)/h$ has a limit as h approaches 0, and that its approximate value is 0.7, so that

$$g'(x) \approx (0.7)2^x.$$

In other words, the derivative is a constant times the function value

2^x. Similarly, for $t(x) = 3^x$,

$$t'(x) = 3^x \cdot \lim_{h \to 0} \frac{3^h - 1}{h}.$$

Again we can find an approximation for the limit which does not depend on the value of x at which we are evaluating the derivative.

DO EXERCISE 40.

The margin exercise suggests that $(3^h - 1)/h$ has a limit as h approaches 0, and that its approximate value is 1.1, so that

$$t'(x) \approx (1.1)3^x.$$

In other words, the derivative is a constant times the function value 3^x.

DO EXERCISES 41 AND 42.

In Fig. 1 we have graphed $g(x) = 2^x$ and $g'(x) \approx (0.7)2^x$. Note that the graph of g' lies *below* the graph of g.

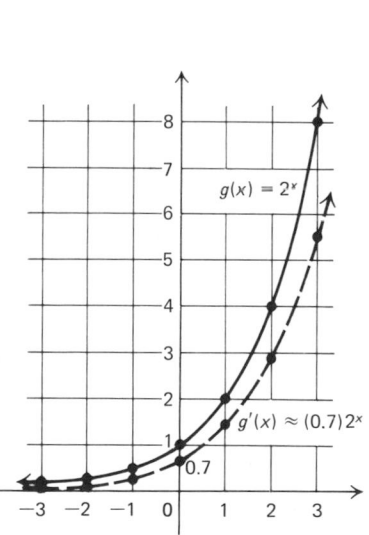

Figure 1

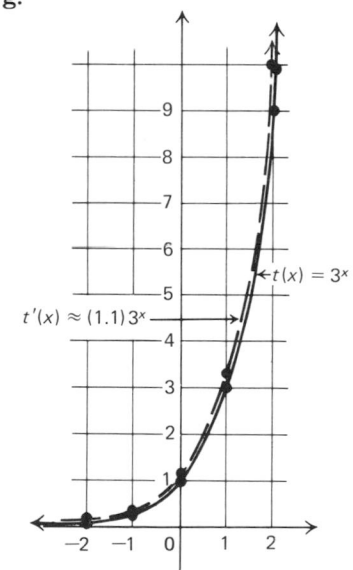

Figure 2

In Fig. 2 we have graphed $t(x) = 3^x$ and $t'(x) \approx (1.1)3^x$. Note that the graph of t' lies *above* the graph of t.

40. a) Complete this table.

h	$\dfrac{3^h - 1}{h}$
$\frac{1}{2}$	
$\frac{1}{4}$	
$\frac{1}{8}$	
$\frac{1}{16}$	
$\frac{1}{32}$	

b) To the nearest tenth, what is the value of

$$\lim_{h \to 0} \frac{3^h - 1}{h}?$$

41. a) Complete this table.

x	-3	-2	-1	0	1	2	3
2^x			0.5				
$(0.7)2^x$			0.35				

b) Using the same set of axes, graph $g(x) = 2^x$, with a solid curve; and $g'(x) \approx (0.7)2^x$ with a dashed curve.

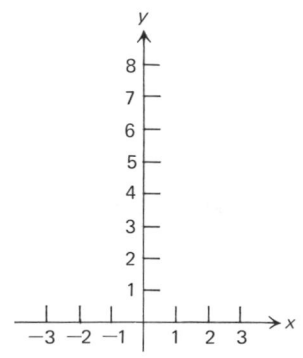

42. a) Complete this table.

x	-2	-1	0	1	2
3^x					
$(1.1)3^x$					

b) Using the same set of axes, graph $h(x) = 3^x$, with a solid line, and $h'(x) \approx (1.1)3^x$ with a dashed line.

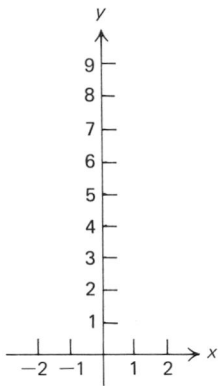

43. (▦ with a y^x key.)

The compound interest formula which we developed in Chapter 1 is

$$P = P_0\left(1 + \frac{i}{n}\right)^{nt},$$

where P is the amount an initial investment P_0 will be worth after t years at interest rate i, compounded n times per year.

Suppose $1 is an initial investment at 100% interest for 1 year (no bank would pay this). The above formula becomes

$$P = \left(1 + \frac{1}{n}\right)^n.$$

We might expect that there is exactly one base a between 2 and 3 for which a^x and its derivative have the same graph. This conjecture can be proved (though we will not do it here). We define the number e to be the unique positive real number for which

$$\lim_{h \to 0} \frac{e^h - 1}{h} = 1.$$

It follows that for the exponential function $f(x) = e^x$,

$$f'(x) = e^x \cdot \lim_{h \to 0} \frac{e^h - 1}{h} = e^x \cdot 1 = e^x.$$

In Margin Exercise 43 you will not only consider an application of e, but also find decimal approximations.

DO EXERCISE 43.

Suppose we were to have the compounding periods n increase indefinitely. The amount in the investment of the margin exercise would be growing at interest compounded continuously, and would approach about $2.718. It can be shown that the number e can be described by a limit:

$$e = \lim_{n \to \infty} \left(1 + \frac{1}{n}\right)^n.$$

That is, e is that number which

$$\left(1 + \frac{1}{n}\right)^n$$

approaches as n gets larger without bound. To ten decimal places e is given by

$$e = 2.7182818284 \cdots.$$

We have established that for the function $f(x) = e^x$, we also have $f'(x) = e^x$. Or, simply,

$$\frac{d}{dx} e^x = e^x.$$

Note that this says that the derivative (the slope of the tangent line) at any x is the same as the function value. Let us find some other derivatives.

Example 1 $\dfrac{d}{dx} 3e^x = 3e^x$

Example 2 $\dfrac{d}{dx} x^2 e^x = x^2 \cdot e^x + 2x \cdot e^x$ (Product Rule)

$= e^x(x^2 + 2x)$, or $xe^x(x + 2)$ (Factoring)

Example 3 $\dfrac{d}{dx}\left(\dfrac{e^x}{x^3}\right) = \dfrac{x^3 \cdot e^x - e^x(3x^2)}{x^6}$ (Quotient Rule)

$= \dfrac{x^2 e^x(x - 3)}{x^6}$ (Factoring)

$= \dfrac{e^x(x - 3)}{x^4}$ (Simplifying)

DO EXERCISES 44 THROUGH 46.

The following rule (a form of the Chain Rule) allows us to find many other derivatives.

$$\frac{d}{dx} e^{f(x)} = f'(x)e^{f(x)}, \quad \text{or} \quad \frac{d}{dx} e^{\square} = \boxed{}' \cdot e^{\square}$$

The following gives us a way to remember this rule.

$$(2x - 5)e^{x^2-5x} \quad \overset{e^{x^2-5x}}{\nearrow} \quad \begin{array}{l}\text{Multiply the original function by the}\\ \text{derivative of the exponent.}\end{array}$$

Example 4 $\dfrac{d}{dx} e^{3x} = 3e^{3x}$

Example 5 $\dfrac{d}{dx} e^{-x^2+4x-7} = (-2x + 4)e^{-x^2+4x-7}$

Example 6 $\dfrac{d}{dx} e^{\sqrt{x^2-3}} = \frac{1}{2}(x^2 - 3)^{-1/2} \cdot 2x \cdot e^{\sqrt{x^2-3}}$

$= x(x^2 - 3)^{-1/2} \cdot e^{\sqrt{x^2-3}}$

$= \dfrac{xe^{\sqrt{x^2-3}}}{\sqrt{x^2 - 3}}$

DO EXERCISES 47 THROUGH 49.

Graphs of e^x, e^{-x}, and $1 - e^{-kx}$

We can use a calculator or Table 4 (in the back of the book) to find approximate values of e^x and e^{-x}. With these we can draw graphs

Complete this table. Round to six decimal places.

n	$\left(1 + \dfrac{1}{n}\right)^n$
1 (compounding annually)	
2 (compounding semiannually)	
3	
4 (compounding quarterly)	
5	
100	
365 (compounding daily)	
8760 (compounding hourly)	

Differentiate.

44. $6e^x$

45. $x^3 e^x$

46. $\dfrac{e^x}{x^2}$

Differentiate.

47. e^{-4x}

48. e^{x^3+8x}

49. $e^{\sqrt{x^2+5}}$

50. Graph $f(x) = 2e^{-x}$. Use Table 4. For example, for $x = 3$, $f(3) = 2e^{-3} = 2(0.0498) \approx 0.1$.

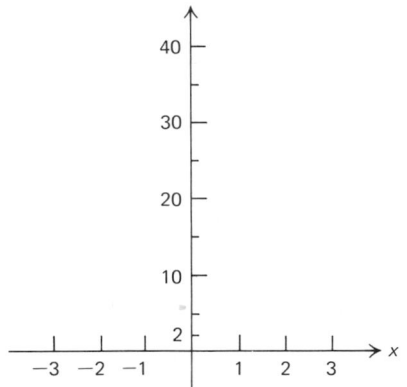

(Figures 3 and 4) of the functions.

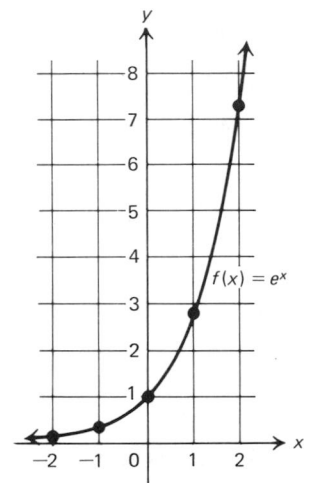

Figure 3

Figure 4

Note that the graph of e^{-x} is a reflection or mirror image of the graph of e^x across the y-axis.

DO EXERCISE 50.

Example 7   Graph $f(x) = 1 - e^{-2x}$, for nonnegative values of x.

Solution   We obtain these values using a calculator or Table 4 at the back of the book.

x	0	$\frac{1}{2}$	1	2	3
$f(x)$	0	0.63	0.86	0.98	0.998

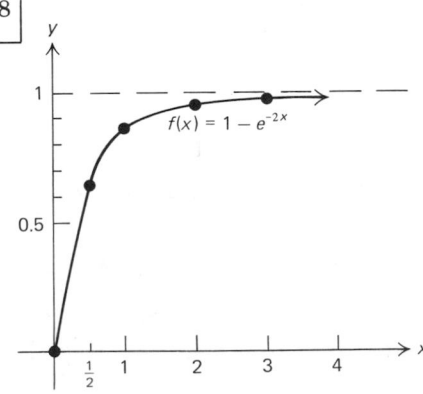

For example,

$$f(1) = 1 - e^{-2 \cdot 1}$$
$$= 1 - e^{-2}$$
$$= 1 - 0.135335 \approx 0.86.$$

DO EXERCISE 51.

In general, the graph of $f(x) = 1 - e^{-kx}$, for $k > 0$, increases from 0, since $f'(x) = ke^{-kx} > 0$, and approaches 1 as x gets larger; that is, $\lim_{x \to \infty} (1 - e^{-kx}) = 1$.

Application

Example 8 *Business*. A company begins a radio advertising campaign in New York City to market a new product. The percentage of the "target market" that buys a product is normally a function of the length of the advertising campaign. The radio station estimates this precentage as $(1 - e^{-0.04t})$ for this type of product, where $t =$ number of days of the campaign. The target market is estimated to be 1,000,000 people and the price per unit is \$0.50. The costs of advertising are \$1000 per day. Find the length of the advertising campaign that will result in maximum profit.

Solution That the percentage of the target market that buys the product can be modeled by $f(t) = 1 - e^{-0.04t}$ is justified by looking at its graph.

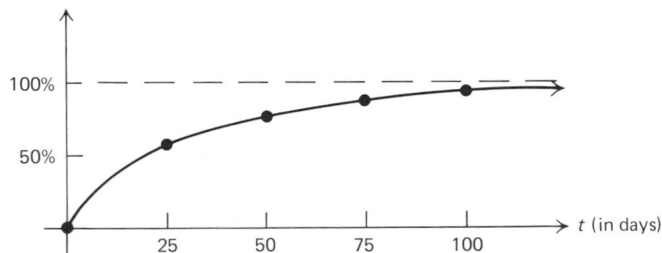

The function increases from 0(0%) towards 1(100%). The longer the advertising campaign, the larger the percentage of the market that has bought the product.

Recall the profit function (here expressed in terms of time, t):

$$\text{Profit} = \text{Revenue} - \text{Cost}$$
$$P(t) = R(t) - C(t).$$

51. a) Complete this table for
$$f(x) = 1 - e^{-x}.$$

x	0	$\frac{1}{2}$	1	2	3	4
$f(x)$						

b) Graph $f(x) = 1 - e^{-x}$.

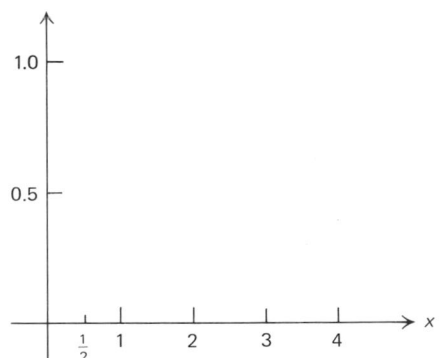

52. Solve the problem in the example when the price per unit is \$0.80.

a) Find $R(t)$.

$R(t)$ = (Price per unit) $\cdot$ (Target market) $\cdot$ (Percentage buying)
$R(t) = 0.5(1{,}000{,}000)(1 - e^{-0.04t}) = 500{,}000 - 500{,}000e^{-0.04t}$

b) Find $C(t)$

$C(t)$ = (Advertising costs per day) $\cdot$ (Number of days)
$C(t) = 1000t.$

c) Find $P(t)$, and take its derivative.

$$P(t) = R(t) - C(t)$$
$$P(t) = 500{,}000 - 500{,}000e^{-0.04t} - 1000t$$
$$P'(t) = (-0.04)(-500{,}000e^{-0.04t}) - 1000$$
$$P'(t) = 20{,}000e^{-0.04t} - 1000.$$

d) Set the first derivative equal to 0 and solve.

$$20{,}000e^{-0.04t} - 1000 = 0$$
$$20{,}000e^{-0.04t} = 1{,}000$$
$$e^{-0.04t} = \frac{1{,}000}{20{,}000} = 0.05 \qquad\qquad (1)$$

$$\ln e^{-0.04t} = \ln 0.05$$
$$-0.04t = \ln 0.05$$
$$t = \frac{\ln 0.05}{-0.04}$$
$$t = \frac{-2.995732}{-0.04} \qquad (\text{▦ or Table 3.})$$
$$t \approx 75.$$

e) We have only one critical point. So we can use the second derivative to determine if we have a maximum.

$$P''(t) = -0.04(20{,}000e^{-0.04t}) = -800e^{-0.04t}$$

Now since exponential functions are positive, $e^{-0.04t} > 0$ for all numbers t. Thus $-800e^{-0.04t} < 0$ for all numbers t. Thus $P''(t)$ is less than 0 for $t = 75$ and we have a maximum.

The length of the advertising campaign must be 75 days to result in maximum profit.

DO EXERCISE 52.

A word of caution! Functions of the type a^x (for example, 2^x, 3^x, and e^x) are different from functions of the type x^a (for example, x^2, x^3, $x^{1/2}$). For a^x the variable is in the exponent. For x^a the variable is in the base. The derivative of a^x is not xa^{x-1}. In particular,

$$\frac{d}{dx}e^x \neq xe^{x-1}, \quad \text{but} \quad \frac{d}{dx}e^x = e^x.$$

EXERCISE SET 10.2

Differentiate.

1. e^{3x} **2.** e^{2x} **3.** $5e^{-2x}$ **4.** $4e^{-3x}$ **5.** $3 - e^{-x}$ **6.** $2 - e^{-x}$

7. $-7e^x$ **8.** $-4e^x$ **9.** $\frac{1}{2}e^{2x}$ **10.** $\frac{1}{4}e^{4x}$ **11.** $x^4 e^x$ **12.** $x^5 e^x$

13. $\dfrac{e^x}{x^4}$ **14.** $\dfrac{e^x}{x^5}$ **15.** e^{-x^2+7x} **16.** e^{-x^2+8x} **17.** $e^{-x^2/2}$ **18.** $e^{x^2/2}$

19. $e^{\sqrt{x-7}}$ **20.** $e^{\sqrt{x-4}}$ **21.** $\sqrt{e^x - 1}$ **22.** $\sqrt{e^x + 1}$ **23.** $xe^{-2x} + e^{-x} + x^3$

24. $e^x + x^3 - xe^x$ **25.** $1 - e^{-x}$ **26.** $1 - e^{-3x}$ **27.** $1 - e^{-kx}$ **28.** $1 - e^{-mx}$

Graph, using Table 4.

29. $f(x) = e^{2x}$ **30.** $f(x) = e^{(1/2)x}$ **31.** $f(x) = e^{-2x}$ **32.** $f(x) = e^{-(1/2)x}$

33. $f(x) = 1 - e^{-x}$, for nonnegative values of x.   **34.** $f(x) = 2(1 - e^{-x})$, for nonnegative values of x.

Applied Problems

35. *Business.* Solve the advertising problem (Example 8) where the costs of advertising are \$2000 per day.

36. *Business.* Solve the advertising problem where the costs of advertising are \$4000 per day.

37. *Business.* A company's total cost, in millions of dollars, is given by

$$C(t) = 100 - 50e^{-t},$$

where $t = $ time. Find
a) the marginal cost $C'(t)$,
b) $C'(0)$,
c) $C'(4)$.

38. *Business.* A company's total cost, in millions of dollars, is given by

$$C(t) = 200 - 40e^{-t},$$

where $t = $ time. Find
a) the marginal cost $C'(t)$,
b) $C'(0)$,
c) $C'(5)$.

39. *Biomedical—Acceptances of a new medicine.* The percentage P of doctors who accept a new medicine is given by

$$P(t) = 1 - e^{-0.2t},$$

where $t = $ time, in months.

a) Find $P(1)$ and $P(6)$.
b) Find $P'(t)$.
c) How many months will it take for 90% of the doctors to become aware of the new medicine?

40. *Psychology—Hullian learning model.* A typist learns to type W words per minute after t weeks of practice, where W is given by

$$W(t) = 100(1 - e^{-0.3t}).$$

a) Find $W(1)$ and $W(8)$.
b) Find $W'(t)$.
c) After how many weeks will the typist's speed be 95 words per minute?

41. Business—*Growth of a stock*. The value of a stock is modeled by

$$V(t) = \$58(1 - e^{-1.1t}) + \$20,$$

where V is the value of the stock after time t, in months.

a) Find $V(1)$ and $V(12)$.
b) Find $V'(t)$.
c) After how many months will the value of the stock be $75?

42. Business—*Marginal revenue*. The demand function for a certain product is given by

$$p = D(x) = 800e^{-0.125x}.$$

Recall that total revenue is given by $R(x) = xD(x)$.

a) Find $R(x)$.
b) Find the marginal revenue, $R'(x)$.
c) At what value of x will the revenue be maximum?

▶───────────────────────────────────

Differentiate.

43. $(e^{3x} + 1)^5$

44. $(e^{x^2} - 2)^4$

45. $\dfrac{e^{3t} - e^{7t}}{e^{4t}}$

46. $\sqrt[3]{e^{3t} + t}$

47. $\dfrac{e^x}{x^2 + 1}$

48. $\dfrac{e^x}{1 - e^x}$

49. $e^{\sqrt{x}} + \sqrt{e^x}$

50. $\dfrac{1}{e^x} + e^{1/x}$

51. $e^{x/2} \cdot \sqrt{x - 1}$

52. $\dfrac{xe^{-x}}{1 + x^2}$

53. $\dfrac{e^x - e^{-x}}{e^x + e^{-x}}$

▦ with y^x key. Each of the following is an expression for e. Find the function values that are approximations for e. Round to five decimal places.

54. $e = \lim_{t \to 0} f(t)$; $f(t) = (1 + t)^{1/t}$. Find $f(1)$, $f(0.5)$, $f(0.2)$, $f(0.1)$, and $f(0.001)$.

55. $e = \lim_{t \to 1} g(t)$; $g(t) = t^{1/(t-1)}$. Find $g(0.5)$, $g(0.9)$, $g(0.99)$, $g(0.999)$, and $g(0.9998)$.

───────────────────────────────────

OBJECTIVES

You should be able to

a) Differentiate functions involving natural logarithms.
b) Graph functions like $y = \ln x$ and $y = 2 + \ln x$, using Table 3.
c) Solve maximum–minimum problems involving natural logarithm functions.

10.3 THE NATURAL LOGARITHM FUNCTION

Recall the definition of logarithms:

$$\text{"}y = \log_e x\text{"} \quad \text{means} \quad \text{"}x = e^y.\text{"}$$

Thus, for natural logarithms, $\log_e x$ is that number y such that $x = e^y$. The number $\log_e x$ is called the *natural logarithm* of x and is abbreviated $\ln x$. That is,

$$\ln x = \log_e x.$$

There are two ways we might obtain the graph of $y = \ln x$. One is by writing its equivalent equation $x = e^y$.

Then we select values for y, and use a calculator or Table 4 to find the corresponding values of e^y. We then plot points, remembering that x still is the first coordinate.

x(or e^y)	0.4	0.1	1	2.7	7.4	20
y	−1	−2	0	1	2	3

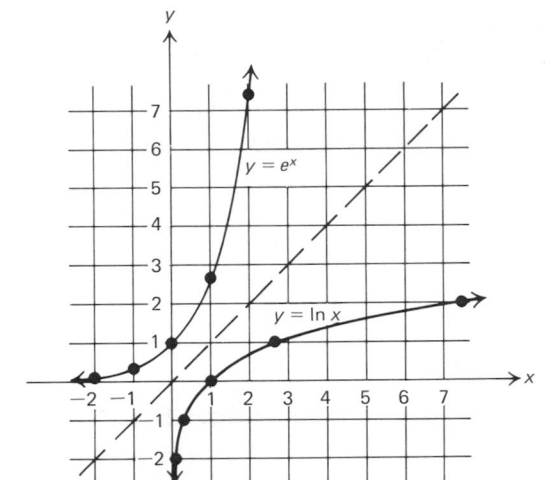

Figure 5

53.
a) Complete, using ▦ or Table 3.

x	0.5	1	2	3	4
$\ln x$			0.7		

b) Graph $y = \ln x$.

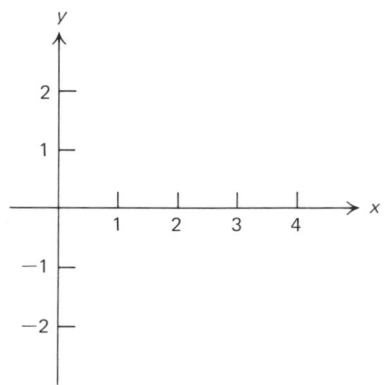

(How does this procedure in Fig. 5 compare with that used in plotting Fig. 3?) Note that the graph of $y = \ln x$ is a reflection, or mirror image, across the line $y = x$, of the graph of $y = e^x$.

The second way of graphing $y = \ln x$ is by using a calculator or Table 3, which is a table of natural logarithms at the back of the book. For example, $\ln 2 = 0.6931 \approx 0.7$.

DO EXERCISE 53.

These properties follow.

$\ln x$ exists only for positive number x.
$\ln x < 0$ for $0 < x < 1$.
$\ln x > 0$ for $x > 1$.

The Derivative of ln x

Consider $f(x) = \ln x$.

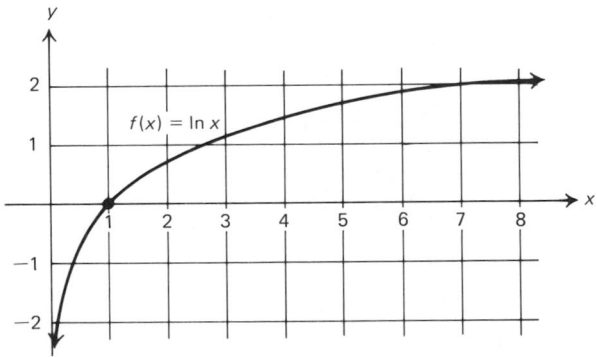

Differentiate.

54. $5 \ln x$

55. $x^3 \ln x + 4x$

56. $\dfrac{\ln x}{x^2}$

We can show that $f'(x) = 1/x$ (the slope of the tangent line at x is just the reciprocal of x). We are trying to find the derivative of

$$f(x) = \ln x. \tag{1}$$

We first write its equivalent exponential equation

$$e^{f(x)} = x. \tag{2}$$

Now we differentiate both sides of this equation.

$$\frac{d}{dx} e^{f(x)} = \frac{d}{dx} x$$

$$f'(x) \cdot e^{f(x)} = 1$$

$$f'(x) \cdot x = 1 \qquad \text{(Substituting } x \text{ for } e^{f(x)} \text{ from Eq. (2))}$$

$$f'(x) = \frac{1}{x}$$

Thus

$$\frac{d}{dx} \ln x = \frac{1}{x}.$$

This is true only for positive values of x, since $\ln x$ is defined only for positive numbers. Let us find some derivatives.

Example 1 $\dfrac{d}{dx} 3 \ln x = \dfrac{3}{x}$

Example 2

$$\frac{d}{dx}(x^2 \ln x + 5x) = x^2 \cdot \frac{1}{x} + 2x \cdot \ln x + 5 \quad \text{(Product Rule on } x^2 \ln x)$$

$$= x + 2x \cdot \ln x + 5$$

$$= x(1 + 2 \ln x) + 5. \qquad \text{(Simplifying)}$$

Example 3

$$\frac{d}{dx}\left(\frac{\ln x}{x^3}\right) = \frac{x^3 \cdot \frac{1}{x} - (\ln x)(3x^2)}{x^6} \qquad \text{(Quotient Rule)}$$

$$= \frac{x^2 - 3x^2 \ln x}{x^6}$$

$$= \frac{x^2(1 - 3 \ln x)}{x^6} \qquad \text{(Factoring)}$$

$$= \frac{1 - 3 \ln x}{x^4}. \qquad \text{(Simplifying)}$$

DO EXERCISES 54 THROUGH 56.

The following rule (a form of the Chain Rule) allows us to find many other derivatives.

$$\frac{d}{dx}\ln f(x) = f'(x)\cdot\frac{1}{f(x)}, \quad \text{or} \quad \frac{d}{dx}\ln \boxed{} = \boxed{}'\cdot\frac{1}{\boxed{}}.$$

The following gives us a way of remembering this rule.

$$\ln(x^2 - 8x)$$

①   ②

$$(2x - 8)\cdot\frac{1}{x^2 - 8x}$$

1. Differentiate the "inside" function.

2. Multiply by the reciprocal of the "inside" function.

Example 4 $\dfrac{d}{dx}\ln 3x = 3\cdot\dfrac{1}{3x} = \dfrac{1}{x}.$

Note that we could have done this another way using Property 1:

$$\ln 3x = \ln 3 + \ln x;$$

then

$$\frac{d}{dx}\ln 3x = \frac{d}{dx}\ln 3 + \frac{d}{dx}\ln x = 0 + \frac{1}{x} = \frac{1}{x}.$$

Example 5 $\dfrac{d}{dx}\ln(x^2 - 5) = 2x\cdot\dfrac{1}{x^2 - 5} = \dfrac{2x}{x^2 - 5}.$

Example 6 $\dfrac{d}{dx}\ln(\ln x) = \dfrac{1}{x}\cdot\dfrac{1}{\ln x} = \dfrac{1}{x\ln x}.$

Example 7

$$\frac{d}{dx}\ln\left(\frac{x^3 + 4}{x}\right) = \frac{d}{dx}[\ln(x^3 + 4) - \ln x] \qquad \text{(Property 2. This avoids using the Quotient Rule.)}$$

$$= 3x^2\cdot\frac{1}{x^3 + 4} - \frac{1}{x} = \frac{3x^2}{x^3 + 4} - \frac{1}{x}$$

$$= \frac{3x^2}{x^3 + 4}\cdot\frac{x}{x} - \frac{1}{x}\cdot\frac{x^3 + 4}{x^3 + 4}$$

$$= \frac{(3x^2)x - (x^3 + 4)}{x(x^3 + 4)} = \frac{3x^3 - x^3 - 4}{x(x^3 + 4)} = \frac{2x^3 - 4}{x(x^3 + 4)}$$

DO EXERCISES 57 THROUGH 60.

Differentiate.

57. $\ln 5x$

58. $\ln(3x^2 + 4)$

59. $\ln(\ln 5x)$

60. $\ln\left(\dfrac{x^5 - 2}{x}\right)$

Application

Example 8 *Forgetting.* In a psychological experiment students were shown a set of nonsense syllables, such as *POK*, and asked to recall them every second thereafter. The percentage $R(t)$ who retained the syllables after t seconds was found to be given by

$$R(t) = 80 - 27 \ln t, \quad \text{for} \quad t \geq 1.$$

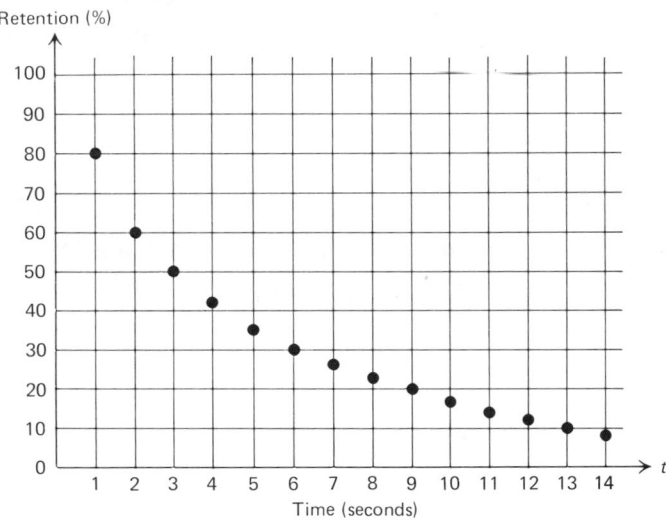

Figure 6

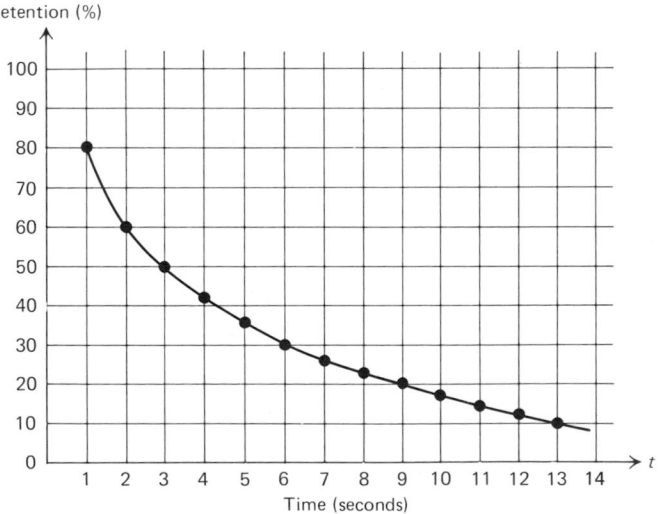

Figure 7

(Strictly speaking, the function is not continuous and has a graph as shown in Fig. 6. But in order to use calculus, we "fill in" the graph with a smooth curve (Fig. 7), considering $R(t)$ to be defined for any number $t \geqslant 1$. This is not unreasonable, since we are now able to find the percentage who retained the syllables after $t = 3.417$ seconds, instead of just after integer values such as 1, 2, 3, 4, and so on.)

a) What percentage retained the syllables after 1 second?

b) Find $R'(t)$, the rate of change of R with respect to t.

c) Find maximum and minimum values, if they exist.

Conduct your own memory experiment. Study this photograph carefully. Then put it aside and write down as many items as you can. Wait a half-hour and again write down as many as you can. Do this ˚ve more times. Make a graph of the number of items you remember versus time. Does the graph appear to be logarithmic? (*Clif Garboden: Stock, Boston*)

Solution

a) $R(1) = 80 - 27 \cdot \ln 1 = 80 - 27 \cdot 0 = 80\%$

b) $R'(t) = -27 \cdot \dfrac{1}{t} = -\dfrac{27}{t}$

c) Now $R'(t)$ exists for all values of t in the interval $[1, \infty)$. Note that for $t \geqslant 1$, $-27/t < 0$. Thus there are no critical points and R is decreasing. Then R has a maximum value at the endpoint 1. This maximum value is $R(1)$, or 80%.

61. *Advertising.* A model for advertising response is given by

$$N(a) = 500 + 200 \ln a, \quad a \geq 1,$$

where

$N(a)$ = number of units sold

a = amount spent on advertising, in thousands of dollars.

a) How many units were sold after spending 1 thousand dollars? (Substitute 1 for a, not 1000.)

b) Find $N'(a)$.

c) Find maximum and minimum values, if they exist.

DO EXERCISE 61.

Example 8 and the problem in the margin exercise are presented in reverse of how they might have come up if we were constructing models with the aid of calculus. That is, we might begin our reasoning about the rate of forgetting. It might be reasoned that the rate of forgetting $R'(t)$ is inversely proportional to time; that is, $R'(t) = -A/t$. Then we would reason backward to determine the function $R(t) = B - A \ln t$. Models constructed with the aid of calculus often grow out of assumptions about rates of change of quantities connected with the phenomenon to be modeled.

EXERCISE SET 10.3

Differentiate.

1. $-6 \ln x$

2. $-4 \ln x$

3. $x^4 \ln x - \frac{1}{2}x^2$

4. $x^5 \ln x - \frac{1}{4}x^4$

5. $\dfrac{\ln x}{x^4}$

6. $\dfrac{\ln x}{x^5}$

7. $\ln \dfrac{x}{4}$ $\left[\textit{Hint:} \ln \dfrac{x}{4} = \ln x - \ln 4\right]$

8. $\ln \dfrac{x}{2}$

9. $\ln (5x^2 - 7)$

10. $\ln (7x^3 + 4)$

11. $\ln (\ln 4x)$

12. $\ln (\ln 3x)$

13. $\ln \left(\dfrac{x^2 - 7}{x}\right)$

14. $\ln \left(\dfrac{x^2 + 5}{x}\right)$

15. $e^x \ln x$

16. $e^{2x} \ln x$

17. $\ln (e^x + 1)$

18. $\ln (e^x - 2)$

19. $(\ln x)^2$

20. $(\ln x)^3$
[*Hint:* The Extended Power Rule]

Applied Problems

21. *Psychology—Forgetting.* Students in college botany took a final exam. They took equivalent forms of the exam in monthly intervals thereafter. The average score, $S(t)$ in percent, after t months was found to be given by

$$S(t) = 68 - 20 \ln (t + 1), \quad t \geq 0.$$

a) What was the average score when they initially took the test, $t = 0$?

b) What was the average score after 4 months?

c) What was the average score after 24 months?

d) What percentage of the initial score did they retain after 2 years (24 months)?

e) Find $S'(t)$.

f) Find maximum and minimum values, if they exist.

22. *Psychology—Forgetting.* Students in college zoology took a final exam. They took equivalent forms of the exam in monthly intervals thereafter. The average score, $S(t)$ in percent, after t months was found to be given by

$$S(t) = 78 - 15 \ln (t + 1), \quad t \geq 0.$$

a) What was the average score when they initially took the test, $t = 0$?

b) What was the average score after 4 months?

c) What was the average score after 24 months?

d) What percentage of the initial score did they retain after 2 years (24 months)?

e) Find $S'(t)$.

f) Find maximum and minimum values, if they exist.

23. *Business—Advertising.* A model for advertising response is given by

$$N(a) = 1000 + 200 \ln a, \qquad a \geqslant 1,$$

where

$N(a)$ = number of units sold

$\quad a$ = amount spent on advertising
$\quad\quad$ in thousands of dollars.

a) How many units were sold after spending 1 thousand dollars ($a = 1$) on advertising?
b) Find $N'(a)$, $N'(10)$.
c) Find maximum and minimum values, if they exist.

25. *Psychology—Walking speed.* Bornstein and Bornstein found in a study that the average walking speed v of a person living in a city of population p, in thousands, is given by

$$v(p) = 0.86 \ln p + 0.05,$$

where v is in feet per second.
a) The population of Seattle is 531,000. What is the average walking speed of a person living in Seattle? Find $v(531)$.
b) The population of New York is 7,900,000. What is the average walking speed of a person living in New York?
c) Find $v'(p)$. Interpret $v'(p)$.

24. *Business—Advertising.* A model for advertising response is given by

$$N(a) = 2000 + 500 \ln a, \qquad a \geqslant 1,$$

where

$N(a)$ = number of units sold

$\quad a$ = amount spent on advertising,
$\quad\quad$ in thousands of dollars.

a) How many units were sold after spending 1 thousand dollars ($a = 1$) on advertising?
b) Find $N'(a)$, $N'(10)$.
c) Find maximum and minimum values, if they exist.

26. *Biomedical—The Reynolds number.* For many kinds of animals the Reynolds number R is given by

$$R = A \ln r - Br,$$

where A and B are positive constants and r is the radius of the aorta. Find the maximum value of R.

▶ ──

Differentiate.

27. $(\ln x)^{-4}$

28. $(\ln x)^n$

29. $\ln (t^3 + 1)^5$

30. $\ln (t^2 + t)^3$

31. $[\ln (x + 5)]^4$

32. $\ln [\ln (\ln 3x)]$

33. $\ln (t^3 + 3)(t^2 - 1)$

34. $\ln \dfrac{1 - t}{1 + t}$

35. $\ln \dfrac{x^5}{(8x + 5)^2}$

36. $\ln \sqrt{5 + x^2}$

37. $\dfrac{\ln t^2}{t^2}$

38. $\frac{1}{5}x^5(\ln x - \frac{1}{5})$

39. $\dfrac{x^{n+1}}{n + 1}\left(\ln x - \dfrac{1}{n + 1}\right)$

40. $\dfrac{x \ln x - x}{x^2 + 1}$

41. $\ln (t + \sqrt{1 + t^2})$

42. Find: $\displaystyle\lim_{h \to 0} \dfrac{\ln (1 + h)}{h}$

43. ▦ Which is larger, e^π or π^e?

44. ▦ Find $\sqrt[e]{e}$. Compare it to other expressions of the type $\sqrt[x]{x}$, $x > 0$. What can you conclude?

OBJECTIVES

You should be able to

a) State the solution of an equation

$$\frac{dP}{dt} = kP, \text{ as } P(t) = P_0 e^{kt}.$$

b) Given a growth rate find the doubling time.

c) Given the doubling time, find the growth rate.

d) Solve applied problems involving exponential growth.

62. a) Differentiate $y = 5e^{4x}$.

b) Express $\frac{dy}{dx}$ in terms of y.

10.4 APPLICATIONS: THE UNINHIBITED GROWTH MODEL, $\frac{dP}{dt} = kP$

What will the world population be in 1986? (*Peter Vandermark: Stock, Boston*)

Consider the function

$$f(x) = 2e^{3x}.$$

Differentiating, we get

$$f'(x) = 3 \cdot 2e^{3x} = 3 \cdot f(x).$$

This, graphically, says that the derivative, or slope of the tangent line, is simply the constant 3 times the function value.

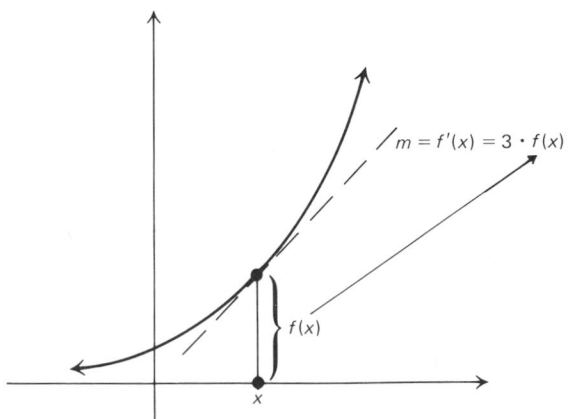

DO EXERCISE 62.

In general,

A function $y = f(x)$ satisfies the equation

$$\frac{dy}{dx} = ky \qquad [f'(x) = k \cdot f(x)]$$

if and only if

$$y = ce^{kx} \qquad [f(x) = ce^{kx}]$$

for some constant c.

No matter what the variables, you should be able to write the solution.

Example 1   The solution of $\dfrac{dA}{dt} = kA$ is $A = ce^{kt}$, or $A(t) = ce^{kt}$.

Example 2   The solution of $\dfrac{dP}{dt} = kP$ is $P = ce^{kt}$, or $P(t) = ce^{kt}$.

Example 3   The solution of $f'(Q) = k \cdot f(Q)$ is $f(Q) = ce^{kQ}$.

DO EXERCISE 63.

The equation

$$\frac{dP}{dt} = kP, \quad k > 0 \qquad [P'(t) = k \cdot P(t), \quad k > 0]$$

is the basic model of uninhibited population growth, whether it be a population of humans, a bacteria culture, or money invested at interest compounded continuously. Neglecting special inhibiting and stimulating factors, a population normally reproduces itself at a rate proportional to its size, and this is exactly what the equation $dP/dt = kP$ says. The solution of the equation is

$$P(t) = ce^{kt}, \tag{1}$$

where $t =$ time. At $t = 0$, we have some "initial" population $P(0)$ that we will represent by P_0. We can rewrite Eq. (1) in terms of P_0 as follows:

$$P_0 = P(0) = ce^{k \cdot 0} = ce^0 = c \cdot 1 = c.$$

Thus $P_0 = c$, so we can express $P(t)$ as

$$P(t) = P_0 e^{kt}.$$

63. a) State the solution of

$$\frac{dN}{dt} = kN.$$

b) State the solution of

$$f'(t) = k \cdot f(t).$$

64. *Exploratory exercises—Growth.* Use a sheet of $8\frac{1}{2} \times 11$ paper. Cut it into two equal pieces. Then cut these into four equal pieces. Then cut these into eight equal pieces, and so on, performing five cutting steps.

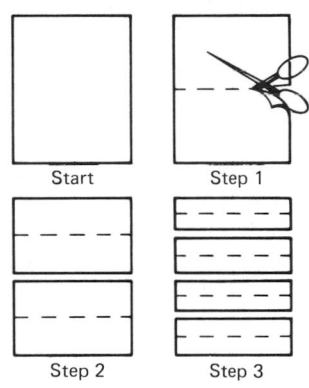

Start   Step 1

Step 2   Step 3

a) Place all the pieces in a stack and measure the thickness.

b) A piece of paper is typically 0.004 in. thick. Check the calculation in (a) by completing this table.

	t	$0.004 \cdot 2^t$
Start	0	$0.004 \cdot 2^0$, or 0.004
Step 1	1	$0.004 \cdot 2^1$, or 0.008
Step 2	2	$0.004 \cdot 2^2$, or 0.016
Step 3	3	
Step 4	4	
Step 5	5	

c) Compute the thickness of the paper (in miles) after 25 steps.

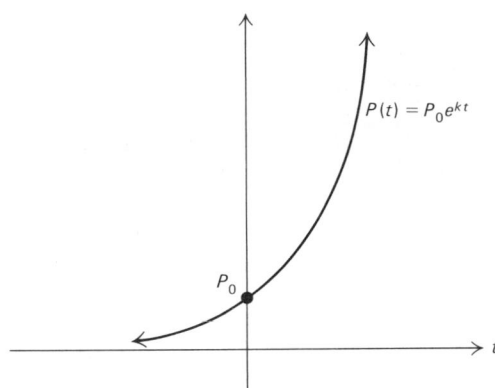

$P(t) = P_0 e^{kt}$

P_0

t

Its graph is the curve shown here, which shows how uninhibited growth results in a "population explosion."

DO EXERCISE 64.

The constant k is called the *rate of exponential growth*, or simply the *growth rate*. This is not the rate of change of the population size, which is

$$\frac{dP}{dt} = kP,$$

but the constant that P must be multiplied by to get its rate of change. It is thus a different use of the word *rate*. It is like the *interest rate* paid by a bank. If the interest rate is 7%, or 0.07, we do not mean that your bank balance P is growing at the rate of 0.07 dollars per year, but at the rate of $0.07P$ dollars per year. We therefore express the rate as 7% per year, rather than 0.07 dollars per year. We could say that the rate is 0.07 dollars *per dollar* per year. When interest is compounded continuously, the interest rate *is* a true exponential growth rate.

Example 4 *Business—Interest compounded continuously.* Suppose an amount P_0 is invested in a savings account where interest is compounded continuously at 7% per year. That is, the balance P grows at the rate given by

$$\frac{dP}{dt} = 0.07P.$$

a) Find the solution of the equation in terms of P_0 and 0.07.
b) Suppose $100 is invested. What is the balance after 1 year?
c) When will an investment of $100 double itself?

Solution

a) $P(t) = P_0 e^{0.07t}$

b) $P(1) = 100 e^{0.07(1)} = 100 e^{0.07} = 100(1.072508)$

$$\approx \$107.25 \quad \text{(⌨ or Table 4.)}$$

c) We are asking at what time T does $P(T) = \$200$. The number T is called the *doubling time*. To find T we solve the equation

$$200 = 100 e^{0.07 \cdot T}$$

$$2 = e^{0.07T} \quad \text{(Multiplying by } \tfrac{1}{100})$$

We use natural logarithms to solve this equation.

$$\ln 2 = \ln e^{0.07T}$$

$$\ln 2 = 0.07T \quad \text{(Property 5: } \ln e^k = k.)$$

$$\frac{\ln 2}{0.07} = T$$

$$\frac{0.693147}{0.07} = T \quad \text{(⌨ or Table 3.)}$$

$$9.9 \approx T$$

Thus \$100 will double itself in 9.9 years.

DO EXERCISE 65.

We can find a general expression relating the growth rate k and the doubling time T by solving the equation

$$2P_0 = P_0 e^{kT}$$

$$2 = e^{kT} \quad \text{(Multiplying by } 1/P_0)$$

$$\ln 2 = \ln e^{kT}$$

$$\ln 2 = kT.$$

The growth rate k and the doubling time T are related by

$$kT = \ln 2 = 0.693147,$$

or

a) $k = \dfrac{\ln 2}{T} = \dfrac{0.693147}{T}$ b) $T = \dfrac{\ln 2}{k} = \dfrac{0.693147}{k}$

Note that this relationship between k and T does not depend on P_0.

65. *Business.* Suppose an amount P_0 is invested in a savings account where interest is compounded continuously at 8% per year. That is, the balance P grows at the rate given by

$$\frac{dP}{dt} = 0.08P.$$

a) Find the solution of the equation in terms of P_0 and 0.08.

b) Suppose \$1000 is invested. What is the balance after 1 year?

c) When will an investment of \$1000 double itself?

66. Complete this table relating growth rate k and doubling time T.

Growth rate k (% per year)	Doubling time T (in years)
2%	
	10
14%	
	15
1%	

67. The population of the United States in 1976 was 216 million. It was estimated that the population P was growing exponentially at the rate of 0.8% per year. That is,

$$\frac{dP}{dt} = 0.008P,$$

where t = time in years.

a) Find the solution of the equation assuming $P_0 = 216$ and $k = 0.008$.

b) Estimate U.S. population in 1981. ($t = 5$).

c) When will the population be double that in 1976?

Example 5   A bank advertises that it will double your money in 4.6 years. What is the interest rate on such an account, assuming interest to be compounded continuously?

Solution $k = \dfrac{\ln 2}{T} = \dfrac{0.693147}{4.6} = 0.151 = 15.1\%$

DO EXERCISE 66.

Example 6   *Ecology—World population growth.* The population of the world passed 4 billion on March 28, 1976. On the basis of data available at that time it was estimated that the population P was growing exponentially at the rate of 1.9% per year. That is, $dP/dt = 0.019P$, where t = time, in years, from 1976. (To facilitate computations we assume the population was 4 billion at the start of 1976).

a) Find the solution of the equation assuming $P_0 = 4$ and $k = 0.019$.

b) Estimate the world population in 1986 ($t = 10$).

c) When will the population be double that in 1976?

Solution

a) $P(t) = 4e^{0.019t}$

b) $P(10) = 4e^{0.019(10)} = 4e^{0.19} = 4(1.209250)$ (▦ or Table 4.)

$$\approx 4.8\ \text{billion}$$

c) $T = \dfrac{\ln 2}{k} = \dfrac{0.693147}{0.019} \approx 36.5\ \text{yr}$

Thus, according to this model, the 1976 population will double by the year 2012. No wonder ecologists are alarmed!

DO EXERCISE 67

The Rule of 70   The relationship between doubling time T and interest rate k is the basis of a rule often used in the investment world, called the *Rule of 70*: To estimate how long it will take to double your money at varying rates of return, divide 70 by the rate of return. To see how this works, let the interest rate $k = r\%$. Then,

$$T = \frac{\ln 2}{k} = \frac{0.693147}{r\%} = \frac{0.693147}{r \times 0.01} = \frac{0.693147}{r \times 0.01} \cdot \frac{100}{100} = \frac{69.3147}{r} \approx \frac{70}{r}$$

Why Would We Expect Population P to Obey the Law $\dfrac{dP}{dt} = kP$?

Suppose, for example, that a growing colony of bacteria has size $P(t)$. Our measurements of P suggest that P grows smoothly; that is, that

P is a differential function of t. Let k be its growth rate when its size is 1. That is,

$$\frac{dP}{dt} = k, \quad \text{when} \quad P = 1.$$

Now we assume, or observe, that the colony grows uniformly. That is, when the colony has grown to size n, we can picture it as composed of n identical colonies, each of size 1, and each growing (at that moment) at the rate k. So

$$\frac{dP}{dt} = kn, \quad \text{when} \quad P = n.$$

That is,

$$\frac{dP}{dt} = kP.$$

Populations grow exponentially under certain conditions. If bacteria were confined to a petri dish, their growth curve would be much different than $P = P_0 e^{kt}$. This is because there is only a limited amount of food, and eventually the waste products cause the population to level off. Human population growth is exponential over relatively short periods of time, say 10 to 50 years.

Strictly speaking, population is an integer-valued and, hence, a discontinuous function of time. Look at this small portion of a graph of $P = P_0 e^{kt}$.

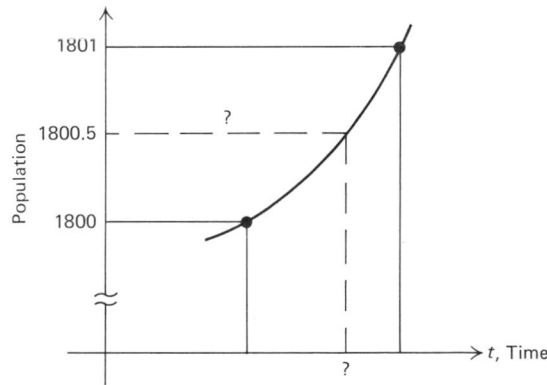

We have assumed that the graph rises from 1800 to 1801. But this implies that we have a value of t for which the population is 1800.5, which is not possible unless you count fractional parts of a preg-

(*Donald C. Dietz: Stock, Boston*)

*Some myths about alcohol.** It's a fact, the blood alcohol concentration (BAC) in the human body is measurable. And there's no cure for its effect on the central nervous system except time. It takes time for the body's metabolism to recover.

That means a cup of coffee, a cold shower, and fresh air can't erase the effect of several drinks.

There are variables, of course: a person's body weight, how many drinks have been consumed in a given time, how much has been eaten, etc. These account for different BAC levels. But the myth that some people can "handle their liquor" better than others is a gross rationalization—especially when it comes to driving. Some people can act more sober than others. But an automobile doesn't act, it reacts.

* *Indianapolis Alcohol Safety Action Project.*

nancy. Nevertheless, estimates of population are usually all we require, and an answer like 1800.5 should cause no difficulty.

Modeling Other Phenomena

Example 7 *Biomedical—Alcohol absorption and the risk of having an accident.* Extensive research has provided data relating the risk $R(\%)$ of having an automobile accident to the blood alcohol level b (%). Note that these data are not a perfect fit (see the part between $b = 0$ and $b = 0.05$), but we shall approximate the data with an exponential function. The modeling assumption is that the rate of change of the risk R with respect to the blood alcohol level b is given by

$$\frac{dR}{db} = kR.$$

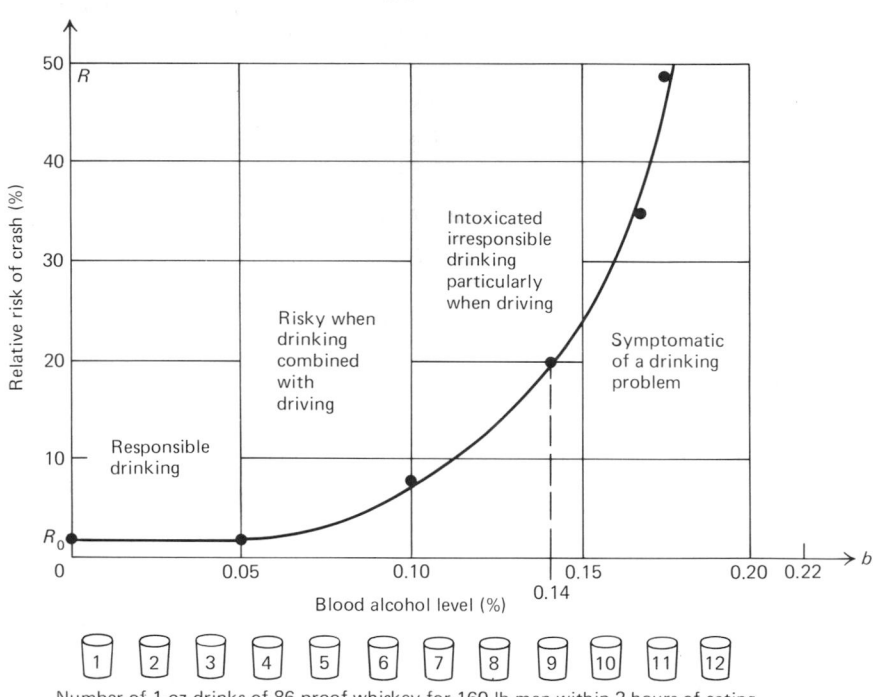

Number of 1-oz drinks of 86 proof whiskey for 160 lb man within 2 hours of eating.

a) Find the solution of the equation, assuming $R_0 = 1\%$.
b) Find k, using the data point $R(0.14) = 20$. (This is how one might fit the data to an exponential equation.)
c) Rewrite $R(b)$ in terms of k.

d) At what blood alcohol level will the risk of having an accident be 100%?

Solution

a) Since both R and b are percents, we omit the % symbol for ease of computation. The solution is

$$R(b) = e^{kb}, \quad \text{since} \quad R_0 = 1.$$

b) We solve this equation for k.

$$20 = e^{k(0.14)} = e^{0.14k}$$

We use natural logarithms to solve this equation.

$$\ln 20 = \ln e^{0.14k}$$

$$\ln 20 = 0.14k$$

$$\frac{\ln 20}{0.14} = k$$

$$\frac{2.995732}{0.14} = k \quad (\text{▦ or Table 3.})$$

$$21.4 = k \quad (\text{Rounding to the nearest tenth})$$

c) $R(b) = e^{21.4b}$

d) We solve this equation for b.

$$100 = e^{21.4b}$$

$$\ln 100 = \ln e^{21.4b}$$

$$\ln 100 = 21.4b$$

$$\frac{\ln 100}{21.4} = b$$

$$\frac{4.605170}{21.4} = b \quad (\text{▦ or Table 3.})$$

$$0.22 = b$$

Thus when the blood alcohol level is 0.22%, according to this model, the risk of an accident is 100%. From the graph, this would occur after 12 1-oz drinks of 86 proof whiskey. "Theoretically" the model tells us that after 12 drinks of whiskey one is "sure" to have an accident. This might be questioned in actuality, since a person who has had 12 drinks might not be able to drive at all.

DO EXERCISE 68.

68. *Ecology—Electrical energy demand.* Past data on electrical energy demand in the U.S. are shown in the graph.

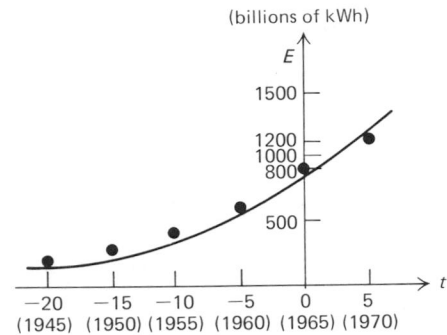
(billions of kWh)

It appears that we can fit an exponential function to the data. We accept the modeling assumption that the rate of change of electrical energy need E (in billion kilowatthours, kWh) with respect to time is given by

$$\frac{dE}{dt} = kE.$$

a) Find the solution of the equation, assuming $E_0 = 800$ billion kWh. That is, at $t = 0(1965)$, $E = 800$.

b) Find k using the data point $E(5) = 1200$ billion kWh. That is, in 1970, 1200 billion kWh were used. Round to the nearest hundredth.

c) Rewrite $E(t)$ in terms of k.

d) How much electrical energy will be needed in 1995?

EXERCISE SET 10.4

1. State the solution of $\dfrac{dQ}{dt} = kQ$ in terms of Q_0.

2. State the solution of $\dfrac{dR}{dt} = kR$ in terms of R_0.

3. *Business—Compound interest.* Suppose P_0 is invested in a savings account where interest is compounded continuously at 9% per year. That is, the balance P grows at the rate given by

$$\frac{dP}{dt} = 0.09P.$$

a) Find the solution of the equation in terms of P_0 and 0.09.
b) Suppose $1000 is invested. What is the balance after 1 year? 2 years?
c) When will an investment of $1000 double itself?

4. *Business—Compound interest.* Suppose P_0 is invested in a savings account where interest is compounded continuously at 10% per year. That is, the balance P grows at the rate given by

$$\frac{dP}{dt} = 0.10P.$$

a) Find the solution of the equation in terms of P_0 and 0.10.
b) Suppose $20,000 is invested. What is the balance after 1 year? 2 years?
c) When will an investment of $20,000 double itself?

5. *Ecology—Population growth.* The growth rate of the population of Central America is 3.5% per year (one of the highest in the world). What is the doubling time?

6. *Ecology—Population growth.* The growth rate of the population of Europe is 1% per year (one of the lowest in the world). What is the doubling time?

7. *Business—Annual interest rate.* A bank advertises that it compounds interest continuously and that it will double your money in 10 years. What is its annual interest rate?

8. *Business—Annual interest rate.* A bank advertises that it compounds interest continuously and that it will double your money in 12 years. What is its annual interest rate?

9. *Ecology—Population growth.* The population of USSR was 209 million in 1959. It was estimated that the population P was growing exponentially at the rate of 1% per year. That is,

$$\frac{dP}{dt} = 0.01P.$$

a) Find the solution of the equation assuming $P_0 = 209$ and $k = 0.01$.
b) Estimate the population of the USSR in 1999.
c) When will the population be double that of 1959?

10. *Ecology—Population growth.* The population of Europe west of the USSR was 430 million in 1961. It was estimated that the population was growing exponentially at the rate of 1% per year. That is,

$$\frac{dP}{dt} = 0.01P.$$

a) Find the solution of the equation assuming $P_0 = 430$ and $k = 0.01$.
b) Estimate the population of Europe in 1991.
c) When will the population be double that of 1961?

11. *Biomedical—Blood alcohol level.* In Example 7 (on alcohol absorption), at what blood alcohol level will the risk of an accident be 80%?

12. *Biomedical—Blood alcohol level.* In Example 7 (on alcohol absorption), at what blood alcohol level will the risk of an accident be 90%?

13. *Business—Franchise expansion.* A national hamburger firm is selling franchises throughout the country. The president estimates that the number of franchises N will increase at a rate of 10% per year. That is,

$$\frac{dN}{dt} = 0.10N.$$

a) Find the solution of the equation, assuming the number of franchises at $t = 0$ is 50.
b) How many franchises will there be in 20 years?
c) When will the initial number of 50 franchises double?

14. *Business—Franchise expansion.* Pizza, Unltd., a national pizza firm, is selling franchises throughout the country. The president estimates that the number of franchises N will increase at a rate of 15% per year. That is,

$$\frac{dN}{dt} = 0.15N.$$

a) Find the solution of the equation, assuming the number of franchises at $t = 0$ is 40.
b) How many franchises will there be in 20 years?
c) When will the initial number of 40 franchises double?

15. *Ecology—Oil demand.* The growth rate of the demand for oil in the United States is 10% per year. When will the demand be double that of 1980?

16. *Ecology—Coal demand.* The growth rate of the demand for coal in the world is 4% per year. When will the demand be double that of 1980?

17. *Ecology—Population growth.* The population of Tempe, Arizona was 25 thousand in 1960. In 1969 it was 52 thousand. Assuming the exponential model,
a) Find the value $k(P_0 = 25)$. Use natural logarithms. Write the equation.
b) Estimate the population of Tempe in 1980.

18. *Ecology—Population growth.* The population of Kansas City was 475 thousand in 1960. In 1970 it was 507 thousand. Assuming the exponential model,
a) Find the value $k(P_0 = 475)$. Use natural logarithms. Write the equation.
b) Estimate the population of Kansas City in 2000.

19. *Business—Wine sales in the U.S.* The total number of dollars spent on wine in the U.S. in 1934 was $90 million. In 1974 the amount spent was $1480 million. Assuming the exponential model,
a) Find the value $k(P_0 = 90)$. Use natural logarithms. Write the equation.
b) Estimate the amount spent on wine in 1984.
c) When will the amount spent on wine be double that spent in 1974?

20. *Business—Cost of a double-dip ice cream cone.* In 1970 the cost of a double dip ice cream cone was 52¢. In 1978 it was 66¢. Assuming the exponential model,
a) Find the value $k(P_0 = 52)$. Use natural logarithms. Write the equation.
b) Estimate the cost of a cone in 1986.
c) When will the cost of a cone be twice that of 1978?

21. *Business—Consumer price index.* The *consumer price index* compares the costs of goods and services over various years. 1967 is used as a base (P_0). The same goods and services that cost $100 in 1967 cost $184.50 in 1977. Assuming the exponential model,
a) Find the value $k(P_0 = \$100)$, and write the equation.
b) Estimate what the same goods and services will cost in 1987.
c) When will the same goods and services cost double that of 1967?

22. *Business—Job opportunities.* It is estimated that there were 714,000 accountants employed in 1972 and it is projected that there will be 935,000 accountants needed in 1985. Assuming the exponential model,
a) Find the value $k(P_0 = 714,000)$, and write the equation.
b) Estimate the number of accountants needed in 1990.
c) When will the need for accountants be double that of 1972?

▶ ─────────────────────────────────────

Business—Effective annual yield. Suppose $100 is invested at 7% compounded continuously for 1 year. We know from Example 4 that the balance will be $107.25. This is the same as if $100 were invested at 7.25% and compounded once a year (simple interest). The 7.25% is called the "effective annual yield." In general, if P_0 is invested at $k(\%)$ compounded continuously, then the effective annual yield is that number i satisfying $P_0(1 + i) = P_0 e^k$. Then $1 + i = e^k$, or

Effective annual yield $= i = e^k - 1$.

23. An amount is invested at 6% per year compounded continuously. What is the effective annual yield?

24. An amount is invested at 8% per year compounded continuously. What is the effective annual yield?

25. The effective annual yield on an investment compounded continuously is 9.42%. At what rate was it invested?

26. The effective annual yield on an investment compounded continuously is 10.52%. At what rate was it invested?

27. Find an expression relating the growth rate k and the *tripling* time T_3.

28. Find an expression relating the growth rate k and the *quadrupling* time T_4.

29. Gather data concerning population growth in your city. Estimate its population in 1984, in 2000.

30. A quantity Q_1 grows exponentially with a doubling time of 1 year. A quantity Q_2 grows exponentially with a doubling time of 2 years. If the initial amounts of Q_1 and Q_2 are the same, when will Q_1 be twice the size of Q_2?

31. ▦ *Business—Value of Manhattan Island.* Peter Minuit, of the Dutch West India Company, purchased Manhattan Island from the Indians in 1626 for $24 worth of merchandise. Assuming an exponential rate of inflation of 8%, how much would Manhattan be worth in 1984?

32. ▦ *Ecology—Population growth in the Virgin Islands.* The U.S. Virgin Islands have one of the highest growth rates in the world, 9.6%. In 1970 the population was 75,150. The land area of the Virgin Islands is 3,097,600 square yards. Assuming this growth rate continues and is exponential, when will the population of the Virgin Islands be such that there is 1 person for every square yard of land?

33. ▦ *Ecology—Bicentennial growth of U.S.* The population of the U.S. in 1776 was about 2,508,000. In its bicentennial year the population was about 216,000,000. Assuming the exponential model, what was the growth rate of the U.S. through its bicentennial years?

34. ▦ *Business—Cost of a first-class postage stamp.* The cost of a first-class postage stamp in 1962 was 4¢. In 1978 it was 15¢. This was exponential growth. What was the growth rate? What will be the cost of a first-class postage stamp in 1987? 1997?

35. ▦ *Business—Cost of a prime-rib dinner.* The average cost of a prime rib dinner in 1962 was $4.65, and was increasing at an exponential growth rate of 5.1%. What will the cost of such a dinner be in 1987? 1997?

36. ▦ *Business—Cost of a Hershey Bar.* The cost of a Hershey Bar in 1962 was $0.05, and was increasing at an exponential growth rate of 9.7%. What will the cost of a Hershey Bar be in 1987? 1997?

OBJECTIVES

You should be able to

a) State the solution of an equation
$$\frac{dP}{dt} = -kP, \text{ as } P(t) = P_0 e^{-kt}.$$

b) Given a decay rate, find the half-life.

c) Given a half-life, find the decay rate.

d) Solve applied problems involving decay.

e) Solve applied problems involving Newton's law of cooling.

69. Using the same set of axes on the opposite page, graph $y = e^{2x}$ and $y = e^{-2x}$.

10.5 APPLICATIONS: DECAY

DO EXERCISE 69.

In the equation of population growth $dP/dt = kP$ the constant k is actually given by

$$k = (\text{Birth rate}) - (\text{Death rate}).$$

Thus a population "grows" only when the *birth rate* is greater than the *death rate*. When the birth rate is less than the death rate, k will be negative so the population will be decreasing, or "decaying," at a rate proportional to its size. The equation

$$\frac{dP}{dt} = -kP \qquad (\text{where } k > 0)$$

shows P to be *decreasing* as a function of time, and the solution

$$P(t) = P_0 e^{-kt}$$

shows it to be decreasing exponentially. This is exponential *decay*. The amount present initially at $t = 0$ is again P_0.

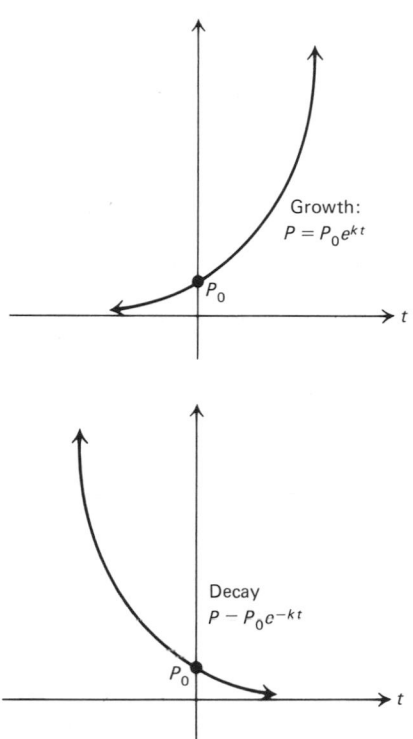

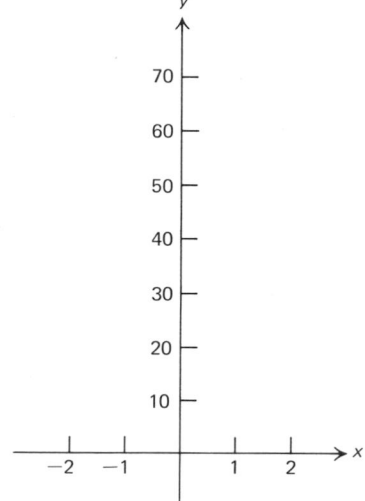

Radioactive Decay

Radioactive elements decay exponentially; that is, they disintegrate at a rate which is proportional to the amount present.

Example 1 Strontium-90 has a decay rate of 2.8% per year. The rate of change of an amount N is given by

$$\frac{dN}{dt} = -0.028N.$$

a) Find the solution of the equation in terms of N_0 (the amount present at $t = 0$).
b) Suppose 1000 grams of strontium-90 is present at $t = 0$. How much will remain after 100 years?
c) After what amount of time will half of the 1000 grams remain?

70. Xenon-133 has a decay rate of 14% per day. The rate of change of an amount N is given by

$$\frac{dN}{dt} = -0.14N.$$

a) Find the solution of the equation in terms of N_0.
b) Suppose 1000 grams of xenon-133 is present at $t = 0$. How much will remain after 10 days?
c) After what time will half of 1000 grams remain?

Solution

a) $N(t) = N_0 e^{-0.028t}$
b) $N(100) = 1000e^{-0.028(100)} = 1000e^{-2.8}$

$$= 1000(0.060810) \qquad (\text{▦ or Table 4.})$$

$$\approx 60.8 \text{ grams}$$

c) We are asking at what time T will $N(T) = 500$. The number T is called the *half-life*. To find T we solve the equation:

$$500 = 1000e^{-0.028T}$$

$$\tfrac{1}{2} = e^{-0.028T}$$

$$\ln \tfrac{1}{2} = \ln e^{-0.028T}$$

$$\ln 1 - \ln 2 = -0.028T$$

$$0 - \ln 2 = -0.028T$$

$$\frac{-\ln 2}{-0.028} = T$$

$$\frac{\ln 2}{0.028} = T$$

$$\frac{0.693147}{0.028} = T$$

$$25 \approx T$$

Thus the half-life of strontium-90 is 25 years.

DO EXERCISE 70.

We can find a general expression relating the decay rate k and the half-life T by solving the equation

$$\tfrac{1}{2}P_0 = P_0 e^{-kT}.$$

$$\tfrac{1}{2} = e^{-kT}$$

$$\ln \tfrac{1}{2} = \ln e^{-kT}$$

$$\ln 1 - \ln 2 = -kT$$

$$0 - \ln 2 = -kT$$

$$-\ln 2 = -kT$$

$$\ln 2 = kT$$

Again,

The *decay rate* k and the *half-life* T are related by

$$kT = \ln 2 = 0.693147,$$

or

$$\text{a) } k = \frac{\ln 2}{T} \qquad \text{b) } T = \frac{\ln 2}{k}.$$

Thus the half-life T depends only on the decay rate k. In particular, it is independent of the initial population size.

The effect of half-life is shown in this radioactive decay curve.

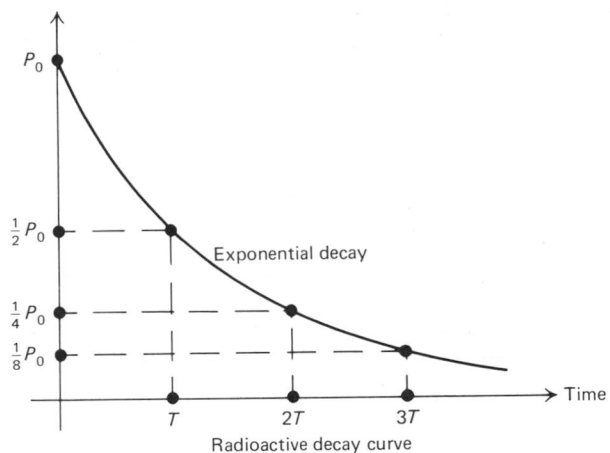

Radioactive decay curve

The exponential function gets closer to 0 as t gets larger, but never reaches 0. Thus, in theory, a radioactive substance never completely decays.

Example 2 Plutonium, a common product and ingredient of nuclear reactors, is of great concern to those who are against the building of nuclear reactors. Its decay rate is 0.003% per year. What is its half-life?

Solution $T = \dfrac{\ln 2}{k} = \dfrac{0.693147}{0.00003} = 23{,}105 \text{ years.}$

DO EXERCISES 71 AND 72.

Example 3 *Carbon dating.* The radioactive element carbon-14 has a half-life of 5750 years. The percentage of carbon-14 present in the remains of plants and animals can be used to determine age. How old is an animal bone that has lost 30% of its carbon-14?

Solution

a) Find the decay rate k.

$$k = \frac{\ln 2}{T} = \frac{0.693147}{5750} = 0.00012, \qquad \text{or } 0.012\% \text{ per year}$$

71. The decay rate of cesium-137 is 2.3% per year. What is its half-life?

72. The half-life of barium-140 is 13 days. What is its decay rate?

How can scientists determine that an animal bone has lost 30% of its carbon-14? The assumption is that the percentage of carbon-14 in the atmosphere and in *living* plants and animals is the same. When a plant or animal dies, the amount of carbon-14 decays exponentially. The scientist burns the animal bone and uses a geiger counter to determine the percentage of the smoke which is carbon-14. It is the amount this varies from the percentage in the atmosphere that tells how much carbon-14 has been lost.

73. How old is a skeleton that has lost 80% of its carbon-14?

b) Find the exponential equation for the amount $N(t)$ that remains from an initial amount N_0 after t years.

$$N(t) = N_0 e^{-0.00012t}$$

[*Note:* This equation can be used for any subsequent carbon dating problem.]

c) If an animal bone has lost 30% of its carbon-14 from an initial amount P_0, then 70% P_0 is the amount present. To find the age t of the bone we solve the following equation for t.

$$70\% \ P_0 = P_0 e^{-0.00012t}$$
$$0.7 = e^{-0.00012t}$$
$$\ln 0.7 = \ln e^{-0.00012t}$$
$$\ln 0.7 = -0.00012t$$
$$-0.356675 = -0.00012t \qquad (\text{⊞ or Table 3.})$$
$$\frac{0.356675}{0.00012} = t$$
$$2973 \approx t$$

Thus, an animal bone that has lost 30% of its carbon-14 is about 2973 years old.

DO EXERCISE 73.

EXERCISE SET 10.5

1. The decay rate of iodine-131 is 9.6% per day. What is its half-life?

3. The half-life of polonium is 3 minutes. What is its decay rate?

5. Of an initial amount of 1000 grams of polonium, how much will remain after 20 minutes? See Exercise 3 for the value of k.

7. *Carbon dating.* How old is a piece of wood that has lost 90% of its carbon-14?

9. *Carbon dating.* How old is a Chinese artifact that has lost 60% of its carbon-14?

2. The decay rate of krypton-85 is 6.3% per year. What is its half-life?

4. The half-life of lead is 22 yr. What is its decay rate?

6. Of an initial amount of 1000 grams of lead, how much will remain after 100 yr? See Exercise 4 for the value of k.

8. *Carbon dating.* How old is an ivory tusk that has lost 40% of its carbon-14?

10. *Carbon dating.* How old is a skeleton that has lost 50% of its carbon-14?

11. In a *chemical reaction* substance A decomposes at a rate proportional to the amount of A present.

a) Write an equation relating A to the amount left of an initial amount A_0 after time t.

b) It is found that 8 grams of A will reduce to 4 grams in 3 hours. At what time will there be only 1 gram left?

13. *Weight loss.* The initial weight of a starving animal is W_0. Its W after t days is given by

$$W = W_0 e^{-0.008t}.$$

a) What percentage of its weight does it lose each day?

b) What percentage of its initial weight remains after 30 days?

15. *Satellite power.* The power supply of a satellite is a radioisotope. The power output P, in watts, decreases at a rate proportional to the amount present. P is given by

$$P = 50e^{-0.004t},$$

where t = time in days.

a) How much power will be available after 375 days?

b) What is the half-life of the power supply?

c) The satellite's equipment cannot operate on less than 10 watts of power. How long can the satellite stay in operation?

d) How much power did the satellite have to begin with?

17. *Salvage value.* A business estimates that the salvage value V of a piece of machinery after t years is given by

$$V(t) = \$40{,}000e^{-t}$$

a) What did the machinery cost initially?

b) What is the salvage value after 2 years?

12. In a *chemical reaction* substance A decomposes at a rate proportional to the amount of A present.

a) Write an equation relating A to the amount left of an initial amount A_0 after time t.

b) It is found that 10 lb of A will reduce to 5 lb in 3.3 hr. At what time will there be only 1 lb left?

14. *Weight loss.* The initial weight of a starving animal is W_0. Its weight after t days is given by

$$W = W_0 e^{-0.009t}.$$

a) What percentage of its weight does it lose each day?

b) What percentage of its initial weight remains after 30 days?

16. *Atmospheric pressure.* Atmospheric pressure P at altitude a is given by

$$P = P_0 e^{-0.00005a},$$

where P_0 = pressure at sea level. Assume $P_0 = 14.7 \text{ lb/in}^2$ (pounds per square inch).

a) Find the pressure at an altitude of 1000 ft.

b) Find the pressure at 20,000 ft.

c) At what altitude is the pressure 1.47 lb/in^2.

18. *Supply and demand.* The supply and demand for the sale of stereos by a sound company are given by

$$S(x) = e^x, \qquad D(x) = 163{,}000e^{-x},$$

where $S(x)$ = price at which the company is willing to supply x stereos, and $D(x)$ = demand price for a quantity of x stereos. Find the equilibrium point.

BEER-LAMBERT LAW. A beam of light enters a medium, such as water or smoky air, with initial intensity I_0. Its intensity is decreased depending on the thickness (or concentration) of the medium. The intensity I at a depth (or concentration) of x units is given by

$$I = I_0 e^{-\mu x}.$$

The constant μ ("mu"), called the *coefficient of absorption*, varies with the medium.

19. *Light through sea water* has $\mu = 1.4$ when x is measured in meters (m).

a) What percentage of I_0 remains at a depth of sea water that is 1 m? 2 m? 3 m?

b) Plant life cannot exist below 10 meters. What percentage of I_0 remains at 10 meters?

20. *Light through smog.* Particulate concentrations of pollution reduce sunlight. In a smoggy area $\mu = 0.01$ and x = concentration of particulates measured in micrograms per cubic meter. What percentage of an initial amount I_0 of sunlight passes through smog that has a concentration of 100 micrograms per cubic meter?

CHAPTER 10 TEST

Differentiate.

1. e^x

2. $\ln x$

3. e^{-x^2}

4. $\ln \dfrac{x}{7}$

5. $e^x - 5x^3$

6. $3e^x \ln x$

7. $\ln(e^x - x^3)$

8. $\dfrac{\ln x}{e^x}$

Given $\ln 2 = 0.6931$ and $\ln 7 = 1.9459$, find:

9. $\ln 14$.

10. $\ln \frac{2}{7}$

11. $\ln 7e$.

12. State the solution of $\dfrac{dM}{dt} = kM$, in terms of M_0.

13. The doubling time of a certain bacteria culture is 4 hours. What is the growth rate?

14. An investment is made at 6.931% per year compounded continuously. What is the doubling time?

15. The demand by airlines for fuel is increasing at the rate of 12% per year. That is,

$$\frac{dF}{dt} = 0.12F,$$

where F = amount of fuel used, and t = time in years.

a) The airlines used 3 billion gallons of fuel in 1960. Find the solution of the equation, assuming $F_0 = 3$ and $k = 0.12$.
b) How much fuel will be needed in 1980?
c) When will the demand be double that in 1960?

16. The half-life of tellurium is 1,000,000 years. What is its decay rate?

17. The decay rate of zirconium is 1.1% per day. What is its half-life?

18. A dose of a drug is injected into the body of a patient. The drug amount in the body decreases at the rate of 10% per hour. That is,

$$\frac{dA}{dt} = -0.1A,$$

where A = amount in body, t = time in hours.

a) A dose of 3 cubic centimeters (cc) is administered. Assuming $A_0 = 3$ and $k = 0.1$, find the solution to the equation.
b) How much of the initial dose of 3 cc will remain after 10 hours?
c) At what time does half the original dose remain?

CHAPTER ELEVEN

The area under a curve can be approximated by a sum of rectangular areas. (*Clif Garboden: Stock, Boston*)

Integration and Applications

OBJECTIVES

You should be able to

a) Find the indefinite integral (anti-derivative) of a given function.
b) Find a function f with a given derivative and function value.
c) Solve applied problems involving antiderivatives.

1. Find three antiderivatives.

$$\frac{dy}{dx} = 7$$

2. Find three antiderivatives.

$$\frac{dy}{dx} = -2$$

Find the general form of each anti-derivative.

3. x

4. x^3

5. e^x

6. $\frac{1}{x}$

11.1 THE ANTIDERIVATIVE

In Chapters 9 and 10 we have considered several interpretations of the derivative. Some are listed below.

Function	Derivative
Distance	Velocity
Revenue	Marginal revenue
Cost	Marginal cost
Population	Rate of growth of population

For population we actually considered the derivative first and then the function. Many problems can be solved by doing the reverse of differentiation, called *antidifferentiation*.

The Antiderivative

Suppose that y is a function of x and that the derivative is the constant 8. Can we find y? It is easy to see that one such function is $8x$. That is, $8x$ is a function whose derivative is 8. Are there other functions whose derivative is 8? Yes. Here are some examples:

$$8x + 3, \quad 8x - 10, \quad 8x + \sqrt{2}.$$

All of these functions are $8x$ plus some constant. There are no other functions having a derivative of 8 other than those of the form $8x + C$. Another way of saying this is that any two functions having a derivative of 8 must differ by a constant. This is true in general.

If two functions F and G have the same derivative on an interval, then

$$F(x) = G(x) + C, \quad \text{where } C \text{ is a constant.}$$

The reverse of differentiating is called *antidifferentiating*. The result of antidifferentiating is called an *antiderivative*. Above we found antiderivatives of the function 8. There are several of them, but they are all $8x$ plus some constant.

Example 1. Antidifferentiate (find the antiderivatives of) x^2.

Solution One antiderivative is $x^3/3$. All other antiderivatives differ from this by a constant, so we can denote them as follows:

$$\frac{x^3}{3} + C.$$

This is the *general form* of the antiderivative.

DO EXERCISES 1 THROUGH 6.

Integrals and Integration

The process of antidifferentiation is, in some contexts, called *integration*, and the general form of the antiderivative is referred to as an *indefinite integral*. A common notation for the indefinite integral, from Leibniz, is as follows:

$$\int f(x)\, dx.$$

The symbol $\int$ is called an *integral sign*. The symbol dx plays no apparent role at this point in our development, but will be useful later. In this context, $f(x)$ is called the *integrand*. We illustrate this notation using the preceding example.

Example 2   Integrate $\int x^2\, dx.$

Solution $\int x^2\, dx = x^3/3 + C$

The symbol on the left is read "the integral of x^2, dx." (The "dx" is often omitted in the reading.) In this case the integrand is x^2. The constant C is called the *constant of integration*.

Example 3   Integrate $\int e^x\, dx.$

Solution $\int e^x\, dx = e^x + C$

DO EXERCISES 7 THROUGH 9.

To integrate (or antidifferentiate) we make use of differentiation formulas, in effect reading them in reverse. Below are some of these, stated in reverse, as integration formulas. These can be checked by differentiating the right-hand side and noting that the result is, in each case, the integrand.

1. $\displaystyle\int k\, dx,\ (k \text{ a constant}) = kx + C$

2. $\displaystyle\int x^r\, dx = \frac{x^{r+1}}{r+1} + C \text{ (provided } r \neq -1\text{), or}$

   $\displaystyle\int (r+1)x^r\, dx = x^{r+1} + C$

(To integrate a power of x, other than -1, increase the power by 1 and divide by the increased power.)

3. $\displaystyle\int x^{-1}\, dx = \int \frac{1}{x}\, dx = \ln x + C$

4. $\displaystyle\int e^x\, dx = e^x + C$

Integrate. Don't forget the constant of integration!

7. $\displaystyle\int x^3\, dx$

8. $\displaystyle\int x\, dx$

9. $\displaystyle\int \frac{1}{x}\, dx$

10. Integrate. Don't forget the constant of integration!

$$\int (7x^4 + 2x)\, dx$$

The following rules allow us to find many other integrals. They are obtainable by reversing two familiar differentiation rules.

RULE A. $\int kf(x)\, dx = k \int f(x)\, dx$
(The integral of a constant times a function is the constant times the integral.)

RULE B. $\int [f(x) + g(x)]\, dx = \int f(x)\, dx + \int g(x)\, dx$
(The integral of a sum is the sum of the integrals.)

Example 4

$$\int (5x + 4x^3)\, dx = \int 5x\, dx + \int 4x^3\, dx \qquad \text{(Rule B)}$$

$$= 5 \int x\, dx + \int 4x^3\, dx \qquad \text{(Rule A)}$$

(Note that we did not factor the 4 out of the second integral. This is because we can find the antiderivative of $4x^3$ directly as x^4, as shown in the second part of formula 2.)

$$= 5 \cdot \frac{x^2}{2} + x^4 + C = \frac{5}{2} x^2 + x^4 + C.$$

(Don't forget the constant of integration!)

Note:

We can always check by differentiating.

Thus, in Example 4,

$$\frac{d}{dx}\left(\frac{5}{2} x^2 + x^4 + C\right) = 2 \cdot \frac{5}{2} \cdot x + 4x^3 = 5x + 4x^3.$$

DO EXERCISE 10.

Example 5 $\displaystyle\int (e^x - \sqrt{x})\, dx = \int e^x\, dx - \int \sqrt{x}\, dx$

$$= \int e^x\, dx - \int x^{1/2}\, dx$$

$$= e^x - \frac{x^{(1/2)+1}}{\frac{1}{2} + 1} + C$$

$$= e^x - \frac{x^{3/2}}{\frac{3}{2}} + C$$

$$= e^x - \tfrac{2}{3} x^{3/2} + C$$

Example 6 $\displaystyle\int\left(1 - \frac{3}{x} + \frac{1}{x^4}\right)dx = \int 1\,dx - 3\int\frac{dx}{x} + \int x^{-4}\,dx$

$$= x - 3\ln x + \frac{x^{-4+1}}{-4+1} + C$$

$$= x - 3\ln x - \frac{x^{-3}}{3} + C$$

DO EXERCISES 11 AND 12.

Another Look at Antiderivatives

The graphs of the antiderivatives of x^2 are the graphs of the functions

$$y = \int x^2\,dx = \frac{x^3}{3} + C$$

for the various values of the constant C.

DO EXERCISE 13.

As shown in the following graphs, x^2 is the derivative of each function. That is, the tangent line at the point

$$\left(a, \frac{a^3}{3} + C\right)$$

has slope a^2. The curves $(x^3/3) + C$ fill up the plane, exactly one curve going through any given point (x_0, y_0).

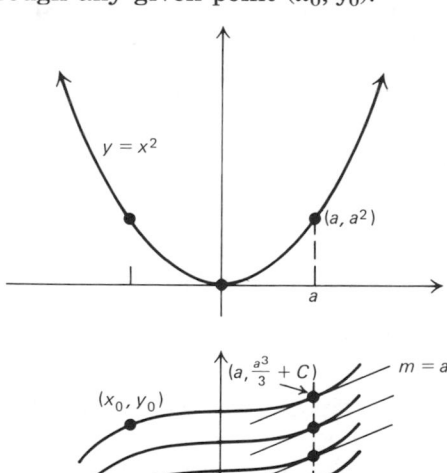

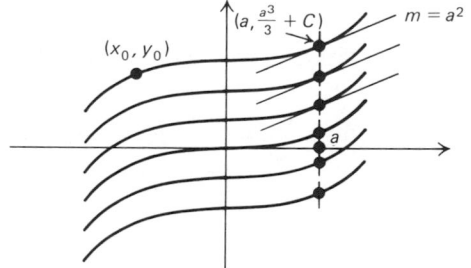

Integrate. Don't forget the constant of integration.

11. $\displaystyle\int (e^x - x^{2/5})\,dx$

12. $\displaystyle\int\left(\frac{5}{x} - 7 + \frac{1}{x^6}\right)dx$

13. Using the same set of axes, graph

$$y = \frac{x^3}{3}, \quad y = \frac{x^3}{3} + 1, \quad \text{and} \quad y = \frac{x^3}{3} - 1.$$

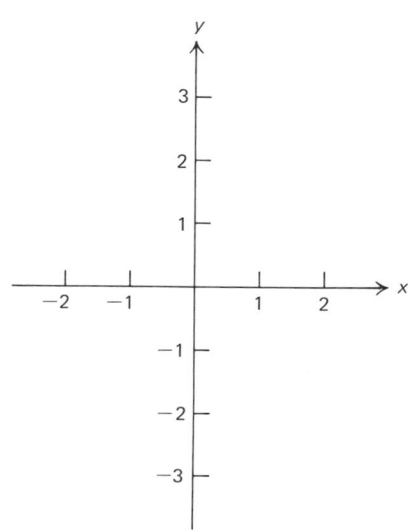

14. Find f such that

$$f'(x) = x^2, \quad \text{and} \quad f(-2) = 5.$$

15. Find g such that

$$g'(x) = 2x - 4, \quad \text{and} \quad g(2) = 9.$$

16. A company determines that the marginal cost, C', of producing the xth unit of a certain product is given by

$$C'(x) = x^2 + 5x.$$

Find the total cost function C, assuming fixed costs to be \$35.

Suppose we look for an antiderivative of x^2 having a specified value at a certain point, say $f(-1) = 2$. We find that there is only one such function.

Example 7. Find the function f such that

$$f'(x) = x^2, \quad \text{and} \quad f(-1) = 2.$$

Solution

a) We find $f(x)$ by integrating.

$$f(x) = \int x^2 \, dx = \frac{x^3}{3} + C.$$

b) The condition $f(-1) = 2$ allows us to find C.

$$f(-1) = \frac{(-1)^3}{3} + C = 2, \quad \text{and solving for } C \text{ we get:}$$

$$-\tfrac{1}{3} + C = 2,$$

$$C = 2 + \tfrac{1}{3}, \quad \text{or } \tfrac{7}{3}.$$

Thus $f(x) = (x^3/3) + (7/3)$.

DO EXERCISES 14 AND 15.

Applied Problems

Example 8 A company determines that the marginal cost, C', of producing the xth unit of a certain product is given by

$$C'(x) = x^3 + 2x.$$

Find the total cost function C, assuming fixed costs (costs when 0 units are produced) are \$45.

Solution

a) We integrate to find $C(x)$, using K for the integration constant to avoid confusion with the cost function C.

$$C(x) = \int C'(x) \, dx = \int (x^3 + 2x) \, dx = \frac{x^4}{4} + x^2 + K.$$

b) Fixed costs are \$45. This means $C(0) = 45$. This allows us to determine the value of K.

$$C(0) = \frac{0^4}{4} + 0^2 + K = 45,$$

$$K = 45.$$

Thus $C(x) = \frac{x^4}{4} + x^2 + 45.$

DO EXERCISE 16.

Recall that the position coordinate, at time t, of an object moving along a number line is $s(t)$. Then

$$s'(t) = v(t) = \text{the } velocity \text{ at time } t,$$
$$v'(t) = a(t) = \text{the } acceleration \text{ at time } t.$$

Example 9   Suppose $v(t) = 5t^4$ and $s(0) = 9$. Find $s(t)$.

Solution
a) We find $s(t)$ by integrating.

$$s(t) = \int v(t)\, dt = \int 5t^4\, dt = t^5 + C$$

b) The condition $s(0) = 9$ allows us to determine C.

$$s(0) = 0^5 + C = 9,$$
$$C = 9$$

Thus $s(t) = t^5 + 9$.

DO EXERCISE 17.

Example 10   Suppose $a(t) = 12t^2 - 6, v(0) = $ initial velocity $= 5$, and $s(0) = 10$. Find $s(t)$.

Solution
a) We find $v(t)$ by integrating $a(t)$.

$$v(t) = \int a(t)\, dt = \int (12t^2 - 6)\, dt = 4t^3 - 6t + C_1$$

b) The condition $v(0) = 5$ allows us to find C_1.

$$v(0) = 4 \cdot 0^3 - 6 \cdot 0 + C_1 = 5, \quad C_1 = 5$$

Thus $v(t) = 4t^3 - 6t + 5$.
c) We find $s(t)$ by integrating $v(t)$.

$$s(t) = \int v(t)\, dt = \int (4t^3 - 6t + 5)\, dt = t^4 - 3t^2 + 5t + C_2$$

d) The condition $s(0) = 10$ allows us to find C_2.

$$s(0) = 0^4 - 3 \cdot 0^2 + 5 \cdot 0 + C_2 = 10, \quad C_2 = 10$$

Thus $s(t) = t^4 - 3t^2 + 5t + 10$.

DO EXERCISE 18.

17. Suppose $v(t) = 4t^3$ and $s(0) = 13$. Find $s(t)$.

18. Suppose $a(t) = 24t^2 - 12, v(0) = 7$, and $s(0) = 8$. Find $s(t)$.

EXERCISE SET 11.1

Integrate.

1. $\int x^6 \, dx$ 2. $\int x^7 \, dx$ 3. $\int 2 \, dx$ 4. $\int 4 \, dx$ 5. $\int x^{1/4} \, dx$ 6. $\int x^{1/3} \, dx$

7. $\int (x^2 + x - 1) \, dx$ 8. $\int (x^2 - x + 2) \, dx$ 9. $\int (t^2 - 2t + 3) \, dt$ 10. $\int (3t^2 - 4t + 7) \, dt$

11. $\int 5e^x \, dx$ 12. $\int 3e^x \, dx$ 13. $\int (x^3 - x^{8/7}) \, dx$ 14. $\int (x^4 - x^{6/5}) \, dx$

15. $\int \frac{1000}{x} \, dx$ 16. $\int \frac{500}{x} \, dx$ 17. $\int \frac{dx}{x^2} \left(\text{or} \int \frac{1}{x^2} \, dx \right)$ 18. $\int \frac{dx}{x^3}$

Find f.

19. $f'(x) = x - 3, f(2) = 9$ 20. $f'(x) = x - 5, f(1) = 6$ 21. $f'(x) = x^2 - 4, f(0) = 7$ 22. $f'(x) = x^2 + 1, f(0) = 8$

Applied Problems

23. A company determines that the marginal cost, C', of producing the xth unit of a certain product is given by

$$C'(x) = x^3 - 2x.$$

Find the total cost function C, assuming fixed costs are $100.

24. A company determines that the marginal cost, C', of producing the xth unit of a certain product is given by

$$C'(x) = x^3 - x.$$

Find the total cost function C, assuming fixed costs are $200.

25. A company determines that the marginal revenue R', from selling the xth unit of a certain product is given by

$$R'(x) = x^2 - 3.$$

a) Find the total revenue function R, assuming $R(0) = 0$.
b) Why is $R(0) = 0$ a reasonable assumption?

Find $s(t)$.

27. $v(t) = 3t^2, s(0) = 4$ 28. $v(t) = 2t, s(0) = 10$

Find $s(t)$.

31. $a(t) = -2t + 6, v(0) = 6,$ and $s(0) = 10$.

33. For a freely falling object, $a(t) = -32 \text{ ft/sec}^2, v(0) = $ initial velocity $= v_0$, and $s(0) = $ initial height $= s_0$. Find a general expression for $s(t)$ in terms of v_0 and s_0.

35. A car with constant acceleration goes from 0 to 60 mph in $\frac{1}{2}$ minute. How far does the car travel during that time?

26. A company determines that the marginal revenue R', from selling the xth unit of a certain product, is given by

$$R'(x) = x^2 - 1.$$

a) Find the total revenue function R, assuming $R(0) = 0$.
b) Why is $R(0) = 0$ a reasonable assumption?

Find $v(t)$.

29. $a(t) = 4t, v(0) = 20$ 30. $a(t) = 6t, v(0) = 30$

32. $a(t) = -6t + 7, v(0) = 10,$ and $s(0) = 20$.

34. A ball is thrown from a height of 10 ft, $s(0) = 10$, at an initial velocity of 80 ft/sec, $v(0) = 80$. How long will it take to hit the ground? (See Exercise 33.)

36. *Efficiency of a machine operator.* The rate at which a machine operator's efficiency E (expressed as a percentage) changes with respect to time is given by

$$\frac{dE}{dt} = 40 - 10t,$$

where t = the number of hours the operator has been at work.

a) Find $E(t)$, given that the operator's efficiency after working 2 hr is 72%. That is, $E(2) = 72$.

b) Use the answer to (a) to find the operator's efficiency after 4 hr, after 8 hr.

38. *Psychology—Memory.* In a certain memory experiment the rate of memorizing is given by

$$M'(t) = 0.2t - 0.003t^2,$$

where $M(t)$ is the number of Spanish words memorized in t minutes.

a) Find $M(t)$ if it is known that $M(0) = 0$.

b) How many words are memorized in 8 minutes?

37. *Efficiency of a machine operator.* The rate at which a machine operator's efficiency E (expressed as a percentage) changes with respect to time is given by

$$\frac{dE}{dt} = 30 - 10t,$$

where t = the number of hours the operator has been at work.

a) Find $E(t)$, given that the operator's efficiency after working 2 hr is 72%. That is, $E(2) = 72$.

b) Use the answer to (a) to find operator's efficiency after 3 hr, after 5 hr.

39. *Biomedical.* The area A of a healing wound is decreasing at the rate given by

$$A'(t) = -43.4t^{-2}, \qquad 1 \leqslant t \leqslant 7,$$

where t is the time in days and A is in square centimeters.

a) Find $A(t)$ if $A(1) = 39.7$.

b) Find the area of the wound after 7 days.

Find f.

40. $f'(t) = \sqrt{t} + \dfrac{1}{\sqrt{t}}, \qquad f(4) = 0$ **41.** $f'(t) = t^{\sqrt{3}}, \qquad f(0) = 8$

Integrate.

42. $\displaystyle\int (5t + 4)^2 \, dt$

43. $\displaystyle\int (x - 1)^2 x^3 \, dx$

44. $\displaystyle\int (1 - t)\sqrt{t} \, dt$

45. $\displaystyle\int \frac{(t + 3)^2}{\sqrt{t}} \, dt$

46. $\displaystyle\int \frac{x^4 - 6x^2 - 7}{x^3} \, dx$

47. $\displaystyle\int (t + 1)^3 \, dt$

48. $\displaystyle\int \frac{1}{\ln 10} \frac{dx}{x}$

49. $\displaystyle\int b e^{ax} \, dx$

50. $\displaystyle\int (3x - 5)(2x + 1) \, dx$

51. $\displaystyle\int \sqrt[3]{64x^4} \, dx$

52. $\displaystyle\int \frac{x^2 - 1}{x + 1} \, dx$

53. $\displaystyle\int \frac{t^3 + 8}{t + 2} \, dt$

11.2 THE DEFINITE INTEGRAL: AREA

In this section we consider the application of integration to finding areas of certain regions. Consider a function whose outputs are positive in an interval (the function might be 0 at one of the endpoints). We wish to find the area of the region between the graph of the function and the x-axis on that interval.

OBJECTIVES

You should be able to

a) Find the area under a curve on a given closed interval.

b) Interpret the area under a curve in two other ways.

19. Consider the constant function $f(x) = 3$.

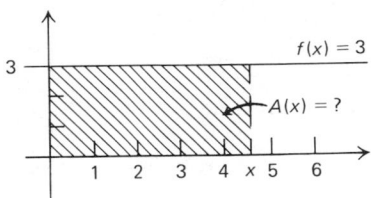

a) Find $A(x)$.
b) Find $A(1)$, $A(2)$, and $A(5)$.
c) Graph $A(x)$.
d) How do $f(x)$ and $A(x)$ compare?

20. A clothing firm, Raggs, Ltd., determines that the marginal cost of each suit it produces is $50.

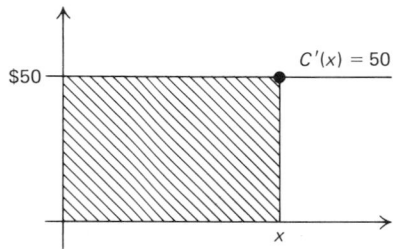

a) Find the total cost, $C(x)$, of producing x suits, assuming fixed costs are $0 (ignore fixed costs).
b) Find the area of the shaded rectangle. Compare your answer to (a).
c) Graph $C(x)$. Why is this an increasing function?

21. Assuming better management Raggs Ltd., of Margin Exercise 20, is able to decrease its production costs by

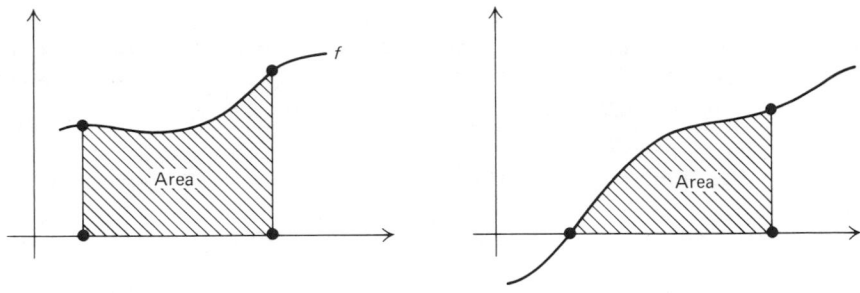

Let us first consider a constant function $f(x) = m$ on the interval from 0 to x, $[0, x]$.

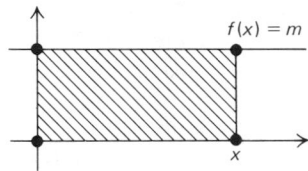

The figure formed is a rectangle, and its area is mx. Suppose we allow x to vary, giving us rectangles of different areas. The area of each rectangle is still mx. We have an area *function*,

$$A(x) = mx.$$

Its graph is shown below.

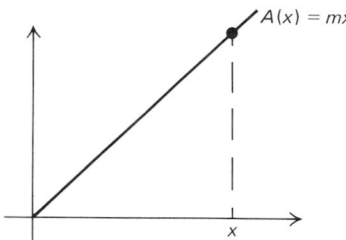

DO EXERCISES 19 THROUGH 21.

Let us next consider the linear function $f(x) = mx$ on the interval from 0 to x, $[0, x]$.

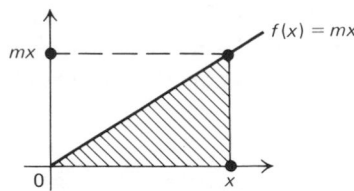

The figure formed this time is a triangle, and its area is $\frac{1}{2}$ the base times the height, $\frac{1}{2} \cdot x \cdot (mx)$, or $\frac{1}{2}mx^2$. If we allow x to vary, we again get an area function

$$A(x) = \tfrac{1}{2}mx^2.$$

Its graph is as shown below.

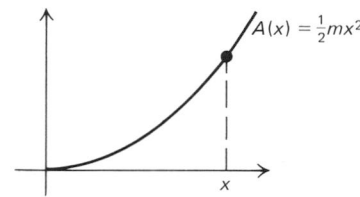

DO EXERCISE 22.

Now consider the linear function $f(x) = mx + b$ on the interval from 0 to x, $[0, x]$.

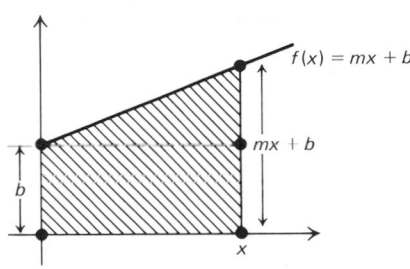

The figure formed this time is a trapezoid, and its area is $\frac{1}{2}$ the height times the sum of the lengths of its parallel sides (or, noting the dashed line, the area of the triangle plus the rectangle),

$$\tfrac{1}{2} \cdot x \cdot [b + (mx + b)], \quad \text{or } \tfrac{1}{2} \cdot x \cdot (mx + 2b), \quad \text{or } \tfrac{1}{2}mx^2 + bx.$$

If we allow x to vary, we again get an area function

$$A(x) = \tfrac{1}{2}mx^2 + bx.$$

Its graph is as shown below.

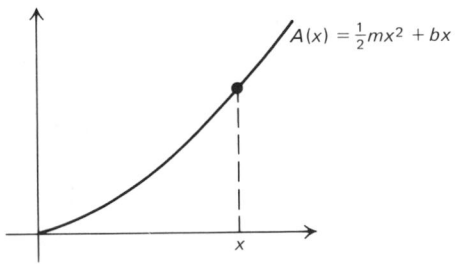

$10 per suit for every hundred suits it produces. This is shown below.

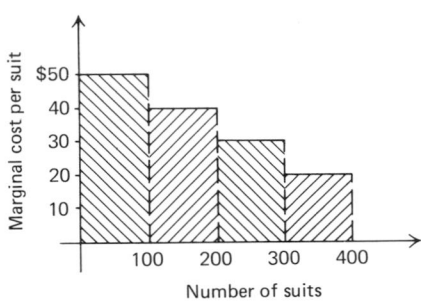

a) Find the total cost of producing 400 suits.
b) Find the total area of the rectangles. Compare your answer with (a).

22. Consider the function $f(x) = 3x$.

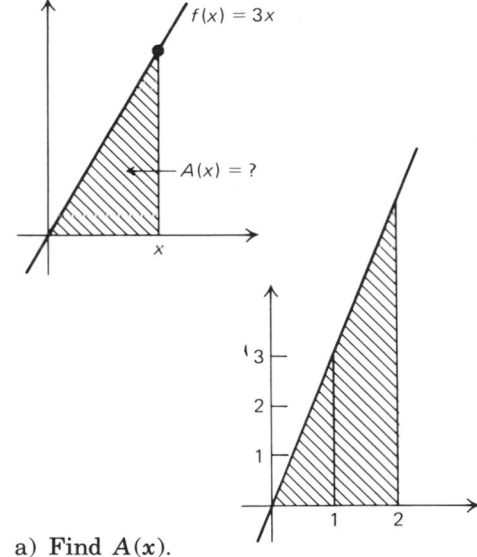

a) Find $A(x)$.

b) Find $A(1)$.

c) Find $A(2)$.

d) Find $A(3.5)$.

e) Graph $A(x)$.

f) How do $f(x)$ and $A(x)$ compare?

23. Raggs Ltd., of Margin Exercise 21, installs new sewing machines. This allows the marginal cost per suit to decrease continually in such a way that

$$C'(x) = -0.1x + 50.$$

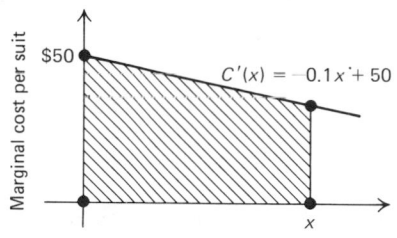

a) Find the total cost of producing x suits, ignoring fixed costs.

b) Find the area of the shaded trapezoid.

c) Find the total cost of producing 400 suits. Compare this answer with that of Margin Exercise 21.

DO EXERCISE 23.

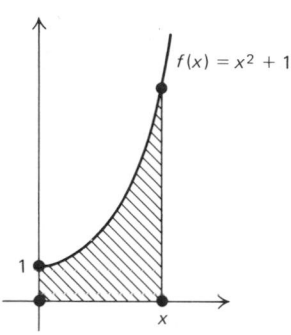

Now we consider the function $f(x) = x^2 + 1$ on the interval from 0 to x, $[0, x]$. The graph of the region in question is as shown, but it is not so easy this time to find the area function because the graph of $f(x)$ is not a straight line. Let us tabulate our previous results and look for a pattern.

$f(x)$	$A(x)$
$f(x) = 3$	$A(x) = 3x$
$f(x) = m$	$A(x) = mx$
$f(x) = 3x$	$A(x) = \frac{3}{2}x^2$
$f(x) = mx$	$A(x) = \frac{1}{2}mx^2$
$f(x) = mx + b$	$A(x) = \frac{1}{2}mx^2 + bx$

You may have conjectured that the area function $A(x)$ is an antiderivative of $f(x)$. In the following exploratory exercises you will investigate further.

Exploratory Exercises: Finding Areas

1. The region under the graph of $f(x) = x^2 + 1$, on the interval $[0, 2]$ is shown to the right.

a) Make a copy of the shaded region on thin paper.

b) Cut up the shaded region in any way you wish in order to fill up squares in the grid below. Make an estimate of the total area.

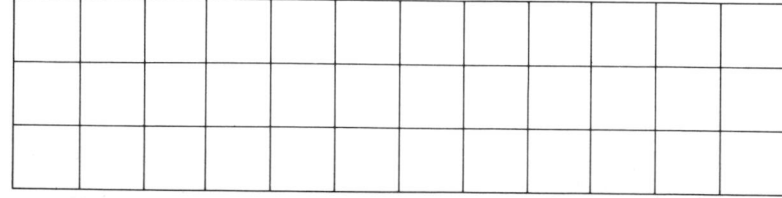

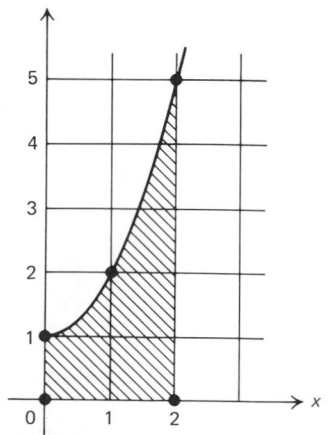

c) Using the antiderivative

$$F(x) = \frac{x^3}{3} + x,$$

find $F(2)$.

2. Repeat Exercise 1(a) and (b) for the shaded region of the graph at the right.

c) Using the antiderivative

$$F(x) = \frac{x^3}{3} + x,$$

find $F(3)$.

d) Compare your answers to (b) and (c).

The conjecture concerning areas and antiderivatives (or integrals), is true; It is expressed as follows.

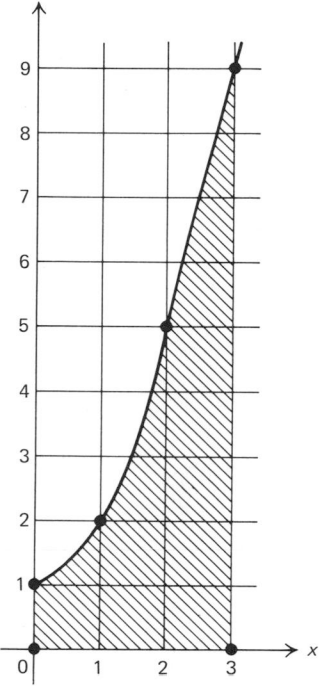

THEOREM. **Let f be a positive, continuous function on an interval $[a, b]$ and let $A(x)$ be the area of the region between the graph of f and the x-axis on the interval $[a, x]$. Then $A(x)$ is a differentiable function of x and**

$$A'(x) = f(x).$$

Proof The situation described in the theorem is shown here.

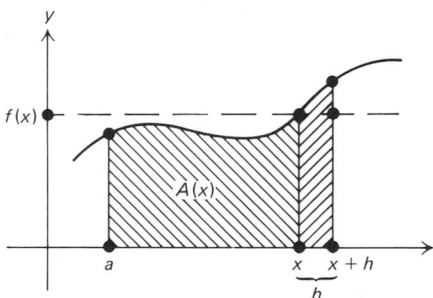

The derivative of $A(x)$ is, by definition of derivative,

$$A'(x) = \lim_{h \to 0} \frac{A(x + h) - A(x)}{h}.$$

Note, from the drawing, that $A(x + h) - A(x)$ is the area of the small, shaded, vertical strip. The area of this small strip is approximately that of a rectangle of base h and height $f(x)$, especially for

small values of h. Thus we have

$$A(x + h) - A(x) \approx f(x) \cdot h.$$

Now

$$A'(x) = \lim_{h \to 0} \frac{A(x + h) - A(x)}{h} = \lim_{h \to 0} \frac{f(x) \cdot h}{h} = \lim_{h \to 0} f(x) = f(x),$$

since $f(x)$ does not involve h.

The theorem above also holds if $f(x) = 0$ at one or both endpoints of the interval $[a, b]$.

Since the area function A is an antiderivative of f, and since any two antiderivatives differ by a constant, we easily conclude that the area function and any antiderivative differ by a constant.

We can think of the function A as given by

$$A(x) = \text{the area on the interval } [a, x],$$

where a is some fixed point and x varies.

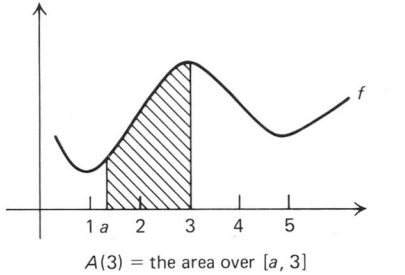

$A(3) =$ the area over $[a, 3]$

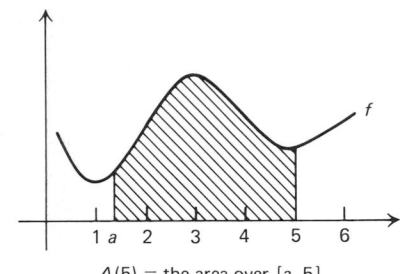

$A(5) =$ the area over $[a, 5]$

Now let us find some areas.

Example 1 Find the area under the graph of $y = x^2 + 1$ on the interval $[-1, 2]$.

Solution
a) We first make a drawing. This includes a graph of the function and the region in question.

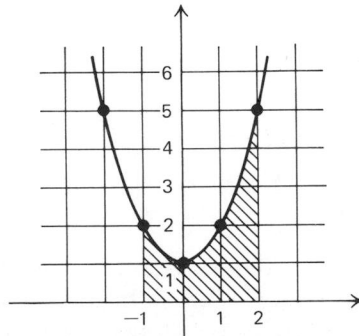

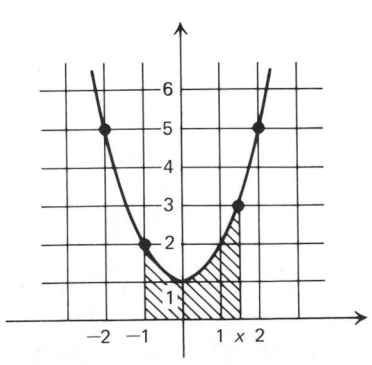

b) Second, we make a drawing showing a portion of the region from -1 to x. Now $A(x)$ is the area of this portion, that is, in the interval $[-1, x]$.

c) Now

$$A(x) = \int (x^2 + 1)\, dx = \frac{x^3}{3} + x + C,$$

where C has to be determined. Since we know that $A(-1) = 0$ (there is no area above the number -1), we can substitute for x in $A(x)$, as follows:

$$A(-1) = \frac{(-1)^3}{3} + (-1) + C = 0,$$

$$-\tfrac{1}{3} - 1 + C = 0,$$

$$C = \tfrac{4}{3}.$$

This determines that $C = \tfrac{4}{3}$, so we have

$$A(x) = \frac{x^3}{3} + x + \frac{4}{3}.$$

Then the area in the interval $[-1, 2]$ is $A(2)$. We compute $A(2)$ as follows:

$$A(2) = \frac{2^3}{3} + 2 + \frac{4}{3} = \frac{8}{3} + 2 + \frac{4}{3} = \frac{12}{3} + 2 = 6.$$

DO EXERCISE 24.

Example 2 Find the area under the graph of $y = x^3$ on the interval $[0, 5]$.

Solution

a) We first make a drawing which includes a graph of the function and the region in question.

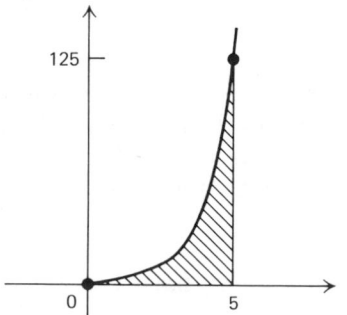 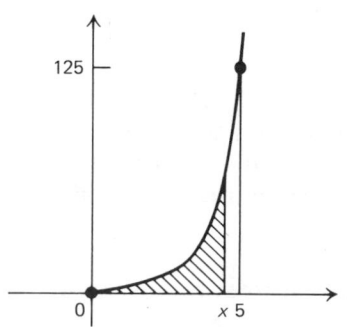

24. Find the area under the graph of $y = x^2 + 3$ on the interval $[1, 2]$.

25. Find the area under the graph of $y = x^2 + x$ on the interval $[0, 3]$.

b) Second, we make a drawing showing a portion of that region from 0 to x. Now $A(x)$ is the area of this portion, that is, on the interval $[0, x]$.

c) Now

$$A(x) = \int x^3 \, dx = \frac{x^4}{4} + C,$$

where C has to be determined. Since we know that $A(0) = 0$, we can substitute 0 for x in $A(x)$, as follows:

$$A(0) = \frac{0^4}{4} + C = 0,$$

$$C = 0.$$

This determines C. So

$$A(x) = \frac{x^4}{4}.$$

Then the area in the interval $[0, 5]$ is $A(5)$. We can compute $A(5)$ as follows:

$$A(5) = \frac{5^4}{4} = \frac{625}{4} = 156\tfrac{1}{4}.$$

DO EXERCISE 25.

Since the area under a curve, as in the preceding examples, is an antiderivative, area can also be associated with various kinds of functions. If, for example, we have a velocity function over an interval $[0, b]$, then the area under the curve in that interval is the total distance. Suppose the velocity function is

$$v(t) = t^3.$$

In 5 hours the total distance covered is $156\tfrac{1}{4}$. This can be seen in the preceding Example 2, simply by changing the variable from x to t. For a marginal cost function over the interval $[0, x]$ the area under the curve is the total cost of producing x units, or the accumulated cost.

Example 3 Raggs, Ltd., goes even further to reduce production costs. In addition to purchasing new sewing machines, air conditioning is installed and the president takes a calculus course. These cause the marginal cost per suit to decrease rapidly in such a way that

$$C'(x) = 0.0003x^2 - 0.2x + 50.$$

Find the total cost of producing 400 suits. (Ignore fixed costs.)

Solution

a) First, we make a drawing. This includes a graph of the function and the region in question.

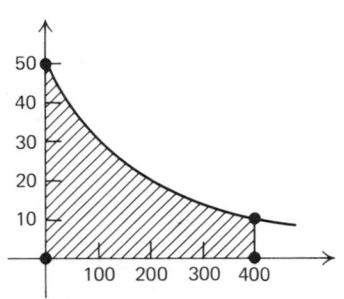 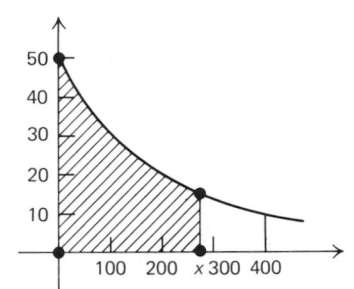

b) Second, we make a drawing showing a portion of that region from 0 to x. Now $A(x)$ is the area of that portion, that is, on the interval $[0, x]$.

c) Now,

$$C(x) = A(x) = \int (0.0003x^2 - 0.2x + 50)\, dx$$
$$= 0.0001x^3 - 0.1x^2 + 50x + K,$$

where K has to be determined. We are ignoring fixed costs, $K = 0$, and we have

$$C(x) = 0.0001x^3 - 0.1x^2 + 50x.$$

Then the area in the interval $[0, 400]$ is $A(400)$ or $C(400)$. We can compute $C(400)$ as follows:

$$C(400) = 0.0001 \cdot 400^3 - 0.1 \cdot 400^2 + 50 \cdot 400, \text{ or } \$10{,}400.$$

DO EXERCISE 26.

26. Referring to Example 3,

a) Compare \$10,400 to your answer for Margin Exercise 23. Has the company reduced total costs?

b) Find the total cost of producing 100 suits.

EXERCISE SET 11.2

Find the area under the given curve on the interval indicated.

1. $y = 4; [1, 3]$ **2.** $y = 5; [1, 3]$ **3.** $y = 2x; [1, 3]$ **4.** $y = x^2; [0, 3]$

5. $y = x^2; [0, 5]$ **6.** $y = x^3; [0, 2]$ **7.** $y = x^3; [0, 1]$ **8.** $y = 1 - x^2; [-1, 1]$

9. $y = 4 - x^2; [-2, 2]$ **10.** $y = e^x; [0, 2]$ **11.** $y = e^x; [0, 3]$ **12.** $y = \dfrac{1}{x}; [1, 2]$

13. $y = \dfrac{1}{x}; [1, 3]$ **14.** $y = x^2 - 4x; [-4, -2]$ **15.** $y = x^2 - 4x; [-4, -1]$

In each case give two interpretations of the shaded region.

16. Velocity

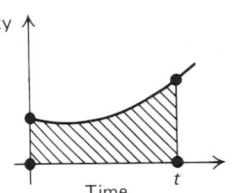

Time

17. Acceleration

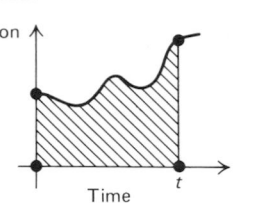

Time

18. Divorce rate (per unit of time)

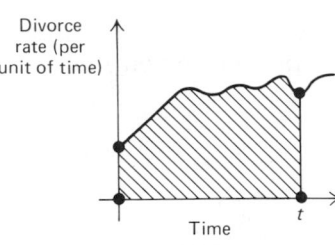

Time

19. Rate of energy use (per unit of time)

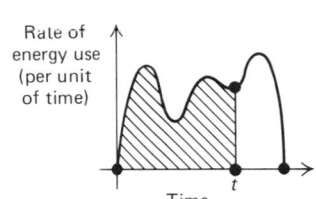

Time

20. Marginal cost (per unit)

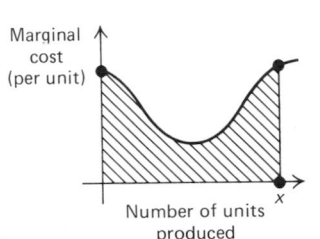

Number of units produced

21. Marginal revenue (per unit)

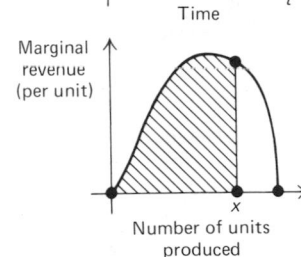

Number of units produced

22. Sales on tth day

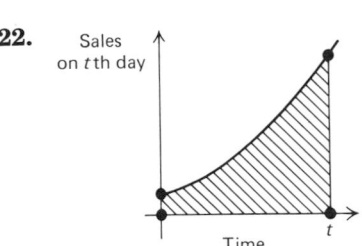

Time

23. Concentration of a drug (in milligrams per cubic centimeter)

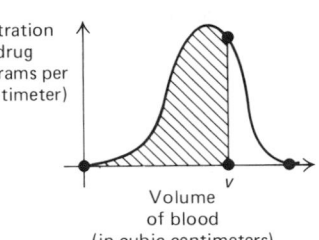

Volume of blood (in cubic centimeters)

24. Rate of memory (in words per minute)

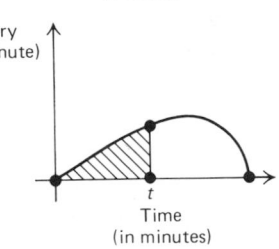

Time (in minutes)

25. A particle starts out from the origin. Its velocity at time t is given by

$$v(t) = 3t^2 + 2t.$$

a) Find the distance the particle has traveled after t hours.
b) Find the distance the particle has traveled after 5 hours.

27. A sound company determines that the marginal cost of producing the xth stereo is given by

$$C'(x) = 100 - 0.2x, \quad C(0) = 0.$$

It also determines that its marginal revenue from the sale of the xth stereo is given by

$$R'(x) = 100 + 0.2x, \quad R(0) = 0.$$

a) Find the total cost of producing x stereos.
b) Find the total revenue of selling x stereos.
c) Find the total profit from the production and sale of x stereos.
d) Find the total profit from the production and sale of 1000 stereos.

26. A particle starts out from the origin. Its velocity at time t is given by

$$v(t) = 4t^3 + 2t.$$

a) Find the distance the particle has traveled after t hours.
b) Find the distance the car has traveled after 2 hours.

28. A refrigeration company determines that the marginal cost of producing the xth refrigerator is given by

$$C'(x) = 50 - 0.4x, \quad C(0) = 0.$$

It also determines that its marginal revenue from the sale of the xth refrigerator is given by

$$R'(x) = 50 + 0.4x, \quad R(0) = 0.$$

a) Find the total cost of producing x refrigerators.
b) Find the total revenue of selling x refrigerators.
c) Find the total profit from the production and sale of x refrigerators.
d) Find the total profit from the production and sale of 1000 refrigerators.

▶

Find the area under the curve on the interval indicated.

29. $y = \dfrac{x^2 - 1}{x - 1}$; $[2, 3]$

30. $y = \dfrac{x^5 - x^{-1}}{x^2}$; $[1, 5]$

31. $y = (x - 1)\sqrt{x}$; $[4, 16]$

32. $y = (x + 2)^3$; $[0, 1]$

33. $y = \dfrac{\sqrt[3]{x^2} - 1}{\sqrt[3]{x}}$; $[1, 8]$

34. $y = \dfrac{x^3 + 8}{x + 2}$; $[0, 1]$

11.3 INTEGRATION ON AN INTERVAL: THE DEFINITE INTEGRAL

Let f be a positive continuous function on an interval $[a, b]$. We know that f has an antiderivative, namely $A(x)$. Let F and G be any two antiderivatives of f. Then

$$F(b) - F(a) = G(b) - G(a).$$

To understand this, recall that F and G differ by a constant. That is, $F(x) = G(x) + C$. Then

$$F(b) - F(a) = [G(b) + C] - [G(a) + C] = G(b) - G(a).$$

Thus the difference $F(b) - F(a)$ has the same value for all anti-derivatives of f. It is called the *definite integral* of f from a to b.

Definite integrals are usually symbolized as follows:

$$\int_a^b f(x)\, dx.$$

This is read "the integral from a to b of $f(x)\, dx$" (the dx is sometimes omitted from the reading). From the preceding development we see that to find a definite integral $\int_a^b f(x)\, dx$ we first find an antiderivative $F(x)$. The simplest one is the one for which the constant of integration is 0. We evaluate F at b and at a and subtract.

$\int_a^b f(x)\, dx$ **is defined to be** $F(b) - F(a)$, **where F is any antiderivative of f.**

Evaluating definite integrals is called *integrating*. The numbers a and b are known as the *limits of integration*.

Example 1 Integrate $\displaystyle\int_a^b x^2\, dx$.

Solution Using the antiderivative $F(x) = x^3/3$, we have

$$\int_a^b x^2\, dx = \frac{b^3}{3} - \frac{a^3}{3}.$$

DO EXERCISES 27 AND 28.

OBJECTIVES

You should be able to

a) Evaluate a definite integral.

b) Find the area under a graph on $[a, b]$.

c) Solve applied problems involving definite integrals.

Integrate.

27. $\displaystyle\int_a^b 2x\, dx$

28. $\displaystyle\int_a^b e^x\, dx$

Integrate.

29. $\displaystyle\int_1^3 2x\,dx$

30. $\displaystyle\int_{-2}^0 e^x\,dx$

31. $\displaystyle\int_0^1 (2x - x^2)\,dx$

32. $\displaystyle\int_1^e \left(1 + 3x^2 - \frac{1}{x}\right)dx$

It is convenient to use an intermediate notation

$$\int_a^b f(x)\,dx = [F(x)]_a^b = F(b) - F(a).$$

We now evaluate several definite integrals.

Example 2 $\displaystyle\int_{-1}^2 x^2\,dx = \left[\frac{x^3}{3}\right]_{-1}^2 = \frac{2^3}{3} - \frac{(-1)^3}{3}$

$$= \frac{8}{3} - \left(-\frac{1}{3}\right) = \frac{8}{3} + \frac{1}{3} = 3$$

Example 3 $\displaystyle\int_0^3 e^x\,dx = [e^x]_0^3 = e^3 - e^0 = e^3 - 1$

Example 4 $\displaystyle\int_1^4 (x^2 - x)\,dx = \left[\frac{x^3}{3} - \frac{x^2}{2}\right]_1^4 = \left(\frac{4^3}{3} - \frac{4^2}{2}\right) - \left(\frac{1^3}{3} - \frac{1^2}{2}\right)$

$$= \left(\frac{64}{3} - \frac{16}{2}\right) - \left(\frac{1}{3} - \frac{1}{2}\right)$$

$$= \frac{64}{3} - 8 - \frac{1}{3} + \frac{1}{2} = 13\tfrac{1}{2}$$

Example 5 $\displaystyle\int_1^e \left(1 + 2x - \frac{1}{x}\right)dx = [x + x^2 - \ln x]_1^e$

$$= (e + e^2 - \ln e) - (1 + 1^2 - \ln 1)$$
$$= (e + e^2 - 1) - (1 + 1 - 0)$$
$$= e + e^2 - 1 - 1 - 1$$
$$= e + e^2 - 3$$

It is important to note that in $\int_a^b f(x)\,dx, a < b$. That is, the largest number is on the top!

DO EXERCISE 29 THROUGH 32.

The area under a curve can be expressed by a definite integral.

THEOREM. Let f be a positive continuous function over the closed interval $[a, b]$. The area under the graph of f on the interval $[a, b]$ is

$$\int_a^b f(x)\,dx.$$

Proof

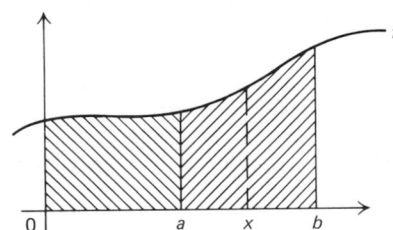

Let

$$A(x) = \text{the area of the region over } [0, x].$$

Then

$$A'(x) = f(x),$$

so $A(x)$ is an antiderivative of $f(x)$. Then

$$\int_a^b f(x)\, dx = A(b) - A(a).$$

But $A(b) - A(a)$ is the area over $[0, b]$ minus the area over $[0, a]$, which is the area over $[a, b]$.

Let us now find some areas.

Example 6 Find the area under $y = x^2 + 1$ on $[-1, 2]$.

Solution $\displaystyle \int_{-1}^{2} (x^2 + 1)\, dx = \left[\frac{x^3}{3} + x\right]_{-1}^{2}$

$$= \left(\frac{2^3}{3} + 2\right) - \left(\frac{(-1)^3}{3} + (-1)\right)$$

$$= \left(\frac{8}{3} + 2\right) - \left(-\frac{1}{3} - 1\right)$$

$$= \frac{8}{3} + 2 + \frac{1}{3} + 1 = 6$$

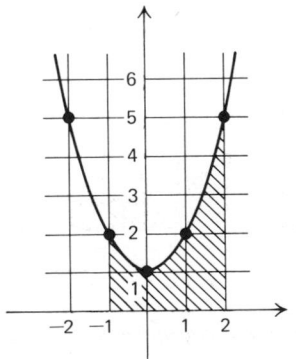

33. Find the area under $y = x^2 + 3$ on $[1, 2]$. Don't forget that it helps to draw the graph. (Compare the result with Margin Exercise 24.)

Compare this with Example 1 on p. 542.

DO EXERCISE 33.

Example 7 Find the area under $y = x^3$ on $[0, 5]$.

Solution
$$\int_0^5 x^3 \, dx = \left[\frac{x^4}{4}\right]_0^5 = \frac{5^4}{4} - \frac{0^4}{4} = \frac{625}{4}$$

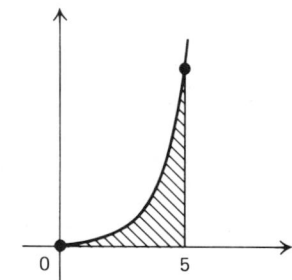

34. Find the area under $y = x^2 + x$ on $[0, 3]$. (Compare the result with Margin Exercise 25.)

Compare this with Example 2 on p. 543.

DO EXERCISE 34.

Example 8 Find the area under $y = \frac{1}{x}$ on $[1, 4]$.

Solution
$$\int_1^4 \frac{dx}{x} = [\ln x]_1^4 = \ln 4 - \ln 1$$
$$= \ln 4$$
$$\approx 1.3863 \qquad (\blacksquare \text{ or Table 3.})$$

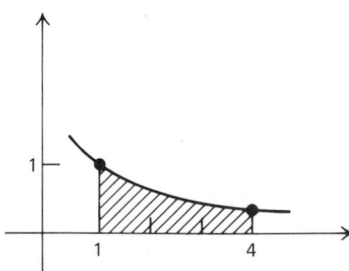

Example 9 Find the area under $y = \frac{1}{x^2}$ on $[1, b]$.

Solution $\displaystyle\int_1^b \frac{dx}{x^2} = \int_1^b x^{-2}\,dx = \left[\frac{x^{-2+1}}{-2+1}\right]_1^b$

$$= \left[\frac{x^{-1}}{-1}\right]_1^b = \left[-\frac{1}{x}\right]_1^b = \left(-\frac{1}{b}\right) - \left(-\frac{1}{1}\right)$$

$$= 1 - \frac{1}{b}.$$

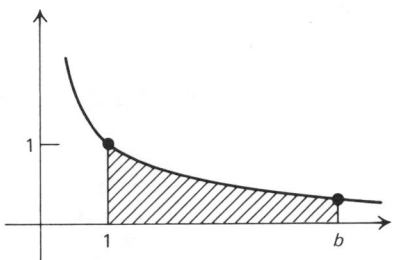

DO EXERCISES 35 AND 36.

The following properties of definite integrals can be derived rather easily from the definition of definite integral and from the properties of the indefinite integral:

PROPERTY 1. $\displaystyle\int_a^b k \cdot f(x)\,dx = k \cdot \int_a^b f(x)\,dx.$

(The integral of a constant times a function is the constant times the integral of the function. That is, we can "factor out" a constant from the integrand.)

Example 10

$$\int_0^5 100e^x\,dx = 100\int_0^5 e^x\,dx = 100[e^x]_0^5 = 100(e^5 - e^0) = 100(e^5 - 1)$$

DO EXERCISE 37.

PROPERTY 2. $\displaystyle\int_a^b [f(x) + g(x)]\,dx = \int_a^b f(x)\,dx + \int_a^b g(x)\,dx.$

(The integral of a sum is the sum of the integrals.)

PROPERTY 3. For $a < c < b$, $\displaystyle\int_a^b f(x)\,dx = \int_a^c f(x)\,dx + \int_c^b f(x)\,dx.$

(For any number c between a and b, the integral from a to b is the integral from a to c plus the integral from c to b.)

35. Find the area under $y = \frac{1}{x}$ on $[1, 7]$.

36. Find the area under $y = \dfrac{1}{x^4}$ on $[1, b]$.

37. Integrate $\displaystyle\int_1^2 20x^3\,dx.$

38. Find the area under the graph of $y = f(x)$ from -3 to 2.

$$f(x) = \begin{cases} 4, & \text{if} \quad x < 0 \\ 4 - x^2, & \text{if} \quad x \geq 0. \end{cases}$$

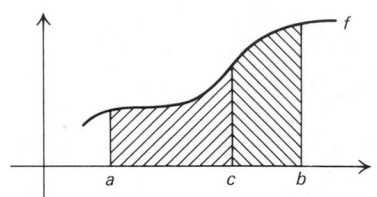

Property 3 has particular application when a function is defined in different ways over subintervals.

Example 11 Find the area under the graph of $y = f(x)$ from -4 to 5, where

$$f(x) = \begin{cases} 9, & \text{if } x < 3 \\ x^2 & \text{if } x \geq 3. \end{cases}$$

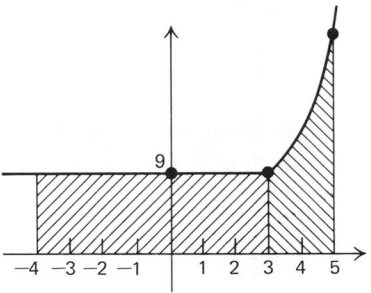

Solution
$$\int_{-4}^{5} f(x)\, dx = \int_{-4}^{3} f(x)\, dx + \int_{3}^{5} f(x)\, dx$$

$$= \int_{-4}^{3} 9\, dx + \int_{3}^{5} x^2\, dx$$

$$= 9 \int_{-4}^{3} dx + \int_{3}^{5} x^2\, dx$$

$$= 9[x]_{-4}^{3} + \left[\frac{x^3}{3}\right]_{3}^{5}$$

$$= 9[3 - (-4)] + (5^3/3) - (3^3/3)$$

$$= 95\tfrac{2}{3}$$

DO EXERCISE 38.

Applied Problem

Example 12.   *Business—Accumulated sales.* The sales of a company are expected to grow continuously at a rate given by the function

$$S'(t) = 100e^t,$$

where $S'(t)$ = sales rate, in dollars per day, at time t.

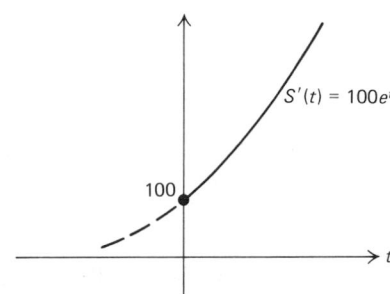

a) Find the accumulated sales for the first 7 days.
b) On what day will accumulated sales exceed \$810,000?

Solution

a) Accumulated sales through day 7 are

$$\int_0^7 S'(t)\,dt = \int_0^7 100e^t\,dt = 100\int_0^7 e^t\,dt$$

$$= 100[e^t]_0^7 = 100(e^7 - e^0)$$

$$= 100(1096.633158 - 1)$$

$$= 100(1095.633158) \approx \$109{,}563.32.$$

b) Accumulated sales through day k are

$$\int_0^k S'(t)\,dt = \int_0^k 100e^t\,dt = 100\int_0^k e^t\,dt$$

$$= 100[e^t]_0^k = 100(e^k - e^0) = 100(e^k - 1).$$

We set this equal to \$810,000 and solve for k.

$$100(e^k - 1) = 810{,}000$$

$$e^k - 1 = 8100$$

$$e^k = 8101$$

39. The sales of a company are expected to grow continuously at a rate given by the function

$$S'(t) = 200e^t,$$

where $S'(t)$ = sales rate, in dollars per day, at time t.

a) Find the accumulated sales for the first 8 days.

b) On what day will accumulated sales exceed \$300,000?

c) The accumulated sales from the 8th through the 10th day is given by

$$\int_8^{10} S'(t)\, dt.$$

Find this.

We solve this equation for k using natural logarithms:

$$e^k = 8101$$

$$\ln e^k = \ln 8101$$

$$k = 8.999743 \quad (\blacksquare \text{ or Table 3})$$

$$k \approx 9$$

DO EXERCISE 39 (starting on page 553).

EXERCISE SET 11.3

Integrate.

1. $\int_0^1 (x - x^2)\, dx$ **2.** $\int_1^2 (x^2 - x)\, dx$ **3.** $\int_{-1}^1 (x^2 - x^4)\, dx$ **4.** $\int_0^b e^x\, dx$ **5.** $\int_a^b e^t\, dt$

6. $\int_0^a (ax - x^2)\, dx$ **7.** $\int_a^b 3t^2\, dt$ **8.** $\int_a^b 4t^3\, dt$ **9.** $\int_1^e \left(x + \frac{1}{x}\right) dx$ **10.** $\int_1^e \left(x - \frac{1}{x}\right) dx$

11. $\int_0^1 \sqrt{x}\, dx$ **12.** $\int_0^1 3\sqrt{x}\, dx$ **13.** $\int_0^1 \frac{10}{17} t^3\, dt$ **14.** $\int_0^1 \frac{12}{13} t^2\, dt$

Find the area under the graph on the interval indicated.

15. $y = x^3;\ [0, 2]$ **16.** $y = x^4;\ [0, 1]$ **17.** $y = x^2 + x + 1;\ [2, 3]$

18. $y = 2 - x - x^2;\ [-2, 1]$ **19.** $y = 5 - x^2;\ [-1, 2]$ **20.** $y = e^x;\ [-2, 3]$

21. $y = e^x;\ [-1, 5]$ **22.** $y = 2x + \frac{1}{x^2};\ [1, 4]$ **23.** $y = 2x - \frac{1}{x^2};\ [1, 3]$

Find the area under the graph on $[-2, 3]$.

24. $f(x) = \begin{cases} x^2, & \text{if } x < 1 \\ 1, & \text{if } x \geq 1 \end{cases}$ **25.** $f(x) = \begin{cases} 4 - x^2, & \text{if } x < 0 \\ 4, & \text{if } x \geq 0 \end{cases}$

26. *Business—Accumulated sales.* Raggs, Ltd. estimates that its sales will grow continuously at a rate given by the function

$$S'(t) = 10e^t,$$

where $S'(t) =$ sales rate, in dollars per day, at time t.

a) Find the accumulated sales for the first 5 days.

b) Find the sales from the second through the fifth day. This is the integral from 1 to 5.

c) On what day will accumulated sales exceed \$40,000?

27. *Business—Accumulated sales.* A company estimates that its sales will grow continuously at a rate given by the function

$$S'(t) = 20e^t,$$

where $S'(t) =$ sales rate, in dollars per day, at time t.

a) Find the accumulated sales for the first 5 days.

b) Find the sales from the second through the fifth day. This is the integral from 1 to 5.

c) On what day will accumulated sales exceed \$20,000?

28. A particle starts out from the origin. Its velocity at time t is given by

$$v(t) = 3t^2 + 2t.$$

How far does it travel from the second through the fifth hour (from $t = 1$ to $t = 5$)?

29. A particle starts out from the origin. Its velocity at time t is given by

$$v(t) = 4t^3 + 2t.$$

How far does it travel from the first through the third hour (from $t = 0$ to $t = 3$)?

30. *Business.* Raggs, Ltd. determines that the marginal cost per suit is given by

$$C'(x) = 0.0003x^2 - 0.2x + 50.$$

Ignoring fixed costs, find the total cost of producing the 101st through the 400th suit (integrate from $x = 100$ to $x = 400$).

31. *Business.* In Exercise 30, find the cost of producing the 201st through the 400th suit (integrate from $x = 200$ to $x = 400$).

Psychology—Memorizing. In the psychological process of memorizing, the rate of memorizing (say in words per minute) increases with respect to time, but eventually a maximum rate of memorizing is reached from which the memory rate decreases.

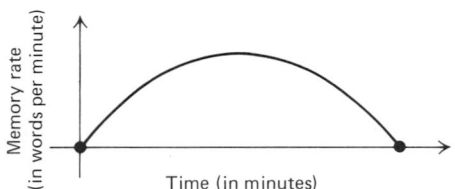

32. Suppose in a certain memory experiment the rate of memorizing is given by

$$M'(t) = -0.009t^2 + 0.2t,$$

where $M'(t)$ = memory rate in words per minute. How many words are memorized the first 10 minutes (from $t = 0$ to $t = 10$)?

34. *Psychology—Business.* A company is producing a new product. However, due to the nature of the product, it is felt that the time required to produce each unit will decrease as the workers become more familiar with the production procedure. It is determined that the function for the learning process is

$$T(x) = ax^b,$$

where

$\quad T(x)$ = cumulative average time to produce x units,
$\quad\quad x$ = number of units produced,
$\quad\quad a$ = hours required to produce 1st unit,
$\quad\quad b$ = slope of the learning curve.

33. Suppose in a certain memory experiment the rate of memorizing is given by

$$M'(t) = -0.003t^2 + 0.2t,$$

where $M'(t)$ = memory rate in words per minute. How many words are memorized in the first 10 minutes ($t = 0$ to $t = 10$)?

a) Find an expression for the total time required to produce 100 units.
b) Suppose $a = 100$ hr and $b = -0.322$. Find the total time required to produce 100 units. [*Hint:* $100^{0.678} \approx 22.7$.]

▶

Integrate.

35. $\displaystyle\int_1^2 (4x + 3)(5x - 2)\, dx$

36. $\displaystyle\int_2^5 (t + \sqrt{3})(t - \sqrt{3})\, dt$

37. $\displaystyle\int_0^1 (t + 1)^3\, dt$

38. $\displaystyle\int_1^3 \left(x - \frac{1}{x}\right)^2 dx$

39. $\displaystyle\int_1^3 \frac{t^5 - t}{t^3}\, dt$

40. $\displaystyle\int_4^9 \frac{t + 1}{\sqrt{t}}\, dt$

41. $\displaystyle\int_3^5 \frac{x^2 - 4}{x - 2}\, dx$

42. $\displaystyle\int_0^1 \frac{t^3 + 1}{t + 1}\, dt$

OBJECTIVES

You should be able to
a) Integrate using substitution.
b) Solve applied problems involving
   integration by substitution.

40. For $y = f(x) = 6x^2 + x$, find dy.

41. For $u = g(x) = x + 3$, find du.

11.4 INTEGRATION TECHNIQUES: SUBSTITUTION

The following formulas provide a basis for an integration technique called *substitution*.

A) $\displaystyle\int n \cdot u^{n-1} \, du = u^n + C$

B) $\displaystyle\int e^u \, du = e^u + C$

C) $\displaystyle\int \frac{1}{u} \, du = \ln u + C$

In the Leibniz notation dy/dx we did not give specific definitions of dy and dx. Nevertheless it will be convenient to treat dy/dx as a quotient. Thus, from

$$\frac{dy}{dx} = f'(x)$$

we can derive

$$dy = f'(x) \, dx.$$

It is possible to define dy and dx, but it is not necessary for our purposes.

Example 1   For $y = f(x) = x^3$,   find dy.

Solution $\dfrac{dy}{dx} = f'(x) = 3x^2$,   so $dy = f'(x) \, dx = 3x^2 \, dx$.

Example 2   For $u = g(x) = \ln x$,   find du.

Solution $\dfrac{du}{dx} = g'(x) = \dfrac{1}{x}$,   so $du = g'(x) \, dx = \dfrac{1}{x} \, dx$,   or $\dfrac{dx}{x}$.

DO EXERCISES 40 AND 41.

So far the dx in

$$\int f(x) \, dx$$

has played no role in integrating, other than indicating the variable of integration. Now it will be convenient to make use of dx. Consider the integral

$$\int 2x \cdot e^{x^2} \, dx.$$

If we set

$$u = x^2,$$

then

$$du = 2x \, dx.$$

If we substitute u for x^2, and du for $2x \, dx$, the integral takes on the form

$$\int e^u \, du.$$

Since

$$\int e^u \, du = e^u + C,$$

it follows that

$$\int 2x \cdot e^{x^2} \, dx = \int e^u \, du = e^u + C = e^{x^2} + C.$$

The result can be checked by differentiation. This procedure is referred to as *substitution*, or *change of variable*. It is a *trial-and-error* procedure; that is, if we try a substitution that doesn't result in an integrand that can be easily integrated, we try another. It will not always work! It *will* work if the integrand fits one of the rules A, B, or C, considered on p. 556.

DO EXERCISE 42.

Let us consider some further examples.

Example 3 $\displaystyle \int \frac{2x \, dx}{1 + x^2} = \int \frac{du}{u}$ Substitution $\boxed{\begin{array}{l} \text{Let } u = 1 + x^2, \\ \text{then } du = 2x \, dx. \end{array}}$

$$= \ln u + C$$
$$= \ln (1 + x^2) + C$$

Remember that this is a trial and error process. Suppose we had made the substitution

$$u = x^2.$$

Then

$$du = 2x \, dx,$$

and the integral becomes

$$\int \frac{du}{1 + u}.$$

42. Integrate.

$$\int 3x^2 e^{x^3} \, dx$$

43. Integrate.

$$\int \frac{2x\,dx}{5 + x^2}.$$

This is still not easily integrated, so we would try another substitution.

DO EXERCISE 43.

Example 4 $\int \frac{2x\,dx}{(1 + x^2)^2} = \int \frac{du}{u^2}$ Substitution
$$\boxed{\begin{array}{l} u = 1 + x^2, \\ du = 2x\,dx \end{array}}$$

$$= -\frac{1}{u} + C = -\frac{1}{1 + x^2} + C$$

DO EXERCISE 44.

44. Integrate.

$$\int \frac{2x\,dx}{(3 + x^2)^2}$$

Example 5 $\int \frac{\ln 3x\,dx}{x} = \int u\,du$ Substitution
$$\boxed{\begin{array}{l} u = \ln 3x, \\ du = \frac{1}{x}\,dx \end{array}}$$

$$= \frac{u^2}{2} + C$$

$$= \frac{(\ln 3x)^2}{2} + C$$

DO EXERCISE 45.

45. Integrate.

$$\int \frac{\ln x\,dx}{x}$$

Example 6 Integrate $\int xe^{x^2}dx$.

Solution Suppose we try

$$u = x^2,$$

then

$$du = 2x\,dx.$$

We don't quite have $2x\,dx$. We have only $x\,dx$ and will need to supply a 2. We do this by multiplying by $\frac{1}{2} \cdot 2$ as follows.

46. Integrate.

$$\int x^2 \cdot e^{x^3}\,dx$$

$$\frac{1}{2} \cdot 2 \cdot \int xe^{x^2}\,dx = \frac{1}{2}\int 2xe^{x^2}\,dx$$

$$= \frac{1}{2}\int e^{x^2}(2x\,dx)$$

$$= \frac{1}{2}\int e^u\,du$$

$$= \frac{1}{2}e^u + C$$

$$= \frac{1}{2}e^{x^2} + C$$

DO EXERCISE 46.

Example 7 $\displaystyle\int e^{ax}\,dx = \frac{1}{a}\int ae^{ax}\,dx \Bigg\}$ Substitution $\boxed{\begin{array}{l} u = ax, \\ du = a\,dx \end{array}}$

$$= \frac{1}{a}\int e^{u}\,du$$

$$= \frac{1}{a}e^{u} + C = \frac{1}{a}e^{ax} + C$$

Note that this gives us a formula for integrating e^{ax}.

DO EXERCISES 47 THROUGH 49.

Example 8 $\displaystyle\int\frac{dx}{x+3} = \int\frac{du}{u}$ Substitution $\boxed{\begin{array}{l} u = x + 3, \\ du = dx \end{array}}$

$$= \ln u + C = \ln(x+3) + C$$

With practice, you will make certain substitutions mentally and just write down the answer. Examples 7 and 8 are good illustrations.

Example 9

$$\int x^{2}(x^{3}+1)^{10}\,dx = \tfrac{1}{3}\int 3x^{2}(x^{3}+1)^{10}\,dx \Bigg\}$$ Substitution $\boxed{\begin{array}{l} u = x^{3} + 1 \\ du = 3x^{2}\,dx \end{array}}$

$$= \tfrac{1}{3}\int u^{10}\,du$$

$$= \frac{1}{3}\cdot\frac{u^{11}}{11} + C = \tfrac{1}{33}(x^{3}+1)^{11} + C.$$

DO EXERCISE 50.

Example 10 Evaluate.

$$\int_{0}^{1} x^{2}(x^{3}+1)^{10}\,dx.$$

Solution
a) First find the indefinite integral (shown in Example 9).
b) Then evaluate the definite integral on $[0, 1]$.

$$\int_{0}^{1} x^{2}(x^{3}+1)^{10}\,dx = [\tfrac{1}{33}(x^{3}+1)^{11}]_{0}^{1}$$

$$= \tfrac{1}{33}[(1^{3}+1)^{11} - (0^{3}+1)^{11}]$$

$$= \tfrac{1}{33}(2^{11} - 1^{11})$$

$$= \frac{2^{11}-1}{33}$$

DO EXERCISE 51.

Integrate.

47. $\displaystyle\int e^{5x}\,dx$

48. $\displaystyle\int e^{0.02x}\,dx$

49. $\displaystyle\int e^{-x}\,dx$

Integrate.

50. $\displaystyle\int x^{3}(x^{4}+5)^{19}\,dx$

51. Evaluate.

$$\int_{1}^{e}\frac{\ln x\,dx}{x}$$

(See Margin Exercise 45.)

EXERCISE SET 11.4

Integrate. (Be sure to check by differentiating!)

1. $\int \dfrac{3x^2\, dx}{7 + x^3}$

2. $\int \dfrac{3x^2\, dx}{1 + x^3}$

3. $\int e^{4x}\, dx$

4. $\int e^{3x}\, dx$

5. $\int e^{x/2}\, dx$

6. $\int e^{x/3}\, dx$

7. $\int x^3 e^{x^4}\, dx$

8. $\int x^4 e^{x^5}\, dx$

9. $\int t^2 e^{-t^3}\, dt$

10. $\int t e^{-t^2}\, dt$

11. $\int \dfrac{\ln 4x\, dx}{x}$

12. $\int \dfrac{\ln 5x\, dx}{x}$

13. $\int \dfrac{dx}{1 + x}$

14. $\int \dfrac{dx}{5 + x}$

15. $\int \dfrac{dx}{4 - x}$

16. $\int \dfrac{dx}{1 - x}$

17. $\int t^2(t^3 - 1)^7\, dt$

18. $\int t(t^2 - 1)^5\, dt$

19. $\int (x^4 + x^3 + x^2)^7(4x^3 + 3x^2 + 2x)\, dx$

20. $\int (x^3 - x^2 - x)^9(3x^2 - 2x - 1)\, dx$

21. $\int \dfrac{e^x\, dx}{4 + e^x}$

22. $\int \dfrac{e^t\, dt}{3 + e^t}$

23. $\int \dfrac{\ln x^2}{x}\, dx$

24. $\int \dfrac{(\ln x)^2}{x}\, dx$

25. $\int \dfrac{dx}{x \ln x}$

26. $\int \dfrac{dx}{x \ln x^2}$

27. $\int \sqrt{ax + b}\, dx$

28. $\int x\sqrt{ax^2 + b}\, dx$

29. $\int b e^{ax}\, dx$

30. $\int P_0 e^{kt}\, dt$

Integrate.

31. $\displaystyle\int_0^1 2x e^{x^2}\, dx$

32. $\displaystyle\int_0^1 3x^2 e^{x^3}\, dx$

33. $\displaystyle\int_0^1 x(x^2 + 1)^5\, dx$

34. $\displaystyle\int_1^2 x(x^2 - 1)^7\, dx$

35. $\displaystyle\int_1^3 \dfrac{dt}{1 + t}$

36. $\displaystyle\int_1^3 e^{2x}\, dx$

37. $\displaystyle\int_1^4 \dfrac{2x + 1}{x^2 + x - 1}\, dx$

38. $\displaystyle\int_1^3 \dfrac{2x + 3}{x^2 + 3x}\, dx$

39. $\displaystyle\int_0^b e^{-x}\, dx$

40. $\displaystyle\int_0^b 2e^{-2x}\, dx$

41. $\displaystyle\int_0^b m e^{-mx}\, dx$

42. $\displaystyle\int_0^b k e^{-kx}\, dx$

43. $\displaystyle\int_0^4 (x - 6)^2\, dx$

44. $\displaystyle\int_0^3 (x - 5)^2\, dx$

45. ▦ *Sociology—Divorce.* The U.S. divorce rate is approximated by

$$D(t) = 100{,}000 e^{0.025t},$$

where
$D(t)$ = number of divorces occurring at time t,
   t = number of years measured from 1900.
That is, $t = 0$ corresponds to 1900, $t = 84\frac{9}{365}$ corresponds to January 9, 1984, and so on.

a) Find the total number of divorces from 1900 to 1984. Note that this is

$$\int_0^{84} D(t)\, dt.$$

b) Find the total number of divorces from 1980 to 1984. Note that this is $\int_{80}^{84} D(t)\, dt$.

46. *Business—Value of an investment.* A company buys a new machine for $250,000. The marginal revenue from the sale of products produced by the machine is projected to be

$$R'(t) = 4000t.$$

The salvage value of the machine decreases at the rate of

$$V(t) = 25{,}000 e^{-0.1t}.$$

The total profit from the machine after T yrs is given by

$$P(T) = \begin{pmatrix}\text{Revenue from}\\ \text{sale of prod.}\end{pmatrix} + \begin{pmatrix}\text{Revenue from}\\ \text{sale of mach.}\end{pmatrix} - \begin{pmatrix}\text{Cost of}\\ \text{machine}\end{pmatrix}$$

$$= \int_0^T R'(t)\, dt \;+\; \int_0^T V(t)\, dt \;-\; \$250{,}000.$$

a) Find $P(T)$.

b) Find $P(10)$.

Integrate.

47. $\int 5x\sqrt{1 - 4x^2}\, dx$ **48.** $\int \frac{dx}{ax + b}$ **49.** $\int \frac{x^2}{e^{x^3}}\, dx$ **50.** $\int \frac{e^{\sqrt{t}}}{\sqrt{t}}\, dt$ **51.** $\int \frac{e^{1/t}}{t^2}$

52. $\int \frac{(\ln x)^2}{x}\, dx$ **53.** $\int \frac{dx}{x(\ln x)^4}$ **54.** $\int (e^t + 2)e^t\, dt$ **55.** $\int x^2\sqrt{x^3 + 1}\, dx$ **56.** $\int \frac{t^2}{\sqrt[4]{2 + t^3}}\, dt$

57. $\int \frac{x - 3}{(x^2 - 6x)^{1/3}}\, dx$ **58.** $\int \frac{[(\ln x)^2 + 3(\ln x) + 4]}{x}\, dx$ **59.** $\int \frac{t^3 \ln(t^4 + 8)}{t^4 + 8}\, dt$

60. $\int \frac{t^2 + 2t}{(t + 1)^2}\, dt$ *Hint:* $\dfrac{t^2 + 2t}{(t + 1)^2} = \dfrac{t^2 + 2t + 1 - 1}{t^2 + 2t + 1} = 1 - \dfrac{1}{(t + 1)^2}$ **61.** $\int \frac{x^2 + 6x}{(x + 3)^2}\, dx$

62. $\int \frac{x + 3}{x + 1}\, dx$ *Hint:*   Divide, $\dfrac{x + 3}{x + 1} = 1 + \dfrac{2}{x + 1}$ **63.** $\int \frac{t - 5}{t - 4}\, dt$ **64.** $\int \frac{dx}{x(\ln x)^n}$

65. $\int \frac{dx}{e^x + 1}$ *Hint:* $\dfrac{1}{e^x + 1} = \dfrac{e^{-x}}{1 + e^{-x}}$ **66.** $\int \frac{e^x - e^{-x}}{e^x + e^{-x}}\, dx$ **67.** $\int \frac{(\ln x)^n}{x}\, dx$

11.5 INTEGRATION TECHNIQUES: INTEGRATION BY PARTS — TABLES

Recall the product rule for derivatives:

$$\frac{d}{dx}\, uv = \frac{du}{dx}\, v + \frac{dv}{dx}\, u = u\frac{dv}{dx} + v\frac{du}{dx}.$$

Integrating both sides, we get

$$uv = \int u\frac{dv}{dx}\, dx + \int v\frac{du}{dx}\, dx = \int u\, dv + \int v\, du.$$

Solving for $\int u\, dv$, we get

$$\int u\, dv = uv - \int v\, du.$$

This equation can be used as a formula for integrating in certain situations. These are situations in which an integrand is a product of two functions; one of the functions can be integrated by using the techniques we have already developed.

For example,

$$\int xe^x\, dx$$

can be considered as follows:

$$\int x(e^x\, dx) = \int u\, dv, \quad \text{where } u = x \text{ and } dv = e^x\, dx.$$

OBJECTIVES

You should be able to

a) Integrate using integration by parts.

b) Solve applied problems involving integration by parts.

c) Integrate using a table of integration formulas.

52. Integrate.

$$\int 3x \cdot e^{3x}\, dx$$

We already know how to integrate $e^x\, dx$, or dv. The simplest anti-derivative is e^x. This is v. Now since $du = dx$, the formula gives us

$$\overset{u}{}\quad \overset{dv}{} \qquad \overset{u}{}\,\,\overset{v}{} \qquad \overset{v}{}\,\,\overset{du}{}$$

$$\int (x)(e^x\, dx) = (x)(e^x) - \int (e^x)(dx)$$

$$= xe^x - e^x + C.$$

This way of integrating is called *integration by parts*, and the following formula provides the basis for it.

INTEGRATION BY PARTS FORMULA

$$\int u\, dv = uv - \int v\, du$$

Note that integration by parts is a trial-and-error process, as is substitution. In the preceding example, suppose we had reversed the roles of x and e^x. We would have obtained

$$u = e^x \qquad dv = x\, dx,$$

$$du = e^x\, dx \qquad v = \frac{x^2}{2},$$

and

$$\int (e^x)(x\, dx) = (e^x)\left(\frac{x^2}{2}\right) - \int \left(\frac{x^2}{2}\right)(e^x\, dx).$$

Now the integrand on the right is more difficult to integrate than the one we started with. When we can integrate *both* factors of an integrand, and thus have a choice as to how to apply the integration by parts formula, it can happen that only one (and maybe none) of the possibilities will work.

DO EXERCISE 52.

Let us consider some further examples.

Example 1   Integrate $\int \ln x\, dx$.

Solution   Note that $\int (dx/x) = \ln x + C$, but we do not yet know how to find $\int \ln x\, dx$. Let

$$u = \ln x \qquad \text{and} \qquad dv = dx.$$

Then

$$du = \frac{1}{x}\, dx \qquad \text{and} \qquad v = x.$$

Using the integration by parts formula gives

$$\overset{u}{\int} \overset{dv}{(\ln x)(dx)} = \overset{u}{(\ln x)}\overset{v}{x} - \int \overset{v}{x}\overset{du}{\left(\frac{1}{x}\,dx\right)}$$

$$= x \ln x - \int dx = x \ln x - x + C.$$

DO EXERCISE 53.

Example 2 Integrate $\int x\sqrt{x + 1}\,dx$.

Solution We let

$$u = x \qquad \text{and} \qquad dv = (x + 1)^{1/2}\,dx.$$

Then

$$du = dx \qquad \text{and} \qquad v = \tfrac{2}{3}(x + 1)^{3/2}.$$

Note that we had to use substitution to integrate dv. Using the integration by parts formula gives

$$\int x\sqrt{x + 1}\,dx = x \cdot \tfrac{2}{3}(x + 1)^{3/2} - \tfrac{2}{3}\int (x + 1)^{3/2}\,dx$$

$$= \tfrac{2}{3}x(x + 1)^{3/2} - \tfrac{2}{3} \cdot \tfrac{2}{5}(x + 1)^{5/2} + C$$

$$= \tfrac{2}{3}x(x + 1)^{3/2} - \tfrac{4}{15}(x + 1)^{5/2} + C.$$

DO EXERCISE 54.

Example 3 Integrate $\int_1^2 \ln x\,dx$.

Solution
a) First find the indefinite integral (Example 1).
b) Then evaluate the definite integral.

$$\int_1^2 \ln x\,dx = [x \ln x - x]_1^2$$

$$= (2 \ln 2 - 2) - (1 \cdot \ln 1 - 1)$$

$$= 2 \ln 2 - 2 + 1$$

$$= 2 \ln 2 - 1$$

DO EXERCISE 55.

Tables of Integration Formulas

You have probably noticed that, generally speaking, integration is more difficult and "tricky" than differentiation. Because of this,

53. Integrate.

$$\int x \cdot \ln x\,dx$$

54. Integrate.

$$\int x\sqrt{x + 3}\,dx$$

55. Integrate $\displaystyle\int_1^2 x \ln x\,dx$.

(See Margin Exercise 63.)

56. Using Table 5, integrate

$$\int \frac{1}{x^2 - 25} \, dx.$$

integral formulas that are reasonable and/or important have been gathered into tables. Table 5 at the back of the book, though quite brief, is such an example. Entire books of integration formulas are available in libraries, and lengthy tables are also available in mathematics handbooks. Such tables are usually classified by the form of the integrand. The idea is to properly match the integral in question with a formula in the table.

Example Integrate $\int \frac{dx}{x(3 - x)}$.

Solution This integral fits *Formula 20* in Table 5:

$$\int \frac{1}{x(ax + b)} \, dx = \frac{1}{b} \ln \left(\frac{x}{ax + b} \right) + C.$$

In our integral, $a = -1$ and $b = 3$, so we have, by the formula,

$$\int \frac{1}{x(3 - x)} \, dx = \int \frac{dx}{x(-1 \cdot x + 3)} = \frac{1}{3} \ln \left(\frac{x}{-1 \cdot x + 3} \right) + C$$

$$= \frac{1}{3} \ln \left(\frac{x}{3 - x} \right) + C.$$

DO EXERCISE 56.

EXERCISE SET 11.5

Integrate. Use integration by parts. Do not use Table 5. Check by differentiating.

1. $\int 5xe^{5x} \, dx$

2. $\int 2xe^{2x} \, dx$

3. $\int x^3(3x^2 \, dx)$

4. $\int x^2(2x \, dx)$

5. $\int xe^{2x} \, dx$

6. $\int xe^{3x} \, dx$

7. $\int xe^{-2x} \, dx$

8. $\int xe^{-x} \, dx$

9. $\int x^2 \ln x \, dx$

10. $\int x^3 \ln x \, dx$

11. $\int x \ln x^2 \, dx$

12. $\int x^2 \ln x^3 \, dx$

13. $\int \ln (x + 3) \, dx$

14. $\int \ln (x + 1) \, dx$

15. $\int (x + 2) \ln x \, dx$

16. $\int (x + 1) \ln x \, dx$

17. $\int (x - 1) \ln x \, dx$

18. $\int (x - 2) \ln x \, dx$

19. $\int x\sqrt{x + 2} \, dx$

20. $\int x\sqrt{x + 4} \, dx$

21. $\int x^3 \ln 2x \, dx$

22. $\int x^2 \ln 5x \, dx$

23. $\int x^2 e^x \, dx$

24. $\int (\ln x)^2 \, dx$

25. $\int x^2 e^{2x} \, dx$

26. $\int x^{-5} \ln x \, dx$

Integrate. Use integration by parts. Do not use Table 5.

27. $\int_1^2 x^2 \ln x \, dx$

28. $\int_1^2 x^3 \ln x \, dx$

29. $\int_2^6 \ln (x + 3) \, dx$

30. $\int_0^5 \ln (x + 1) \, dx$

31. $\int_0^1 xe^x \, dx$

32. $\int_0^1 xe^{-x} \, dx$

Integrate. Use Table 5.

33. $\int xe^{-3x}\, dx$ **34.** $\int xe^{4x}\, dx$ **35.** $\int 5^x\, dx$ **36.** $\int \frac{1}{\sqrt{x^2 - 9}}\, dx$ **37.** $\int \frac{1}{16 - x^2}\, dx$

38. $\int \frac{1}{x\sqrt{4 + x^2}}\, dx$ **39.** $\int \frac{x}{5 - x}\, dx$ **40.** $\int \frac{x}{(1 - x)^2}\, dx$ **41.** $\int \frac{1}{x(5 - x)^2}\, dx$ **42.** $\int \sqrt{x^2 + 9}\, dx$

43. *Ecology—Electrical energy use.* The rate of electrical energy used by a family in kilowatt hours per day is given by

$$K(t) = 10te^{-t},$$

where t is the time, in hours. That is, t is in the interval $[0, 24]$.
a) How many kilowatt hours does the family use in the first T hours of a day ($t = 0$ to $t = T$)?
b) How many kilowatt hours does the family use in the first 4 hours of the day?

44. *Biomedical—Drug dosage.* Suppose an oral dose of a drug is taken. From that time, the drug is assimilated in the body and excreted through the urine. The total amount of the drug that has passed through the body in time T is given by

$$\int_0^T E(t)\, dt,$$

where R is the rate of excretion of the drug through the urine. A typical rate of excretion function is

$$E(t) = te^{-kt},$$

where $k > 0$ and t is time, in hours.
a) Use integration by parts to find a formula for $\int_0^T E(t)\, dt$.
b) ▦ Find $\int_0^{10} E(t)\, dt$, when $k = 0.2$ mg/hr.

▶───

Integrate by parts. Do not use Table 5.

45. $\int \sqrt{x}\, \ln x\, dx$ **46.** $\int x^n \ln x\, dx$ **47.** $\int \frac{te^t}{(t + 1)^2}\, dt$ **48.** $\int x^2 (\ln x)^2\, dx$ **49.** $\int \frac{\ln x}{\sqrt{x}}\, dx$ **50.** $\int x^n (\ln x)^2\, dx$

51. a) Verify that, for any positive integer n,

$$\int x^n e^x\, dx = x^n e^x - n \int x^{n-1} e^x\, dx.$$

b) Apply (a) repeatedly to integrate

$$\int x^3 e^x\, dx.$$

52. a) Verify that, for any positive integer n,

$$\int (\ln x)^n\, dx = x(\ln x)^n - n \int (\ln x)^{n-1}\, dx.$$

b) Apply (a) repeatedly to integrate

$$\int (\ln x)^3\, dx.$$

───

11.6 *(OPTIONAL) THE DEFINITE INTEGRAL AS A LIMIT OF SUMS

We now consider approximating the area of a region by dividing it into subregions that are almost rectangles. In the next drawing $[a, b]$ has been divided into 4 subintervals, each having width Δx, or $(b - a)/4$.

The heights of the rectangles shown are

$$f(x_1), \qquad f(x_2), \qquad f(x_3), \qquad \text{and} \quad f(x_4).$$

OBJECTIVES

You should be able to
a) Approximate

$$\int_a^b f(x)\, dx$$

by adding areas of rectangles.
b) Find the average value of a function over a given interval.

Write summation notation.

57. $1 + 4 + 9 + 16 + 25 + 36$

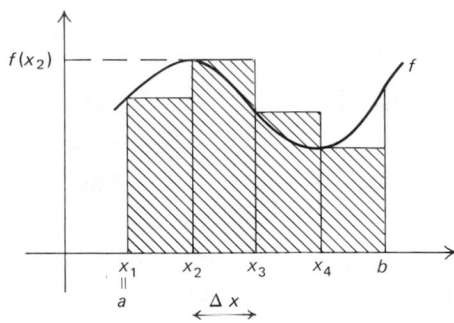

58. $e + e^2 + e^3 + e^4$

59. $P(x_1)\,\Delta x + P(x_2)\,\Delta x + \cdots + P(x_{38})\,\Delta x.$

The area of the region under the curve is approximately the sum of the areas of the four rectangles,

$$f(x_1)\,\Delta x + f(x_2)\,\Delta x + f(x_3)\,\Delta x + f(x_4)\,\Delta x.$$

We can name this sum using *summation notation* which utilizes the Greek capital letter sigma, Σ,

$$\sum_{i=1}^{4} f(x_i)\,\Delta x.$$

This is read "the sum of the numbers $f(x_i)\,\Delta x$ from $i = 1$ to $i = 4$." To recover the original expression substitute the numbers 1 through 4 successively into $f(x_i)\,\Delta x$ and write plus signs between the results.

Example 1 Write summation notation for $2 + 4 + 6 + 8 + 10$.

Solution $\quad 2 + 4 + 6 + 8 + 10 = \displaystyle\sum_{i=1}^{5} 2i.$

Example 2 Write summation notation for

$$g(x_1)\,\Delta x + g(x_2)\,\Delta x + \cdots + g(x_{19})\,\Delta x.$$

Solution $\quad g(x_1)\,\Delta x + g(x_2)\,\Delta x + \cdots + g(x_{19})\,\Delta x = \displaystyle\sum_{i=1}^{19} g(x_i)\,\Delta x.$

DO EXERCISES 57 THROUGH 59.

Example 3 Express $\displaystyle\sum_{i=1}^{4} 3^i$ without using summation notation.

Solution $\quad \displaystyle\sum_{i=1}^{4} 3i = 3^1 + 3^2 + 3^3 + 3^4$, or 120.

Example 4   Express $\sum_{i=1}^{30} h(x_i)\,\Delta x$ without using summation notation.

Solution $\sum_{i=1}^{30} h(x_i)\,\Delta x = h(x_1)\,\Delta x + h(x_2)\,\Delta x + \cdots + h(x_{30})\,\Delta x.$

DO EXERCISES 60 THROUGH 62.

Approximation of area by rectangles becomes better as we use more rectangles and smaller subintervals, as we show in the following drawings.

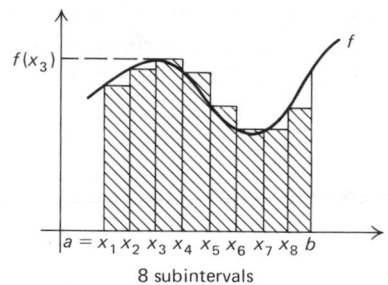

8 subintervals

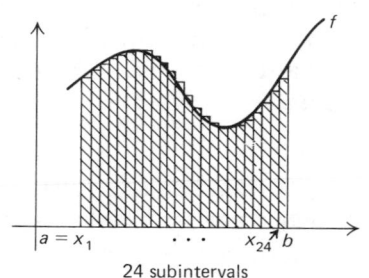

24 subintervals

In general, the interval $[a, b]$ is divided into n equal subintervals, each of width $\Delta x = (b - a)/n$.

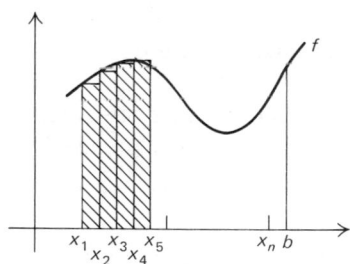

The heights of the rectangles are

$$f(x_1), f(x_2), \cdots, f(x_n).$$

The area of the region under the curve is approximated by the sum of the areas of the rectangles,

$$\sum_{i=1}^{n} f(x_i)\,\Delta x.$$

Express without using summation notation.

60. $\sum_{i=1}^{3} 4^i$

61. $\sum_{i=1}^{5} ie^i$

62. $\sum_{i=1}^{20} t(x_i)\,\Delta x$

We now obtain the actual area by letting the number of intervals increase indefinitely and by taking the limit. The area is thus given by

$$A = \lim_{n \to \infty} \sum_{i=1}^{n} f(x_i) \, \Delta x.$$

The area is also given by a definite integral:

$$\int_a^b f(x) \, dx = \lim_{n \to \infty} \sum_{i=1}^{n} f(x_i) \, \Delta x.$$

The fact that we can so express the integral of a function (positive or otherwise) as a limit of a sum or in terms of an antiderivative is so important that it has a name, *The Fundamental Theorem of Integral Calculus*.

THE FUNDAMENTAL THEOREM OF INTEGRAL CALCULUS. If a function f has an antiderivative F on $[a, b]$, then

$$\int_a^b f(x) \, dx = F(b) - F(a) = \lim_{n \to \infty} \sum_{i=1}^{n} f(x_i) \, \Delta x.$$

It is interesting to envision that, as we take the limit on the right, the summation sign stretches into something reminiscent of an S (the integral sign) and the Δx becomes dx. This is also a motivation for the use of dx in the integral notation.

This result allows us to approximate the value of a definite integral by a sum, making it as good as we please by taking n sufficiently large.

Example 5 Raggs, Ltd., determines that the marginal cost per suit is

$$C'(x) = 0.0003x^2 - 0.2x + 50.$$

Approximate the total cost of producing 400 suits by computing the sum $\sum_{i=1}^{4} C'(x_i) \, \Delta x$.

Solution The interval $[0, 400]$ is divided into 4 subintervals, each of length $\Delta x = (400 - 0)/4 = 100$. Now x_i is varying from $x_1 = 0$ to $x_5 = 400$.

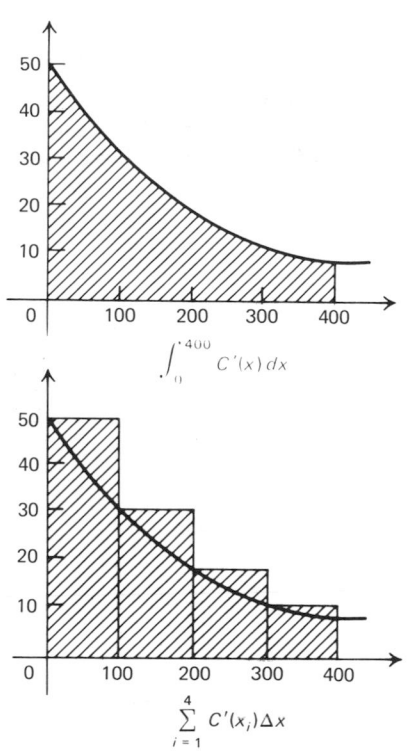

$$\int_0^{400} C'(x)\, dx$$

$$\sum_{i=1}^{4} C'(x_i)\Delta x$$

$$\sum_{i=1}^{4} C'(x_i)\,\Delta x = C'(0)\cdot 100 + C'(100)\cdot 100 + C'(200)\cdot 100$$

$$+\, C'(300)\cdot 100$$

$$= 50\cdot 100 + 33\cdot 100 + 22\cdot 100 + 17\cdot 100 = \$12{,}200$$

Now

$$\int_0^{400} C'(x)\, dx = \$10{,}400.$$

Thus this approximation is not too far off, even though the number of subintervals is small. In Margin Exercise 63 you will obtain a better approximation using 8 subintervals.

DO EXERCISES 63 AND 64.

The fact that an integral can be approximated by a sum is useful when the antiderivative of a function does not have an elementary formula. For example, for the function $e^{-x^2/2}$, important in probability, there is no formula for the antiderivative. So, tables of approximate values of its integral have been computed using summation methods.

63. Referring to Example 5, find

$$\sum_{i=1}^{8} C'(x_i)\,\Delta x,$$

where the interval $[0, 400]$ is divided into 8 equal subintervals of length

$$\Delta x = \frac{400 - 0}{8} = 50.$$

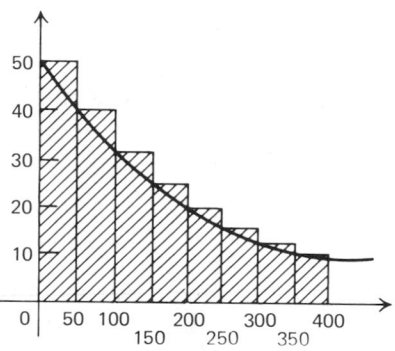

64. In graphs (a) and (b) compute the areas of each rectangle to four decimal places. Then add them to approximate the area under the curve $y = 1/x$ over $[1, 7]$.

a)

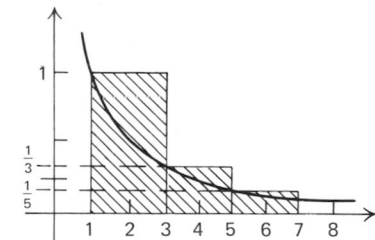

b)

EXERCISE SET 11.6

1. a) Approximate $\int_1^7 (dx/x^2)$ by computing the area of each rectangle to four decimal places and adding.

b) Evaluate $\int_1^7 (dx/x^2)$. Compare the answer to (a).

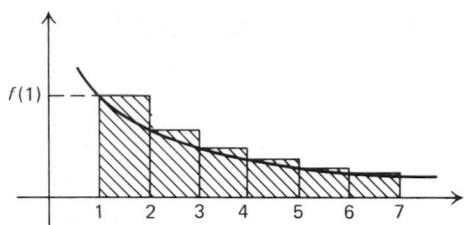

2. a) Approximate $\int_0^5 (x^2 + 1)\, dx$ by computing the area of each rectangle and adding.

b) Evaluate $\int_0^5 (x^2 + 1)\, dx$. Compare the answer to (a).

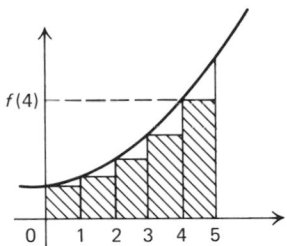

The Trapezoidal Rule. Another way to approximate an integral is to replace each rectangle in the sum (see the figure at the right) by a trapezoid, as shown in the second figure. The area of a trapezoid is $h(c_1 + c_2)/2$, where c_1 and c_2 are the lengths of the parallel sides. Thus, in the second figure,

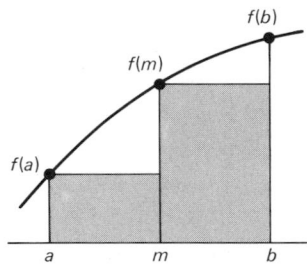

$$\int_a^b f(x)\, dx = \int_a^m f(x)\, dx + \int_m^b f(x)\, dx$$

$$\approx \Delta x \frac{f(a) + f(m)}{2} + \Delta x \frac{f(m) + f(b)}{2}$$

$$\approx \Delta x \left[\frac{f(a)}{2} + f(m) + \frac{f(b)}{2} \right]$$

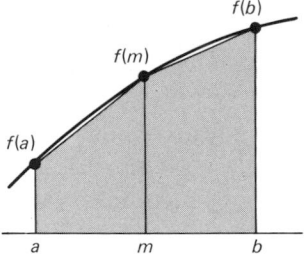

For an interval $[a, b]$ subdivided into n equal subintervals of length $\Delta x = (b - a)/n$, we get the approximation

$$\int_a^b f(x)\, dx = \Delta x \left[\frac{f(a)}{2} + f(x_2) + f(x_3) + \cdots + f(x_n) + \frac{f(b)}{2} \right],$$

where $x_1 = a$. This is called the *Trapezoidal Rule*.

3. Use the Trapezoidal Rule and the interval subdivision of Exercise 1 to approximate

$$\int_1^7 \frac{dx}{x^2}$$

4. Use the Trapezoidal Rule and the interval subdivision of Exercise 2 to approximate

$$\int_0^5 (x^2 + 1)\, dx$$

11.7 IMPROPER INTEGRALS

Let us try to find the area of the region under the graph of $y = 1/x^2$ on the interval $[1, \infty)$.

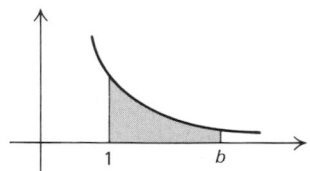 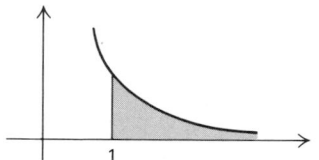

Note that this region is of infinite extent. We have not yet considered how to find the area of such a region. Let us find the area under the curve on the interval from 1 to b, and then see what happens as b gets very large. The area on $[1, b]$ is

$$\int_1^b \frac{dx}{x^2} = \left[-\frac{1}{x}\right]_1^b = \left(-\frac{1}{b}\right) - \left(-\frac{1}{1}\right) = -\frac{1}{b} + 1 = 1 - \frac{1}{b}.$$

Then

$$\lim_{b\to\infty} [\text{area from 1 to } b] = \lim_{b\to\infty}\left(1 - \frac{1}{b}\right).$$

Let us investigate this limit.

DO EXERCISE 65.

Note that as $b \to \infty$, $1/b \to 0$, so $[1 - (1/b)] \to 1$. Thus

$$\lim_{b\to\infty} [\text{area from 1 to } b] = \lim_{b\to\infty}\left(1 - \frac{1}{b}\right) = 1.$$

We *define* the area from 1 to ∞ to be this limit. Here we have an example of an infinitely long region with a finite area.

DO EXERCISE 66.

Such areas may not always be finite. Let us try to find the area of the region under the graph of $y = 1/x$ on the interval $[1, \infty)$.

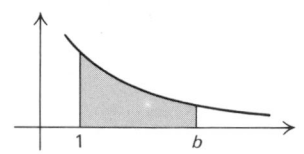 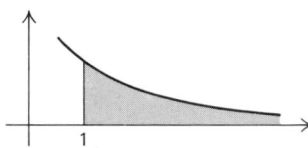

OBJECTIVES

You should be able to determine whether an improper integral is convergent or divergent, and calculate its value if it is convergent.

65. Complete.

b	$1 - \dfrac{1}{b}$
2	$1 - \frac{1}{2}$, or $\frac{1}{2}$
3	
10	
100	
200	

66. Find the area of the region under the graph of

$$y = \frac{1}{x^2}$$

on the interval $[2, \infty)$.

67. Find the area under the graph of

$$y = \frac{1}{x}$$

from $x = 2$ to $x = \infty$.

By definition, the area A from 1 to ∞ is the limit as $b \to \infty$ of the area from 1 to b, so

$$A = \lim_{b \to \infty} \int_1^b \frac{dx}{x} = \lim_{b \to \infty} [\ln x]_1^b = \lim_{b \to \infty} (\ln b - \ln 1) = \lim_{b \to \infty} \ln b.$$

On p. 505 of Chapter 10 we can see that since $\ln b$ increases indefinitely as b increases, this limit does not exist.

Thus we have an infinitely long region with an infinite area. Note that the graphs of $y = 1/x^2$ and $y = 1/x$ have similar shapes, but the region under one of them has a finite area and the other does not.

DO EXERCISE 67.

An integral such as

$$\int_a^\infty f(x)\, dx,$$

with an upper limit of ∞, is called an *improper integral*. Its value is defined to be the following limit.

$$\int_a^\infty f(x)\, dx = \lim_{b \to \infty} \int_a^b f(x)\, dx$$

If the limit exists, then we say that the improper integral *converges*. If the limit does not exist, we say that the improper integral *diverges*. Thus

$$\int_1^\infty \frac{dx}{x^2} = 1 \quad converges; \quad \text{and} \quad \int_1^\infty \frac{dx}{x} = \infty \quad diverges.$$

Example 1.

$$\int_0^\infty 2e^{-2x}\, dx = \lim_{b \to \infty} \int_0^b 2e^{-2x}\, dx = \lim_{b \to \infty} [2(-\tfrac{1}{2})e^{-2x}]_0^b$$

$$= \lim_{b \to \infty} [-e^{-2x}]_0^b$$

$$= \lim_{b \to \infty} [-e^{-2b} - (-e^{-2 \cdot 0})]$$

$$= \lim_{b \to \infty} (-e^{-2b} + 1)$$

$$= \lim_{b \to \infty} \left(1 - \frac{1}{e^{2b}}\right).$$

Now as $b \to \infty$, $e^{2b} \to \infty$ (from Chapter 10), so

$$\frac{1}{e^{2b}} \to 0 \quad \text{and} \quad \left(1 - \frac{1}{e^{2b}}\right) \to 1.$$

Thus

$$\int_0^\infty 2e^{-2x}\, dx = 1. \qquad \text{(The integral is convergent.)}$$

DO EXERCISES 68 AND 69.

The following are definitions of two types of improper integrals.

$$\int_{-\infty}^b f(x)\, dx = \lim_{a \to -\infty} \int_a^b f(x)\, dx$$

$$\int_{-\infty}^\infty f(x)\, dx = \int_{-\infty}^c f(x)\, dx + \int_c^\infty f(x)\, dx$$

For $\int_{-\infty}^\infty f(x)\, dx$ to converge, both integrals on the right above must converge.

Application

When an amount P of radioactive material is being released into the atmosphere annually, the amount present at time T is given by

$$A_T = \int_0^T Pe^{-kt}\, dt = \frac{P}{k}(1 - e^{-kT}).$$

As $T \to \infty$ (the radioactive material is to be released forever),

$$A_T \to P/k.$$

That is, the buildup of radioactive material approaches a limiting value P/k.

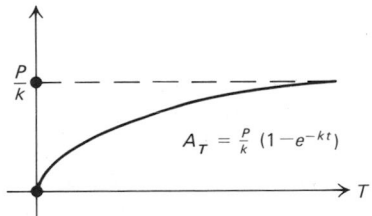

$$A_T = \frac{P}{k}(1 - e^{-kt})$$

Determine whether each of the following improper integrals is convergent or divergent, and calculate its value if it is convergent.

68. $\displaystyle\int_0^\infty 5e^{-5x}\, dx$

69. $\displaystyle\int_0^\infty 2x\, dx$

EXERCISE SET 11.7

Determine whether each of the following improper integrals is convergent or divergent, and calculate its value if it is convergent.

1. $\int_3^\infty \frac{dx}{x^2}$ **2.** $\int_4^\infty \frac{dx}{x^2}$ **3.** $\int_3^\infty \frac{dx}{x}$ **4.** $\int_4^\infty \frac{dx}{x}$

5. $\int_0^\infty 3e^{-3x}\,dx$ **6.** $\int_0^\infty 4e^{-4x}\,dx$ **7.** $\int_1^\infty \frac{dx}{x^3}$ **8.** $\int_1^\infty \frac{dx}{x^4}$

9. $\int_0^\infty \frac{dx}{1+x}$ **10.** $\int_0^\infty \frac{4\,dx}{1+x}$ **11.** $\int_1^\infty 5x^{-2}\,dx$ **12.** $\int_1^\infty 7x^{-2}\,dx$

13. $\int_0^\infty e^x\,dx$ **14.** $\int_0^\infty e^{2x}\,dx$ **15.** $\int_3^\infty x^2\,dx$ **16.** $\int_5^\infty x^4\,dx$

17. $\int_0^\infty xe^x\,dx$ **18.** $\int_0^\infty \ln x\,dx$ **19.** $\int_0^\infty me^{-mx}\,dx,\quad m>0$ **20.** $\int_0^\infty Qe^{-kt}\,dt,\quad k>0$

21. *Radioactive buildup.* Plutonium has a decay rate of 0.003% per year. Suppose 1 lb of plutonium is released into the atmosphere each year. What is the limiting value of the radioactive buildup?

22. *Radioactive buildup.* Cesium-137 has a decay rate of 2.3% per year. Suppose 1 lb of cesium-137 is released into the atmosphere each year. What is the limiting value of the radioactive buildup?

Accumulated present values of stock dividends paid perpetually. The accumulation of all present values of dividends that are assumed to be paid perpetually is given by

$$V = \int_0^\infty d(t)e^{-mt}\,dt,$$

where $d(t)$ is the instantaneous dividend payment and m is the current interest rate.

23. Find V when $d(t) = e^{-t}$ and $m = 7\%$.   **24.** Find V when $d(t) = \$1000$ and $m = 8\%$.

OBJECTIVES

You should be able to
a) Verify that a given function satisfies the property

$$\int_a^b f(x)\,dx = 1,$$

for being a probability density function.
b) Find k such that a function like

$$f(x) = kx^2$$

is a probability density function over an interval $[a, b]$.
c) Solve applied problems involving probability density functions.

11.8 PROBABILITY

Here we continue the study of probability started in Chapter 5 and show the role played by the definite integral in the theory of probability.

Continuous Random Variables

In Chapter 6 we reconsidered random variables with discrete values. Now we let the random variable take on a *continuous* range of values. While in Section 6.1 we defined a random variable X as a rule which assigns a numerical value x to the outcome of an experiment, for simplicity of notation we will only use lower case x throughout this chapter.

Suppose we throw a dart at a number line in such a way that it always lands in the interval $[1, 3]$.

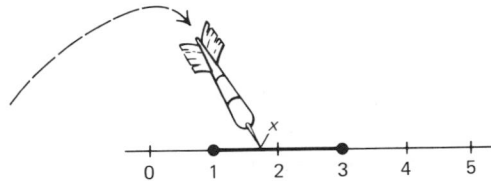

Let x be the number that the dart hits. Note that x is a quantity which can be observed (or measured) repeatedly and whose possible values consist of an entire interval of real numbers. Such a variable is called a *continuous random variable*. Suppose we throw the dart a large number of times and it lands in the subinterval $[1.6, 2.8]$ 43% of the time; the probability, then, that the dart lands in that interval is 0.43.

Let us consider some other examples of continuous random variables.

Example 1 Suppose x is the arrival time of buses at a bus stop in a three-hour period from 2 P.M. to 5 P.M. The interval is $[2, 5]$.

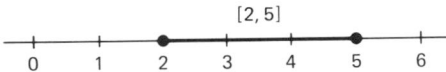

Then x is a continuous random variable distributed over the interval $[2, 5]$.

Example 2 Suppose x is the corn acreage of each farm in the U.S. and Canada. The interval is $[0, a]$, where a is the highest acreage. Or, not knowing what the highest acreage might be, the interval might be $[0, \infty)$ to allow for all possibilities.

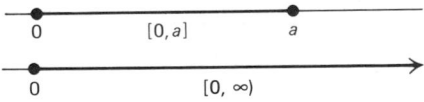

[*Note:* It might be argued that there is a value in $[0, a]$ or $[0, \infty)$ for which no farm has that acreage, but for practical purposes these values are often disregarded.]

70. Suppose

a) The dosage x of a drug is from 15 milligrams to 25 milligrams. What interval is determined?

b) The distance x is the distance between successive cars on a highway. What interval is determined?

Then x is a continuous random variable distributed over the interval $[0, a]$, or $[0, \infty)$.

DO EXERCISE 70.

Suppose, considering Example 1 on the arrival times of buses, that we wanted to know the probability that a bus will arrive between 4 P.M. and 5 P.M., as represented by

$$P([4, 5]), \quad \text{or} \quad P(4 \leqslant x \leqslant 5).$$

In some cases it is possible to find a function over $[2, 5]$ such that areas over subintervals give the probabilities that a bus will arrive during these subintervals. For example, suppose we had a constant function $f(x) = \frac{1}{3}$ that will give us these probabilities. Look at its graph.

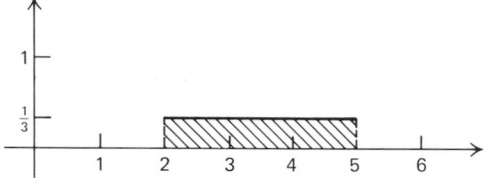

The area under the curve is $3 \cdot \frac{1}{3}$, or 1. The probability that a bus will arrive between 4 P.M. and 5 P.M. is that fraction of the large area which lies over the interval $[4, 5]$. That is,

$$P([4, 5]) = \frac{1}{3} = 33\frac{1}{3}\%.$$

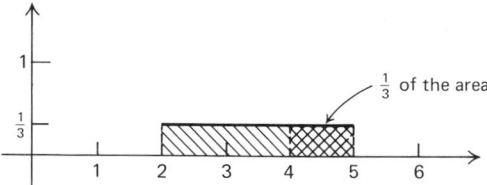

The probability that a bus will arrive between 2:00 P.M. and 4:30 P.M. is $\frac{5}{6}$ or $83\frac{1}{3}\%$.

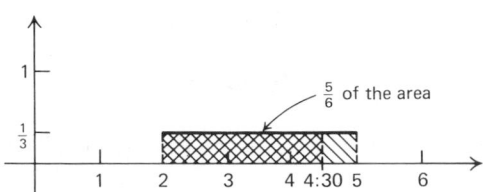

DO EXERCISE 71.

71. Find the probability that a bus will arrive between 2:30 P.M. and 4:30 P.M.

Note that any interval of length 1 has probability $\frac{1}{3}$. This may not always happen. Suppose we have a function

$$f(x) = \frac{3}{117} x^2$$

whose definite integral over the interval $[4, 5]$ would yield the probability that a bus will arrive between 4 P.M. and 5 P.M. Then

$$P([4, 5]) = \int_4^5 f(x)\, dx = \int_4^5 \frac{3}{117} x^2\, dx$$

$$= \left[\frac{3}{117} \cdot \frac{1}{3} x^3 \right]_4^5$$

$$= \frac{1}{117} (5^3 - 4^3)$$

$$= \frac{61}{117} \approx 0.52.$$

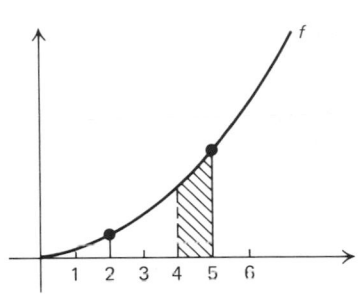

Thus 52% of the time you will be able to catch a bus between 4 P.M. and 5 P.M. The function f is called a *probability density function*. Its integral over *any* subinterval gives the probability that x "lands" in that subinterval.

Let x be a continuous random variable distributed over some interval $[a, b]$. A function f is said to be a *probability density function* for x if
1. f is nonnegative over $[a, b]$, that is, $f(x) \geqslant 0$ for all x in $[a, b]$;
2. for any subinterval $[c, d]$ of $[a, b]$, the probability $P([c, d])$, or $P(c \leqslant x \leqslant d)$, that x lands in that subinterval is given by

$$P([c, d]) = \int_c^d f(x)\, dx;$$

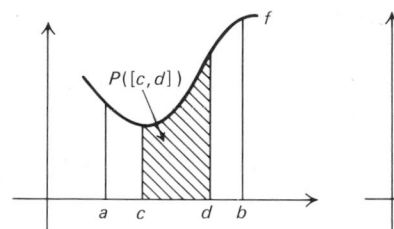 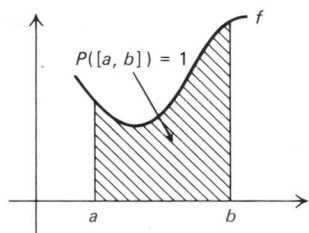

72. Verify property 3 of the definition of a probability density function for

$$f(x) = \tfrac{2}{3}x \quad \text{over } [1, 2]$$

3. the probability that x lands in $[a, b]$ is 1:

$$\int_a^b f(x)\, dx = 1.$$

That is, we are "certain" that x is in the interval $[a, b]$.

Example 3   Verify property 3 of the above definition for

$$f(x) = \frac{3}{117}\, x^2.$$

Solution   The "big" interval under consideration is $[2, 5]$. So

$$\int_2^5 \frac{3}{117}\, x^2\, dx = \left[\frac{3}{117} \cdot \frac{1}{3}\, x^3\right]_2^5 = \frac{1}{117}(5^3 - 2^3) = \frac{117}{117} = 1.$$

DO EXERCISE 72.

73. In reference to Example 6,

a) Verify property 3 of the definition of a probability density function.

Example 4   A company produces transistors. It determines that the life t of a transistor is from 3 to 6 years and that the probability density function for t is given by

$$f(t) = \frac{24}{t^3}, \quad \text{for} \quad 3 \leqslant t \leqslant 6.$$

a) Find the probability that a transistor will last no more than 4 years.
b) Find the probability that a transistor will last from 4 to 5 years.

Solution

a) The probability that a transistor will last no more than 4 years is

b) Find the probability that a transistor will last no more than 5 years.

$$P(3 \leqslant t \leqslant 4) = \int_3^4 \frac{24}{t^3}\, dt = \left[24\left(-\frac{1}{2}\, t^{-2}\right)\right]_3^4 = \left[-\frac{12}{t^2}\right]_3^4 = -12\left(\frac{1}{4^2} - \frac{1}{3^2}\right)$$

$$= -12\left(\frac{1}{16} - \frac{1}{9}\right) = -12\left(-\frac{7}{144}\right) = \frac{7}{12} \approx 0.58.$$

c) Find the probability that a transistor will last from 4 to 6 years.

b) The probability that a transistor will last from 4 to 5 years is

$$P(4 \leqslant t \leqslant 5) = \int_4^5 \frac{24}{t^3}\, dt = \left[24\left(-\frac{1}{2}\, t^{-2}\right)\right]_4^5 = \left[-\frac{12}{t^2}\right]_4^5 = -12\left(\frac{1}{5^2} - \frac{1}{4^2}\right)$$

$$= -12\left(\frac{1}{25} - \frac{1}{16}\right) = -12\left(-\frac{9}{400}\right) = \frac{27}{100} = 0.27.$$

DO EXERCISE 73.

Constructing Probability Density Functions

Suppose you have an arbitrary nonnegative function $f(x)$ whose definite integral over some interval $[a, b]$ is K. Then

$$\int_a^b f(x)\,dx = K.$$

Now multiply on both sides by $\dfrac{1}{K}$.

$$\frac{1}{K}\int_a^b f(x)\,dx = \frac{1}{K}\cdot K = 1 \quad \text{or} \quad \int_a^b \frac{1}{K}\cdot f(x)\,dx = 1.$$

Thus when we multiply the function $f(x)$ by $1/K$ we have a function whose area over the given interval is 1.

Example 5 Find k such that

$$f(x) = kx^2$$

is a probability density function over the interval $[2, 5]$.

Solution

$$\int_2^5 x^2\,dx = \left[\frac{x^3}{3}\right]_2^5 = \frac{5^3}{3} - \frac{2^3}{5} = \frac{125}{3} - \frac{8}{3} = \frac{117}{3}$$

Thus $k = \dfrac{1}{(117/3)} = \dfrac{3}{117}$, and $f(x) = \dfrac{3}{117}x^2$.

DO EXERCISES 74 AND 75.

Uniform Distributions

Suppose the probability density function of a continuous random variable is constant. How is it described? Consider the following graph.

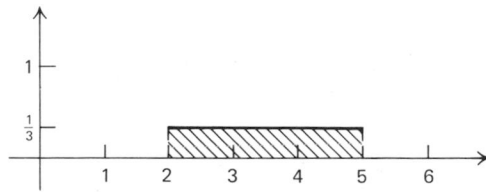

The length of the shaded rectangle is the length of the interval $[2, 5]$ which is 3. For the shaded area to be 1, the height of the rectangle

74. Find k such that

$$f(x) = kx^2$$

is a probability density function over the interval $[1, 3]$.

75. Find k such that

$$f(x) = kx^3$$

is a probability density function over the interval $[0, 1]$.

76. A number x is selected at random from the interval $[7, 15]$. The probability density function for x is given by

$$f(x) = \tfrac{1}{8}, \quad \text{for} \quad 7 \leqslant x \leqslant 15.$$

Find the probability that a number selected is in the subinterval $[11, 13]$.

must be $\tfrac{1}{3}$. Thus $f(x) = \tfrac{1}{3}$. For the general case consider the following graph.

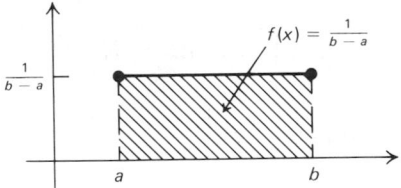

The length of the shaded rectangle is the length of the interval $[a, b]$, which is $b - a$. For the shaded area to be 1, the height of the rectangle must be $1/(b - a)$. Thus $f(x) = 1/(b - a)$.

A continuous random variable x is said to be *uniformly distributed* over an interval $[a, b]$ if it has a probability density function f given by

$$f(x) = \frac{1}{b - a}, \quad \text{for} \quad a \leqslant x \leqslant b.$$

Example 6 A number x is selected at random from the interval $[40, 50]$. The probability density function for x is given by

$$f(x) = \tfrac{1}{10}, \quad \text{for} \quad 40 \leqslant x \leqslant 50.$$

Find the probability that a number selected is in the subinterval $[42, 48]$.

Solution The probability is

$$P(42 \leqslant x \leqslant 48) = \int_{42}^{48} \tfrac{1}{10}\, dx = \tfrac{1}{10}[x]_{42}^{48} = \tfrac{1}{10}(48 - 42) = \tfrac{6}{10} = 0.6.$$

DO EXERCISE 76.

Example 7 A company produces guitars for a rock concert. The maximum loudness L of the guitars ranges from 70 to 100 decibels. The probability density for L is

$$f(L) = \tfrac{1}{30}, \quad \text{for} \quad 70 \leqslant L \leqslant 100.$$

A guitar is selected at random off the assembly line. Find the probability that its maximum loudness is from 70 to 92 decibels.

A rock concert. (*Jeff Albertson: Stock, Boston*)

77. A person arrives at a bus stop. The waiting time t for a bus is 0 to 20 minutes. The probability density function for t is

$$f(t) = \tfrac{1}{20}, \quad \text{for} \quad 0 \leqslant t \leqslant 20.$$

What is the probability that the person will have to wait no more than 5 minutes for a bus?

Solution   The probability is

$$P(70 \leqslant L \leqslant 92) = \int_{70}^{92} \tfrac{1}{30}\, dL = \tfrac{1}{30}[L]_{70}^{92} = \tfrac{1}{30}(92 - 70) = \tfrac{22}{30} = \tfrac{11}{15} \approx 0.73.$$

DO EXERCISE 77.

Exponential Distributions

The duration of a phone call, the distance between successive cars on a highway, and the amount of time required to learn a task are all examples of exponentially distributed random variables. That is, their probability density functions are exponential.

A continuous random variable is *exponentially distributed* if it has a probability density function given by

$$f(x) = ke^{-kx}, \quad \text{over the interval } [0, \infty).$$

The function $f(x) = 2e^{-2x}$ is such a probability density function. That

$$\int_0^\infty 2e^{-2x}\, dx = 1$$

is shown in Section 6.4 The general case

$$\int_0^\infty ke^{-kx}\, dx = 1$$

can be verified in a similar way.

Why is it reasonable to assume that distance between cars is exponentially distributed? This is because there are many more cases in which distances are small. The same argument holds for the duration of a phone call. That is, there are more short calls than long ones.

Example 8   *Transportation planning.* The distance x, in feet, between successive cars on a certain stretch of highway has probability density function

$$f(x) = ke^{-kx}, \quad \text{for} \quad 0 \leqslant x < \infty,$$

where $k = 1/a$ and a = average distance between successive cars over some period of time.

A transportation planner determines that the average distance between cars on a certain stretch of highway is 166 ft. What is the probability that the distance between cars is 50 ft or less?

A transportation planner can determine probabilities that cars are certain distances apart. (*Ellis Herwig: Stock, Boston*)

Solution   We first determine k.

$$k = \tfrac{1}{166} \approx 0.006$$

The probability density function for x is

$$f(x) = 0.006e^{-0.006x}, \quad \text{for} \quad 0 \leqslant x < \infty.$$

The probability that the distance between cars is 50 ft or less is

$$P(0 \leq x \leq 50) = \int_0^{50} 0.006 e^{-0.006x} \, dx$$

$$= \left[\frac{0.006}{-0.006} e^{-0.006x} \right]_0^{50}$$

$$= [-e^{-0.006x}]_0^{50}$$

$$= (-e^{-0.006 \cdot 50}) - (-e^{0.006 \cdot 0})$$

$$= -e^{-0.3} + 1$$

$$= 1 - e^{-0.3}$$

$$= 1 - 0.740818$$

$$\approx 0.2592$$

DO EXERCISE 78.

78. A transportation planner determines that the average distance between cars on a certain stretch of highway is 125 ft. What is the probability that the distance between cars is 50 ft or less?

EXERCISE SET 11.8

Verify property 3 of the definition of a probability density function over the given interval.

1. $f(x) = 2x$, $[0, 1]$ **2.** $f(x) = \frac{1}{4}x$, $[1, 3]$ **3.** $f(x) = \frac{1}{3}$, $[4, 7]$ **4.** $f(x) = \frac{1}{4}$, $[9, 13]$

5. $f(x) = \frac{3}{26}x^2$, $[1, 3]$ **6.** $f(x) = \frac{3}{64}x^2$, $[0, 4]$ **7.** $f(x) = \frac{1}{x}$, $[1, e]$ **8.** $f(x) = \frac{1}{e - 1} e^x$, $[0, 1]$

9. $f(x) = \frac{3}{2}x^2$, $[-1, 1]$ **10.** $f(x) = \frac{1}{3}x^2$, $[-2, 1]$ **11.** $f(x) = 3e^{-3x}$, $[0, \infty)$ **12.** $f(x) = 4e^{-4x}$, $[0, \infty)$

Find k such that each function is a probability density function over the given integral.

13. $f(x) = kx$, $[1, 3]$ **14.** $f(x) = kx$, $[1, 4]$ **15.** $f(x) = kx^2$, $[-1, 1]$ **16.** $f(x) = kx^2$, $[-2, 2]$

17. $f(x) = k$, $[2, 7]$ **18.** $f(x) = k$, $[3, 9]$ **19.** $f(x) = k(2 - x)$, $[0, 2]$ **20.** $f(x) = k(4 - x)$, $[0, 4]$

21. $f(x) = \frac{k}{x}$, $[1, 3]$ **22.** $f(x) = \frac{k}{x}$, $[1, 2]$ **23.** $f(x) = ke^x$, $[0, 3]$ **24.** $f(x) = ke^x$, $[0, 2]$

25. A dart is thrown at a number line in such a way that it always lands in the interval $[0, 10]$. Let $x =$ the number the dart hits. Suppose the probability density function for x is given by

$$f(x) = \frac{1}{50}x, \quad \text{for} \quad 0 \leq x \leq 10.$$

Find $P(2 \leq x \leq 6)$, the probability that it lands in $[2, 6]$.

26. Suppose the situation of Exercise 25, but that the dart always lands in the interval $[0, 5]$, and that the probability density function for x is given by

$$f(x) = \frac{3}{125}x^2, \quad \text{for} \quad 0 \leq x \leq 5.$$

Find $P(1 \leq x \leq 4)$, the probability that it lands in $[1, 4]$.

27. A number x is selected at random from the interval $[4, 20]$. The probability density function for x is given by

$$f(x) = \frac{1}{16}, \quad \text{for} \quad 4 \leq x \leq 20.$$

Find the probability that a number selected is in the subinterval $[9, 17]$.

28. A number x is selected at random from the interval $[5, 29]$. The probability density function for x is given by

$$f(x) = \frac{1}{24}, \quad \text{for} \quad 5 \leq x \leq 29.$$

Find the probability that a number selected is in the subinterval $[13, 29]$.

29. A transportation planner determines that the average distance between cars on a certain highway is 100 ft. What is the probability that the distance between cars is 40 ft or less?

31. A telephone company determines the duration t of a phone call is an exponentially distributed random variable with probability density function

$$f(t) = 2e^{-2t}, \quad 0 \leqslant t < \infty.$$

Find the probability that a phone call will last no more than 5 minutes.

33. In a psychology experiment, the time t, in seconds, that it takes a rat to learn its way through a maze is an exponentially distributed random variable with probability density function

$$f(t) = 0.02e^{-0.02t}, \quad 0 \leqslant t < \infty.$$

Find the probability that a rat will learn its way through a maze in 150 seconds, or less.

30. A transportation planner determines that the average distance between cars on a certain highway is 200 ft. What is the probability that the distance between cars is 10 ft or less?

32. Referring to the data in Exercise 31, find the probability that a phone call will last no more than 2 minutes.

34. Assume the situation and equation in Exercise 33, but find the probability that a rat will learn its way through a maze in 50 seconds or less.

The time it takes a rat to learn its way through a maze is an exponentially distributed random variable. (*Sol Schwartz from Monkmeyer*)

35. The *time to failure* t, in hours, of a certain machine can often be assumed to be exponentially distributed with probability density function

$$f(t) = ke^{-kt}, \quad 0 \leqslant t < \infty$$

where $k = 1/a$, and a = average time that will pass before a failure occurs. Suppose the average time that will pass before a failure occurs is 100 hours. What is the probability that a failure will occur in 50 hours or less?

36. The *reliability* of the machine (probability that it will work) in Exercise 35 is defined as

$$R(T) = 1 - \int_0^T 0.01e^{-0.01t} \, dt,$$

where $R(T)$ is the reliability at time T. Find $R(T)$.

37. The function $f(x) = x^3$ is a probability density on $[0, b]$. What is b?

38. The function $f(x) = 12x^2$ is a probability density on $[-a, a]$. What is a?

11.9 PROBABILITY: EXPECTED VALUE; THE NORMAL DISTRIBUTION

Expected Value

Let us again consider throwing a dart at a number line in such a way that it always lands in the interval $[1, 3]$.

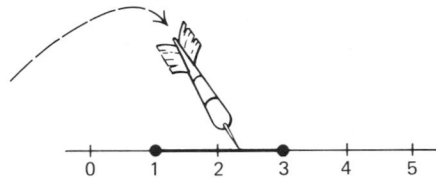

Suppose we throw the dart at the line 100 times and keep track of the numbers it hits. Then suppose we calculate the arithmetic mean (or average) $\bar{x}$ of all these numbers.

$$\bar{x} = \frac{x_1 + x_2 + x_3 + \cdots + x_{100}}{100} = \frac{\sum_{t=1}^{100} x_i}{100} = \sum_{i=1}^{100} x_i \cdot \frac{1}{100}$$

The expression

$$\sum_{i=1}^{n} x_i \cdot \frac{1}{n}$$

is analogous to the integral

$$\int_1^3 x \cdot f(x)\, dx,$$

where f is the probability density function for x. That is, $1/n$ gives a weight to x_i, and similarly $f(x)$ gives a weight to x. We add all the $x_i \cdot (1/n)$ values when we find $\sum_{i=1}^n x_i \cdot (1/n)$; and similarly we add all the $x \cdot f(x)$ values when we find $\int_1^3 x \cdot f(x)\, dx$. Suppose $f(x) = \frac{1}{4}x$. Then

$$\int_1^3 x \cdot f(x)\, dx = \int_1^3 x \cdot \frac{1}{4} x\, dx = \left[\frac{1}{4} \cdot \frac{x^3}{3}\right]_1^3 = \left[\frac{x^3}{12}\right]_1^3$$

$$= \frac{1}{12}(3^3 - 1^3) = \frac{26}{12} \approx 2.17.$$

Suppose we keep throwing the dart and computing averages. The more times we throw the dart, the closer we expect the averages to come to 2.17.

Let x be a continuous random variable over the interval $[a, b]$ with probability density function f.

OBJECTIVES

You should be able to

a) Given a probability density, find $E(x)$, $E(x^2)$, the mean, variance, and standard deviation.
b) Use Table 6 to evaluate probabilities involving a normal distribution.

79. Given the probability density function

$$f(x) = 2x, \quad \text{over } [0, 1],$$

find $E(x)$ and $E(x^2)$.

The expected value of x is defined by

$$E(x) = \int_a^b x \cdot f(x)\, dx.$$

The notion of expected value generalizes to other functions of x. Suppose $y = g(x)$. Then

The expected value of $g(x)$ is defined by

$$E(g(x)) = \int_a^b g(x) \cdot f(x)\, dx.$$

For example,

$$E(x) = \int_a^b x f(x)\, dx, \quad E(x^2) = \int_a^b x^2 f(x)\, dx, \quad E(e^x) = \int_a^b e^x f(x)\, dx,$$

and

$$E(2x + 3) = \int_a^b (2x + 3) f(x)\, dx.$$

Example 1 Given the probability density function

$$f(x) = \tfrac{1}{2}x, \quad \text{over } [0, 2],$$

find $E(x)$ and $E(x^2)$.

Solution

$$E(x) = \int_0^2 x \cdot \frac{1}{2} x\, dx = \int_0^2 \frac{1}{2} x^2\, dx = \frac{1}{2}\left[\frac{x^3}{3}\right]_0^2$$

$$= \frac{1}{2}\left(\frac{2^3}{3} - \frac{0^3}{3}\right) = \frac{1}{2} \cdot \frac{8}{3} = \frac{4}{3}$$

$$E(x^2) = \int_0^2 x^2 \cdot \frac{1}{2} x\, dx = \int_0^2 \frac{1}{2} x^3\, dx = \frac{1}{2}\left[\frac{x^4}{4}\right]_0^2$$

$$= \frac{1}{2}\left(\frac{2^4}{4} - \frac{0^4}{4}\right) = \frac{1}{2} \cdot \frac{16}{4} = 2.$$

DO EXERCISE 79.

The *mean* μ of a continuous random variable is defined to be $E(x)$. That is,

$$\mu = E(x) = \int_a^b x f(x)\, dx.$$

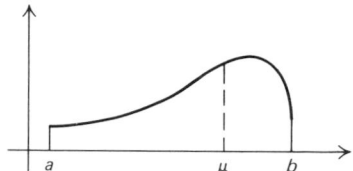 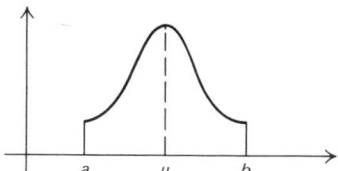

If we could imagine cutting out the region under the curve, the mean is the balance point. Note that the mean can be thought of as an average on the x-axis in contrast to the "average value of a function" that lies on the y-axis.

The *variance* σ^2 of a continuous random variable is defined

$$\sigma^2 = E(x^2) - \mu^2 = E(x^2) - [E(x)]^2$$
$$= \int_a^b x^2 f(x)\, dx - \left[\int_a^b x f(x)\, dx \right]^2.$$

The *standard deviation* σ of a continuous random variable is defined

$$\sigma = \sqrt{\text{variance}}.$$

Example 2   Given the probability density function

$$f(x) = \tfrac{1}{2}x, \quad \text{over } [0, 2],$$

find the mean, variance, and standard deviation.

Solution   From Example 1.

$$E(x) = \tfrac{4}{3} \quad \text{and} \quad E(x^2) = 2.$$

Then

$$\text{The mean} = \mu = E(x) = \tfrac{4}{3};$$
$$\text{The variance} = \sigma^2 = E(x^2) - [E(x)]^2$$
$$= 2 - (\tfrac{4}{3})^2$$
$$= 2 - \tfrac{16}{9}$$
$$= \tfrac{18}{9} - \tfrac{16}{9}$$
$$= \tfrac{2}{9};$$
$$\text{The standard deviation} = \sigma = \sqrt{\tfrac{2}{9}} = \tfrac{1}{3}\sqrt{2} \approx 0.47.$$

Loosely speaking, the standard deviation is a measure of how close the graph of f is to the mean. Note these examples.

80. Given the probability density function

$$f(x) = 2x, \quad \text{over } [0, 1],$$

find the mean, variance, and standard deviation.

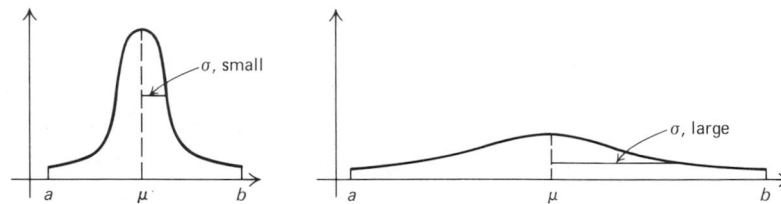

DO EXERCISE 80.

The Normal Distribution

Suppose the average on a test is 70. Usually there are about as many scores above the average as there are below; and the further away from the average, the fewer people there are who get a given score. For example, more people would score in the 80s than in the 90s; and more people would score in the 60s than in the 50s. Test scores, heights of human beings, and weights of human beings are all examples of random variables that may be *normally* distributed.

Consider the function

$$g(x) = e^{-x^2/2}, \quad \text{over the interval } (-\infty, \infty).$$

This function has the entire set of real numbers as domain. Its graph is the bell-shaped curve which follows. Function values are found by using a calculator or Table 4.

$$y = e^{-x^2/2}$$

x	0	1	2	3	-1	-2	-3
y	1	0.6	0.1	0.01	0.6	0.1	0.01

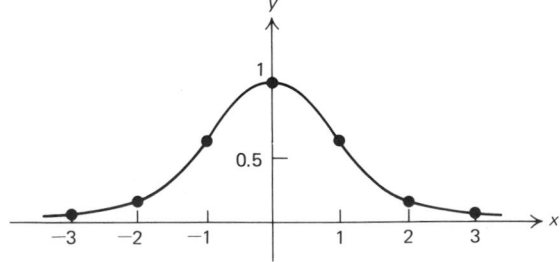

This function has an antiderivative, but that antiderivative has no elementary formula. Nevertheless, it has been shown that its improper integral converges over the interval $(-\infty, \infty)$, and

$$\int_{-\infty}^{\infty} e^{-x^2/2}\, dx = \sqrt{2\pi}.$$

That is, while an expression for the antiderivative cannot be found, there is a numerical value for the improper integral evaluated over the set of real numbers. Note that since the area is not 1, the function g is not a probability density function; but the following is:

$$\frac{1}{\sqrt{2\pi}}\, e^{-x^2/2}.$$

A continuous random variable x has a *standard normal distribution* if its probability density function is

$$f(x) = \frac{1}{\sqrt{2\pi}}\, e^{-x^2/2}, \quad \text{over } (-\infty, \infty).$$

This distribution has a mean of 0 and standard deviation 1. Its graph is shown below.

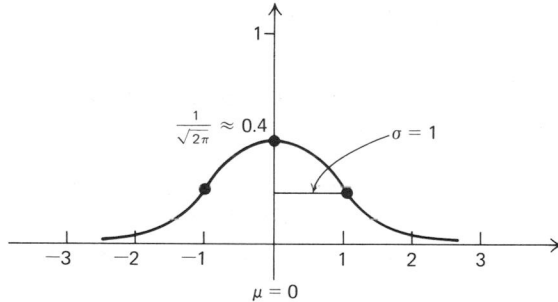

The general case is defined as follows.

A continuous random variable x is *normally distributed* with mean μ and standard deviation σ if its probability density function is given by

$$f(x) = \frac{1}{\sigma\sqrt{2\pi}} \cdot e^{-(1/2)[(x-\mu)/\sigma]^2}, \quad \text{over } (-\infty, \infty).$$

The graph is a transformation of the graph of the standard density. This is done by translating the graph along the x-axis and changing the way the graph is clustered about the mean. Some examples are shown on the following page.

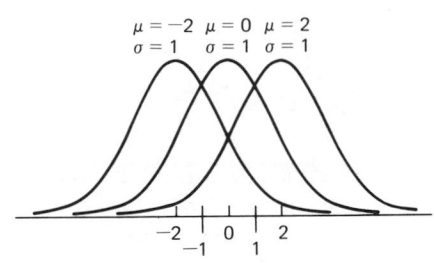

Normal distributions with same standard deviations but different means.

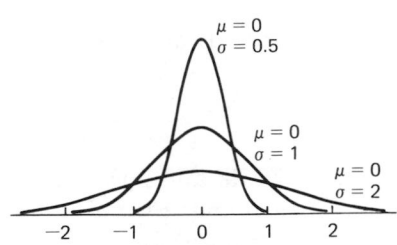

Normal distributions with same means but different standard deviations.

The normal distribution is extremely important in statistics; it underlies much of the research in the behavioral and social sciences. Because of this, tables of approximate values of the definite integral of the standard density functions have been prepared. Table 6, at the back of the book, is such a table. It contains values of

$$P(0 \leq x \leq t) = \int_0^t \frac{1}{\sqrt{2\pi}} e^{-x^2/2} \, dx.$$

The symmetry of the graph about the mean allows many types of probabilities to be computed from the table.

Example 3 Let x be a continuous random variable with standard normal density. Using Table 6, find:

a) $P(0 \leq x \leq 1.68)$,   b) $P(-0.97 \leq x \leq 0)$,

c) $P(-2.43 \leq x \leq 1.01)$,   d) $P(1.90 \leq x \leq 2.74)$,

e) $P(-2.98 \leq x \leq -0.42)$,   f) $P(x \geq 0.61)$.

Solution

a) $P(0 \leq x \leq 1.68)$ is the area bounded by the standard normal curve and the lines $x = 0$ and $x = 1.68$. We look this up in Table 6 by going down the left column to 1.6, then moving to the right to the column headed 0.08. There we read 0.4535. Thus

$$P(0 \leq x \leq 1.68) = 0.4535.$$

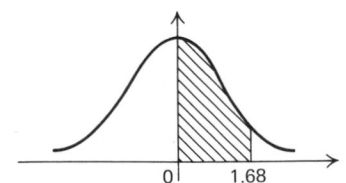

b) Due to the symmetry of the graph,

$P(-0.97 \leqslant x \leqslant 0)$
$= P(0 \leqslant x \leqslant 0.97) = 0.3340.$

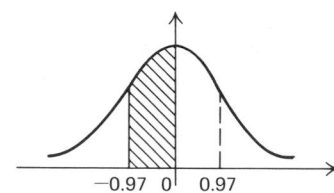

81. Let x be a continuous random variable with standard normal density. Using Table 6, find:

a) $P(0 \leqslant x \leqslant 2.17)$

c) $P(-2.43 \leqslant x \leqslant 1.01)$
$= P(-2.43 \leqslant x \leqslant 0) + P(0 \leqslant x \leqslant 1.01)$
$= P(0 \leqslant x \leqslant 2.43) + P(0 \leqslant x \leqslant 1.01)$
$= 0.4925 + 0.3438$
$= 0.8363$

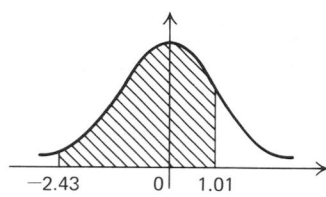

b) $P(-1.76 \leqslant x \leqslant 0)$

d) $P(1.90 \leqslant x \leqslant 2.74)$
$= P(0 \leqslant x \leqslant 2.74) - P(0 \leqslant x \leqslant 1.90)$
$= 0.4969 - 0.4713$
$= 0.0256$

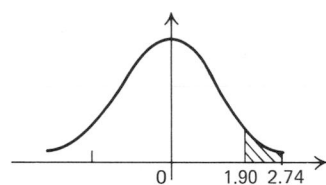

c) $P(-1.77 \leqslant x \leqslant 2.53)$

e) $P(-2.98 \leqslant x \leqslant -0.42)$
$= P(0.42 \leqslant x \leqslant 2.98)$
$= P(0 \leqslant x \leqslant 2.98) - P(0 \leqslant x \leqslant 0.42)$
$= 0.4986 - 0.1628$
$= 0.3358$

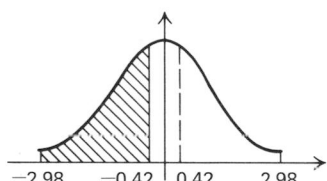

d) $P(0.49 \leqslant x \leqslant 1.75)$

e) $P(-1.66 \leqslant x \leqslant -1.00)$

f) $P(x \geqslant 0.61)$
$= P(x \geqslant 0) - P(0 \leqslant x \leqslant 0.61)$
$= 0.5000 - 0.2291$

(Because of the symmetry about the line $x = 0$, half the area is on each side of the line, and since the entire area is 1, $P(x \geqslant 0) = 0.5000$).

$= 0.2709$

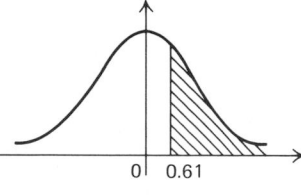

f) $P(x \geqslant 1.87)$

DO EXERCISE 81.

In many applications, a normal distribution is not standard. It would be a hopeless task to make tables for all values of the mean, μ, and the standard deviation, σ. In such cases, the transformation

$$z = \frac{x - \mu}{\sigma}$$

standardizes the distribution, permitting the use of Table 6 at the back of the book. That is,

$$P(a \le x \le b) = P\left(\frac{a - \mu}{\sigma} \le z \le \frac{b - \mu}{\sigma}\right),$$

and the probability on the right can be found using Table 6. To see this, consider

$$P(a \le x \le b) = \int_a^b \frac{1}{\sigma\sqrt{2\pi}} e^{-(1/2)[(x - \mu)/\sigma]^2} \, dx,$$

and make the substitution

$$z = \frac{x - \mu}{\sigma} = \frac{x}{\sigma} - \frac{\mu}{\sigma}.$$

Then

$$dz = \frac{1}{\sigma} \, dx.$$

When $x = a$, $z = (a - \mu)/\sigma$; and when $x = b$, $z = (b - \mu)/\sigma$. Then

$$P(a \le x \le b) = \int_a^b \frac{1}{\sigma\sqrt{2\pi}} e^{-(1/2)[(x - \mu)/\sigma]^2} \, dx$$

$$= \int_{(a-\mu)/\sigma}^{(b-\mu)/\sigma} \frac{1}{\sqrt{2\pi}} e^{-(1/2)z^2} \, dz \qquad \text{(The integrand is now in the form of the standard density.)}$$

$$= P\left(\frac{a - \mu}{\sigma} \le z \le \frac{b - \mu}{\sigma}\right). \qquad \text{(We can look this up in Table 6.)}$$

Example 4 The weights w of the students in a calculus class are normally distributed with mean 150 lb and standard deviation 25 lb. Find the probability that a student's weight is from 160 lb to 180 lb.

Solution We first standardize the weights:

$$180 \text{ is standardized to } \frac{b - \mu}{\sigma} = \frac{180 - 150}{25} = 1.2.$$

$$160 \text{ is standardized to } \frac{a - \mu}{\sigma} = \frac{160 - 150}{25} = 0.4.$$

Then

$$P(160 \leqslant w \leqslant 180) = P(0.4 \leqslant z \leqslant 1.2) \quad \text{(Now we can use Table 6.)}$$

$$= P(0 \leqslant z \leqslant 1.2) - P(0 \leqslant z \leqslant 0.4)$$

$$= 0.3849 - 0.1554$$

$$= 0.2295$$

Thus the probability that a student's weight is from 160 lb to 180 lb is 0.2295. That is, about 23% of the students have weights from 160 lb to 180 lb.

DO EXERCISE 82.

82. The daily profits p of a small firm are normally distributed with mean $200 and standard deviation $40. Find the probability that the daily profit will be from $230 to $250.

EXERCISE SET 11.9

For each probability density function, over the given interval, find $E(x)$, $E(x^2)$, the mean, variance, and standard deviation.

1. $f(x) = \frac{1}{3}$, $[2, 5]$

2. $f(x) = \frac{1}{4}$, $[3, 7]$

3. $f(x) = \frac{2}{9}x$, $[0, 3]$

4. $f(x) = \frac{1}{8}x$, $[0, 4]$

5. $f(x) = \frac{2}{3}x$, $[1, 2]$

6. $f(x) = \frac{1}{4}x$, $[1, 3]$

7. $f(x) = \frac{1}{3}x^2$, $[-2, 1]$

8. $f(x) = \frac{3}{2}x^2$, $[-1, 1]$

9. $f(x) = \frac{1}{\ln 3} \cdot \frac{1}{x}$, $[1, 3]$

10. $f(x) = \frac{1}{\ln 2} \cdot \frac{1}{x}$, $[1, 2]$

Let x be a continuous random variable with standard normal density. Using Table 6, find:

11. $P(0 \leqslant x \leqslant 2.69)$

12. $P(0 \leqslant x \leqslant 0.04)$

13. $P(-1.11 \leqslant x \leqslant 0)$

14. $P(-2.61 \leqslant x \leqslant 0)$

15. $P(-1.89 \leqslant x \leqslant 0.45)$

16. $P(-2.94 \leqslant x \leqslant 2.00)$

17. $P(1.76 \leqslant x \leqslant 1.86)$

18. $P(0.76 \leqslant x \leqslant 1.45)$

19. $P(-1.45 \leqslant x \leqslant -0.69)$

20. $P(-2.45 \leqslant x \leqslant -1.69)$

21. $P(x \geqslant 3.01)$

22. $P(x \geqslant 1.01)$

23. a) $P(-1 \leqslant x \leqslant 1)$
b) What percentage of the area is from -1 to 1?

24. a) $P(-2 \leqslant x \leqslant 2)$
b) What percentage of the area is from -2 to 2?

Let x be a continuous random variable which is normally distributed with mean $\mu = 22$ and standard deviation $\sigma = 5$. Using Table 6, find:

25. $P(24 \leqslant x \leqslant 30)$

26. $P(22 \leqslant x \leqslant 27)$

27. $P(19 \leqslant x \leqslant 25)$

28. $P(18 \leqslant x \leqslant 26)$

29. The heights h of the students in a calculus class are normally distributed with mean 65 in. and standard deviation 10 in.
a) Find the probability that a student's height is from 67 to 72 in.
b) Find the probability that a student's height is from 60 to 70 in.
c) Find the probability that a student's height is more than 6 ft (72 in.).

30. The daily production N of stereos by a recording company is normally distributed with mean 1000 and standard deviation 50. The company promises to pay bonuses to its employees on those days when the production of stereos is 1100 or more. What percentage of the days will the company have to pay a bonus?

31. The number of daily orders N received by a mail order firm is normally distributed with mean 250 and standard deviation 20. The company has to hire extra help or pay overtime on those days when the number of orders received is 300 or higher. What percentage of the days will a company have to hire extra help or pay overtime?

32. The scores S on a psychology test are normally distributed with mean 65 and standard deviation 20. A score of 80 to 89 is a B. What is the probability of getting a B?

▶ ───

For each probability density function over the given interval, find $E(x)$, $E(x^2)$, the mean, variance, and standard deviation.

33. The uniform probability density

$$f(x) = \frac{1}{b - a}, \quad \text{over } [a, b].$$

34. The exponential probability density

$$f(x) = ke^{-kx}, \quad \text{over } [0, \infty).$$

Median. Let x be a continuous random variable over $[a, b]$ with probability density function f. Then the *median* of x is that number m for which

$$\int_a^m f(x)\, dx = \tfrac{1}{2}.$$

Find the median.

35. $f(x) = \tfrac{1}{2}x$, $[0, 2]$

36. $f(x) = \tfrac{3}{2}x^2$, $[-1, 1]$

37. $f(x) = ke^{-kx}$, $[0, \infty)$.

CHAPTER 11 TEST

Integrate.

1. $\int dx$ **2.** $\int 1000x^4\, dx$ **3.** $\int \left(e^x + \dfrac{1}{x} + x^{3/8} \right) dx$

Find the area under the curve on the interval indicated.

4. $y = x - x^2;\, [0, 1]$ **5.** $y = \dfrac{4}{x};\, [1, 3]$

6. Give two interpretations of the shaded area.

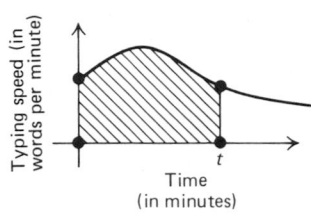

Integrate.

7. $\int_{-1}^{2} (2x + 3x^2)\, dx$ **8.** $\int_{0}^{1} e^{-2x}\, dx$ **9.** $\int_{a}^{b} \dfrac{dx}{x}$

Integrate. Use substitution. Do not use Table 5.

10. $\int \dfrac{dx}{x + 8}$ **11.** $\int e^{-0.5x}\, dx$ **12.** $\int t^3 (t^4 + 1)^9\, dt$

Integrate. Use integration by parts. Do not use Table 5.

13. $\int x e^{5x}\, dx$ **14.** $\int x^3 \ln x^4\, dx$

Integrate. Use Table 5.

15. $\int 2^x\, dx$

16. $\int \dfrac{dx}{x(7 - x)}$

17. Find the area of the region bounded by $y = x$, $y = x^5$, $x = 0$, $x = 1$.

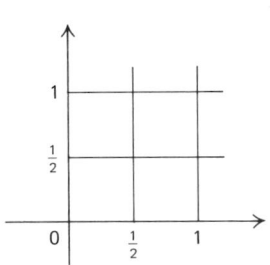

18. Approximate $\int_0^5 (25 - x^2)\, dx$, by computing the area of each rectangle and adding.

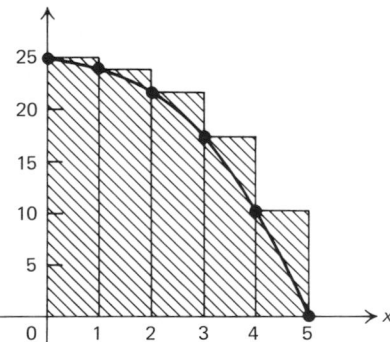

Determine if each of the following improper integrals is convergent or divergent, and calculate its value if convergent.

19. $\int_{1}^{\infty} \dfrac{dx}{x^5}$

20. $\int_{0}^{\infty} \dfrac{3}{1 + x}\, dx$

21. Find k such that $f(x) = kx^3$ is a probability density function over the interval $[0, 2]$.

22. A telephone company determines that the length of time t of a phone call is an exponentially distributed random variable with probability density function

$$f(t) = 2e^{-2t}, \quad 0 \leqslant t \leqslant \infty.$$

Find the probability that a phone call will last no more than 1 minute.

Given the probability density function $f(x) = 3x^2$, over $[0, 1]$, find:

23. $E(x)$ **24.** $E(x^2)$ **25.** the mean.   **26.** the variance.   **27.** the standard deviation.

Let x be a continuous random variable with standard normal density. Using Table 6, find

28. $P(0 \leqslant x \leqslant 1.5)$ **29.** $P(0.12 \leqslant x \leqslant 2.32)$ **30.** $P(-1.61 \leqslant x \leqslant 1.76)$

31. The price per pound p of T-bone steak at various stores in a certain city is normally distributed with mean \$3.75 and standard deviation \$0.25. What is the probability that the price per pound is \$3.80 or more?

CHAPTER TWELVE

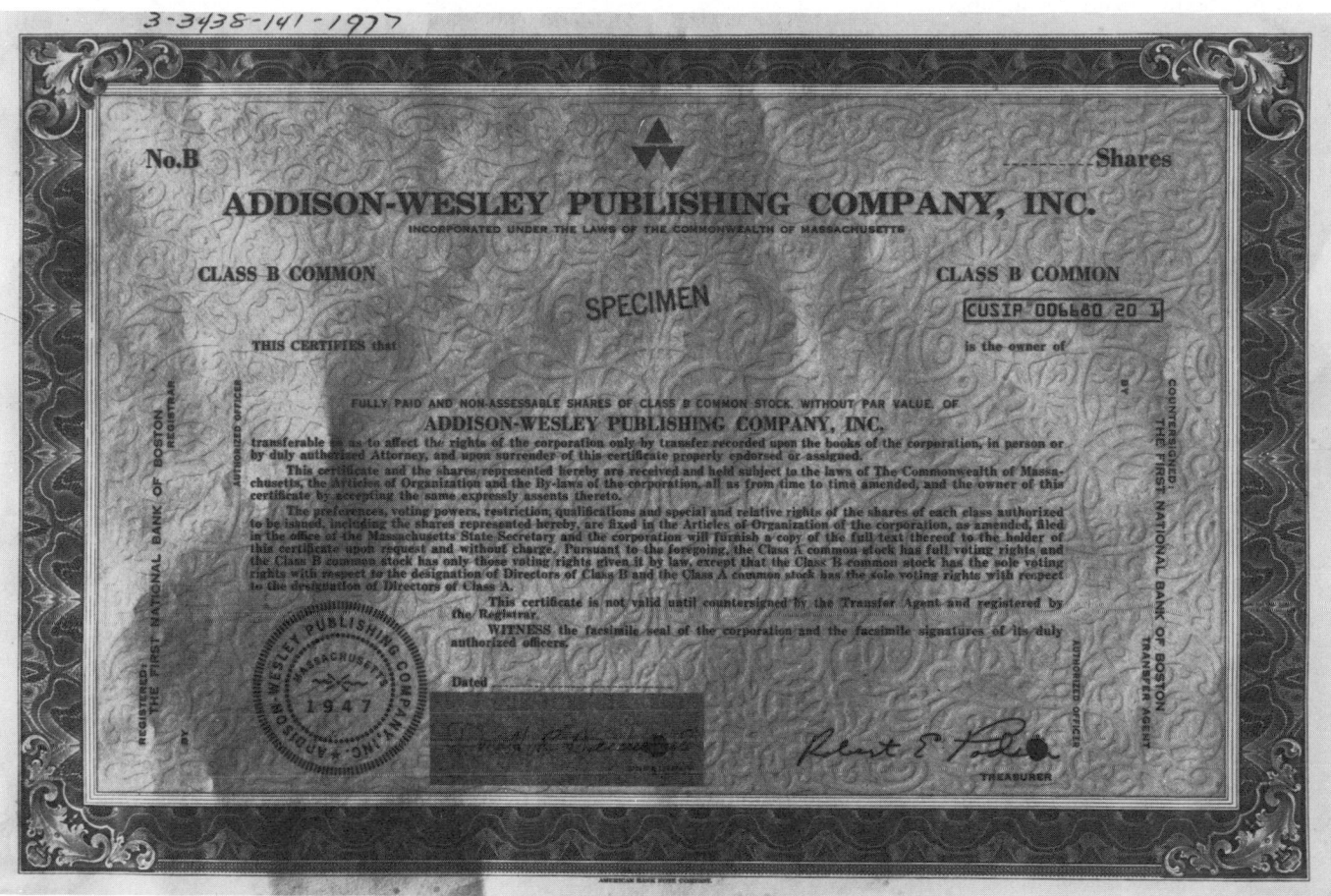

The *yield* and *price-earnings* ratio of a stock are functions of two variables.

Functions of Several Variables

OBJECTIVES

You should be able to

a) Find a function value for a function of several variables.
b) Find the partial derivatives of a given function.
c) Evaluate the partial derivatives of a function at a given point.

1. For $P(x, y) = 4x + 6y$,

a) find $P(14, 12)$ and interpret its meaning.

b) find $P(0, 8)$ and interpret its meaning.

2. For
$$A(n, i) = 9 \cdot \frac{n}{i},$$
find $A(4, 5)$, $A(7, \frac{2}{3})$, and $A(1, 15)$.

12.1 PARTIAL DERIVATIVES

Functions of Several Variables

Suppose a one-product firm produces x items of its product at a profit of \$4 per item. Then its total profit $P(x)$ is given by
$$P(x) = 4x.$$
This is a function of *one* variable.

Suppose a two-product firm produces x items of one product at a profit of \$4 per item, and y items of a second at a profit of \$6 per item. Then its total profit P is a function of the *two* variables, x and y, and is given by
$$P(x, y) = 4x + 6y.$$
This function assigns to the input pair (x, y) a unique output number $4x + 6y$.

Example 1 For $P(x, y) = 4x + 6y$, find $P(25, 10)$.

Solution $P(25, 10)$ is defined to be the value of the function found by substituting 25 for x and 10 for y:
$$P(25, 10) = 4 \cdot 25 + 6 \cdot 10 = 100 + 60 = 160.$$
This means that the two-product firm, by selling 25 items of the first product and 10 of the second, will make a profit of \$160.

DO EXERCISE 1.

The following are further examples of functions of several variables, that is, functions of two or more variables.

Example 2 A pitcher's *earned-run average* is given by
$$A(n, i) = 9 \cdot \frac{n}{i},$$
where n is the total number of earned runs given up in i innings of pitching. In a recent year Tom Seaver, of the Cincinnati Reds, gave up 75 earned runs in 261 innings. Find his earned-run average.

Solution We substitute 75 for n and 261 for i.
$$A(75, 261) = 9 \cdot \frac{75}{261} = 2.59$$

DO EXERCISE 2.

Example 3   The volume of a rectangular solid is given by

$$V(x, y, z) = xyz,$$

where x is the length, y the width, and z the height. This is a function of three variables.

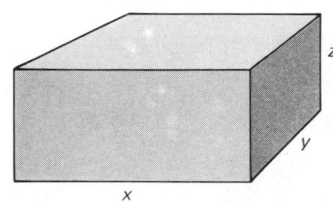

DO EXERCISE 3.

Example 4   The production of a company is given by

$$f(x, y, z, w) = 4x^2 + 5y + z - w,$$

where x dollars are spent for labor, y dollars are spent for raw materials, z dollars are spent for advertising, and w dollars are spent for machinery. This is a function of four variables.

Find $f(3, 2, 0, 10)$.

Solution   We substitute 3 for x, 2 for y, 0 for z, and 10 for w.

$$f(3, 2, 0, 10) = 4 \cdot 3^2 + 5 \cdot 2 + 0 \quad 10$$
$$= 4 \cdot 9 + 10 + 0 - 10 = 36$$

DO EXERCISE 4.

Example 5   Suppose an amount P_0 is invested in a savings account, where interest is compounded continuously at interest rate k. The balance after t years is given by

$$P(P_0, k, t) = P_0 e^{kt}.$$

Find $P(\$100, 0.07, 1)$.

Solution

$$P(\$100, 0.07, 1) = 100e^{0.07(1)} = 100(1.072508) \quad (\text{▦ or Table 4})$$
$$\approx \$107.25$$

DO EXERCISE 5.

3. For $V(x, y, z) = xyz$, find $V(5, 10, 40)$.

4. For the function f of Example 4, find

a) $f(9, 5, 3, 0)$

b) $f(1, 2, 3, 4)$

5. For the function of Example 5, find

$$P(\$1000, 0.08, 2).$$

6. The constant function g is given by

$$g(x, y) = 4$$

for all inputs x and y. Find

a) $g(-9, 10)$ b) $g(560, 43)$.

Example 6 *The gravity model*. The number of telephone calls between two cities is given by

$$N(d, P_1, P_2) = \frac{2.8 P_1 P_2}{d^{2.4}},$$

where d is the distance between the cities, and P_1 and P_2 are their populations.

Sociologists say that as two cities merge the communication between them increases. (*USGS EROS Data Center*)

A constant can also be thought of as a function of several variables.

Example 7 The constant function f is given by

$$f(x, y) = -3, \quad \text{for all inputs } x \text{ and } y.$$

Find $f(5, 7)$ and $f(-2, 0)$.

Solution Since this is a constant function, it has the value -3 for any x and y.

So
$$f(5, 7) = -3 \quad \text{and} \quad f(-2, 0) = -3.$$

DO EXERCISE 6.

Partial Derivatives

Consider the function f given by

$$z = f(x, y) = x^2 y^3 + xy + 4y^2.$$

Suppose for the moment that we fix y at 3. Then

$$f(x, 3) = x^2 3^3 + x3 + 4 \cdot 3^2 = 27x^2 + 3x + 36.$$

Note that we now have a function of only one variable. Taking the first derivative with respect to x, we have

$$54x + 3.$$

DO EXERCISE 7.

Now, without replacing y by a specific number, let us consider y fixed. Then f becomes a function of x alone and we can calculate its derivative with respect to x. This derivative is called the *partial derivative of f with respect to x*. Notation for this partial derivative is

$$\frac{\partial f}{\partial x} \quad \text{or} \quad \frac{\partial z}{\partial x}.$$

Thus fixing y (treating it as a constant) and calculating the derivative with respect to x, we have

$$\frac{\partial f}{\partial x} = \frac{\partial z}{\partial x} = 2xy^3 + y.$$

DO EXERCISE 8.

Similarly, we find $\partial f/\partial y$ or $\partial z/\partial y$ by fixing x (treating it as a constant) and calculating the derivative with respect to y. We have

$$\frac{\partial f}{\partial y} = \frac{\partial z}{\partial y} = 3x^2y^2 + x + 8y.$$

DO EXERCISE 9.

Partial differentiation can be done for any number of variables.

Example 8   For $w = x^2 - xy + y^2 + 2yz + 2z^2 + z$, find

$$\frac{\partial w}{\partial x}, \quad \frac{\partial w}{\partial y}, \quad \text{and} \quad \frac{\partial w}{\partial z}.$$

Solution

$$\frac{\partial w}{\partial x} = 2x - y$$

$$\frac{\partial w}{\partial y} = -x + 2y + 2z$$

$$\frac{\partial w}{\partial z} = 2y + 4z + 1$$

DO EXERCISE 10.

7. Consider

$$f(x, y) = 1 - x^2 - y^2.$$

a) Fix y at 4 and find $f(x, 4)$.

b) The answer to (a) could be interpreted as a function of one variable x. Find the first derivative.

8. For

$$f(x, y) = 1 - x^2 - y^2,$$

find $\dfrac{\partial f}{\partial x}$.

9. For $z = 3x^2y + 5x^3$, find

a) $\dfrac{\partial z}{\partial x}$,

b) $\dfrac{\partial z}{\partial y}$.

10. For $t = xy + xz + x^2 + y^3$, find

a) $\dfrac{\partial t}{\partial x}$,

b) $\dfrac{\partial t}{\partial y}$,

c) $\dfrac{\partial t}{\partial z}$.

11. For $f(x, y) = 3x^3y + 2xy$, find

a) f_x, b) $f_x(-4, 1)$,

c) f_y, d) $f_y(2, 6)$.

12. For

$$f(x, y) = \ln(xy) + ye^x,$$

find f_x and f_y.

We will often make use of a simpler notation f_x for the partial derivative of f with respect to x, and f_y for the partial derivative of f with respect to y.

Example 9 For $f(x, y) = 3x^2y + xy$, find f_x and f_y.

Solution

$$f_x = 6xy + y, \qquad f_y = 3x^2 + x$$

For the function in the preceding example, let us evaluate f_x at $(2, -3)$.

$$f_x(2, -3) = 6 \cdot 2 \cdot (-3) + (-3) = -39$$

Using the notation $\partial z/\partial x = 6xy + y$, where $z = 3x^2y + xy$, the value of the partial derivative at $(2, -3)$ is given by

$$\left. \frac{\partial z}{\partial x} \right|_{(2, -3)} = 6 \cdot 2 \cdot (-3) + (-3) = -39,$$

but this notation is not as convenient as $f_x(2, -3)$.

DO EXERCISE 11.

Example 10 For $f(x, y) = e^{xy} + y \ln x$, find f_x and f_y.

Solution

$$f_x = y \cdot e^{xy} + y \cdot \frac{1}{x} = ye^{xy} + \frac{y}{x},$$

$$f_y = x \cdot e^{xy} + 1 \cdot \ln x = xe^{xy} + \ln x$$

DO EXERCISE 12.

Geometric Interpretations

Consider a function of two variables

$$z = f(x, y).$$

Recall the mapping interpretation of function that we considered in Chapter 1. As a mapping, a function of two variables can be thought of as mapping a point (x_1, y_1) in an xy-plane onto a point z_1 on a number line:

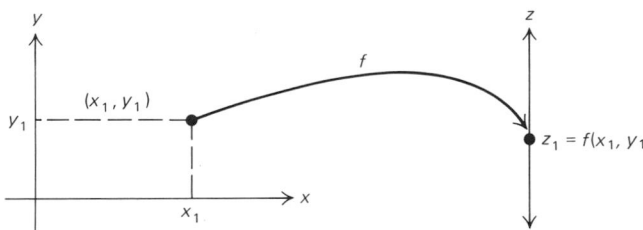

To graph a function of two variables, we need a three-dimensional coordinate system. The axes are usually placed as follows. The line z, called the z-axis, is placed perpendicular to the xy-plane at the origin.

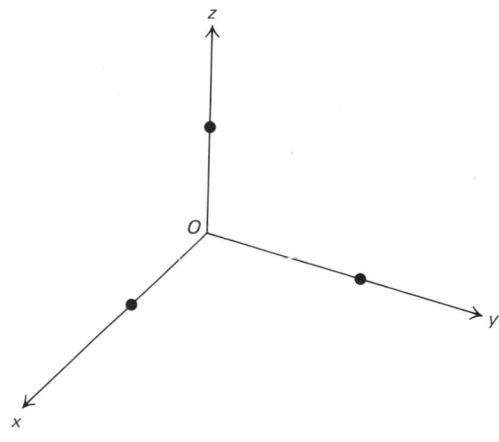

To help visualize this, think of looking into the corner of a room, where the floor is the xy-plane and the z-axis is the intersection of two walls. To plot a point (x_1, y_1, z_1) we locate the point (x_1, y_1) in the xy-plane, and move up or down in space according to the value of z_1.

Example 11 Plot these points: $P_1(2, 3, 5)$, $P_2(2, -2, -4)$, and $P_3(0, 5, 2)$.

13. Using the axes shown below, graph $P_1(3, 2, 5)$, $P_2(2, 3, 1)$, and $P_3(-3, 2, 0)$.

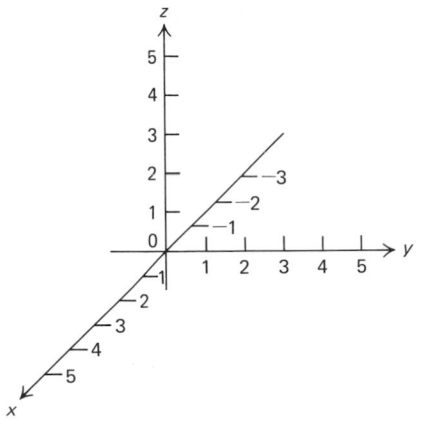

Solution

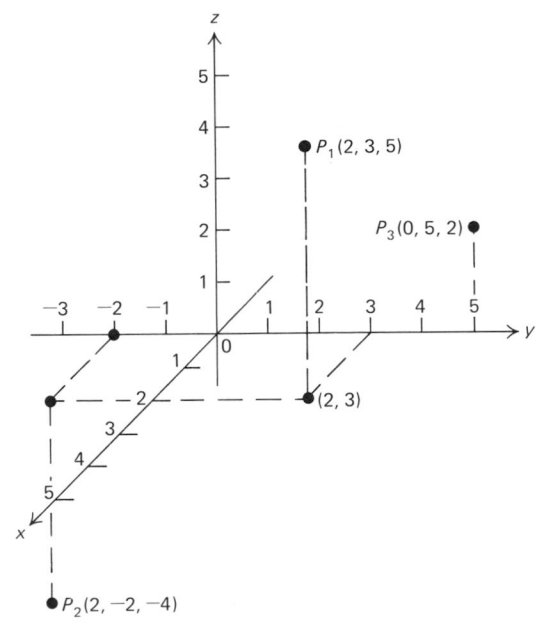

DO EXERCISE 13.

The *graph* of a function of two variables

$$z = f(x, y)$$

consists of ordered triples (x_1, y_1, z_1), where $z_1 = f(x_1, y_1)$. The domain of f is a region D in the xy-plane, and the graph of f is a surface S, as shown below.

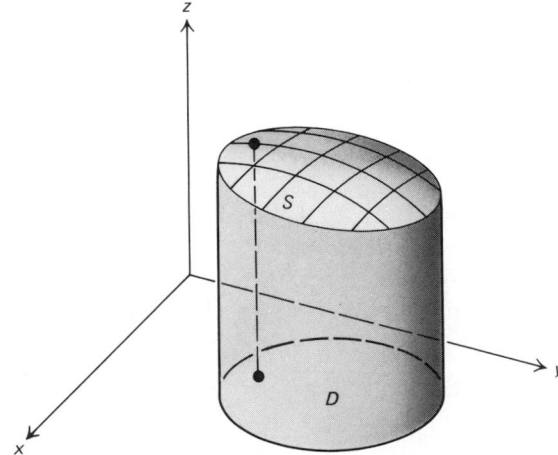

Here are some equations and their graphs.

Plane: $x = a$.

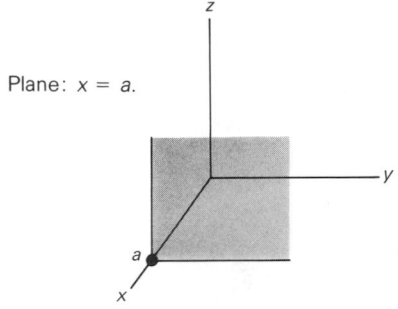

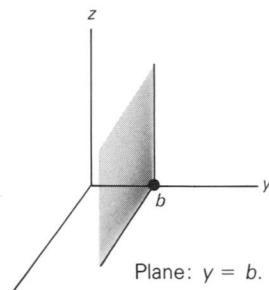

Plane: $y = b$.

Elliptic paraboloid: $z = x^2 + y^2$.

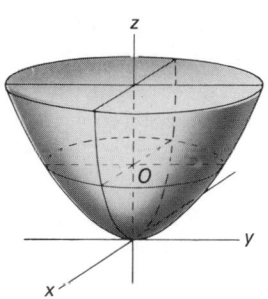

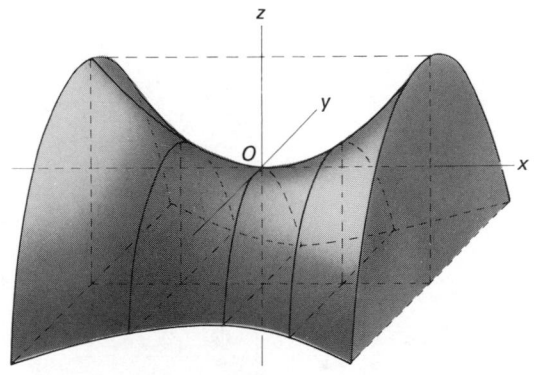

Hyperbolic paraboloid: $z = x^2 - y^2$.

Now suppose we hold x fixed, say at the value a. The set of all points for which $x = a$ is a plane parallel to the yz-plane, so when x

is fixed at a, y and z vary along the plane as shown in the following figure.

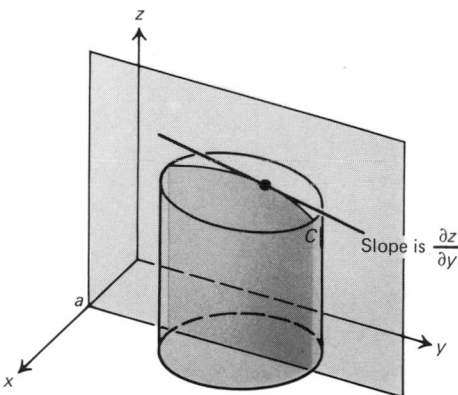

The plane shown cuts the surface S in some curve C as shown. The partial derivative f_y gives the slopes of tangent lines to this curve. Similarly, if we hold y fixed, say at the value b, we obtain a curve C' as shown in the following figure. The partial derivative f_x gives the slopes of tangent lines to this curve.

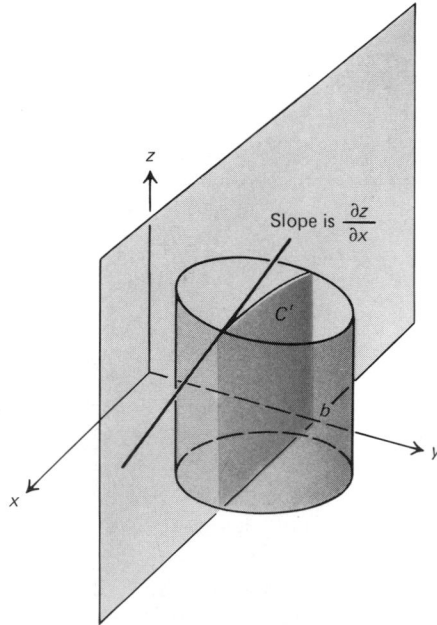

EXERCISE SET 12.1

1. For $f(x, y) = x^2 - 2xy$, find $f(0, -2)$, $f(2, 3)$, and $f(10, -5)$.

2. For $f(x, y) = y^2 + 3xy$, find $f(-2, 0)$, $f(3, 2)$, and $f(-5, 10)$.

3. For $f(x, y) = 3^x + 7xy$, find $f(0, -2)$, $f(-2, 1)$, and $f(2, 1)$.

4. For $f(x, y) = \log_{10} x - 5y^2$, find $f(10, 2)$, $f(1, -3)$, and $f(100, 4)$.

5. For $f(x, y) = \ln x + y^3$, find $f(e, 2)$, $f(e^2, 4)$, and $f(e^3, 5)$.

6. For $f(x, y) = 2^x - 3^y$, find $f(0, 0)$, $f(1, 1)$, and $f(2, 2)$.

7. For $f(x, y, z) = x^2 - y^2 + z^2$, find $f(-1, 2, 3)$ and $f(2, -1, 3)$.

8. For $f(x, y, z) = 2^x + 5zy - x$, find $f(0, 1, -3)$ and $f(1, 0, -3)$.

9. *Psychology—Intelligence quotient.* The intelligence quotient in psychology is given by

$$Q(m, c) = 100 \cdot \frac{m}{c},$$

where m is the mental age of a person and c is the chronological, or actual age. Find $Q(21, 20)$ and $Q(19, 20)$.

10. *Business—Price-earnings ratio.* The *price-earnings ratio* of a stock is given by

$$R(P, E) = \frac{P}{E},$$

where P is the price per share of the stock and E is the earnings per share. Recently, the price per share of IBM stock was $\$287\frac{3}{8}$ and the earnings per share were $\$23.30$. Find the price-earnings ratio. Give decimal notation to the nearest tenth.

11. *Business—Yield of a stock.* The *yield* of a stock is given by

$$Y(D, P) = \frac{D}{P},$$

where D is the dividends per share of a stock and P is the price per share. Recently, the price per share of Goodyear stock was $\$16\frac{7}{8}$ and the dividends per share were $\$1.30$. Find the yield. Give percent notation to the nearest tenth of a percent.

12. *Biomedical—Poiseuille's Law.* The speed of blood in a vessel is given by

$$V(L, p, R, r, v) = \frac{p}{4Lv}(R^2 - r^2),$$

where R is the radius of the vessel, r is the distance of the blood from the center of the vessel, L is the length of the blood vessel, p is pressure, and v is viscosity. Find $V(1, 100, 0.0075, 0.0025, 0.05)$.

Find $\dfrac{\partial z}{\partial x}$, $\dfrac{\partial z}{\partial y}$, $\dfrac{\partial z}{\partial x}\bigg|_{(-2, -3)}$, and $\dfrac{\partial z}{\partial y}\bigg|_{(0, -5)}$.

13. $z = 2x - 3xy$ **14.** $z = 5y + 2xy$

15. $z = 3x^2 - 2xy + y$ **16.** $z = 2x^3 + 3xy - x$

Find f_x, f_y, $f_x(-2, 4)$, and $f_y(4, -3)$.

Find f_x, f_y, $f_x(-2, 1)$, and $f_y(-3, -2)$.

17. $f(x, y) = 2x - 3y$ **18.** $f(x, y) = 5x + 7y$

19. $f(x, y) = \sqrt{x^2 + y^2}$ **20.** $f(x, y) = \sqrt{x^2 - y^2}$

Find f_x and f_y.

21. $f(x, y) = e^{2x+3y}$ **22.** $f(x, y) = e^{3x-2y}$

23. $f(x, y) = e^{xy}$ **24.** $f(x, y) = e^{2xy}$

25. $f(x, y) = y \ln(x + y)$ **26.** $f(x, y) = x \ln(x + y)$

27. $f(x, y) = x \ln(xy)$ **28.** $f(x, y) = y \ln(xy)$

29. $f(x, y) = \dfrac{x}{y} - \dfrac{y}{x}$ **30.** $f(x, y) = \dfrac{x}{y} + \dfrac{y}{x}$

31. $f(x, y) = 3(2x + y - 5)^2$ **32.** $f(x, y) = 4(3x + y - 8)^2$

Find $\dfrac{\partial f}{\partial b}$ and $\dfrac{\partial f}{\partial m}$.

33. $f(b, m) = (m + b - 4)^2 + (2m + b - 5)^2 + (3m + b - 6)^2$

34. $f(b, m) = (m + b - 6)^2 + (2m + b - 8)^2 + (3m + b - 9)^2$

Find f_x, f_y, and f_λ.

35. $f(x, y, \lambda) = 3xy - \lambda(2x + y - 8)$

36. $f(x, y, \lambda) = 4xy - \lambda(3x - y + 7)$

37. $f(x, y, \lambda) = x^2 + y^2 - \lambda(10x + 2y - 4)$

38. $f(x, y, \lambda) = x^2 - y^2 - \lambda(4x - 7y - 10)$

---

39. ▦ *Wind chill temperature.* Wind speed affects the actual temperature, making a person colder due to extra heat loss from the skin. The *wind chill temperature* is what the temperature would have to be with no wind to give the same chilling effect. The wind chill temperature, W, is given by

$$W(v, T) = 91.4 - \frac{(10.45 + 6.68\sqrt{v} - 0.447v)(457 - 5T)}{110},$$

where T is the actual temperature as given by a thermometer, in degrees Fahrenheit, and v is the speed of the wind, in mph. Find the wind chill temperature in each case. Round to the nearest one degree.

a) $T = 30°F$, $v = 25$ mph b) $T = 20°F$, $v = 20$ mph
c) $T = 20°F$, $v = 40$ mph d) $T = -10°F$, $v = 30$ mph

40. ▦ For W in Exercise 39, find W_v and W_T.

Find f_x and f_t.

41. $f(x, t) = \dfrac{x^2 + t^2}{x^2 - t^2}$

42. $f(x, t) = \dfrac{x^2 - t}{x^3 + t}$

43. $f(x, t) = \dfrac{2\sqrt{x} - 2\sqrt{t}}{1 + 2\sqrt{t}}$

44. $f(x, t) = \sqrt[4]{x^3 t^5}$

45. $f(x, t) = 6x^{2/3} - 8x^{1/4}t^{1/2} - 12x^{-1/2}t^{3/2}$

46. $f(x, t) = \left(\dfrac{x^2 + t^2}{x^2 - t^2}\right)^5$

OBJECTIVES

You should be able to find the four second partial derivatives of a function.

14. Consider

$$z = 3xy^2 + 2xy + x^2.$$

a) Find $\dfrac{\partial z}{\partial y}$.

b) For the function in (a), find the first partial derivative with respect to x.

c) For the function in (a), find the first partial derivative with respect to y; that is, differentiate "twice" with respect to y.

12.2 HIGHER-ORDER PARTIAL DERIVATIVES

Consider

$$z = f(x, y) = 3xy^2 + 2xy + x^2. \tag{1}$$

Then

$$\frac{\partial z}{\partial x} = \frac{\partial f}{\partial x} = 3y^2 + 2y + 2x. \tag{2}$$

Suppose we find the first partial derivative of function (2) with respect to y. This will be a *second-order partial derivative*. Notation for it is as follows.

$$\frac{\partial}{\partial y}\left(\frac{\partial z}{\partial x}\right) = \frac{\partial}{\partial y}\left(\frac{\partial f}{\partial x}\right) = \frac{\partial^2 z}{\partial y\, \partial x} = \frac{\partial^2 f}{\partial y\, \partial x} = 6y + 2.$$

DO EXERCISE 14.

We could also denote the preceding partial derivative using the notation f_{xy}. Then

$$f_{xy} = 6y + 2.$$

Note that in the notation f_{xy}, x and y are in the order (left to right) in which the differentiation is done. In the other symbolisms that order is reversed, but the meaning is not.

DO EXERCISE 15.

Notation for the four second-order partial derivatives is as follows:

$$\frac{\partial^2 z}{\partial x\, \partial x} = \frac{\partial^2 f}{\partial x\, \partial x} = \frac{\partial^2 z}{\partial x^2} = \frac{\partial^2 f}{\partial x^2} = f_{xx} \qquad \text{(Take the partial with respect to } x \text{, and then with respect to } x \text{ again.)}$$

$$\frac{\partial^2 z}{\partial y\, \partial x} = \frac{\partial^2 f}{\partial y\, \partial x} = f_{xy} \qquad \text{(Take the partial with respect to } x \text{, and then with respect to } y \text{.)}$$

$$\frac{\partial^2 z}{\partial x\, \partial y} = \frac{\partial^2 f}{\partial x\, \partial y} = f_{yx} \qquad \text{(Take the partial with respect to } y \text{, and then with respect to } x \text{.)}$$

$$\frac{\partial^2 z}{\partial y\, \partial y} = \frac{\partial^2 f}{\partial y\, \partial y} = \frac{\partial^2 z}{\partial y^2} = \frac{\partial^2 f}{\partial y^2} = f_{yy} \qquad \text{(Take the partial with respect to } y \text{, and then with respect to } y \text{ again.)}$$

Example 1   For

$$z = f(x, y) = x^2 y^3 + x^4 y + x e^y$$

find the four second-order partial derivatives.

Solution

a) $\dfrac{\partial^2 f}{\partial x^2} = \dfrac{\partial}{\partial x}(2xy^3 + 4x^3 y + e^y)$

       (Differentiate twice with respect to x.)

    $= 2y^3 + 12x^2 y$

b) $\dfrac{\partial^2 f}{\partial y\, \partial x} = \dfrac{\partial}{\partial y}(2xy^3 + 4x^3 y + e^y)$

       (Differentiate with respect to x, then with respect to y.)

    $= 6xy^2 + 4x^3 + e^y$

c) $\dfrac{\partial^2 f}{\partial x\, \partial y} = \dfrac{\partial}{\partial x}(3x^2 y^2 + x^4 + x e^y)$

       (Differentiate with respect to y, then with respect to x.)

    $= 6xy^2 + 4x^3 + e^y$

d) $\dfrac{\partial^2 f}{\partial y^2} = \dfrac{\partial}{\partial y}(3x^2 y^2 + x^4 + x e^y)$

       (Differentiate twice with respect to y.)

    $= 6x^2 y + x e^y$

DO EXERCISE 16.

---

15. Consider

$$f(x, y) = 3xy^2 + 2xy + x^2.$$

a) Find f_y.

b) For the function in (a), find the first partial derivative with respect to x. Denote this f_{yx}.

c) For the function in (a), find the first partial derivative with respect to y. Denote this f_{yy}.

16. For

$$z = f(x, y) = 3xy^2 + 2xy + x^2 + x \ln y,$$

find the four second-order partial derivatives.

Note by comparing (b) and (c) above that

$$\frac{\partial^2 f}{\partial y\,\partial x} = \frac{\partial^2 f}{\partial x\,\partial y} \qquad \text{(And similarly, } f_{xy} = f_{yx}.)$$

This will be true for all functions that we consider in this text, but is *not* true for all functions.

EXERCISE SET 12.2

Find the four second-order partial derivatives.

1. $f(x, y) = 3x^2 - xy + y$ **2.** $f(x, y) = 5x^2 + xy - x$ **3.** $f(x, y) = 3xy$

4. $f(x, y) = 4xy$ **5.** $f(x, y) = x^5y^4 + x^3y^2$ **6.** $f(x, y) = x^4y^3 - x^2y^3$

Find f_{xx}, f_{yx}, f_{xy}, and f_{yy}. (Remember, f_{yx} means differentiate with respect to y, then x.)

7. $f(x, y) = 2x - 3y$ **8.** $f(x, y) = 3x + 5y$ **9.** $f(x, y) = e^{2xy}$

10. $f(x, y) = e^{xy}$ **11.** $f(x, y) = x + e^y$ **12.** $f(x, y) = y - e^x$

13. $f(x, y) = y \ln x$ **14.** $f(x, y) = x \ln y$

▶ ───────────────────────────────────

Find f_{xx}, f_{yx}, f_{xy}, and f_{yy}:

15. $f(x, y) = \dfrac{x}{y^2} - \dfrac{y}{x^2}$ **16.** $f(x, y) = \dfrac{xy}{x - y}$

17. Consider $f(x, y) = \ln(x^2 + y^2)$. Show that f is a solution of the partial differential equation

$$\frac{\partial^2 f}{\partial x^2} + \frac{\partial^2 f}{\partial y^2} = 0.$$

18. Consider $f(x, y) = x^3 - 5xy^2$. Show that f is a solution of the partial differential equation

$$xf_{xy} - f_y = 0.$$

OBJECTIVES

You should be able to find maximum and minimum values of functions of two variables.

12.3 MAXIMUM–MINIMUM PROBLEMS

In this section we shall find maximum and minimum values of functions of two variables.

DEFINITION. A function f of two variables

i) has a relative maximum at (a, b) if

$$f(x, y) \leqslant f(a, b)$$

for all points in a circular region containing (a, b).

ii) has a relative minimum at (a, b) if

$$f(x, y) \geqslant f(a, b)$$

for all points in a circular region containing (a, b).

This definition is illustrated in Fig. 1.

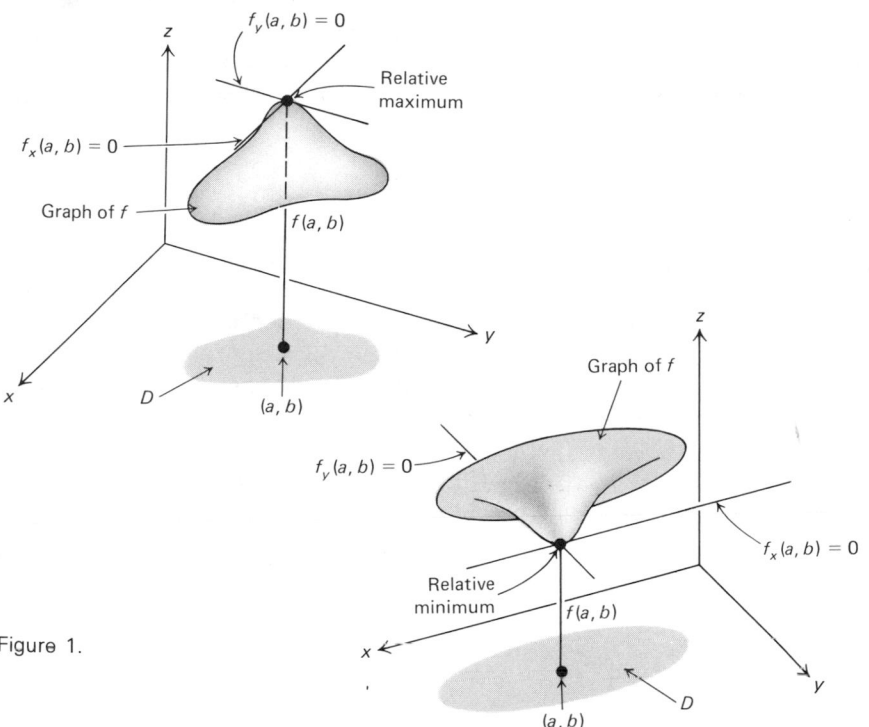

Figure 1.

A relative maximum (minimum) may not be an "absolute" maximum (minimum) as illustrated below in Fig. 2.

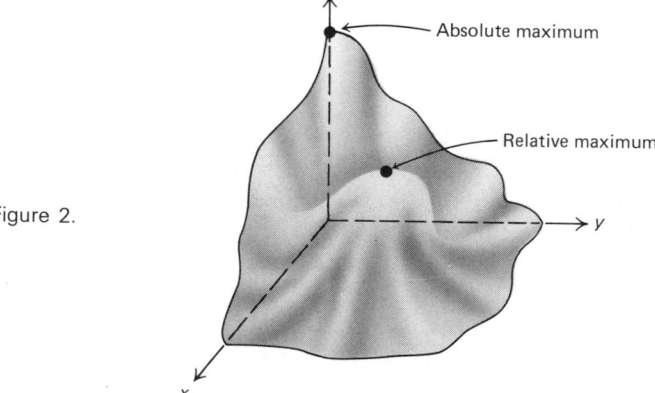

Figure 2.

Determining Maximum and Minimum Values

Suppose a function f assumes a relative maximum (or minimum) value at some point (a, b) inside its domain. If we hold y constant at the value b, then $f(x, b)$ is a function of one variable x having its relative maximum value at $x = a$, so its derivative must be 0 there. That is, $f_x = 0$ at the point (a, b). Similarly, $f_y = 0$ at (a, b). The equations

$$f_x = 0, \qquad f_y = 0$$

are thus satisfied by the point (a, b) at which the relative maximum occurs. We call a point (a, b) where both partial derivatives are 0 a *critical point*. This is comparable to the earlier definition for functions of one variable. Thus one strategy for finding relative maximum or minimum values is to solve the above system of equations to find critical points. Just as for functions of one variable, this strategy does *not* guarantee that we will have a relative maximum or minimum value. We have argued only that *if* f has a maximum or minimum value at (a, b), *then* both its partial derivatives must be 0 at that point. Look at Fig. 1. That this does not hold in all cases is shown in Fig. 3.

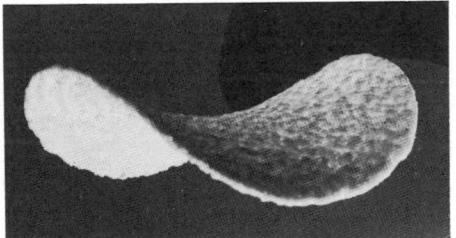

Where is the saddle point? (*Marshall Henrichs*)

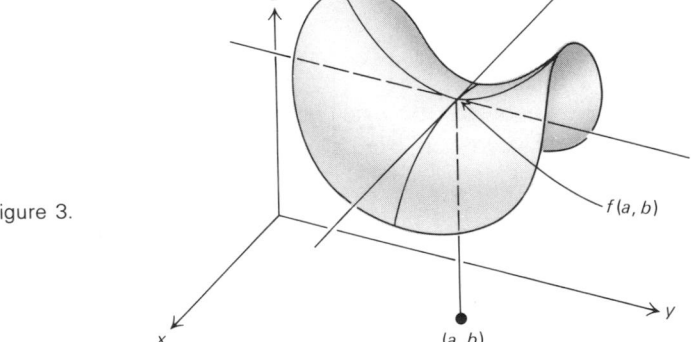

Figure 3.

Now suppose we fix y at a point b. Then $f(x, b)$, considered as a function of one variable, has a maximum at a, but f does not. Similarly, if we fix x at a, then $f(a, y)$, considered as a function of one variable, has a minimum at b, but f does not. The point $f(a, b)$ is called a *saddle point*. In other words, $f_x(a, b) = 0$ and $f_y(a, b) = 0$ [the point (a, b) is a critical point], but f does not attain a relative maximum or minimum at (a, b).

A test for finding relative maximum and minimum values that involves the use of first- and second-order partial derivatives is stated below. We shall not prove this theorem.

THEOREM.   *The D-test.* **To find the relative maximum and minimum values of** f**,**
1. **Find** f_x**,** f_y**,** f_{xx}**,** f_{yy}**, and** f_{xy}**.**
2. **Solve the system of equations** $f_x = 0$**,** $f_y = 0$**.**
   Let (a, b) **represent a solution.**
3. **Evaluate** D **where** $D = f_{xx}(a, b) \cdot f_{yy}(a, b) - [f_{xy}(a, b)]^2$**.**
4. **Then**
   i) f **has a maximum at** (a, b) **if** $D > 0$ **and** $f_{xx}(a, b) < 0$**.**
   ii) f **has a minimum at** (a, b) **if** $D > 0$ **and** $f_{xx}(a, b) > 0$**.**
   iii) f **has neither a maximum nor a minimum at** (a, b) **if** $D < 0$**.**
       The function has a *saddle point* **at** (a, b)**. See Fig. 3.**
   iv) **This test is not applicable if** $D = 0$**.**

A relative maximum or minimum *may not be an absolute maximum or minimum.* Tests for absolute maximum or minimum are rather complicated. We shall restrict our attention to finding *relative* maximum or minimum values. Fortunately, in most applications relative maximum and minimum values turn out to be absolute maximum and minimum values.

Example 1   Find the relative maximum and minimum values of

$$f(x, y) = x^2 + xy + y^2 - 3x.$$

Solution

1. Find f_x, f_y, f_{xx}, f_{yy}, and f_{xy}.

$$f_x = 2x + y - 3, \qquad f_y = x + 2y,$$
$$f_{xx} = 2, \qquad\qquad f_{yy} = 2,$$
$$f_{xy} = 1.$$

2. Solve the system of equations $f_x = 0$, $f_y = 0$:

$$2x + y - 3 = 0 \qquad\qquad (1)$$
$$x + 2y = 0. \qquad\qquad (2)$$

Solving Eq. (2) for x we get $x = -2y$. Substituting $-2y$ for x in Eq. (1) and solving we get

$$2(-2y) + y - 3 = 0$$
$$-4y + y - 3 = 0$$
$$-3y - 3 = 0$$
$$y = -1.$$

17. Find the relative maximum and minimum values of

$$f(x, y) = x^2 + 2xy + 2y^2 - 6y.$$

To find x when $y = -1$, we substitute -1 for y in either Eq. (1) or Eq. (2). We use Eq. (2):

$$x + 2(-1) = 0$$
$$x = 2.$$

Thus $(2, -1)$ is our candidate for a maximum or minimum.

3. We have to check to see if $f(2, -1)$ is a maximum or minimum.

$$D = f_{xx}(2, -1) \cdot f_{yy}(2, -1) - [f_{xy}(2, -1)]^2$$
$$= 2 \cdot 2 - [1]^2$$
$$= 3.$$

4. Thus $D = 3$ and $f_{xx}(2, -1) = 2$. Since $D > 0$ and $f_{xx}(2, -1) > 0$, it follows that f has a relative minimum at $(2, -1)$ and that the minimum is found as follows:

$$f(2, -1) = 2^2 + 2(-1) + (-1)^2 - 3 \cdot 2$$
$$= 4 - 2 + 1 - 6 = -3.$$

DO EXERCISE 17.

Example 2 Find the relative maximum and minimum values of

$$f(x, y) = xy - x^3 - y^2.$$

Solution

1. Find f_x, f_y, f_{xx}, f_{yy}, and f_{xy}:

$$f_x = y - 3x^2, \qquad f_y = x - 2y,$$
$$f_{xx} = -6x, \qquad f_{yy} = -2,$$
$$f_{xy} = 1.$$

2. Solve the system of equations $f_x = 0$, $f_y = 0$:

$$y - 3x^2 = 0 \tag{1}$$
$$x - 2y = 0. \tag{2}$$

Solving Eq. (1) for y, we get $y = 3x^2$. Substituting $3x^2$ for y in Eq. (2) and solving, we get

$$x - 2(3x^2) = 0$$
$$x - 6x^2 = 0$$
$$x(1 - 6x) = 0. \qquad \text{(Factoring)}$$

Setting each factor equal to 0 and solving, we have

$$x = 0 \qquad \text{or} \qquad 1 - 6x = 0$$
$$x = 0 \qquad \text{or} \qquad x = \tfrac{1}{6}.$$

To find y when $x = 0$ we substitute 0 for x in either Eq. (1) or Eq. (2). We use Eq. (2):

$$0 - 2y = 0$$
$$-2y = 0$$
$$y = 0.$$

Thus $(0, 0)$ is one critical value (candidate for a maximum or minimum). To find the other critical value we substitute $\frac{1}{6}$ for x in either Eq. (1) or (2). We use Eq. (2).

$$\tfrac{1}{6} - 2y = 0$$
$$-2y = -\tfrac{1}{6}$$
$$y = \tfrac{1}{12}.$$

Thus $(\frac{1}{6}, \frac{1}{12})$ is another critical point.

3. We have to check both $(0, 0)$ and $(\frac{1}{6}, \frac{1}{12})$ as to whether they yield maximum or minimum values.

$$\text{For } (0, 0): \quad D = f_{xx}(0, 0) \cdot f_{yy}(0, 0) - [f_{xy}(0, 0)]^2$$
$$= [-6 \cdot 0] \cdot [-2] - [1]^2$$
$$= -1$$

Since $D < 0$, it follows that $f(0, 0)$ is neither a maximum nor a minimum, but a saddle point.

$$\text{For } (\tfrac{1}{6}, \tfrac{1}{12}): \quad D = f_{xx}(\tfrac{1}{6}, \tfrac{1}{12}) f_{yy}(\tfrac{1}{6}, \tfrac{1}{12}) - [f_{xy}(\tfrac{1}{6}, \tfrac{1}{12})]^2$$
$$= [-6 \cdot \tfrac{1}{6}] \cdot [-2] - [1]^2$$
$$= -1(-2) - 1$$
$$= 1$$

4. Thus $D = 1$ and $f_{xx}(\frac{1}{6}, \frac{1}{12}) = -1$. Since $D > 0$ and $f_{xx}(\frac{1}{6}, \frac{1}{12}) < 0$, it follows that f has a relative maximum at $(\frac{1}{6}, \frac{1}{12})$ and that maximum is found as follows:

$$f(\tfrac{1}{6}, \tfrac{1}{12}) = \tfrac{1}{6} \cdot \tfrac{1}{12} - (\tfrac{1}{6})^3 - (\tfrac{1}{12})^2$$
$$= \tfrac{1}{72} - \tfrac{1}{216} - \tfrac{1}{144} = \tfrac{1}{432}.$$

DO EXERCISE 18.

Example 3 *Business—Maximizing profit.* A firm produces two kinds of golf balls, one that sells for \$3 each and the other for \$2 each. The total revenue from the sale of x thousand balls at \$3 each and y thousand at \$2 each is given by

$$R(x, y) = 3x + 2y.$$

18. Find the relative maximum and minimum values of

$$f(x, y) = 2xy - 4x^3 - y^2.$$

19. *Business—Maximizing profit.* A firm produces two kinds of calculators, one that sells for $15 each and the other for $20 each. The total revenue from the sale of x thousand calculators at $15 each and y thousand at $20 each is given by

$$R(x, y) = 15x + 20y$$

The company determines that the total cost, in thousands of dollars, of producing x thousand of the $15 calculator and y thousand of the $20 calculator is given by

$$C(x, y) = 3x^2 - 3xy + \tfrac{3}{2}y^2 + 6x$$
$$+ 14y - 50.$$

Find the amount of each type of calculator that must be produced and sold to maximize profit.

The company determines that the total cost, in thousands of dollars, of producing x thousand of the $3 ball and y thousand of the $2 ball is given by

$$C(x, y) = 2x^2 - 2xy + y^2 - 9x + 6y + 7.$$

Find the amount of each type of ball that must be produced and sold to maximize profit.

Solution Total profit, $P(x, y)$ is given by

$$P(x, y) = R(x, y) - C(x, y)$$
$$= 3x + 2y - (2x^2 - 2xy + y^2 - 9x + 6y + 7)$$
$$P(x, y) = -2x^2 + 2xy - y^2 + 12x - 4y - 7.$$

1. Find P_x, P_y, P_{xx}, P_{yy}, and P_{xy}.

$$P_x = -4x + 2y + 12, \qquad P_y = 2x - 2y - 4,$$
$$P_{xx} = -4, \qquad\qquad\qquad P_{yy} = -2,$$
$$P_{xy} = 2.$$

2. Solve the system of equations $P_x = 0$, $P_y = 0$:

$$-4x + 2y + 12 = 0 \tag{1}$$
$$2x - 2y - 4 = 0. \tag{2}$$

Adding these equations, we get

$$-2x + 8 = 0.$$

Then

$$-2x = -8$$
$$x = 4.$$

To find y when $x = 4$, we substitute 4 for x in either Eq. (1) or Eq. (2). We use Eq. (2):

$$2 \cdot 4 - 2y - 4 = 0$$
$$-2y + 4 = 0$$
$$-2y = -4$$
$$y = 2.$$

Thus, $(4, 2)$ is our candidate for a maximum or minimum.

3. We have to check to see if $P(4, 2)$ is a maximum or minimum.

$$D = P_{xx}(4, 2)P_{yy}(4, 2) - [P_{xy}(4, 2)]^2$$
$$= (-4)(-2) - 2^2$$
$$= 4$$

4. Thus $D = 4$ and $P_{xx} = -4$. Since $D > 0$ and $P_{xx}(4, 2) < 0$, it follows that P has a relative maximum at $(4, 2)$. Thus to maximize profit, the company must produce and sell 4 thousand of the $3 golf balls and 2 thousand of the $2 golf balls.

DO EXERCISE 19. (on preceding page).

EXERCISE SET 12.3

Find the relative maximum and minimum values.

1. $f(x, y) = x^2 + xy + y^2 - y$

2. $f(x, y) = x^2 + xy + y^2 - 5y$

3. $f(x, y) = 2xy - x^3 - y^2$

4. $f(x, y) = 4xy - x^3 - y^2$

5. $f(x, y) = x^3 + y^3 - 3xy$

6. $f(x, y) = x^3 + y^3 - 6xy$

7. $f(x, y) = x^2 + y^2 - 2x + 4y - 2$

8. $f(x, y) = x^2 + 2xy + 2y^2 - 6y + 2$

9. $f(x, y) = x^2 + y^2 + 2x - 4y$

10. $f(x, y) = 4y + 6x - x^2 - y^2$

11. $f(x, y) = 4x^2 - y^2$

12. $f(x, y) = x^2 - y^2$

In these problems assume that relative maximum and minimum values are absolute maximum and minimum values.

13. *Business—Maximizing profit.* A firm produces two kinds of radios, one that sells for $17 each and the other for $21 each. The total revenue from the sale of x thousand radios at $17 each and y thousand at $21 each is given by

$$R(x, y) = 17x + 21y.$$

The company determines that the total cost, in thousands of dollars, of producing x thousand of the $17 radio and y thousand of the $21 radio is given by

$$C(x, y) = 4x^2 - 4xy + 2y^2 - 11x + 25y - 3.$$

Find the amount of each type of radio that must be produced and sold to maximize profit.

15. A one-product company found that its profit in millions of dollars is a function P given by
$$P(a, p) = 2ap + 80p - 15p^2 - \tfrac{1}{10}a^2p - 100,$$

where a = amount spent on advertising, in millions of dollars, and p = price charged per item of the product, in dollars. Find the maximum value of P and the values of a and p at which it is attained.

14. *Business—Maximizing profit.* A firm produces two kinds of baseball gloves, one that sells for $18 each and the other for $25 each. The total revenue from the sale of x thousand gloves at $18 each and y thousand at $25 each is given by

$$R(x, y) - 18x + 25y.$$

The company determines that the total cost, in thousands of dollars, of producing x thousand of the $18 glove and y thousand of the $25 glove is given by

$$C(x, y) = 4x^2 - 6xy + 3y^2 + 20x + 19y - 12.$$

Find the amount of each type of glove that must be produced and sold to maximize profit.

16. A one-product company finds that its profit in millions of dollars is a function P given by
$$P(a, n) = -5a^2 - 3n^2 + 48a - 4n + 2an + 300,$$

where a = amount spent on advertising, in millions of dollars, and n = number of items sold. Find the maximum value of P and the values of a and n at which it is attained.

Find the relative maximum and minimum values.

17. $f(x, y) = e^x + e^y - e^{x+y}$

18. $f(x, y) = xy + \dfrac{2}{x} + \dfrac{4}{y}$

19. $S(b, m) = (m + b - 72)^2 + (2m + b - 73)^2$
$\qquad\qquad + (3m + b - 75)^2$

20. An open-top rectangular box is to be made with a 20-meter^2 surface area. Find the dimensions that will yield the maximum volume.

OBJECTIVES

You should be able to find the regression line for a given set of data points and use the regression line to make predictions regarding further data.

20. a) Use Fig. 5 to predict life expectancy of the female in 1980.
  b) Use Fig. 6 to predict life expectancy of the female in 1980.
  c) Compare your answers.

12.4 APPLICATION: THE LEAST SQUARES TECHNIQUE

The problem of fitting an equation to a set of data occurs frequently. We considered one procedure for doing this in Section 1.6. Such an equation provides a model of the phenomena from which predictions can be made. For example, in business one might want to predict future sales based on past data. In ecology, one might want to predict future demands for natural gas based on past need. Suppose we are trying to determine a linear equation

$$y = mx + b$$

to fit the data. To determine this equation is to determine the values of m and b. But how? Let us consider some factual data on life expectancy of females in the U.S.

ear, x	1. 1950	2. 1960	3. 1970	4. 1980
Life expectancy of female (in years), y	72	73	75	?

Suppose we plot these points and try to draw a line that fits. Note that there are several ways this might be done (see Figs. 4 through 6). Each would give a different estimate of life expectancy in 1980.

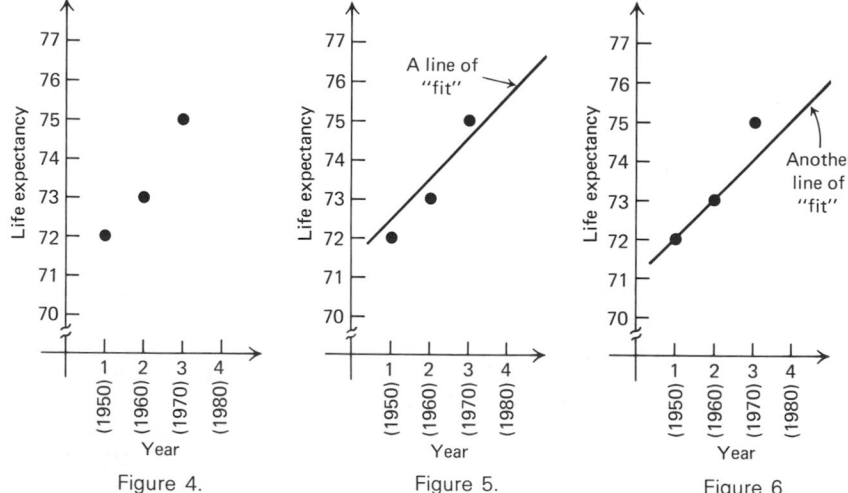

Figure 4.　　　　Figure 5.　　　　Figure 6.

DO EXERCISE 20.

Note that time is incremented in tens of years, making computation easier. Consider the data points $(1, 72)$, $(2, 73)$, and $(3, 75)$ as plotted in Fig. 7.

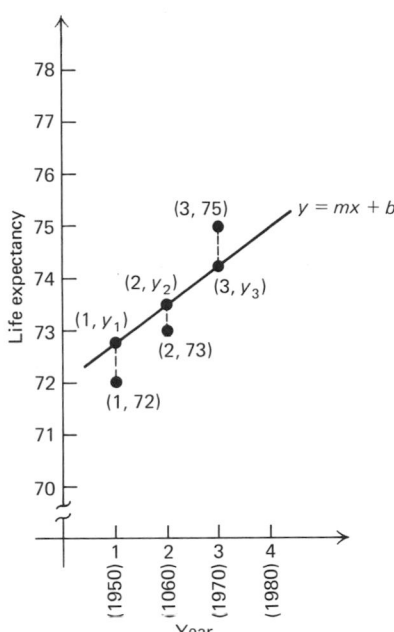

Figure 7.

We will try to fit this data with a line

$$y = mx + b$$

by determining values of m and b. Note the y-errors, or y-deviations, $y_1 - 72$, $y_2 - 73$, and $y_3 - 75$ between the observed points $(1, 72)$, $(2, 73)$, and $(3, 75)$ and the points $(1, y_1)$, $(2, y_2)$, and $(3, y_3)$ on the line. We would like, somehow, to minimize these deviations, in order to have a good fit. One way of minimizing the deviations is based on the so-called *least squares assumption*, as follows.

LEAST SQUARES ASSUMPTION. **The line of best fit is the line for which the sum of squares of the y-deviations is a minimum. This is called the *regression line*.**

Using the least squares assumption for the life expectancy data, we would minimize

$$(y_1 - 72)^2 + (y_2 - 73)^2 + (y_3 - 75)^2 \tag{1}$$

21. Consider this factual data on life expectancy of males in the United States.

Year, x	1 1950	2 1960	3 1970	4 1980
Life expectancy of male (in years), y	65	67	68	?

$$y = mx + b.$$

a) Find the regression line

b) Use the regression line to predict life expectancy of the male in 1980.

and since the points $(1, y_1)$, $(2, y_2)$, and $(3, y_3)$ must be solutions of $y = mx + b$, it follows that

$$y_1 = m1 + b = m + b$$
$$y_2 = m2 + b = 2m + b$$
$$y_3 = m3 + b = 3m + b.$$

Substituting $m + b$ for y_1, $2m + b$ for y_2, and $3m + b$ for y_3 in (1), we have

$$(m + b - 72)^2 + (2m + b - 73)^2 + (3m + b - 75)^2.$$

Thus, to find the regression line for the given set of data, we must find the values of m and b that minimize the function S given by $S(b, m) = (m + b - 72)^2 + (2m + b - 73)^2 + (3m + b - 75)^2$.

To apply the D-test, we first find the partial derivatives $\dfrac{\partial S}{\partial b}$ and $\dfrac{\partial S}{\partial m}$:

$$\frac{\partial S}{\partial b} = 2(m + b - 72) + 2(2m + b - 73) + 2(3m + b - 75)$$
$$= 12m + 6b - 440;$$

$$\frac{\partial S}{\partial m} = 2(m + b - 72) + 2(2m + b - 73)2 + 2(3m + b - 75)3$$
$$= 28m + 12b - 886.$$

We set these derivatives equal to 0 and solve the resulting system.

$$\begin{array}{ccc} 12m + 6b - 440 = 0 & & 6m + 3b = 220 \quad (2) \\ & \text{or} & \\ 28m + 12b - 886 = 0 & & 14m + 6b = 443 \quad (3) \end{array}$$

The solution of this system is

$$b = \tfrac{211}{3} = 70\tfrac{1}{3}, \qquad m = \tfrac{3}{2}.$$

We leave it to the reader to complete the D-test to verify that $(70\tfrac{1}{3}, \tfrac{3}{2})$ does, in fact, yield the minimum of S. We need not bother to compute $S(70\tfrac{1}{3}, \tfrac{3}{2})$. The values of m and b are all we need to determine $y = mx + b$. The regression line is

$$y = \frac{3}{2}x + 70\tfrac{1}{3}, \qquad \text{or} \qquad y = 1.5x + 70.3 \quad \text{(rounded)}.$$

We can extrapolate from the data to find a predicted life expectancy in 1980.

$$y = 1.5 \cdot 4 + 70.3$$
$$= 6 + 70.3$$
$$= 76.3$$

Thus, life expectancy in 1980 is 76.3 years.

DO EXERCISE 21.

The method of least squares is a statistical process illustrated here with only three data points to ease the explanation. Most statistical researchers would warn that many more than three data points should be used to get a "good" regression line. Furthermore, making predictions too far in the future from any linear model may be suspect. It can be done, but the further into the future the prediction is made, the more dubious one should be about the prediction.

The Regression Line for an Arbitrary Collection of Data Points $(c_1, d_1), (c_2, d_2), \cdots, (c_n, d_n)$ (Optional)

Look again at the regression line

$$y = \frac{3}{2}x + \frac{211}{3}$$

for the data points $(1, 72)$, $(2, 73)$, and $(3, 75)$. Let us consider the arithmetic averages, or means, of the x-coordinates, denoted $\bar{x}$; and the y-coordinates, denoted $\bar{y}$.

$$\bar{x} = \frac{1 + 2 + 3}{3} = 2, \qquad \bar{y} = \frac{72 + 73 + 75}{3} = \frac{220}{3}$$

It turns out that the point $(\bar{x}, \bar{y})$, or $(2, \frac{220}{3})$, is on the regression line for

$$\frac{220}{3} = \frac{3}{2} \cdot 2 + \frac{211}{3}.$$

Thus the regression line is as follows

$$y - \bar{y} = m(x - \bar{x}), \qquad \text{or} \qquad y - \frac{220}{3} = m(x - 2).$$

All that remains, in general, is to determine m. Suppose we wanted to find the regression line for an arbitrary number of points (c_1, d_1), $(c_2, d_2), \cdots, (c_n, d_n)$.

To do so, find the values m and b that minimize the function S given by

$$S(b, m) = (y_1 - d_1)^2 + (y_2 - d_2)^2 + \cdots + (y_n - d_n)^2 = \sum_{i=1}^{n} (y_i - d_i)^2,$$

where $y_i = mc_i + b$.

22. Repeat Margin Exercise 21(a) using the procedure outlined in this optional part of the section.

Using a procedure like the one we used earlier to minimize S, we can show that $y = mx + b$ takes the form

$$y - \bar{y} = m(x - \bar{x}),$$

where

$$\bar{x} = \frac{\sum_{i=1}^{n} c_i}{n}, \qquad \bar{y} = \frac{\sum_{i=1}^{n} d_i}{n},$$

and

$$m = \frac{\sum_{i=1}^{n} (c_i - \bar{x})(d_i - \bar{y})}{\sum_{i=1}^{n} (c_i - \bar{x})^2}.$$

Let us see how this works out for the life expectancy example done previously.

c_i	d_i	$c_i - \bar{x}$	$(c_i - \bar{x})^2$	$(d_i - \bar{y})$	$(c_i - \bar{x})(d_i - \bar{y})$
1	72	-1	1	-1.3	1.3
2	73	0	0	-0.3	0
3	75	1	1	1.7	1.7

$$\sum_{i=1}^{3} c_i = 6, \quad \sum_{i=1}^{3} d_i = 220 \qquad \sum_{i=1}^{3} (c_i - \bar{x})^2 = 2 \qquad \sum_{i=1}^{3} (c_i - \bar{x})(d_i - \bar{y}) = 3$$

$$\bar{x} = 2 \qquad \bar{y} = 73.3 \qquad\qquad\qquad\qquad m = \tfrac{3}{2}$$

Thus the regression line is

$$y - 73.3 = \frac{3}{2}(x - 2)$$

which simplifies to

$$y = \frac{3}{2}x + 70.3.$$

DO EXERCISE 22.

EXERCISE SET 12.4

1. Use the regression line $y = 1.5x + 70.3$ to predict life expectancy of the female in the year 1990.

2. Use the regression line $y = 1.5x + 63.7$ to predict life expectancy of the male in the year 2000.

3. Consider the following factual data on natural gas demand.

Year, x	1. 1950	2. 1960	3. 1970	4. 1980
Demand, y (in quadrillion BTU	19	21	22	?

a) Find the regression line $y = mx + b$.
b) Use the regression line to predict gas demand in 1980.
c) Use the regression line to predict gas demand in 2000.

5. Consider this data relating cricket chirps per minute to Fahrenheit temperature.

Chirps per minute, x	60	76	88	100
Fahrenheit temperature, y	55°	59°	62°	65°

a) Find the regression line $y = mx + b$. [*Hint:* The y-deviations are $60m + b - 55$, $76m + b - 59$, and so on.]
b) One night the crickets chirped 84 times per minute. Use the regression line to determine the temperature that night.

4. A student wanted to predict his final examination score, based on what his midterm test score was. He decided to base his data (see below) on scores of three students who took the same course with the same instructor the previous semester.

Midterm score (%), x	70	60	85
Final exam score (%), y	75	62	89

a) Find the regression line $y = mx + b$. [*Hint:* The y-deviations are $70m + b - 75$, $60m + b - 62$, and so on.]
b) The midterm score of the student was 81. Use the regression line to predict his final exam score.

6. Consider the following total sales data of a company during the first 4 years of operation.

Year, x	1	2	3	4
Sales (in millions), y	$22	$34	$44	$60

a) Find the regression line $y = mx + b$.
b) Use the regression line to predict sales in the 5th year.

---

7. ▦
a) Find the regression line $y = mx + b$ that fits the set of data in the table.

b) Use the regression line to predict the world record in the mile in 1984.

c) In July 1979 Sebastian Coe set a new world record of 3:49.0 for the mile. How does this compare with what can be predicted by the regression?

Year, x	World record in mile, y (min:sec)
1875 (Walter Slade)	4:24.5
1894 (Fred Bacon)	4:18.2
1923 (Paavo Nurmi)	4:10.4
1937 (Sidney Wooderson)	4:06.4
1942 (Gunder Haegg)	4:06.2
1945 (Gunder Haegg)	4:01.4
1954 (Roger Bannister)	3:59.4
1964 (Peter Snell)	3:54.4
1967 (Jim Ryun)	3:51.1
1975 (John Walker)	3:49.4

Hint: Convert each time to decimal notation; for example, $4{:}24.5 = 4\dfrac{24.5}{60} = 4.4083$.

OBJECTIVES

You should be able to

a) Find a maximum or minimum value of a given function subject to a given constraint, using the Method of LaGrange Multipliers.

b) Solve applied problems involving LaGrange Multipliers.

12.5 CONSTRAINED MAXIMUM AND MINIMUM VALUES—LAGRANGE MULTIPLIERS

Before we get into detail let us look at a problem we considered in Chapter 3.

Example 1 A hobby store has 20 ft of fencing to fence off a rectangular electric-train area in one corner of its display room. What dimensions of the rectangle will maximize the area?

We maximize the function

$$A = xy$$

subject to the condition or *constraint* $x + y = 20$. Note that A is a function of two variables:

$$A(x, y) = xy.$$

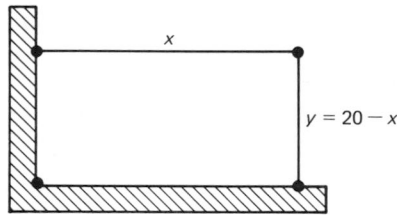

When we solved this earlier, we first solved the constraint for y:

$$y = 20 - x.$$

We then substituted $20 - x$ for y to obtain

$$A(x, 20 - x) = x(20 - x) = 20x - x^2,$$

which is a function of one variable. We then found a maximum value using Maximum–Minimum Principle 1 (see p. 161). By itself, the function of two variables

$$A(x, y) = xy$$

has no maximum value. This can be checked using the D-test. But, with the constraint $x + y = 20$, the function does have a maximum. We see this pictorially in the following figure.

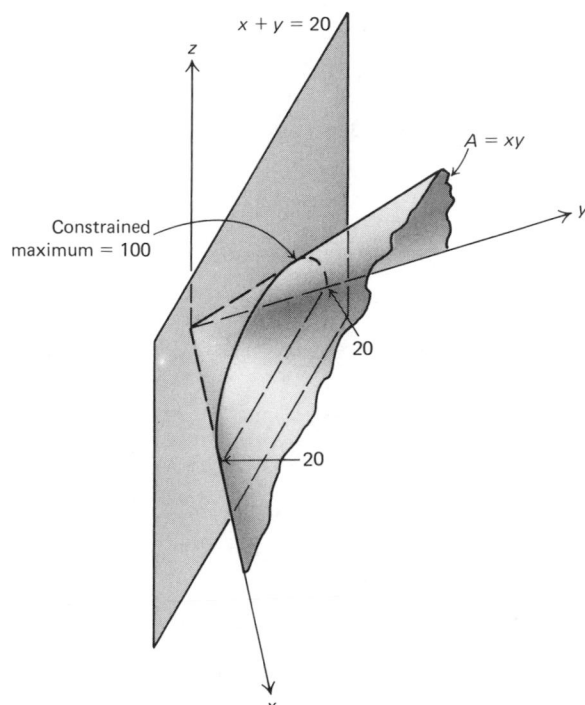

It may be quite difficult to solve a constraint for one variable. The procedure outlined below allows us to proceed without solving a constraint for one variable.

METHOD OF LAGRANGE MULTIPLIERS. **To find a maximum or minimum value of a function $f(x, y)$ subject to the constraint $g(x, y) = 0$,**

1. Form a new function:

$$F(x, y, \lambda) = f(x, y) - \lambda g(x, y).$$

2. Find the partial derivatives F_x, F_y, and F_λ.
3. Solve the system

$$F_x = 0, \quad F_y = 0, \quad \text{and} \quad F_\lambda = 0.$$

Let (a, b) represent a solution. We still must determine whether (a, b) yields a maximum or minimum, but we will assume one or the other in the problems considered here.

The variable λ (lambda) is called a *LaGrange Multiplier*. We first illustrate the Method of LaGrange Multipliers in Example 2.

23. A rancher has 50 ft of fencing to fence off a rectangular animal pen in the corner of a barn. What dimensions of the rectangle will yield the maximum area?

a) Express the area as a function of two variables with a constraint.

b) Find the maximum value of the function in (a) above, using the Method of LaGrange Multipliers.

Example 2   Find the maximum value of

$$A(x, y) = xy$$

subject to the constraint $x + y = 20$.

Solution

1. We form the new function F given by

$$F(x, y, \lambda) = xy - \lambda \cdot (x + y - 20).$$

Note that we first had to express $x + y = 20$ as $x + y - 20 = 0$.

2. We find the first partial derivatives.

$$F_x = y - \lambda,$$
$$F_y = x - \lambda,$$
$$F_\lambda = -(x + y - 20)$$

3. We set these derivatives equal to 0 and solve the resulting system.

$$y - \lambda = 0 \tag{1}$$
$$x - \lambda = 0 \tag{2}$$
$$x + y - 20 = 0 \tag{3}$$

[If $-(x + y - 20) = 0$, then $x + y - 20 = 0$.]

From Eqs. (1) and (2) it follows that

$$x = y = \lambda.$$

Substituting λ for x and y in Eq. (3) we get

$$\lambda + \lambda - 20 = 0,$$
$$2\lambda = 20,$$
$$\lambda = 10.$$

Thus $x = \lambda = 10$, and $y = \lambda = 10$. The maximum occurs at $(10, 10)$ and is

$$A(10, 10) = 10 \cdot 10 = 100.$$

DO EXERCISE 23.

Example 3 Find the maximum value of

$$f(x, y) = 3xy$$

subject to the constraint

$$2x + y = 8.$$

[*Note:* f could be interpreted as a production function with budget constraint $2x + y = 8$.]

Solution

1. We form the new function F given by

$$F(x, y, \lambda) = 3xy - \lambda(2x + y - 8).$$

Note that we had to express $2x + y = 8$ as $2x + y - 8 = 0$.

2. We find the first partial derivatives.

$$F_x = 3y - 2\lambda$$
$$F_y = 3x - \lambda$$
$$F_\lambda = -(2x + y - 8)$$

3. We set these derivatives equal to 0 and solve the resulting system

$$3y - 2\lambda = 0 \qquad (1)$$
$$3x - \lambda = 0 \qquad (2)$$
$$-(2x + y - 8) = 0, \quad \text{or} \quad 2x + y - 8 = 0 \qquad (3)$$

Solving Eq. (1) for y, we get

$$y = \frac{2}{3}\lambda.$$

Solving Eq. (2) for x, we get

$$x = \frac{\lambda}{3}.$$

Substituting $(2/3)\lambda$ for y and $(\lambda/3)$ for x in Eq. (3), we get

$$2\left(\frac{\lambda}{3}\right) + \left(\frac{2}{3}\lambda\right) - 8 = 0$$
$$\tfrac{4}{3}\lambda = 8$$
$$\lambda = \tfrac{3}{4} \cdot 8 = 6.$$

24. Find the maximum value of

$$f(x, y) = 5xy$$

subject to the constraint

$$4x + y = 20.$$

Then

$$x = \frac{\lambda}{3} = \frac{6}{3} = 2 \quad \text{and} \quad y = \frac{2}{3}\lambda = \frac{2}{3} \cdot 6 = 4.$$

The maximum of f subject to the constraint occurs at $(2, 4)$ and is

$$f(2, 4) = 3 \cdot 2 \cdot 4 = 24.$$

DO EXERCISE 24.

Example 4 *The beverage can problem.* The standard beverage can has a volume of 12 oz, or 26 in³. What dimensions yield the minimum surface area? Find the minimum surface area.

Solution We want to minimize the function s given by

$$s(h, r) = 2\pi rh + 2\pi r^2$$

subject to the volume constraint

$$\pi r^2 h = 26, \quad \text{or} \quad \pi r^2 h - 26 = 0.$$

Note that s does not have a minimum without the constraint.

1. We form the new function S given by

$$S(h, r, \lambda) = 2\pi rh + 2\pi r^2 - \lambda(\pi r^2 h - 26).$$

2. We find the first partial derivatives.

$$\frac{\partial S}{\partial h} = 2\pi r - \lambda \pi r^2$$

$$\frac{\partial S}{\partial r} = 2\pi h + 4\pi r - 2\lambda \pi r h$$

$$\frac{\partial S}{\partial \lambda} = -(\pi r^2 h - 26)$$

3. We set these derivatives equal to 0 and solve the resulting system.

$$2\pi r - \lambda \pi r^2 = 0 \qquad (1)$$

$$2\pi h + 4\pi r - 2\lambda \pi r h = 0 \qquad (2)$$

$$-(\pi r^2 h - 26) = 0, \quad \text{or} \quad \pi r^2 h - 26 = 0 \qquad (3)$$

Note that Eq. (1) can be solved for r:

$$\pi r(2 - \lambda r) = 0$$

$$\pi r = 0 \quad \text{or} \quad 2 - \lambda r = 0$$

$$r = 0 \qquad \text{or} \qquad r = \frac{2}{\lambda}.$$

Now $r = 0$ cannot be a solution to the original problem, so we continue by substituting $2/\lambda$ for r in Eq. (2):

$$2\pi h + 4\pi \cdot \frac{2}{\lambda} - 2\lambda \pi \cdot \frac{2}{\lambda} \cdot h = 0$$

$$2\pi h + \frac{8\pi}{\lambda} - 4\pi h = 0$$

$$\frac{8\pi}{\lambda} - 2\pi h = 0$$

$$-2\pi h = -\frac{8\pi}{\lambda},$$

so

$$h = \frac{4}{\lambda}.$$

25. Repeat Example 4 for a can of 16 oz, or 35 in³.

Since $h = 4/\lambda$ and $r = 2/\lambda$, it follows that $h = 2r$. Substituting $2r$ for h in Eq. (3) yields

$$\pi r^2(2r) - 26 = 0$$
$$2\pi r^3 - 26 = 0$$
$$2\pi r^3 = 26$$
$$\pi r^3 = 13$$
$$r^3 = \frac{13}{\pi}$$
$$r = \sqrt[3]{\frac{13}{\pi}} \approx 1.6 \text{ in.} \qquad (\text{▦ or Table 1})$$

So when $r = 1.6$ in., $h = 3.2$ in., the surface area is a minimum and is about $2\pi(1.6)(3.2) + 2\pi(1.6)^2$, or 48.3 in².

DO EXERCISE 25.

EXERCISE SET 12.5

Find the maximum value of f subject to the given constraint.

1. $f(x, y) = xy;$ $2x + y = 8$

2. $f(x, y) = 2xy;$ $4x + y = 16$

3. $f(x, y) = 4 - x^2 - y^2;$ $x + 2y = 10$

4. $f(x, y) = 3 - x^2 - y^2;$ $x + 6y = 37$

Find the minimum value of f subject to the given constraint.

5. $f(x, y) = x^2 + y^2;$ $2x + y = 10$

6. $f(x, y) = x^2 + y^2;$ $x + 4y = 17$

7. $f(x, y) = 2y^2 - 6x^2;$ $2x + y = 4$

8. $f(x, y) = 2x^2 + y^2 - xy;$ $x + y = 8$

9. $f(x, y, z) = x^2 + y^2 + z^2;$ $y + 2x - z = 3$

10. $f(x, y, z) = x^2 + y^2 + z^2;$ $x + y + z = 1$

Use the *Method of LaGrange Multipliers* to solve these problems.

11. Of all numbers whose sum is 70, find the two that have the maximum product.

12. Of all numbers whose sum is 50, find the two that have the maximum product.

13. Of all numbers whose difference is 6, find the two that have the minimum product.

14. Of all numbers whose difference is 4, find the two that have the minimum product.

15. A standard piece of typing paper has a perimeter of 39 in. Find the dimensions of the paper that will give the most typing area, subject to the perimeter constraint of 39 in. What is its area? Does the standard $8\frac{1}{2} \times 11$ paper have maximum area?

16. A carpenter is building a room with a fixed perimeter of 80 ft. What are the dimensions of the largest room that can be built? What is its area?

17. An oil drum of standard size has a volume of 200 gal or 27 ft^3. What dimensions yield the minimum surface area? Find the minimum surface area.

▶ ───

Find the indicated maximum or minimum value of f, subject to the given constraint.

18. Minimum: $f(x, y) = xy$; $x^2 + y^2 = 4$

19. Minimum: $f(x, y) = 2x^2 + y^2 + 2xy + 3x + 2y$; $y^2 = x + 1$

20. Maximum: $f(x, y, z) = x + y + z$; $x^2 + y^2 + z^2 = 1$

21. Maximum: $f(x, y, z) = x^2 y^2 z^2$; $x^2 + y^2 + z^2 = 1$

22. Maximum: $f(x, y, z) = x + 2y - 2z$; $x^2 + y^2 + z^2 = 4$

23. Maximum: $f(x, y, z, t) = x + y + z + t$; $x^2 + y^2 + z^2 + t^2 = 1$

24. Suppose $p(x, y)$ represents the production of a two product firm. We give no formula for p. The company produces x items of the first product at a cost c_1 of each, and y items of the second product at a cost c_2 of each. The budget constraint B is given by

$$B = c_1 x + c_2 y.$$

Find the value of λ in the LaGrange Multiplier method in terms of p_x, p_y, c_1, and c_2. The resulting equation is called the *Law of Equimarginal Productivity*.

CHAPTER 12 TEST

Given $f(x, y) = e^x + 2x^3y + y$, find

1. $\dfrac{\partial f}{\partial x}$ **2.** $\dfrac{\partial f}{\partial y}$ **3.** $\dfrac{\partial^2 f}{\partial x^2}$ **4.** $\dfrac{\partial^2 f}{\partial x \, \partial y}$ **5.** $\dfrac{\partial^2 f}{\partial y \, \partial x}$ **6.** $\dfrac{\partial^2 f}{\partial y^2}$

7. Find the relative maximum and minimum values.

$$f(x, y) = x^2 - xy + y^3 - x$$

8. Find the relative maximum and minimum values.

$$f(x, y) = y^2 - x^2$$

9. Consider this data regarding total sales of a company during the first three years of operation.

Year, x	1	2	3
Sales (in millions), y	$10	$15	$19

a) Find the regression line $y = mx + b$.
b) Use the regression line to predict sales in the 4th year.

10. Find the maximum value of

$$f(x, y) = 6xy - 4x^2 - 3y^2$$

subject to the constraint $x + 3y = 19$.

TABLES

TABLE 1 POWERS, ROOTS, AND RECIPROCALS

TABLE 2 COMMON LOGARITHMS

TABLE 3 NATURAL LOGARITHMS (In x)

TABLE 4 EXPONENTIAL FUNCTIONS

TABLE 5 INTEGRATION FORMULAS

TABLE 6 AREAS FOR A STANDARD NORMAL DISTRIBUTION

TABLE 1 POWERS, ROOTS, AND RECIPROCALS

n	n^2	n^3	$\sqrt{n}$	$\sqrt[3]{n}$	$\sqrt{10n}$	$\frac{1}{n}$	n	n^2	n^3	$\sqrt{n}$	$\sqrt[3]{n}$	$\sqrt{10n}$	$\frac{1}{n}$
1	1	1	1.000	1.000	3.162	1.0000	51	2,601	132,651	7.141	3.708	22.583	.0196
2	4	8	1.414	1.260	4.472	.5000	52	2,704	140,608	7.211	3.733	22.804	.0192
3	9	27	1.732	1.442	5.477	.3333	53	2,809	148,877	7.280	3.756	23.022	.0189
4	16	64	2.000	1.587	6.325	.2500	54	2,916	157,464	7.348	3.780	23.238	.0185
5	25	125	2.236	1.710	7.071	.2000	55	3,025	166,375	7.416	3.803	23.452	.0182
6	36	216	2.449	1.817	7.746	.1667	56	3,136	175,616	7.483	3.826	23.664	.0179
7	49	343	2.646	1.913	8.367	.1429	57	3,249	185,193	7.550	3.849	23.875	.0175
8	64	512	2.828	2.000	8.944	.1250	58	3,364	195,112	7.616	3.871	24.083	.0172
9	81	729	3.000	2.080	9.487	.1111	59	3,481	205,379	7.681	3.893	24.290	.0169
10	100	1,000	3.162	2.154	10.000	.1000	60	3,600	216,000	7.746	3.915	24.495	.0167
11	121	1,331	3.317	2.224	10.488	.0909	61	3,721	226,981	7.810	3.936	24.698	.0164
12	144	1,728	3.464	2.289	10.954	.0833	62	3,844	238,328	7.874	3.958	24.900	.0161
13	169	2,197	3.606	2.351	11.402	.0769	63	3,969	250,047	7.937	3.979	25.100	.0159
14	196	2,744	3.742	2.410	11.832	.0714	64	4,096	262,144	8.000	4.000	25.298	.0156
15	225	3,375	3.873	2.466	12.247	.0667	65	4,225	274,625	8.062	4.021	25.495	.0154
16	256	4,096	4.000	2.520	12.648	.0625	66	4,356	287,496	8.124	4.041	25.690	.0152
17	289	4,913	4.123	2.571	13.038	.0588	67	4,489	300,763	8.185	4.062	25.884	.0149
18	324	5,832	4.243	2.621	13.416	.0556	68	4,624	314,432	8.246	4.082	26.077	.0147
19	361	6,859	4.359	2.668	13.784	.0526	69	4,761	328,509	8.307	4.102	26.268	.0145
20	400	8,000	4.472	2.714	14.142	.0500	70	4,900	343,000	8.367	4.121	26.458	.0143
21	441	9,261	4.583	2.759	14.491	.0476	71	5,041	357,911	8.426	4.141	26.646	.0141
22	484	10,648	4.690	2.802	14.832	.0455	72	5,184	373,248	8.485	4.160	26.833	.0139
23	529	12,167	4.796	2.844	15.166	.0435	73	5,329	389,017	8.544	4.179	27.019	.0137
24	576	13,824	4.899	2.884	15.492	.0417	74	5,476	405,224	8.602	4.198	27.203	.0135
25	625	15,625	5.000	2.924	15.811	.0400	75	5,625	421,875	8.660	4.217	27.386	.0133
26	676	17,576	5.099	2.962	16.125	.0385	76	5,776	438,976	8.718	4.236	27.568	.0132
27	729	19,683	5.196	3.000	16.432	.0370	77	5,929	456,533	8.775	4.254	27.749	.0130
28	784	21,952	5.292	3.037	16.733	.0357	78	6,084	474,552	8.832	4.273	27.928	.0128
29	841	24,389	5.385	3.072	17.029	.0345	79	6,241	493,039	8.888	4.291	28.107	.0127
30	900	27,000	5.477	3.107	17.321	.0333	80	6,400	512,000	8.944	4.309	28.284	.0125
31	961	29,791	5.568	3.141	17.607	.0323	81	6,561	531,441	5.000	4.327	28.460	.0123
32	1,024	32,768	5.657	3.175	17.889	.0312	82	6,724	551,368	9.055	4.344	28.636	.0122
33	1,089	35,937	5.745	3.208	18.166	.0303	83	6,889	571,787	9.110	4.362	28.810	.0120
34	1,156	39,304	5.831	3.240	18.439	.0294	84	7,056	592,704	9.165	4.380	28.983	.0119
35	1,225	42,875	5.916	3.271	18.708	.0286	85	7,225	614,125	9.220	4.397	28.155	.0118
36	1,296	46,656	6.000	3.302	18.974	.0278	86	7,396	636,056	9.274	4.414	29.326	.0116
37	1,369	50,653	6.083	3.332	19.235	.0270	87	7,569	658,503	9.327	4.431	29.496	.0115
38	1,444	54,872	6.164	3.362	19.494	.0263	88	7,744	681,472	9.381	4.448	29.665	.0114
39	1,521	59,319	6.245	3.391	19.748	.0256	89	7,921	704,969	9.434	4.465	29.833	.0112
40	1,600	64,000	6.325	3.420	20.000	.0250	90	8,100	729,000	9.487	4.481	30.000	.0111
41	1,681	68,921	6.403	3.448	20.248	.0244	91	8,281	753,571	9.539	4.498	30.166	.0110
42	1,764	74,088	6.481	3.476	20.494	.0238	92	8,464	778,688	9.592	4.514	30.332	.0109
43	1,849	79,507	6.557	3.503	20.736	.0233	93	8,649	804,357	9.644	4.531	30.496	.0108
44	1,936	85,184	6.633	3.530	20.976	.0227	94	8,836	830,584	9.695	4.547	30.659	.0106
45	2,025	91,125	6.708	3.557	21.213	.0222	95	9,025	857,375	9.747	4.563	30.822	.0105
46	2,116	97,336	6.782	3.583	21.448	.0217	96	9,216	884,736	9.798	4.579	30.984	.0104
47	2,209	103,823	6.856	3.609	21.679	.0213	97	9,409	912,673	9.849	4.595	31.145	.0103
48	2,304	110,592	6.928	3.634	21.909	.0208	98	9,604	941,192	9.899	4.610	31.305	.0102
49	2,401	117,649	7.000	3.659	22.136	.0204	99	9,801	970,299	9.950	4.626	31.464	.0101
50	2,500	125,000	7.071	3.684	22.361	.0200	100	10,000	1,000,000	10.000	4.642	31.623	.0100

TABLE 2 **635**

TABLE 2 COMMON LOGARITHMS

x	0	1	2	3	4	5	6	7	8	9
1.0	.0000	.0043	.0086	.0128	.0170	.0212	.0253	.0294	.0334	.0374
1.1	.0414	.0453	.0492	.0531	.0569	.0607	.0645	.0682	.0719	.0755
1.2	.0792	.0828	.0864	.0899	.0934	.0969	.1004	.1038	.1072	.1106
1.3	.1139	.1173	.1206	.1239	.1271	.1303	.1335	.1367	.1399	.1430
1.4	.1461	.1492	.1523	.1553	.1584	.1614	.1644	.1673	.1703	.1732
1.5	.1761	.1790	.1818	.1847	.1875	.1903	.1931	.1959	.1987	.2014
1.6	.2041	.2068	.2095	.2122	.2148	.2175	.2201	.2227	.2253	.2279
1.7	.2304	.2330	.2355	.2380	.2405	.2430	.2455	.2480	.2504	.2529
1.8	.2553	.2577	.2601	.2625	.2648	.2672	.2695	.2718	.2742	.2765
1.9	.2788	.2810	.2833	.2856	.2878	.2900	.2923	.2945	.2967	.2989
2.0	.3010	.3032	.3054	.3075	.3096	.3118	.3139	.3160	.3181	.3201
2.1	.3222	.3243	.3263	.3284	.3304	.3324	.3345	.3365	.3385	.3404
2.2	.3424	.3444	.3464	.3483	.3502	.3522	.3541	.3560	.3579	.3598
2.3	.3617	.3636	.3655	.3674	.3692	.3711	.3729	.3747	.3766	.3784
2.4	.3802	.3820	.3838	.3856	.3874	.3892	.3909	.3927	.3945	.3962
2.5	.3979	.3997	.4014	.4031	.4048	.4065	.4082	.4099	.4116	.4133
2.6	.4150	.4166	.4183	.4200	.4216	.4232	.4249	.4265	.4281	.4298
2.7	.4314	.4330	.4346	.4362	.4378	.4393	.4409	.4425	.4440	.4456
2.8	.4472	.4487	.4502	.4518	.4533	.4548	.4564	.4579	.4594	.4609
2.9	.4624	.4639	.4654	.4669	.4683	.4698	.4713	.4728	.4742	.4757
3.0	.4771	.4786	.4800	.4814	.4829	.4843	.4857	.4871	.4886	.4900
3.1	.4914	.4928	.4942	.4955	.4969	.4983	.4997	.5011	.5024	.5038
3.2	.5051	.5065	.5079	.5092	.5105	.5119	.5132	.5145	.5159	.5172
3.3	.5185	.5198	.5211	.5224	.5237	.5250	.5263	.5276	.5289	.5307
3.4	.5315	.5328	.5340	.5353	.5366	.5378	.5391	.5403	.5416	.5428
3.5	.5441	.5453	.5465	.5478	.5490	.5502	.5514	.5527	.5539	.5551
3.6	.5563	.5575	.5587	.5599	.5611	.5623	.5635	.5647	.5658	.5670
3.7	.5682	.5694	.5705	.5717	.5729	.5740	.5752	.5763	.5775	.5786
3.8	.5798	.5809	.5821	.5832	.5843	.5855	.5866	.5877	.5888	.5899
3.9	.5911	.5922	.5933	.5944	.5955	.5966	.5977	.5988	.5999	.6010
4.0	.6021	.6031	.6042	.6053	.6064	.6075	.6085	.6096	.6107	.6117
4.1	.6128	.6138	.6149	.6160	.6170	.6180	.6191	.6201	.6212	.6222
4.2	.6232	.6243	.6253	.6263	.6274	.6284	.6294	.6304	.6314	.6325
4.3	.6335	.6345	.6355	.6365	.6375	.6385	.6395	.6405	.6415	.6425
4.4	.6435	.6444	.6454	.6464	.6474	.6484	.6493	.6503	.6513	.6522
4.5	.6532	.6542	.6551	.6561	.6571	.6580	.6590	.6599	.6609	.6618
4.6	.6628	.6637	.6646	.6656	.6665	.6675	.6684	.6693	.6702	.6712
4.7	.6721	.6730	.6739	.6749	.6758	.6767	.6776	.6785	.6794	.6803
4.8	.6812	.6821	.6830	.6839	.6848	.6857	.6866	.6875	.6884	.6893
4.9	.6902	.6911	.6920	.6928	.6937	.6946	.6955	.6964	.6972	.6981
5.0	.6990	.6998	.7007	.7016	.7024	.7033	.7042	.7050	.7059	.7067
5.1	.7076	.7084	.7093	.7101	.7110	.7118	.7126	.7135	.7143	.7152
5.2	.7160	.7168	.7177	.7185	.7193	.7202	.7210	.7218	.7226	.7235
5.3	.7243	.7251	.7259	.7267	.7275	.7284	.7292	.7300	.7308	.7316
5.4	.7324	.7332	.7340	.7348	.7356	.7364	.7372	.7380	.7388	.7396
x	0	1	2	3	4	5	6	7	8	9

(*Continued on next page*)

TABLE 2 COMMON LOGARITHMS (*continued*)

x	0	1	2	3	4	5	6	7	8	9
5.5	.7404	.7412	.7419	.7427	.7435	.7443	.7451	.7459	.7466	.7474
5.6	.7482	.7490	.7497	.7505	.7513	.7520	.7528	.7536	.7543	.7551
5.7	.7559	.7566	.7574	.7582	.7589	.7597	.7604	.7612	.7619	.7627
5.8	.7634	.7642	.7649	.7657	.7664	.7672	.7679	.7686	.7694	.7701
5.9	.7709	.7716	.7723	.7731	.7738	.7745	.7752	.7760	.7767	.7774
6.0	.7782	.7789	.7796	.7803	.7810	.7818	.7825	.7832	.7839	.7846
6.1	.7853	.7860	.7868	.7875	.7882	.7889	.7896	.7903	.7910	.7917
6.2	.7924	.7931	.7938	.7945	.7952	.7959	.7966	.7973	.7980	.7987
6.3	.7993	.8000	.8007	.8014	.8021	.8028	.8035	.8041	.8048	.8055
6.4	.8062	.8069	.8075	.8082	.8089	.8096	.8102	.8109	.8116	.8122
6.5	.8129	.8136	.8142	.8149	.8156	.8162	.8169	.8176	.8182	.8189
6.6	.8195	.8202	.8209	.8215	.8222	.8228	.8235	.8241	.8248	.8254
6.7	.8261	.8267	.8274	.8280	.8287	.8293	.8299	.8306	.8312	.8319
6.8	.8325	.8331	.8338	.8344	.8351	.8357	.8363	.8370	.8376	.8382
6.9	.8388	.8395	.8401	.8407	.8414	.8420	.8426	.8432	.8439	.8445
7.0	.8451	.8457	.8463	.8470	.8476	.8482	.8488	.8494	.8500	.8506
7.1	.8513	.8519	.8525	.8531	.8537	.8543	.8549	.8555	.8561	.8567
7.2	.8573	.8579	.8585	.8591	.8597	.8603	.8609	.8615	.8621	.8627
7.3	.8633	.8639	.8645	.8651	.8657	.8663	.8669	.8675	.8681	.8686
7.4	.8692	.8698	.8704	.8710	.8716	.8722	.8727	.8733	.8739	.8745
7.5	.8751	.8756	.8762	.8768	.8774	.8779	.8785	.8791	.8797	.8802
7.6	.8808	.8814	.8820	.8825	.8831	.8837	.8842	.8848	.8854	.8859
7.7	.8865	.8871	.8876	.8882	.8887	.8893	.8899	.8904	.8910	.8915
7.8	.8921	.8927	.8932	.8938	.8943	.8949	.8954	.8960	.8965	.8971
7.9	.8976	.8982	.8987	.8993	.8998	.9004	.9009	.9015	.9020	.9025
8.0	.9031	.9036	.9042	.9047	.9053	.9058	.9063	.9069	.9074	.9079
8.1	.9085	.9090	.9096	.9101	.9106	.9112	.9117	.9122	.9128	.9133
8.2	.9138	.9143	.9149	.9154	.9159	.9165	.9170	.9175	.9180	.9186
8.3	.9191	.9196	.9201	.9206	.9212	.9217	.9222	.9227	.9232	.9238
8.4	.9243	.9248	.9253	.9258	.9263	.9269	.9274	.9279	.9284	.9289
8.5	.9294	.9299	.9304	.9309	.9315	.9320	.9325	.9330	.9335	.9340
8.6	.9345	.9350	.9555	.9360	.9365	.9370	.9375	.9380	.9385	.9390
8.7	.9395	.9400	.9405	.9410	.9415	.9420	.9425	.9430	.9435	.9440
8.8	.9445	.9450	.9455	.9460	.9465	.9469	.9474	.9479	.9484	.9489
8.9	.9494	.9499	.9504	.9509	.9513	.9518	.9523	.9528	.9533	.9538
9.0	.9542	.9547	.9552	.9557	.9562	.9566	.9571	.9576	.9581	.9586
9.1	.9590	.9595	.9600	.9605	.9609	.9614	.9619	.9624	.9628	.9633
9.2	.9638	.9643	.9647	.9652	.9657	.9661	.9666	.9671	.9675	.9680
9.3	.9685	.9689	.9694	.9699	.9703	.9708	.9713	.9717	.9722	.9727
9.4	.9731	.9736	.9741	.9745	.9750	.9754	.9759	.9763	.9768	.9773
9.5	.9777	.9782	.9786	.9791	.9795	.9800	.9805	.9809	.9814	.9818
9.6	.9823	.9827	.9832	.9836	.9841	.9845	.9850	.9854	.9859	.9863
9.7	.9868	.9872	.9877	.9881	.9886	.9890	.9894	.9899	.9903	.9908
9.8	.9912	.9917	.9921	.9926	.9930	.9934	.9939	.9943	.9948	.9952
9.9	.9956	.9961	.9965	.9969	.9974	.9978	.9983	.9987	.9991	.9996
x	0	1	2	3	4	5	6	7	8	9

TABLE 3   **637**

TABLE 3 NATURAL LOGARITHMS (ln x)

x	0.00	0.01	0.02	0.03	0.04	0.05	0.06	0.07	0.08	0.09
1.0	0.0000	0.0100	0.0198	0.0296	0.0392	0.0488	0.0583	0.0677	0.0770	0.0862
1.1	0.0953	0.1044	0.1133	0.1222	0.1310	0.1398	0.1484	0.1570	0.1655	0.1740
1.2	0.1823	0.1906	0.1989	0.2070	0.2151	0.2231	0.2311	0.2390	0.2469	0.2546
1.3	0.2624	0.2700	0.2776	0.2852	0.2927	0.3001	0.3075	0.3148	0.3221	0.3293
1.4	0.3365	0.3436	0.3507	0.3577	0.3646	0.3716	0.3784	0.3853	0.3920	0.3988
1.5	0.4055	0.4121	0.4187	0.4253	0.4318	0.4383	0.4447	0.4511	0.4574	0.4637
1.6	0.4700	0.4762	0.4824	0.4886	0.4947	0.5008	0.5068	0.5128	0.5188	0.5247
1.7	0.5306	0.5365	0.5423	0.5481	0.5539	0.5596	0.5653	0.5710	0.5766	0.5822
1.8	0.5878	0.5933	0.5988	0.6043	0.6098	0.6152	0.6206	0.6259	0.6313	0.6366
1.9	0.6419	0.6471	0.6523	0.6575	0.6627	0.6678	0.6729	0.6780	0.6831	0.6881
2.0	0.6931	0.6981	0.7031	0.7080	0.7130	0.7178	0.7227	0.7275	0.7324	0.7372
2.1	0.7419	0.7467	0.7514	0.7561	0.7608	0.7655	0.7701	0.7747	0.7793	0.7839
2.2	0.7885	0.7930	0.7975	0.8020	0.8065	0.8109	0.8154	0.8198	0.8242	0.8286
2.3	0.8329	0.8372	0.8416	0.8459	0.8502	0.8544	0.8587	0.8629	0.8671	0.8713
2.4	0.8755	0.8796	0.8838	0.8879	0.8920	0.8961	0.9002	0.9042	0.9083	0.9123
2.5	0.9163	0.9203	0.9243	0.9282	0.9322	0.9361	0.9400	0.9439	0.9478	0.9517
2.6	0.9555	0.9594	0.9632	0.9670	0.9708	0.9746	0.9783	0.9821	0.9858	0.9895
2.7	0.9933	0.9969	1.0006	1.0043	1.0080	1.0116	1.0152	0.0188	1.0225	1.0260
2.8	1.0296	1.0332	1.0367	1.0403	1.0438	1.0473	1.0508	1.0543	1.0578	1.0613
2.9	1.0647	1.0682	1.0716	1.0750	1.0784	1.0818	1.0852	1.0886	1.0919	1.0953
3.0	1.0986	1.1019	1.1053	1.1086	1.1119	1.1151	1.1184	1.1217	1.1249	1.1282
3.1	1.1314	1.1346	1.1378	1.1410	1.1442	1.1474	1.1506	1.1537	1.1569	1.1600
3.2	1.1632	1.1663	1.1694	1.1725	1.1756	1.1787	1.1817	1.1848	1.1878	1.1909
3.3	1.1939	1.1970	1.2000	1.2030	1.2060	1.2090	1.2119	1.2149	1.2179	1.2208
3.4	1.2238	1.2267	1.2296	1.2326	1.2355	1.2384	1.2413	1.2442	1.2470	1.2499
3.5	1.2528	1.2556	1.2585	1.2613	1.2641	1.2669	1.2698	1.2726	1.2754	1.2782
3.6	1.2809	1.2837	1.2865	1.2892	1.2920	1.2947	1.2975	1.3002	1.3029	1.3056
3.7	1.3083	1.3110	1.3137	1.3164	1.3191	1.3218	1.3244	1.3271	1.3297	1.3324
3.8	1.3350	1.3376	1.3403	1.3429	1.3455	1.3481	1.3507	1.3533	1.3558	1.3584
3.9	1.3610	1.3635	1.3661	1.3686	1.3712	1.3737	1.3762	1.3788	1.3813	1.3838
4.0	1.3863	1.3888	1.3913	1.3938	1.3962	1.3987	1.4012	1.4036	1.4061	1.4085
4.1	1.4110	1.4134	1.4159	1.4183	1.4207	1.4231	1.4255	1.4279	1.4303	1.4327
4.2	1.4351	1.4375	1.4398	1.4422	1.4446	1.4469	1.4493	1.4516	1.4540	1.4563
4.3	1.4586	1.4609	1.4633	1.4656	1.4679	1.4702	1.4725	1.4748	1.4770	1.4793
4.4	1.4816	1.4839	1.4861	1.4884	1.4907	1.4929	1.4952	1.4974	1.4996	1.5019
4.5	1.5041	1.5063	1.5085	1.5107	1.5129	1.5151	1.5173	1.5195	1.5217	1.5239
4.6	1.5261	1.5282	1.5304	1.5326	1.5347	1.5369	1.5390	1.5412	1.5433	1.5454
4.7	1.5476	1.5497	1.5518	1.5539	1.5560	1.5581	1.5602	1.5623	1.5644	1.5665
4.8	1.5686	1.5707	1.5728	1.5748	1.5769	1.5790	1.5810	1.5831	1.5851	1.5872
4.9	1.5892	1.5913	1.5933	1.5953	1.5974	1.5994	1.6014	1.6034	1.6054	1.6074
5.0	1.6094	1.6114	1.6134	1.6154	1.6174	1.6194	1.6214	1.6233	1.6253	1.6273
5.1	1.6292	1.6312	1.6332	1.6351	1.6371	1.6390	1.6409	1.6429	1.6448	1.6467
5.2	1.6487	1.6506	1.6525	1.6544	1.6563	1.6582	1.6601	1.6620	1.6639	1.6658
5.3	1.6677	1.6696	1.6715	1.6734	1.6752	1.6771	1.6790	1.6808	1.6827	1.6845
5.4	1.6864	1.6882	1.6901	1.6919	1.6938	1.6956	1.6974	1.6993	1.7011	1.7029
5.5	1.7047	1.7066	1.7084	1.7102	1.7120	1.7138	1.7156	1.7174	1.7192	1.7210
5.6	1.7228	1.7246	1.7263	1.7281	1.7299	1.7317	1.7334	1.7352	1.7370	1.7387
5.7	1.7405	1.7422	1.7440	1.7457	1.7475	1.7492	1.7509	1.7527	1.7544	1.7561
5.8	1.7579	1.7596	1.7613	1.7630	1.7647	1.7664	1.7682	1.7699	1.7716	1.7733
5.9	1.7750	1.7766	1.7783	1.7800	1.7817	1.7834	1.7851	1.7867	1.7884	1.7901

(*Continued on next page*)

TABLE 3 NATURAL LOGARITHMS (*continued*)

Examples.

$$\ln 96{,}700 = \ln 9.67 + 4 \ln 10$$
$$= 2.2690 + 9.2103$$
$$= 11.4793.$$

$$\ln 0.00967 = \ln 9.67 - 3 \ln 10$$
$$= 2.2690 - 6.9078$$
$$= -4.6388.$$

x	0.00	0.01	0.02	0.03	0.04	0.05	0.06	0.07	0.08	0.09
6.0	1.7918	1.7934	1.7951	1.7967	1.7984	1.8001	1.8017	1.8034	1.8050	1.8066
6.1	1.8083	1.8099	1.8116	1.8132	1.8148	1.8165	1.8181	1.8197	1.8213	1.8229
6.2	1.8245	1.8262	1.8278	1.8294	1.8310	1.8326	1.8342	1.8358	1.8374	1.8390
6.3	1.8406	1.8421	1.8437	1.8453	1.8469	1.8485	1.8500	1.8516	1.8532	1.8547
6.4	1.8563	1.8579	1.8594	1.8610	1.8625	1.8641	1.8656	1.8672	1.8687	1.8703
6.5	1.8718	1.8733	1.8749	1.8764	1.8779	1.8795	1.8810	1.8825	1.8840	1.8856
6.6	1.8871	1.8886	1.8901	1.8916	1.8931	1.8946	1.8961	1.8976	1.8991	1.9006
6.7	1.9021	1.9036	1.9051	1.9066	1.9081	1.9095	1.9110	1.9125	1.9140	1.9155
6.8	1.9169	1.9184	1.9199	1.9213	1.9228	1.9242	1.9257	1.9272	1.9286	1.9301
6.9	1.9315	1.9330	1.9344	1.9359	1.9373	1.9387	1.9402	1.9416	1.9430	1.9445
7.0	1.9459	1.9473	1.9488	1.9502	1.9516	1.9530	1.9544	1.9559	1.9573	1.9587
7.1	1.9601	1.9615	1.9629	1.9643	1.9657	1.9671	1.9685	1.9699	1.9713	1.9727
7.2	1.9741	1.9755	1.9769	1.9782	1.9796	1.9810	1.9824	1.9838	1.9851	1.9865
7.3	1.9879	1.9892	1.9906	1.9920	1.9933	1.9947	1.9961	1.9974	1.9988	2.0001
7.4	2.0015	2.0028	2.0042	2.0055	2.0069	2.0082	2.0096	2.0109	2.0122	2.0136
7.5	2.0149	2.0162	2.0176	2.0189	2.0202	2.0215	2.0229	2.0242	2.0255	2.0268
7.6	2.0282	2.0295	2.0308	2.0321	2.0334	2.0347	2.0360	2.0373	2.0386	2.0399
7.7	2.0412	2.0425	2.0438	2.0451	2.0464	2.0477	2.0490	2.0503	2.0516	2.0528
7.8	2.0541	2.0554	2.0567	2.0580	2.0592	2.0605	2.0618	2.0631	2.0643	2.0665
7.9	2.0669	2.0681	2.0694	2.0707	2.0719	2.0732	2.0744	2.0757	2.0769	2.0782
8.0	2.0794	2.0807	2.0819	2.0832	2.0844	2.0857	2.0869	2.0882	2.0894	2.0906
8.1	2.0919	2.0931	2.0943	2.0956	2.0968	2.0980	2.0992	2.1005	2.1017	2.1029
8.2	2.1041	2.1054	2.1066	2.1078	2.1090	2.1102	2.1114	2.1126	2.1133	2.1150
8.3	2.1163	2.1175	2.1187	2.1199	2.1211	2.1223	2.1235	2.1247	2.1258	2.1270
8.4	2.1282	2.1294	2.1306	2.1318	2.1330	2.1342	2.1353	2.1365	2.1377	2.1389
8.5	2.1401	2.1412	2.1424	2.1436	2.1448	2.1459	2.1471	2.1483	2.1494	2.1506
8.6	2.1518	2.1529	2.1541	2.1552	2.1564	2.1576	2.1587	2.1599	2.1610	2.1622
8.7	2.1633	2.1645	2.1656	2.1668	2.1679	2.1691	2.1702	2.1713	2.1725	2.1736
8.8	2.1748	2.1759	2.1770	2.1782	2.1793	2.1804	2.1815	2.1827	2.1838	2.1849
8.9	2.1861	2.1872	2.1883	2.1894	2.1905	2.1917	2.1928	2.1939	2.1950	2.1961
9.0	2.1972	2.1983	2.1994	2.2006	2.2017	2.2028	2.2039	2.2050	2.2061	2.2072
9.1	2.2083	2.2094	2.2105	2.2116	2.2127	2.2138	2.2148	2.2159	2.2170	2.2181
9.2	2.2192	2.2203	2.2214	2.2225	2.2235	2.2246	2.2257	2.2268	2.2279	2.2289
9.3	2.2300	2.2311	2.2322	2.2332	2.2343	2.2354	2.2364	2.2375	2.2386	2.2396
9.4	2.2407	2.2418	2.2428	2.2439	2.2450	2.2460	2.2471	2.2481	2.2492	2.2502
9.5	2.2513	2.2523	2.2534	2.2544	2.2555	2.2565	2.2576	2.2586	2.2597	2.2607
9.6	2.2618	2.2628	2.2638	2.2649	2.2659	2.2670	2.2680	2.2690	2.2701	2.2711
9.7	2.2721	2.2732	2.2742	2.2752	2.2762	2.2773	2.2783	2.2793	2.2803	2.2814
9.8	2.2824	2.2834	2.2844	2.2854	2.2865	2.2875	2.2885	2.2895	2.2905	2.2915
9.9	2.2925	2.2935	2.2946	2.2956	2.2966	2.2976	2.2986	2.2996	2.3006	2.3016

$\ln 10 = 2.3026$	$7 \ln 10 = 16.1181$	
$2 \ln 10 = 4.6052$	$8 \ln 10 = 18.4207$	
$3 \ln 10 = 6.9078$	$9 \ln 10 = 20.7233$	
$4 \ln 10 = 9.2103$	$10 \ln 10 = 23.0259$	
$5 \ln 10 = 11.5129$	$11 \ln 10 = 25.3284$	
$6 \ln 10 = 13.8155$	$12 \ln 10 = 27.6310$	

Note: Adapted from *Functional Approach to Precalculus*, 2nd ed., Mustafa A. Munem and James P. Yizze (New York, NY: Worth Publishers, Inc., © 1974), pp. 500–501. Reproduced by permission of the publisher.

TABLE 4 **639**

TABLE 4 EXPONENTIAL FUNCTIONS

x	e^x	e^{-x}	x	e^x	e^{-x}	x	e^x	e^{-x}
0.00	1.0000	1.0000	0.55	1.7333	0.5769	3.6	36.598	0.0273
0.01	1.0101	0.9900	0.60	1.8221	0.5488	3.7	40.447	0.0247
0.02	1.0202	0.9802	0.65	1.9155	0.5220	3.8	44.701	0.0224
0.03	1.0305	0.9704	0.70	2.0138	0.4966	3.9	49.402	0.0202
0.04	1.0408	0.9608	0.75	2.1170	0.4724	4.0	54.598	0.0183
0.05	1.0513	0.9512	0.80	2.2255	0.4493	4.1	60.340	0.0166
0.06	1.0618	0.9418	0.85	2.3396	0.4274	4.2	66.686	0.0150
0.07	1.0725	0.9324	0.90	2.4596	0.4066	4.3	73.700	0.0136
0.08	1.0833	0.9231	0.95	2.5857	0.3867	4.4	81.451	0.0123
0.09	1.0942	0.9139	1.0	2.7183	0.3679	4.5	90.017	0.0111
0.10	1.1052	0.9048	1.1	3.0042	0.3329	4.6	99.484	0.0101
0.11	1.1163	0.8958	1.2	3.3201	0.3012	4.7	109.95	0.0091
0.12	1.1275	0.8869	1.3	3.6693	0.2725	4.8	121.51	0.0082
0.13	1.1388	0.8781	1.4	4.0552	0.2466	4.9	134.29	0.0074
0.14	1.1503	0.8694	1.5	4.4817	0.2231	5	148.41	0.0067
0.15	1.1618	0.8607	1.6	4.9530	0.2019	6	403.43	0.0025
0.16	1.1735	0.8521	1.7	5.4739	0.1827	7	1096.6	0.0009
0.17	1.1853	0.8437	1.8	6.0496	0.1653	8	2981.0	0.0003
0.18	1.1972	0.8353	1.9	6.6859	0.1496	9	8103.1	0.0001
0.19	1.2092	0.8270	2.0	7.3891	0.1353	10	22026	0.00005
0.20	1.2214	0.8187	2.1	8.1662	0.1225	11	59874	0.00002
0.21	1.2337	0.8106	2.2	9.0250	0.1108	12	162,754	0.000006
0.22	1.2461	0.8025	2.3	9.9742	0.1003	13	442,413	0.000002
0.23	1.2586	0.7945	2.4	11.023	0.0907	14	1,202,604	0.0000008
0.24	1.2712	0.7866	2.5	12.182	0.0821	15	3,269,017	0.0000003
0.25	1.2840	0.7788	2.6	13.464	0.0743			
0.26	1.2969	0.7711	2.7	14.880	0.0672			
0.27	1.3100	0.7634	2.8	16.445	0.0608			
0.28	1.3231	0.7558	2.9	18.174	0.0550			
0.29	1.3364	0.7483	3.0	20.086	0.0498			
0.30	1.3499	0.7408	3.1	22.198	0.0450			
0.35	1.4191	0.7047	3.2	24.533	0.0408			
0.40	1.4918	0.6703	3.3	27.113	0.0369			
0.45	1.5683	0.6376	3.4	29.964	0.0334			
0.50	1.6487	0.6065	3.5	33.115	0.0302			

TABLE 5 INTEGRATION FORMULAS

(Whenever $\ln X$ is used it is assumed that $X > 0$.)

1. $\int x^n \, dx = \dfrac{x^{n+1}}{n+1} + C, n \neq -1$

2. $\int \dfrac{dx}{x} = \ln x + C$

3. $\int u \, dv = uv - \int v \, du$

4. $\int e^x \, dx = e^x + C$

5. $\int e^{ax} \, dx = \dfrac{1}{a} \cdot e^{ax} + C$

6. $\int x e^{ax} \, dx = \dfrac{1}{a^2} \cdot e^{ax}(ax - 1) + C$

7. $\int x^n e^{ax} \, dx = \dfrac{x^n e^{ax}}{a} - \dfrac{n}{a} \int x^{n-1} e^{ax} \, dx$

8. $\int \ln x \, dx = x \ln x - x + C$

9. $\int (\ln x)^n \, dx = x(\ln x)^n - n \int (\ln x)^{n-1} \, dx, n \neq -1$

10. $\int x^n \ln x \, dx = x^{n+1} \left[\dfrac{\ln x}{n+1} - \dfrac{1}{(n+1)^2} \right] + C, n \neq -1$

11. $\int a^x \, dx = \dfrac{a^x}{\ln a} + C, a > 0, a \neq 1$

12. $\int \dfrac{1}{\sqrt{x^2 + a^2}} \, dx = \ln(x + \sqrt{x^2 + a^2}) + C$

13. $\int \dfrac{1}{\sqrt{x^2 - a^2}} \, dx = \ln(x + \sqrt{x^2 - a^2}) + C$

14. $\int \dfrac{1}{x^2 - a^2} \, dx = \dfrac{1}{2a} \ln\left(\dfrac{x - a}{x + a} \right) + C$

15. $\int \dfrac{1}{a^2 - x^2} \, dx = \dfrac{1}{2a} \ln\left(\dfrac{a + x}{a - x} \right) + C$

16. $\int \dfrac{1}{x\sqrt{a^2 + x^2}} \, dx = -\dfrac{1}{a} \ln\left(\dfrac{a + \sqrt{a^2 + x^2}}{x} \right) + C$

17. $\int \dfrac{1}{x\sqrt{a^2 - x^2}} \, dx = -\dfrac{1}{a} \ln\left(\dfrac{a + \sqrt{a^2 - x^2}}{x} \right) + C, 0 < x < a$

18. $\int \dfrac{x}{ax + b} \, dx = \dfrac{b}{a^2} + \dfrac{x}{a} - \dfrac{b}{a^2} \ln(ax + b) + C$

19. $\int \dfrac{x}{(ax + b)^2} \, dx = \dfrac{b}{a^2(ax + b)} + \dfrac{1}{a^2} \ln(ax + b) + C$

20. $\int \dfrac{1}{x(ax + b)} \, dx = \dfrac{1}{b} \ln\left(\dfrac{x}{ax + b} \right) + C$

21. $\int \dfrac{1}{x(ax + b)^2} \, dx = \dfrac{1}{b(ax + b)} + \dfrac{1}{b^2} \ln\left(\dfrac{x}{ax + b} \right) + C$

22. $\int \sqrt{x^2 \pm a^2} \, dx$

$\quad = \frac{1}{2}[x\sqrt{x^2 \pm a^2} \pm a^2 \ln(x + \sqrt{x^2 \pm a^2})] + C$

TABLE 6   **641**

TABLE 6 AREAS FOR A STANDARD NORMAL DISTRIBUTION

Entries in the table represent area under the
curve between $t = 0$ and a positive value of t.
Because of the symmetry of the curve, area under
the curve between $t = 0$ and a negative value of
t would be found in a like manner.

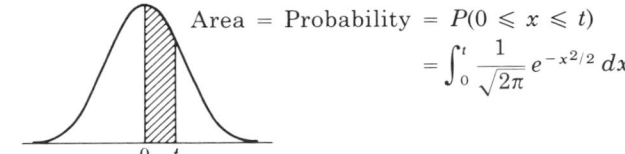

Area = Probability = $P(0 \leqslant x \leqslant t)$
$$= \int_0^t \frac{1}{\sqrt{2\pi}} e^{-x^2/2} \, dx$$

t	0.00	0.01	0.02	0.03	0.04	0.05	0.06	0.07	0.08	0.09
0.0	.0000	.0040	.0080	.0120	.0160	.0199	.0239	.0279	.0319	.0359
0.1	.0398	.0438	.0478	.0517	.0557	.0596	.0636	.0675	.0714	.0753
0.2	.0793	.0832	.0871	.0910	.0948	.0987	.1026	.1064	.1103	.1141
0.3	.1179	.1217	.1255	.1293	.1331	.1368	.1406	.1443	.1480	.1517
0.4	.1554	.1591	.1628	.1664	.1700	.1736	.1772	.1808	.1844	.1879
0.5	.1915	.1950	.1985	.2019	.2054	.2088	.2123	.2157	.2190	.2224
0.6	.2257	.2291	.2324	.2357	.2389	.2422	.2454	.2486	.2517	.2549
0.7	.2580	.2611	.2642	.2673	.2704	.2734	.2764	.2794	.2823	.2852
0.8	.2881	.2910	.2939	.2967	.2995	.3023	.3051	.3078	.3106	.3133
0.9	.3159	.3186	.3212	.3238	.3264	.3289	.3315	.3340	.3365	.3389
1.0	.3413	.3438	.3461	.3485	.3508	.3531	.3554	.3577	.3599	.3621
1.1	.3643	.3665	.3686	.3708	.3729	.3749	.3770	.3790	.3810	.3830
1.2	.3849	.3869	.3888	.3907	.3925	.3944	.3962	.3980	.3997	.4015
1.3	.4032	.4049	.4066	.4082	.4099	.4115	.4131	.4147	.4162	.4177
1.4	.4192	.4207	.4222	.4236	.4251	.4265	.4279	.4292	.4306	.4319
1.5	.4332	.4345	.4357	.4370	.4382	.4394	.4406	.4418	.4429	.4441
1.6	.4452	.4463	.4474	.4484	.4495	.4505	.4515	.4525	.4535	.4545
1.7	.4554	.4564	.4573	.4582	.4591	.4599	.4608	.4616	.4625	.4633
1.8	.4641	.4649	.4656	.4664	.4671	.4678	.4686	.4693	.4699	.4706
1.9	.4713	.4719	.4726	.4732	.4738	.4744	.4750	.4756	.4761	.4767
2.0	.4772	.4778	.4783	.4788	.4793	.4798	.4803	.4808	.4812	.4817
2.1	.4821	.4826	.4830	.4834	.4838	.4842	.4846	.4850	.4854	.4857
2.2	.4861	.4864	.4868	.4871	.4875	.4878	.4881	.4884	.4887	.4890
2.3	.4893	.4896	.4898	.4901	.4904	.4906	.4909	.4911	.4913	.4916
2.4	.4918	.4920	.4922	.4925	.4927	.4929	.4931	.4932	.4934	.4936
2.5	.4938	.4940	.4941	.4943	.4945	.4946	.4948	.4949	.4951	.4952
2.6	.4953	.4955	.4956	.4957	.4959	.4960	.4961	.4962	.4963	.4964
2.7	.4965	.4966	.4967	.4968	.4969	.4970	.4971	.4972	.4973	.4974
2.8	.4974	.4975	.4976	.4977	.4977	.4978	.4979	.4979	.4980	.4981
2.9	.4981	.4982	.4982	.4983	.4984	.4984	.4985	.4985	.4986	.4986
3.0	.4987	.4987	.4987	.4988	.4988	.4989	.4989	.4989	.4990	.4990

Answers

CHAPTER 1

MARGIN EXERCISES

1. $3 \cdot 3 \cdot 3 \cdot 3$, or 81   **2.** $(-3)(-3)$, or 9   **3.** $1.02 \times 1.02 \times 1.02$, or 1.061208 **4.** $\frac{1}{4} \cdot \frac{1}{4}$, or $\frac{1}{16}$ **5.** 1   **6.** $5t$ **7.** 1

8. m **9.** $\frac{1}{4}$ **10.** 1   **11.** $\frac{1}{2 \cdot 2 \cdot 2 \cdot 2}$, or $\frac{1}{16}$ **12.** $\frac{1}{10 \cdot 10}$, or $\frac{1}{100}$, or 0.01 **13.** 64   **14.** $\frac{1}{t^7}$ **15.** $\frac{1}{e^t}$ **16.** $\frac{1}{M}$

17. $\frac{1}{(x+1)^2}$ **18.** t^9 **19.** t^{-3} **20.** $50e^{-13}$ **21.** t^{-6} **22.** $24b^3$ **23.** x^4 **24.** x^{-4} **25.** 1   **26.** e^{2-k} **27.** e^{12}

28. e^2 **29.** x^{-12} **30.** e^4 **31.** e^{3x} **32.** $25x^6y^{10}$ **33.** $\frac{1}{256}x^{20}y^{24}z^{-8}$, or $\frac{x^{20}y^{24}}{256z^8}$ **34.** $2x + 14$ **35.** $P - Pi$

36. $x^2 + 3x - 28$ **37.** $a^2 - 2ab + b^2$ **38.** $a^2 - b^2$ **39.** $x^2 - 2xh + h^2$ **40.** $9x^2 + 6xt + t^2$ **41.** $25t^2 - m^2$

42. $P(1-i)$ **43.** $(x + 5y)^2$ **44.** $4(x + 5)(x + 2)$ **45.** $(5c - d)(5c + d)$ **46.** $h(3x^2 + 3xh + h^2)$ **47.** 1.01 **48.** $1144.90

49. $1302.26   **50.** $\frac{56}{9}$ **51.** $725   **52.** $0, -2, \frac{3}{2}$ **53.** $-4, 3$ **54.** $0, -1, 1$ **55.** $x < \frac{11}{5}$ **56.** $\frac{20}{17} \leq x$ **57.** More than 19,975 suits.   **58.** a) $(-1, 3)$ b) $(1, 4)$ **59.** a) $(-1, 4)$ b) $(-\frac{1}{4}, \frac{1}{4})$ **60.** a) $[-1, 4]$ b) $(-1, 4]$ c) $[-1, 4)$ d) $(-1, 4)$

61. a) $(-\sqrt{2}, \sqrt{2})$ b) $[0, 1)$ c) $(-6.7, -4.2]$ d) $[3, 7\frac{1}{2}]$ **62.** a) $(-\infty, 5]$ b) $(4, \infty)$ c) $(-\infty, 4.8)$ d) $(-\infty, 5]$ **63.** a) $[8, \infty)$
b) $(-\infty, -7)$ c) $(10, \infty)$ d) $(-\infty, -0.78]$

64.

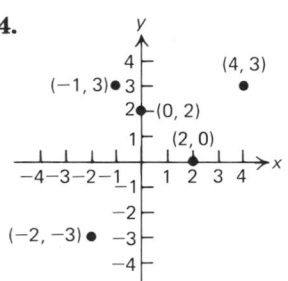

65. a) Yes b) No   **66.**

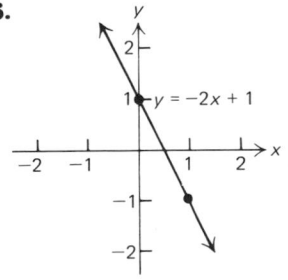

67.

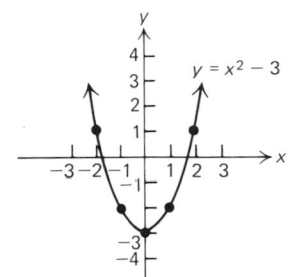

68.

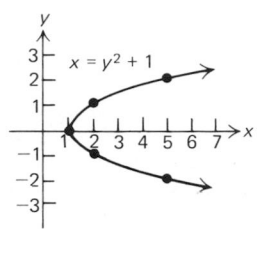

69.

Inputs	Outputs
5	$\frac{1}{5}$
$-\frac{2}{3}$	$-\frac{3}{2}$
$\frac{1}{4}$	4
$\frac{1}{a}$	a
k	$\frac{1}{k}$
$1 + t$	$\dfrac{1}{1 + t}$

70. $f(5) = \frac{1}{5}$, $f(-2) = -\frac{1}{2}$, $f(\frac{1}{4}) = 4$, $f(\frac{1}{a}) = a$, $f(k) = \frac{1}{k}$, $f(1 + t) = \dfrac{1}{1 + t}$, $f(x + h) = \dfrac{1}{x + h}$ **71.** $t(5) = 30$, $t(-5) = 20$,

$t(x + h) = x + h + x^2 + 2xh + h^2$ **72.** a) All real numbers except 3, since an input of 3 would result in division by 0;

b) $f(5) = \frac{1}{2}$, $f(4) = 1$, $f(2.5) = -2$, $f(x + h) = \dfrac{1}{x + h - 3}$. **73.** Same as margin exercise 66, only labeled $f(x) = -2x + 1$.

74. Same as margin exercise 67, only labeled $g(x) = x^2 - 3$. **75.** c, d. **76.** a) f b) g, h c) i

77. a) $[1, 3]$ b) $[-1, 1]$ **78.** a) Horizontal line through $(0, 3)$. b) Yes. **79.** a) Vertical line through $(1, 0)$. b) No.

80. a)

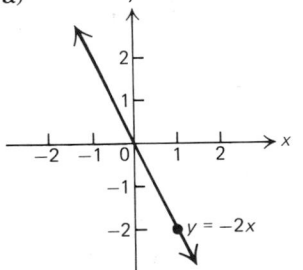

b) Yes c) -2 **81.** a) A, b) B, c) C, d) E, e) A, f) E. **82.** a) $T = \frac{1}{36}h$ b) 4.5

83. a)

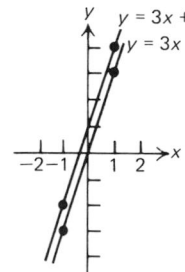

b) By moving it upward 1 unit **84.** $m = -\frac{2}{3}$, y-intercept: $(0, 2)$. **85.** $y = -4x + 1$

86. $y + 7 = -4(x - 2)$, or $y = -4x + 1$ **87.** 2 **88.** $\frac{1}{8}$ **89.** $-\frac{17}{2}$ **90.** 0 **91.** 0 **92.** No slope. **93.** a) Decreasing,
b) Increasing, c) Neither.

94. a)

y-axis labels: $20,000, $10,000; x-axis labels: 100, 300, 500

$C(x) = 30x + 15{,}000$

Variable costs = $30x$

Fixed costs = $15,000

b) $C(100) = \$18{,}000$, $C(400) = \$27{,}000$. **c)** $C(400) - C(100) = \$9000$.

95. a), b)

y-axis labels: $30,000, $20,000, $10,000, −$10,000, −$20,000; x-axis labels: 100, 300, 500

$R(x) = 90x$

$C(x) = 30x + 15{,}000$

$P(x) = R(x) - C(x)$
$= 60x - 15{,}000$

c) Break even is 250.

96. Yes   **97.** No   **98.** Consistent, dependent   **99.** Inconsistent, independent   **100.** Consistent, independent   **101. a)** $(1, 2)$ **b)** $(-2, 2)$ **102.** $(\frac{4}{3}, \frac{14}{3})$ **103.** $(4, -2)$ **104.** $(\frac{1}{2}, \frac{1}{3})$ **105.** $(-2, 5)$ **106.** $(-7, 10)$ **107.** No solution   **108.** An infinite number of solutions

109.

$y = x^2$

$y = -x^2$

110. a)

$y = x^2$ $y = (x - 6)^2$

b) By moving it to the right 6 units.   **111.** $\dfrac{-1 \pm \sqrt{22}}{3}$ **112.**

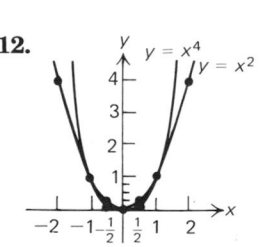

$y = x^4$ $y = x^2$

113. a) All real numbers except -4. **b)** All real numbers except $-5, 1$. **c)** All real numbers except 5.

114.

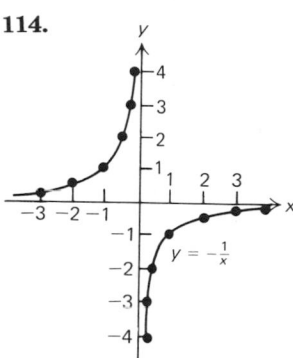

115.

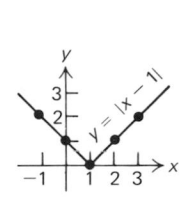

116.

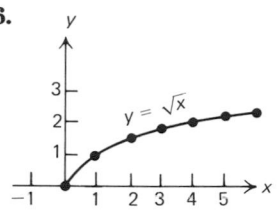

117. $[-\frac{3}{2}, \infty)$

118. a) $t^{3/4}$ b) $\dot{y}^{1/5}$ c) $x^{-2/5}$ d) $t^{-1/3}$ e) x^3 f) $x^{7/2}$ **119.** a) $\sqrt[7]{y}$ b) $\sqrt{x^3}$ c) $\dfrac{1}{\sqrt{t^3}}$ d) $\dfrac{1}{\sqrt{b}}$ **120.** 32   **121.** 9

122. (2, $9)   **123.** 3.77744 ≈ 3:46.6   **124.** 1984   **125.** a) $T = 0.5x - 946$ b) 46.5¢; 1.5¢ more than in (b) of Example 1.

126. a) $A = \frac{131}{300}x^2 - 39\frac{7}{10}x + 1039\frac{1}{3}$, or $A = 0.437x^2 - 39.7x + 1039.333$ b) 516

EXERCISE SET 1.1, p. 8

1. $5 \cdot 5 \cdot 5$, or 125   **3.** $(-7)(-7)$, or 49   **5.** 1.0201   **7.** $\frac{1}{16}$ **9.** 1   **11.** t **13.** 1   **15.** $\dfrac{1}{3^2}$, or $\dfrac{1}{9}$ **17.** 8   **19.** 0.1

21. $\dfrac{1}{e^b}$ **23.** $\dfrac{1}{b}$ **25.** x^5 **27.** x^{-6}, or $\dfrac{1}{x^6}$ **29.** $35x^5$ **31.** x^4 **33.** 1   **35.** x^3 **37.** x^{-3}, or $\dfrac{1}{x^3}$ **39.** 1   **41.** e^{t-4}

43. t^{14} **45.** t^2 **47.** t^{-6}, or $\dfrac{1}{t^6}$ **49.** e^{4x} **51.** $8x^6y^{12}$ **53.** $\frac{1}{81}x^8y^{20}z^{-16}$, or $\dfrac{x^8y^{20}}{81z^{16}}$ **55.** $9x^{-16}y^{14}z^4$, or $\dfrac{9y^{14}z^4}{x^{16}}$ **57.** $5x - 35$

59. $x - xt$ **61.** $x^2 - 7x + 10$ **63.** $x^2 + 3x - 10$ **65.** $2x^2 + 3x - 5$ **67.** $a^2 - 4$ **69.** $25x^2 - 4$ **71.** $a^2 - 2ah + h^2$

73. $25x^2 + 10xt + t^2$ **75.** $5x^5 + 30x^3 + 45x$ **77.** $a^3 + 3a^2b + 3ab^2 + b^3$ **79.** $x^3 - 15x^2 + 75x - 125$ **81.** $x(1 - t)$

83. $(x + 3y)^2$ **85.** $(x - 5)(x + 3)$ **87.** $(x - 5)(x + 4)$ **89.** $(7x - t)(7x + t)$ **91.** $4(3t - 2m)(3t + 2m)$

93. $ab(a + 4b)(a - 4b)$ **95.** $(a^4 + b^4)(a^2 + b^2)(a + b)(a - b)$ **97.** $10x(a + 2b)(a - 2b)$ **99.** $2(1 + 4x^2)(1 + 2x)(1 - 2x)$

101. a) 0.81 b) 0.0801 c) 0.008001   **103.** a) 1.261 b) 0.120601 c) 0.012006001   **105.** a) $1080 b) $1081.60

c) $1082.43 d) $1083.278 assuming 365 days in a year. e) $1083.283   **107.** $444.96

EXERCISE SET 1.2, p. 17

1. $\frac{7}{4}$ **3.** -8 **5.** 120   **7.** 200   **9.** 480 lb   **11.** $650   **13.** 810,000   **15.** $0, -3, \frac{4}{5}$ **17.** 0, 2   **19.** 0, 3   **21.** 0, 7

23. $0, \frac{1}{3}, -\frac{1}{3}$ **25.** 1   **27.** $-\frac{4}{5} \leq x$ **29.** $x > -\frac{1}{12}$ **31.** $x > -\frac{4}{7}$ **33.** $x \leq -3$ **35.** $x > \frac{2}{3}$ **37.** $x < -\frac{2}{5}$ **39.** $2 < x < 4$

41. $\frac{3}{2} \leq x \leq \frac{11}{2}$ **43.** $-1 \leq x \leq \frac{14}{5}$ **45.** More than 7000 units.   **47.** $60\% \leq x < 100\%$ **49.** $(0, 5)$ **51.** $[-9, -4]$

53. $[x, x + h]$ **55.** (p, ∞) **57.** $[-3, 3]$ **59.** $[-14, -11)$ **61.** $(-\infty, -4]$

EXERCISE SET 1.3, p. 30

1. a)

Inputs	Outputs
4.1	11.2
4.01	11.02
4.001	11.002
4	11

b) $f(5) = 13$, $f(-1) = 1$, $f(k) = 2k + 3$, $f(1 + t) = 2t + 5$, $f(x + h) = 2x + 2h + 3$

3. $g(-1) = -2$, $g(0) = -3$, $g(1) = -2$, $g(5) = 22$, $g(u) = u^2 - 3$, $g(a + h) = a^2 + 2ah + h^2 - 3$, $g(1 - h) = h^2 - 2h - 2$ **5.** a) $f(4) = 1$, $f(-2) = 25$, $f(0) = 9$, $f(a) = a^2 - 6a + 9$, $f(t + 1) = t^2 - 4t + 4$, $f(t + 3) = t^2$, $f(x + h) = x^2 + 2xh + h^2 - 6x - 6h + 9$ b) Take an input, square it, subtract 6 times the input, add 9.

7.

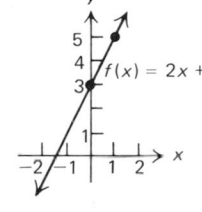

9.

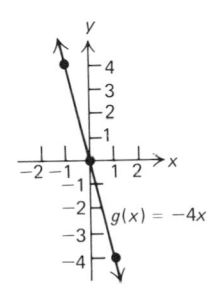

11.

13.

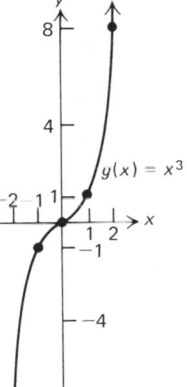

15. Yes   **17.** Yes   **19.** No   **21.** No   **23.** a)

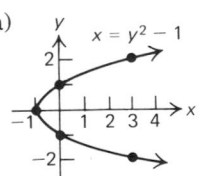

b) No

25. $f(x + h) = x^2 + 2xh + h^2 - 3x - 3h$ **27.** $R(10) = \$70$, $R(100) = \$250$ **29.** Increasing   **31.** Decreasing
33. Neither   **35.** Increasing on $[-3, -1]$; decreasing on $[-1, 1]$.   **37.** $y = 5$; a function.   **39.** $y = \pm\sqrt{x}$; not a function.

EXERCISE SET 1.4, p. 44

1. Horizontal line through $(0, -4)$　**3.** Vertical line through $(4.5, 0)$

5.
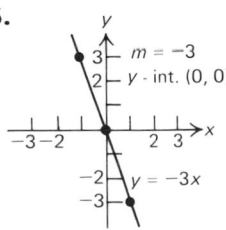

7.
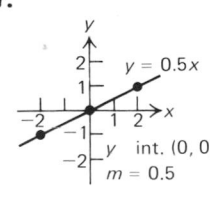

9.
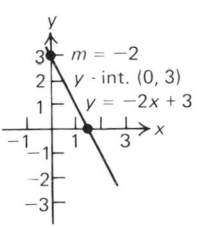

11.
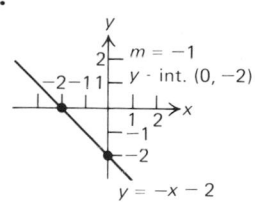

13. $m = -2$, y-int.: $(0, 2)$　**15.** $m = -1$, y-int.: $(0, -\frac{5}{2})$　**17.** $y + 5 = -5(x - 1)$, or $y = -5x$　**19.** $y - 3 = -2(x - 2)$, or $y = -2x + 7$　**21.** $y = \frac{1}{2}x - 6$　**23.** $y = 3$　**25.** $\frac{3}{2}$　**27.** $\frac{1}{2}$　**29.** No slope　**31.** 0　**33.** 3　**35.** 2　**37.** $y - 1 = \frac{3}{2}(x + 2)$, or $y + 2 = \frac{3}{2}(x + 4)$, or $y = \frac{3}{2}x + 4$　**39.** $y + 4 = \frac{1}{2}(x - 2)$, or $y = \frac{1}{2}x - 5$　**41.** $x = 3$　**43.** $y = 3$　**45.** $y = 3x$
47. $y = 2x + 3$　**49.** Neither　**51.** Decreasing　**53.** Increasing　**55.** a) $R = 4.17T$　b) $R = 25.02$
57. a) $B = 0.025W$　b) $B = 2.5\%W$. The weight of the brain is 2.5% of the body weight.　c) 3 lb.
59. a) $A = P + 8\%P = P + 0.08P = 1.08P$　b) $108　c) $240
61. a) $D(0°) = 115$ ft, $D(-20°) = 75$ ft, $D(10°) = 135$ ft, $D(32°) = 179$ ft

b)
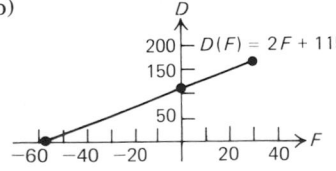

c) Temperatures below $-57.5°$ would yield a negative stopping distance, which has no meaning here. For temperatures above $32°$ there would be no ice.

63. a) $A(0) = 2$, $A(1) = 3.1$, $A(4) = 6.4$, $A(10) = 13$　b) Straight line through $(0, 2)$ and $(10, 13)$　c) The area is measured only from the time the organism is released. Thus only nonnegative values of t would be used as inputs.
65. a) $C(x) = 20x + 100{,}000$　b) $R(x) = 45x$　c) $P(x) = R(x) - C(x) = 25x - 100{,}000$　d) $3,650,000, a profit.　e) 4000
67. a) 200.69 cm　b) 195.23 cm

EXERCISE SET 1.5, p. 51

1. No　**3.** Consistent, independent　**5.** Consistent, dependent　**7.** Inconsistent, independent　**9.** $(2, -1)$

11. No solution　**13.** $(-5, 4)$

EXERCISE SET 1.6, pp. 55–56

1. $(2, -3)$ **3.** $(2, -2)$ **5.** $(6, 2)$ **7.** $(\frac{9}{19}, \frac{51}{38})$ **9.** $(\frac{3}{2}, \frac{5}{2})$ **11.** $(-\frac{1}{3}, -4)$ **13.** An infinite number of solutions

15. No solution   **17.** 13 and 16   **19.** 5 pairs of cloth gloves and 15 pairs of pigskin gloves   **21.** 42 L of 2%,

18 L of 6%   **23.** \$2000 at 12%, \$2800 at 13%   **25.** 20 km/h, 3 km/h   **27.** $\left(\dfrac{\pi + 3\sqrt{2}}{\pi^2 + 2}, \dfrac{3\pi - \sqrt{2}}{\pi^2 + 2} \right)$ **29.** $(0.924, -0.833)$

EXERCISE SET 1.7, p. 67

1.
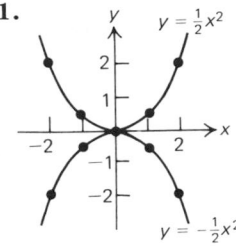

3. See Margin Exercise 109 for the graph of $y = x^2$. Move it to the right 1 unit to get the graph of $y = (x - 1)^2$.

5. See Margin Exercise 109 for $y = x^2$. Move it to the left 1 unit to get $y = (x + 1)^2$.

7.
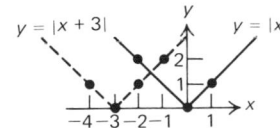

9. See Exercise Set 1.3, Ex. 13 for $y = x^3$. Move it up 1 unit for $y = x^3 + 1$.

11. See Margin Exercise 116 for $y = \sqrt{x}$. Move it to the left 1 unit for $y = \sqrt{x + 1}$.

13.
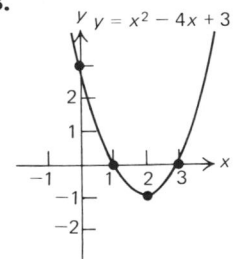

15.
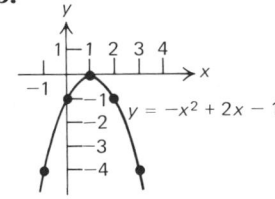

17.

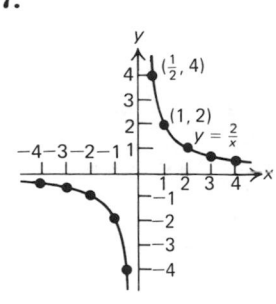

19.

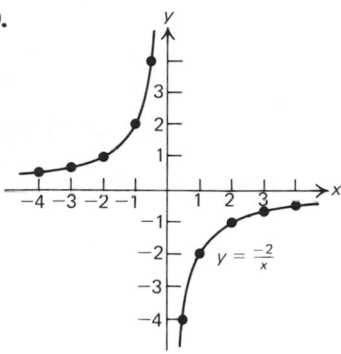

21.

x	-2	-1	$-\frac{1}{2}$	$\frac{1}{2}$	1	2
y	$\frac{1}{4}$	1	4	4	1	$\frac{1}{4}$

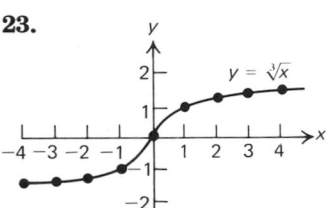

23.

25. $1 \pm \sqrt{3}$ **27.** $-3 \pm \sqrt{10}$ **29.** $\dfrac{1 \pm \sqrt{2}}{2}$ **31.** $\dfrac{-4 \pm \sqrt{10}}{3}$ **33.** $x^{3/2}$ **35.** $a^{3/5}$ **37.** $t^{1/7}$

39. $t^{-4/3}$ **41.** $t^{-1/2}$ **43.** $(x^2 + 7)^{-1/2}$ **45.** $\sqrt[5]{x}$ **47.** $\sqrt[3]{y^2}$ **49.** $\dfrac{1}{\sqrt[5]{t^2}}$ **51.** $\dfrac{1}{\sqrt[3]{b}}$ **53.** $\dfrac{1}{\sqrt[6]{e^{17}}}$ **55.** $\dfrac{1}{\sqrt{x^2 - 3}}$ **57.** 27   **59.** 16

61. 8   **63.** All real numbers except 5.   **65.** All real numbers except 2, 3.   **67.** $[-\frac{4}{5}, \infty)$ **69.** (2, \$4)   **71.** (1, \$4)

73. (2, \$4)   **75.**

W	0	10	20	30	40	50	100	150
T	0	20	51	86	126	168	417	709

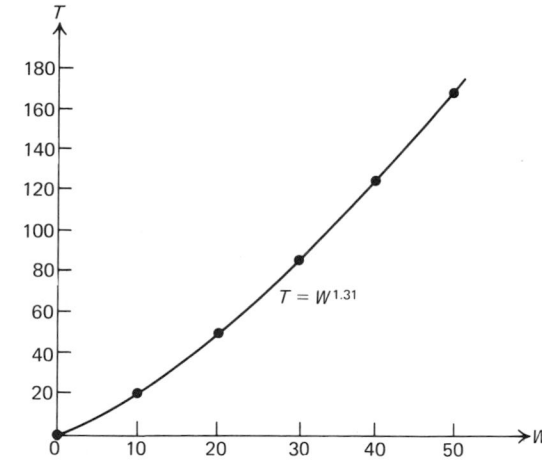

EXERCISE SET 1.8, p. 75

1. b) Linear c) $y = 32x + 9998$ d) \$10,126; \$10,318 e) $C = 29.5x + 10,000$ f) \$10,118; \$10,295   **3.** b) Quadratic c) $D = 0.9875x^2 - 121.5x + 3756.25$ d) 21.25   **5.** b) Constant, linear c) $S = -20x + 100,330$ d) Let $S = b$, where b is average of sales totals: \$100,296.25.   **7.** b) Quadratic c) $y = 104.5x^2 - 1501.5x + 6016$ d) 1682, 769, 1451

CHAPTER 1, TEST, p. 77

1. $\dfrac{1}{e^k}$ **2.** e^{-13} or $\dfrac{1}{e^{13}}$ **3.** $x^2 + 2xh + h^2$ **4.** $(5x - t)(5x + t)$ **5.** 920 **6.** $x > -4$ **7. a)** $f(-3) = 5$

b) $x^2 + 2xh + h^2 - 4$ **8.** $m = -3$; y-int.: $(0, 2)$ **9.** $y + 5 = \frac{1}{4}(x - 8)$, or $y = \frac{1}{4}x - 7$ **10.** $m = 6$ **11.** $F = \frac{2}{3}W$

12. a) $C(x) = 0.5x + 10,000$ **b)** $R(x) = 1.3x$ **c)** $P(x) = R(x) - C(x) = 0.8x - 10,000$ **d)** $12,500$ **13.** Decreasing

14. $(3, \$16)$ **15.**

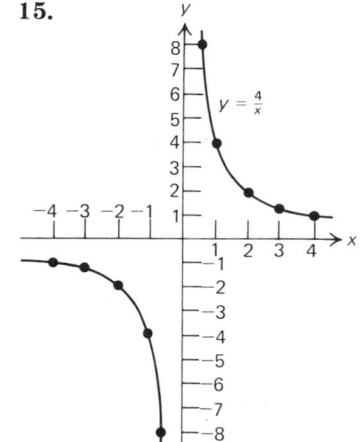

$y = \dfrac{4}{x}$

16. $t^{-1/2}$ **17.** $\dfrac{1}{\sqrt[5]{t^3}}$ **18.** All real numbers except $2, -7$.   **19.** $[-2, \infty)$

20. $y = 4x - 1$ **21.** $y = -4.5x^2 + 17.5x - 8$ **22.** $[c, d)$

23. Inconsistent, independent   **24.** Yes

25. $(3, -1)$ **26.** $(4, 1)$ **27.** $(\frac{3}{2}, \frac{5}{2})$

28. An infinite number of solutions

CHAPTER 2

MARGIN EXERCISES

1. $2x_1 - 6x_2 = \frac{1}{4}$
$-3x_1 + x_2 = 8$

2.
x_1	x_2	1
2	5	-24
5	-2	-2

3. $(-2, -4)$ **4.** $(-\frac{1}{3}, \frac{1}{2})$

5. $x_1 = 2$, $x_2 = \frac{1}{2}$, $x_3 = -2$, or $(2, \frac{1}{2}, -2)$ **6.** 1800 @ 7%, 1900 @ 9% **7.** No solution   **8.** $x_1 = \frac{3}{2} + \frac{1}{2}x_2$, $x_2 =$ any number; $x_2 = 0$, $x_1 = \frac{3}{2}$; $x_2 = 1$, $x_1 = 2$; $x_2 = -4$, $x_1 = -\frac{1}{2}$

9. a)
x_1	x_2	x_3	x_4	1
1	4	0	0	-2
0	0	1	0	8
0	0	0	1	-3

b) $x_1 = -2 - 4x_2$, $x_2 =$ any number, $x_3 = 8$, $x_4 = -3$ **c)** $x_1 = -2$, $x_2 = 0$, $x_3 = 8$, $x_4 = -3$; $x_1 = -6$, $x_2 = 1$, $x_3 = 8$, $x_4 = -3$; $x_1 = 6$, $x_2 = -2$, $x_3 = 8$, $x_4 = -3$

10. No solution   **11.** $x_1 = -10$, $x_2 = \frac{11}{3}$, $x_3 = \frac{1}{6}$, or $(-10, \frac{11}{3}, \frac{1}{6})$ **12.** 2×3 **13.** 1×3 **14.** 3×1 **15.** 1×1

16. 2×2 **17. a)** $a_{12} = 0$ **b)** $a_{22} = -6$ **c)** $a_{21} = 4$ **d)** $a_{32} = -2$ **18. a)** No **b)** No   **19.** $a = -6$, $b = \frac{1}{2}$

20. Row vectors are B and C, Column vectors are A and D **21.** $A^T = \begin{bmatrix} -8 & -4 & 6 \\ 1 & 0 & 7 \\ -2 & -1 & 8 \end{bmatrix}$, $B^T = \begin{bmatrix} -7 \\ 9 \\ 10 \\ \frac{1}{4} \end{bmatrix}$,

$C^T = [-20 \quad 41]$, $D^T = \begin{bmatrix} -4 & 1 & 0 \\ 5 & 0 & 1 \end{bmatrix}$ **22.** a) $\begin{bmatrix} -3 & -6 \\ 9 & 0 \end{bmatrix}$ b) Same as (a) c) Yes   **23.** $\begin{bmatrix} 0 & 15 \\ 1 & 12 \\ 1 & \frac{1}{2} \end{bmatrix}$

24. a) A b) A c) Yes   **25.** a) $\begin{bmatrix} 36 & 6 & 0 \\ 12 & 6 & -30 \\ -18 & 54 & 3 \end{bmatrix}$ b) $\begin{bmatrix} -6 & -1 & 0 \\ -2 & -1 & 5 \\ 3 & -9 & -\frac{1}{2} \end{bmatrix}$ c) O d) O e) $\begin{bmatrix} -\frac{1}{5} & -\frac{1}{30} & 0 \\ -\frac{1}{15} & -\frac{1}{30} & \frac{1}{6} \\ \frac{1}{10} & -\frac{3}{10} & -\frac{1}{60} \end{bmatrix}$

f) $\begin{bmatrix} 6t & t & 0 \\ 2t & t & -5t \\ -3t & 9t & \frac{1}{2}t \end{bmatrix}$ **26.** a) $\begin{bmatrix} 13 & 2 \\ 5 & -8 \end{bmatrix}$ b) $\begin{bmatrix} -13 & -2 \\ -5 & 8 \end{bmatrix}$ c) No   **27.** $\sum_{i=1}^{6} i^2$ **28.** $\sum_{i=1}^{4} t^i$ **29.** $\sum_{i=1}^{38} p_i$ **30.** $2 + 2^2 + 2^3$

31. $p_1 q_1 + p_2 q_2 + \cdots + p_{20} q_{20}$ **32.** $t + 2t^2 + 3t^3 + 4t^4 + 5t^5$ **33.** $[4a + 5b]$, or $4a + 5b$ **34.** $[7]$, or 7

35. $[-36]$, or -36 **36.** $\begin{bmatrix} 11 & 2 \\ 7 & -13 \end{bmatrix}\begin{bmatrix} x_1 \\ x_2 \end{bmatrix} = \begin{bmatrix} -1 \\ 8 \end{bmatrix}$ **37.** $\begin{aligned} 3y_1 - 7y_2 &= 4 \\ -2y_1 + y_2 &= 5 \end{aligned}$

38. a) $[5c + 6d \quad 5g + 6h]$ b) $[-3 \quad -24]$ **39.** a) $\begin{bmatrix} 5c + 6d & 5g + 6h \\ 3c - d & 3g - h \end{bmatrix}$ b) $\begin{bmatrix} -3 & -24 \\ -11 & 4 \end{bmatrix}$

40. a) $AB = \begin{bmatrix} -23 & 16 & 39 \\ 6 & 4 & 14 \end{bmatrix}$, BA not possible b) $AB = [31]$, or 31; $BA = \begin{bmatrix} 12 & 18 & -6 & 6 \\ 16 & 24 & -8 & 8 \\ 0 & 0 & 0 & 0 \\ -10 & -15 & 5 & -5 \end{bmatrix}$

41. a) $AB = \begin{bmatrix} -32 & -3 \\ -20 & -1 \end{bmatrix}$ b) $BA = \begin{bmatrix} -1 & 5 \\ 12 & -32 \end{bmatrix}$ c) No d) No, we have a counterexample with AB and BA

42. a) $(AB)C = A(BC) = \begin{bmatrix} -43 & -19 \\ 99 & 39 \end{bmatrix}$ b) $A(B + C) = AB + AC = \begin{bmatrix} 5 & 1 \\ 30 & 6 \end{bmatrix}$

43. a) $AI = A$ b) $IA = A$ c) Same $= A$ d) $IX = X$ e) XI cannot be found f) Same   **44.** $A^{-1}A = AA^{-1} = I$; that is, A and A^{-1} commute and their product is the identity.

45. a) $\begin{bmatrix} 3 & 5 \\ 1 & 2 \end{bmatrix}\begin{bmatrix} x_1 \\ x_2 \end{bmatrix} = \begin{bmatrix} -1 \\ 4 \end{bmatrix}$ b) $A = \begin{bmatrix} 3 & 5 \\ 1 & 2 \end{bmatrix}$ c) $x_1 = -22$, $x_2 = 13$

46. a) $[X \quad Y] = \begin{bmatrix} x_1 & y_1 \\ x_2 & y_2 \\ x_3 & y_3 \end{bmatrix}$ b) $\begin{bmatrix} X \\ Y \end{bmatrix} = \begin{bmatrix} x_1 \\ x_2 \\ x_3 \\ y_1 \\ y_2 \\ y_3 \end{bmatrix}$ **47.** $[A \quad I] = \begin{bmatrix} -2 & 3 & 1 & 0 \\ 4 & 5 & 0 & 1 \end{bmatrix}$, $\begin{bmatrix} A \\ I \end{bmatrix} = \begin{bmatrix} -2 & 3 \\ 4 & 5 \\ 1 & 0 \\ 0 & 1 \end{bmatrix}$

48. a) $[A \quad I]^T = \begin{bmatrix} 5 & -2 \\ 7 & 0 \\ \hdashline 1 & 0 \\ 0 & 1 \end{bmatrix}$ b) $\begin{bmatrix} A \\ I \end{bmatrix}^T = \begin{bmatrix} 5 & -2 & \vdots & 1 & 0 \\ 7 & 0 & \vdots & 0 & 1 \end{bmatrix}$ **49.** $A^{-1} = \begin{bmatrix} \frac{5}{22} & \frac{3}{22} \\ -\frac{2}{11} & \frac{1}{11} \end{bmatrix}$

EXERCISE SET 2.1, pp. 88–89

1. $(-3, 2)$ **3.** $(7, 3)$ **5.** $(\frac{5}{2}, -1)$ **7.** $(-3, 5, 7)$ **9.** $(0, 2, 1)$ **11.** $(2, \frac{1}{2}, -2)$ **13.** $(4, 5, 6, 7)$ **15.** \$4100 @ 7%,
\$4700 @ 8% **17.** \$30,000 @ 5%, \$40,000 @ 6% **19.** 8 white, 22 yellow **21.** 150 lb soybean meal; 200 lb corn meal
23. \$400 @ 7%, \$500 @ 8%, \$1600 @ 9% **25.** $y = 2x^2 + 3x - 1$ **27.** a) $y = 2500x^2 - 6500x + 5000$ b) \$19,000

EXERCISE SET 2.2, p. 95

1. $x_1 = 2 + 3x_2$, $x_2 =$ any number **3.** No solution **5.** $x_1 = 1 - x_3$, $x_2 = -x_3$, $x_3 =$ any number **7.** $x_1 = \frac{1}{5}x_3$,
$x_2 = -\frac{7}{5}x_3$, $x_3 =$ any number **9.** $x_1 = -\frac{7}{2}x_3$, $x_2 = -\frac{19}{2}x_3$, $x_3 =$ any number **11.** $x_1 = 1 + 3x_3$, $x_2 = 4 - 5x_3$,
$x_3 =$ any number, $x_4 = -2$ **13.** No solution **15.** $x_1 = 19 - 2x_3 - 16x_4$, $x_2 = -6 + 3x_3 + 4x_4$, $x_3 =$ any number,
$x_4 =$ any number **17.** $x_1 = -1 + 4x_3$, $x_2 = 4 + 2x_3$, $x_3 =$ any number **19.** $x_1 = 6 - 8x_3 + 3x_4$, $x_2 = 4 - 4x_3 - 2x_4$,
$x_3 =$ any number, $x_4 =$ any number, $x_5 = -5$

EXERCISE SET 2.3, pp. 100–101

1. 2×2 **3.** 2×3 **5.** $\begin{bmatrix} -1 & 3 \\ 2 & 1 \end{bmatrix}$ **7.** $\begin{bmatrix} -1 & 6 & -7 \\ 4 & 2 & 0 \end{bmatrix}$ **9.** $\begin{bmatrix} 3 & 9 \\ 12 & 6 \end{bmatrix}$ **11.** $\begin{bmatrix} 5 & 10 & 15 \\ -15 & -10 & -5 \end{bmatrix}$ **13.** $\begin{bmatrix} 3 & 6 \\ 6 & 3 \end{bmatrix}$

15. $\begin{bmatrix} -1 & -10 & 1 \\ 2 & 2 & 2 \end{bmatrix}$ **17.** Not possible **19.** $\begin{bmatrix} -k & -2k & -3k \\ 3k & 2k & k \end{bmatrix}$ **21.** A **23.** $\begin{bmatrix} 1 & 4 \\ 3 & 2 \end{bmatrix}$ **25.** $\begin{bmatrix} -1 & 3 \\ -2 & 2 \\ -3 & 1 \end{bmatrix}$ **27.** $\begin{bmatrix} -1 & 2 \\ 3 & 1 \end{bmatrix}$

29. $a_{11} = -4$, $a_{12} = 5$, $a_{31} = 1$, $a_{22} = 9$, $a_{32} = 3$, $a_{21} = 0$ **31.** $X^T = [x_1 \quad x_2 \quad x_3 \quad x_4]$

EXERCISE SET 2.4, pp. 112–114

1. $AB = [-10]$, or -10; $BA = \begin{bmatrix} -6 & 3 \\ 8 & -4 \end{bmatrix}$ **3.** $AB = [17]$, or 17; $BA = \begin{bmatrix} 18 & 0 & -36 \\ -10 & 0 & 20 \\ \frac{1}{2} & 0 & -1 \end{bmatrix}$

5. $AB = \begin{bmatrix} 7 & -6 & 1 \\ -15 & 12 & 3 \\ -2 & -1 & 8 \end{bmatrix}$, $BA = \begin{bmatrix} 9 & 11 & -10 \\ 3 & 4 & 4 \\ -3 & -1 & 14 \end{bmatrix}$ **7.** $AB = [35 \quad -35]$, BA not possible **9.** $AB = 0$

11. $\begin{bmatrix} 11 & 3 \\ 7 & 2 \end{bmatrix}\begin{bmatrix} x_1 \\ x_2 \end{bmatrix} = \begin{bmatrix} -4 \\ 5 \end{bmatrix}$ **13.** $\begin{bmatrix} 3 & 1 & 0 \\ 1 & -1 & 2 \\ 1 & 1 & 1 \end{bmatrix}\begin{bmatrix} x_1 \\ x_2 \\ x_3 \end{bmatrix} = \begin{bmatrix} 2 \\ -4 \\ 5 \end{bmatrix}$ **15.** $\begin{aligned} x_1 + 2x_2 &= -1 \\ 4x_1 - 3x_2 &= 2 \end{aligned}$

17. $x_1 = -23$, $x_2 = 83$ **19.** $x_1 = \frac{1}{11}$, $x_2 = -\frac{6}{11}$ **21.** $x_1 = -1$, $x_2 = 5$, $x_3 = 1$

23. $(A + B)(A + B) = \begin{bmatrix} 1 & 0 \\ 4 & 4 \end{bmatrix}$, $A^2 + 2AB + B^2 = \begin{bmatrix} -3 & -3 \\ 9 & 8 \end{bmatrix}$ **25.** $\begin{bmatrix} X \\ Y \end{bmatrix}^T = [a \quad b \quad c \quad e \quad f \quad g]$

27. $[A \quad B] - \begin{bmatrix} 2 & -1 & 3 & | & 0 & 1 & -2 \\ 4 & 1 & 0 & | & 1 & -3 & 7 \end{bmatrix}$, $[A \quad B]^T = \begin{bmatrix} 2 & 4 \\ -1 & 1 \\ 3 & 0 \\ \hline 0 & 1 \\ 1 & -3 \\ -2 & 7 \end{bmatrix}$ **29.** Yes

---

EXERCISE SET 2.5, p. 116

1. $A^{-1} = \begin{bmatrix} -3 & -2 \\ 5 & -3 \end{bmatrix}$ **3.** $A^{-1} = \begin{bmatrix} 2 & -3 \\ -7 & 11 \end{bmatrix}$ **5.** $A^{-1} = \begin{bmatrix} \frac{2}{11} & \frac{3}{11} \\ -\frac{1}{11} & \frac{4}{11} \end{bmatrix}$ **7.** $A^{-1} = \begin{bmatrix} \frac{3}{8} & \frac{1}{8} & -\frac{1}{4} \\ -\frac{1}{8} & -\frac{3}{8} & \frac{3}{4} \\ -\frac{1}{4} & \frac{1}{4} & \frac{1}{2} \end{bmatrix}$ **9.** $A^{-1} = \begin{bmatrix} \frac{1}{3} & 0 & \frac{1}{3} \\ -\frac{2}{5} & \frac{2}{5} & \frac{1}{5} \\ \frac{2}{15} & \frac{1}{5} & -\frac{1}{15} \end{bmatrix}$

CHAPTER 2 TEST, p. 117

1. $\begin{bmatrix} 7 & 4 \\ 3 & 1 \end{bmatrix} \begin{bmatrix} x_1 \\ x_2 \end{bmatrix} = \begin{bmatrix} -21 \\ -9 \end{bmatrix}$, $x_1 = -3, x_2 = 0$ **2.** $\begin{bmatrix} 3 & -2 & 3 \\ 1 & 1 & -1 \\ 2 & 3 & -5 \end{bmatrix} \begin{bmatrix} x_1 \\ x_2 \\ x_3 \end{bmatrix} = \begin{bmatrix} 24 \\ -7 \\ -32 \end{bmatrix}$, $x_1 = 1, x_2 = -3, x_3 = 5$

3. $\begin{array}{l} 4x_1 - 8x_2 = -20 \\ 3x_1 - 6x_2 = -15 \end{array}$, $x_1 = -5 + 2x_2, x = $ any number **4.** $\begin{array}{l} 8x_1 - 4x_2 = 20 \\ 6x_1 - 3x_2 = 16 \end{array}$, no solution

5. $x_1 = 5 - 6x_3, x_2 = 3 + 2x_3, x_3 = $ any number, $x_4 = 2$ **6.** No solution **7.** $\begin{bmatrix} -3 & 1 \\ -4 & 1 \end{bmatrix}$ **8.** $\begin{bmatrix} 2 & 3 \\ 1 & 5 \end{bmatrix}$

9. $\begin{bmatrix} -3 & 3 \\ -6 & 1 \end{bmatrix}$ **10.** $\begin{bmatrix} 12 & -8 \\ 20 & -4 \end{bmatrix}$ **11.** $C^T = [2 \quad -3 \quad 4]$ **12.** AB not possible, $BA = \begin{bmatrix} 5 \\ 7 \end{bmatrix}$ **13.** $x_1 = \frac{1}{2}, x_2 = \frac{2}{3}$

14. $x_1 = -2, x_2 = 3, x_3 = 9$ **15.** $\begin{bmatrix} X \\ Y \end{bmatrix}^T = [p \quad q \quad t \quad u]$ **16.** \$4000 @ 8%, \$5300 @ 10% **17.** $A^{-1} = \begin{bmatrix} 2 & 0 & 1 \\ 3 & 1 & 2 \\ 1 & 0 & 1 \end{bmatrix}$

CHAPTER 3

MARGIN EXERCISES

1.

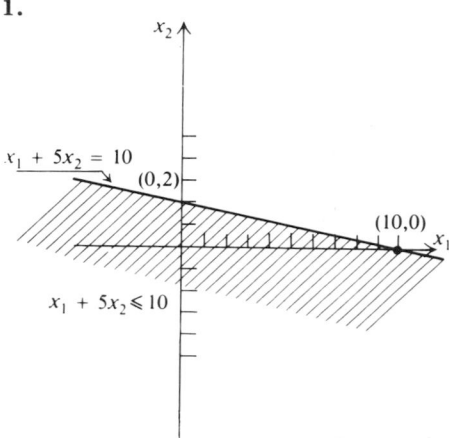

2.

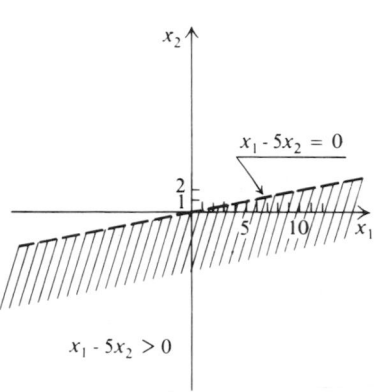

3.

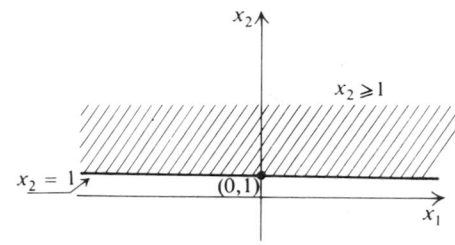

4, 5, 6, 7, 8.

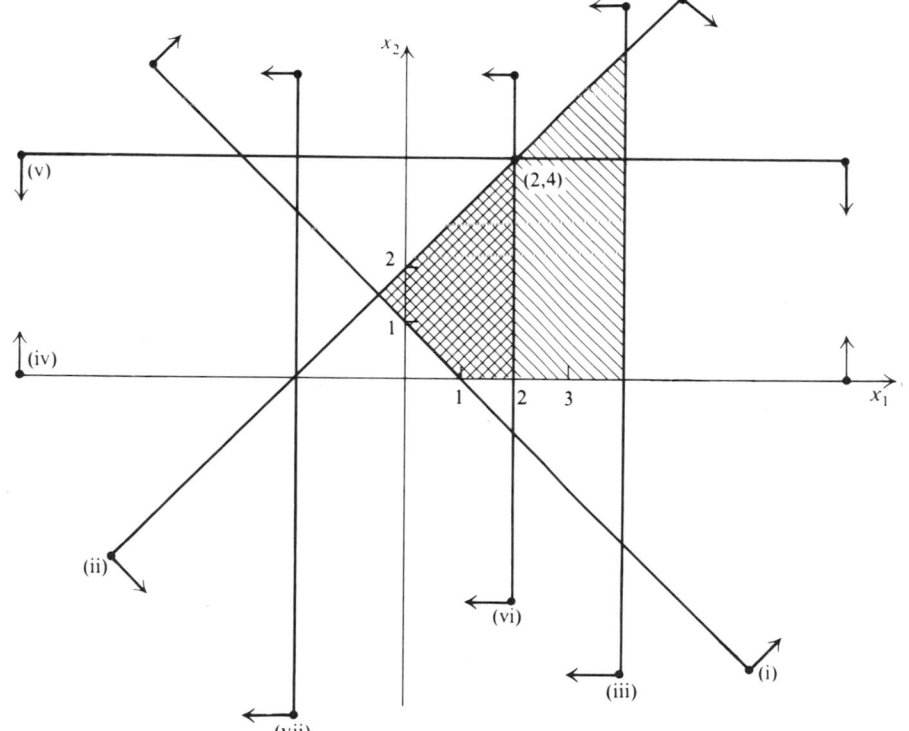

7. a) $(2, 4)$ degenerate; intersection of ii, v, and vi   b) Constraints (iii) and (v) redundant   **8.** Solution set empty

9. a, b, e **10.** $(\frac{20}{7}, \frac{12}{7})$ **11.** min: $f(-\frac{4}{3}, -\frac{2}{3}) = -\frac{14}{3}$,   max: $f(2, 2) = 10$ **12.** min: $f(-\frac{4}{3}, -\frac{2}{3}) = f(2, 1) = 0$ [any

point on the line segment between $(-\frac{4}{3}, -\frac{2}{3})$ and $(2, 1)$],   max: $f(0, 2) = 4$.

13.

	Composition		Supply available
	Chairs	Sofas	
Number of units	x_1	x_2	
Wood (feet)	20	100	1900
Foam (lbs)	1	50	500
Material (yds)	2	20	240
Unit price ($)	20	300	Maximize income

$$20x_1 + 100x_2 \leqslant 1900,$$
$$x_1 + 50x_2 \leqslant 500,$$
$$2x_1 + 20x_2 \leqslant 240,$$
$$\max f: f = 20x_1 + 300x_2; \quad x_1, x_2 \geqslant 0.$$

14. $\begin{bmatrix} 20 & 100 \\ 1 & 50 \\ 2 & 50 \end{bmatrix} \begin{bmatrix} x_1 \\ x_2 \end{bmatrix} \leqslant \begin{bmatrix} 1900 \\ 500 \\ 240 \end{bmatrix}$ $\max f: f = \begin{bmatrix} 20 & 300 \end{bmatrix} \begin{bmatrix} x_1 \\ x_2 \end{bmatrix}; \quad \begin{bmatrix} x_1 \\ x_2 \end{bmatrix} \geqslant \begin{bmatrix} 0 \\ 0 \end{bmatrix}.$ **15.** $x_1 = 25$, $x_2 = \frac{19}{2}$, $f = 3350$

16. a)

	Composition (tons)		Amount required (tons)
	Ore A	Ore B	
Number of units (tons)	y_1	y_2	
Iron	0.10	0	200
Aluminum	0	0.20	500
Copper	0.02	0.01	100
Cost ($) per ton	10	15	Minimize cost

$$0.1y_1 + 0y_2 \geqslant 200,$$
$$0y_1 + 0.2y_2 \geqslant 500,$$
$$0.02y_1 + 0.01y_2 \geqslant 100,$$
$$\min f: f = 10y_1 + 15y_2; \quad y_1, y_2 \geqslant 0.$$

b) $\begin{bmatrix} 0.1 & 0 \\ 0 & 0.2 \\ 0.02 & 0.01 \end{bmatrix} \begin{bmatrix} y_1 \\ y_2 \end{bmatrix} \geqslant \begin{bmatrix} 200 \\ 500 \\ 100 \end{bmatrix}$

$\min f: f = \begin{bmatrix} 10 & 15 \end{bmatrix} \begin{bmatrix} y_1 \\ y_2 \end{bmatrix}; \quad \begin{bmatrix} y_1 \\ y_2 \end{bmatrix} \geqslant \begin{bmatrix} 0 \\ 0 \end{bmatrix}.$

17. $f = 75,000$ at $(y, y_2) = (3750, 2500)$

18.

$$20x_1 + 100x_2 + y_1 \qquad\qquad = 1900$$
$$x_1 + 50x_2 \qquad + y_2 \qquad = 500,$$
$$2x_1 + 20x_2 \qquad\qquad + y_3 = 240,$$

max x_0: $x_0 = 20x_1 + 300x_2 + 0y_1 + 0 \cdot y_2 + 0 \cdot y_3$; $x_i \geqslant 0$ for all i, $y_i \geqslant 0$ for all i.

$$\begin{bmatrix} 20 & 100 & 1 & 0 & 0 \\ 1 & 50 & 0 & 1 & 0 \\ 2 & 20 & 0 & 0 & 1 \end{bmatrix} \begin{bmatrix} x_1 \\ x_2 \\ y_1 \\ y_2 \\ y_3 \end{bmatrix} = \begin{bmatrix} 1900 \\ 500 \\ 240 \end{bmatrix},$$

max x_0: $x_0 = [20\ \ 300\ \ 0\ \ 0\ \ 0][x_1\ x_2\ y_1\ y_2\ y_3]^T$, $[x_1\ x_2\ y_1\ y_2\ y_3]^T = [0\ 0\ 0\ 0\ 0\ 0]^T$.

19.

x_1	x_2	y_1	y_2	y_3	1
20	100	1	0	0	1900
1	50	0	1	0	500
2	20	0	0	1	240
−20	−300	0	0	0	0

20. $x_1 = x_2 = 0$; $y_1 = 1900$, $y_2 = 500$, $y_3 = 240$; $x_0 = 0$.

21. Second column, second row

22. $x_1 = 25$, $x_2 = \frac{19}{2}$, $y_1 = 450$, $y_2 = y_3 = 0$; $x_0 = 3350$

23.

$$20y_1 + y_2 + 2y_3 \geqslant 20,$$
$$100y_1 + 50y_2 + 20y_3 \geqslant 300,$$
$$y_1, y_2, y_3 \geqslant 0,$$

min y_0: $y_0 = 1900y_1 + 500y_2 + 240y_3$

24.

$$0.1x_1 + 0 \cdot x_2 + 0.02x_3 \leqslant 10,$$
$$0 \cdot x_1 + 0.2x_2 + 0.01x_3 \leqslant 15,$$
$$x_1, x_2, x_3 \geqslant 0,$$

max x_0: $x_0 = 200x_1 + 500x_2 + 100x_3$

25.

x_1	x_2	y_1	y_2	y_3	1
20	100	1	0	0	1900
1	50	0	1	0	500
2	20	0	0	1	240
−20	−300	0	0	0	0

Initial primal sol.: $x_0 = x_1 = x_2 = 0$, $y_1 = 1900$,
$y_2 = 500$, $y_3 = 240$
Initial dual sol.: $y_0 = y_1 = y_2 = y_3 = 0$, $x_1 = -20$,
$x_2 = -300$

26.

$$0.1x_1 + 0 \cdot x_2 + 0.02x_3 + y_1 \qquad = 10,$$
$$0 \cdot x_1 + 0.2x_2 + 0.01x_3 \qquad + y_2 = 15,$$
$$x_1, x_2, x_3, y_1, y_2 \geqslant 0,$$

max x_0: $x_0 = 200x_1 + 500x_2 + 100x_3 + 0 \cdot y_1 + 0 \cdot y_2$

x_1	x_2	x_3	y_1	y_2	1
0.1	2	0.02	1	0	10
0	0.2	0.01	0	1	15
−200	−500	−100	0	0	0

Initial primal sol.: $x_0 = x_1 = x_2 = x_3 = 0$, $y_1 = 10$, $y_2 = 15$
Initial dual sol.: $y_0 = y_1 = y_2 = 0$,
$x_1 = -200$, $x_2 = -500$, $x_3 = -100$

27. Dual (min) solution: $y_1 = 3750$, $y_2 = 2500$, $x_1 = 175$, $x_2 = x_3 = 0$; $y_0 = 75,000$

28. Primal (max) solution: $x_1 = 0$, $x_2 = 75$, $x_3 = 500$, $y_1 = y_2 = 0$; $x_0 = 75,000$

EXERCISE SET 3.1, pp. 129–130

1.

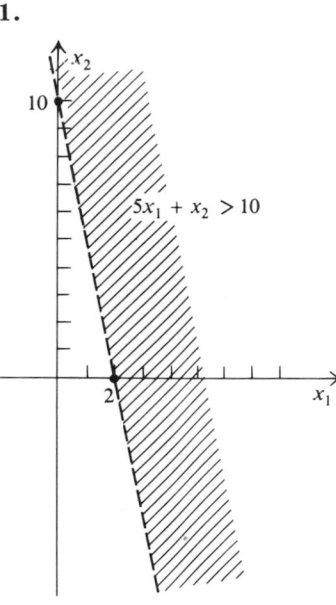

3.

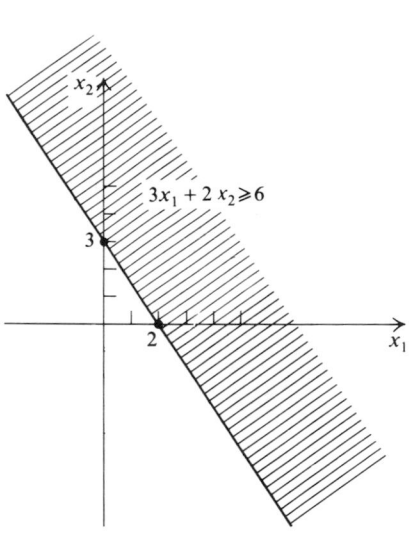

5.

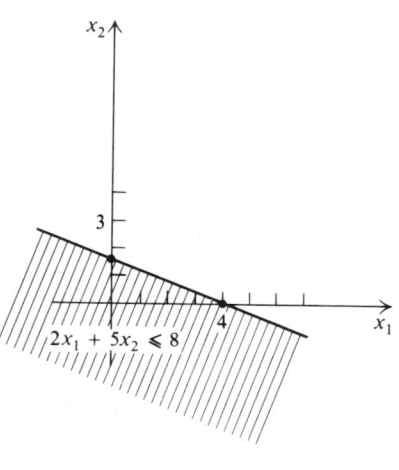

7.

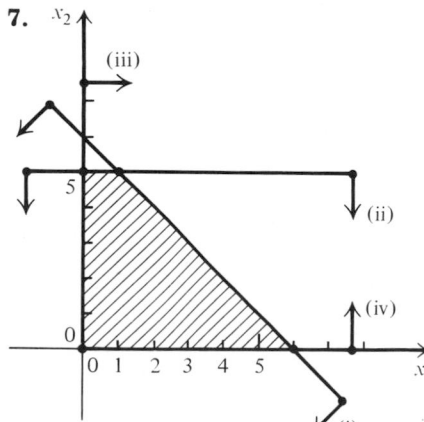

9.

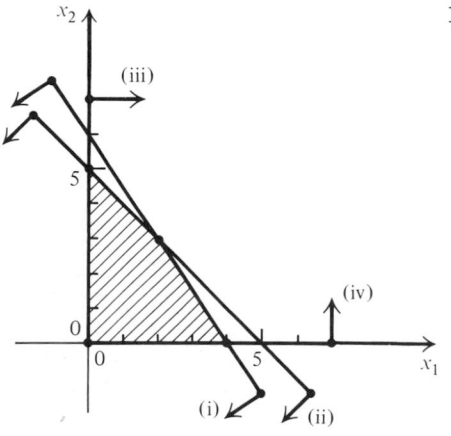

11.

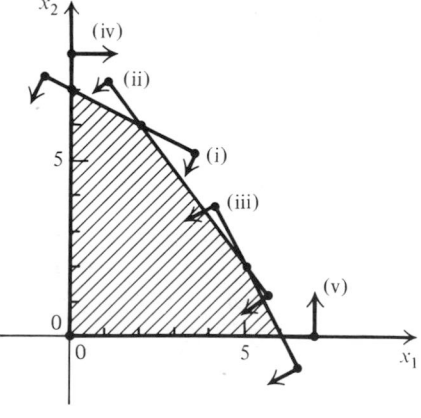

13.

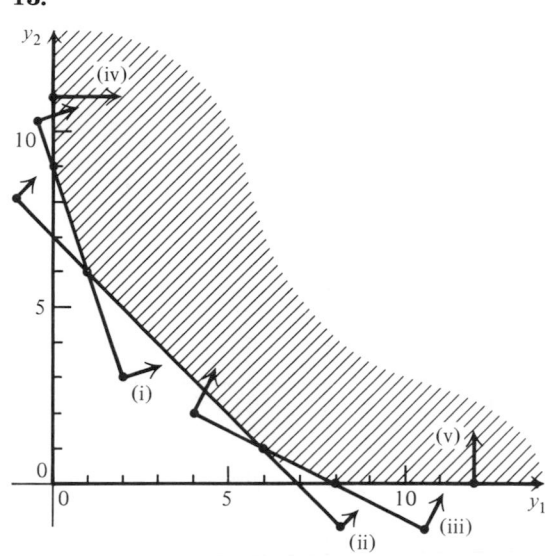

15.

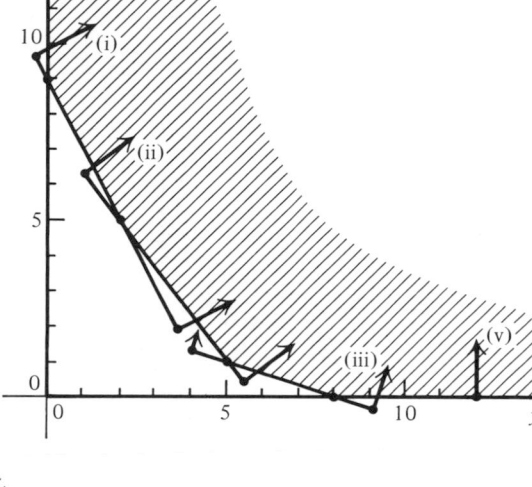

17.

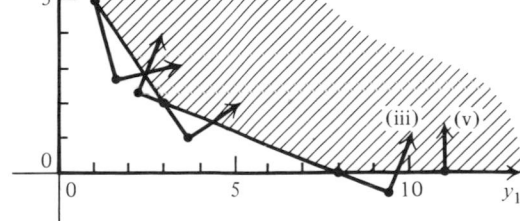

19. a)   i) $x_1 + 2x_2 \leq 6$
     ii) $0 \leq x_1 \leq 5$
     iii) $x_2 \leq -2$
  b) Nonempty
  c) Bounded
  d) No redundancies
  e) No degeneracies

21. a)   i) $x_1 \geq -3$
     ii) $x_1 - 2x_2 \leq 4$
     iii) $x_2 - 3x_1 \leq 9$
     iv) $3x_1 + x_2 \leq 10$
  b) Nonempty
  c) Bounded
  d) No redundancies
  e) No degeneracies

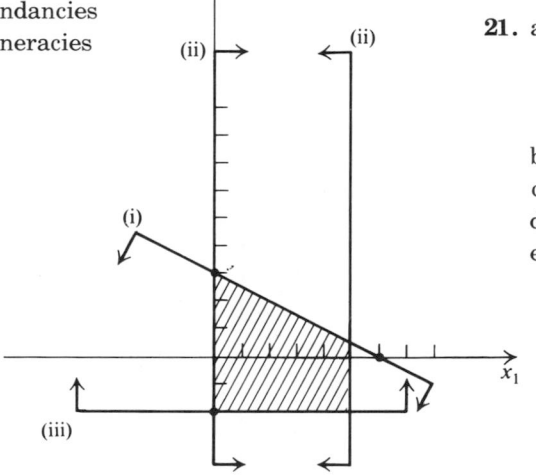

23. a) i) $-3x_1 + 2x_2 \geqslant 6$
ii) $2x_1 + x_2 \leqslant -2$
iii) $x_1 + x_2 \geqslant 4$
iv) $2x_1 + 7x_2 \leqslant 21$
b) Empty
c) Bounded
d) (i) or (ii)
e) No degeneracies

25. a) i) $x_1 \geqslant 0$
ii) $x_2 \geqslant 0$
iii) $x_1 + x_2 \geqslant 2$
iv) $x_1 - x_2 \leqslant 2$
v) $x_2 \leqslant 6$
b) Nonempty
c) Bounded
d) (ii)
e) $(2, 0)$

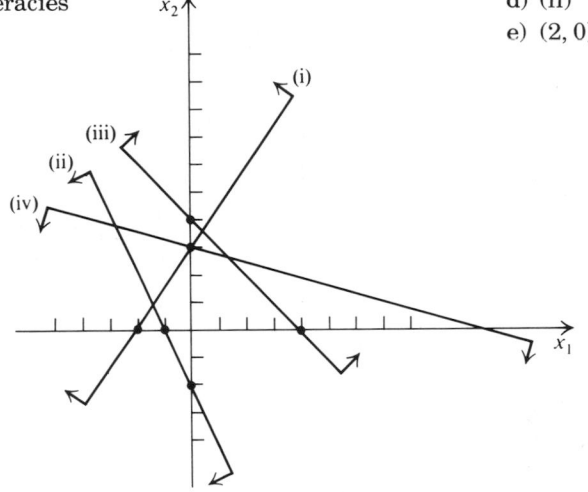

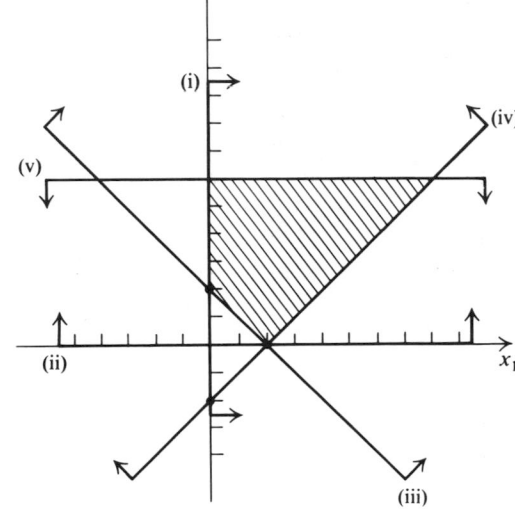

27. a) i) $x_1 + x_2 \leqslant 0$
ii) $2x_1 - 3x_2 \leqslant 15$
iii) $x_2 \leqslant 5$
iv) $x_1 \geqslant 0$
v) $2x_1 + x_2 \geqslant 3$
b) Nonempty ⎤ one point,
c) Bounded ⎦ $(3, -3)$
d) (iii), (iv)
e) $(3, -3)$

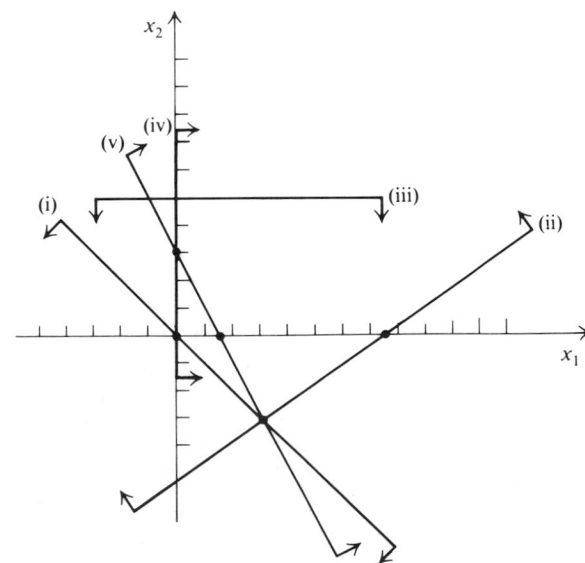

29. a) i) $x_1 \geqslant 0$
    ii) $x_2 \geqslant 0$
    iii) $5x_2 - 3x_1 \leqslant 15$
    iv) $x_1 \leqslant 4x_2$
    v) $2x_1 - 5x_2 \leqslant 10$
 b) Nonempty
 c) Unbounded
 d) (ii)
 e) $(0, 0)$

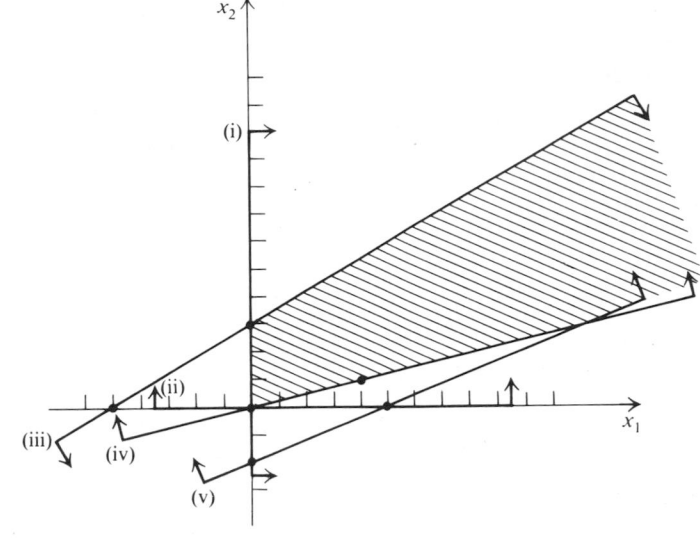

EXERCISE SET 3.2, pp. 134–136

1. $x_1 = 1, x_2 = 5; f = 11$ **3.** $x_1 = 2, x_2 = 3; f = 22$ **5.** $x_1 = 2, x_2 = 6; f = 30$

7. $y_1 = 6, y_2 = 1; f = 22$ **9.** $y_1 = 5, y_2 = 1; f = 15$ **11.** $y_1 = 3, y_2 = 2; f = 19$

13. max: $f = 7$ at $(5, -2)$; min: $f = -3$ at $(0, 3)$ **15.** max: $f = \frac{26}{7}$ at $\left(\frac{24}{7}, \frac{-2}{7}\right)$;   min: $f = -\frac{28}{3}$ at $\left(\frac{1}{6}, \frac{19}{2}\right)$ **17.** max: $f = \frac{64}{3}$

at $(8, \frac{8}{3})$; min: $f = 0$ for any point on the line segment between $\left(\frac{24}{11}, \frac{72}{11}\right)$ and $\left(\frac{5}{4}, \frac{15}{4}\right)$ **19.** No feasible solution, $\therefore$ no

optimum feasible solution, max or min.   **21.** max: unbounded; min: $f = 0$ at $(0, 0)$ **23.** max and min: both have

$f = 3$ at $(3, -3)$

EXERCISE SET 3.3, pp. 141–142

1. Let x_1 = number of suits to be made,
    x_2 = number of dresses to be made,
    f = income.
  Then
$$x_1 + 2x_2 \leqslant 60,$$
$$4x_1 + 3x_2 \leqslant 120,$$
$$\max f: f = 120x_1 + 75x_2; \quad x_1, x_2 \geqslant 0.$$

b) $\begin{bmatrix} 1 & 2 \\ 4 & 3 \end{bmatrix} \begin{bmatrix} x_1 \\ x_2 \end{bmatrix} \leqslant \begin{bmatrix} 60 \\ 120 \end{bmatrix}$

$\max f: f = [120 \quad 75] \begin{bmatrix} x_1 \\ x_2 \end{bmatrix}; \quad \begin{bmatrix} x_1 \\ x_2 \end{bmatrix} \geqslant \begin{bmatrix} 0 \\ 0 \end{bmatrix}$

c) max income = \$3600 for $x_1 = 30$ suits and
   $x_2 = 0$ dresses.

3. a) Let x_1 = number of lbs of Mixture I,
    x_2 = number of lbs of Mixture II,
    f = income.
  Then
$$0.6x_1 + 0.2x_2 \leqslant 1800,$$
$$0.3x_1 + 0.5x_2 \leqslant 1500,$$
$$0.1x_1 + 0.3x_2 \leqslant 750.$$
$$\max f: f = 0.75x_1 + 2x_2; \quad x_1, x_2 \geqslant 0.$$

b) $\begin{bmatrix} 0.6 & 0.2 \\ 0.3 & 0.5 \\ 0.1 & 0.3 \end{bmatrix} \begin{bmatrix} x_1 \\ x_2 \end{bmatrix} \leqslant \begin{bmatrix} 1800 \\ 1500 \\ 750 \end{bmatrix}$

$\max f: f = [0.75 \quad 2] \begin{bmatrix} x_1 \\ x_2 \end{bmatrix}; \quad \begin{bmatrix} x_1 \\ x_2 \end{bmatrix} \geqslant \begin{bmatrix} 0 \\ 0 \end{bmatrix}$

c) max income = \$5,156.25 for $x_1 = 1875$ lbs and
   $x_2 = 1875$ lbs.

5. a) Let x_1 = number of animals of species A1,
x_2 = number of animals of species A2,
f = total number of animals;

F1: $x_1 + 1.2x_2 \leqslant 600$,
F2: $2x_1 + 1.8x_2 \leqslant 960$,
F3: $2x_1 + 0.6x_2 \leqslant 720$.
max $f: f = x_1 + x_2$; $x_1, x_2 \geqslant 0$.

b) $\begin{bmatrix} 1 & 1.2 \\ 2 & 1.8 \\ 2 & 0.6 \end{bmatrix} \begin{bmatrix} x_1 \\ x_2 \end{bmatrix} \leqslant \begin{bmatrix} 600 \\ 960 \\ 720 \end{bmatrix}$

max $f: f = [1 \quad 1]\begin{bmatrix} x_1 \\ x_2 \end{bmatrix}$; $\begin{bmatrix} x_1 \\ x_2 \end{bmatrix} \geqslant \begin{bmatrix} 0 \\ 0 \end{bmatrix}$

c) max number = 520 for $x_1 = 120$ and $x_2 = 400$.

7. a) F1: $x_1 + 1.2x_2 \leqslant 720$,
F2: $2x_1 + 1.8x_2 \leqslant 960$,
F3: $2x_1 + 0.6x_2 \leqslant 600$.
max $f: f = x_1 + x_2$; $x_1 x_2 \geqslant 0$.

c) max number = $\frac{1600}{3}$ (ignore fractions) for $x_1 = 0$ and $x_2 = \frac{1600}{3}$. Species A1 would become extinct (in that area).

EXERCISE SET 3.4, p. 146–147

1. a) Let y_1 = number of sacks of soybean meal,
y_2 = number of sacks of oats,
f = cost.

Then $50y_1 + 15y_2 \geqslant 120$,
$8y_1 + 5y_2 \geqslant 24$,
$5y_1 + y_2 \geqslant 10$.
min $f: f = 15y_1 + 5y_2$; $y_1, y_2 \geqslant 0$.

b) $\begin{bmatrix} 50 & 15 \\ 8 & 5 \\ 5 & 1 \end{bmatrix} \begin{bmatrix} y_1 \\ y_2 \end{bmatrix} \geqslant \begin{bmatrix} 120 \\ 24 \\ 10 \end{bmatrix}$

min $f: f = [15 \quad 5]\begin{bmatrix} y_1 \\ y_2 \end{bmatrix}$; $\begin{bmatrix} y_1 \\ y_2 \end{bmatrix} \geqslant \begin{bmatrix} 0 \\ 0 \end{bmatrix}$

c) min cost = $\$\frac{480}{13}$ for $y_1 = \frac{24}{13}$ and $y_2 = \frac{24}{13}$.

3. a) Third constraint $5y_1 + 8y_3 \geqslant 10$, becomes $5y_1 + 8y_3 \geqslant 20$. **c)** min cost = $\$\frac{276}{7}$ for $y_1 = \frac{12}{7}$ and $y_3 = \frac{12}{7}$.

5. a) Let y_1 = number of P1 airplanes, y_2 = number of P2 airplanes, f = cost in $ thousands. Then

$40y_1 + 80y_2 \geqslant 2000$,
$40y_1 + 30y_2 \geqslant 1500$,
$120y_1 + 40y_2 \geqslant 2400$,
min $f: f = 12y_1 + 10y_2$; $y_1, y_2 \geqslant 0$.

b) $\begin{bmatrix} 40 & 80 \\ 40 & 30 \\ 120 & 40 \end{bmatrix} \begin{bmatrix} y_1 \\ y_2 \end{bmatrix} \geqslant \begin{bmatrix} 2000 \\ 1500 \\ 240 \end{bmatrix}$,

min $f: f = [12 \quad 10]\begin{bmatrix} y_1 \\ y_2 \end{bmatrix}$; $\begin{bmatrix} y_1 \\ y_2 \end{bmatrix} \geqslant \begin{bmatrix} 0 \\ 0 \end{bmatrix}$.

c) min cost = $460 thousand for $y_1 = 30$ and $y_2 = 10$.

7. a) $40y_1 + 40y_3 \geqslant 2000$,
$40y_1 + 80y_3 \geqslant 1500$,
$120y_1 + 80y_3 \geqslant 2400$,
min $f: f = 12y_1 + 15y_3$; $y_1, y_3 \geqslant 0$.

b) $\begin{bmatrix} 40 & 40 \\ 40 & 80 \\ 120 & 80 \end{bmatrix} \begin{bmatrix} y_1 \\ y_3 \end{bmatrix} \geqslant \begin{bmatrix} 2000 \\ 1500 \\ 2400 \end{bmatrix}$,

min $f: f = [12 \quad 15]\begin{bmatrix} y_1 \\ y_2 \end{bmatrix}$; $\begin{bmatrix} y_1 \\ y_3 \end{bmatrix} \geqslant \begin{bmatrix} 0 \\ 0 \end{bmatrix}$.

c) min cost = $600 thousand for $y_1 = 50$ and $y_3 = 0$.

9. Replace $\begin{bmatrix} 2000 \\ 1500 \\ 2400 \end{bmatrix}$ by $\begin{bmatrix} 1600 \\ 2100 \\ 2400 \end{bmatrix}$. **a)** Set $y_3 = 0$. **c)** min cost = $630 thousand for $y_1 = 52.5$ and $y_2 = 0$.

11. a) Set $y_2 = 0$. **c)** min cost = $517.5 thousand for $y_i = \frac{55}{2}$ and $y_3 = \frac{25}{2}$.

EXERCISE SET 3.5, pp. 163–164

1. $x_1 = 1$, $x_2 = 5$, $y_1 = y_2 = 0$; $x_0 = 11$ **3.** $x_1 = 2$, $x_2 = 3$, $y_1 = y_2 = 0$; $x_0 = 22$ **5.** $x_1 = 2$, $x_2 = 6$, $y_1 = y_2 = 0$, $y_3 = 2$; $x_0 = 30$ **7.** $x_1 = 30$, $x_2 = y_1 = 0$, $y_2 = 20$, $x_0 = 3600$ **9.** $x_1 = 0$, $x_2 = 4$, $y_1 = 1$, $y_2 = y_3 = 0$; $x_0 = 90$
11. $x_1 = 8$, $x_2 = 3$, $y_1 = 2$, $y_2 = y_3 = 0$; $x_0 = 19$ **13.** $x_1 = 9$, $x_2 = 4$, $y_1 = 5$, $y_2 = y_3 = 0$; $x_0 = 79$ **15.** $x_1 = \frac{6}{5}$, $x_2 = 0$,
$x_3 = \frac{22}{5}$, $y_1 = y_2 = 0$, $y_3 = \frac{144}{5}$; $x_0 = 22$ **17.** $x_1 = 0$, $x_2 = \frac{9}{4}$, $x_3 = \frac{19}{4}$, $y_1 = \frac{11}{4}$, $y_2 = y_3 = 0$; $x_0 = \frac{75}{4}$
19. No bookcases, 12 desks, 65 tables; max sales = $6400.

EXERCISE SET 3.6, pp. 174–177

For solutions to Exercise 1 through 17, see solutions to Exercises 1 through 17 of Set 3.2.

19. $y_1 = 4$, $y_2 = 6$, $x_1 = 4$, $x_2 = x_3 = 0$; $y_0 = 76$ $(P: x_1 = 0, x_2 = \frac{4}{7}, x_3 = \frac{11}{7}, y_1 = y_2 = 0)$
21. $y_1 = 5$, $y_2 = 7$, $x_1 = x_2 = 0$, $x_3 = 1$; $y_0 = 53$ $(P: x_1 = \frac{9}{7}, x_2 = \frac{2}{7}, x_3 = y_1 = y_2 = 0)$
23. $x_1 = 0$, $x_2 = \frac{7}{4}$, $x_3 = \frac{5}{4}$, $y_1 = \frac{59}{4}$, $y_2 = y_3 = 0$; $x_0 = \frac{19}{4}$ $(P: y_1 = 0, y_2 = \frac{1}{8}, y_3 = \frac{5}{4}, x_1 = \frac{27}{8}, x_2 = x_3 = 0)$
25. $y_1 = \frac{8}{11}$, $y_2 = \frac{2}{11}$, $y_3 = x_1 = x_2 = 0$, $x_3 = \frac{13}{11}$; $y_0 = \frac{82}{11}$ $(P: x_1 = \frac{29}{11}, x_2 = \frac{6}{11}, x_3 = y_1 = y_2 = 0, y_3 = 4)$
27. Cost $0.29 with all soybeans = 288 gms   **29.** Cost $0.91 with 130 gms cheese and 138 gms beef.

CHAPTER 3 TEST, p. 178

1. Let x_1 = number of lbs of mixture I,

x_2 = number of lbs of mixture II,

f = income.

Then

$$0.8x_1 + 0.6x_2 \leqslant 100,$$
$$0.01x_1 + 0.03x_2 \leqslant 10,$$
$$0x_1 + 0.04x_2 \leqslant 5,$$
$$0.12x_1 + 0.24x_2 \leqslant 25,$$
$$0.07x_1 + 0.09x_2 \leqslant 15,$$

max f: $f = 0.95x_1 + 1.35x_2$; $x_1, x_2 \geqslant 0$.

$$\begin{bmatrix} 0.8 & 0.6 \\ 0.01 & 0.03 \\ 0 & 0.04 \\ 0.12 & 0.24 \\ 0.07 & 0.09 \end{bmatrix} \begin{bmatrix} x_1 \\ x_2 \end{bmatrix} \leqslant \begin{bmatrix} 100 \\ 10 \\ 5 \\ 25 \\ 15 \end{bmatrix},$$

max f: $f = \begin{bmatrix} 0.95 & 1.35 \end{bmatrix} \begin{bmatrix} x_1 \\ x_2 \end{bmatrix}$; $\begin{bmatrix} x_1 \\ x_2 \end{bmatrix} \geqslant \begin{bmatrix} 0 \\ 0 \end{bmatrix}$.

2.

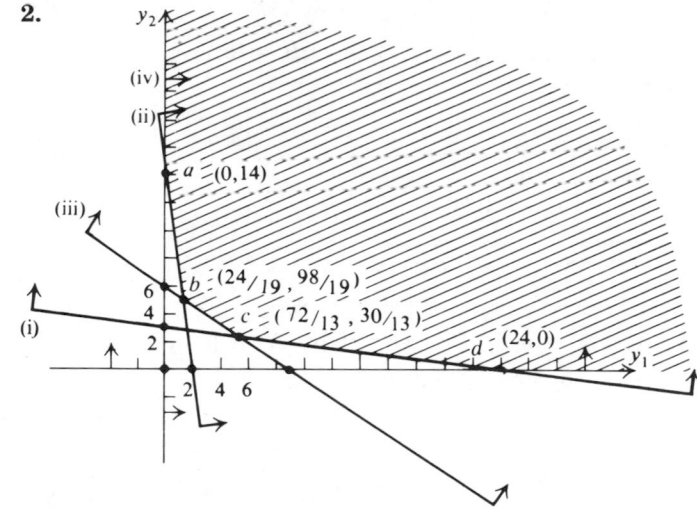

3. Minimum is $6\frac{8}{19}$ when $y_1 = \frac{24}{19}$ and $y_2 = \frac{98}{19}$.

4. $x_1 = 11$, $x_2 = 5$, $y_1 = 5$, $y_2 = y_3 = 0$; $x_0 = 59$.

5. a)
$$y_1 + y_2 + 5y_3 \geq 4,$$
$$2y_1 + y_2 + 3y_3 \geq 3,$$
$$\min y_0: y_0 = 26y_1 + 16y_2 + 70y_3;$$
$$y_1, y_2, y_3 \leq 0.$$

b) $y_1 = 0$, $y_2 = \frac{3}{2}$, $y_3 = \frac{1}{2}$, $x_1 = x_2 = 0$; $y_0 = 59$

c)
$$x_1 + 2x_2 + y_1 = 26,$$
$$x_1 + x_2 + y_2 = 16,$$
$$5x_1 + 3x_2 + y_3 = 70,$$
$$x_0 = 4x_1 + 3x_2;$$
$$x_1, x_2; y_1, y_2, y_3 \geq 0.$$

d)
$$y_1 + y_2 + 5y_3 - x_1 = 4,$$
$$2y_1 + y_2 + 3y_3 - x_2 = 3,$$
$$y_0 = 26y_1 + 16y_2 + 70y_3 = x_0;$$
$$y_1, y_2, y_3; x_1, x_2 \geq 0.$$

CHAPTER 4

MARGIN EXERCISES

1. a) Yes, no b) No, yes   **2.** a) $D = \{x \mid x = 2n + 1, n \text{ is an integer}, n \geq 0\}$ b) $F = \{1, 4, 7, 10, 13, \ldots\}$

3. $\{a\}, \{b\}, \{c\}, \{d\}, \{a, b\}, \{a, c\}, \{a, d\}, \{b, c\}, \{b, d\}, \{c, d\}, \{a, b, c\}, \{a, b, d\}, \{b, c, d\}, \{a, c, d\}, \{a, b, c, d\}, \emptyset$

4. a) Yes b) No, $0 \notin A$ **5.** a) 4 b) 0   **6.** *Before* he buys, the universal set is the set of articles which can be bought for \$10, or less. *After* he buys, it is that set of purchases.   **7.** $A^c = \{2, 4, 6, 8\}$ **8.** $A \times B =$ $\{(a, 1), (a, 2), (a, 3), (a, 4), (b, 1), (b, 2), (b, 3), (b, 4), (c, 1), (c, 2), (c, 3), (c, 4)\}$, $\mathcal{N}(A \times B) = 12$

9. $A \cup B = \{a, b, c, e, g, f, s\}$, $A \cap B = \{b, c, e\}$.

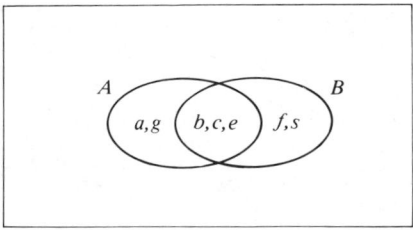

10. a) $\{\{1, 2, 3, 4\}, \{5\}\}$ | 1, 2, 3, 4 | 5 |

b) $\{\{1, 2, 3\}, \{4, 5\}\}$ | 1, 2, 3 | 4, 5 |

c) $\{\{1\}, \{2\}, \{3\}, \{4, 5\}\}$ | 1 | 2 | 3 | 4, 5 |

There are other possibilities.

11. $A - B = \{a, g\}$, $B - A = \{f, h\}$, $A^c = \{f, h, i, j, k, \ldots, z\}$, $B^c = \{a, g, i, j, k, l, \ldots, z\}$, $B^c - A^c = \{a, g\}$, shaded as in the accompanying figure.

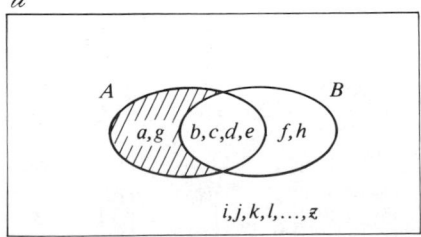

12. 60   **13.** 53   **14.** $S = \{(0, 3), (1, 2), (2, 1), (3, 0)\}$ where each ordered pair $=$ (no. G, no. S)

15.

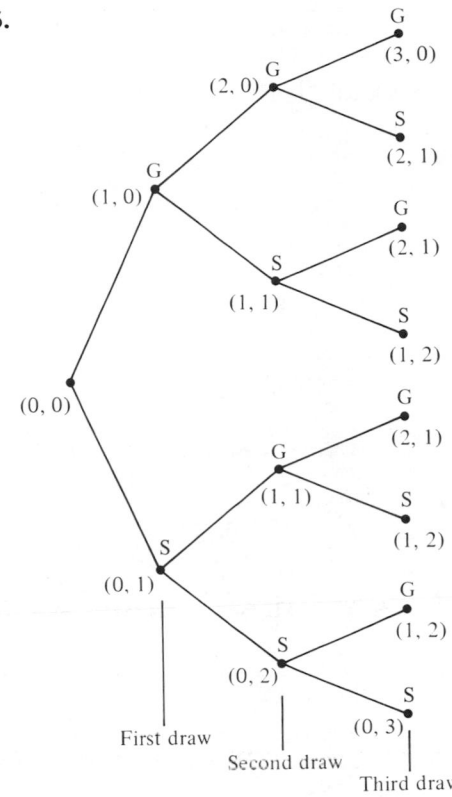

G
(3, 0)

G
(2, 0)

S
(2, 1)

G
(1, 0)

G
(2, 1)

S
(1, 1)

S
(1, 2)

(0, 0)

G
(2, 1)

G
(1, 1)

S
(1, 2)

S
(0, 1)

G
(1, 2)

S
(0, 2)

S
(0, 3)

First draw

Second draw

Third draw

16. $3 \cdot 2 \cdot 1$, or 6   **17.** $5 \cdot 4 \cdot 3 \cdot 2 \cdot 1$, or 120   **18.** $4 \cdot 3 \cdot 2 \cdot 1$, or 24

19. 2^{10}, or 1024   **20.** $5 \cdot 40 \cdot 8$, or 1600   **21.** $5 \cdot 4 \cdot 3 \cdot 2 \cdot 1$, or 120

22. 120   **23.** 720, 120   **24.** 40,320   **25.** 362,880   **26.** 6!

27. a) $8! = 8 \cdot 7!$ b) $38! = 38 \cdot 37!$ **28.** 360   **29.** a) 5040

b) 362,880   **30.** a) 5040 b) 2520 c) 840 d) 210

31. $(6 - 1)!$, or 120   **32.** a) $52 \cdot 51$, or 2652 b) $52 \cdot 52$, or 2704

33. a) 720 b) 240 c) 144 d) 576 **34.** a) 5 b) 1 c) 10 d) 5

e) 1   **35.** a) $P(4, 3) = 4 \cdot 3 \cdot 2 = 24$ b) *ABC, ACB, ABD, ADB,*

ACD, ADC, BAC, BCA, BAD, BDA, BCD, BDC, CAB, CBA, CAD,

CDA, CBD, CDB, DAB, DBA, DAC, DCA, DBC, DCB

c) $\binom{4}{3} = \dfrac{4 \cdot 3 \cdot 2}{3 \cdot 2 \cdot 1} = 4$ d) $\{A, B, C\}, \{A, C, D\}, \{B, C, D\}, \{A, B, D\},$

36. a) 120 b) 120 c) 126 d) 126   **37.** a) 56 b) $\binom{8}{3}$

38. $\binom{10}{8} = 45$ **39.** $\binom{7}{4} \cdot \binom{5}{3} = 350$ **40.** $\binom{3}{1}\binom{4}{1} + \binom{4}{2}$, or 18

41. $\binom{7}{1}\binom{6}{2}\binom{4}{2}\binom{2}{2} = \dfrac{7!}{1!2!2!2!} = 630$ **42.** a) 9 b) 1 c) 4 d) 2

e) 2 f) 3780 g) 3780 **43.** $-1512x^5$ **44.** $5670x^4$ **45.** $8064y^5$

46. $243x^5$ **47.** $x^{10} - 5x^8 + 10x^6 - 10x^4 + 5x^2 - 1$

48. $16x^4 + \dfrac{32x^3}{y} + \dfrac{24x^2}{y^2} + \dfrac{8x}{y^3} + \dfrac{1}{y^4}$

49. $x^6 - 6\sqrt{2}x^5 + 30x^4 - 40\sqrt{2}x^3 + 60x^2 - 24\sqrt{2}x + 8$

50. $\binom{n}{0}2^n + \binom{n}{1}2^{n-1} + \cdots + \binom{n}{n}2^0$

51. 1 6 15 20 15 6 1

EXERCISE SET 4.1, pp. 185–186

1. $A = \{0, 1, 2, 3, 4, 5, 6, 7, 8, 9, 10\}, B = \{1, 2, 3, 4, 5, 6, 7, 8, 9\}, C = \{0, 2, 4, 6, 8, 10, 12\}$ **3.** $A^c = \{11, 12\},$

$B^c = \{0, 10, 11, 12\}, C^c = \{1, 3, 5, 7, 9, 11\}$ **5.** True   **7.** False   **9.** True   **11.** True   **13.** False   **15.** False

17. $A^c = \{m, n, r, t, c\}, B^c = \{a, e, i, o, u\}, C^c = \{i, o, u, m, n, r, t\}$ **19.** False   **21.** True   **23.** False   **25.** False

27. $E \times F = \{(e, r), (e, t), (i, r), (i, t)\}$ **29.** Yes   **31.** $C \times E = \{(a, e), (a, i), (c, e), (c, i), (e, e), (e, i)\}$ **33.** 6

35. a) 0 b) $\emptyset$ c) 1   **37.** a) 2 b) $\emptyset, \{a\}, \{b\}, \{a, b\}$ c) 4   **39.** $1, 2, 4, 8, 2^4, 2^n$

EXERCISE SET 4.2, pp. 197–198

1. a) A b) B c) $\{0, 10\}$ d) $\emptyset$ e) A f) $\{2, 4, 8, 10\}$ g) $\{1, 5, 7, 11\}$ h) $\{0, 2, 4, 8, 10, 12\}$ i) $\{10\}$ j) $\{1, 5, 7\}$

3. $\mathscr{U}$

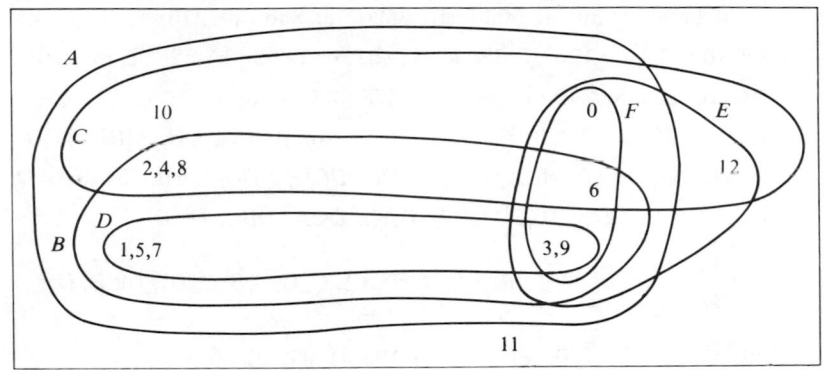

5. a) $\{e, i, r, t\}$ b) $\emptyset$ c) $\{c, r\}$ d) $\{i\}$ e) $\emptyset$ f) $\{m, n, r, t\}$ g) $\emptyset$ h) $\{a, e\}$ **7.** 5   **9.** 4   **11.** 85   **13.** 47%, 36%, 5%

EXERCISE SET 4.3, pp. 201–202

1. $S = \{I, II, III, \text{none}\}$

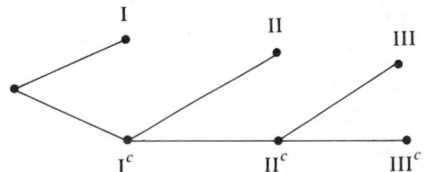

3. $S = \{(AN, CLR), (AN, CLD), (AN, PR), (N, CLR), (N, CLD), (N, PR),$
$(BN, CLR), (BN, CLD), (BN, PR)\}$

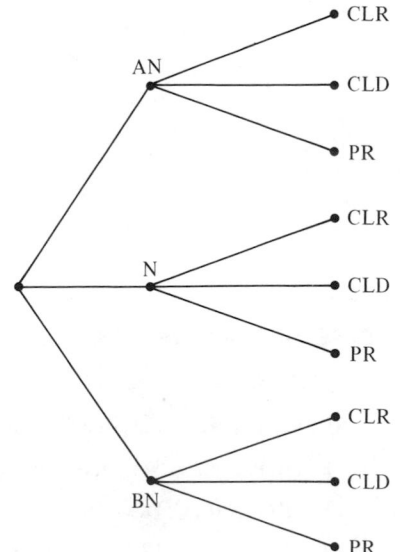

5. $S = \{(M, S, <18), (M, S, \geqslant18), (M, M, <18), (M, M, \geqslant18),$
$\quad (F, S, <18), (F, S, \geqslant18), (F, M, <18), (F, M, \geqslant 18)\}$

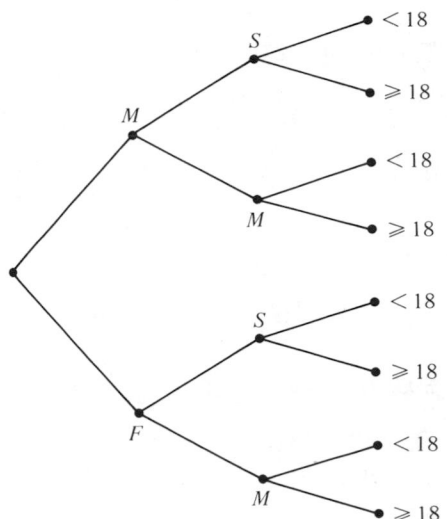

7.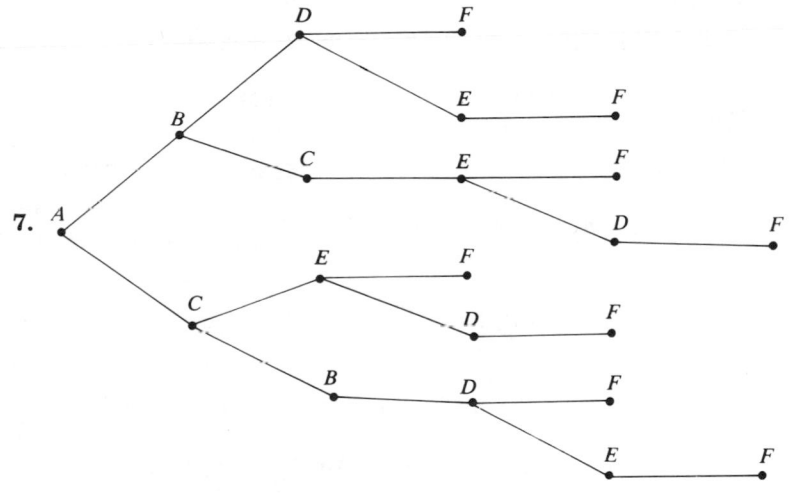

9. $n - 1$

EXERCISE SET 4.4, pp. 210–211

1. 120   **3.** 1   **5.** 720   **7.** 380   **9.** 2520   **11.** $n(n - 1)(n - 2)$ **13.** $3 \cdot 4 \cdot 2$, or 24   **15.** $P(4, 4) = 4! = 24$

17. $P(4, 2) = 4 \cdot 3 = 12$ **19.** $P(4, 4) = 24$ **21.** $(4 - 1)! = 3! = 6$; no   **23.** $5 \cdot 5 \cdot 5 \cdot 5 = 625, 5 \cdot 4 \cdot 3 \cdot 2 = 120$

25. a) $5 \cdot 5 \cdot 5 = 125$ b) $5 \cdot 4 \cdot 3 = 60$ **27.** $7^4 = 2401, 7 \cdot 6 \cdot 5 \cdot 4 = 840$ **29.** $7^3 = 343$ **31.** 5040, 144, 1440

33. $6! = 720, 2 \cdot 2 \cdot 2 \cdot 4 \cdot 3! = 192$

EXERCISE SET 4.5, pp. 216–220

1. 78 **3.** 78 **5.** 7 **7.** $\dfrac{n(n-1)}{2}$ **9.** $\dbinom{6}{4} = 15$ **11.** $\dbinom{10}{7} \cdot \dbinom{8}{5} = 6720$ **13. a)** $\dbinom{14}{5} = 2002$

b) $\dbinom{6}{2}\dbinom{8}{3} + \dbinom{6}{1}\dbinom{8}{4} + \dbinom{8}{5} = 1316$ **c)** $\dbinom{6}{1}\dbinom{8}{4} + \dbinom{8}{5} = 476$ **d)** $\dbinom{6}{4}\dbinom{8}{1} + \dbinom{6}{3}\dbinom{8}{2} + \dbinom{6}{2}\dbinom{8}{3} + \dbinom{6}{1}\dbinom{8}{4} + \dbinom{8}{5} = 1996$

15. $\dfrac{11!}{3!\,4!\,2!\,2!} = 69{,}300$ **17.** $\dfrac{10!}{2!\,3!\,3!\,1!\,1!} = 50{,}400$ **19.** $\dfrac{20!}{2!\,5!\,8!\,3!\,2!} = 20{,}951{,}330{,}400$ **21.** $\dfrac{9!}{3!\,2!\,4!} = 1260$

23. a) $\dbinom{20}{3} = 1140$ **b)** $\dbinom{20}{3}\dbinom{3}{1} = 3420$ **25.** $5 \cdot 4 \cdot 3 = 60,\ \dfrac{5 \cdot 4 \cdot 3}{3 \cdot 2 \cdot 1} = 10$ **27. a)** $\dbinom{30}{6} = 593{,}775$

b) $\dbinom{10}{4}\dbinom{20}{2} = 39{,}900$ **29.** $\dbinom{12}{4}\dbinom{8}{4}\dbinom{4}{4} = 34{,}650$ **31.** $\dfrac{34{,}650}{3!} = 5775$ **33.** 2520, 2520, 1320

35. 240 **37. a)** $\dbinom{8}{0} + \dbinom{8}{1} + \cdots + \dbinom{8}{8} = \displaystyle\sum_{i=0}^{8} \dbinom{8}{i}$ **b)** $\displaystyle\sum_{i=0}^{8} \dbinom{8}{i} = 2^8 = 256$ **c)** $\displaystyle\sum_{i=0}^{9} \dbinom{9}{i} = 2^9 = 512$

39. 4 **41.** $13 \cdot 48 = 624$ **43.** $13 \cdot \dbinom{4}{2}\dbinom{12}{3}\dbinom{4}{1}\dbinom{4}{1}\dbinom{4}{1} = 1{,}098{,}240$ **45.** $\dbinom{13}{5}4 - 36 - 4 = 5108$

47. $10 \cdot 4^5 - 36 - 4 = 10{,}200$ **49.** $P(10, 6) = 10 \cdot 9 \cdot 8 \cdot 7 \cdot 6 \cdot 5 = 151{,}200$ **51.** $C(8, 2) = 28,\ 2C(8, 2) = 56$

53. Math: $4! = 24$; Business: $\dfrac{8!}{3!} = 6720$; Philosophical: $\dfrac{13!}{2!2!2!2!2!} = 194{,}594{,}400$ **55.** $2^7 - 1 = 127$

57. $80 \cdot 26 \cdot 9999 = 20{,}797{,}920$ **59.** $C(n, 2) - n = \dfrac{n^2 - 3n}{2}$ **61. a)** $P(6, 5) = 6 \cdot 5 \cdot 4 \cdot 3 \cdot 2 = 720$ **b)** $6^5 = 7776$

c) $P(5, 4) = 5 \cdot 4 \cdot 3 \cdot 2 = 120$ **d)** $P(3, 2) = 3 \cdot 2 = 6$ **63.** $\dbinom{n}{4} = \dfrac{n(n-1)(n-2)(n-3)}{24}$ **65.** 5 **67.** 8 **69.** 6

71. 11

EXERCISE SET 4.6, pp. 224–225

1. $15a^4b^2$ **3.** $-745{,}472a^3$ **5.** $1120x^{12}y^2$ **7.** $-1{,}959{,}552u^5v^{10}$ **9.** $m^5 + 5m^4n + 10m^3n^2 + 10m^2n^3 + 5mn^4 + n^5$

11. $x^{10} - 15x^8y + 90x^6y^2 - 270x^4y^3 + 405x^2y^4 - 243y^5$ **13.** $x^{-8} + 4x^{-4} + 6 + 4x^4 + x^8$ **15.** $\dbinom{n}{0} - \dbinom{n}{1} + \dbinom{n}{2} - \dbinom{n}{3} +$

$\cdots + (-1)^n\dbinom{n}{n}$ **17.** $140\sqrt{2}$ **19.** $9 - 12\sqrt{3}t + 18t^2 - 4\sqrt{3}t^3 + t^4$ **21.** -3 **23.** $-5 + \sqrt{8}$

CHAPTER 4, TEST p. 226

1. B **2.** A **3.** $\{(a, d), (a, e), (b, d), (b, e), (c, d), (c, e)\}$ **4.** C **5.** Yes **6.** No **7.** 840 **8.** 35 **9.** 720

10. 1 **11.** $6 \cdot 6 \cdot 6 = 216$ **12.** $\dbinom{52}{2} = 1326$ **13.** $5! = 120$ **14.** $4 \cdot 4 \cdot 4 = 64,\ 4 \cdot 3 \cdot 2 = 24$ **15.** $2 \cdot 4! \cdot 4! = 1152$

16. $\dfrac{8!}{1!2!2!1!1!1!} = 10{,}080$ **17.** $\dbinom{12}{3} + \dbinom{9}{1}\dbinom{12}{2} = 814$ **18.** 35 **19.** $x^8 + 12x^6y + 54x^4y^2 + 108x^2y^3 + 81y^4$

CHAPTER 5

MARGIN EXERCISES

1. a) $\frac{1}{6}$ b) Answers may vary c) Answers may vary **2.** a) $\frac{4}{52}$, or $\frac{1}{13}$ b) $\frac{2}{52}$, or $\frac{1}{26}$ c) $\frac{26}{52}$, or $\frac{1}{2}$ d) $\frac{12}{52}$, or $\frac{3}{13}$ **3.** $\frac{5}{8}$

4. 0 **5.** 1 **6.** $\frac{1}{17}$ **7.** $\frac{4}{13}$ **8.** $\frac{6}{13}$ **9.** a) $\frac{5}{36}$ b) $\frac{6}{36}$ c) $\frac{2}{36}$ d) 0 **10.** a) $\frac{23}{34}$ b) 0.37 **11.** a) $\frac{1}{26}$ b) $\frac{25}{26}$

c) $\frac{1}{4}$ d) $\frac{3}{4}$ **12.** a) 1:25 b) 25:1 **13.** 0.63 **14.** 0.76 **15.** $\frac{45}{91}$ **16.** a) $\frac{1}{6}$ b) $\frac{1}{2}$ c) $\frac{1}{12}$ d) Yes

17. a) $\frac{1}{2}$ b) $\frac{1}{2}$ c) $\frac{1}{4}$ **18.** a) $\frac{1}{6}$ b) 0.132 **19.** $\frac{1}{2} \cdot \frac{1}{2} \cdot \frac{1}{6}$, or $\frac{1}{24}$ **20.** $\frac{1}{6} \cdot \frac{1}{6} \cdot \frac{1}{6}$, or $\frac{1}{216}$ **21.** $(\frac{1}{2})^4$, $\frac{1}{16}$

22. a) $\frac{4}{5} \cdot \frac{2}{3}$, or $\frac{8}{15}$ b) $(1 - \frac{4}{5})(1 - \frac{2}{3})$, or $\frac{1}{15}$ c) $\frac{6}{15}$, or $\frac{2}{5}$ **23.** Same answers **24.** $\frac{5}{32}$ **25.** Answers may vary **26.** 0.9

27. a) 0.37 b) 0.63 **28.** 0.06 **29.** a) $p_4 = 0.625$, $p_5 = 0.63333$, $p_6 = 0.63194$, $p_7 = 0.63213$ **30.** 0.2

31. $\frac{2}{20} \div \frac{8}{20} = \frac{1}{4}$ **32.** a) 0.5 b) 0.47 c) 0.56 d) No; $0.47 \neq 0.56$ e) 0.38 **33.** No; $\frac{4}{5} \cdot \frac{2}{3} \neq \frac{3}{5}$ **34.** $\frac{45}{91}$ **35.** $\frac{26}{91}$, or $\frac{2}{7}$

36. a) $\frac{200}{450}(0.03) + \frac{150}{450}(0.04) + \frac{100}{450}(0.05)$, or $\frac{17}{450}$ b) $\frac{6}{17}, \frac{6}{17}, \frac{5}{17}$

EXERCISE 5.1, pp. 234–235

1. $\frac{4}{52}$, or $\frac{1}{13}$ **3.** $\frac{13}{52}$, or $\frac{1}{4}$ **5.** $\frac{4}{52}$, or $\frac{1}{13}$ **7.** $\frac{1}{2}$ **9.** $\frac{8}{52}$, or $\frac{2}{13}$ **11.** $\frac{6}{16}$, or $\frac{3}{8}$ **13.** 0 **15.** $\dfrac{\binom{7}{2} \cdot \binom{8}{2}}{\binom{15}{4}} = \dfrac{28}{65}$

17. $\frac{4}{36}$, or $\frac{1}{9}$ **19.** $\frac{1}{36}$ **21.** 0 **23.** $\dfrac{\binom{7}{3}\binom{8}{2}\binom{10}{2}}{\binom{25}{7}} = \dfrac{441}{4807}$ **25.** $\dfrac{\binom{4}{3} \cdot \binom{4}{2}}{\binom{52}{5}} = \dfrac{1}{108,290}$ **27.** $\dfrac{\binom{4}{4}\binom{4}{1}}{\binom{52}{5}} = \dfrac{1}{649,740}$

29. 0 **31.** $\dfrac{\binom{10}{2}\binom{10}{2}}{\binom{20}{4}} = \dfrac{135}{323}$ **33.** a) $\dfrac{\binom{5}{1}\binom{3}{1}}{\binom{8}{2}} = \dfrac{15}{28}$ b) $\dfrac{\binom{5}{2}}{\binom{8}{2}} = \dfrac{5}{14}$ c) $\dfrac{\binom{3}{2}}{\binom{8}{2}} = \dfrac{3}{28}$ d) $\frac{5}{14} + \frac{3}{28} = \frac{13}{28}$

35. a) $\frac{28}{45}$ b) 17:28 c) 28:17 **37.** a) $\frac{1}{201}$ b) $\frac{3}{250,003}$ c) $\frac{3}{500,003}$ d) $\frac{2}{201}$

EXERCISE SET 5.2, p. 241

1. a) 0.77 b) $\frac{9}{14}$ **3.** $\frac{1}{6}$ **5.** a) $\frac{68}{95}$ b) $\frac{27}{95}$ c) $\frac{51}{190}$ d) $\frac{3}{190}$ **7.** a) $\frac{5}{9}$ b) $\frac{4}{9}$ c) $\frac{1}{9}$ **9.** a) $\frac{1}{221}$ b) $\frac{1}{17}$ c) $\frac{4}{17}$ d) $\frac{12}{17}$

11. $\frac{1}{3}$ **13.** $\frac{1}{30}$

EXERCISE SET 5.3, pp. 255–257

1. $\frac{11}{18}$ **3.** 0.1584 **5.** a) $\frac{1}{4}$ b) $\frac{1}{13}$ c) $\frac{1}{52}$ **7.** a) $\frac{1}{4}$ b) $\frac{13}{51}$ c) $\frac{13}{204}$ **9.** $\frac{1}{3} \cdot \frac{2}{3} \cdot \frac{2}{3} \cdot \frac{1}{3} \cdot \frac{2}{3}$, or $\frac{8}{243}$ **11.** 0.2

13. $(0.90)^6$, or 0.531441; $(0.10)^6$ or 0.000001 **15.** 0.2 **17.** a) $\frac{15}{32}$ b) $\frac{3}{4} \cdot \frac{3}{8} + \frac{1}{4} \cdot \frac{5}{8}$, or $\frac{7}{16}$ c) $\frac{3}{32}$ **19.** $\frac{37}{64}$

21. Left to the student **23.** $\frac{4}{15}$, $\frac{23}{45}$ **25.** a) $\dfrac{\binom{13}{1}\binom{4}{2}\binom{12}{3}\binom{4}{1}\binom{4}{1}}{\binom{52}{5}} = \dfrac{1760}{4165}$ b) $\dfrac{\binom{13}{1}\binom{4}{3}\binom{12}{2}\binom{4}{1}\binom{4}{1}}{\binom{52}{5}} = \dfrac{88}{4165}$

(Continued on next page.)

25. (*Continued*)

c) $\dfrac{\binom{13}{2}\binom{4}{2}\binom{4}{2}\binom{44}{1}}{\binom{52}{5}} = \dfrac{198}{4165}$ d) $\dfrac{\binom{13}{1}\binom{4}{2}\binom{12}{1}\binom{4}{3}}{\binom{52}{5}} = \dfrac{6}{4165}$ e) $\dfrac{\binom{13}{1}\binom{4}{4}\binom{48}{1}}{\binom{52}{5}} = \dfrac{1}{4165}$ **27.** $\frac{29}{54}$ **29.** $\frac{20}{27}$

31. $p_n^T(2) = p_{n-1}(1)[1 + A_n^T(1)(1 + A_n^T(2)(1 + \cdots (1 + A_n^T(m_L)) \cdots], \text{ where } A_n^T(m) = \dfrac{(n - 2m)(n - 2m - 1)}{2m(365 + m + 1 - n)} \text{ and}$

$m_L = \text{the largest integer} \leqslant \dfrac{(n - 2)}{2}.$

EXERCISE SET 5.4, p. 264

1. 0.71   **3.** 0.8   **5.** $\frac{11}{15}$ **7.** 0.70018   **9.** 51% (100% − 49%)   **11.** (0.4)(0.4) = 0.16, (0.6)(0.4) + (0.4)(0.6) = 0.48
13. 5%

EXERCISE SET 5.5, pp. 273–275

1. 0.5   **3.** $\frac{12}{13}$ **5.** $\frac{1}{13}$ **7.** a) $\frac{3}{7}$ b) $\frac{4}{7}$ **9.** $\frac{3}{4}$; no   **11.** 51% should show up, ∴ not independent; O.K.
13. Yes   **15.** $\frac{7}{30}$ **17.** 0.0215   **19.** 14%, 14.5%   **21.** $\frac{3}{4}$ **23.** $\frac{2}{3}$ **25.** $\frac{281}{480}, \frac{7}{96}$ **27.** 0.875   **29.** 0.19

EXERCISE SET 5.6, pp. 279–281

1. $\frac{3}{43}$ **3.** $\frac{4}{7}, \frac{8}{29}$ **5.** $\frac{2}{3}$ **7.** $\frac{49}{89}$ **9.** $\frac{1}{32}$ **11.** 0.16, $\frac{5}{16}$ **13.** $\frac{1}{2}$ **15.** $\frac{17}{30}, \frac{5}{8}$ **17.** $\frac{1}{2}, \frac{297}{400}, \frac{13}{33}$ **19.** $\frac{5}{16}, \frac{1}{2}, \frac{1}{8}$

CHAPTER 5, TEST, p. 282

1. a) $p(E^c) = \frac{16}{39}$ b) 23:16 c) 16:23   **2.** 0.9   **3.** $\frac{2}{5}$ **4.** 0.4838   **5.** a) 0.0344 b) No   **6.** $\frac{1}{3}$
7. a) $\frac{1}{775}$ b) $\frac{60}{775}$ c) $\frac{714}{775}$ **8.** $\frac{3}{5} \cdot \frac{2}{5} \cdot \frac{2}{5} \cdot \frac{2}{5} \cdot \frac{3}{5} = \frac{72}{3125}$ **9.** 0.31   **10.** 0.8; no   **11.** a) $\frac{39}{1000}$ b) $\frac{10}{39}, \frac{21}{39}, \frac{8}{39}$

CHAPTER 6

MARGIN EXERCISES

1.

x	0	1	2	3	4	5
p	$\frac{6}{36}$	$\frac{10}{36}$	$\frac{8}{36}$	$\frac{6}{36}$	$\frac{4}{36}$	$\frac{2}{36}$

2.

x	0	1	2	3
p	$\frac{24}{91}$	$\frac{45}{91}$	$\frac{20}{91}$	$\frac{2}{91}$

3. 78.08   **4.** a) 2.277 b) Easy   **5.** $\frac{70}{36}$ **6.** 1   **7.** $E(\bar{X}) = \$1.76$; to charity $8.20 per ticket

8. Example 6.   **9.** Betting *with* shooter $E(\bar{X}) = -\frac{7}{495} = -\frac{28}{1980}$ to win. Betting *against* shooter $E(\bar{X}) = -\frac{3}{220} = -\frac{27}{1980}$
to win. ∴ Better to bet *against* shooter.   **10.** 64.91   **11.** $\frac{52}{91}$ **12.** 8.06   **13.** 0.75593

14. $\binom{6}{2} \cdot \frac{1}{2^6} = \frac{15}{64}$

TTHHHH	HHTTHH	HHHHTT	HTHHTH
HTTHHH	HHTHTH	HHHTTH	THHHHT
THTHHH	HTHHHT	HHHTHT	THHHTH
HTHTHH	THHTHH	HHTHHT	

15. $\binom{6}{4}(0.8)^4(0.2)^2 = 0.24576$

16. K is the number cured.

k	p_k
0	0.000064
1	0.001536
2	0.015360
3	0.081920
4	0.245760
5	0.393216
6	0.262144

17. 4.8 **18.** $\sigma^2 = 0.96$, $\sigma = 0.97980$ **19.** $p = \frac{20}{91}$ ($= p_2$ in Margin Exercise 2). Now $p_3 = \binom{4}{3}\left(\frac{20}{91}\right)^3\left(\frac{71}{91}\right)^1 = 0.033132$

EXERCISE SET 6.1, pp. 287–288

1.

n	0	1	2	3	4
p	$\frac{14}{323}$	$\frac{80}{323}$	$\frac{135}{323}$	$\frac{80}{323}$	$\frac{14}{323}$

3.

n	0	1	2	3
p	$\frac{1}{14}$	$\frac{6}{14}$	$\frac{6}{14}$	$\frac{1}{14}$

5.

n	0	1	2	3	4	5
p	$\frac{1001}{7752}$	$\frac{3003}{7752}$	$\frac{2730}{7752}$	$\frac{910}{7752}$	$\frac{105}{7752}$	$\frac{3}{7752}$

7.

n	0	1	2	3
p	$\frac{1}{8}$	$\frac{3}{8}$	$\frac{3}{8}$	$\frac{1}{8}$

9.

n_{White}	0	1	2
n_{Red}	3	2	1
p	$\frac{5}{12}$	$\frac{6}{12}$	$\frac{1}{12}$

11.

n	0	1	2	3
p	$\frac{1}{20}$	$\frac{9}{20}$	$\frac{9}{20}$	$\frac{1}{20}$

13.

n	0	1	2	3
p	$\frac{8}{100}$	$\frac{54}{100}$	$\frac{36}{100}$	$\frac{2}{100}$

15.

n	4	5	6	7
p	$\frac{1}{35}$	$\frac{4}{35}$	$\frac{10}{35}$	$\frac{20}{35}$

17.

n	1	2	3	4	5	6	$\cdots$
p	0	$\frac{1}{2}$	$\frac{1}{4}$	$\frac{1}{8}$	$\frac{1}{16}$	$\frac{1}{32}$	$\cdots$

EXERCISE SET 6.2, pp. 296–297

1. 86.533 mph **3.** \$5.89 **5.** 2 women, 1 man **7.** 2 **9.** $\frac{3}{5}$ **11.** $\frac{3}{2}$ **13.** $\frac{3}{4}$, 1 **15.** $\frac{3}{2}$ **17.** $\frac{132}{100}$ **19.** $\frac{224}{35}$

EXERCISE SET 6.3, pp. 299–300

1. 11.530, 3.396 mph **3.** 1.53, \$1.24 **5.** $\frac{114}{190} = 0.6$, 0.77460 **7.** 0.44, 0.66332

EXERCISE SET 6.4, pp. 306–307

1.

n	0	1	2	3	4	5
p	$\frac{1}{32}$	$\frac{5}{32}$	$\frac{10}{32}$	$\frac{10}{32}$	$\frac{5}{32}$	$\frac{1}{32}$

$\frac{5}{2}, \frac{5}{4}, 1.11803$ **3.** $3, 3, \frac{11}{16}$ **5.** $\frac{3}{16}, \frac{5}{2}, \frac{5}{4}, 1.11803$ **7.** 0.885735, 0.6, 0.54, 0.73485

9.

n	0	1	2	3	4	5
p	0.00032	0.00640	0.05120	0.20480	0.40960	0.32768

; 4, 0.67232

11.

n	0	1	2	3	4	5
10%p	0.59049	0.32805	0.07290	0.00810	0.00045	0.00001
20%p	0.32768	0.40960	0.20480	0.05120	0.00640	0.00032
30%p	0.16807	0.36015	0.30870	0.13230	0.02835	0.00243

10%, 20%, 30%, 30%

13. $\frac{3}{16}, \frac{15}{16}$ **15.** 0.26272, 0.15053

CHAPTER 6, TEST, p. 308

1.

n	0	1	2	3	4	5
p	$\binom{5}{0}\left(\frac{3}{4}\right)^5\left(\frac{1}{4}\right)^0$ $=\frac{243}{1024}$	$\binom{5}{1}\left(\frac{3}{4}\right)^4\left(\frac{1}{4}\right)^1$ $=\frac{405}{1024}$	$\binom{5}{2}\left(\frac{3}{4}\right)^3\left(\frac{1}{4}\right)^2$ $=\frac{270}{1024}$	$\binom{5}{3}\left(\frac{3}{4}\right)^2\left(\frac{1}{4}\right)^3$ $=\frac{90}{1024}$	$\binom{5}{4}\left(\frac{3}{4}\right)\left(\frac{1}{4}\right)^4$ $=\frac{15}{1024}$	$\binom{5}{5}\left(\frac{3}{4}\right)^0\left(\frac{1}{4}\right)^5$ $=\frac{1}{1024}$

2. $\binom{5}{4}\left(\frac{1}{4}\right)^1\left(\frac{3}{4}\right)^4 + \binom{5}{5}\left(\frac{1}{4}\right)^0\left(\frac{3}{4}\right)^5 = \frac{648}{1024} = \frac{81}{128}$ **3.** $5 \cdot \frac{3}{4} = \frac{15}{4}$ **4.** $\mu = 2$ **5.** $\sigma^2 = \frac{4}{3}$ **6.** $\sigma = \frac{2}{3}\sqrt{3} \approx 1.155$

CHAPTER 7

MARGIN EXERCISES

1. a)

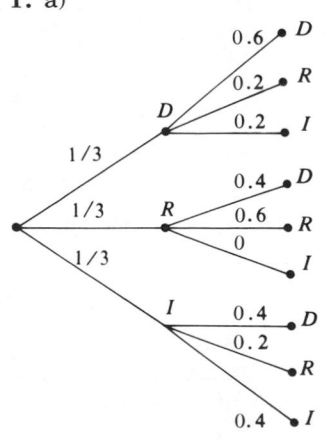

b)

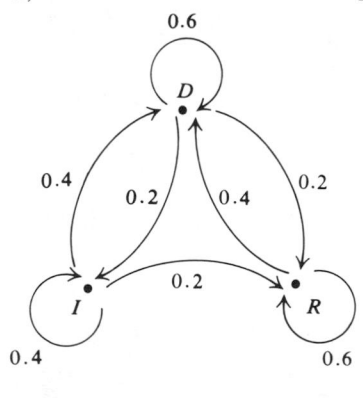

2.

$$\begin{array}{c} \\ D \\ R \\ I \end{array}\begin{array}{ccc} D & R & I \\ \begin{bmatrix} 0.6 & 0.2 & 0.2 \\ 0.4 & 0.6 & 0 \\ 0.4 & 0.2 & 0.4 \end{bmatrix} \end{array}$$

3. a) Qualifies

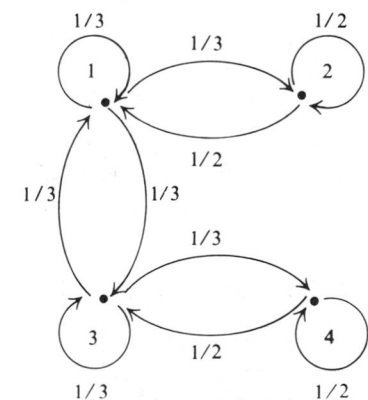

b) Does not qualify. Sum of elements in second row now is less than one.

4. $P_0 = [\frac{1}{3} \quad \frac{1}{3} \quad \frac{1}{3}]$

$P_1 = [\frac{7}{15} \quad \frac{5}{15} \quad \frac{3}{15}]$

$P_2 = [\frac{37}{75} \quad \frac{25}{75} \quad \frac{13}{75}]$

5. Regular

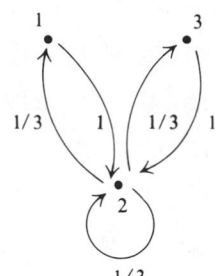

6. Not regular, second state absorbing

7. Regular, There is a nonzero element in the main diagonal.

8. Regular

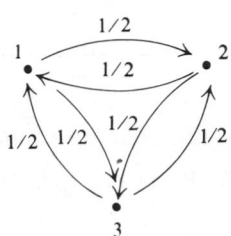

9. $P_8 = [0.235349 \quad 0.305860 \quad 0.458791]$ for $P_0 = [1 \quad 0 \quad 0]$;

$\bar{P} = [0.230769 \quad 0.307692 \quad 0.461538]$

10. $\bar{P} = [\frac{1}{5} \quad \frac{3}{5} \quad \frac{1}{5}]$

11. Chain is cyclic: $T = T^3 = T^5 = \cdots$, $\therefore$ not regular. Since chain is ergodic, we can obtain $\bar{P} = [\frac{1}{2} \quad \frac{1}{3} \quad \frac{1}{6}]$.
But this is *not* a "long-run" probability vector since chain is cyclic.

12. Answers may vary **13.** $\alpha_3\beta_2$, $v = 4$ **14.** Answers may vary **15.** $\alpha_4\beta_2$, $v = 5$

16. $X = [\frac{3}{8} \quad \frac{5}{8}]$, $Y = [\frac{1}{4} \quad \frac{3}{4}]$, $v = \frac{13}{4}$ **17.** $\phi = [2, 2, 1, 1]/6$ **18.** $\phi = \beta = [1, 1, 1, 0]/3$

EXERCISE SET 7.1, pp. 315–316

1.

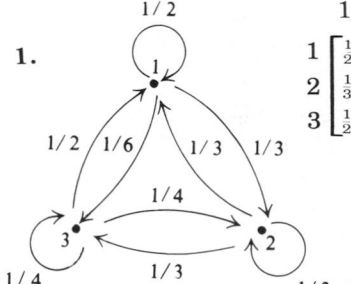

$$\begin{array}{c c} & \begin{array}{c c c} 1 & 2 & 3 \end{array} \\ \begin{array}{c} 1 \\ 2 \\ 3 \end{array} & \begin{bmatrix} \frac{1}{2} & \frac{1}{3} & \frac{1}{6} \\ \frac{1}{3} & \frac{1}{3} & \frac{1}{3} \\ \frac{1}{2} & \frac{1}{4} & \frac{1}{4} \end{bmatrix} \end{array}$$

3. No; negative element in first row.

5.

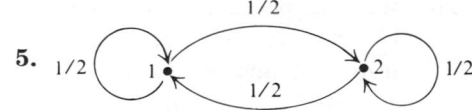

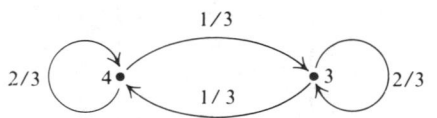

7. 1 (○) 2 (○) 1

9.

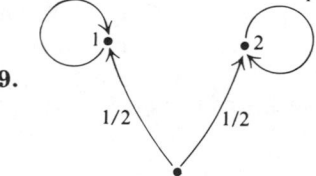

11. 1 •——→ 2 •——→ 3 • 1

13.

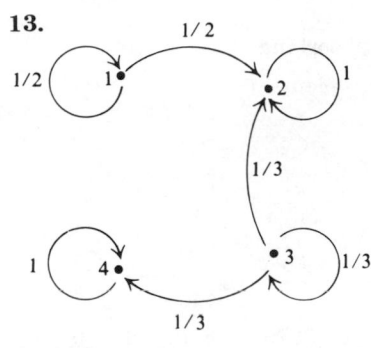

15. First zone: $P_0 = [0 \quad 1 \quad 0]$, $P_1 = [\frac{1}{3} \quad \frac{1}{3} \quad \frac{1}{3}]$, $P_2 = [\frac{16}{36} \quad \frac{11}{36} \quad \frac{9}{36}]$

17. $P_0 = [0 \quad 0 \quad 1]$ $P_1 = [\frac{1}{2} \quad \frac{1}{4} \quad \frac{1}{4}]$ $P_2 = [\frac{22}{48} \quad \frac{15}{48} \quad \frac{11}{48}]$

19. $P_2 = [\frac{9}{25} \quad \frac{14}{25} \quad \frac{2}{25}]$ **21.** $P_2 = [\frac{1}{4} \quad \frac{1}{4} \quad \frac{1}{4} \quad \frac{1}{4}]$ **23.** $P_1 = [\frac{1}{2} \quad \frac{1}{2}]$

25. $P_2 = [1 \quad 0]$ **27.** $P_2 = [\frac{1}{2} \quad \frac{1}{2} \quad 0]$ **29.** $P_3 = [0 \quad 0 \quad 1]$

31. $P_2 = [\frac{1}{16} \quad \frac{3}{16} \quad \frac{7}{16} \quad \frac{5}{16}]$ **33.** $P_2 = [\frac{9}{144} \quad \frac{79}{144} \quad \frac{4}{144} \quad \frac{52}{144}]$

35. $P_2 = [\frac{5}{36} \quad \frac{13}{36} \quad \frac{13}{36} \quad \frac{5}{36}]$

EXERCISE SET 7.2, pp. 320–321

1. Not regular; second state absorbing **3.** Not regular; absorbing states **5.** Not regular; absorbing states

7. Not regular; absorbing state **9.** Not regular; absorbing states **11.** Regular **13.** Regular **15.** Not regular

17. Regular **19.** Not regular

EXERCISE SET 7.3, pp. 325–326

1. $[\frac{1}{3}, \quad \frac{2}{3}]$ **3.** $[\frac{2}{12} \quad \frac{5}{12} \quad \frac{5}{12}]$ **5.** $[\frac{2}{10} \quad \frac{3}{10} \quad \frac{5}{10}]$ **7.** $[\frac{5}{17} \quad \frac{2}{17} \quad \frac{5}{17} \quad \frac{5}{17}]$ **9.** $[\frac{1}{10} \quad \frac{2}{10} \quad \frac{4}{10} \quad \frac{3}{10}]$ **11.** $[\frac{30}{67} \quad \frac{21}{67} \quad \frac{16}{67}]$

13.
$$T = \begin{bmatrix} 1 & 0 & 0 & 0 & 0 \\ \frac{1}{2} & 0 & \frac{1}{2} & 0 & 0 \\ 0 & \frac{1}{2} & 0 & \frac{1}{2} & 0 \\ 0 & 0 & \frac{1}{2} & 0 & \frac{1}{2} \\ 0 & 0 & 0 & 0 & 1 \end{bmatrix}$$

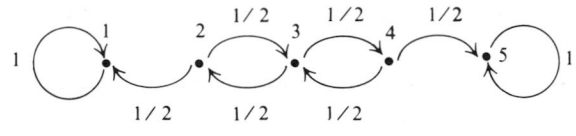

The two fixed points are his home and the bar.
$P_0 = [0 \quad 0 \quad 1 \quad 0 \quad 0]$, $P_1 = [0 \quad \frac{1}{2} \quad 0 \quad \frac{1}{2} \quad 0]$, $P_2 = [\frac{1}{4} \quad 0 \quad \frac{1}{2} \quad 0 \quad \frac{1}{4}]$, $P_3 = [\frac{1}{4} \quad \frac{1}{4} \quad 0 \quad \frac{1}{4} \quad \frac{1}{4}]$, $P_4 = [\frac{3}{8} \quad 0 \quad \frac{2}{8} \quad 0 \quad \frac{3}{8}]$ $\cdots$
For large n, P_n approaches $[\frac{1}{2} \quad 0 \quad 0 \quad 0 \quad \frac{1}{2}]$; p (home) $= \frac{1}{2}$.

15.
$$T = \begin{bmatrix} 0 & 1 & 0 & 0 & 0 \\ \frac{1}{2} & 0 & \frac{1}{2} & 0 & 0 \\ 0 & \frac{1}{2} & 0 & \frac{1}{2} & 0 \\ 0 & 0 & \frac{1}{2} & 0 & \frac{1}{2} \\ 0 & 0 & 0 & 0 & 1 \end{bmatrix}$$

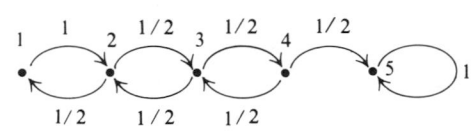

The fixed point is his home.
$P_0 = [0 \quad 0 \quad 1 \quad 0 \quad 0]$
$P_1 = [0 \quad \frac{1}{2} \quad 0 \quad \frac{1}{2} \quad 0]$
$P_2 = [\frac{1}{4} \quad 0 \quad \frac{1}{2} \quad 0 \quad \frac{1}{4}]$
$P_3 = [0 \quad \frac{1}{2} \quad 0 \quad \frac{1}{4} \quad \frac{1}{4}]$
$P_4 = [\frac{2}{8} \quad 0 \quad \frac{3}{8} \quad 0 \quad \frac{3}{8}]$

For large n, P_n approaches $[0 \quad 0 \quad 0 \quad 0 \quad 1]$; p (home) $= 1$.

17. $T = \begin{bmatrix} 0 & \frac{2}{3} & 0 & \frac{1}{3} \\ \frac{2}{3} & 0 & \frac{1}{3} & 0 \\ 0 & \frac{1}{2} & 0 & \frac{1}{2} \\ \frac{1}{2} & 0 & \frac{1}{2} & 0 \end{bmatrix}$ $\bar{P} = [\frac{3}{10} \quad \frac{3}{10} \quad \frac{2}{10} \quad \frac{2}{10}]$. Thus, the mouse keeps moving.

EXERCISE SET 7.4, p. 333

1. (α, β_1), 3   **3.** (α_2, β_1), 2   **5.** (α_2, β_2), 4   **7.** (α_4, β_4), 3

EXERCISE SET 7.5, pp. 337–338

1. $X = [\frac{1}{2} \quad \frac{1}{2}]$, $Y = [\frac{1}{4} \quad \frac{3}{4}]$, $v = \frac{5}{2}$ **3.** $X = [0 \quad \frac{6}{7} \quad \frac{1}{7}]$, $Y = [\frac{6}{7} \quad 0 \quad \frac{1}{7} \quad 0]$, $v = \frac{20}{7}$ **5.** $X = [0 \quad 0 \quad \frac{1}{2} \quad \frac{1}{2}]$, $Y = [\frac{3}{5} \quad 0 \quad \frac{2}{5} \quad 0]$, $v = 4$ **7.** Plant: $[\frac{9}{10} \quad \frac{1}{10}]$; inspector: $[\frac{9}{10} \quad \frac{1}{10}]$; $v = -\frac{7}{10}$, where first strategy is "far out" and second is "local stream."   **9.** 5   **11.** $X = [\frac{1}{11} \quad \frac{10}{11}]$, $Y = [\frac{5}{11} \quad \frac{6}{11}]$, $v = \frac{5}{11}$; $\therefore$ for *one* trial: take $X = [0 \quad 1]$, $Y = [0 \quad 1]$; then $v = 0$.

EXERCISE SET 7.6, p. 344

1. $\phi = \beta = [3, 1, 1, 1]/6$ **3.** $\phi = \beta = [1, 0, 0, 0]$; no one else can influence a decision   **5.** [51; 49, 48, 3], $\phi = \beta = [1, 1, 1]/3$ **7.** $\phi = \beta = [2, 2, 1, 1]/6$ for [9; 5, 5, 3, 2] so that quota must be raised from 8 to 9   **9.** [4; 3, 1, 1, 1]; $\phi = [9, 1, 1, 1]/12$, $\beta = [7, 1, 1, 1]/10$

CHAPTER 7, TEST, p. 345

1.

2. a) No, has absorbing states) b) No, cyclic c) No, not ergodic) d) Yes, taking powers

3. $P_2 = [\frac{3}{12} \quad \frac{7}{12} \quad \frac{2}{12}]$, $\bar{P} = [\frac{2}{9} \quad \frac{3}{9} \quad \frac{4}{9}]$

4. Minimax $\neq$ maximin, $\therefore$ no pure-strategy solution

5. β_1 dominates β_2
   α_1 dominates α_2
   α_4 dominates α_3

$$\begin{array}{cc} & \begin{array}{cc} \beta_1 & \beta_3 \end{array} \\ \begin{array}{c} \alpha_1 \\ \alpha_4 \end{array} & \begin{bmatrix} 1 & 5 \\ 3 & 2 \end{bmatrix} \end{array}$$

6. $X = [\frac{1}{5} \quad 0 \quad 0 \quad \frac{4}{5}]$, $Y = [\frac{3}{5} \quad 0 \quad \frac{2}{5}]$, $v = \frac{13}{5}$ **7.** $\phi = [4, 1, 1, 0]/6$ **8.** $\beta = [3, 1, 1, 0]/5$

CHAPTER 8

MARGIN EXERCISES

1. 4, 7, 12, 19; 228   **2.** 2; 3   **3.** 19; −5   **4.** $6300; −$346.75   **5.** 2; $\frac{1}{2}$ **6.** 70   **7.** $2832.50   **8.** 20,100

9. 256   **10.** $22,100   **11.** 5   **12.** −3   **13.** 0.85   **14.** 1.09   **15.** $\frac{1}{2}$ **16.** 256   **17.** $\frac{1}{81}$ **18.** 510

19. $21,470,000 approx..   **20.** No   **21.** No   **22.** 1   **23.** $\frac{3125}{3}$, or $1041\frac{2}{3}$ **24.** $20,000   **25.** $7100, $1420, 20%

26.

Year	Rate of depreciation	Annual depreciation	Book value	Total depreciation
0			$8700	
1	$\frac{1}{5}$ or 20%	$1420	7280	$1420
2	20%	1420	5860	2840
3	20%	1420	4440	4260
4	20%	1420	3020	5680
5	20%	1420	1600	7100

27. a) $V_n = \$8700 - (\$1420)n$ b) −$1420

28.

Year	Rate of depreciation	Annual depreciation	Book value	Total depreciation
0			$8700	
1	$\frac{2}{5}$ or 40%	$3480	5220	$3480
2	40%	2088	3132	5568
3	40%	1252.80	1879.20	6820.80
4		279.20	1600	7100
5		0	1600	7100

29. $V_n = \$8700(0.60)^n$

30. a) $\frac{5}{15}, \frac{4}{15}, \frac{3}{15}, \frac{2}{15}, \frac{1}{15}$ b) $2366.67, $6333.33; $1893.33, $4440.00; $1420.00, $3020.00

31.

Year	Rate of depreciation	Annual depreciation	Book value	Total depreciation
0			$8700	
1	$\frac{5}{15}$	$2366.67	6333.33	$2366.67
2	$\frac{4}{15}$	1893.33	4440.00	4260.00
3	$\frac{3}{15}$	1420.00	3020.00	5680.00
4	$\frac{2}{15}$	946.67	2073.33	6626.67
5	$\frac{1}{15}$	473.33	1600.00	7100.00

32. $1015   **33.** $1725.50
34. $2524.95
35. a) $1120   b) $1123.60
c) $1125.51   d) $1126.49
e) $1127.49
36. $2969.88   **37.** 9.203%
38. 15.2%   **39.** $5134.51
40. $21,241.93   **41.** $723.78
42. $19,636.29   **43.** $13,677.74
44. $404.08

EXERCISE SET 8.1, pp. 352–353

1. $\frac{1}{2}, \frac{2}{3}, \frac{3}{4}, \frac{4}{5}; \frac{15}{16}$ **3.** $0, \frac{1}{3}, \frac{2}{7}, \frac{3}{13}; \frac{14}{211}$ **5.** 2; 5   **7.** $1.06; $0.06   **9.** 5; $-\frac{2}{3}$ **11.** 47   **13.** $-$1628.16   **15.** 45,150

17. 690   **19.** $\dfrac{n(n+1)}{2}$ **21.** $4.96   **23.** $18,450

EXERCISE SET 8.2, pp. 358–359

1. 2   **3.** $-\frac{1}{3}$ **5.** 0.95   **7.** 2187   **9.** $\dfrac{1}{5^6}$, or $\dfrac{1}{15,625}$ **11.** $2331.64   **13.** 1016   **15.** $5866.60

17. $2,684,000 approx.   **19.** No   **21.** $12\frac{1}{2}$ **23.** 486   **25.** $12,500   **27.** $4,000,000,000   **29.** $\approx$$3,333,333; $66\frac{2}{3}$%

EXERCISE SET 8.3, p. 367

1. a)

Year	Rate of depreciation	Annual depreciation	Book value	Total depreciation
0			$8000	
1	$\frac{1}{4}$ or 25%	$1500	6500	$1500
2	25%	1500	5000	3000
3	25%	1500	3500	4500
4	25%	1500	2000	6000

b) $V_n = $8000 - ($1500)n$

c) $-$1500

3. a)

Year	Rate of depreciation	Annual depreciation	Book value	Total depreciation
0			$450	
1	$\frac{1}{8}$ or 12.5%	$56.25	393.75	$56.25
2	12.5%	56.25	337.50	112.50
3	12.5%	56.25	.281.25	168.75
4	12.5%	56.25	225.00	225.00
5	12.5%	56.25	168.75	281.25
6	12.5%	56.25	112.50	337.50
7	12.5%	56.25	56.25	393.75
8	12.5%	56.25	0	450.00

b) $V_n = \$450 - (\$56.25)n$ c) $-\$56.25$

5. a)

Year	Rate of depreciation	Annual depreciation	Book value	Total depreciation
0			$8000	
1	$\frac{2}{4}$ or 50%	$4000	4000	$4000
2	50%	2000	2000	6000
3		0	2000	6000
4		0	2000	6000

b) $V_n = \$8000(0.50)^n$

7. a)

Year	Rate of depreciation	Annual depreciation	Book value	Total depreciation
0			$450	
1	$\frac{2}{8}$ or 25%	$112.50	337.50	$112.50
2	25%	84.38	252.12	196.88
3	25%	63.28	189.84	260.16
4	25%	47.46	142.38	307.62
5	25%	35.60	106.78	343.22
6	25%	26.70	80.08	369.92
7	25%	20.02	60.06	389.94
8	25%	15.02	45.04	404.96

b) $V_n = \$450(0.75)^n$

9. a) $\frac{4}{10}, \frac{3}{10}, \frac{2}{10}, \frac{1}{10}$

b)

Year	Rate of depreciation	Annual depreciation	Book value	Total depreciation
0			$8000	
1	$\frac{4}{10}$	$2400	5600	$2400
2	$\frac{3}{10}$	1800	3800	4200
3	$\frac{2}{10}$	1200	2600	5400
4	$\frac{1}{10}$	600	2000	6000

11. a) $\frac{8}{36}, \frac{7}{36}, \frac{6}{36}, \frac{5}{36}, \frac{4}{36}, \frac{3}{36}, \frac{2}{36}, \frac{1}{36}$ b)

Year	Rate of depreciation	Annual depreciation	Book value	Total depreciation
0			$450	
1	$\frac{8}{36}$	$100	350	$100.00
2	$\frac{7}{36}$	87.50	262.50	187.50
3	$\frac{6}{36}$	75.00	187.50	262.50
4	$\frac{5}{36}$	62.50	125.00	325.00
5	$\frac{4}{36}$	50.00	75.00	375.00
6	$\frac{3}{36}$	37.50	37.50	412.50
7	$\frac{2}{36}$	25.00	12.50	437.50
8	$\frac{1}{36}$	12.50	0	450.00

EXERCISE SET 8.4, p. 373

1. $2060   **3.** $2560   **5.** a) $2280 b) $2289.80 c) $2295.05 d) $2297.76 e) $2300.52   **7.** $793.83
9. $788.49   **11.** $3450.32   **13.** $1144.90   **15.** $2.41   **17.** 6.25%

EXERCISE SET 8.5, p. 376

1. 8.16%   **3.** 9.308%   **5.** 8.271%   **7.** 8.328%   **9.** 8.329%   **11.** 13.7%   **13.** 16.8%

EXERCISE SET 8.6, pp. 379–380

1. $4439.94   **3.** $13,816.45   **5.** $48,594.74   **7.** $1228.29   **9.** $103,399.40   **11.** $3673.84   **13.** $992.37
15. $100.33   **17.** $P = \dfrac{Vi}{[(1 + i)^N - 1]}$

EXERCISE SET 8.7, pp. 383–384

1. $3387.21   **3.** $7023.58   **5.** $32,702.87   **7.** $760.95   **9.** $139.50   **11.** $548.84   **13.** $P = \dfrac{Si}{[1 - (1 + i)^{-N}]}$
15 $12,500; you want $1000 of interest each year, so just solve $I = Pit$ for P when $I = $1000, $i = 0.08$, and $t = 1$.

CHAPTER 8 TEST, p. 385

1. 5; 3   **2.** 62   **3.** $31.40   **4.** 1.05   **5.** $155.13   **6.** $1257.79   **7.** Yes, $1086.96

8. a)

Year	Rate of depreciation	Annual depreciation	Book value	Total depreciation
0			$8500	
1	$\frac{1}{4}$ or 25%	$1487.50	7012.50	$1487.50
2	25%	1487.50	5525	2975
3	25%	1487.50	4037.50	4462.50
4	25%	1487.50	2550	5950

b) $V_n = \$8500 - (\$1487.50)n$

c) $-\$1487.50$

9. a)

Year	Rate of depreciation	Annual depreciation	Book value	Total depreciation
0			$8500	
1	$\frac{2}{4}$ or 50%	$4250	4250	$4250
2		1700	2550	5950
3		0	2550	5950
4		0	2550	5950

b) $V_n = \$8500(0.50)^n$

10. a) $\frac{4}{10}, \frac{3}{10}, \frac{2}{10}, \frac{1}{10}$ b)

Year	Rate of depreciation	Annual depreciation	Book value	Total depreciation
0			$8500	
1	$\frac{4}{10}$	$2380	6120	$2380
2	$\frac{3}{10}$	1785	4335	4165
3	$\frac{2}{10}$	1190	3145	5355
4	$\frac{1}{10}$	595	2550	5950

11. a) $1180   b) $1191.02   c) $1194.05   d) $1195.62   **12.** 9.308%   **13.** 18.3%   **14.** $37,089.80   **15.** $1090.52
16. $103,796.58   **17.** $723.77

CHAPTER 9

MARGIN EXERCISES

1. a, b **2.** a) No, Yes, b) Yes, No, **3.** a) 20, 19.7, 18.2, 17.3, 17.03, 17.003; 14, 16.1, 16.4, 16.7, 16.97, 16.997
b) 17 c) 17 **4.** a) 7, 6.6, 6.2, 6.1, 6.01, 6.001; 4, 5, 5.4, 5.9, 5.99, 5.999 b) 6 **5.** L: 3, 3.4, 3.8, 3.98, 3.998; R: 1.6,
1.2, 0.2, 0.02, 0.002 **6.** a) Yes, 17; b) Yes, 17; c) Yes; d) Yes. **7.** $\sqrt[3]{x}$ is continuous by ii; 7 is continuous by i, and
x^2 is continuous by ii, so $7x^2$ is continuous by iii. Then $\sqrt[3]{x} - 7x^2$ is continuous by iii. Now x is continuous by ii, and 2
is continuous by i, so $x - 2$ is continuous by iii. Thus the quotient $\dfrac{\sqrt[3]{x} - 7x^2}{x - 2}$ is continuous by iv. **8.** 53 **9.** $\sqrt{8}$
10. a) 4.00, 4.50, 5.14, 6.00, 7.20, 9.00, 12.00, 18.00, 36.00, 54.00, 108.00; b) ∞; c) ∞. **11.** a) -0.3333, -0.2778,
-0.2632, -0.2564, -0.251, -0.2501; -0.2, -0.2222, -0.2273, -0.2439, -0.2494, -0.2499; b) -0.25, or $-\frac{1}{4}$. **12.** -8
13. $\frac{1}{6}$ **14.** a) $2x + 1$, $2x + 0.7$, $2x + 0.4$, $2x + 0.1$, $2x + 0.01$, $2x + 0.001$; b) $2x$. **15.** a) 3.25, 2.25, 2.0625, 2.025,
2.005, 2.0005; b) 2. **16.** a) $20\dfrac{\text{suits}}{\text{hr}}$, $35\dfrac{\text{suits}}{\text{hr}}$, $9\dfrac{\text{suits}}{\text{hr}}$, $36\dfrac{\text{suits}}{\text{hr}}$; b) 11 A.M. to 12 P.M. c) Workers were anticipating
the lunch break; answers may vary. d) 10 A.M. to 11 A.M. e) Workers were fatigued or took longer breaks than they
should have; answers may vary. f) $25\dfrac{\text{suits}}{\text{hr}}$. **17.** a) 21 b) 7 c) 28 d) 21 **18.** a) $\frac{1}{2}$ b) $\frac{1}{2}$ c) $\frac{1}{2}$

19.

x	h	$x + h$	$f(x)$	$f(x + h)$	$f(x + h) - f(x)$	$\dfrac{f(x + h) - f(x)}{h}$
3	2	5	36	100	64	32
3	1	4	36	64	28	28
3	0.1	3.1	36	38.44	2.44	24.4
3	0.01	3.01	36	36.2404	0.2404	24.04
3	0.001	3.001	36	36.024004	0.024004	24.004

20. a) $f(x + h) = 4x^2 + 8xh + 4h^2$; b) $f(x + h) - f(x) = 8xh + 4h^2$; c) $\dfrac{f(x + h) - f(x)}{h} = 4(2x + h)$; d) 36, 40, 44, 47.6,
47.96, 47.996. **21.** a) $4(3x^2 + 3xh + h^2)$; b) 28, 45.64, 47.7604, 47.976004. **22.** a) $\dfrac{-1}{x(x + h)}$; b) -0.1, -0.1667,
-0.2381, -0.2488, -0.2499. **23.** a) L_2, L_3, L_4, L_6; b) -2, -1, 0, 1, 2; c) $m(x) = 2x$. **24.** $f'(5) = 10$ **25.** $f'(x) = 8x$,
$f'(5) = 40 = $ the slope of the tangent line at $(5, f(5))$, or $(5, 100)$. **26.** $f'(x) = 12x^2$, $f'(-5) = 300$, $f'(0) = 0$.

27. $f'(x) = \dfrac{-1}{x^2}$, $f'(-10) = -\frac{1}{100} = -0.01$, $f'(-2) = -\frac{1}{4} = -0.25$ **28.** x_2, x_4, x_5, x_6 **29.** a) $3x^2$ b) $3x^2$ c) 48 **30.** $\dfrac{dy}{dx} = 6x^5$

31. $\dfrac{dy}{dx} = -7x^{-8}$, or $-\dfrac{7}{x^8}$ **32.** $\dfrac{dy}{dx} = \frac{1}{3}x^{-2/3}$, or $\dfrac{1}{3\sqrt[3]{x^2}}$ **33.** $\dfrac{dy}{dx} = -\frac{1}{4}x^{-5/4}$, or $\dfrac{-1}{4\sqrt[4]{x^5}}$ **34.** $g'(x) = 0$ **35.** $\dfrac{dy}{dx} = 100x^{19}$

36. $\dfrac{dy}{dx} = \dfrac{3}{x^2}$ **37.** $\dfrac{dy}{dx} = -4x^{-1/2}$, or $\dfrac{-4}{\sqrt{x}}$ **38.** $\dfrac{dy}{dx} = x^{5.25}$ **39.** $\dfrac{dy}{dx} = -\frac{1}{4}$ **40.** $\dfrac{dy}{dx} = 28x^3 + 12x$ **41.** $\dfrac{dy}{dx} = 30x - \dfrac{4}{x^2} + \dfrac{1}{2\sqrt{x}}$

42. $(2, \frac{8}{3})$ **43.** $(2 + \sqrt{3}, \frac{8}{3} + \sqrt{3})$, $(2 - \sqrt{3}, \frac{8}{3} - \sqrt{3})$ **44.** a) $70\,\dfrac{\text{mi}}{\text{hr}}$ b) $100\,\dfrac{\text{mi}}{\text{hr}}$ **45.** a) $v(t) = 32t$;

b) $v(2) = 64\,\text{ft/sec}$; c) $v(10) = 320\,\text{ft/sec}$. **46.** $a(t) = 32\,\text{ft/sec}^2$ **47.** a) $V'(s) = 3s^2$; b) $V'(10) = 300\,\text{ft}^2$.

48. a) $P'(t) = 9700 + 20{,}000t$. b) $308{,}500$; $109{,}700\,\dfrac{\text{bacteria}}{\text{hr}}$. c) $428{,}200$; $129{,}700\,\dfrac{\text{bacteria}}{\text{hr}}$. **49.** a) $P(x) = 40x - 0.5x^2 - 3$;

b) $R(40) = \$1200$, $C(40) = \$403$, $P(40) = \$797$; c) $R'(x) = 50 - x$, $C'(x) = 10$, $P'(x) = 40 - x$; d) $R'(40) = \$10$ per

unit, $C'(40) = \$10$ per unit, $P'(40) = \$0$ per unit; e) No. **50.** $f'(x) = 54x^{17}$

51. $f'(x) = (9x^3 + 4x^2 + 10)(-14x + 4x^3) + (27x^2 + 8x)(-7x^2 + x^4)$. **52.** $f'(x) = 4x^3$ **53.** $\dfrac{3x^2 - 5}{x^6}$ **54.** $\dfrac{-x^4 + 3x^2 + 2x}{(x^3 + 1)^2}$

55. a) $R(x) = x(200 - x) = 200x - x^2$; b) $R'(x) = 200 - 2x$. **56.** $20x(1 + x^2)^9$ · **57.** $\dfrac{-x}{\sqrt{1 - x^2}}$ **58.** $-2x(1 + x^2)(1 + 3x^2)$

59. $2(x - 4)^4(6 - x)^2(21 - 4x)$ **60.** $\left(\dfrac{x + 5}{x - 4}\right)^{-2/3} \cdot \dfrac{-3}{(x - 4)^2}$ **61.** $3(x^2 - 1)$, $9x^2 - 1$ **62.** $4 \cdot \sqrt[3]{x} + 5$, $\sqrt[3]{4x + 5}$

63. $f'(x) = 12x^5 - 5x^4$, $f''(x) = 60x^4 - 20x^3$, $f'''(x) = 240x^3 - 60x^2$, $f^{(4)}(x) = 720x^2 - 120x$, $f^{(5)}(x) = 1440x - 120$,

$f^{(6)}(x) = 1440$. **64.** a) $\dfrac{dy}{dx} = 7x^6 - 3x^2$, b) $\dfrac{d^2y}{dx^2} = 42x^5 - 6x$, c) $\dfrac{d^3y}{dx^3} = 210x^4 - 6$, d) $\dfrac{d^4y}{dx^4} = 840x^3$. **65.** $\dfrac{d^2y}{dx^2} = \dfrac{4}{x^3}$

66. $a(t) = 12t^2$ **67.** a) $[a, b], [c, d]$; b) $[a, b), (c, d)$; c) $[b, c], [d, e]$; d) $(b, c), (d, e)$. **68.** a) $[b, c], [d, e]$; b) $[a, b]$,

$[c, d]$. **69.** a) $(0, b]$; b) $[a, 0)$. **70.** R, T, V. **71.** a) At $x = 3$ the function is increasing and the graph is concave up.

b) At $x = 1$, the function has a horizontal tangent and is concave up. **72.** a) $x_1, x_3, x_5, x_6, x_8, x_{10}$; b) x_4, x_7, x_9;

c) $x_1, x_3, x_4, x_5, x_6, x_7, x_8, x_9, x_{10}$.

73. Answers will vary on the graph, but it must have at least one critical point.

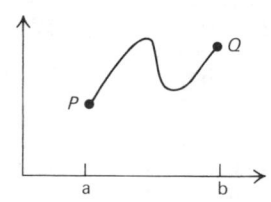

74. It is not possible.

75. a) Maximum at c_8, minimum at b;

b) Maximum at c_1, minimum at c_2.

76. Maximum $= 4$ at $x = 2$;

minimum $= 1$ at $x = 1$ and $x = -1$.

77. Maximum $= 176$ at $x = 6$;

minimum $= 97$ at $x = 5$.

78. Minimum $= -4$ at $x = 2$. There is no maximum **79.** Minimum $= -4$ at $x = 2$; maximum $= 0$ at $x = 0$ and

$x = 4$. **80.** a) Maximum $= 8$ at $x = 2$; minimum $= -8$ at $x = -2$. b) There are none.

81. Minimum $= \dfrac{10}{\sqrt{10}} + \sqrt{10}$ at $x = \dfrac{1}{\sqrt{10}}$.

82. a) y: 20, 16, 13.5, 12, 10, 8, 6.8, 0; A: 0, 64, 87.75, 96, 100, 96, 89.76, 0.

b) 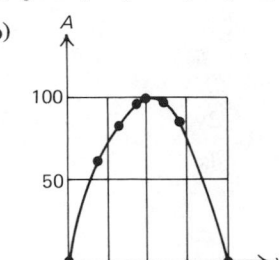   c) Yes.   d) Max. $= 100\,\text{ft}^2$ at $x = 10\,\text{ft}$.

83. 25 ft by 25 ft; $625\,\text{ft}^2$.

84. a) $R(x) = x(200 - x) = 200x - x^2$,

b) $P(x) = -x^2 + 192x - 5000$,

c) 96,   d) \$4216,

e) $200 - 96$, or \$104 per unit.

85. a) h: 4, 3.5, 3, 2.5, 2, 1.7, 1.5, 1, 0.6, 0.5, 0; V: 0, 3.5, 12, 22.5, 32, 35.972, 37.5, 36, 27.744, 24.5, 0.

b) 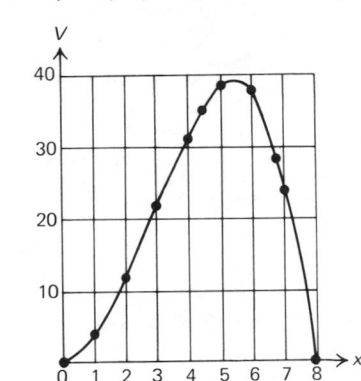   c) About 38; between 5 and 6, maybe 5.2.

86. Max. volume $= 74\frac{2}{27}\,\text{in}^3$.;
dimensions $6\frac{2}{3}$ in. by $6\frac{2}{3}$ in. by $1\frac{2}{3}$ in.

87. Min. area $= 300\,\text{cm}^2$;
dimensions 10 cm by 10 cm by 5 cm

88. 40¢

89.

x	$\dfrac{2500}{x}$	$\dfrac{x}{2}$	$10 \cdot \dfrac{x}{2}$	$20 + 9x$	$(20 + 9x)\dfrac{2500}{x}$	$10 \cdot \dfrac{x}{2} + (20 + 9x)\dfrac{2500}{x}$
2500	1	1250	\$12,500	\$22,520	\$22,520	\$35,020
1250	2	625	\$ 6250	\$11,270	\$22,540	\$28,790
500	5	250	\$ 2500	\$ 4520	\$22,600	\$25,100
250	10	125	\$ 1250	\$ 2270	\$22,700	\$23,950
167	15	84	\$ 840	\$ 1523	\$22,845	\$23,685
125	20	63	\$ 630	\$ 1145	\$22,900	\$23,530
100	25	50	\$ 500	\$ 920	\$23,000	\$23,500
90	28	45	\$ 450	\$ 830	\$23,240	\$23,690
50	50	25	\$ 250	\$ 470	\$23,500	\$23,750

90. 15 times per year at lot size 40.   **91.** 19 times per year at lot size 31.

EXERCISE SET 9.1, pp. 398–390

1. No   **3.** Yes   **5.** a) Yes b) No   **7.** a) Yes b) Yes   **9.** No, Yes, No, Yes.   **11.** Does not exist.   **13.** 41¢

15. a) 1, -0.75, -1.56, -1.79, -1.9799, -1.997999; b) Answers may vary; c) -2.   **17.** -10 **19.** Does not exist.

21. $-\frac{3}{2}$ **23.** 12   **25.** $\frac{2}{3}$ **27.** $\frac{1}{2}$ **29.** -0.25 **31.** $2x + 1$ **33.** a) 0.775, 0.753125, 0.75125, 0.7505, 0.75003125,

0.75000625; b) 0.75 or $\frac{3}{4}$.   **35.** 0   **37.** 5   **39.** a) \$800, \$736, \$677.12, \$622.95, \$573.11; b) \$5656.12; c) \$10,000.

41. 0, No.

EXERCISE SET 9.2, pp. 406–408

1. a) 70, 39, 29, 23 pleasure units/unit of product; b) The more you get, the less pleasure you get from each

additional unit.   **3.** a) $1.25\dfrac{\text{words}}{\text{min}}$, $1.25\dfrac{\text{words}}{\text{min}}$, $0.625\dfrac{\text{words}}{\text{min}}$, $0\dfrac{\text{words}}{\text{min}}$, $0\dfrac{\text{words}}{\text{min}}$. b) You have reached a saturation point;

you cannot memorize any more.   **5.** a) $125\dfrac{\text{million people}}{\text{yr}}$ for each; b) No; c) A: $290\dfrac{\text{million people}}{\text{yr}}$,

$-40\dfrac{\text{million people}}{\text{yr}}$, $-50\dfrac{\text{million people}}{\text{yr}}$, $300\dfrac{\text{million people}}{\text{yr}}$; B: $125\dfrac{\text{million people}}{\text{yr}}$ in all intervals; d) A.   **7.** a) \$93.99,

b) \$100, c) $-\$6.01$, d) $-\$6.01$.   **9.** a) 144 ft, b) 256 ft, c) $128\dfrac{\text{ft}}{\text{sec}}$.   **11.** $72,500\dfrac{\text{marriages}}{\text{yr}}$.   **13.** \$275,780,000 per

day.   **15.** a) $7(2x + h)$; b) 70, 63, 56.7, 56.07.   **17.** a) $-7(2x + h)$; b) $-70, -63, -56.7, -56.07$.

19. a) $7(3x^2 + 3xh + h^2)$, b) 532, 427, 344.47, 336.8407.   **21.** a) $\dfrac{-5}{x(x + h)}$; b) $-0.2083, -0.25, -0.3049, -0.3117$.

23. a) -2, b) All -2.   **25.** a) $2x + h - 1$, b) 9, 8, 7.1, 7.01.   **27.** $2ax + b + ah$.   **29.** $\dfrac{\sqrt{x + h} - \sqrt{x}}{h}$.

EXERCISE SET 9.3, p. 418

1. $f'(x) = 10x$, $f'(-2) = -20$, $f'(-1) = -10$, $f'(0) = 0$, $f'(1) = 10$, $f'(2) = 20$.   **3.** $f'(x) = -10x$, $f'(-2) = 20$, $f'(-1) = 10$,

$f'(0) = 0$, $f'(1) = -10$, $f'(2) = -20$.   **5.** $f'(x) = 15x^2$, $f'(-2) = 60$, $f'(-1) = 15$, $f'(0) = 0$, $f'(1) = 15$, $f'(2) = 60$.

7. $f'(x) = 2$, all 2.   **9.** $f'(x) = -4$, all -4.   **11.** $f'(x) = 2x + 1$, $f'(-2) = -3$, $f'(-1) = -1$, $f'(0) = 1$, $f'(1) = 3$, $f'(2) = 5$.

13. $f'(x) = \dfrac{-4}{x^2}$; $f'(-2) = -1$; $f'(-1) = -4$; $f'(0)$ does not exist; $f'(1) = -4$; $f'(2) = -1$.   **15.** $f'(x) = m$, all m.   **17.** $x_0, x_3,$

x_4, x_6, x_{12}.   **19.** $x = -3$

EXERCISE SET 9.4, pp. 425–426

1. $7x^6$ **3.** 0 **5.** $600x^{149}$ **7.** $3x^2 + 6x$ **9.** $\dfrac{4}{\sqrt{x}}$ **11.** $0.07x^{-0.93}$ **13.** $\dfrac{2}{5 \cdot \sqrt[5]{x}}$ **15.** $\dfrac{-3}{x^4}$ **17.** $6x - 8$ **19.** $\dfrac{1}{4\sqrt[4]{x^3}} + \dfrac{1}{x^2}$

21. $1.6x^{1.5}$ **23.** $\dfrac{-5}{x^2} - 1$ **25.** 4 **27.** 4 **29.** x^3 **31.** $-0.02x - 0.5$ **33.** $-2x^{-5/3} + \frac{3}{4}x^{-1/4} + \frac{6}{5}x^{1/5} - 24x^{-4}$ **35.** $(0,0)$

37. $(0,0)$ **39.** $(\frac{5}{6}, \frac{23}{12})$ **41.** $(-25, 76.25)$ **43.** There are none. **45.** Tangent is horizontal at all points on graph.

47. $(\frac{5}{3}, \frac{148}{27})$, $(-1, -4)$ **49.** $(\sqrt{3}, 2 - 2\sqrt{3})$, $(-\sqrt{3}, 2 + 2\sqrt{3})$ **51.** $(\frac{19}{2}, \frac{399}{4})$ **53.** $(60, 150)$ **55.** $(-2 + \sqrt{3}, \frac{4}{3} - \sqrt{3})$,

$(-2 - \sqrt{3}, \frac{4}{3} + \sqrt{3})$ **57.** $(0, -4)$, $(\frac{\sqrt{2}}{3}, -\frac{40}{9})$, $(-\frac{\sqrt{2}}{3}, -\frac{40}{9})$ **59.** $2x - 1$ **61.** $2x + 1$ **63.** $3x^2 - \dfrac{1}{x^2}$ **65.** $-192x^2$

67. $\dfrac{2}{3 \cdot \sqrt[3]{x^2}}$ **69.** $3x^2 + 6x + 3$

EXERCISE SET 9.5, pp. 431–432

1. a) $v(t) = 3t^2 + 1$; b) $a(t) = 6t$; c) $v(4) = 49\,\text{ft/sec}$, $a(4) = 24\,\text{ft/sec}^2$. **3.** a) $\dfrac{dV}{dh} = 0.61/\sqrt{h}$, b) 244 miles, c) 0.00305

miles per foot. **5.** $P'(t) = 1.25$ **7.** $\dfrac{dC}{dr} = 2\pi$ **9.** a) $T'(t) = -0.2t + 1.2$; b) 100.175°; c) 0.9 degrees/day. **11.** $\dfrac{dB}{dx} = $

$0.1x - 0.9x^2$ **13.** $\dfrac{dT}{dW} = 1.31W^{0.31}$ **15.** a) $P(x) = -0.001x^2 + 3.8x - 60$; b) $R(100) = \$500$, $C(100) = \$190$,

$P(100) = \$310$; c) $R'(x) = 5$, $C'(x) = 0.002x + 1.2$, $P'(x) = -0.002x + 3.8$; d) $R'(100) = \$5$ per unit, $C'(100) = \$1.4$

per unit, $P'(100) = \$3.6$ per unit.

EXERCISE SET 9.6, pp. 436–437

1. $11x^{10}$ **3.** $\dfrac{1}{x^2}$ **5.** $3x^2$ **7.** $(8x^5 - 3x^2 + 20)\left(32x^3 - \dfrac{3}{2\sqrt{x}}\right) + (40x^4 - 6x)(8x^4 - 3\sqrt{x})$ **9.** $300 - 2x$ **11.** $\dfrac{300}{(300 - x)^2}$

13. $\dfrac{17}{(2x + 5)^2}$ **15.** $\dfrac{-x^4 - 3x^2 - 2x}{(x^3 - 1)^2}$ **17.** $\dfrac{1}{(1 - x)^2}$ **19.** $\dfrac{2}{(x + 1)^2}$ **21.** $\dfrac{-1}{(x - 3)^2}$ **23.** $\dfrac{-2x^2 + 6x + 2}{(x^2 + 1)^2}$ **25.** $\dfrac{-18x + 35}{x^8}$

27. a) $R(x) = x(400 - x) = 400x - x^2$; b) $R'(x) = 400 - 2x$. **29.** a) $R(x) = 4000 + 3x$; b) $R'(x) = 3$.

31. $A'(x) = \dfrac{xC'(x) - C(x)}{x^2}$ **33.** $\dfrac{5x^3 - 30x^2\sqrt{x}}{2\sqrt{x}(\sqrt{x} - 5)^2}$ **35.** $\dfrac{-3(1 + 2v)}{(1 + v + v^2)^2}$ **37.** $\dfrac{2t^3 - t^2 + 1}{(1 - t + t^2 - t^3)^2}$ **39.** $\dfrac{5x^3 + 15x^2 + 2}{2x\sqrt{x}}$

41. $[x(9x^2 + 6) + (3x^3 + 6x - 2)](3x^4 + 7) + 12x^4(3x^3 + 6x - 2)$ **43.** $\dfrac{6t^2(t^5 + 3)}{(t^3 + 1)^2} + \dfrac{5t^4(t^3 - 1)}{t^3 + 1}$

45. $\dfrac{(x^7 - 2x^6 + 9)[(2x^2 + 3)(12x^2 - 7) + 4x(4x^3 - 7x + 2)] - (7x^6 - 12x^5)(2x^2 + 3)(4x^3 - 7x + 2)}{(x^7 - 2x^6 + 9)^2}$

EXERCISE SET 9.7, p. 442

1. $-55(1 - x)^{54}$ **3.** $\dfrac{4}{\sqrt{1 + 8x}}$; **5.** $\dfrac{3x}{\sqrt{3x^2 - 4}}$ **7.** $-240x(3x^2 - 6)^{-41}$ **9.** $\sqrt{2x + 3} + \dfrac{x}{\sqrt{2x + 3}}$, or $\dfrac{3(x + 1)}{\sqrt{2x + 3}}$

11. $2x\sqrt{x - 1} + \dfrac{x^2}{2\sqrt{x - 1}}$, or $\dfrac{5x^2 - 4x}{2\sqrt{x - 1}}$ **13.** $\dfrac{-6}{(3x + 8)^3}$ **15.** $(1 + x^3)^2(-3x^2 - 12x^5)$, or $-3x^2(1 + x^3)^2(1 + 4x^3)$ **17.** $4x - 400$

19. $2(x + 6)^9(x - 5)^3(7x - 13)$ **21.** $4(x - 4)^7(3 - x)^3(10 - 3x)$ **23.** $4(2x - 3)^2(3 - 8x)$ **25.** $\left(\dfrac{1 - x}{1 + x}\right)^{-1/2} \cdot \dfrac{-1}{(x + 1)^2}$

27. a) $\dfrac{2x - 3x^2}{(1 + x)^6}$, b) $\dfrac{2x - 3x^2}{(1 + x)^6}$, c) Same. **29.** $C'(x) = \dfrac{1500x^2}{\sqrt{x^3 + 2}}$ **31.** $\dfrac{x^2 - 2}{\sqrt[3]{(x^3 - 6x + 1)^2}}$ **33.** $\dfrac{x - 2}{2(x - 1)^{3/2}}$

35. $\dfrac{-4(1 + 2v)^3}{v^5}$ **37.** $\dfrac{1}{\sqrt{1 - x^2}(1 - x)}$ **39.** $3\left(\dfrac{x^2 - x - 1}{x^2 + 1}\right)^2 \cdot \dfrac{x^2 + 4x - 1}{(x^2 + 1)^2}$ **41.** $\dfrac{1}{\sqrt{t}(1 + \sqrt{t})^2}$

EXERCISE SET 9.8, pp. 444–445

1. 0 **3.** $-\dfrac{2}{x^3}$ **5.** $-\dfrac{3}{16}x^{-7/4}$ **7.** $12x^2 + \dfrac{8}{x^3}$ **9.** $\dfrac{12}{x^5}$ **11.** $n(n - 1)x^{n-2}$ **13.** $12x^2 - 2$ **15.** $-\dfrac{1}{4}(x - 1)^{-3/2}$, or

$\dfrac{-1}{4\sqrt{(x - 1)^3}}$ **17.** $2a$ **19.** 24 **21.** $720x$ **23.** $n(n - 1)(n - 2)(n - 3)(n - 4)(n - 5)x^{n-6}$ **25.** $a(t) = 6t + 2$

27. $P''(t) = 200{,}000$ **29.** $y' = -x^{-2} - 2x^{-3}$, $y'' = 2x^{-3} + 6x^{-4}$, $y''' = -6x^{-4} - 24x^{-5}$ **31.** $y' = \dfrac{1 + 2x^2}{\sqrt{1 + x^2}}$, $y'' = \dfrac{2x^3 + 3x}{(1 + x^2)^{3/2}}$,

$y''' = \dfrac{3}{(1 + x^2)^{5/2}}$ **33.** $y' = \dfrac{11}{(2x + 3)^2}$, $y'' = \dfrac{-44}{(2x + 3)^3}$, $y''' = \dfrac{264}{(2x + 3)^4}$ **35.** $y' = \dfrac{x - 2}{2(x - 1)^{3/2}}$, $y'' = \dfrac{4 - x}{4(x - 1)^{5/2}}$,

$y''' = \dfrac{3(x - 6)}{8(x - 1)^{7/2}}$ **37.** $\dfrac{dy}{dx} - \dfrac{1 + y}{2 - x}$, $\dfrac{d^2y}{dx^2} - \dfrac{2 + 2y}{(2 - x)^2}$ **39.** $\dfrac{dy}{dx} = \dfrac{x}{y}$, $\dfrac{d^2y}{dx^2} = \dfrac{y^2 - x^2}{y^3}$

EXERCISE SET 9.9, pp. 459–460

1. a) 41 mph b) 80 mph c) 13.5 mpg d) 16.5 mph e) About 22% **3.** Max. $= 5\frac{1}{4}$ at $x = \frac{1}{2}$; min. $= 3$ at $x = 2$.

5. Max. $= 4$ at $x = 2$; min. $= 1$ at $x = 1$. **7.** Max. $= \frac{59}{27}$ at $x = -\frac{1}{3}$; min. $= 1$ at $x = -1$. **9.** Max $= 1$ at $x = 1$;

min. $= -5$ at $x = -1$. **11.** None. **13.** Max. $= 1225$ at $x = 35$. **15.** Min. $= 200$ at $x = 10$. **17.** Max. $= \frac{1}{3}$ at $x = \frac{1}{2}$.

19. Max. $= \frac{289}{4}$ at $x = \frac{17}{2}$ **21.** Max. $= 2\sqrt{3}$ at $x = -\sqrt{3}$; min. $= -2\sqrt{3}$ at $x = \sqrt{3}$. **23.** Max. $= 5700$ at $x = 2400$

25. Min. $= -55\frac{1}{3}$ at $x = 1$. **27.** Max. $= 2000$ at $x = 20$; min. $= 0$ at $x = 0$ and $x = 30$. **29.** Min. $= 24$ at $x = 6$.

31. Min. $= 108$ at $x = 6$. **33.** Max. $= 3$ at $x = -1$; min. $= -\frac{3}{8}$ at $x = \frac{1}{2}$. **35.** Max. $= 2$ at $x = 8$; min. $= 0$ at $x = 0$.

37. None. **39.** 22506; \$150,000. **41.** 61.25 mph. **43.** Max. $= 3\sqrt{6}$ at $x = 3$; min. $= -2$ at $x = -2$.

45. Max. $= 1$ at $x = -1$ and $x = 1$; min. $= 0$ at $x = 0$. **47.** None. **49.** Max. $= -\frac{10}{3} + 2\sqrt{3}$ at $x = 2 - \sqrt{3}$;

min. $= -\frac{10}{3} - 2\sqrt{3}$ at $x = 2 + \sqrt{3}$. **51.** Min. $= -1$ at $x = -1$ and $x = 1$. **53.** 7.

EXERCISE SET 9.10, pp. 469–474

1. 25 and 25. Max. prod. = 625 **3.** No. $Q = x(50 - x)$ has no minimum. **5.** 2 and -2. Min. prod. = -4 **7.** $x = \frac{1}{2}$, $y = \sqrt{\frac{1}{2}}$; max. = $\frac{1}{4}$ **9.** $x = 10$, $y = 10$; min. = 200 **11.** $x = 2$, $y = \frac{32}{3}$; max. = $\frac{64}{3}$ **13.** $x = 30$ yd, $y = 60$ yd; max area = 1800 yd^2 **15.** 13.5 ft by 13.5 ft; 182.25 ft^2 **17.** 46 units; max. profit = $1048 **19.** 70 units; max. profit = $19 **21.** Approx. 1667 units; max. profit $\approx$ $5500 **23.** a) $R(x) = 150x - 0.5x^2$, b) $P(x) = -0.75x^2 + 150x - 4000$, c) 100, d) $3500, e) $100 **25.** 20 in. by 20 in. by 5 in.; max. = 2000 in^3 **27.** 5 in. by 5 in. by 2.5 in.; min. = 75 in^2 **29.** $5.75, 72,500 (Will the stadium hold that many?) **31.** 25 **33.** 14 in. by 14 in. by 28 in. **35.** $\sqrt[3]{\frac{1}{10}}$ **37.** 4 ft by 4 ft by 20 ft. **39.** $x = y = \dfrac{24}{4 + \pi}$ **41.** 9% **43.** Min. at $x = \dfrac{24\pi}{\pi + 4} \approx 10.55$, $24 - x = \dfrac{96}{\pi + 4} \approx 13.45$; there is no maximum if the string is to be cut. One would interpret the maximum to be at the endpoint, with the string uncut and used to form a circle. **45.** S should be about 4.25 miles down shore from A. **47.** a) $C'(x) = 8 + \dfrac{3x^2}{100}$; b) $A(x) = 8 + \dfrac{20}{x} + \dfrac{x^2}{100}$; c) $A'(x) = \dfrac{x}{50} - \dfrac{20}{x^2}$; d) Min. = 11 at $x_0 = 10$, $C(10) = 11$; e) $A(10) = 11$, $c'(10) = 11$. **49.** Min. = $6 - 4\sqrt{2}$ at $x = 2 - \sqrt{2}$ and $y = -1 + \sqrt{2}$.

EXERCISE SET 9.11, p. 478

1. Reorder 5 times per year; lot size = 20. **3.** Reorder 12 times per year; lot size = 30. **5.** Reorder about 13 times per year; lot size = 28. **7.** $x = \sqrt{\dfrac{2bQ}{a}}$

CHAPTER 9, TEST, pp. 479–480

1. Yes **2.** No **3.** a) 19, 15.25, 12.61, 12.0601, 12.006001; b) 12. **4.** $3(2x + h)$ **5.** a) 3.4, 3.9625, 3.985, 3.9997; b) 4. **6.** $(0, 0)$, $(2, -4)$. **7.** $84x^{83}$ **8.** $\dfrac{5}{\sqrt{x}}$ **9.** $\dfrac{10}{x^2}$ **10.** $\frac{5}{4}x^{1/4}$, or $\frac{5}{4} \cdot \sqrt[4]{x}$. **11.** $-x + 0.61$ **12.** $x^2 - 2x + 2$ **13.** $\dfrac{-6x + 20}{x^5}$ **14.** $\dfrac{5}{(5 - x)^2}$ **15.** $(x + 3)^3(7 - x)^4(-9x + 13)$ **16.** $-5(x^5 - 4x^3 + x)^{-6}(5x^4 - 12x^2 + 1)$ **17.** $\sqrt{x^2 + 5} + \dfrac{x^2}{\sqrt{x^2 + 5}}$, or $\dfrac{2x^2 + 5}{\sqrt{x^2 + 5}}$ **18.** a) $P(x) = -0.001x^2 + 48.8x - 60$; b) $R(10) = 500, $C(10) = 72.10, $P(10) = $427.90; c) $R'(x) = 50$, $C'(x) = 0.002x + 1.2$, $P'(x) = -0.002x + 48.8$; d) $R'(10) = $50 per unit, $C'(10) = $1.22 per unit, $P'(10) = $48.78. **19.** a) $\dfrac{dM}{dt} = -0.003t^2 + 0.2t$; b) 9 c) 1.7 words/min. **20.** $\dfrac{d^3y}{dx^3} = 24x$ **21.** Max. = 9 at $x = 3$. **22.** Max. = 2 at $x = -1$; min. = -1 at $x = -2$. **23.** Max. = 28.49 at $x = 4.3$. **24.** Max. = 7 at $x = -1$; min. = 3 at $x = 1$. **25.** None. **26.** Min. = $-\frac{13}{12}$ at $x = \frac{1}{6}$. **27.** Min. = 48 at $x = 4$. **28.** 4 and -4. **29.** $x = 5$, $y = -5$; min. = 50. **30.** Max. profit = $24,980; 500 units. **31.** 40 in. by 40 in. by 10 in.; max. volume = 16,000 in.^{3} **32.** 35 times at lot size 35.

CHAPTER 10

MARGIN EXERCISES

1. a)

x	0	$\frac{1}{2}$	1	2	-1	-2
3^x	1	1.7	3	9	$\frac{1}{3}$	$\frac{1}{9}$

b)

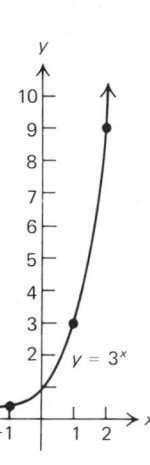

2. a)

x	0	$\frac{1}{2}$	1	2	-1	-2
3^x	1	0.6	$\frac{1}{3}$	$\frac{1}{9}$	3	9

b)

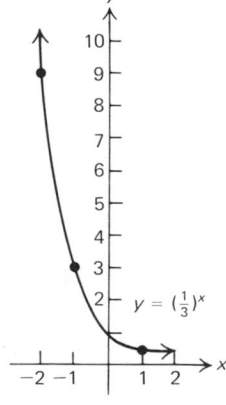

3. a) $b^T = P$, b) $9^{1/2} = 3$, c) $10^3 = 1000$, d) $10^{-1} = 0.1$. **4.** a) $\log_e T = k$, b) $\log_{16} 2 = \frac{1}{4}$, c) $\log_{10} 10{,}000 = 4$, d) $\log_{10} 0.001 = -3$.

5.

![graph of y = log_3 x]

6. 1 **7.** -0.398 **8.** 0.398 **9.** 0.699 **10.** $\frac{3}{2}$ **11.** 1.699 **12.** 1.204 **13.** 2.3025 **14.** -0.9163 **15.** 0.9163
16. 2.7724 **17.** 2.6094 **18.** $\frac{1}{2}$ **19.** -1.6094 **20.** 0.693147 **21.** 2.995732 **22.** 4.605170 **23.** -2.599375
24. 0.076961 **25.** -0.000100 **26.** 2.0956 **27.** 11.3059 **28.** -2.5096 **29.** 7.6009 **30.** -9.2103 **31.** a) 4,
b) -4. **32.** a) 4.4977, b) -0.0419. **33.** a) 3.8954, b) 2.8954, c) 1.8954, d) 0.8954, e) -0.1046, f) -1.1046,
g) -2.1046. **34.** a) 1.8831, b) 3.3674, c) -2.0605. **35.** a) $t \approx 4$, b) $t \approx 44$. **36.** 90 decibels **37.** 23 decibels
38. 8, 8.574188, 8.815241, 8.821353, 8.824411; $2^\pi \approx 8.82$. **39.** a) 0.8284, 0.7568, 0.7241, 0.7083, 0.7008; b) 0.7.
40. a) 1.4641, 1.2642, 1.1776, 1.1372, 1.1177; b) 1.1.

41. a)

x	-3	-2	-1	0	1	2	3
2^x	0.125	0.25	0.5	1	2	4	8
$(0.7)2^x$	0.09	0.18	0.35	0.7	1.4	2.8	5.6

b) See later text figure.

42. a)

x	-2	-1	0	1	2
3^x	0.11	0.33	1	3	9
$(1.1)3^x$	0.12	0.36	1.1	3.3	9.9

b) See later text figure.

43. $2, $2.25, $2.370370, $2.441406, $2.488320, $2.704814, $2.714567, $2.718121. **44.** $6e^x$ **45.** $e^x(x^3 + 3x^2)$, or

$x^2 e^x(x + 3)$ **46.** $\dfrac{e^x(x - 2)}{x^3}$ **47.** $-4e^{-4x}$ **48.** $(3x^2 + 8)e^{x^3+8x}$ **49.** $\dfrac{xe^{\sqrt{x^2+5}}}{\sqrt{x^2 + 5}}$

50.

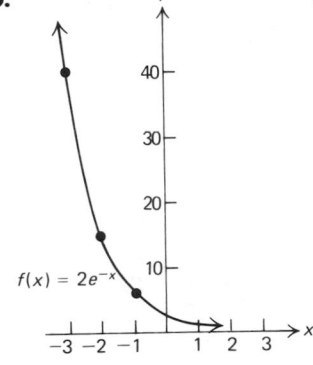

$f(x) = 2e^{-x}$

51. a)

x	0	$\frac{1}{2}$	1	2	3	4
$f(x)$	0	0.39	0.63	0.86	0.95	0.98

b)

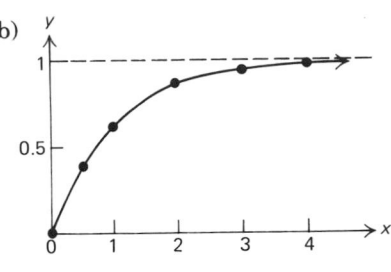

52. 87 days

53. a)

x	0.5	1	2	3	4
$\ln x$	-0.7	0	0.7	1.1	1.4

b) Same as text figure.

54. $\dfrac{5}{x}$ **55.** $x^2(1 + 3 \ln x) + 4$ **56.** $\dfrac{1 - 2 \ln x}{x^3}$ **57.** $\dfrac{1}{x}$ **58.** $\dfrac{6x}{3x^2 + 4}$ **59.** $\dfrac{1}{x \ln 5x}$ **60.** $\dfrac{4x^5 + 2}{x(x^5 - 2)}$ **61.** a) 500,

b) $N'(a) = \dfrac{200}{a}$, c) Min. = 500 at $a = 1$. **62.** a) $\dfrac{dy}{dx} = 20e^{4x}$, b) $\dfrac{dy}{dx} = 4y$. **63.** a) $N(t) = ce^{kt}$, b) $f(t) = ce^{kt}$.

64. a) Should be about $\frac{1}{8}$ in. or 0.125 in. b) 0.032, 0.064, 0.128. c) About 2 mi. **65.** a) $P(t) = P_0 e^{0.08t}$, b) $1083.29,

c) 8.7 yr. **66.** 2%, 34.7; 6.9%, 10; 14%, 5.0; 4.6%, 15; 1%, 69.3 **67.** a) $P(t) = 216e^{0.008t}$, b) 225 million, c) 86.6 yr

(2062). **68.** a) $E(t) = 800e^{kt}$ b) $k = 0.08$ c) $E(t) = 800e^{0.08t}$ d) 8818.5 billion kWh. **69.** See Exercises 29 and 31 of

Exercise Set 10.2. **70.** a) $N(t) = N_0 e^{-0.14t}$ b) 246.6 gr c) 5 days **71.** 30 yr **72.** 5.3% **73.** 13,412 yr

---

EXERCISE SET 10.1, pp. 492–494

1.

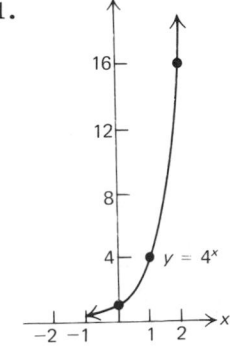

$y = 4^x$

3.

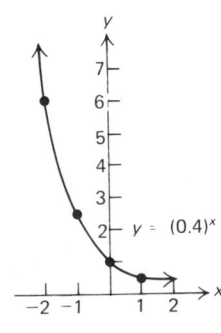

$y = (0.4)^x$

5.

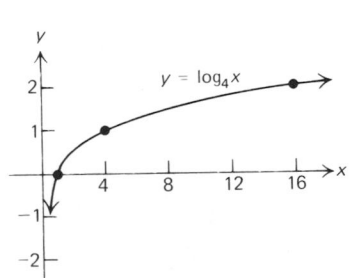

$y = \log_4 x$

7. $2^3 = 8$ **9.** $8^{1/3} = 2$

11. $a^J = K$ **13.** $b^v = T$ **15.** $\log_e b = M$ **17.** $\log_{10} 100 = 2$ **19.** $\log_{10} 0.1 = -1$ **21.** $\log_M V = p$ **23.** 2.708

25. 0.51   **27.** -1.609 **29.** $\frac{3}{2}$ **31.** 2.609   **33.** 3.218   **35.** 2.9957   **37.** 0.2231   **39.** -1.3863 **41.** 2.3863   **43.** 4

45. 2.7726   **47.** 8.681690   **49.** -4.006334 **51.** 0.6313   **53.** -3.9739 **55.** 6.8091   **57.** -4.5099 **59.** 5.9428

61. -0.9539 **63.** 0.3284   **65.** -0.6716 **67.** 3.6532   **69.** -3.0969 **71.** $t \approx 5$ **73.** $t \approx 4$ **75.** $t \approx 2$

77. $t \approx 141$ **79.** 64 decibels.   **81.** 120 decibels.   **83.** 7.85   **85.** 7.8   **87.** $t = \dfrac{\ln P - \ln P_0}{k}$ **89.** $\propto$

91. Let $a = \ln x$; then $e^a = x$, so $\log x = \log e^a = a \log e$; then $a = \ln x = \dfrac{\log x}{\log e} \approx \dfrac{\log x}{0.4343} \approx 2.3026 \log x.$

EXERCISE SET 10.2, pp. 503–504

1. $3e^{3x}$ **3.** $-10e^{-2x}$ **5.** e^{-x} **7.** $-7e^x$ **9.** e^{2x} **11.** $x^3 e^x (x + 4)$ **13.** $\dfrac{e^x(x - 4)}{x^5}$ **15.** $(-2x + 7)e^{-x^2 + 7x}$

17. $-xe^{-x^2/2}$ **19.** $\dfrac{e^{\sqrt{x-7}}}{2\sqrt{x-7}}$ **21.** $\dfrac{e^x}{2\sqrt{e^x - 1}}$ **23.** $(1 - 2x)e^{-2x} - e^{-x} + 3x^2$ **25.** e^{-x} **27.** ke^{-kx}

29.   **31.**   **33.**   **35.** 58 days

37. a) $C'(t) = 50e^{-t}$, b) $50 million, c) $.916 million.   **39.** a) 18.1%, 69.9%; b) $P'(t) = 0.2e^{-0.2t}$; c) 11.5.

41. a) $58.69, $78.00; b) $63.80 $e^{-1.1t}$; c) 2.7.   **43.** $15(e^{3x} + 1)^4 e^{3x}$ **45.** $-e^{-t} - 3e^{3t}$ **47.** $\dfrac{e^x(x - 1)^2}{(x^2 + 1)^2}$

49. $\dfrac{e^{\sqrt{x}}}{2\sqrt{x}} + \frac{1}{2}\sqrt{e^x}$ **51.** $\frac{1}{2}e^{x/2}\left[\dfrac{x}{\sqrt{x} - 1}\right]$ **53.** $\dfrac{4}{(e^x + e^{-x})^2}$ **55.** 4, 2.86797, 2.73200, 2.71964, 2.71855

EXERCISE SET 10.3, pp. 510–511

1. $-\dfrac{6}{x}$ **3.** $x^3(1 + 4 \ln x) - x$ **5.** $\dfrac{1 - 4 \ln x}{x^5}$ **7.** $\dfrac{1}{x}$ **9.** $\dfrac{10x}{5x^2 - 7}$ **11.** $\dfrac{1}{x \ln 4x}$ **13.** $\dfrac{x^2 + 7}{x(x^2 - 7)}$ **15.** $e^x\left(\dfrac{1}{x} + \ln x\right)$

17. $\dfrac{e^x}{e^x + 1}$ **19.** $\dfrac{2\ln x}{x}$ **21.** a) 68% b) 36% c) 3.6% d) 5% e) $-\dfrac{20}{t + 1}$ f) Max. = 68% at $t = 0$.   **23.** a) 1000;

b) $N'(a) = \dfrac{200}{a}$, $N'(10) = 20$; c) Min. = 1000 at $a = 1$.   **25.** a) 5.4 ft/sec, b) 7.8 ft/sec, c) $v'(p) = \dfrac{0.86}{p}$, $v'(p) =$ the

acceleration of the walker.   **27.** $\dfrac{-4(\ln x)^{-5}}{x}$ **29.** $\dfrac{15t^2}{t^3 + 1}$ **31.** $\dfrac{4[\ln(x + 5)]^3}{x + 5}$ **33.** $\dfrac{5t^4 - 3t^2 + 6t}{(t^3 + 3)(t^2 - 1)}$ **35.** $\dfrac{24x + 25}{8x^2 + 5x}$

37. $\dfrac{2(1 - \ln t^2)}{t^3}$ **39.** $x^n \ln x$ **41.** $\dfrac{1}{\sqrt{1 + t^2}}$ **43.** e^π

EXERCISE SET 10.4, pp. 520–522

1. $Q(t) = Q_0 e^{kt}$ **3.** a) $P(t) = P_0 e^{0.09t}$ b) \$1094.17, \$1197.22 c) 7.7 yr   **5.** 19.8 yr   **7.** 6.9%   **9.** a) $P(t) = 209e^{0.01t}$
b) 312 million c) 69.3 yr   **11.** 0.20%   **13.** a) $N(t) = 50e^{0.1t}$ b) 369 c) 6.9 yr   **15.** 6.9 yr (1986)   **17.** a) $k = 0.08$,
$P(t) = 25e^{0.08t}$ b) 124 thousand   **19.** a) $k = 0.07$, $P(t) = \$90e^{0.07t}$; b) \$2980 million; c) 9.9 yr (1983).   **21.** a) $k = 0.06$,
$P(t) = \$100e^{0.06t}$; b) \$332.01; c) 11.6 yr (1978).   **23.** 6.18%   **25.** 9%   **27.** $T_3 = \dfrac{\ln 3}{k}$ **29.** Answers depend on
particular data.   **31.** $\approx \$66,000,000,000,000$ **33.** $\approx 2.2\%$ **35.** \$16.64, \$27.71.

EXERCISE SET 10.5, pp. 526–527

1. 7.2 days   **3.** 23% per min   **5.** 10.1 g   **7.** 19,188 yr   **9.** 7636 yr   **11.** a) $A = A_0 e^{-kt}$ b) 9 hr   **13.** a) 0.8%
b) 79% W_0 **15.** a) 11 watts b) 173 days c) 402 days d) 50 watts   **17.** a) \$40,000 b) \$5412   **19.** a) 25% I_0, 6% I_0,
1.5% I_0 b) 0.00008%

CHAPTER 10, TEST, p. 528

1. e^x **2.** $\dfrac{1}{x}$ **3.** $-2xe^{-x^2}$ **4.** $\dfrac{1}{x}$ **5.** $e^x - 15x^2$ **6.** $3e^x\left(\dfrac{1}{x} + \ln x\right)$ **7.** $\dfrac{e^x - 3x^2}{e^x - x^3}$ **8.** $\dfrac{\frac{1}{x} - \ln x}{e^x}$, or $\dfrac{1 - x\ln x}{xe^x}$ **9.** 2.639
10. -1.2528 **11.** 2.9459   **12.** $M(t) = M_0 e^{kt}$ **13.** 17.3% per hr   **14.** 10 yr   **15.** a) $F(t) = 3e^{0.12t}$, b) 33 billion gal,
c) 5.8 yr (1965).   **16.** 0.000069% per yr   **17.** 63 days   **18.** a) $A(t) = 3e^{-0.1t}$, b) 1.1 cc, c) 6.9 hr.

CHAPTER 11

MARGIN EXERCISES

1. $y = 7x$, $y = 7x - \frac{1}{2}$, $y = 7x + C$ (Answers can vary.)   **2.** $y = -2x$, $y = -2x + 27$, $y = -2x + C$ (Answers can
vary.)   **3.** $\dfrac{x^2}{2} + C$ **4.** $\dfrac{x^4}{4} + C$ **5.** $e^x + C$ **6.** $\ln x + C$ **7.** $\dfrac{x^4}{4} + C$ **8.** $\dfrac{x^2}{2} + C$ **9.** $\ln x + C$ **10.** $\frac{7}{5}x^5 + x^2 + C$

11. $e^x - \frac{5}{7}x^{7/5} + C$ **12.** $5\ln x - 7x - \frac{1}{5}x^{-5} + C$ **13.**

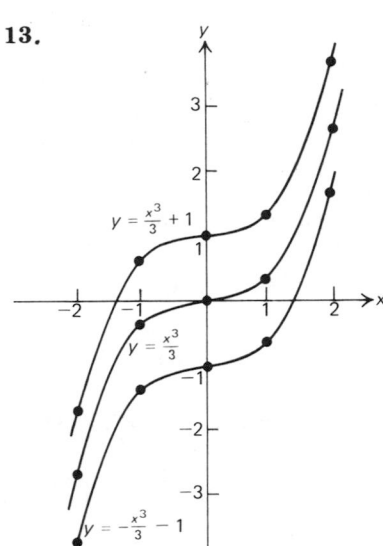

14. $f(x) = \dfrac{x^3}{3} + \dfrac{23}{3}$

15. $g(x) = x^2 - 4x + 13$ **16.** $C(x) = \frac{1}{3}x^3 + \frac{5}{2}x^2 + 35$ **17.** $s(t) = t^4 + 13$ **18.** $s(t) = 2t^4 - 6t^2 + 7t + 8$ **19.** a) $A(x) = 3x$;
b) $A(1) = 3, A(2) = 6, A(5) = 15$;   c)   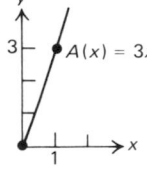   d) $A(x)$ is an antiderivative of $f(x)$.

20. a) $C(x) = 50x$;   b) $A(x) = 50x$;   c) As x increases, the area over $[0, x]$ increases.   **21.** a) $5000 + 4000 + 3000 +$
2000, or \$14,000;   b) 14,000, same.   **22.** a) $A(x) = \frac{3}{2}x^2$;   b) $A(1) = \frac{3}{2}$;   c) $A(2) = 6$;   d) $A(3.5) = 18.375$;

e)

f) $A(x)$ is an antiderivative of $f(x)$.

23. a) $C(x) = -0.05x^2 + 50x$. b) $A(x) = -0.05x^2 + 50x$. c) \$12,000; less than in margin exercise 21.   **24.** $5\frac{1}{3}$.   **25.** $13\frac{1}{2}$

26. a) Yes. b) \$4100.   **27.** $\displaystyle\int_a^b 2x\,dx = b^2 - a^2$ **28.** $\displaystyle\int_a^b e^x\,dx = e^b - e^a$ **29.** 8   **30.** $1 - \dfrac{1}{e^2}$ **31.** $\frac{2}{3}$ **32.** $e + e^3 - 3$

33. $5\frac{1}{3}$ **34.** $13\frac{1}{2}$ **35.** $\ln 7$ **36.** $-\frac{1}{3}b^{-3} + \frac{1}{3}$ **37.** 75   **38.** $17\frac{1}{3}$ **39.** a) \$596,991.60;   b) $k = 7.3$, so it is the 8th day;

c) \$3,809,101.56.　　**40.** $dy = (12x + 1)\,dx$　　**41.** $du = dx$　　**42.** $e^{x^3} + C$　　**43.** $\ln(5 + x^2) + C$

44. $-\dfrac{1}{3 + x^2} + C$　　**45.** $\dfrac{(\ln x)^2}{2} + C$　　**46.** $\frac{1}{3}e^{x^3} + C$　　**47.** $\frac{1}{5}e^{5x} + C$　　**48.** $\dfrac{1}{0.02}e^{0.02x} + C$, or $50e^{0.02x} + C$　　**49.** $-e^{-x} + C$

50. $\frac{1}{80}(x^4 + 5)^{20} + C$　　**51.** $\displaystyle\int \dfrac{\ln x\,dx}{x} = \dfrac{(\ln x)^2}{2} + C$, so $\displaystyle\int_1^e \dfrac{\ln x\,dx}{x} = \left[\dfrac{(\ln x)^2}{2}\right]_1^e = \dfrac{(\ln e)^2}{2} - \dfrac{(\ln 1)^2}{2} = \frac{1}{2} - 0 = \frac{1}{2}$

52. $xe^{3x} - \frac{1}{3}e^{3x} + C$　　**53.** $\dfrac{x^2}{2}\ln x - \dfrac{x^2}{4} + C$; let $u = \ln x$, $dv = x\,dx$.　　**54.** $\frac{2}{3}x(x + 3)^{3/2} - \frac{4}{15}(x + 3)^{5/2} + C$　　**55.** $2\ln 2 - \frac{3}{4}$

56. $\frac{1}{10}\ln\left(\dfrac{x - 5}{x + 5}\right) + C$　　**57.** $\displaystyle\sum_{i=1}^{6} i^2$　　**58.** $\displaystyle\sum_{i=1}^{4} e^i$　　**59.** $\displaystyle\sum_{i=1}^{38} P(x_i)\,\Delta x$　　**60.** $4^1 + 4^2 + 4^3$, or 84　　**61.** $e^1 + 2e^2 + 3e^3 + 4e^4 + 5e^5$

62. $t(x_1)\,\Delta x + t(x_2)\,\Delta x + \cdots + t(x_{20})\,\Delta x$　　**63.** \$11,250　　**64.** a) 3.0667　b) 2.4501　c) $\displaystyle\int_1^7 \dfrac{1}{x}\,dx = \ln 7 \approx 1.9459$

65. $\frac{2}{3}, \frac{9}{10}, \frac{99}{100}, \frac{199}{200}$　　**66.** $\frac{1}{2}$　　**67.** ∞　　**68.** Convergent, 1　　**69.** Divergent　　**70.** a) $[15, 25]$　b) $[0, \infty)$　　**71.** $\frac{2}{3}$

72. $\displaystyle\int_1^2 \dfrac{2}{3}x\,dx = \left[\dfrac{2}{3}\cdot\dfrac{1}{2}x^2\right]_1^2 = \dfrac{1}{3}(2^2 - 1) = 1$　　**73.** a) $\displaystyle\int_3^6 \dfrac{24}{t^3}\,dt = \left[24\left(-\dfrac{t^{-2}}{2}\right)\right]_3^6 = -12\left(\dfrac{1}{6^2} - \dfrac{1}{3^2}\right) = 1$,　b) $\frac{64}{75}$,　c) $\frac{5}{12}$.

74. $\frac{3}{26}$　　**75.** 4　　**76.** $\frac{1}{4}$　　**77.** $\frac{1}{4}$　　**78.** 0.3297　　**79.** $E(x) = \frac{2}{3}$, $E(x^2) = \frac{1}{2}$　　**80.** $\mu = \frac{2}{3}$, $\sigma^2 = \frac{1}{18}$, $\sigma = \sqrt{\frac{1}{18}} = \frac{1}{3}\sqrt{\frac{1}{2}}$

81. a) 0.4850　b) 0.4608　c) 0.9559　d) 0.2720　e) 0.1102　f) 0.0307　　**82.** 0.1210

EXERCISE SET 11.1, pp. 536–537

1. $\dfrac{x^7}{7} + C$　　**3.** $2x + C$　　**5.** $\frac{4}{5}x^{5/4} + C$　　**7.** $\dfrac{x^3}{3} + \dfrac{x^2}{2} - x + C$　　**9.** $\dfrac{t^3}{3} - t^2 + 3t + C$　　**11.** $5e^x + C$　　**13.** $\dfrac{x^4}{4} - \frac{7}{15}x^{15/7} + C$

15. $1000\ln x + C$　　**17.** $-x^{-1} + C$　　**19.** $f(x) = \dfrac{x^2}{2} - 3x + 13$　　**21.** $f(x) = \dfrac{x^3}{3} - 4x + 7$　　**23.** $C(x) = \dfrac{x^4}{4} - x^2 + 100$

25. a) $R(x) = \dfrac{x^3}{3} - 3x$; b) If you sell no products you make no money.　　**27.** $s(t) = t^3 + 4$　　**29.** $v(t) = 2t^2 + 20$

31. $s(t) = -\frac{1}{3}t^3 + 3t^2 + 6t + 10$　　**33.** $s(t) = -16t^2 + v_0t + s_0$　　**35.** $\frac{1}{4}$ mi　　**37.** a) $E(t) = 30t - 5t^2 + 32$　b) $E(3) = 77\%$,

$E(5) = 57\%$　　**39.** a) $A(t) = 43.4t^{-1} - 3.7$; b) 2.5 cm².　　**41.** $f(t) = \dfrac{t^{\sqrt{3}+1}}{\sqrt{3} + 1} + 8$　　**43.** $\dfrac{x^6}{6} - \frac{2}{5}x^5 + \frac{1}{4}x^4 + C$

45. $\frac{2}{5}t^{5/2} + 4t^{3/2} + 18\sqrt{t} + C$　　**47.** $\dfrac{t^4}{4} + t^3 + \frac{3}{2}t^2 + t + C$　　**49.** $\dfrac{b}{a}e^{ax} + C$　　**51.** $\frac{12}{7}x^{7/3} + C$　　**53.** $\dfrac{t^3}{3} - t^2 + 4t + C$

EXERCISE SET 11.2, pp. 545–547

1. 8　　**3.** 8　　**5.** $41\frac{2}{3}$　　**7.** $\frac{1}{4}$　　**9.** $10\frac{2}{3}$　　**11.** $e^3 - 1 \approx 19.086$　　**13.** $\ln 3 \approx 1.0986$　　**15.** 51　　**17.** An antiderivative, velocity.　**19.** An antiderivative, energy used in time t.　**21.** An antiderivative, total revenue.　**23.** An antiderivative, amount of drug in blood.　　**25.** a) $s(t) = t^3 + t^2$, b) 150.　　**27.** a) $C(x) = 100x - 0.1x^2$, b) $R(x) = 100x + 0.1x^2$, c) $P(x) = 0.2x^2$, d) $P(1000) = \$200,000$.　　**29.** $3\frac{1}{2}$　　**31.** $359\frac{7}{15}$　　**33.** $6\frac{3}{4}$

EXERCISE SET 11.3, pp. 554–555

1. $\frac{1}{6}$ **3.** $\frac{4}{15}$ **5.** $e^b - e^a$ **7.** $b^3 - a^3$ **9.** $\frac{e^2}{2} + \frac{1}{2}$ **11.** $\frac{2}{3}$ **13.** $\frac{5}{34}$ **15.** 4   **17.** $9\frac{5}{6}$ **19.** 12   **21.** $e^5 - \frac{1}{e}$ **23.** $7\frac{1}{3}$

25. $17\frac{1}{3}$ **27.** a) \$2948.26 b) \$2913.90 c) $k = 6.9$, so it will be on the 7th day.   **29.** 90   **31.** \$3600   **33.** 9

35. $\frac{307}{6}$ **37.** $\frac{15}{4}$ **39.** 8   **41.** 12

EXERCISE SET 11.4, pp. 560–561

1. $\ln(7 + x^3) + C$ **3.** $\frac{1}{4}e^{4x} + C$ **5.** $2e^{x/2} + C$ **7.** $\frac{1}{4}e^{x^4} + C$ **9.** $-\frac{1}{3}e^{-t^3} + C$ **11.** $\frac{(\ln 4x)^2}{2} + C$ **13.** $\ln(1 + x) + C$

15. $-\ln(4 - x) + C$ **17.** $\frac{1}{24}(t^3 - 1)^8 + C$ **19.** $\frac{1}{8}(x^4 + x^3 + x^2)^8 + C$ **21.** $\ln(4 + e^x) + C$ **23.** $\frac{1}{4}(\ln x^2)^2 + C$, or $(\ln x)^2 + C$

25. $\ln(\ln x) + C$ **27.** $\frac{2}{3a}(ax + b)^{3/2} + C$ **29.** $\frac{b}{a}e^{ax} + C$ **31.** $e - 1$ **33.** $\frac{21}{4}$ **35.** $\ln 4 - \ln 2 = \ln\frac{4}{2} = \ln 2$

37. $\ln 19$ **39.** $1 - \frac{1}{e^b}$ **41.** $1 - \frac{1}{e^{mb}}$ **43.** $\frac{208}{3}$ **45.** a) $\frac{100{,}000}{0.025}(e^{2.1} - 1) \approx 28{,}664{,}680$ b) $\frac{100{,}000}{0.025}(e^{2.1} - e^2) \approx 3{,}108{,}455$

47. $-\frac{5}{12}(1 - 4x^2)^{3/2} + C$ **49.** $-\frac{1}{3}e^{-x^3} + C$ **51.** $-e^{1/t} + C$ **53.** $-\frac{1}{3}(\ln x)^{-3} + C$ **55.** $\frac{2}{9}(x^3 + 1)^{3/2} + C$ **57.** $\frac{3}{4}(x^2 - 6x)^{2/3} + C$

59. $\frac{1}{8}[\ln(t^4 + 8)]^2 + C$ **61.** $x + \frac{1}{x + 3} + C$ **63.** $t - \ln(t - 4) + C$ **65.** $-\ln(1 + e^{-x}) + C$ **67.** $\frac{1}{n + 1}(\ln x)^{n+1} + C$

EXERCISE SET 11.5, pp. 564–565

1. $xe^{5x} - \frac{1}{5}e^{5x} + C$ **3.** $\frac{1}{2}x^6 + C$ **5.** $\frac{x}{2}e^{2x} - \frac{1}{4}e^{2x} + C$ **7.** $-\frac{x}{2}e^{-2x} - \frac{1}{4}e^{-2x} + C$ **9.** $\frac{x^3}{3}\ln x - \frac{x^3}{9} + C$ **11.** $\frac{x^2}{2}\ln x^2 - \frac{x^2}{2} + C$

13. $(x + 3)\ln(x + 3) - x + C$. Let $u = \ln(x + 3)$, $dv = dx$, and choose $v = x + 3$ for an antiderivative of v.

15. $\left(\frac{x^2}{2} + 2x\right)\ln x - \frac{x^2}{4} - 2x + C$ **17.** $\left(\frac{x^2}{2} - x\right)\ln x - \frac{x^2}{4} + x + C$ **19.** $\frac{2}{3}x(x + 2)^{3/2} - \frac{4}{15}(x + 2)^{5/2} + C$

21. $\frac{x^4}{4}\ln 2x - \frac{x^4}{16} + C$ **23.** $x^2e^x - 2xe^x + 2e^x + C$ **25.** $\frac{1}{2}x^2e^{2x} + \frac{1}{4}e^{2x} - \frac{1}{2}xe^{2x} + C$ **27.** $\frac{8}{3}\ln 2 - \frac{7}{9}$ **29.** $9\ln 9 - 5\ln 5 - 4$

31. 1   **33.** $\frac{1}{9}e^{-3x}(-3x - 1) + C$ **35.** $\frac{5^x}{\ln 5} + C$ **37.** $\frac{1}{8}\ln\left(\frac{4 + x}{4 - x}\right) + C$ **39.** $5 - x - 5\ln(5 - x) + C$

41. $\frac{1}{5(5 - x)} + \frac{1}{25}\ln\left(\frac{x}{5 - x}\right) + C$ **43.** a) $10[e^{-T}(-T - 1) + 1]$ b) ≈ 9.085 **45.** $\frac{2}{9}x^{3/2}[3\ln x - 2] + C$ **47.** $\frac{e^t}{t + 1} + C$

49. $4\sqrt{x}(\ln\sqrt{x}) - 4\sqrt{x} + C$.   **51.** a) Let $u = x^n$ and $dv = e^x\,dx$. Then $du = nx^{n-1}\,dx$ and $v = e^x$. Then use integration by parts.

b) $x^3e^x - 3\int x^2e^x\,dx = x^3e^x - 3[x^2e^x - 2\int xe^x\,dx] = x^3e^x - 3x^2e^x + 6[xe^x - \int x^0e^x\,dx] = x^3e^x - 3x^2e^x + 6xe^x - 6e^x + C$

EXERCISE SET 11.6, p. 570

1. a) 1.4914 b) 0.8571   **3.** 1.0016

EXERCISE SET 11.7, pp. 573–574

1. $\frac{1}{3}$ **3.** Divergent **5.** 1 **7.** $\frac{1}{2}$ **9.** Divergent **11.** 5 **13.** Divergent **15.** Divergent **17.** Divergent **19.** 1

21. 33,333 lb **23.** $0.93

EXERCISE SET 11.8, pp. 582–583

1. $\int_0^1 2x\,dx = [x^2]_0^1 = 1^2 - 0^2 = 1$ **3.** $\int_4^7 \frac{1}{3}\,dx = \left[\frac{1}{3}x\right]_4^7 = \frac{1}{3}(7-4) = 1$ **5.** $\int_1^3 \frac{3}{26}x^2\,dx = \left[\frac{3}{26}\cdot\frac{x^3}{3}\right]_1^3 = \frac{1}{26}(3^3 - 1^3) = 1$

7. $\int_1^e \frac{1}{x}\,dx = [\ln x]_1^e = \ln e - \ln 1 = 1 - 0 = 1$ **9.** $\int_{-1}^1 \frac{3}{2}x^2\,dx = \left[\frac{3}{2}\cdot\frac{1}{3}x^3\right]_{-1}^1 = \frac{1}{2}(1^3 - (-1)^3) = \frac{1}{2}(1+1) = 1$

11. $\int_0^\infty 3e^{-3x}\,dx = \lim_{b\to\infty}\int_0^b 3e^{-3x}\,dx = \lim_{b\to\infty}\left[\frac{3}{-3}e^{-3x}\right]_0^b = \lim_{b\to\infty}[-e^{-3x}]_0^b = \lim_{b\to\infty}[-e^{-3b} - (-e^{-3\cdot 0})] = \lim_{b\to\infty}\left(1 - \frac{1}{3^b}\right) = 1$

13. $k = \frac{1}{4}$ **15.** $k = \frac{3}{2}$ **17.** $k = \frac{1}{5}$ **19.** $k = \frac{1}{2}$ **21.** $k = \frac{1}{\ln 3}$ **23.** $k = \frac{1}{e^3 - 1}$ **25.** $\frac{8}{25}$ **27.** $\frac{1}{2}$ **29.** 0.3297

31. 0.99995 **33.** 0.9502 **35.** 0.3935 **37.** $b = \sqrt[4]{4}$, or $\sqrt{2}$.

EXERCISE SET 11.9, pp. 592–593

1. $\mu = E(x) = \frac{7}{2}$, $E(x^2) = 13$, $\sigma^2 = \frac{3}{4}$, $\sigma = \frac{1}{2}\sqrt{3}$. **3.** $\mu = E(x) = 2$, $E(x^2) = \frac{9}{2}$, $\sigma^2 = \frac{1}{2}$, $\sigma = \sqrt{\frac{1}{2}}$. **5.** $\mu = E(x) = \frac{14}{9}$,

$E(x^2) = \frac{5}{2}$, $\sigma^2 = \frac{13}{162}$, $\sigma = \sqrt{\frac{13}{162}}$. **7.** $\mu = E(x) = -\frac{5}{4}$, $E(x^2) = \frac{11}{5}$, $\sigma^2 = \frac{51}{80}$, $\sigma = \sqrt{\frac{51}{80}} = \frac{1}{4}\sqrt{\frac{51}{5}}$. **9.** $\mu = E(x) = \frac{2}{\ln 3}$, $E(x^2) =$

$\frac{4}{\ln 3}$, $\sigma^2 = \frac{4\ln 3 - 4}{(\ln 3)^2}$, $\sigma = \frac{2}{\ln 3}\sqrt{\ln 3 - 1}$. **11.** 0.4964 **13.** 0.3665 **15.** 0.6442 **17.** 0.0078 **19.** 0.1716

21. 0.0013 **23.** a) 0.6826 b) 68.26% **25.** 0.2898 **27.** 0.4514 **29.** a) 0.2088 b) 0.3830 c) 0.2420 **31.** 0.62%

33. $\mu = E(x) = \frac{a+b}{2}$, $E(x^2) = \frac{b^3 - a^3}{3(b-a)}$, or $\frac{b^2 + ba + a^2}{3}$, $\sigma^2 = \frac{(b-a)^2}{12}$, $\sigma = \frac{b-a}{2\sqrt{3}}$ **35.** $\sqrt{2}$ **37.** $\frac{\ln 2}{k}$

CHAPTER 11, TEST, pp. 594–595

1. $x + C$ **2.** $200x^5 + C$ **3.** $e^x + \ln x + \frac{8}{11}x^{11/8} + C$ **4.** $\frac{1}{6}$ **5.** $4\ln 3$ **6.** An antiderivative, total number of words

typed in t minutes. **7.** 12 **8.** $-\frac{1}{2}\left(\frac{1}{e^2} - 1\right)$ **9.** $\ln b - \ln a$ **10.** $\ln(x+8) + C$ **11.** $-2e^{-0.5x} + C$

12. $\frac{1}{40}(t^4 + 1)^{10} + C$ **13.** $\frac{x}{5}e^{5x} - \frac{e^{5x}}{25} + C$ **14.** $\frac{x^4}{4}\ln x^4 - \frac{x^4}{4} + C$ **15.** $\frac{2^x}{\ln 2} + C$ **16.** $\frac{1}{7}\ln\left(\frac{x}{7-x}\right) + C$ **17.** $\frac{1}{3}$

18. 95 **19.** Convergent, $\frac{1}{4}$ **20.** Divergent **21.** $k = \frac{1}{4}$ **22.** 0.8647 **23.** $E(x) = \frac{3}{4}$ **24.** $E(x^2) = \frac{3}{5}$ **25.** $\mu = \frac{3}{4}$ **26.** $\sigma^2 = \frac{3}{80}$

27. $\sigma = \frac{1}{4}\sqrt{\frac{3}{5}}$ **28.** 0.4332 **29.** 0.4420 **30.** 0.9071 **31.** 0.4207

CHAPTER 12

MARGIN EXERCISES

1. a) 128; the profit from selling 14 items of the first product and 12 of the second is $128. b) 48; the profit from selling none of the first product and 8 items of the second is $48. **2.** 7.2, 94.5, 0.60 **3.** 2000 **4.** a) 352 b) 13

5. $1173.51 **6.** a) 4, b) 4. **7.** a) $f(x, 4) = -x^2 - 15$, b) $-2x$. **8.** $\dfrac{\partial f}{\partial x} = -2x$. **9.** a) $\dfrac{\partial z}{\partial x} = 6xy + 15x^2$ b) $\dfrac{\partial z}{\partial y} = 3x^2$

10. a) $\dfrac{\partial t}{\partial x} = y + z + 2x$, b) $\dfrac{\partial t}{\partial y} = x + 3y^2$, c) $\dfrac{\partial t}{\partial x} = x$. **11.** a) $f_x = 9x^2y + 2y$, b) $f_x(-4, 1) = 146$, c) $f_y = 3x^3 + 2x$,

d) $f_y(2, 6) = 28$. **12.** $f_x = \dfrac{1}{x} + ye^x$, $f_y = \dfrac{1}{y} + e^x$. **13.**

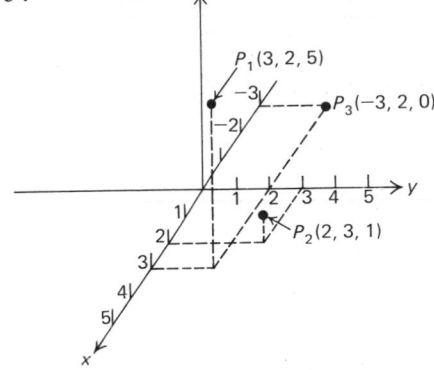

14. a) $\dfrac{\partial z}{\partial y} = 6xy + 2x$, b) $\dfrac{\partial}{\partial x}\left(\dfrac{\partial z}{\partial y}\right) = 6y + 2$, c) $\dfrac{\partial}{\partial y}\left(\dfrac{\partial z}{\partial y}\right) = 6x$. **15.** a) $f_y = 6xy + 2x$, b) $f_{yx} = 6y + 2$, c) $f_{yy} = 6x$.

16. $\dfrac{\partial^2 f}{\partial x^2} = 2$, $\dfrac{\partial^2 f}{\partial y\,\partial x} = 6y + 2 + \dfrac{1}{y}$, $\dfrac{\partial^2 f}{\partial x\,\partial y} = 6y + 2 + \dfrac{1}{y}$, $\dfrac{\partial^2 f}{\partial y^2} = 6x - \dfrac{x}{y^2}$ **17.** Min. $= -9$ at $(-3, 3)$ **18.** Max. $= \frac{1}{108}$ at $(\frac{1}{6}, \frac{1}{6})$

19. 5 thousand of $15 calculator; 7 thousand of $20 calculator. **20.** a) 75.7 yr; answers will vary. b) 75 yr; answers will vary. c) 0.7 yr difference; answers will vary. **21.** a) $y = 1.5x + 63.7$ b) 69.7 yr **22.** $y = 1.5x + 63.7$

23. a) $A(x, y) = xy$, subject to $x + y = 50$. b) Max. $= 625$ at $(25, 25)$. **24.** Max. $= 125$ at $(2.5, 10)$. **25.** $r \approx 1.8$ in., $h \approx 3.6$ in. Surface area is about 61.04 in.^{2}

EXERCISE SET 12.1, pp. 607–608

1. $f(0, -2) = 0$, $f(2, 3) = -8$, $f(10, -5) = 200$. **3.** $f(0, -2) = 1$, $f(-2, 1) = -13\frac{8}{9}$, $f(2, 1) = 23$. **5.** $f(e, 2) = \ln e + 2^3 =$

$1 + 8 = 9$, $f(e^2, 4) = 66$, $f(e^3, 5) = 128$. **7.** $f(-1, 2, 3) = 6$, $f(2, -1, 3) = 12$. **9.** 105, 95 **11.** 7.7% **13.** $\dfrac{\partial z}{\partial x} = 2 - 3y$,

$\dfrac{\partial z}{\partial y} = -3x$, $\dfrac{\partial z}{\partial x}\Big|_{(-2, -3)} = 11$, $\dfrac{\partial z}{\partial y}\Big|_{(0, -5)} = 0$ **15.** $\dfrac{\partial z}{\partial x} = 6x - 2y$, $\dfrac{\partial z}{\partial y} = -2x + 1$, $\dfrac{\partial z}{\partial x}\Big|_{(-2, -3)} = -6$, $\dfrac{\partial z}{\partial y}\Big|_{(0, -5)} = 1$ **17.** $f_x = 2$,

$f_y = -3$, $f_x(-2, 4) = 2$, $f_y(4, -3) = -3$ **19.** $f_x = \dfrac{x}{\sqrt{x^2 + y^2}}$, $f_y = \dfrac{y}{\sqrt{x^2 + y^2}}$, $f_x(-2, 1) = \dfrac{-2}{\sqrt{5}}$, $f_y(-3, -2) = \dfrac{-2}{\sqrt{13}}$

21. $f_x = 2e^{2x+3y}$, $f_y = 3e^{2x+3y}$ **23.** $f_x = ye^{xy}$, $f_y = xe^{xy}$ **25.** $f_x = \dfrac{y}{x+y}$, $f_y = \dfrac{y}{x+y} + \ln(x+y)$ **27.** $f_x = 1 + \ln xy$, $f_y = \dfrac{x}{y}$

29. $f_x = \dfrac{1}{y} + \dfrac{y}{x^2}$, $f_y = -\dfrac{x}{y^2} - \dfrac{1}{x}$ **31.** $f_x = 12(2x + y - 5)$, $f_y = 6(2x + y - 5)$ **33.** $\dfrac{\partial f}{\partial b} = 12m + 6b - 30$, $\dfrac{\partial f}{\partial m} = 28m +$

$12b - 64$ **35.** $f_x = 3y - 2\lambda$, $f_y = 3x - \lambda$, $f_\lambda = -(2x + y - 8)$ **37.** $f_x = 2x - 10\lambda$, $f_y = 2y - 2\lambda$, $f_\lambda = -(10x + 2y - 4)$

39. a) $0°$ b) $-10°$ c) $-22°$ d) $-64°$ **41.** $f_x = \dfrac{-4xt^2}{(x^2 - t^2)^2}$, $f_t = \dfrac{4x^2 t}{(x^2 - t^2)^2}$ **43.** $f_x = \dfrac{1}{\sqrt{x}(1 + 2\sqrt{t})}$, $f_t = \dfrac{-1 - 2\sqrt{x}}{\sqrt{t}(1 + 2\sqrt{t})^2}$.

45. $f_x = 4x^{-1/3} - 2x^{-3/4}t^{1/2} + 6x^{-3/2}t^{3/2}$, $f_t = -4x^{1/4}t^{-1/2} - 18x^{-1/2}t^{1/2}$.

EXERCISE SET 12.2, p. 610

1. $\dfrac{\partial^2 f}{\partial x^2} = 6$, $\dfrac{\partial^2 f}{\partial y\,\partial x} = \dfrac{\partial^2 f}{\partial x\,\partial y} = -1$, $\dfrac{\partial^2 f}{\partial y^2} = 0$. **3.** $\dfrac{\partial^2 f}{\partial x^2} = 0$, $\dfrac{\partial^2 f}{\partial y\,\partial x} = \dfrac{\partial^2 f}{\partial x\,\partial y} = 3$, $\dfrac{\partial^2 f}{\partial y^2} = 0$. **5.** $\dfrac{\partial^2 f}{\partial x^2} = 20x^3 y^4 + 6xy^2$,

$\dfrac{\partial^2 f}{\partial y\,\partial x} = \dfrac{\partial^2 f}{\partial x\,\partial y} = 20x^4 y^3 + 6x^2 y$, $\dfrac{\partial^2 f}{\partial y^2} = 12x^5 y^2 + 2x^3$. **7.** $f_{xx} = 0$, $f_{yx} = 0$, $f_{xy} = 0$, $f_{yy} = 0$. **9.** $f_{xx} = 4y^2 e^{2xy}$,

$f_{yx} = f_{xy} = 4xye^{2xy} + 2e^{2xy}$, $f_{yy} = 4x^2 e^{2xy}$. **11.** $f_{xx} = 0$, $f_{yx} = f_{xy} = 0$, $f_{yy} = e^y$. **13.** $f_{xx} = -\dfrac{y}{x^2}$, $f_{yx} = f_{xy} = \dfrac{1}{x}$, $f_{yy} = 0$.

15. $f_{xx} = \dfrac{-6y}{x^4}$, $f_{yx} = f_{xy} = \dfrac{2(y^3 - x^3)}{x^3 y^3}$, $f_{yy} = \dfrac{6x}{y^4}$. **17.** $\dfrac{\partial^2 f}{\partial x^2} = \dfrac{2y^2 - 2x^2}{(x^2 + y^2)^2}$, $\dfrac{\partial^2 f}{\partial y^2} = \dfrac{2x^2 - 2y^2}{(x^2 + y^2)^2}$, so the sum is 0.

EXERCISE SET 12.3, pp. 617–618

1. Min. $= -\frac{1}{3}$ at $(-\frac{1}{3}, \frac{2}{3})$. **3.** Max. $= \frac{4}{27}$ at $(\frac{2}{3}, \frac{2}{3})$. **5.** Min. $= -1$ at $(1, 1)$. **7.** Min. $= -7$ at $(1, -2)$. **9.** Min. $= -5$ at

$(-1, 2)$. **11.** None. **13.** 6 (thousand) of the $17 radio and 5 (thousand) of the $21 radio. **15.** Max. of $P = 35$

(million dollars) when $a = 10$ (million dollars) and $p = \$3$. **17.** a) $R(p_1, p_2) = 78p_1 - 6p_1^2 - 6p_1 p_2 + 66p_2 - 6p_2^2$;

b) $p_1 = 5$ ($\$50$), $p_2 = 3$ ($\$30$); c) $q_1 = 78 - 6 \cdot 5 - 3 \cdot 3 = 39$ (hundreds), $q_2 = 33$ (hundreds); d) $R = 50 \cdot 3900 +$

$30 \cdot 3300 = \$294{,}000$. **19.** None. **21.** Min. $= \frac{1}{6}$ at $(\frac{211}{3}, \frac{3}{2})$.

EXERCISE SET 12.4, pp. 622–623

1. 77.8 yr **3.** a) $y = 1.5x + 17.7$, b) 23.7, c) 26.7. **5.** a) $y = \frac{1}{4}x + 40$, b) $61°$. **7.** a) $y = -0.00582x + 15.3476$,

b) 3:48.0, c) Letting $x = 1979\frac{7}{12}$, $y = 3:49.6$.

EXERCISE SET 12.5, pp. 630–631

1. Max. $= 8$ at $(2, 4)$.   **3.** Max. $= -16$ at $(2, 4)$.   **5.** Min. $= 20$ at $(4, 2)$.   **7.** Min. $= -96$ at $(8, -12)$.   **9.** Min. $= \frac{3}{2}$ at

$(1, \frac{1}{2}, -\frac{1}{2})$.   **11.** 35 and 35.   **13.** 3 and -3.   **15.** $9\frac{3}{4}$ in., $9\frac{3}{4}$ in.; $95\frac{1}{16}$ in.^{2}; No.   **17.** $r = \sqrt[3]{\dfrac{27}{2\pi}} \approx 1.6\,\text{ft}$; $h = 2 \cdot r \approx 3.2\,\text{ft}$;

min. surface area $\approx 48.3\,\text{ft}^2$.   **19.** Min. $= -\frac{155}{128}$ at $(-\frac{7}{16}, -\frac{3}{4})$.   **21.** Max. $= \frac{1}{27}$ at $\left(\dfrac{1}{\sqrt{3}}, \dfrac{1}{\sqrt{3}}, \dfrac{1}{\sqrt{3}}\right)$ and $\left(-\dfrac{1}{\sqrt{3}}, -\dfrac{1}{\sqrt{3}}, -\dfrac{1}{\sqrt{3}}\right)$.

23. Max. $= 2$ at $(\frac{1}{2}, \frac{1}{2}, \frac{1}{2}, \frac{1}{2})$.

CHAPTER 12, TEST, p. 632

1. $\dfrac{\partial f}{\partial x} = e^x + 6x^2 y$ **2.** $\dfrac{\partial f}{\partial y} = 2x^3 + 1$ **3.** $\dfrac{\partial^2 f}{\partial x^2} = e^x + 12xy$ **4.** $\dfrac{\partial^2 f}{\partial x\,\partial y} = 6x^2$ **5.** $\dfrac{\partial^2 f}{\partial y\,\partial x} = 6x^2$ **6.** $\dfrac{\partial^2 f}{\partial y^2} = 0$

7. Min. $= -\frac{7}{16}$ at $(\frac{3}{4}, \frac{1}{2})$ **8.** None   **9.** a) $y = \frac{9}{2}x + \frac{17}{3}$, b) \$23.7.   **10.** Max. $= -19$ at $(4, 5)$

Index

Absolute complement of a set, 184
Absolute maximum, 453, 613
Absolute minimum, 453, 613
Absolute value function(s), 60–61
Absorbing state, 318
Absorption, coefficient of, 527
Acceleration, 427
  as higher derivative, 444
  integral of, 546
  and integration, 535
Accidents, and alcohol absorption, 518
  automotive, 74, 75, 460
Accumulated sales, 553, 554
Accumulations, as integrals, 544, 553
Addition Principle, 10, 52
Addition Theorem, 237, 257–263
  three events, 262
  two events, 257
Advertising, 406, 432, 501, 511
Aircraft, 493
Airlines, use of fuel, 528
Alcohol absorption, and the risk of having
    an accident, 518
Algebra review, 2–78
Algorithm, 148n
  echelon, 80ff
  simplex, 147ff
Amortization, 382
Annuities and sinking funds, 376
  amortization, 382
  amount, 378
  annuities, 376
  future value, 376
  ordinary, 377
  present value, 380, 382
  sinking funds, 378
Answers, 651

Anthropology applications, estimating
    heights, 46
Antiderivatives, 529–537
  area as, 541
  general form of, 531
  see also Integrals
Applications, of exponential functions; see
    Exponential functions
  of linear functions; see Linear functions
  of logarithmic functions; see Logarithmic
    functions
  using differentiation, 443–480
  using integration, 571–596
  See also Acceleration, Automotive
    applications, Biology, Biomedical,
    Business applications, Chemistry,
    Craps, Ecology, Economic
    applications, Games, Linear
    programs, Markov chains,
    Optimization, Probability,
    Psychology, Sociology, Statistics,
    Velocity, Voting coalitions
Archaeological applications; see Carbon
    dating
Arcs, trees, 200
Area, as antiderivative, 541
  as definite integral, 548–555
  and integration, 537–547
  maximizing, 460–462
  surface, minimizing, 467, 468, 471
Area function, 537–547
Arithmetic mean, 585
Arithmetic sequences, 349
  sum of first n terms, 350
Arrangements (*See* Permutations)
Associative law, matrices, addition, 99
    multiplication, 107

Atmospheric pressure, 527
Automotive applications, accidents, 74, 75,
    460
  cost of operation, 73
  gasoline mileage, 459
  stopping distance on glare ice, 45, 431
  transportation planning, 582, 583
Average, of a set of numbers, 18, 585
Average cost, 430
  derivative of, 437
  marginal, 437, 474
  minimizing, 474
Average rate of change, 400–408, 426
  as difference quotient, 403
  as slope, 401
Average revenue, derivative of, 437
  marginal, 437
Average speed, 408, 426, 427
Average value, 288–295
Average velocity, 408, 426, 427
Axis, x-axis, y-axis, 19

Banzhaf value, 339, 342
Barium, 447, 525
Base, exponential, 2
Basic feasible solution, 148, 154
Basic probability principle, 229
Basic variables, 148, 154–155
Bayes' Theorem, 275–279
Beer–Lambert Law, 527
Bell, Alexander G., 491
Bels, 491
Bernoulli trials, 300–305
Beverage can problem, 628
Binomial, cubing, 9
  squaring, 5
Binomial coefficients, 214, 221

Binomial probability, 300–305
   expected value, 304
   standard deviation, 305
   variance, 305
Binomial Theorem, 220–224
Biology applications, 11, 17, 36, 45, 63, 68,
      432, 473, 522–527, 623
   population growth; see Population
      growth
Biomedical applications, 44, 45, 76, 268,
      407, 428, 431, 432, 503, 511, 518, 520,
      537, 565, 607
Birth rate, 522
Birthday problem, 249–255
   extended, 250
   four people, 250, 253
   leap year, 253
   three people, 250, 252
   twins, 254
   two people, 249–250
Blood alcohol level, 518, 520
Blood pressure, 432
Bounded solution set, 125
Brain weight, 45
Branch of a tree, 200
Break-even point, 43, 463
Break-even value, 43
Build-up of radioactive material, 574
Business applications, 10, 42, 45, 46, 63–67,
      71, 76, 400, 406, 408, 429, 432, 437,
      442, 460, 464, 470, 474–478, 501, 504,
      510, 511, 520–522, 527, 553, 554, 560,
      584, 607, 615, 632
   compound interest; see Compound
      interest
   cost; see Total cost
   revenue; see Total revenue

Calcium content of foods, 176
Calculator, use of, 8
Calculator and finding logs, natural, 487
   common, 489
Calculus, differential; see Differential
      calculus
   Fundamental Theorem of, 568
   integral; see Integrals
Cancer and smoking, 268
Cancer tumor, 428
Carbon dating, 525, 526
Carbon-14, 525
Cardinality of a set, 183
Cards, 229
Carr, H. J., 69n
Carrying costs, 474
Cartesian product, 184
   cardinality, 184
Cesium-137, 525
Chain Rule, 441
   see also Extended Power Rule

Change of variable, 557
Change in x, y, see Slope
Characteristic, 490
Characteristic form of a game, 339–344
Checking solutions, linear programs, 171
Chemistry applications, carbon dating, 525,
      526
   decay, 522–527
   pH, 494
   radioactive decay, 523–526
Circular permutations, 207
Closed interval, 15
Coe, Sebastian, 623
Coefficient of absorption, 527
Combination lock, 212
Combinations, 211–216
   binomial coefficients, 214
   notation, 212–214
Common logarithms, 489, 490
   characteristic of, 490
   mantissa of, 490
Commutative law, 459
   matrices, addition, 99
      multiplication, 107
Complement of a set, 184
Complementary events, 233
Composition of functions, 440
Compound events, 236
Compound interest, 7, 8, 10, 17, 31, 442,
      509, 520, 521
   compounded annually, 7
   compounded continuously, 514, 520, 521
   compounded n times per year, 8
   and the number e, 498, 499
Concave down, 446
Concave up, 446
Concavity, 445–452
Conditional probability, 264–273, 275–279
   Bayes' Theorem, 275–279
Conformability of matrices, 107
   addition, 98, 111
   augmented, 111
   equality, 97, 107
   multiplication, 105, 111
Conjunction, events, 236
   probability, 236
Constant, of integration, 531, 547
   of proportionality, 35
   of variation, 35
Constant function, 37
   continuity of, 393
   derivative of, 421
Constant times a function, derivative of,
      421
Constraints, 120
   explicit, 138
   implicit, 138
   integer, 144
   matrix game, 327

nonnegativity, 138–139
   positivity, 148
   redundant, 126
Consumer price index, 521
Continuity, 388–390
   of a constant function, 393
   on an interval, 393
   at a point, 393
   principles, 393
   of a rational function, 394
Continuous functions, 388–390
   and maximum-minimum values, 453
Continuous money flow, 504
Continuous random variable, 575–594
   expected value of, 586
   and integration, 577
   mean of, 586
   normally distributed, 589
   standard deviation of, 587
   standard normal, 589
   uniformly distributed, 580
   variance of, 587
Continuously compounded interest, 509,
      520, 521
Converges, 572
Convex combination, 128, 134, 322, 336
Convex set, 128
Coordinates, 19
   first, 19
   second, 19
Corner points, 130, 153
Cost, average, 430, 437, 474
   carrying, 474
   derivative of, 429, 442
   fixed, 41, 429
   marginal, 429, 442, 464, 503, 538
   marginal average, 437, 474
   reorder, 474
   total; see Total cost
   variable, 41, 429
Counting techniques,
   combinations, 211–216
   permutations, 202–210
Craps, 228, 294–295
Cricket chirps as a function of temperature,
      623
Critical points, 451, 452, 612
   for a function of one variable, 451
   for a function of two variables, 612
   shape of graph between, 451, 452
Cubic function, 58–59
Cubing a binomial, 9
Curve(s), family of, 533
Curve fitting, 70–74, 618–623

Death rate, 522
Debt, personal, 373
Decay, exponential, 522–527
   radioactive, 523–526

Decay rate, 522
   and half-life, 524
Decibels, 491, 492
Decisions (See Games)
Deck of cards, 219, 229
Decreasing functions, 28, 496, 522
   and derivatives, 445–452
   linear, 40
Definite integral(s), 547–555
   as area, 548–555
   definition of, 547
   as a limit of sums, 565–569
   of a positive function, 547–555
   properties of, 551
Degenerate points, 127
Demand, for coal, 521
   for electrical energy, 519
   for natural gas, 623
   for oil, 521
   and supply, 63–67, 527
Demand functions, 63–67, 527
Density function, probability, 577
Dependent variable, of a function, 29
Depreciation, 400
   declining-balance, 362
   straight-line, 360
   sum-of-year's-digits, 365
Derivative notation, $\frac{d}{dx}f(x)$, 419
   dot, 420
   $\frac{dy}{dx}$, 419
   $f'(x)$, 412
   Leibniz's, 419, 443
Derivatives, 412–442, 496–499, 505–507,
      598–631
   anti, 530–537
   of a constant, 421
   of a constant times a function, 421
   of a difference, 422
   of $e^{f(x)}$, 499
   of e^x, 498
   of exponential functions, 496–504
   fifth, 443
   first, 412–442
   fourth, 443
   higher, 443
   of linear functions, 412, 413
   of ln $f(x)$, 507
   of ln x, 506
   of natural logarithmic functions, 505
   partial, 600–610
   of a power, 420, 437–442
   of a product, 433
   of a quotient, 434
   second, 443
   of a sum, 422
   third, 443
   see also Differentiation
Deviations, 619

Dice, 228
   craps, 294–295
Dictator, 344
Die, singular of dice, $q.v$
Diet problem, 176
Difference, set, 190
Difference quotient(s), 403–408
   as average rate of change, 403
   and differentiation, 412–418
Differential calculus, 412
   see also Differentiation
Differentiation, 388–442, 496–499,
      505–507, 598–631
   anti-, 530
   applications of, 426–432, 435, 443–480
   of a constant, 421
   of a constant times a function, 421
   of a difference, 422
   of $e^{f(x)}$, 499
   of e^x, 498
   of exponential functions, 496–504
   of ln $f(x)$, 507
   of ln x, 506
   and maximum-minimum values, 453–460,
      460–480
   of natural logarithmic functions, 505
   partial, 600–610
   of a power, 420, 437–442
   of a product, 433
   of a quotient, 434
   of a sum, 422
   using limits, 409–418
   see also Derivatives, Differentiation
      techniques
Differentiation notation; see Derivative
   notation
Differentiation techniques, 419–425,
      433–436, 437–442
   chain rule, 441
   difference rule, 422
   extended power rule, 438
   power rule, 420
   product rule, 433
   quotient rule, 434
   sum-difference rule, 422
   sum rule, 422
Direct variation, 35
Directly proportional, 35, 513
Discontinuous function, 389, 391–393, 415
Disjoint sets, 187
Disjunction, events, 236
   probability, 236
Distinguishability, 198–201, 216
Distribution, exponential, 581
   normal, 588–594
   uniform, 580
Distributive law, 5
   matrices, 107
Diverges, 572

Dividing with exponents, 4
Divorce rate, 408, 560
   integral of, 546
Domain, of a function, 22–25
   of square root function, 23, 61
Dominant strategies, 329
Doubling time, 514–516
   and growth rate, 515
Drug dosage, 528, 565
D-Test, 613
Dual solution, 165
Dual variables, 165
Duality, 164ff
   economic interpretation, 172
   games, 336
   Theorem, 165

e, 486, 498
Earthquakes, Anchorage, 494
   magnitude of, 494
   Mexico City, 494
   and the Richter scale, 494
Echelon method, 80ff
   pivot element, 81
   pivot row, 81
   special cases, 89–94
   tableau, 80
   unique solutions, 80–88
Echelon tableau, 80
Ecology applications, 36, 44, 142, 492, 519,
      521, 565
   population growth; see Population
      growth
Economic applications, 406, 429–431
Economic interpretation of duality, 172
Economic multiplier, 358
Economic ordering quantity, 478
Educational applications, grade average,
      18
Effective annual yield, 521
Efficiency, of machine operator, 537
$e^{f(x)}$, differentiating, 499
Electrical energy demand, 519
Electrical energy use, 565
Element, matrix, 96
   set, 180
Emlen, J. M., 63
Empty sack in an empty sack, 182
Empty set, 127, 182
Endpoints, of interval, 15
Energy, electrical, 519, 565
   see also Ecology applications
Energy conservation, 44
Equation(s), 10–12
   addition principle for, 10, 52
   exponential, 484, 490, 491
   graphs of, 20, 21
   linear, 32–46
   multiplication principle for, 10

point-slope, 37
principle of zero products, 12
related, 121
slope–intercept, 37
solution of, 19
systems (*See* Linear equations)
two-point, 44
Equilibrium-point, 66
    matrix game, 333
    supply and demand, 63–67
Equilibrium price, 66
Equimarginal productivity, 631
Equiprobability, 228
Ergodic Markov chain, 317
Estimating heights, 46
Europe, population growth in, 520
Events, cannot occur, 230
    certain to occur, 230
    complementary, 233
    compound, 236
    conjunction, 236
    dependent, 243, 264–273
    disjoint, 236
    distinguishability, 198–201
    independent, 243, 258, 268
    mutually exclusive, 200, 237
    partitioning, 188–189
    probability, 228
e^x, differentiating, 498
    integrating, 531
Executive decision-maker, 228
Expected value, Banzaf, 339, 342
    binomial, 304
    of continuous random variable, 586
    hypergeometric, 291, 297
    Shapley, 339
Experimental probability, 228
Experiments, 198–201
    distinguishability of outcomes, 198–201
    time ordering, 199
Exponent(s), 2–5
    base of, 2
    dividing with, 4
    fractional, 62
    irrational, 495
    multiplying with, 3
    negative, 3
    of one, 2
    properties of, 3–5
    raising a power to a power, 4
    of zero, 2
Exponential decay, 522–527
    exponential distribution, 581
    and telephone calls, 584
    and transportation planning, 582–584
Exponential equations, 484, 490, 491
Exponential functions, 482–528, 494–503
    derivatives of, 496–504
    graphs of, 482, 483, 495, 497, 500, 505

integration of, 531
    and population growth, 512–522
Exponential growth, 512–522
Exponential growth model, 512–522, 599
Exponential growth rate, 514, 515
    and doubling time, 515
    and effective annual yield, 521
Exponential notation; *see* Exponents
Extended Power Rule, 438

Factor, 2
Factorial, 204
    zero, 205
Factoring, 6
Fair game, 292, 333
Family of curves, 533
Feasible solution, 148
    initial, 148, 152
Fever, 432
First coordinate, 19
First derivative, and increasing, decreasing
        functions, 445
    see also Differentiation
Fixed costs, 41, 429
Fixed points, 321–325
Fixed probability vector, 323
Flights of homing pigeons, 473
Fluid weight in body, 77
Forgetting, 508–510
Fork, tree, 200
Formula(s), differentiation; *see*
        Differentiation
    integration by parts, 562
    quadratic, 57
Fractional exponents, 62, 63
Franchise expansion, 520
Fuel, demand by airlines, 528
Function(s), 21–76
    absolute value, 61
    area, 537–547
    composition of, 440
    constant, 37, 58
    continuous, 388–390, 393
    critical point of, 451, 452, 612
    cubic, 58
    decreasing, 28, 445–452, 496, 522
    decreasing linear, 40
    demand, 63–67, 527
    dependent variable of, 29
    derivatives; *see* Differentiation
    differentiation of; *see* Differentiation
    discontinuous, 389, 391–393, 415
    domain of, 22–25, 61
    exponential, 482–528, 494–503
    graphs of, 25–66
    increasing, 28, 445–452, 495
    increasing linear, 40
    independent variable of, 29
    as input–output relations, 22

inputs, 22
integration of; *see* Integration
limits of, 390–400
linear, 32–43, 58
logarithmic, 504–511
as a mapping, 24
natural logarithmic, 504–511
outputs of, 22
polynomial, 58
postage, 398, 418
power, 58
power with fractional exponents, 62
probability, 284ff
probability density, 577
quadratic, 47, 58
range of, 22
rational, 59
reciprocal, 22, 23
of several variables, 598–632
    derivatives of, 600–610
    graphs of, 603, 604
square root, 23, 61
squaring, 22, 24
supply, 65–67
vertical line test, 27, 33
Function notation, 22
Fundamental counting principle, 203
Fundamental Theorem of Calculus, 568
Future value of an annuity, 367

Gambling (*see* Craps, Probability)
Games, 327–344
    Banzaf value, 339, 342
    characteristic form, 339–344
    choices, 328
    coalitions, 339
    craps, 294–295
    dictator, 344
    dominant strategies, 329
    duality, 336
    dummy, 342, 344
    equilibrium point, 333
    fair, 292, 333
    give-away, 235
    individual rationality, 329
    information sets, 328
    lottery, 292
    matrix, 327
    maximin, 333, 335
    minimax, 333, 336
    Minimax Theorem, 336
    mixed strategies, 334–338
    payoffs, 328
    Prisoner's Dilemma problem, 327
    probability distributions, 334
    proper, 339
    pure strategies, 328
    quota, 340
    security level, 333

Shapley value, 339
solution, 329
strategies, mixed, 334
    pure, 328
tree, 327
value, 333, 336
von Neumann Theorem, 336
voting, 340
voting coalitions, 339-344
zero sum, 329
Gas, natural, 623
Geometric sequences, 353
    common ratio, 353
    infinite, 356
    sum of first n terms, 355, 357
Graphs, of equations, 20, 21
    of exponential functions, 482, 483, 495,
        497, 500, 505
    of functions, 25-66
    of functions of several variables, 603,
        604
    and interval notation, 15, 16
    of linear equations, 20, 25, 32-41
    of linear functions, 20, 25, 32-41
    of logarithmic functions, 484, 505
    shape of, 445-452
    single inequality, 121
    system of inequalities, 121-129
Gravity model, 600
Growth, average rate, 407
    exponential, 512-522
    population; see Population growth
    of stock, 504
    uninhibited, 512-522
Growth model, $P' = kP$, 513
Growth rate, 428
    average, 407
    and doubling time, 515
    exponential, 514, 515

Hair growth, 36
Half-life, 524
    and decay rate, 524
Half-open interval, 15
Healing wound, 431, 432, 537
Health food store, 178
Hellin's Law, 254
Herget, P., 69n
Higher derivations, 443
    Leibniz's notation for, 443
Higher-order partial derivatives, 608-610
    notation for, 608-610
Home range of an animal, 63, 432
Horizon, view to the, 431
Horizontal lines, 32, 33, 37
Hugo, Victor, 412
Hullian learning model, 503
Hypergeometric probability, 287, 291,
    302

Improper integrals, 572, 571-574
Increasing functions, 28, 495
    and derivatives, 445-452
    linear, 40
Indefinite integral(s), formulas for, 531
    see Antiderivatives, Integration
Independent events, 243, 258, 268
Independent trials, repeated, 300
Independent variable, of function, 29
Indianapolis Safety Action Project, 518
Indicators, 156, 161
Industrial psychology, 400
Inequalities, 13, 14, 120
    graphing, 121
    properties of solving, 13
    solution, 13
Infinite intervals, 16
Infinite limits, 395
Infinity, 395
    limits at, 397
Inflection point, 447
Information set, 328
    Cartesian product, 184
Initial feasible solution, 148, 152
Input-output tables, and limits, 390-400
Inputs of functions, 22, 29
Instantaneous rates of change, 426-432
Instantaneous velocity, 426
Insulation, R-factor of, 44
Integer constraint, 144
Integer program, 144
Integers, 2
Integral(s), 531-537
    as an accumulation, 544, 553
    definite, 547-555
        properties of, 551
    improper, 571-574
    indefinite, 531
        properties of, 531, 532
    see also Integration
Integral sign, 531
Integration, 530-596
    and antiderivatives, 530-537
    applications of, 573-595
    and area, 537-547
    constant of, 531, 547
    of exponential functions, 531
    Fundamental Theorem of Calculus, 568
    improper, 571-574
    limits of, 547
    of logarithmic functions, 562, 563
    by parts, 561-565
    of powers, 531
    by substitution, 556-561
    using tables, 564
    see also Table 5
Integration by parts formula, 562
Integration tables; see Table 5
Intelligence quotient, 607

Intensity, of earthquake, 494
    of sound, 491, 492
Interest, 86, 367
    add-on, 374
    compound, 368
        annually, 369
        continuously, 371
        n times per year, 370
    simple, 367
Interior point of interval, 16
Intersection of sets, 187-196
Interval(s), 15, 16
    closed, 15
    endpoints of, 15
    half-open, 15
    infinite, 16
    interior points of, 16
    open, 15
Interval notation, 14-16
Intuitive, 388
Inventory cost, minimizing, 474-478
Inverse matrix, 108-116
Inverse variation, 60, 510
Inversely proportional, 60, 510
Investment, 560
Iodine-131, 526
Iron content of foods, 176

Job opportunities, 521
Joint probability, 272

Key points, 452, 453
Key problem, 262
Krypton-85, 526

LaGrange Multipliers, 624-631
    method of, 625-630
Lambert; see Beer-Lambert Law, Law(s)
Law(s), Beer-Lambert, 527
    distributive, 5
    Poiseuille's, 607
Lead, 526
Learning, Hullian model of, 503
    rate in maze, 584
    see also Memory, Psychological
        applications
Least squares assumption, 619
Least squares technique, 618-624
Leibniz, Gottfried W. von, 420
Leibniz's notation, 419, 443
    for first derivatives, 419
    for higher derivatives, 443
Life expectancy, of the female, 618-620
    of the male, 620
Light, through sea water, 527
    through smog, 527
Limit(s), 390-400, 496
    of a constant, 396
    of a difference, 396

differentiation, using, 409–418
finding, using input–output tables, 390–400
finding, using substitution, 394
at infinity, 397
of integration, 547
notation, 391
principles, 396
of a product, 396
of a reciprocal, 396
of a sum, 396
of sums, 565–569
Line(s), horizontal, 32, 33, 37
number, 15
parallel, 37
regression, 619, 620
secant, 401, 411
slope of, 34, 39
straight, 32–46
tangent, 409–411
vertical, 33
see also Linear equations, Linear functions
Line of symmetry, 47
Linear constraints (See Linear inequalities)
Linear equations, 32–46, 80ff
addition principle, 10
algebraic solution by addition method, 52–55
by echelon method, 80ff
by substitution method, 52
consistent, 48
echelon method, q.v., 80ff
graphical solution, 47–50
graphs, 20, 25, 32–46
inconsistent, 48, 93
intercepts, 37
linearly dependent, 48, 54, 91, 93
linearly independent, 48
multiplication principle, 10
point–slope form, 37
slope–intercept form, 37
solution of single equation, 10
two-point form, 44
see also Line(s), Linear functions
Linear functions, 32–43
applications of, 41–46
and curve fitting, 70–72
decreasing, 40
graphs of, 20, 25, 32–46
increasing, 40
slope of, 34–40
y-intercept of, 37
see also Line(s), Linear equation(s)
Linear inequalities, 13–14, 130–133
corners of solution set, 133
graphing single inequality, 121–124
graphing system of inequalities, 124–129
redundant constraint, 126

related equation, 121
solution set, bounded, 125
empty, 127
unbounded, 124–125
vertices of solution set, 130–133
Linear optimization (See Linear programs)
Linear programs, 120ff, 147ff
basic feasible solution, 148, 154
basic variables, 148, 154–155
checking solutions, 171
constraints (See Constraints)
corner points, 153
dual programs, 164ff
dual solution, 165
duality, 164ff
Duality Theorem, 165
economic interpretation, 172
explicit constraints, 138
extended tableau, 151
feasible solution, 148
final tableau, 162
formulating maximum type, 137–141
formulating minimum type, 142–146
graphical solution, 140
implicit constraints, 138
indicators, 156, 161
initial feasible solution, 148, 152
initial simplex tableau, 148, 150
matrix form, 139, 152
maximum value, 132–133
minimum value, 132–133
nonbasic variables, 148, 154–155
nonnegativity constraints, 138–139
objective function, 132–133, 134, 148, 151
optimum feasible solution, 148
optimum value, 130–134
pivot, choice of, 148, 155ff
wrong, 159
pivoting, 149, 157ff
positivity constraint, 148
primal program, 165
primal solution, 165
simplex algorithm, 147ff
slack variables, 148, 149
solution, 148
basic feasible, 148
dual, 168
empty, 127
feasible, 148
graphically, 140
optimum feasible, 148
simplex, 147ff
standard form, maximum type, 147ff
minimum type, 164ff
structural variable, 148
summary, simplex algorithm, 161
termination, 149, 160
ln f(x), derivative of, 507

ln x, derivative of, 506
integration of, 562
Logarithm(s), base a, 483–486
base e, 486–489, 504–511
base 10, 489, 490
basic properties of, 484, 486
common, 489, 490
characteristic of, 490
mantissa of, 490
natural, 486–489, 504–511
definition of, 504
Logarithmic functions, 482–528
derivatives of, 506, 507
graphs of, 484, 505
integration of, 562, 563
see also Logarithm(s)
Long run, 321–325
Lot size, 466
Lottery, 292
Loudness of sound, 491, 492
Lung cancer and smoking, 268

Mantissa, 490
Mapping, 24
Marginal average cost, 437, 474
Marginal average revenue, 437
Marginal cost, 429, 442, 464, 503, 538
and integrating, 534, 536
Marginal profit, 429, 462–464
Marginal revenue, 429, 442, 462–464, 504
approximating, 536
and integrating, 536
Marginalist School of Economic Thought, 429n
Market, target, 501
Markov chains, 310–325
absorbing state, 318
cyclic, 319
ergodic, 317
fixed points, 321–325
probability vector, 323
irregular, 317
long run, 321–325
periodic, 319, 325
power test for regularity, 316
regular, 316, 324–325
transition matrices, 310
Marriage rate, 408
Mathematical model(s), 69–74
applications, 71–74
curve fitting, 70–74, 618–624
gravity, 600
growth, $P' = kP$, 513
least squares technique, 618–624
of uninhibited growth, 513
voting coalitions, 339–344
Matrices, 95ff
addition, 98
associative, 99

commutative, 99
additive identity, 99
additive inverse, 100
Associative Law,
    addition, 99
    multiplication, 107
augmented, 111
coefficient, 109
column vector, 97
conformability, 105–107
definition, 96
dimensions, 97
Distributive Law, 107
division (not possible), 109
dual linear programs, 164ff
elements, 96
equality, 97, 107
formulation of linear programs, 137–146
games, 327
identity, multiplicative, 107
inverse, computed, 114–116
inverse, multiplicative, 112–116
main diagonal, 107
multiplication, 101–112
partitioned, 111
product, matrix, 101–112
    scalar, 99
row vector, 97
scalar product, 99
square, 96
subtraction, 100
transition, 310–315
transpose, 98
Matrix (plur. matrices, q.v.)
Maxima and minima, absolute, 453
    area, 460–462
    constrained, with LaGrange Multipliers,
        624–631
    definition of, 452
    D-Test, 613
    finding, 453–478
    functions of two variables, 610–618,
        624–631
    inventory costs, 474–478
    least squares technique, 618–624
    problems, 460–478
    profit, 410, 462–464, 470, 502, 615, 617
    relative, 453, 610
    revenue, 468, 471
    second derivatives, 454
    surface area, 467, 468, 471
    total revenue, 468, 471
    volume, 465–467
Maximin, 333, 335
Maximum-Minimum Principle 1, 453
Maximum-Minimum Principle 2, 455
Mean, 288–295
    arithmetic, 585
    of a continuous random variable, 586

Medical applications; see Biomedical
    applications
Memory, 407, 462, 537, 555
Minimax, 333–336
Minimax Theorem, 336
Minimum; see Maxima and Minima
Mixed counting, 208
Models; see Mathematical models
Monthly mortgage payment, 10
Motel problem, 189
Multiplication, and the distributive laws, 5
    with exponents, 3
    principle, 10, 52
Multiplication Theorem, 242–255, 264–266
    n events, 271
Multiplier, economic, 358
Muscle weight, 45

National debt, 408
Natural gas, demand for, 623
Natural logarithmic (function), 486–489,
        504–511
    definition, 504
    derivative of, 506
    integral of, 562
    notation for, 504
    properties of, 486, 487
Negative exponent, 3
Nerve impulse speed, 44
Newton, Sir Isaac, 420
Nodes, games, 327
    information sets, 328
Noise pollution, 492
Nonbasic variables, 148, 154–155
Nonnegativity constraint, 138–139
Normal distribution, 588–594
    mean of, 589
    probability with tables, 590
    standard, 589
    standard deviation, 589
Norman window, 472
Notation, $\frac{d}{dx}f(x)$, 419
    dot, 420
    $\frac{dy}{dx}$, 419
    exponential; see Exponents
    for first derivatives, 412, 419
    function, 22
    $f'(x)$, 412
    for higher order derivatives, 443
    for integrals, 547
    interval, 14–16
    Leibniz's, 419, 443
    limit, 391
    for partial derivatives, 600, 601
    radical, 62
    for second-order partial derivatives, 608,
        609
    summation, 566, 567

Null set, 182
Number(s), e, 486, 498
    integers, 2
    real, 14
Nutrition problem, 176

Objective function, 132–133, 134, 148,
        151
Odds, 233–234
    against, 234
    for, 234
Oil, demand for, 521
Open interval, 15
Operating cost of a car, 460
Optimization (See Linear programs, Games)
Optimum, 145
Optimum value, 130ff
Or, 186
    exlusive, 186
    inclusive, 186
Ordered pair, 19, 184, 285, 328
    coordinates of, 19
Ordering quantity, economic, 478
Organism, spread of, 45
Outcomes, 198, 200
Outputs of functions, 22, 29

Pair, ordered, 19, 184, 285, 328
Parallel lines, 37
Partial derivative(s), 600–610
    geometric interpretation, 602–606
    higher order, 608–610
    notation for, 600, 601
    second-order, 608
Partial differentiation; see Partial
    derivatives
Partition, matrix, 111
    set, 188–189
Parts, integration by, 561–565
Pascal's triangle, 223–224
Payoffs of a game, 328
Percentage, 374
    annual rate, 374
Permutations, 202–210
    circular, 207
    factorial, 204
    Fundamental Counting Principle,
        203
    notation, 204, 206
    tree, 203
    with repetition, 208
    with replacement, 208
    without replacement, 202
Petri dish, 517
pH, 494
Pi (π), 494
Pivot element, 81
    echelon method, 80
    simplex algorithm, 147ff

Pivoting, 148
Plutonium, 525
Point(s), break-even, 43, 463
  coordinates of, 19
  critical, 451, 612
  equilibrium, 66
  inflection, 447
  key, 452, 453
  saddle, 612, 613
Point–slope equation, 37
Poiseuille's law, 607
Poker, 218–219
Pollution, noise, 492
Polynomial functions, 49
Population in college, 45, 431, 460
Population growth, 17, 45, 407, 432, 445,
    512–522
  bacteria colony, 428
  Central America, 520
  and derivatives, 444
  Europe, 520
  Kansas City, 521
  Tempe, Arizona, 521
  U.S., 516, 522
  U.S.S.R., 520
  Virgin Islands, 522
  world, 516
Positive functions, integral of,
    547–555
Postage function, 398, 418
Postage stamp, cost of, 522
Power(s), derivatives of, 420, 437–442
  integrating, 531
  see also Exponents
Power functions, 58
  applications of, 63
  with fractional exponents, 62
Power Rule, 420
  extended, 438
Power test for regularity, 316
Present value of an annuity, 371, 373, 389,
    382
Pressure, atmospheric, 527
Price-earnings ratio, 607
Primal program, 165
Primal solution, 165
Principal square root(s), 23
Principle(s), addition, 10
  continuity, 393
  limit, 396
  maximum, minimum, 453, 455
  multiplication, 10, 52
  shape, 452
  of zero products, 12
Principle of individual rationality, 329
Prisoner's Dilemma Problem, 327
Probability, 228–279
  Addition Theorem, 237, 257–263
  basic principle, 229

Bayes' Theorem, 275–279
Bernoulli trials, 300–305
binomial, 300–305
birthday problem, 249–255
complementary events, 233
compound events, 236
conditional, 264–273, 275–279
conjunction of events, 236
continuous random variable; see
    Continuous random variable
craps, 294–295
dependent and independent, 258,
    264–273
dice, 228, 232
disjunction, 236
equiprobability, 228
events, certain to occur, 230
  independent, 243
  mutually exclusive, 237, 239
  which cannot occur, 230
expected value, 288–295, 304, 586
experimental, 228
frequency function, 285
function, 284–305
hypergeometric, 287, 291, 302
independent events, 243, 258, 268
joint, 272
key problem, 262
Markov chains, 310–325
mean, 228, 295, 586
Multiplication Theorem, 242–255,
    264–266
mutually exclusive events, 237, 239
normal distribution, 588–594
numerical value of a random variable,
    285
odds, 233–234
outcomes, 228
partitioning events, 238
random variable, 284–305
reduced sample space, 267
sample space, 228
standard deviation, 297–299, 305, 587
standard normal, 589
theoretical, 228
transition, 311
trees, 247, 259
uniformly distributed, 580
variance, 297–299, 305, 587
vector, 310–315
Venn diagrams, 233, 236–237, 261
Probability density function(s), 577–585
  constructing, 579
  see also Distribution
Product, Cartesian, 184
Product Rule, 433
Profit, 410, 429, 462–464, 471, 502
  marginal; see Marginal profit
  maximizing; see Maxima and minima

Profit-and-loss analysis, 42, 46
Programs (See Linear programs)
Properties, of definite integrals, 551
  of derivatives; see Derivatives
  of exponents, 3–5
  of indefinite integrals, 531
Proportional, directly, 35, 513
  inversely, 61, 510
Protein content of foods, 176
Psychology applications, 400, 407, 503,
    508–511, 555, 584, 607
Pure strategies, 328–333

Quadratic formula, 58
Quadratic functions, 56
  and curve fitting, 71, 73, 74
Quadrupling time, 521
Quality units, 460
Quota games, 340
Quotient rule, 434

Rabbits, doubling time, 515
Radical notation, 62
  see also Roots, Square roots
Radioactive build-up, 574
Radioactive decay, 523–526
Random variable, continuous; see
    Continuous random variable
Random variables, 284–305
Range, of a function, 22
Rate(s), birth, 522
  death, 522
  decay, 522
  divorce, 408, 546, 560
  of exponential growth, 514, 515
  growth, 428; see also Growth
  integral of, 546
  marriage, 408
Rate(s) of change, 426–432
  acceleration, 427, 444
  average, 400–408, 426
  as derivative, 428
  and differentiation, 426–432
  in economics, 429–431
  instantaneous, 426–432
  velocity, 427
Rational exponents, 62, 494, 495
Rational functions, 59
  continuity, 394
  domain of, 59
Rationality, individual principle of, 329
Real numbers, 14
Reciprocal function, 22, 23
Recycling newspaper, 36
Reduced row echelon method, 89
Redundant constraint, 126
Regression line, 619–622
Regular Markov chains, 316
Related equation, 121

Relative complement, 190
Relative maximum, 453, 610, 613
Relative minimum, 453, 610, 613
Reliability, 584
Reorder cost, 474
Repeated independent trials, 300
Revenue, 14, 18, 31, 43, 407, 442,
    462–464
  marginal; *see* Marginal revenue
  marginal average, 437
  total; *see* Total revenue
Revenue maximization, 468, 469, 471
Reynolds number, 511
R-factor of insulation, 44
Richter scale, 494
Rollercoaster, 529
Roots, square, 23, 61
Roster method, 180
Row equivalence, 162
Rules, chain, 441
  extended power, 438
  power, 420
  product, 433
  quotient, 434
  sum–difference, 422
Russia, population growth, 520
Ryder, H. W., 69n

Saddle point, 612, 613
Sales commissions, 46
Salvage value, 527
Sample space, 198, 228
  reduced, 276
Satellite power, 527
Seaver, Tom, 598
Secant line, 401, 411
Second derivatives,
    and concavity, 446
    and maximum-minimum values, 454
Second-order partial derivatives, 608
Security level, 333
Sets, 180–216
  absolute complement, 184
  cardinality, 183
  Cartesian product, 184
  complement, 184
  difference, 190
  disjoint, 187
  distinguishable elements, 198–201
  elements, 180
  empty, 182
  equality, 180
  experiments, 198–201
  information, 328
  intersection, 187–196
  null, 182
  or, 186
  partition of, 188–189
  relative complement, 190

roster method, 180
sample space, 198–201
set-builder notation, 181
subsets, 182, 223
trees, 198–201, 203
union, 186–196
universal, 183
Venn diagrams, 186–196
Several variables, functions of,
    598–632
Shadow prices, 174
Shape, of a graph, 445–452
Shape principle, 452
Sigma, 566
Sigma notation; *see* Summation
    notation
Simple interest, 367
Simplex algorithm, 147ff
  extended tableau, 151
  final tableau, 162
  initial extended tableau, 148
  interchanging, 158
  pivot, choice of, 148, 155ff
    wrong, 159
  pivoting, 149, 157ff
  row equivalence, 160
  termination, 149, 160
Simplex tableau, condensed,
  extended, 148, 150
  final, 162
  initial, 148, 150
  terminal, 162
Sinking fund, 376, 378
Slack variables, 148, 149
Slope, 34, 39
  as average rate of change, 401
  as difference quotient, 403
  of linear function, 34–40
  and partial derivatives, 606
  of tangent line, 412
Slope-intercept equation, 37
Smog, 527
Smoking and cancer, 268
Sociology applications, 45, 408, 431, 503,
    546, 560, 600, 618–620
  population; *see* Population
Solution(s), of equations, 11, 12
  of equations in two variables, 19
  of inequalities, 13, 14
Sound, bels, 491
  decibels, 491
  intensity of, 491, 492
  loudness of, 491, 492
  minimum intensity of, 491
Soybeans, 142
Speed, average, 408, 426
Square root function(s), 23, 61
  continuity of, 393
  domain of, 23, 61

Square roots, 23, 61
  principal, 23
Squaring a binomial, 5
Squaring function, 22, 24
Standard deviation, 297–299, 305, 587
Standard normal distribution, 589
Statistics, 284–305
  average value, 288–295
  expected value, 288–295
  mean, 288–295
  random variable, 284–305
  standard deviation, 297–299
  variance, 297–299
Stopping distance on glare ice, 45, 431
Straight-line depreciation, 46
Straight lines, 32–46
Strategies, 327–344
  dominant, 329
  mixed, 334
  pure, 328
Strontium-90, 529
Structural variables, 148
Subscripts, variables with, 47
Subsets, 182
  combinations, 211
  partitions, 188–189
Substitution, and finding limits, 394
Substitution, and integration, 556–561
Sum–difference rule, 422
Summation notation, 101, 566, 567
Supply and demand, 527
Supply function(s), 65–67
Surface area, minimizing, 467–468, 471
Symmetry, line of, 56

Tableau (*See* Echelon—*or* Simplex—)
Tables, 633ff
  areas for a Standard Normal Distribution,
      641
  common logarithms, 635
  compound interest, 642
  exponential functions, 639
  integration formulas, 640
  natural logarithms (ln x), 637
  powers, roots, reciprocals, 634
Tables, integration by, 564
Tangent lines, 409–411
  definition of, 411
  horizontal, 423
  slope of, 412
  vertical, 417
Target market, 501
Taxes earned from each dollar, 71
Telephone calls, between two cities, 600
  and exponential distribution, 584, 596
Television sets, 474
Tellurium, 528
Temperature, during an illness, 407
Termination, 149, 160

Territory area of an animal, 68, 432
Test, vertical line, 27, 33
Theorems, addition, 237, 257–263
    Bayes, 275–279
    duality, 165, 336
    minimax, 336
    multiplication, 242–255, 264–266
    von Neumann's, 336
Theoretical probability, 228
Ticket prices, 468, 471
Time, doubling, 514–516
    quadrupling, 521
    tripling, 521
Time to failure, 584
Total cost, 41, 43, 75, 407, 429, 442,
        462–464, 503, 538, 546
    derivative of, 429
    see also Cost
Total profit, 410, 429, 462–464, 470, 502,
        546
    derivative of, 429
    maximizing, 462–464
    see also Profit
Total revenue, 14, 18, 31, 43, 407, 442,
        462–464, 471
    and demand, 435
    derivative of, 429
    maximizing, 468, 469, 471
    see also Revenue
Transit fares, 469
Transition diagrams, 310
    matrices, 310
    probability, 311
Transportation planning, 582, 583
Trees, 198–201, 247
    fork, 200
    games, 327–329
    probability, 259, 276
    vertex, 200
Tripling time, 521
Tumor, growth and rate of change of a, 428
Two-point equation, 44

Unbounded solution set, 125
Uniform distribution, 580
Uninhibited population growth, 512–522
Union of sets, 186–196
Universal set, 183
U.S. population growth, 516
U.S.S.R. population growth, 520
Utility, 406

Value, capital, 574
    maximum, 452, 611
    minimum, 452, 611
Value, game, 333, 336
    present, 371, 373
    random variable, 285
Variable(s), basic, 148
    change of, 557
    continuous random, 575–594
    dependent, 29
    dual, 165
    independent, 29
    interchanging, 158
    nonbasic, 148
    primal, 148, 165
    random, 284–305
    slack, 148, 149, 167
    structural, 148
Variable costs, 41, 429
Variance, 297–299, 587
Variation, direct, 35
    inverse, 60
Variation constant, 35
Vector, column, 97
    fixed probability, 323
    probability, 310–315
    product, 103
    row, 97
Velocity, 444
    average, 408, 426
    and derivatives, 444
    instantaneous, 426
    and integration, 535

Venn diagrams, 186–196, 261, 267, 269
    solution with, 192–196
Vertex, trees, 200
Vertical line test, 27, 33
Vertical lines, 32, 33
Volume, maximizing, 465–467, 471
Von Neumann Minimax Theorem, 336
Voting coalitions, 339–344
Voting games, 339–344
    dictator, 344
    dummy, 342, 344
    veto power, 344

Walking speed, 511
Weight gain, 11, 17
Weight loss, 527
Whiskey, and the risk of having an
        accident, 518
Wind-chill temperature, 608
Window, Norman, 472
World population growth, 516
Wrong pivot, 159

x-axis, 19
Xenon-133, 524
x^{-1}, differentiating, 420
    integrating, 531
x^r, differentiating, 420
    integrating, 531

y-axis, 19
Yield, effective annual, 521
y-intercept, 37

Zero products, principle of, 12
Zero-sum games, 329
Zirconium, 528